2nd Annual Technology Applied Science and Engineering Conference (ATASEC 2020)

IOP Conference Series: Materials Science and Engineering
Volume 1073

Online
5 August 2020

Editors:

Ratna Ika Putri
Rosa Andrie Asmara
Ferdian Ronilaya
Sapto Wibowo
Imam Fahrur Rozi
Arie Rachmad Syulistyo
Vivi Nur Wijayaningrum
M. Shulhan Khairy

ISBN: 978-1-7138-4580-5
ISSN: 1757-8981

Printed from e-media with permission by:

Curran Associates, Inc.
57 Morehouse Lane
Red Hook, NY 12571

Some format issues inherent in the e-media version may also appear in this print version.

Additional copies of this publication are available from:

Curran Associates, Inc.
57 Morehouse Lane
Red Hook, NY 12571 USA
Phone: 845-758-0400
Fax: 845-758-2633
Email: curran@proceedings.com
Web: www.proceedings.com

TABLE OF CONTENTS

IOP Conf. Series: Materials Science and Engineering 1073 (2021) 011001 doi:10.1088/1757-899X/1073/1/011001

Preface

Good morning, Ladies and Gentlemen, Dear Colleagues,

It gives me great pleasure to extend to you all a very warm welcome on behalf of Politeknik Negeri Malang and to say how grateful I am to all guests and presenters of The Second Annual Technology, Applied Science and Engineering Conference (ATASEC 2020) and 11th Joint National Seminar on Engineering 2020.

I also would like to convey my special appreciation to Mr. Indrazno and Mr. Rosa Andrie, as the chairman of the conference and all committees, which comes from a combination of all engineering majors in Politeknik Negeri Malang, who are responsible for the implementation of these conference.

We are very grateful to be able to continue to hold this conference at the second time, despite of the limitations due to Covid19 pandemic. Our plan was, to welcome all participants of conference to visit our beautiful city Malang, However, even though we cannot meet face to face we can still carry out this conference virtually with enthusiasm and joy.

The ATASEC 2020 has a theme “Welcoming Industry 4.0 through Advanced Technology, Science and Vocational Engineering Education”. As for SNGBR 2020 theme “Implementasi Penelitian Terapan Multidisipliner Bidang Rekayasa melalui Kolaborasi Pendidikan Tinggi Vokasi dengan Industri”. These theme will be a media for sharing knowledge from researchers and industry from all fields of advanced technology and science.
These conferences will be attended by 5 distinguish keynote speakers:

1. Prof. Nobuo Funabiki, from Dept. of Electrical and Communication Engineering, Okayama University, Japan.
2. Prof. Madya Dr. Rosli bin Omar, from Dept. of Electonic Engineering, Universiti Tun Hussein Onn Malaysia, (UTHM), Malaysia.
3. Dr. Eng. Anggit Murdani, ST., M.Eng., Vice Director on Student Affair and a senior lecturer in Machine engineering department, Politeknik Negeri Malang.
4. Ir. Anang Yahmadi, MengSc, Executive Vice Director, Divisi Perencanaan Pengadaan Strategis, PT. Perusahaan Listrik Negara (National Electrical Company), Indonesia.
5. And Mr. Nana Sutiksna, Practician QA/QC from Oil and Gas Industry.

Please accept my high appreciation to all 5 keynote speakers for joining here with us today.

Ladies and Gentlemen,
Such a conference is an extraordinary opportunity, as a good time to share knowledge from a variety of affiliates and also a variety of concentrations in the scope of the field of Advanced Technology, Science and Vocational Engineering Education.

IOP Conf. Series: Materials Science and Engineering 1073 (2021) 011001 doi:10.1088/1757-899X/1073/1/011001

Thank you also to all experts, researchers, lecturers who come to share their knowledge today.

I am sure you that we will have fruitful and rewarding exchanges today. I wish you all a very success with this important conference and I look forward to learning about the outcome.

So, let me now open these conferences officially by wishing you a delightful and vibrant day.

BISMILLAH HIRROHMAN NIROHIM.

Thank you.

Malang, August 5th 2020
Director of State Polytechnic of Malang
Drs. Awan Setiawan, MMT., M.M.

IOP Conf. Series: Materials Science and Engineering 1073 (2021) 011001 doi:10.1088/1757-899X/1073/1/011001

List of committee

Conference Chair
Indrazno Siradjuddin, Ph.D., Politeknik Negeri Malang, Indonesia

Co- Conference Chair
Erfan Rohadi, Ph.D., Politeknik Negeri Malang, Indonesia

International Advisory Board:

- Prof. Ir. Dr. Marizan bin Sulaiman (Universiti Teknikal Malaysia Melaka)
- Prof. Madya Dr. Wahyu Mulyo Utomo (Universiti Tun Hussein Onn Malaysia – UTHM)
- Prof. Taufik (California Polytechnic State University – Cal Poly)
- Prof. Ming-Shyan Wang (Southern Taiwan University of Science and Technology – STUST)
- Prof. Chi Jo Wang (Southern Taiwan University of Science and Technology – STUST)
- Prof. Dr. Ade Gafar Abdullah (Universitas Pendidikan Indonesia)
- Syeilendra Pramudya, Ph.D. (Institut Teknologi Bandung)
- Dr.Eng. Asep Bayu Dani Nandiyanto (Universitas Pendidikan Indonesia)
- Dr. Isma Widiaty, M.Pd. (Universitas Pendidikan Indonesia)
- Kolonel Lek Dr. Ir. Arwin Datumaya Wahyudi Sumari, S.T., M.T., IPM, ASEAN Eng. (TNI Angkatan Udara/Politeknik Negeri Malang/Universitas Pertahanan)
- Prof. Dr. Wahyu Widada (Universitas Bengkulu)
- Ir. Franky M S Telupere, MP. Ph.D. (Universitas Nusa Cendana Kupang)
- Dr. Dadang Lukman Hakim, MT (Universitas Pendidikan Indonesia)
- Prof. Dr. Muhammad Ali Ramdhani, M.T. (UIN Sunan Gunung Djati Bandung)
- Dr. Hadi Kuncoro, M. Farm., Apt. (Universitas Mulawarman)
- Dr. Astri Rinanti, MT (Universitas Trisakti)
- Dr. Hilmi Aulawi, M.T. (Sekolah Tinggi Teknologi Garut)
- Dr. Juniastel Rajagukguk, M.Si (Universitas Negeri Medan)
- Dr. Ir. Herman S., MBA. (Universitas Komputer Indonesia)
- Dr. Endarto Y Wardhono, (Universitas Sultan Ageng Tirtayasa, Banten)
- Dr. Niken Subekti, MSi (Universitas Negeri Semarang)
- Dr. Sudi Dul Aji (Universitas Kanjuruhan Malang)
- Dr. July Hidayat, M.Sn. (Universitas Pelita Harapan)
- Dr. Jarman Arroisi, M.A. (Universitas Darussalam Gontor Ponorogo Jatim)
- Dr. Ir. Suharsono. (UPN Veteran Yogyakarta)
- Dr. Setia Budi, M.Sc (universitas Negeri jakarta)
- Dr. Ade Yeti Nuryantini, S.Pd., M.Pd., M.Si. (UIN Sunan Gunung Djati Bandung)
- Dr. Bebeh Wahid Nuryadin (UIN Sunan Gunung Djati Bandung)
- Dr. Melati Ferianita Fachrul, MS. (Universitas Trisakti)
- Dr.rer.nat. I Gusti Ngurah Agung Suryaputra, S.T., M.Sc. (Universitas Pendidikan Ganesha)
- Daniel Sutopo Pamungkas, PhD (Politeknik Negeri Batam)
- Cynthia Wuisang, ST, MUrbHabMgt., PhD (Universitas Sam Ratulangi)
- Ir. Isri Ronald Mangangka, M.Eng., Ph.D. (Universitas Sam Ratulangi)

Scientific Committee

- Prof. Dr. Asep Kadarohman, Universitas Pendidikan Indonesia, Indonesia
- Prof. Dr. Didi Sukyadi, M. A., Universitas Pendidikan Indonesia, Indonesia
- Prof. Dr. Anna Permanasari, Universitas Pendidikan Indonesia, Indonesia
- Prof. Dr. Sumarto, Universitas Pendidikan Indonesia, Indonesia
- Prof. Dr. M. Syaom Barliana, Universitas Pendidikan Indonesia, Indonesia
- Prof. Dr. Ratih Hurriyati, Universitas Pendidikan Indonesia, Indonesia
- Dr. Ir. Taufiq Saidi, M. Eng., Universitas Syiah Kuala, Indonesia
- Dr. Erfan Handoko, M.Si., Universitas Negeri Jakarta, Indonesia
- Dr. Hilmi Aulawi, MT., Sekolah Tinggi Teknologi Garut, Indonesia
- Dr. Astri Rinanti, MT., Universitas Trisakti, Indonesia
- Dr. Prantasi H. Tjahjanti, Universitas Muhammadiyah Sidoarjo, Indonesia
- Dr. Rositayanti Hadisoebroto, ST., MT., Universitas Trisakti, Indonesia
- Irwan Alnarus Kautsar, Ph.D., Universitas Muhammadiyah Sidoarjo, Indonesia
- Muhammad Nur Hudha, M.Pd., Universitas Kanjuruhan Malang, Indonesia

Organizing Committee

- Dr. Ratna Ika Putri, Politeknik Negeri Malang, Indonesia
- Dr. Eng. Rosa Andrie Asmara, Politeknik Negeri Malang, Indonesia
- Dr. Ferdian Ronilaya, Politeknik Negeri Malang, Indonesia
- Dr. Sapto Wibowo, Politeknik Negeri Malang, Indonesia
- Imam Fahrur Rozi, M.T., Politeknik Negeri Malang, Indonesia
- Arie Rachmad Syulistyo, M.Kom., Politeknik Negeri Malang, Indonesia
- Vivi Nur Wijayaningrum, M.Kom., Politeknik Negeri Malang, Indonesia
- M. Shulhan Khairy, M.T., Politeknik Negeri Malang, Indonesia

Editors

- Dr. Ratna Ika Putri, Politeknik Negeri Malang, Indonesia
- Dr. Eng. Rosa Andrie Asmara, Politeknik Negeri Malang, Indonesia
- Dr. Ferdian Ronilaya, Politeknik Negeri Malang, Indonesia
- Dr. Sapto Wibowo, Politeknik Negeri Malang, Indonesia
- Imam Fahrur Rozi, M.T., Politeknik Negeri Malang, Indonesia
- Arie Rachmad Syulistyo, M.Kom., Politeknik Negeri Malang, Indonesia
- Vivi Nur Wijayaningrum, M.Kom., Politeknik Negeri Malang, Indonesia
- M. Shulhan Khairy, M.T., Politeknik Negeri Malang, Indonesia

Conference Documentations

Project of Programming Self-Study Tools
August 5, 2020
Nobuo FUNABIKI
Okayama University, Japan
ATASEC 2020, Malang, Indonesia

OKAYAMA UNIVERSITY

Room 1
ATASEC
5 - 6 August 2020

POLITEKNIK NEGERI MALANG
Designing Real-time Research Data Portal of JTI POLINEMA

IOP Conf. Series: Materials Science and Engineering 1073 (2021) 011001 doi:10.1088/1757-899X/1073/1/011001

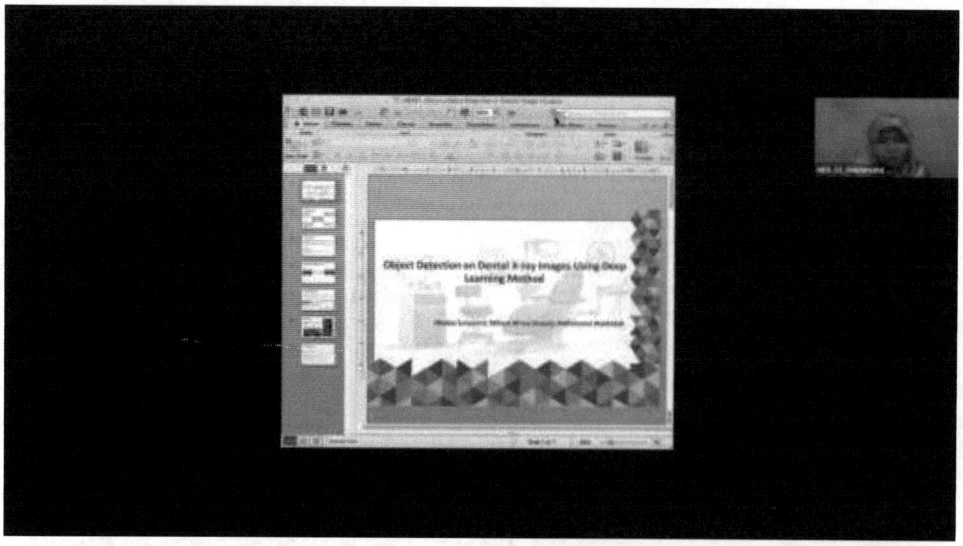

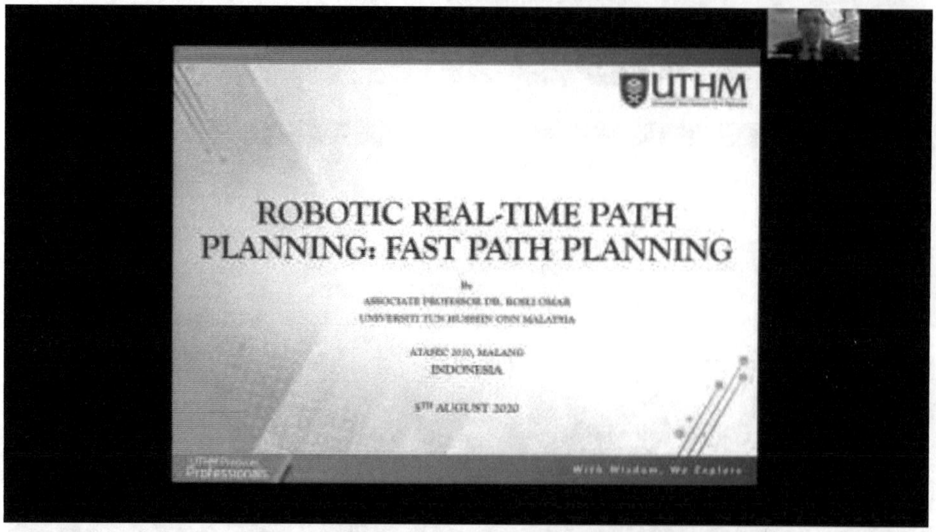

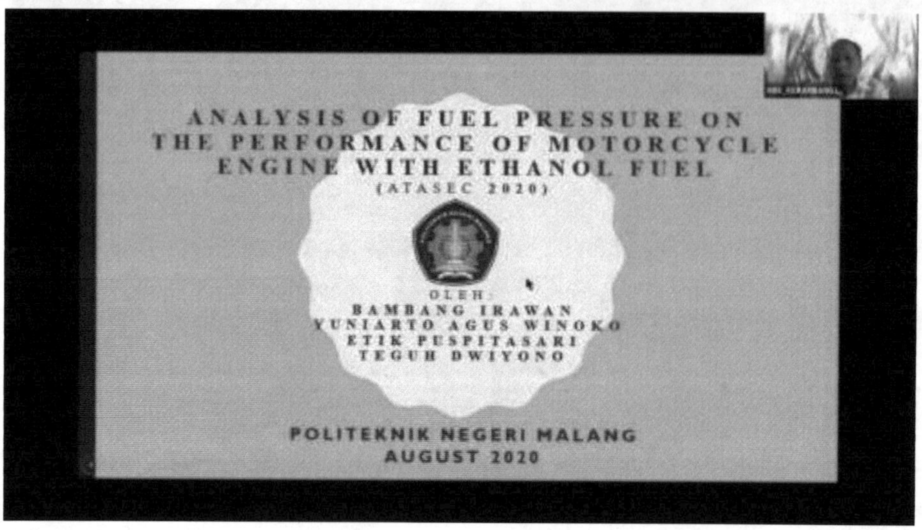

IOP Conf. Series: Materials Science and Engineering 1073 (2021) 011001 doi:10.1088/1757-899X/1073/1/011001

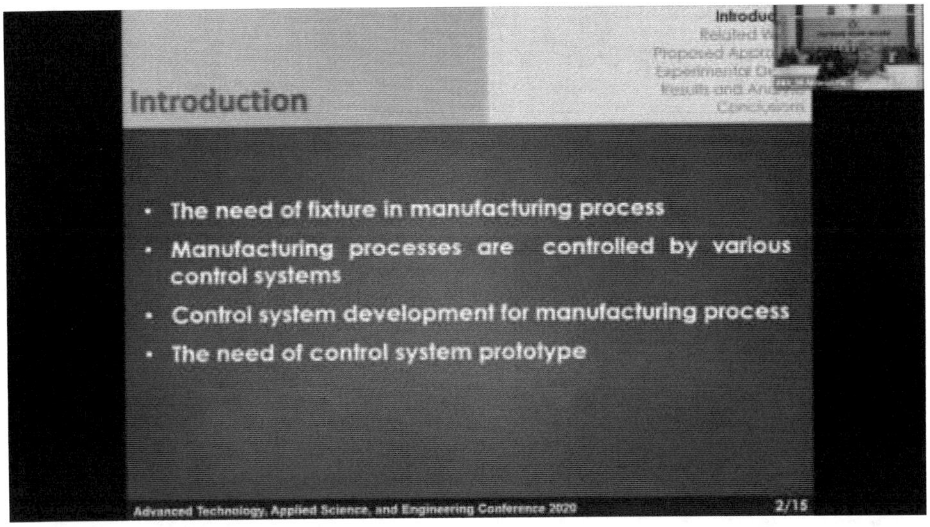

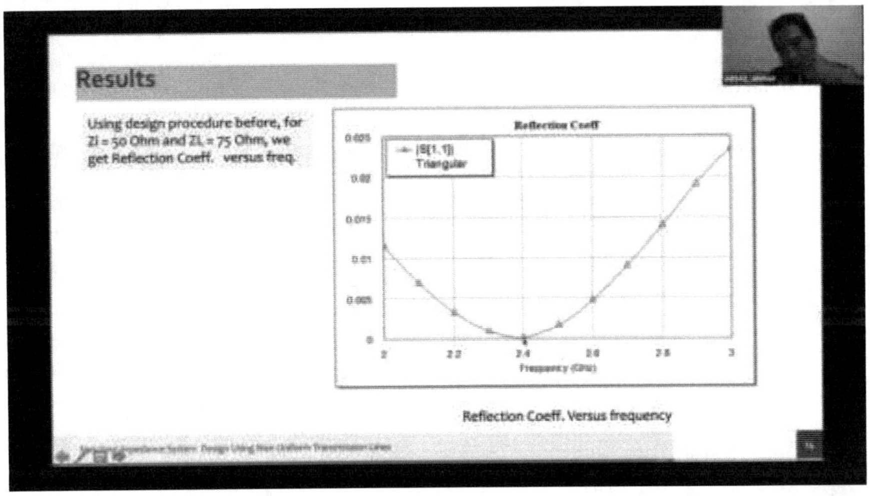

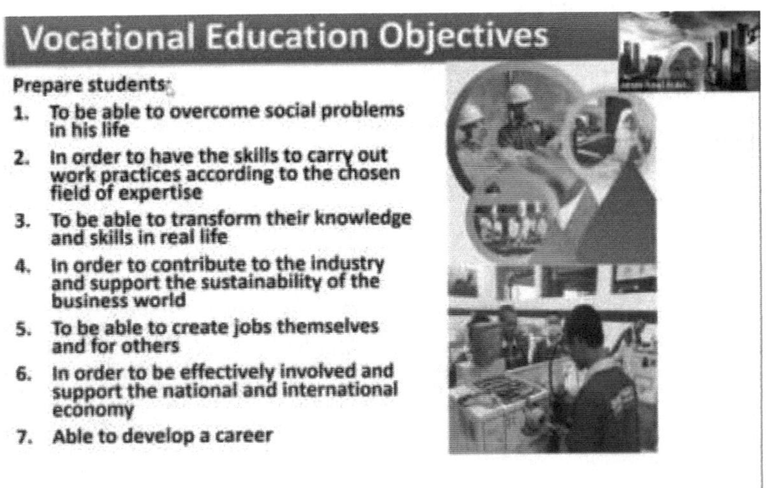

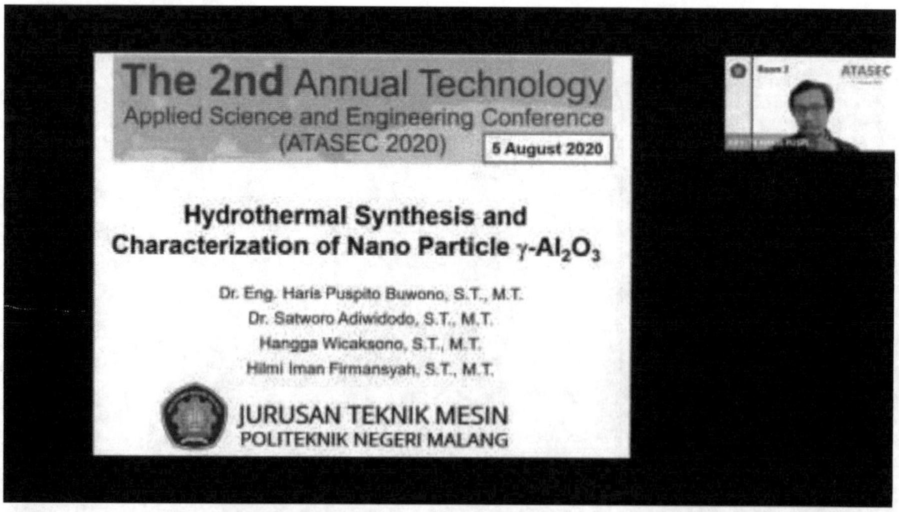
The 2nd Annual Technology
Applied Science and Engineering Conference
(ATASEC 2020)
5 August 2020
Hydrothermal Synthesis and Characterization of Nano Particle γ-Al2O3
Dr. Eng. Haris Puspito Buwono, S.T., M.T.
Dr. Satworo Adiwidodo, S.T., M.T.
Hangga Wicaksono, S.T., M.T.
Hilmi Iman Firmansyah, S.T., M.T.
JURUSAN TEKNIK MESIN
POLITEKNIK NEGERI MALANG
ATASEC

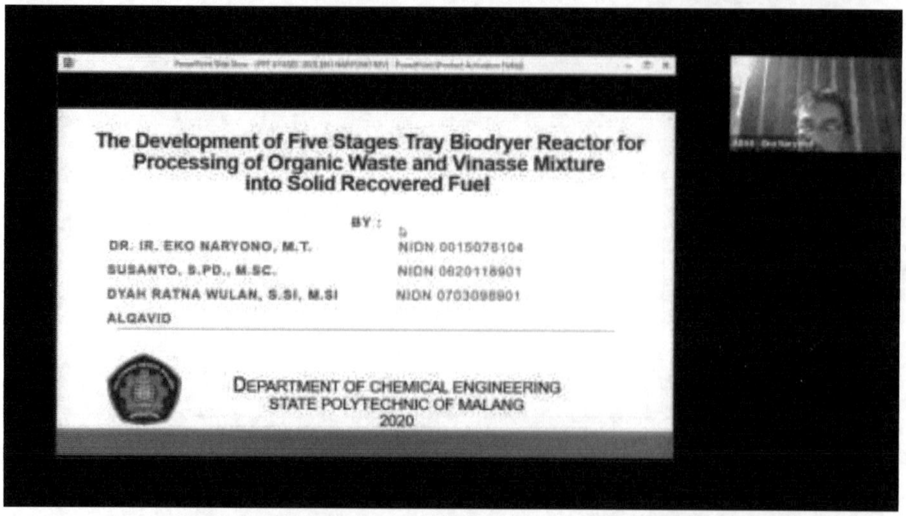
The Development of Five Stages Tray Biodryer Reactor for Processing of Organic Waste and Vinasse Mixture into Solid Recovered Fuel
BY :
DR. IR. EKO NARYONO, M.T. NIDN 0015076104
SUSANTO, S.PD., M.SC. NIDN 0620118901
DYAH RATNA WULAN, S.SI, M.SI NIDN 0703098901
ALQAVID
DEPARTMENT OF CHEMICAL ENGINEERING
STATE POLYTECHNIC OF MALANG
2020

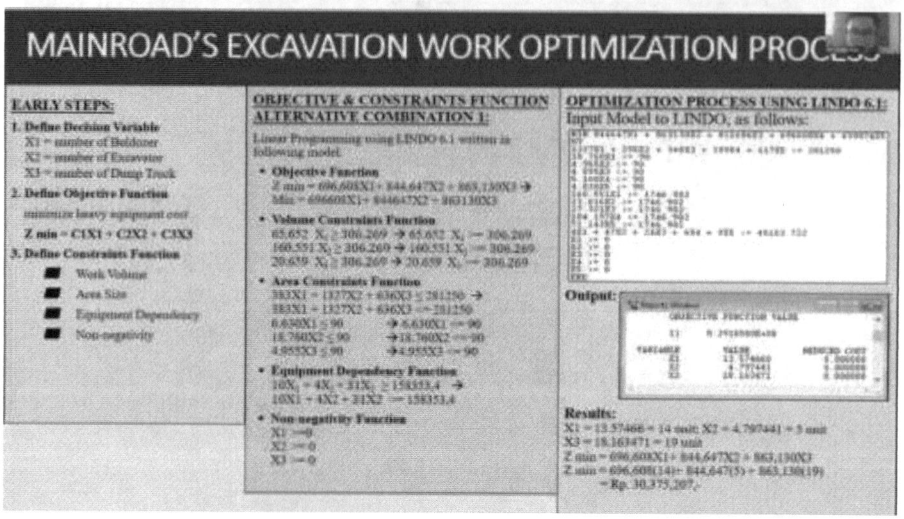
MAINROAD'S EXCAVATION WORK OPTIMIZATION PROC
EARLY STEPS:
1. Define Decision Variable
2. Define Objective Function
Z min = C1X1 + C2X2 + C3X3
3. Define Constraints Function
Work Volume
Area Size
Equipment Dependency
Non-negativity
OBJECTIVE & CONSTRAINTS FUNCTION ALTERNATIVE COMBINATION 1:
Objective Function
Volume Constraints Function
Area Constraints Function
Equipment Dependency Function
Non-negativity Function
OPTIMIZATION PROCESS USING LINDO 6.1:
Input Model to LINDO, as follows:
Output:
Results:
= Rp. 30.375.207,-

IOP Conf. Series: Materials Science and Engineering 1073 (2021) 011002 doi:10.1088/1757-899X/1073/1/011002

Peer review declaration

All papers published in this volume of *IOP Conference Series: Materials Science and Engineering* have been peer reviewed through processes administered by the Editors. Reviews were conducted by expert referees to the professional and scientific standards expected of a proceedings journal published by IOP Publishing.

- **Type of peer review:**

  Single-blind review

  Each paper is distributed to reviewers in which the reviewers know the names of the authors yet the authors do not know. The reviewed papers were sent to the authors through their personal account and they have to send them back to the organizers using their account as well. A second round review will take place if necessary.

- **Describe criteria used by Reviewers when accepting/declining papers. Was there the opportunity to resubmit articles after revisions?**

  First thing the reviewers take into account is whether the scope of the papers they analyze meet that of the intended publisher. Following the process, the reviewers should make sure that the papers meet the standards of international publications in terms of both content and references. Authors are given three decisions; accepted, revision required, and rejected. Once their paper(s) is rejected, the do not have an opportunity to resubmit it. If they need to revise their paper(s) and is not able to submit the revised paper prior to the deadline, they are considered withdrawing their paper.

- **Conference submission management system:**
  Konfrenzi Automatic Conference System

- **Number of submissions received:**
  101

- **Number of submissions sent for review:**
  96

- **Number of submissions accepted:**

  84

- **Acceptance Rate (Number of Submissions Accepted / Number of Submissions Received X 100):**
  101/ 84 x 100 = 83.1%

- **Average number of reviews per paper:**

2 (two)

- **Total number of reviewers involved:**
  10 (two)

- **Any additional info on review process (ie plagiarism check system):**
  None

- **Contact person for queries:**

  ade_gaffar@upi.edu

IOP Conf. Series: Materials Science and Engineering 1073 (2021) 012001 doi:10.1088/1757-899X/1073/1/012001

Enhancement in synthesis of citronellyl laurate flavour by combined effect of ultrasound and immobilized lipase as heterogeneous biocatalyst

D Moentamaria*, S Rulianah, A Chumaidi, R R Bilghis A and R Romadoni

Department of Chemical Engineering, State Polytechnic of Malang, Jalan Soekarno-Hatta No. 9 Malang, Indonesia

*dwina_mnt@yahoo.com

Abstract. This study aims to improve the performance of heterogeneous biocatalysts by using the combination of immobilized and ultrasonic lipases in the esterification reaction for the synthesis of citronellyl lauric flavors. Synthesis was conducted out through the reaction of citronellol with lauric acid by using immobilized lipase and ultrasonic. The results showed that immobilized lipase provided stability for up to 9 weeks and could be reused for 5 cycles of reactions. Immobilized lipase 473 U / gram was applied to the synthesis of citronellyl laurate with a reaction time of 90 minutes, 45^{0}C, with and without ultrasonic. The result of citronellol conversion with ultrasonic is 3.6 x higher than conventional methode. It shows that the immobilization method provided stability to the biocatalyst, but because it had a limited internal mass transfer so the rate of reaction of ester formation reduced. On the other hand, ultrasonic contributed to increase the dispersion and collision of substrate molecules, reducing reaction time, and intensifying catalytic processes. Thus, the combination of immobilized enzyme and ultrasonic methods can be applied to biochemical processes widely.

1. Introduction

Ester is one of the most important natural flavors and can be applied in various fields, including in the food, beverage, jam, candy, and dairy-based, cosmetics industries [1,2]. Short chain esters with fruity flavors have been widely used in the food industry as a flavor component. The short chain of aliphatic ester is produced from short chains of alcohol and short chains of carboxylic acids. Its constituent carbon atoms is less than 10 and produces a fresh and fruity odor [2,3].

However, the industry still requires a lot of flavor components with high molecular weight, because it has a strong and durable odor [1]. The high % conversion of esterification is correlated with the length of the carbon free fatty acid chain [4]. The results showed that lauric acid C_{12} had a higher ester conversion when compared to capric acid (C_{10}), caprylic acid (C_8), 79%, 43%, 25% respectively.

Natural flavors which are conventionally obtained by direct extraction from plant material are low yield and often too rare and expensive for commercial scale. Meanwhile, industry to obtain artificial flavor components using chemical synthesis methods 303K-533K [5]. The market value that is more desirable is the ester of the natural flavor, so it needs to be developed sustainably [2,6].

In a commercial scale, natural flavors can be produced by enzymatic reactions using lipases from various microbes including *Mucor miehei, Candida rugosa, Candida Antarctica B (CALB), Bacillus aerius, Thermomyces lanuginosus, Penicillium crustosum, Aspergillus oryzae*. This enzymatic (free)

IOP Conf. Series: Materials Science and Engineering 1073 (2021) 012001 doi:10.1088/1757-899X/1073/1/012001

lipase reaction is an option because it uses mild conditions, specific reactions and works on the interface area between the insoluble substrate and water in hydraulic reactions and esterification of the hydrophobic matrix [2,4,6-10]. However, the use of this free enzyme is unstable over a wide temperature and pH range and is only used one reaction. This causes the enzymatic synthesis of natural flavor with free lipase to be very expensive [11,12].

Lipase immobilization technique in making natural flavor uses matrix hydrophobic, the enzyme is more stable and can be used repeatedly, so it is cheaper than free lipase. Various matrices, including polyurethane foam (PUF), Amberlite MB-1, Celite, Silica gel, Exfoliated Graphene Oxide (EGO), Silica-epoxy, have been used for flavor making at reaction times of 10 - 72 hours, resulting in a conversion of 30% -90 %. On an industrial scale, the length of reaction time in the enzyme immobilization method becomes an obstacle for large production [2,4,6-9,13].

So comprehensive efforts need to be made to overcome the problem of long reaction time. One solution that can be used is by utilizing ultrasonic technology. Ultrasonic apparatus is a device that can convert electrical energy into mechanics in the form of ultrasonic sound waves above 20kHz. The basic mechanism that occurs in the use of ultrasonic is acoustic cavitation, which after formation occurs, growth and microbubbles collapse in the liquid medium. This occurs due to a microscopic implosion which causes very high local turbulence and releases heat energy. This condition causes a significant increase in temperature and pressure, such as the phenomenon that occurs hydrodynamic cavitation which causes mechanical damage to propellers and pumps [14]. In acoustic cavitation, sound waves form a sinusoidal at various pressures during cavitation in solution (Figure 1).

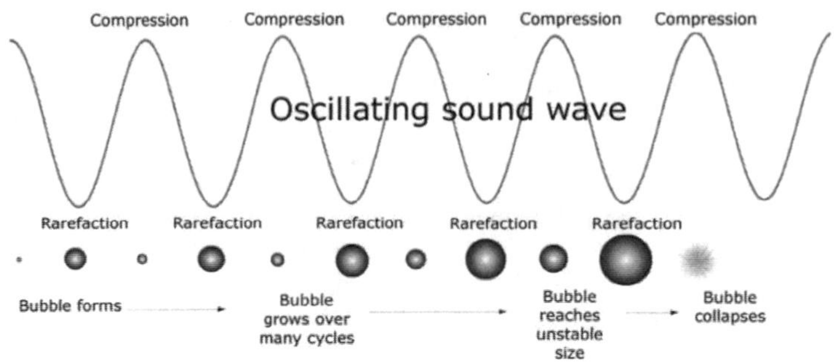

Figure 1. Phenomena of acoustic cavitation cycles in ultrasound.

Phenomena of acoustic cavitation cycles in ultrasound, which is the formation, growth and collapse of microbubbles. The bubbles oscillate through sound waves in the presence of compression followed by rarefaction and eventually will burst and produce energy that can produce vibrations and can emulsify a material [14].

Ultrasonic has been widely used in various industries, as shown in Figure 2.

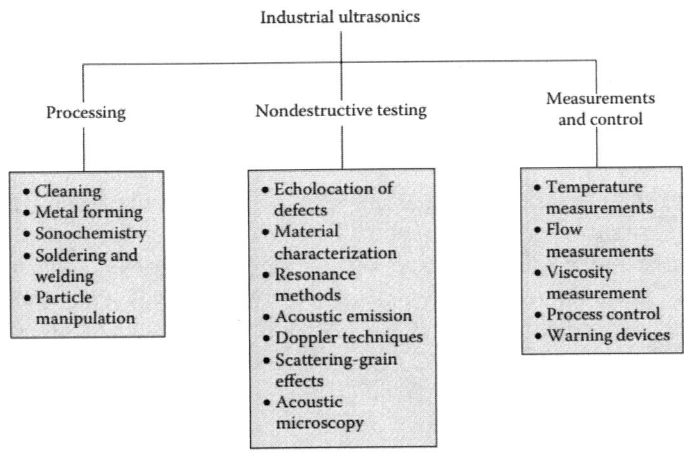

Figure 2. The use of ultrasonic in the industry [15].

In the field of sonochemistry, ultrasonic is used in various industries (Table 1).

Table 1. Sonochemistry application.

Industry	Product	Application	Reference
Flavor	Ethyl butirate	Mixing of butyric acid and ethanol	[4]
Water treatment	Clean water	Mixing, inactivate *Escherichia coli*	[16]
Chemical	Wetting agent	Mixing dichloromethane with ethanolamine	[17]
Cosmetic ester	Skin care	Mixing butyl alcohol and acetic acid	[18]
Medicine	Bioactive compounds of anticancer	Extraction of *Cuminum cyminum*	[19]
Fuel	Biodiesel	Mixing soybean fatty acid and ethanol	[20]
Oil and gas	Oil recovery	Extraction oil dan gas	[21]

Natural flavour- citronellyl laurate is a high –value processed product from long chain carbocyclic and alcohol because it has a strong and long lasting odor [1]. The objective of this study was to know the effect of ultrasound on the reaction between lauric acid and terpene alcohol citronellol, not using solvent, catalyzed by immobilized lipase *Mucor miehei* on polyurethane foam.

The synergy of ultrasound and immobilized enzyme is an attractive solution to reduce reaction time and increase the performance of lipases so that they remain stable and can be used repeatedly. This condition supports the achievement of green technology process, especially the existence of free solvent and washing is not carried out in repeated reactions [10].

2. Material and methods

Lauric acid, citronellol, lipase Mucor miehei are obtained from Sigma. Co-immobilized consisting of gelatin, lecithin, polyethylene glycol (PEG), MgC12 was purchased from Merck (Darmstadt, Germany). PUF is obtained from local market, the isocyanate and polyol reaction with a volume ratio of 1: 1. PUF cut into sized cubes 0.5 x 0.5 x 0.5 cm3 [22].

2.1. *Procedure of immobilization lipase on PUF*

We made a co-immobilized solution from a mixture of lecithin, gelatin, PEG, $MgCl_2$ solutions. PUF was immersed in a co-immobilzed solution in a ratio of 1:20, for 1 hour and then dried to 30^0C. The result is soaked in lipase for 24 hours then dried to 30^0C. drying is carried out in the oven [22-24].

2.2. *Determination of lipase activity*

Olive oil as a substrate was used to determine lipase activity. An amount of 25 ml of olive oil and 75 ml of a 7% solution of gum arabic was emulsified for 2 minutes. Next, 5 ml of emulsified olive oil was added to 2 ml of 0.1 M phosphate buffer (pH 7) and 1 ml of lipase. Then, it was incubated in lipase at 37 ° C for 30 minutes using an orbital shaker. Furthermore, to deactive the lipase activity, 15 ml of acetone-ethanol (1: 1 v / v) was added and free fatty acids were titrated with 0.05 M NaOH. One unit of lipase activity is defined as the number of enzymes capable of releasing 1μmol of fatty acids per minute [22,25,26].

Lipase activity can be calculated by formula.

$$Lipase\ Activity\ (U/ml) = \frac{(A\text{-}B) \times N_{NaOH} \times 1000}{30}$$

Where,
A = The amount of NaOH needed to titrate the sample (ml)
B = The amount of NaOH needed to titrate the blank (ml)
30 = 30 minutes incubation time [27]

2.3. *Determine of storage stability storage*

The immobilized lipase in PUF was stored at 4^0C and its remain activity was examined every day.

2.4. *Esterification*

Lauric acid and citronellol 1:1 mole ratio, were placed in erlenmeyer 100 ml, and being reacted at the temperature of 40°C, 20 h, pH 7 using immobilized lipase on Digital ultrasonic cleaner. The specifications of this equipment are tank size 300 x 240 x 150 mm (L x W x H), transducer 4 pcs, power 240 W, 40 KHZ.

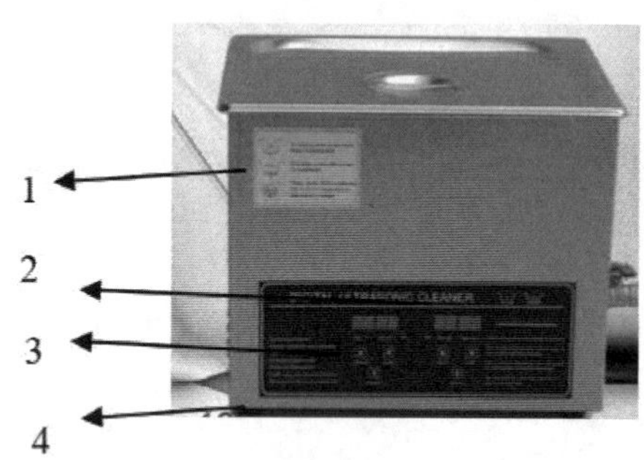

1. Sample in the tank
2. Temperature control
3. Time control
4. Transducers

Figure 3. Digital ultrasonic cleaner.

GC FID (HP 5890, Santa Clara-California, United States) was used to analyze the natural-flavor resulting from the experiment. GC FID uses an HPL 608 column with dimensions of length 30 m x 0.53 mm i.d. x film thickness 0.5 μm. The GC-FID oven temperature was set to 125 ° C for 3 minutes in increments of 7.5 ° C / min to 250 ° C. The injector and detector temperatures were kept at 255 ° C and 275 ° C, respectively. Then the area under the curve is calculated to determine the conversion of citronellol [28-31].

3. Results and discussion

3.1. Storage stability free and immobilized lipase

Storage stabillity was used to test the performance of free and immobilized lipase against long storage time at 4^{0}C. Figure 4 showed, until the storage time of 9 weeks, immobilized lipase on PUF still had high stability of 91.69%, while free lipase dropped to 9.63%. This significant difference was due to i) the surface of the PUF matrix became hydrophobic after being coated with co-immobilized MgCl2, PEG, lechitin, gelatin, ii) the strong covalent bond between PUF and lipase became stronger because it was co-immobilized as a spacer arm that extended the bond. the. Starting from weeks 1-3, free and immobilized lipase were still the same as the initial conditions, with 100% remaining activity. However, after week 4, there was a significant decrease in activity at 45.98% free lipase, while immobilized lipase was still stable as indicated by the remaining activity of 94.81%.

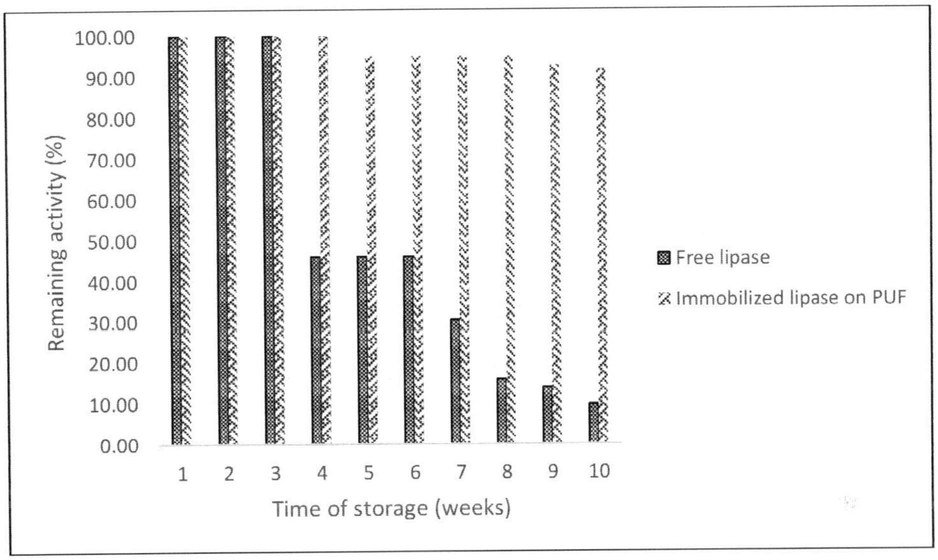

Figure 4. Performance storage stability free and immobilized lipase, 4^{0}C.

The same characteristics were shown in PUF coating using co-immobilized polyethilenimin (PEI) and glutaraldehyde (GA), which had higher stability when compared to free lipase, each remaining activity was 50% and 0%, after incubating 30 minutes, 50^{0}C [28]. The use of co-immobilized in matrix coating can significantly increased enzyme stability, so this method was often used in enzymatic reactions [7,28,31].

3.2. Effect of ultrasound irradiation

Ultrasonic has been used in the esterification reaction of fatty acids and alcohol The results of data analysis showed that ultrasonic could increase enzyme activity and increase substrate transfer on the active side of the enzyme [33,34]. The same phenomenon was shown in the enzymatic reaction of esterification of lauric acid and citronellol to natural-flavor citronelllyl laurate below (Figure 5). Ultrasonic with a power of 240 W, 40 KHZ, was used to produce citronellyl laurate at a reaction time of up to 90 minutes. The citronellol conversion was calculated based on [[initial citronellol – citronellol at t] / [initial citronellol]] x 100% by using GC FID [31].

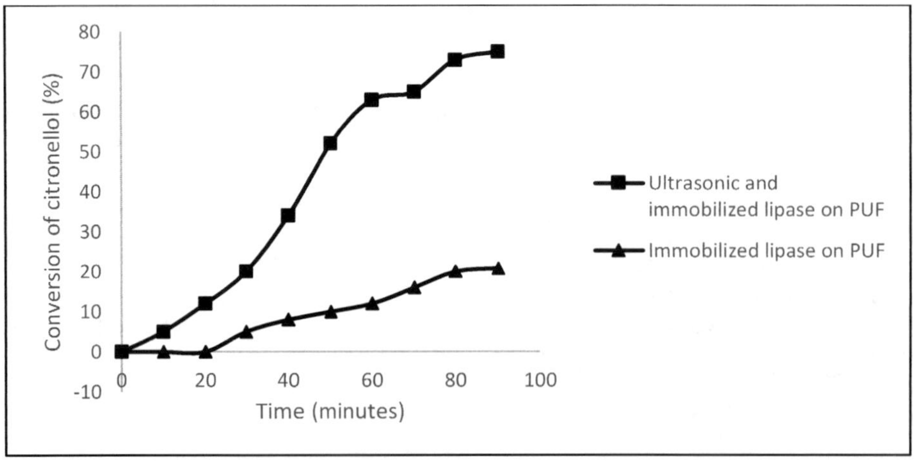

Figure 5. Effect of ultrasound irradiation on conversion of citronellol by using immobilized lipase for esterification reaction in 45^0C.

The esterification reaction has been carried out using immobilized lipase on PUF, resulting in 55% conversion of citronellol, with a reaction time of 20 hours [31]. So by looking at the trend in Figure 5, it can be seen that there were significant differences for each method of immobilized lipase (without ultrasonic) and immobilized lipase with ultrasonic respectively. Without using ultrasonic, the citronellol conversion changes were not very significant, from 30-90 minutes. This was different from the conversion rate of citronellol using ultrasonic, there was a very significant change, so that the remaining citronellol after the esterification reaction was less or the number reacts more, which was influenced by the reaction rate. After esterification reaction time of 90 minutes, the conversion of citronellol with and without ultrasonic was 75% and 20.8%, respectively or 3.6 x. Ultrasonic plays an important role through mixing technique in fast esterification, so it could reduce time. The use of ultrasonic was also applied to carry out the enzymatic synthesis of butyl acetate, which was an increase of 7.5 x times compared to conventional stirring methods. The reaction rate in ultrasonic increased 2.85 times higher than conventional [18]. The use of high frequencies in ultrasonic caused small bubbles to occur, causing the ratio of surface area to volume to increase, so as to increase mass and heat transfer through the interface of the bubble. This would increase the emulsification reaction and reaction rate of soybean, canola and corn oil to produce biodiesel. The reaction results increased with increasing output power from 150W to 450 W, producing 95% fatty acid methyl ester, reaction time of 30 minutes [20].

In this study, a heterogeneous biocatalyst immobilized lipase on PUF was used, with lauric acid and citronellol as a substrate. There was an increase in substrate reactions and mass transfer due to ultrasound irradiation, causing bubble bursts and cavitation which would increase enzymatic reactions, through ultrasonic liquid media. In this reaction, the bubble collapsed near the surface of heterogeneous solids which could cause high intensity shock waves which would increase the substrate enzyme reaction [5].

3.3. Reusability

One of the advantages of immobilized enzyme is the reuse of heterogeneous biocatalysts, thereby reducing operational costs on an industrial scale [5]. In this study, to obtain heterogeneous biocatalysts again, washing was not performed because the co-immobilized ones (lecithin, gelatine, $MgCl_2$, PEG) used were edible. If it is not used immediately for reuse, it can be stored at a temperature of 4^0C.

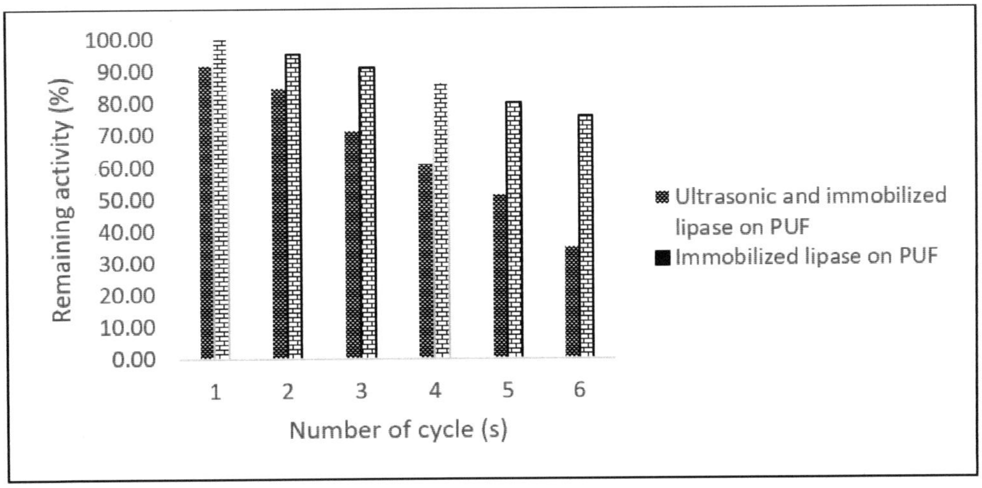

Figure 6. Reusability of the enzyme on % remaining activity of citronellyl laurat as natural flavor.

From Figure 6, using ultrasonic it can be shown that in 1st and 2nd cycle, there has been no change in immobilized lipase activity. Furthermore, in 3rd to 5th cycle, there was a significant decrease. In the 5th cycle, the remaining activity using ultrasonic and without ultrasonic was 80% and 51.18%, respectively. A decrease in remaining activity means that the enzyme activity decreases to carry out the esterification reaction. This is possible due to the release of enzymes from the matrix and the occurrence of deactivation during the process [5].

4. Conclusion

Process productivity can be increased through a combination of ultrasonic and immobilized lipase, into a synergy. Ultrasonic provides a reliable mixing technique on the substrate and reduces the formation of substrate / product layers in the porous PUF matrix. Meanwhile, immobilized lipase in the porous matrix PUF provides stability, so that the substrate enzymes can react properly. The use of ultrasonic can be used in all fields widely.

References

[1] Li C, Sun J, Fu C, Yu B, Liu S Q, Li T and Huang D 2014 Synthesis and evaluation of odour-active methionyl esters of fatty acids via esterification and transesterification of butter oil *Food Chem.* **145** 796–801

[2] Paludo N, Alves J S, Altmann C, Ayub M A Z, Fernandez-Lafuente R and Rodrigues R C 2015 The combined use of ultrasound and molecular sieves improves the synthesis of ethyl butyrate catalyzed by immobilized Thermomyces lanuginosus lipase *Ultrason. Sonochem.* **22** 89–94

[3] Su L, Hong R, Guo X, Wu J and Xia Y 2016 Short-chain aliphatic ester synthesis using Thermobifida fusca cutinase *Food Chem.* **206** 131–6

[4] Shintre M S, Ghadge R S and Sawant S B 2002 Lipolase catalyzed synthesis of benzyl esters of fatty acids *Biochem. Eng. J.* **12** 131–41

[5] Gawas S D and Rathod V K 2018 Enhancement in synthesis of ethyl laurate catalyzed by fermase by combined effect of ultrasound and stage wise addition of ethanol *Chem. Eng. Process. Intensif.* **125** 207–13

[6] Pires-Cabral P, Da Fonseca M M R and Ferreira-Dias S 2010 Esterification activity and operational stability of Candida rugosa lipase immobilized in polyurethane foams in the production of ethyl butyrate *Biochem. Eng. J.* **48** 246–52

[7] Moentamaria D, Muharja M, Widjaja T and Widjaja A 2019 A Performance Study of Home-Made Co-Immobilized Lipase from Mucor miehei in Polyurethane Foam on The Hydrolysis of Coconut Oil to Fatty Acid *Bull. Chem. React. Eng. Catal.* **14** 391–403

[8] Narwal S K, Saun N K, Dogra P and Gupta R 2016 Green synthesis of isoamyl acetate via silica immobilized novel thermophilic lipase from Bacillus aerius *Russ. J. Bioorganic Chem.* **42** 69–73
[9] Patel V, Gajera H, Gupta A, Manocha L and Madamwar D 2015 Synthesis of ethyl caprylate in organic media using Candida rugosa lipase immobilized on exfoliated graphene oxide: Process parameters and reusability studies *Biochem. Eng. J.* **95** 62–70
[10] Nyari N L D, Fernandes I A, Bustamante-Vargas C E, Steffens C, de Oliveira D, Zeni J, Rigo E and Dallago R M 2016 In situ immobilization of Candida antarctica B lipase in polyurethane foam support *J. Mol. Catal. B Enzym.* **124** 52–61
[11] Zhao X, Qi F, Yuan C, Du W and Liu D 2015 Lipase-catalyzed process for biodiesel production: Enzyme immobilization, process simulation and optimization *Renew. Sustain. Energy Rev.* **44** 182–97
[12] Knežević Z D, Šiler-Marinković S S and Mojović L V 2004 Immobilized lipases as practical catalysts *Acta Period. Technol.* 151–64
[13] Mohammadi M, Habibi Z, Dezvarei S, Yousefi M and Ashjari M 2015 Selective enrichment of polyunsaturated fatty acids by hydrolysis of fish oil using immobilized and stabilized Rhizomucor miehei lipase preparations *Food Bioprod. Process.* **94** 414–21
[14] Leong T, Ashokkumar M and Kentish S 2011 The fundamentals of power ultrasound-A review
[15] Ensminger D and Bond L J 2011 *Ultrasonics: fundamentals, technologies, and applications* (CRC press)
[16] Hashim K S, Ali S S M, AlRifaie J K, Kot P, Shaw A, Al Khaddar R, Idowu I and Gkantou M 2020 Escherichia coli inactivation using a hybrid ultrasonic–electrocoagulation reactor *Chemosphere* **247** 125868
[17] Chakraborty N, Rani M and Juglan K C 2020 Ultrasonic Investigation of Dichloromethane With Ethanolamine At Various Temperatures *Plant Arch.* **20** 2825–9
[18] Khan N R and Rathod V K 2015 Enzyme catalyzed synthesis of cosmetic esters and its intensification: A review *Process Biochem.* **50** 1793–806
[19] Nirmala M J, Durai L, Rao K A and Nagarajan R 2020 Ultrasonic Nanoemulsification of Cuminum cyminum Essential Oil and Its Applications in Medicine *Int. J. Nanomedicine* **15** 795
[20] Santin C M T, Scherer R P, Nyari N L D, Dalla Rosa C, Dallago R M, de Oliveira D and Oliveira J V 2014 Batch esterification of fatty acids charges under ultrasound irradiation using Candida antarctica B immobilized in polyurethane foam *Biocatal. Agric. Biotechnol.* **3** 90–4
[21] Wang Z, Fang R and Guo H 2020 Advances in ultrasonic production units for enhanced oil recovery in China *Ultrason. Sonochem.* **60** 104791
[22] Moentamaria D, Muharja M, Widjaja T and Widjaja A 2019 Thermal stability and reusability of home-made co-immobilized lipase from Mucor miehei in polyurethane foam for the production of bio-flavor *IOP Conference Series: Materials Science and Engineering* vol 543 (IOP Publishing) p 12025
[23] Awang R, Ghazuli M R and Basri M 2007 Immobilization of lipase from Candida rugosa on palm-based polyurethane foam as a support material *Am. J. Biochem. Biotechnol.* **3** 163–6
[24] Chua L S, Alitabarimansor M, Lee C T and Mat R 2012 Hydrolysis of virgin coconut oil using immobilized lipase in a batch reactor *Enzyme Res.* **2012**
[25] Ferraz L I R, Possebom G, Alvez E V, Cansian R L, Paroul N, de Oliveira D and Treichel H 2015 Application of home-made lipase in the production of geranyl propionate by esterification of geraniol and propionic acid in solvent-free system *Biocatal. Agric. Biotechnol.* **4** 44–8
[26] Lopes D B, Fraga L P, Fleuri L F and Macedo G A 2011 Lipase and esterase: to what extent can this classification be applied accurately? *Food Sci. Technol.* **31** 603–13
[27] Murty V R, Bhat J and Muniswaran P K A 2002 Hydrolysis of oils by using immobilized lipase enzyme: a review *Biotechnol. Bioprocess Eng.* **7** 57–66
[28] Cui C, Tao Y, Li L, Chen B and Tan T 2013 Improving the activity and stability of Yarrowia

lipolytica lipase Lip2 by immobilization on polyethyleneimine-coated polyurethane foam *J. Mol. Catal. B Enzym.* **91** 59–66
[29] Garlapati V K and Banerjee R 2013 Solvent-free synthesis of flavour esters through immobilized lipase mediated transesterification *Enzyme Res.* **2013**
[30] Sharma A, Chaurasia S P and Dalai A K 2013 Enzymatic hydrolysis of cod liver oil for the fatty acids production *Catal. today* **207** 93–100
[31] Moentamaria D, Dewajani H, Chumaidi A, Nurmahdi H and Sinduwati C 2020 Heterogeneous biocatalyst: Polyurethane foam coating technique with co-immobilized lipase for bio-flavor production *IOP Conference Series: Materials Science and Engineering* vol 732 (IOP Publishing) p 12003
[32] El-Ghaffar M A A and Hashem M S 2010 Chitosan and its amino acids condensation adducts as reactive natural polymer supports for cellulase immobilization *Carbohydr. Polym.* **81** 507–16
[33] K Z, H L, P H and P W 2010 Study of ultrasound-promoted, lipasecatalyzed synthesis of fructose ester *Front. Chem. Eng. China* **4** 367–371
[34] Chisti Y 2003 Sonobioreactors: using ultrasound for enhanced microbial productivity *TRENDS Biotechnol.* **21** 89–93

IOP Conf. Series: Materials Science and Engineering 1073 (2021) 012002 doi:10.1088/1757-899X/1073/1/012002

Effect of temperature and Mg-Zn catalyst ratio on decarboxylation reaction to produce green diesel from kapok oil with saponification pretreatment using NaOH

A Chumaidi*, H Dewajani, M A Sulaiman, F Angestine, A Putri and S A Pravitasari

Jurusan Teknik Kimia Politeknik Negeri Malang, Malang, Indonesia

*achmad.chumaidi@yahoo.com

Abstract. Green diesel is an emerging option to solve the increasing consumption of diesel (fossil based fuel). It involves converting readily available vegetable oils and animal fats to alkane hydrocarbons that can be considered to be drop-in replacements for petroleum-based fuel component. This review journal investigate the results of diesel-like hydrocarbon fuel by catalytic thermal decarboxylation of Mg-Zn basic soap by saponification using NaOH from kapok oil. Specifically, it outlines a comparative assessment of the operating parameters of the temperature and mole ratio of Mg-Zn catalyst to select the appropriate path based on the desired conversion, yield and selectivity. The results of the literature study show that kapok oil can produce diesel equivalent biohydrocarbon products by reviewing the main composition of the kapok oil which containing linoleic acid. An increase in the temperature of the decarboxylation reaction can increase the conversion to green diesel products, n-heptadecane. The result from previous experiment showed that the highest conversion of decarboxylation of castor oil is 65% at 475°C. However, high temperatures can cause decreased selectivity due to the production of undesirable products like in the decarboxylation experiment of soybean oil which showed that the yield at 400°C is decreased to 25%. The mole ratio of Mg-Zn used in the decarboxylation reaction does not have a significant effect on the yield of green diesel products.

1. Introduction

Currently, the world's population is increasing which in turn has resulted rapid increase for the need of petroleum as fuel for transportation. Duc to increased demand of petroleum fuels, there is a rapid depletion in world's petroleum reserves. In addition, the environmental impact of petroleum fuels on global climate change and air quality continues to increase. The various problems above motivate researchers to seek alternative fuels from renewable sources. Biodiesel is a type of biofuel that has been established as a renewable fuel. However, biodiesel is a type of oxygenated fuel [1], therefore its use is very limited (about 15-20% by volume) as a blending agent for petroleum fuels. For example, a blended fuel consisting of 20% biodiesel and 80% petrodiesel is called B20 [2]. Whereas green diesel refers to liquid hydrocarbon fuels that do not contain oxygen in their molecular structure. Therefore, the nature of green diesel is similar to petroleum fuel which can be used in larger quantities, even up to 100% by volume [3]. The advantages of green diesel compared to biodiesel are that it produces a high heating value, a cetane number between 80-90 and has lower NOx emissions. According to Asikin Mijan, et al

[4] green diesel has good fuel properties compared to biodiesel, such as lower viscosity and better fuel stability.

Choosing kapok seed oil as raw material for green diesel is a wise step. This is because the oil is non-edible oil. Using edible oil as a raw material for making green diesel will cause problems, which are a food crisis in the future, an increase in the need for edible oil in the market, which will eventually increase the price of edible oil, and it is possible that many forests will be converted into edible oil field. So, it must used a raw materials that can't be consumed and it is a waste, one of the example is waste palm cooking oil. Kapok seed oil contains about 71.95% unsaturated fatty acids which causes kapok seed oil to go rancid easily, so it is not good to be developed as oil for cooking [5]. Hence, kapok seed oil is used as feed in green diesel production.

Green Diesel can be produced through several methods. The methods often used are thermal cracking, catalytic cracking and deoxygenation. Thermal cracking or pyrolysis and catalytic cracking are methods of breaking down triglyceride molecules with high molecular masses into lower molecular masses [6]. The difference between these two methods is the presence of a catalyst and the operating temperature used. The catalytic cracking method uses a catalyst and an operating temperature of 400-450°C [7]. The catalysts that are often used in this method are zeolite and mesoporous [8]. Meanwhile, the pyrolysis method does not use a catalyst and the temperature used is higher than the catalytic cracking method, which is 850°C [9]. The product of this cracking method is called bio-oil which cannot be used directly as a fuel due to the high level of oxygenated compounds (compounds with oxygen atoms in their molecular formula) in the form of phenols and their derivatives, such as esters, carboxylic acids, ethers, ketones and aldehydes. They are derived from the thermal degradation of cellulose, hemicellulose and lignin in biomass. This causes bio-oil to have undesirable properties, including: corrosive properties, low combustion efficiency and fuel instability which results in re-polymerization during storage [10,11].

Deoxygenation method is believed to be able to remove oxygenated compounds and convert them as carbon dioxide (CO_2) or carbon monoxide (CO) and water (H_2O). However, this method requires an external hydrogen gas source (H_2) to remove oxygenated compounds [12]. Deoxygenation can be done through different routes, namely hydrodeoxygenation, decarboxylation and decarbonylation [13]. The reaction steps that occur in hydrodeoxygenation, decarboxylation, and decarbonylation for saturated fatty acids in general can be illustrated as shown figure 1 below:

Liquid-phase reaction

1. Decarboxylation: $R\text{-}COOH \rightarrow R\text{-}H + CO_2$
2. Decarbonylation: $R\text{-}COOH \rightarrow R'\text{-}H + CO + H_2O$
3. $R\text{-}COOH + H_2 \rightarrow R\text{-}H + CO + H_2O'$
4. Hydrogenation: $R\text{-}COOH + 3H_2 \rightarrow R\text{-}CH_3 + 2H_2O$

R= Saturated alkyl group; R'= unsaturated alkyl group

Figure 1. Deoxygenation reactions on saturated fatty acids in general.

According to Sari [14], hydrodeoxygenation is a reaction between triglycerides and FFA with hydrogen which can eliminate oxygen which then forms water and n-paraffin which is often referred to as n-alkanes. Direct hydrodeoxygenation of vegetable oils and animal fats is a method that is frequently used and has been studied in detail.

However, the hydrodeoxygenation method is an expensive method because it requires a large amount of H_2 gas. Based on Cheah [15], in the hydrodeoxygenation method 1 mole of fatty acids requires 3 moles of H_2 molecules. In addition, this method reacts at high temperatures and pressures [16]. The involvement of H_2 gas under high pressure to suppress O_2 bound to oil and fat feed molecules or fatty acids to be released from its bonds and produce liquid biohydrocarbons as fuel. Based on the explanation above, it can be concluded that the hydrodeoxygenation method is considered less effective for the manufacture of green diesel.

Decarbonylation and decarboxylation methods are alternative methods for making green diesel. The decarboxylation method produces a by-product in the form of CO_2, while for decarbonylation it produces CO and water as a by-product [17]. Decarbonylation can take place well if you use a metal-based catalyst such as Ni/ZrO_2 and with the addition of H_2 gas, which for decarbonylation of one fatty acid requires one mole of H_2. Meanwhile, decarboxylation can take place well under inert conditions when the catalysts used are Pd and Pt. This is because the Pd and Pt catalysts are able to produce hydrogen in situ. However, the use of Pd and Pt catalysts requires a large amount of money [18]. However, currently researchers have conducted research so that green diesel can be produced through the decarboxylation process of metal soap using catalysts in the form of alkaline-earth metals and transition metals so that this method can be more affordable [13]. Therefore, the decarboxylation method is considered as an effective method for producing green diesel.

Decarboxylation of basic soaps (e.g., magnesium soap) into drop-in fuels, illustrated by the equation of reaction (1)

$$(R\text{-}COO)Mg(OH) \rightarrow RH + MgCO_3 \qquad (1)$$

2. Methodology

The general method to produce metal soaps derived from vegetable oil is metathesis process. The kapok seed oil is first mixed with hot ethanol in a glass batch reactor with a stirrer to get a uniform mixture. Then an aqueous solution of sodium hydroxide at 20% by weight is added to the mixture under continued stirring. Once the mixture has reached a firm consistency, a solution consist of magnesium–zinc acetate combination in distilled water is added to the mixture while stirring. The Mg-Zn basic soap produced by ion exchange is insoluble in water and can be easily separated from its soluble sodium counterparts. The saponification process usually takes about 1 hour. The produced basic soap from this process is filtered, washed with hot water and then dried in an oven at 70°C for at least 48 hours. The dried basic soap is subsequently used as the material for the decarboxylation process.

Mg-Zn basic soap decarboxylation was performed, i.e. destructive distillation at 350°C and atmospheric pressure in a glass batch reactor for 5 hours, without catalysts. The heat was supplied and maintained in the reactor by using an automatic system with thermocouples and electronic controls. Nitrogen was flushed into the reactor to remove the remaining air. When the temperature was reached, the reactant evaporated into gaseous phase products. These products were cooled in a condenser tube to get the liquid products, after which they were stored in a glass beaker. This liquid product is called green diesel and then the product is analysed with GC [13].

3. Result and discussion

3.1. Effect of temperature on conversion and yield of green diesel

The deoxygenation studies of triglycerides, fatty acids, and fatty esters are preferably carried out at temperatures ranging from 250 to 360°C.It has been reported by many researchers that the reaction temperature has a considerable effect on the deoxygenation process and the product selevtivity [19]. Research conducted by Setiadi and Susanto [20] shows a decarboxylation reaction through saponification with $Ca(OH)_2$ to produce hydrocarbons equivalent to diesel. The temperature variation of the decarboxylation reaction was carried out from 400°C - 475°C. Based on Table 1, it is known that an increase in reaction temperature can increase the conversion of n-heptadecane. N-heptadecane is the major hydrocarbons in diesel fuel. The largest conversion was obtained at a temperature of 475°C with the conversion of 65%.

In addition, research on the decarboxylation reaction with saponification pretreatment was also carried out by Setiadi and Rosul [21]. Cooking oil waste is reacted with $Ca(OH)_2$ to form a soap with an oil to base ratio of 1: 4.5 and a reaction time of 90 minutes. Temperature variations carried out at

400°C-460°C. Based on Table 1, the conversion increases with increasing temperature, but at 460°C the conversion decreases. The largest conversion was obtained for a temperature of 440°C with the conversion of 32%.

The main composition of kapok oil is linoleic acid. The approach that can be used in this decarboxylation reaction is to use a similar material which has a large linoleic acid content as well. Soybean oil is one of the vegetable oils with a high linoleic acid content with 52%. Research on the decarboxylation of soybean oil has been carried out by Pimenta et al. [22], based on the results of this study, it was found that the higher the temperature, the yield of n-heptadecane products decreased. The high temperature causes the formation of short chain hydrocarbons (C_8-C_{13}). In addition, the production of aromatic and saturated hydrocarbons increases with a higher reaction temperature, whereas the production of unsaturated, but non-aromatic compounds, decreases with increasing reaction temperature. This shows that the hydrogenation reaction, which is responsible for the formation of more saturated compounds and the aromatization reaction is favored by the higher temperature. According to Yang et al [23], an increase in temperature leads to an increase in the α / β cutting rate of carbon, which could explain the larger concentration of the smaller compound produced at 400 ° C.

Table 1. Effect of temperature on conversion and yield of green diesel.

No	*Feed*	Catalyst	Type of Reaction & Reactor	Time (hr)	Temp. (°C)	Result	Reference
1	Castor Oil	$Ca(OH)_2$	Soap Decarboxylation & Batch reactor	1,5	400	C_{17} Conversion = 17%	[20]
					425	C_{17} Conversion = 42%	
					450	C_{17} Conversion = 56%	
					475	C_{17} Conversion = 65%	
2	Waste Palm Cooking Oil	$Ca(OH)_2$	Soap Decarboxylation & Batch reactor	1,5	400	C_{17} Conversion = 17%	[21]
					420	C_{17} Conversion = 21%	
					440	C_{17} Conversion = 32%	
					460	C_{17} Conversion = 20%	
3	Soybean Oil	NiMo	Decarboxylation & Autoclave reactor	1,5	360	Yield C_{17} = 72%	[22]
					380	Yield C_{17} = 56%	
					400	Yield C_{17} = 25%	

3.2. Effect of Mg-Zn catalyst ratio on conversion and yield of green diesel

The amount of the catalyst used can affect a large percentage of the yield composition of the fraction hydrocarbon. Magnesium (Mg) catalyst is a catalyst for the decarboxylation process [24]. Meanwhile, Zinc (Zn) catalyst is a hydrocarbon isomerization catalyst [25]. The effect of the addition of Zn catalyst on the decarboxylation process can affect the order of hydrocarbons of the i-paraffin type. This is because Zn is known as a hydrogenation catalyst for unsaturated compounds [26] which can change the name of iso-alkanes as well [27]. i-paraffin group is a desired hydrocarbon in fuel because it has a lower cetane number than n-paraffin (between 10-80) which is very much needed in fuel. However, I-paraffin has a lower freezing point. The use of a lower amount of catalyst in the DO reaction leads to polymerization, which consequently increases the formation of aromatics and other unwanted products. Therefore, the optimal amount of catalyst plays an important role in determining the highest yield and selectivity of the desired hydrocarbon products [19].

Neonufa [13] has been researching using an Mg-Zn Catalyst (9:1 ratio) with palm stearin (saturated fatty acid, was approximately 63% by weight, while unsaturated fatty acid, was only approximately 26% by weight) at 350 °C and atmospheric pressure for 5 hours. In this study, the result were shown in Figure 1 shows that approximately 41.08 %-mole of liquid bio-hydrocarbons obtained from Mg-Zn basic soap decarboxylation were various 1-alkene molecules. The amount of Mg-Zn used will increase the formation of 1-alkene. That shows that there was a dehydrogenation reaction during the Mg-Zn basic soap decarboxylation or decomposition of long chain paraffin.

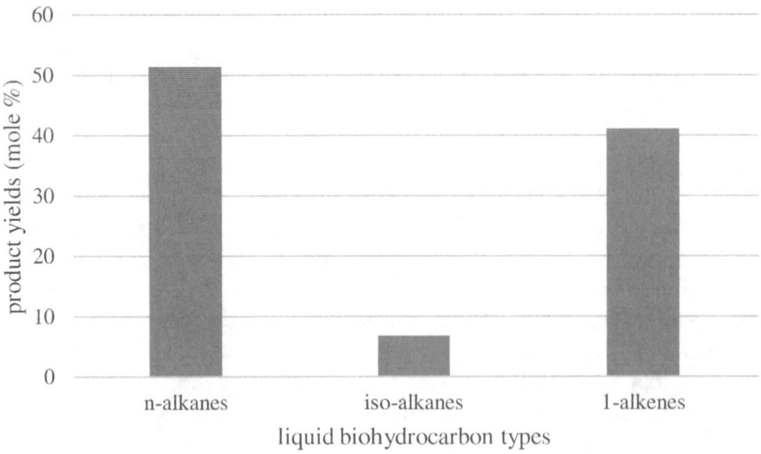

Figure 2. Distribute Biohydrocarbons fractions.

In addition, Neonufa [28] has been researching using an Mg-Zn catalyst (4: 1 ratio) with Pongamia pinnata seed oil (unsaturated fatty acid 47.45% mole, saturated fatty acid 25.42% mole) at 370 °C and atmospheric pressure for five hours. In this study, the results were shown in Figure 2 with the% yield of n-paraffin of 29.87%, while for i-paraffin of 50.78% and olefins of 19.35% mole. The amount of olefins formed proves that the more Mg-Zn mole ratio used will affect the formation of olefins in the decarboxylation reaction. The number of olefins formed at a 4:1 mole ratio is less than the formation of olefins in the nenoufa, 2017 which used the Mg-Zn 9: 1 mole ratio.

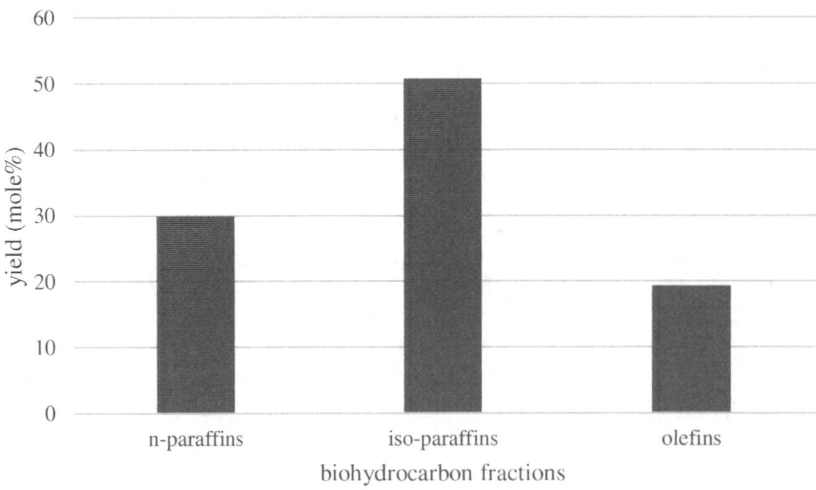

Figure 3. Distribute Biohydrocarbons Fractions.

Based on the two studies that have been done, it is known that the Mg-Zn ratio can affect the hydrocarbons produced. In addition, the raw materials used also play an important role in determining the hydrocarbons produced. Kapok oil contains more unsaturated fatty acids, according to Neonufa et al. [2] it is necessary to hydrogenate palm oil in the biohydrocarbon (renewable fuel) production process (Green diesel), because otherwise it will produce n-heptadecene hydrocarbons and will produce more unwanted products such as alkenes. In addition, the Mg-Zn mole ratio does not really affect the yield of

IOP Conf. Series: Materials Science and Engineering 1073 (2021) 012002 doi:10.1088/1757-899X/1073/1/012002

hydrocarbons because the raw materials used have a high content of unsaturated fatty acids. According to Fu et al. [29], polyunsaturated fatty acids can pass through the decarboxylation reaction pathway as shown in Figure 4 to produce n-heptadecane, a hydrocarbon equivalent to diesel.

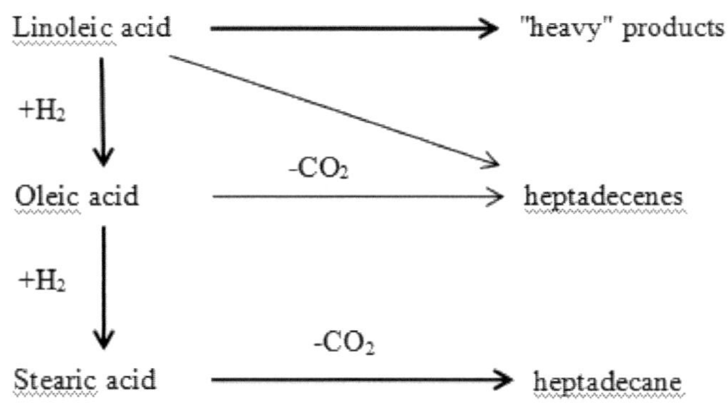

Figure 4. Reaction pathways for linoleic acid.

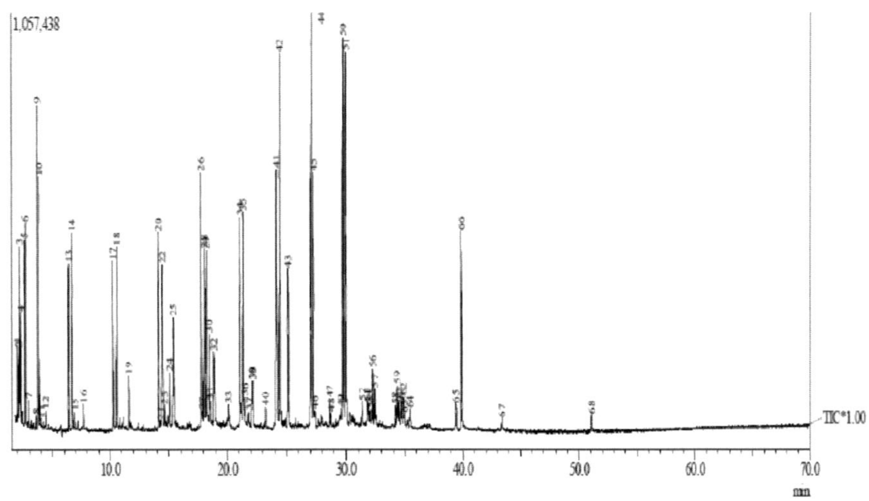

Figure 5. GC analysis green diesel.

4. Conclusion

An increase in the temperature of the decarboxylation reaction can increase the conversion of green diesel products, n-heptadecane and the Mg-Zn catalyst ratio used in the decarboxylation reaction does not have a significant effect on the yield of green diesel products.

Reference

[1] Kubičková I, Snåre M, Eränen K, Mäki-Arvela P and Murzin D Y 2005 Hydrocarbons for diesel fuel via decarboxylation of vegetable oils *Catal. Today* 106 197–200

[2] Neonufa G F, Pratiwi M, Soerawidjaja T H and Prakoso T 2018 High Selectivity of Alkanes Production by Calcium Basic Soap Thermal Decarboxylation *MATEC Web of Conferences* vol 156 (EDP Sciences) p 3035

[3] Neonufa G F. 2018. *Drop-in Fuel Production Technology of Green Diesel and Jet Fuel Types via Catalytic Thermal Decarboxylation of Basic Soap Based on Magnesium and Transition Metal Combination.* PhD Thesis (Bandung: Institut Teknologi Bandung)

[4] Asikin-Mijan N, Lee H V, Abdulkareem-Alsultan G, Afandi A, and Taufiq-Yap Y H 2016 Production of green diesel via cleaner catalytic deoxygenation of Jatropha curcas oil *Journal of Cleaner Production* **167** 1048-1059

[5] Ningsih E, Mirzayanti Y W, Niam A C, Febrianita P and Vangesti W 2019 Biodiesel dari Minyak Biji Kapuk (Ceiba Pentandra) Menggunakan Katalis Ca/Hydrotalcite *Reka Buana J. Ilm. Tek. Sipil dan Tek. Kim.* **4** 16–21

[6] Lestari S, Mäki-Arvela P, Beltramini J, Lu G Q M and Murzin D Y 2009 Transforming triglycerides and fatty acids into biofuels *ChemSusChem Chem. Sustain. Energy Mater.* **2** 1109–19

[7] Sirajudin N, Jusaff K, Yani S, Ifa L, Roesyadi A 2013 Biofuel production from catalytic cracking of palm oil *World Applied Sciences Journal* 26 67-71

[8] Buzetzki E, Sidorová K, Cvengrošová Z, Kaszonyi A and Cvengroš J 2011 The influence of zeolite catalysts on the products of rapeseed oil cracking *Fuel Process. Technol.* **92** 1623–31

[9] Hassan S N, Sani Y M, Aziz A R A, Sulaiman N M N and Daud W M A W 2015 Biogasoline: An out-of-the-box solution to the food-for-fuel and land-use competitions *Energy Convers. Manag.* **89** 349–67

[10] Şenol O İ, Viljava T-R and Krause A O I 2005 Hydrodeoxygenation of methyl esters on sulphided NiMo/γ-Al2O3 and CoMo/γ-Al2O3 catalysts *Catal. Today* **100** 331–5

[11] Bykova M V, Ermakov D Y, Kaichev V V, Bulavchenko O A, Saraev A A, Lebedev M Y and Yakovlev V A 2012 Ni-based sol–gel catalysts as promising systems for crude bio-oil upgrading: Guaiacol hydrodeoxygenation study *Appl. Catal. B Environ.* **113** 296–307

[12] Fisk C A, Morgan T, Ji Y, Crocker M, Crofcheck C, Lewis S A 2009 Bio-oil upgrading: guaiacol hydrodeoxygenation study *Appl Catal A-Gen 2009* 358 150-6

[13] Neonufa G F, Soerawidjaja T H and Prakoso T 2017 Catalytic and thermal decarboxylation of Mg-Zn basic soap to produce drop-in fuel in diesel boiling ranges *J. Eng. Technol. Sci.* **49** 575–86

[14] Sari E 2013 Green diesel production via catalytic hydrogenation/decarboxylation of triglycerides and fatty acids of vegetable oil and brown grease Wayne State University.

[15] Cheah K W, Yusup S, Singh H K G, Uemura Y and Lam H L 2017 Process simulation and techno economic analysis of renewable diesel production via catalytic decarboxylation of rubber seed oil–A case study in Malaysia *J. Environ. Manage.* **203** 950–61

[16] Arun N, Sharma R V and Dalai A K 2015 Green diesel synthesis by hydrodeoxygenation of bio-based feedstocks: Strategies for catalyst design and development *Renew. Sustain. Energy Rev.* **48** 240–55

[17] Abdulkareem-Alsultan G, Asikin-Mijan N, Mansir N, Lee H V, Zainal Z, Islam A and Taufiq-Yap Y H 2019 Pyro-lytic de-oxygenation of waste cooking oil for green diesel production over Ag2O3-La2O3/AC nano-catalyst *J. Anal. Appl. Pyrolysis* **137** 171–84.

[18] Miao C, Marin-Flores O, Davidson S D, Li T, Dong T, Gao D, Wang Y, Garcia-Pérez M and Chen S 2016 Hydrothermal catalytic deoxygenation of palmitic acid over nickel catalyst *Fuel* **166** 302–8

[19] Pattanaik B P and Misra R D 2017 Effect of reaction pathway and operating parameters on the deoxygenation of vegetable oils to produce diesel range hydrocarbon fuels: A review *Renew. Sustain. Energy Rev.* **73** 545–57

[20] Setiadi and Susanto, Andres 2009 *Reaksi Dekarboksilasi Minyak Jarak Pagar Untuk Pembuatan Hidrokarbon Setara Fraksi Diesel dengan Penambahan* $Ca(OH)_2$ (Depok: Universitas Indonesia)

[21] Rosul, Andika Jaya and Setiadi 2013 *Konversi Minyak Jelantah Sawit Menjadi Bahan Bakar Setara Diesel Melalui Reaksi Dekarboksilasi dengan Pretreatment Saponifikasi Menggunakan Kalsium Hidroksida* (Depok: Universitas Indonesia)

[22] Pimenta J L C W, Barreto R D T, dos Santos O A A and de Matos Jorge L M 2020 Effects of reaction parameters on the deoxygenation of soybean oil for the sustainable production of hydrocarbons *Environ. Prog. Sustain. Energy* **39** e13450

[23] Yang Y, Wang Q, Zhang X, Wang L and Li G 2013 Hydrotreating of C18 fatty acids to hydrocarbons on sulphided NiW/SiO2–Al2O3 *Fuel Process. Technol.* **116** 165–74

[24] Zhang A, Ma Q, Wang K, Liu X, Shuler P and Tang Y 2006 Naphthenic acid removal from crude oil through catalytic decarboxylation on magnesium oxide *Appl. Catal. A Gen.* **303** 103–9

[25] Pratiwi M, Neonufa G F, Prakoso T and Soerawidjaja T H 2018 The synthesis of magnesium soaps as feed for biohydrocarbon production *MATEC Web of Conferences* vol 156 (EDP Sciences) p 3001

[26] Matter P H and Ozkan U S 2005 Effect of pretreatment conditions on Cu/Zn/Zr-based catalysts for the steam reforming of methanol to H2 *J. Catal.* **234** 463–75

[27] Danuthai T, Jongpatiwut S, Rirksomboon T, Osuwan S and Resasco D E 2009 Conversion of methylesters to hydrocarbons over Zn-modified H-ZSM-5 zeolite catalyst *Catal. Letters* **132** 197–204

[28] Neonufa G, Prakoso T, and Neonufa N E M 2018 Production of Renewable Fuel Through Decarboxylation of *Pongamia pinnata L.* Seed Oils *International Conference on Climate Change, Biodiversity, Food Security and Local Knowledge* **1**

[29] Fu J, Lu X and Savage P E 2011 Hydrothermal decarboxylation and hydrogenation of fatty acids over Pt/C *ChemSusChem* **4** 481–6

[30] Hermida L, Abdullah A Z and Mohamed A R 2015 Deoxygenation of fatty acid to produce diesel-like hydrocarbons: A review of process conditions, reaction kinetics and mechanism *Renew. Sustain. Energy Rev.* **42** 1223–33

The development of five - level trays type biodryer reactor for processing of organic waste and vinasse mixture into solid recovered fuel

E Naryono*, S Susanto, D R Wulan, A Alqavid, I Nasukha and T I N Auliya

Chemical Engineering Department, State Polytechnic of Malang, Jalan Soekarno-Hatta No. 9 Malang, Indonesia

*eko.naryono@polinema.ac.id

Abstract. At Malang city, municipal solid waste (MSW) composition is mostly an organic waste that can utilize for Solid Recovered Fuel (SRF). Vinasse is a side product of the bioethanol industry, that mixed with MSW and dried by biodrying technique for solid fuel. The purpose of this research is to evaluate SRF characteristics of organic waste and vinasse mixture generate from biodrying. Research performed by drying the mixture of organic waste and vinasse at (0, 10, 20, 30, 40%) b/b ratio using biodrying reactor with five-level trays. This mixture inserted into the reactor tray, aerated at aeration flow rate 2 L/min from the compressor for nine days. The resulting SRF performed then characterized its moisture content, burning rate, and calorific value. The results showed that the biodryer reactor was successfully made with dimensions of 165 x 50 x 70 cm, consisting of five-level tray with a size of each level tray biodryer of 60 x 45 x 5 cm. The best SFR which has the highest heating value obtained from organic waste with vinasse 10%. It has a moisture content of 11.56%, the burning rate of 0.0135 g/sec, and calorific value of 3912.56 cal/g.

1. Introduction

MSW composition of cities in Indonesia is mostly organic waste (80%) [1]. The increasing amount of solid waste can cause environmental pollution problems. As a description in the year of 2012, Malang City generates MSW of 480 tons, while the current has risen to 600 tons every day [2]. Solid waste management mostly by piled on open land and only a small part (2%) used as compost. The open process of organic waste decomposition can contaminate the environment, producing gases (H_2S and CH_4) that cause a foul odour. It also produces leachate, which causes groundwater and surface water pollution [3].

The current organic waste technology consists of sanitary landfill, incineration, garbage briquette, bio-charcoal, biodrying, digestor anaerobic, and livestock feed pellets [3–5]. At Malang, utilization of organic solid waste processing products is mostly for compost. However, because the use of compost is still in small quantities, it is still required other MSW waste processing technology to maximize the amount of waste processed. One alternative method of processing waste environmentally friendly is biodrying because it can imposition the occurrence of leachate in organic waste. The biodrying product is stable dry garbage, and its can be used as solid fuel or SRF. The SRF already used in industrial areas, including power plants as co-firing fuel [6]. The prospective environmental benefits of using SRF as a fuel are enhanced carbon emissions [7], and decrease air pollutants owing to their low sulphur and nitrogen contents [8]. Biodrying (biological drying) is a biological and physical drying technique in the processing of garbage in bioreactors. The process of organic material decomposition using microorganisms is an exothermic biochemical reaction that can raise the temperature of the material [4]. The biodrying method relies on the heat of the biological activity of microorganisms and supports aeration to reduce the moisture content from organic waste [5,9]. The method can reduce water content by 40% to less than 20% [10].

Vinasse, wastewater generated by the bioethanol industry, is made of molasses [11]. Vinasse such as organic waste can pollute the environment so it should be processed first before it is discarded or utilized. Vinasse contains more than 90% of the water, and the rest are dried ingredients containing dead yeast, unfermented carbohydrates, unconverted sugars and various inorganic compounds [12]. Vinasse still contains many microorganisms that can utilize as a bioactivator for the fermentation process [13]. Also, the net calorific value (NCV) Vinasse is still relatively high 7800 J/g so it can apply as fuel [14]. The addition, vinasse in an organic waste can accelerate the fermentation process that impacts the garbage pile temperature, thereby accelerating the evaporation of water. Similarly, the garbage pile can utilize as a bulking agent for drying vinasse. The Bulking agent serves to increase porosity to increase aeration efficiency [15].

Waste processing products are biodrying called SRF with biodrying moisture content less to 20%. The effectiveness of the biodrying process application depends on the equipment, operation and control condition of the process operation and the initial conditioning of the dried material [4]. Preliminary conducted research showing that the waste of food leftovers when dried has a heating value of about 3400 kcal/kg, which is sufficient when used as fuel [2]. The calorific value on fuel is affected by moisture content. In the complete combustion process, the moisture content should be lowered to about 10% so that the calorific value reaches approximately 11600 J/g. The condition achieves by proper drying method [4].

The tray-type biodryer reactor builds from a rectangular-shaped reactor that stands upright. On the inside, five pieces of trays installed serially. This form of design is advantageous as it can use for dry out the garbage mixture and the laid-down vinasse to facilitate the arrangement of aeration airflow with contact material. The purpose of this research is to create biodrying reactor with a five-level tray reactor and evaluate the characteristics of moisture content, calorific value and burning rate from SRF generated from organic waste processing biodrying with a variation of the ratio of vinasse as substrate processing organic waste.

2. Material and methods

2.1. Materials

Organic waste consists of a mixture of leftover vegetables and food then added vinasse by (0, 10, 20, 30, 40%) b/b ratio. Two kg mixture of organic garbage and vinasse, is filled in each tray than dried for nine days using reactor biodrying with aeration flow rate 2L/min. The drying result using five-level biodrying reactor SRF, were analyzed for moisture, burning rate and the calorific value.

2.2. Apparatus

Biodrying Reactor consisting of five-level tray biodryer, bomb calorimeter merk PAAR 1241 EF, oven, and analytical balance.

2.3. Solid recovered fuel analysis procedure

2.3.1. Moisture content analysis

Moisture content determined using gravimetry method. Water content analyzed by weighing 1 g of samples in the porcelain cup then heated in an oven with a temperature of 110°C to a constant weight. Moisture content determined by the formula:

Moisture content (%) = $\frac{m2}{m1} \times 100$; with m1 = mass of samples (g); m2 = mass of samples after heated

2.3.2. *Analysis of burning rate*
Burning rate (g/sec) reported in the form of an SRF mass burned/combustion time. A total of 10 g of SRF weighed, then burned. The burning time and the remaining mass of combustion recorded. The burned SRF mass calculated based on the initial mass minus combustion mass. Besides, the flame burning colour observed.

2.3.3. *Analysis of the calorific value*
A total of 1 g of SRF weighed in order to get an accepted temperature rise of 2 litres of water in the oxygen bomb jacket (30-40 atm) for calorimetry measurements at the ambient temperature of 25°C.

3. Results and discussion
Biodrying is a drying technique assist with aeration that relies on the biological activity of microorganisms to reduce moisture content in organic waste, aims to eliminate as much water content in a short time compared to drying naturally. Organic takes a relatively long time in the process of achieving a dry condition of natural about 30-50 days with regular composting [10]. The biodrying process lasts between two to three weeks [9]. This research, the duration of the proses biodrying was faster for nine days with the addition of vinasse because, on vinasse, may contain microorganisms that can act as a bioactivator [14].

Bioactivator assists the fermentation process of organic waste that can increase the temperature of organic waste due to decomposition of simple organic compounds such as amino acids, glucose and organic acids that produce heat because of its exothermic reaction. A mixture of organic waste and vinasse can be dried successful by biodrying method in tray type reactor. The reactor can be used to dry organic waste and vinasse simultaneously. Vinasse can serve as a bioactivator and organic waste as a bulking agent while the bioreactor type tray increases the effectiveness of aeration. The combination of the fermentation process and the presence of aeration airflow caused the garbage drying process.

3.1. *Specification of biodryer reactor*
The Biodryer Reactor was successfully made with dimensions of 165 x 50 x 70 cm consisting of five trays with a size of each tray of 60 x 45 x 5 cm (Figure 1 B). The Biodyer reactor is connected with the compressor to supply aeration air. The Biodryer reactor is equipped with a flow meter to regulate air flow rate, inlet air tube, outlet air tube, leachate output pipe, five-level iron frame, and five biodryer reactor Racks (Figure 1 A).

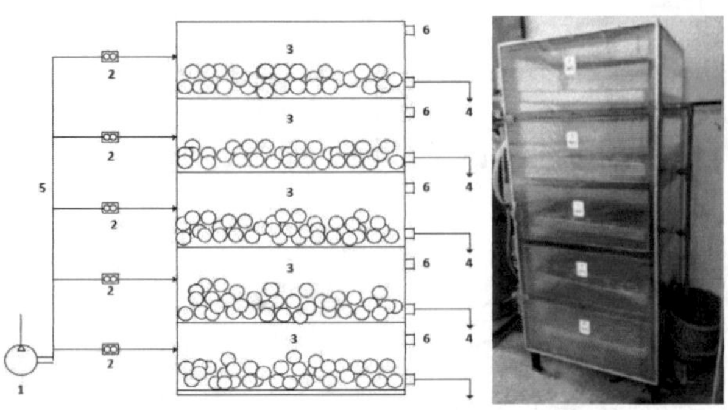

Figure 1. Biodryer reactor: **a**. biodryer reactor design scheme (1. Compressor, 2. Flowmeter, 3. Mix organic waste and vinasse, 4. leachate output pipe, 5. inlet air tube, 6. outlet air tube); **b**. Biodryer reactor with consisting of five-level tray Biodryer.

3.2. *Moisture content*
In this study, the biodrying process for nine days of organic waste with the addition of vinasse produces solid recovered fuel (SRF) with a moisture content of between 9.80 to 36.44% as shown in Figure 2.

This is appropriate because of the biodrying method capable of lowering the moisture content between 40 to ≤ 20% [11]. The higher addition of vinasse used in biodrying, the higher the moisture content of SRF. The most optimal addition of vinasse is at a ratio of 10%. But the moisture content in the SFR (11.65%) higher than that without the addition of vinasse (9.80%). This is because vinasse is a liquid waste that has high water content [12].

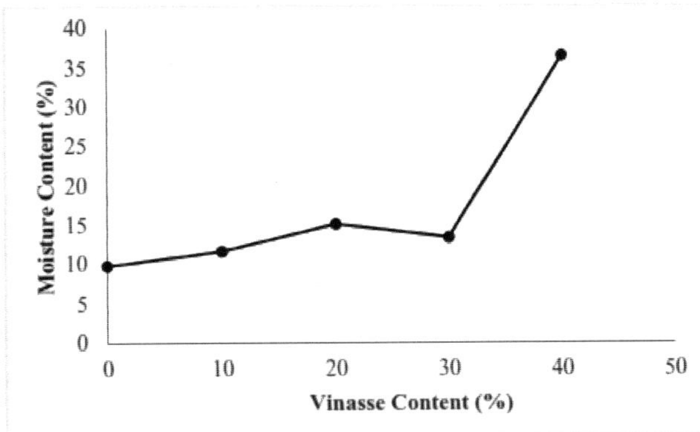

Figure 2. Moisture content of SRF with various vinasse content in biodrying process.

3.3. Burning rate

The rate of combustion is a measure of the energy released by combustion of fuel per unit time per area expressed in kilowatts per square meter [16]. In this study, the burning rate was determined by dividing the SRF burned mass and combustion time and is expressed in mass / time. The burning on SRF shown in figure 3.

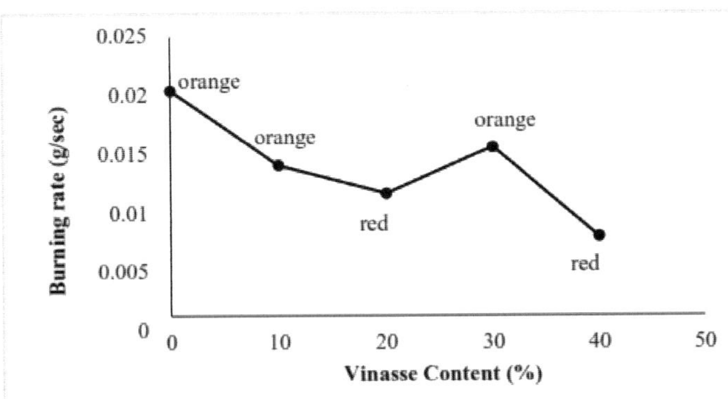

Figure 3. Burning rate of SRF with various vinasse content in biodrying process.

Generated SRF has a value of burning rate between 0. 0071 to 0. 0202 g/sec. The smallest burning rate value is in SRF generated from vinasse 40% with a red flame colour. The largest burning rate value is in SRF without vinasse addition. SRF generates from vinasse 10% has an orange coloured flame having a burning rate value slightly smaller (0.0135 g/sec) than the SRF without the addition of a vinasse (0.0202 g/sec) with an orange flame. The greater the burning rate, the greater the energy released during combustion, as well as the greater the caloric value of SRF.

3.4. Calorific value
Calorific value is the most important parameter to the fuel. Calorific value on SRF resulted from organic waste processing in the biodrying method with variations in the addition of the vinasse ratio shown in figure 4.

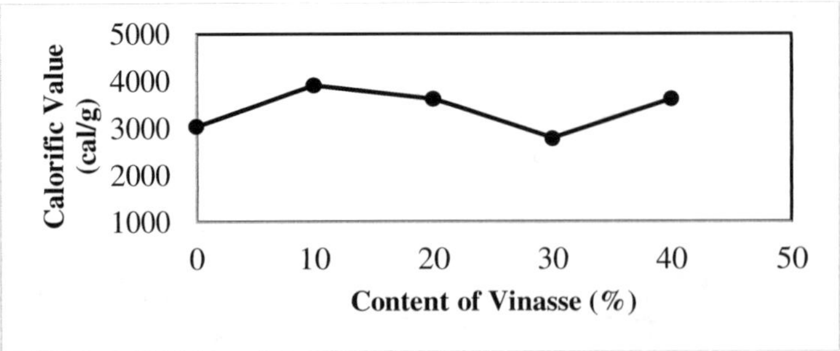

Figure 4. The calorific value of SRF with various vinasse content in biodrying process.

SRF generated has the calorific value between 2783. 43 to 3912. 56 cal/g. The largest calorific value is on SRF generate from vinasse 10%, and the smallest calorific value is on SRF generate from vinasse 30%. The vinasse addition can increase SRF caloric value, due its high caloric value (10-15 MJ / kg on a dry basis) [17]. However the greater the ratio of vinasse increases water content of organic waste, which can reduce the caloric value of the SRF. Thus, the optimum vinasse ratio in organic waste processing biodrying is 10% with SRF calorific value of 3912.56 cal/g or 16.37 MJ/Kg. The calorific value derived from bituminous coal is 26-28 MJ/Kg [18,19], and SRF class 3 calorific value based on the European Committee for Standardization, 2011 is > 15 KJ/Kg, indicate that the generated SRF is very potential to be used as a solid fuel.

4. Conclusion
The Biodryer Reactor was successfully made with five-level tray biodryer. SRF resulting from the processing of organic waste in biodrying method with variations added to the ratio of vinasse. The best SRF which has the highest heating value obtained from organic waste with vinasse 10%. It has a moisture content of 11.56%, the burning rate of 0.0135 g/sec, and heating value of 3912.56 cal/gram.

Acknowledgments
The authors would like to acknowledge The Ministry of Education and Culture of Republic Indonesia for financial support through "Penelitian DIPA Riset Terapan 2020" State Polytechnic of Malang.

References
[1] Sulistyorini L 2005 Pengelolaan Sampah dengan Cara Menjadikannya Kompos *J. Kesehat. Lingkung. Unair* **2** 77–85
[2] Naryono E, Prayito P and Santosa S 2019 *Pembuatan Briket Campuran Sampah Organik dan Sampah Plastik Sebagai Binder untuk Bahan Bakar padat* (Malang)
[3] Wahyono S 2001 Pengolahan Sampah Organik dan Aspek Sanitasi *J. Teknol. Lingkung.* **2** 113–8
[4] Naryono E and Soemarno S 2013 Pengeringan Sampah Organik Rumah Tangga *Indones. Green Technol. J.* **2** 61–9
[5] Dianda P, Munawar E, Briukhanov A Y, Vasilev E V, Himawanto D A and Arifin N A 2017 Bio-charcoal production from municipal organic solid wastes *IOP Conf. Ser. Mater. Sci. Eng.* **227** 1–6
[6] Gehrmann H, Seifert H and Beckmann M 2012 Usage of Solid Recovered Fuels in Power Plants *Chemie Ing. Tech.* **84** 1–15
[7] Séverin M, Velis C A, Longhurst P J and Pollard S J T 2010 The biogenic content of process streams from mechanical-biological treatment plants producing solid recovered fuel . Do the

manual sorting and selective dissolution determination methods correlate? *Biog. content Process streams from Mech. Treat. plants Prod. solid Recover. fuel. Do Man. sorting Sel. dissolution Determ. methods Correl.* **30** 1171–82

[8] Velis C, Wagland S, Longhurst P, Robson B, Sinfield K, Wise S and Pollard S 2012 Solid Recovered Fuel : Influence of Waste Stream Composition and Processing on Chlorine Content and Fuel Quality *Environ. Sci. Technol.* **46** 1923–31

[9] Sadaka S, Costello T, Sharara M and VanDevender K 2013 Partial Composting for Biodrying Organic Materials *Univ. Arkansas Div. Agric.*

[10] Nurul Fadlilah, Yudihanto G, Fadlilah N and Yudihanto G 2013 Pemanfaatan Sampah Makanan Menjadi Bahan Bakar Alternatif dengan Metode Biodrying *Tek. Pomits* **2** 289–93

[11] Velis C A, Longhurst P J, Drew G H, Smith R and Pollard S J T 2009 Biodrying for mechanical-biological treatment of wastes: A review of process science and engineering *Bioresour. Technol.* **100** 2747–61

[12] Amelia J R, Hasanudin U and Suroso E 2019 Potensi Biogas dari Proses Rekayasa Aklimatisasi Bioreaktor Akibat Perubahan Substrat pada Industri Bioethanol *J. Tek. Pertan. Lampung* **8** 224–33

[13] Braunbeck O A and Cortez L A B 2000 Sugarcane culture and use of residues *Industrial Uses of Biomass Energy* ed . Rosillo-Calle, S V Bajay and H Rothma (London: Taylor & Francis)

[14] Mulyanto A 2010 Air Limbah Industri Bioetanol Berbahan Baku Ubi Kayu sebagai Bahan Baku Produksi Biogas *J. Rekayasa Lingkung.* **6** 263–73

[15] Schopf N and Erbino P 2010 Thermal utilisation of vinasse as alternative fuel *Proc. Int. Soc. Sugar Cane Technol.* **27** 1–7

[16] Tran H C and White R H 1992 Burning rate of solid wood measured in a heat release rate calorimeter *Fire Mater.* **16** 197–206

[17] Dirbeba M J, Brink A, Zevenhoven M, DeMartini N, Lindberg D, Hupa L and Hupa M 2019 Characterization of Vinasse for Thermochemical Conversion—Fuel Fractionation, Release of Inorganics, and Ash-Melting Behavior *Energy & Fuels* **33** 5840–8

[18] Teixeira P, Lopes H, Gulyurtlu I, Lapa N and Abelha P 2012 Evaluation of slagging and fouling tendency during biomass co-firing with coal in a fluidized bed *Biomass and Bioenergy* **39** 192–203

[19] Wu H, Glarborg P, Frandsen F J, Kim D, Jensen P A and Sander B 2013 Trace elements in co-combustion of solid recovered fuel and coal Publication date: Trace elements in co-combustion of solid recovered fuel and coal *Fuel Process. Technol.* **105** 212–21

IOP Conf. Series: Materials Science and Engineering 1073 (2021) 012004 doi:10.1088/1757-899X/1073/1/012004

Modeling and simulation of CO_2 gas desorption process in promoted MDEA solution using packed column

A Ariani, A Chalim* and H Hardjono

Chemical Engineering Department, Politeknik Negeri Malang, Indonesia

*ariani.chalim@gmail.com

Abstract. Carbon dioxide (CO_2) gas at atmospheric pressure will tend to freeze, because of low temperature (-78°C), this results in frequent blockages in the piping section. In the process of synthesis of ammonia in the industry in the presence of a catalyst, CO_2 gas will be toxic because it can cover the active-side of the catalyst. The process of CO_2 gas absorption that is often used is reactive absorption using promoted solvent. The most commonly used solvent is a tertiary amine alkanol solution, one of which is a solution of Methyldiethanolamin (MDEA) using a piperazine catalyst. The output solution from the absorption of CO_2 gas is processed by desorption to regenerate the solvent. CO_2 gas desorption process is needed to save the use of solvents and catalysts. The purpose of this research is to do modeling and simulation on the process of desorption of CO_2 gas from a promote MDEA solution using a packed column. Modeling assisted software media with a rate-based model approach using numerical equations with the completion of Ordinary Differential Equation (ODE) - 45. The data obtained was simulated and validated by comparing experimental data.The research parameters are the temperature of the input solution desorber, type and concentration of catalyst. Increasing the temperature of the solution and the concentration of the catalyst will increase the percent removal of CO_2. This is caused by the effect of the temperature of the solution on the value of the reaction rate constant, and the type of catalyst influences the catalyst reactivity level of CO_2 gas. The results of simulation data validation with experimental data show the average deviation of 22.05%, 14.24% and 7.85%.

1. Introduction

Carbon dioxide (CO_2) gas can disrupt processes in the chemical industry because CO_2 gas binds to the air in the form of corrosive acid gas in the piping system. In the Liquified Natural Gas (LNG) industry, CO_2 gas must be removed because at very low temperatures (-78°C, 1 atm) CO_2 will freeze and can clog the piping system and damage the tubing of the main heat exchanger. In the fertilizer industry, CO_2 gas is a byproduct of the process of making ammonia, where CO_2 gas is toxic to the catalyst which can cover the active site of the catalyst. This can inhibit the performance of the catalyst in the ammonia synthesis process [1,2].

The separation process using the absorption method is always integrated with the desorption process because it requires regeneration of the absorbent solvent which is called the desorption process. The process of absorption and desorption is a unitary system.CO_2 gas is absorbed by the solvent in the countercurrent absorber to produce a CO_2-rich solution then goes to the desorber. Before entering the desorber, the solution exchanges heat with a lean solution of CO_2 that comes out of the stripper to the absorber. CO_2 gas is removed from the rich solvent in the desorber column. The

IOP Conf. Series: Materials Science and Engineering 1073 (2021) 012004 doi:10.1088/1757-899X/1073/1/012004

desorption process takes place in three parts, namely in the desorber inlet, the packed column and the reboiler. Optimal conditions in the desorber are very important because the energy demand in the process is 80% of the operational costs in the absorption-desorption system [3].

Energy requirements in the desorption process require industry to optimize operating conditions in order to maximize desorption efficiency. Therefore, modeling and simulating the CO_2 desorption process in a promoted MDEA solution is needed to plan and know the details of the operation of the system. This requires some thermodynamic data and equations including mass and heat transfer, energy requirements according to operating conditions in order to achieve optimal conditions. Rochelle [4] has conducted research on the desorption process using several types of solvents, the best results are using MDEA solution with piperazine promoter. And also in 2009 modeling and simulation of CO_2 gas desorption from potassium carbonate (K_2CO_3) solution promoted by piperazine (PZ) using a rate-based modeling approach [3].

In this research, modeling and simulation of CO_2 gas from methyldiethanolamine (MDEA) promoted by diethanolamine (DEA), Glycine (Gly) and Piperazine (PZ) was carried out using a rate-based modeling approach. Making mathematical models in this desorption process using Matlab with the temperature of the desorber entering solution is 353.15K, 358.15K and 363.15K; the promoter concentration is 3% by volume, 4% by volume and 5% by volume and the types of promoters are DEA, Gly and PZ. The CO_2 gas desorption from MDEA was 20% by volume with the addition of a promoter according to variables using a packed column. Then perform mathematical model validation by comparing the results of the calculation of prediction equation against the experimental data and theoretically study the effect of various process variables such as the inlet rich solution temperature, desorber column pressure, and inlet steam temperature on desorption efficiency.

2. Steam desorption

The feed of a stream of mixed solutions is heated and contacted with steam in a packed tower. Steam is injected through the bottom of the tower to provide heat of a vapor stream. Solution without CO_2 content exits through the bottom of the tower. The mixed solution is introduced through the top of the tower and steam containing organic matter exits through the top of the tower.

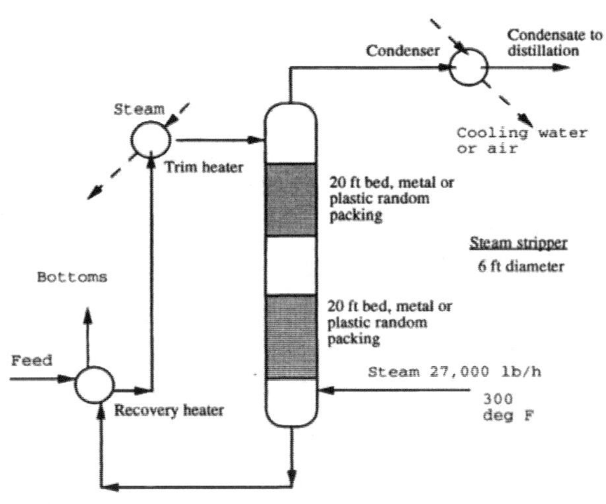

Figure 1. Schematic diagram of desorption proses.

3. Reaction kinetic

The reaction mechanism of CO2 gas desorption from MDEA solution using the PZ promoter is as follows [5-8].

$$R_3NH^+ + HCO_3^- \overset{k_{26}}{\Leftrightarrow} CO_2 + R_3N + H_2O \quad (1)$$

IOP Conf. Series: Materials Science and Engineering 1073 (2021) 012004 doi:10.1088/1757-899X/1073/1/012004

$$HCO_3^- \overset{k_{21}}{\Leftrightarrow} CO_2 + OH^- \tag{2}$$

For the CO_2 (MDEA-PZ-H_2O) system the values of k_{21} and k_{26} can be estimated using the Arhenius equation, so that the following correlation is obtained. This correlation is in accordance with the experimental results carried out by Bishnoi dan Rochelle [1].

$$k_{21} = 1.75 \times 10^4 \, exp\left(\frac{-8.75\times10^4}{R}\left(\frac{1}{T} - \frac{1}{298}\right)\right) \tag{3}$$

$$k_{26} = 1.55 \times 10^4 \, exp\left(\frac{-8.75\times10^4}{R}\left(\frac{1}{T} - \frac{1}{298}\right)\right) \tag{4}$$

Based on research by Pawlak, et al. [5] regarding the study of the absorption kinetics of CO_2 gas in activated MDEA solution, the results show that basically the overall reaction between CO_2 and MDEA is as follows.

$$CO_2 + R_3N + H_2O \overset{k_{26}}{\Leftrightarrow} R_3NH^+ + HCO_3^- \tag{5}$$

So for the kinetics of CO_2 gas desorption in MDEA solution, the overall reaction is as follows.

$$R_3NH^+ + HCO_3^- \overset{K_1,k_{21}}{\Longleftrightarrow} CO_2 + H_2O + R_3N \tag{6}$$

In an activated MDEA solution, the model of the stripping rate can be described as two parallel rapid pseudo-first-order reversible reactions and the overall reaction rate of CO_2 is.

$$r = \left(k_2 C_{am} + k_p C_p\right)\left(C_{CO_2,eq} - C_{CO_2}\right) \tag{7}$$

Where C_{am} and C_p are the concentrations of amine and piperazine. While the Arhenius plot value for k_p is given in the following equation.

$$k_p = 2.98\times10^{11} \exp\left(-6424/T\right) \tag{8}$$

4. Mathematical model

Rochelle, et.al [3] developed a mathematical model for CO_2 gas desorption in a PZ promoted K_2CO_3 solution in a packed column. The mathematical model was adopted and used in the CO_2 gas desorption process in a PZ promoted MDEA. Mass balance in each segment :

$$L_{ij-1} + \left(V_{ij+1} \times y_{i,j+1}\right) = L_{ij} + \left(V_{ij} \times y_{ij}\right) \tag{9}$$

Assuming the evaporation of amines is neglected, where $L_{ij-1} = L_{ij}$; $i = amine$, $j = segmen$

Total pressure on each segment :

$$P_{CO_{2,i}} + P_{H2O,i} = P_T \quad ; \quad P_{\mathbf{CO_2}} + P_{H2O} = P_T \tag{10}$$

CO_2 which transfer in the liquid phase :

$$N_{CO_2} = k_g' \times 1000 \times \left(P_{CO_2}^* - P_{CO_{2,i}}\right) \tag{11}$$

CO_2 which transfer in the gas phase:

$$N_{CO_2} = k_g \times 1000\left(P_{CO_{2,i}} - P_{CO_2}\right) + \left[\left(N_{CO_2} + N_{H_2O}\right)\frac{P_{CO_{2,i}} - P_{CO_2}}{P_T \ln\left(P_{CO_{2,i}}/P_{CO_2}\right)}\right] \tag{12}$$

5. Methods

The research was carried out theoretically by developing a mathematical model of the CO_2 gas desorption process from a PZ promoted MDEA solution on an industrial-scale packed column. The research steps can be seen schematically in Figure 2.

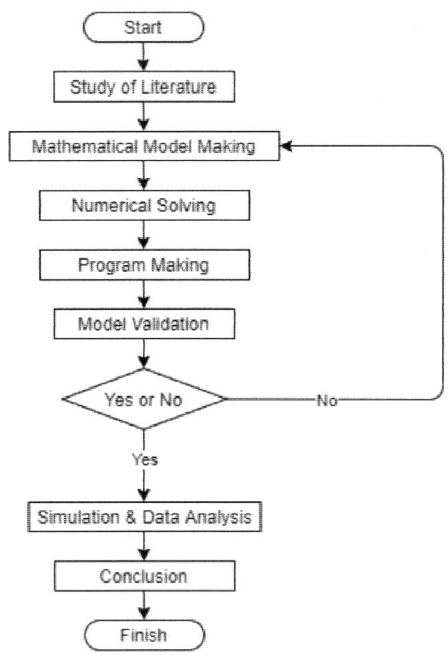

Figure 2. Schematic diagram of the research steps.

5.1 Development of mathematical model

The development of a mathematical model is calculating by making a differential mass balance assuming an isothermal system. Differential mass balance :

$$0 = LC_A - L(C_A + dC_A) - \bar{R}Aadz + rlAdz \tag{13}$$

$$0 = -L\frac{dC_A}{dz} - \bar{R}Aa + rlA \tag{14}$$

$$\overline{R_A}aA = rlA \text{ (assuming no dissolved gases)} \tag{15}$$

The numerical solution used in this study is to use the orthogonal collocation method. This process produces a series of simultaneous algebraic equations. The equation is then solved by the Newton Raphson method :

$$\%\,Efficiency = \left[1 - \frac{C_{A_{in}}}{C_{A_{out}}}\right] \times 100\% \tag{16}$$

Simultaneous solving of mathematical models of differential equations is made in the Matlab program, based on the Ordinary Differential Equation (ODE) – 45 method. With this method, the gas composition exiting the column is iterated until the inlet gas composition meets the predetermined conditions and convergence can be achieved.

Validation is done by comparing the predicted simulation data with the real data from the equipment design.

6. Results and discussion

The CO_2 gas desorption simulation program has been created using Matlab with the formulation of a differential mass balance involving four main parameters, consist of reaction kinetics, reaction equilibrium, gas solubility data, and mass – heat transfer. The system is created using several assumptions, as follows:

- Steady-state system
- Adiabatic operation
- Evaporation of solvent is negligible
- Presure across columns is constant

- The reaction in the liquid phase and component transferred at th interface are only CO_2 and H_2O

Validation of the simulation model that has been created using experimental data from the desorber column with the same system. Comparison of simulation data with experimental data after parameter fitting can be seen in Figure 3.

Validation is done by comparing the removal of each promoter with the same input variable, namely the temperature of the rich solution.

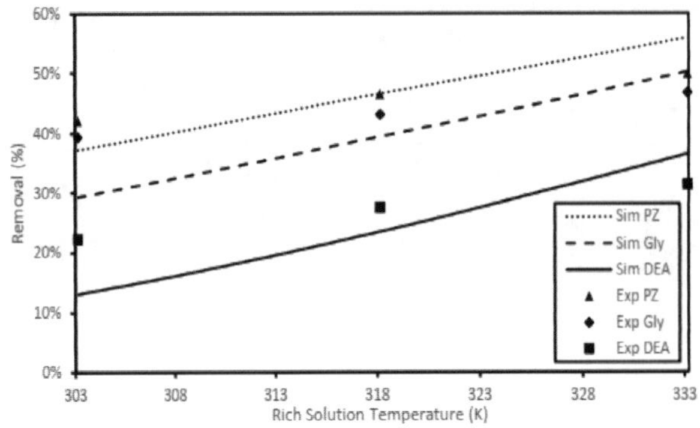

Figure 3. Comparison of simulation data with experimental data on each promoter based on the temperature of the rich solution to CO2 removal

Figure 3 shows the comparison of simulation data, namely lines and experimental data in the form of solid markers. Each promoter is marked with a different marker for PZ (▲), Glycine (♦) and DEA (■).

The determination of the CO2 removal value is based on the CO2 gas content in the liquid. In a system with DEA, Glycine and PZ promoters, the error is 22.05%, 14.24% and 7.85%, respectively. However, the reduction in error of each promoter can be due to the use of empirical correlation to estimate the thermodynamic properties and component displacements and the parameters associated with mass transfer that have been shown to be suitable [9].

6.1. Molar flow distribution of HCO3-, MDEAH +, MDEA, H_2O and CO_2 to column height

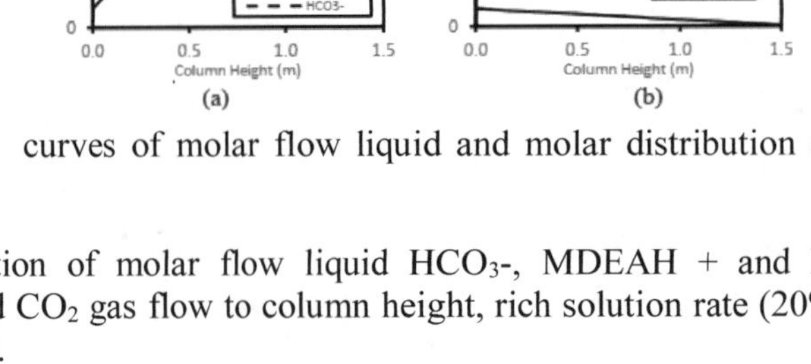

Figure 4. Distribution curves of molar flow liquid and molar distribution of gas phase to column height.

Figure 4 (a) Distribution of molar flow liquid HCO_3-, MDEAH + and MDEA. (b) The molar distribution of H_2O and CO_2 gas flow to column height, rich solution rate (20% MDEA 5% DEA) 0.6 L / min, pressure 2 atm.

Based on the desorption reaction above, MDEA's molar flow in the liquid phase will gradually decrease from 0.1038 kmol / m2s to 0.0102 kmol / m^2s starting from the bottom towards the top of the column and is called lean solution (solution with low CO_2 content).

Figure 4 (b) shows that as the desorption process progresses, the molar flow of CO_2 increases in contrast to H_2O which decreases from the bottom to the top of the column. This shows that high column affects the desorption process [10].

7. Conclusions

Based on the results of the simulation model, it can be concluded as follows:

- A mathematical model for the process of desorption of CO_2 gas from MDEA solution with promoters in the form of Piperazine (PZ), Glycine (Gly) and Diethanolamine (DEA) in a lab-scale packed column equipped with a condenser and a reboiler can be carried out.
- Validation of the mathematical model for the process of desorption of CO_2 gas from MDEA solution with promoters in the form of PZ, Gly and DEA in the packing column has been carried out from experimental data with the same system obtained an error of 22.05%. 14.24% and 7.85%.
- Several operating variables have been studied theoretically for their effect on column performance and it is stated that the model used can be accepted as a real process, although the accuracy still has errors.

Modeling and simulating the stripping process integrated with the absorption process to evaluate the overall performance of the CO2 removal unit.

Acknowledgements

Thanks are conveyed to UPT P2M for funding support for this research, as well as colleagues for their input to improve this writing.

References

[1] Bishnoi S and Rochelle G T 2002 Thermodynamic of Piperazine/Methylalkanolamine/Water/ Carbondioxide *Ind Eng. Chem. Res* **41** 604-612

[2] Zhang, Xiaowen, Zhang X, et al 2017 Reduction of energy requirement of CO_2 desorption from a rich CO_2-loaded MEA solution by using solid acid catalysts *Applied Energy* **202** 673-68

[3] Oyenekan and Rochelle G T 2009 Rate Modeling of CO2 Stripping from Potassium Carbonate Promoted by Piperazine *International Journal of Greenhouse Gas Control 3* 121-132

[4] Rochelle G T, Oyenekan, and Babatunde A 2007 Alternative Stripper Configurations for CO_2 Capture by Aqueous Amines *AIChE Journal* **53** 3144-54

[5] Pawlak H K and Chacuk A 2010 Kinetics of carbon dioxide absorption into aqueous MDEA solution *Ecological Chemistry and Engineering's* **17** 463-475

[6] Cullinane J T and Rochelle G T 2004 Carbondioxide absorbtion with aqueous potassium carbonate with promoter piperazine *Chemical Engineering Science* **59** 3619-30

[7] Cullinane J T 2005 *Thermodynamics and Kinetics of aqueous piperazine with potassium carbonate for carbon dioxide absorption* Ph.D. Dissertation, University of Texas at Austin

[8] Bishnoi S and Rochelle G T 2000 Absorption of carbon dioxide in aqueous piperazine/Methyldiethanolamine *AIChE Journal* **48** 2788–99

[9] Khan F M, Khrisnamoorthi V and Mahmud T 2011 Modeling Reactive absorption of packed columns for post-combustion Carbodioxide capture *Chemical Engineering Research and Design* **89** 1600-08

[10] Mahmoodi L and Darvishi P 2017 Mathematical modeling and optimization of carbondioxide stripping tower in an industrial ammonia plant *International Journal of Greenhouse Gas Control* **58** 42-51

IOP Conf. Series: Materials Science and Engineering 1073 (2021) 012005 doi:10.1088/1757-899X/1073/1/012005

Combination of biodiesel, glycerol, and methanol as liquid fuel

L Rubianto*, H P Sudarminto and S Udjiana

Chemical Engineering Department, State Polytechnic of Malang, Malang, East Java, Indonesia

*luchis.rubianto@polinema.ac.id

Abstract. Waste frying oil is a harmful liquid to human body and environment. However, it can be utilized as fuel. Transesterification of waste frying oil with methanol is presence of potassium hydroxide as catalyst produces fatty acid methyl ester (FAME) known as biodiesel as main product with glycerol as byproduct. Previously, several researches have been carried out concerning about utilization of both substances as renewable liquid fuel. In this research, three materials biodiesel, glycerol, and methanol which involved in transesterification process are combined in order to determine another renewable liquid fuel. Combination with composition 1:1:1 to 1:1:4 (v/v) is burned in an external combustion engine. Viscosity of glycerol is the main problem to form fuel for the engine. The result shows that the higher percentage of the methanol, the easier is the combustion. Composition by volume 1:1:4 of biodiesel, glycerol, and methanol respectively is the best mixture of liquid fuel for boiler burner.

1. Introduction

Environmental sustainability is a major issue recently. Determining a renewable source of energy is another to consider. Production of renewable and sustainable fuel from plants becomes more intensive since fossil fuel reserves continously decreasing. In Indonesia, several researches have already carried out in order to determine renewable fuel especially biodiesel that produced from palm oil. On contrary, the palm oil is also used for domestic and industrial daily purpose such as for frying. It is well known that frying oil is one of nine basic needs for Indonesian people. Consequently, using palm oil as raw material for biodiesel production is not a great option. Avoiding conflict of interest on palm oil utilization, this research attempts to make biodiesel from waste frying oil (WFO).

During frying process, at least 30% WFO is left and it is harmful to human health as well as to environment [1]. Alternatively, WFO must be converted into renewable fuel called biodiesel. Previous experiment showed that transesterification method could reach almost 100% conversion [2,3]. During frying process, palm oil is usually exposed to an extreme heat to more than 190 °C. At this temperature and higher, most organic nutritious substances are decomposed. Chemically, frying also cause acid number tends to increase. This is another problem that must be taken care prior transesterification process. Furthermore, this process may cause formation of several harmful and carcinogenic substances, for example aldehydes, acroleines, and peroxides [3]. Therefore, WFO must not be consumed. It is better to convert WFO into a liquid fuel by transesterification.

Transesterification of WFO and methanol using basic catalyst, such as potassium hydroxide (KOH), producing biodiesel as main product and glycerol as byproduct, with a chemical reaction as figure 1 follows.

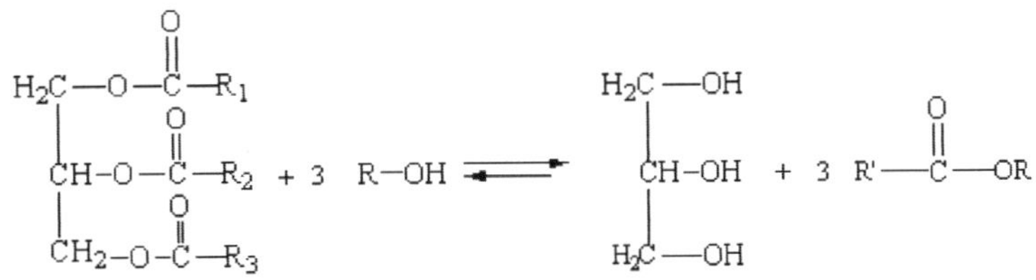

Figure 1. Transesterification of WFO.

Commonly, this biodiesel is called fatty acid methyl ester (FAME) and used as fuel to replace conventional diesel fuel that produced from fossils [3]. Unfortunely, glycerol produced at this point is slippery dark brown viscous liquid with strong odour. Therefore, this byproduct mostly discharged into environment through water system. Since glycerol is soluble in water (hydrophilic), it is potentially may cause water pollution [4]. Consequently, it is not in accordance with principle of environmentally friendly. Moreover, it is aginst sutainabilty development goals as stated by the UN on October 21st, 2015 [5].

In order to increase environment quality and determine sustainable renewable fuel, production of biodiesel must not create another waste. A biofuel production with zero waste must be taken into account. In other words, glycerol as byproduct of transesterificaton must be processed to create more valuable material. During transesterification, there are three major liquids involved, namely biodiesel, glycerol, and methanol. Objectives of this research is to replace ordinary diesel fuel comes from fossil, which almost runs out and unrenewable, by combining and mixing these liquids to create a renewable fuel.

2. Methodology

Transesterification is carried out by mixing waste frying oil, that has been filtered to separate solid contents, with methanol containing a strong base potassium hydroxide (KOH) as catalyst in a three necked roundbottom flask. Pretreatment is required in order to keep waste frying oil clear and solid free [6,7]. The treatment is also intended to keep acid number of waste frying oil less than 3 [2]. The flask was also equipped with an electric stirrer operated at 70 rpm to ensure the mixture reacts thoroughly. The mixture then was heated using an electric water bath with temperature was automatically kept around 64 °C, just below the boiling point of methanol. This process was carried out for at least one hour for complete reaction. After that, the mixture was allowed to cool and reach room temperature. In order to speed up cooling process, the flask was immersed in a cool water bath. Next, the mixture was transferred into a conical flask for separation. In the flask, two layers were formed naturally due to difference in density. Top layer was biodiesel and the bottom layer was glycerol [2]. The layers were separated by opening the valve near the bottom of the flask as shown on following Figure 2.

Figure 2. Layers of biodiesel and glycerol by transesterification.

IOP Conf. Series: Materials Science and Engineering 1073 (2021) 012005 doi:10.1088/1757-899X/1073/1/012005

Volume of each liquid layer, biodiesel and glycerol, was measured by means of a measuring cylinder. Physical properties of both liquid were separately analysed and calculated. They were viscosity, combustion energy, density, flash point, and molar mass [8]. Density was analyzed using a digital densitometer while viscosity was analyzed using viscometer refer to ASTM. Boiling point was analyzed using a distillation methods. Finally, the colour was analysed by means of a UV-Vis spectrophotometer.

Afterwards, a series of composition were prepared by mixing three liquids, namely biodiesel, glycerol, and methanol in four flasks. The compositions were 1:1:1, 1:1:2, 1:1:3, and 1:1:4 based on volume. Densities of these compositions were measured. Each composition then applied to a boiler burner replacing ordinary diesel fuel. Pressurized air was also added to the composition and controlled to create optimum flame. Appearance of the flame was closely monitored and compared to common flame that usually produced when burning ordinary diesel fuel. This procedure was repeated to all of the compositions.

3. Results and discussion

According to measurement at laboratory and calculation based on technical data, values of their physical properties are shown on Table 1.

Table 1. Physical properties.

Substance	Viscosity (Pa.s)	Energy (kJ/mol)	Density (kg/liter)	Auto Ignition (°C)	Molar Mass (gram/mol)
Biodiesel	36-50	37,270	0.880	>130	varies
Glycerol	1,412	1,654	1.261	393	92
Methanol	0.545	715	0.792	470	32

According to the experiment, every liter mixture of waste frying oil and methanol produces around 700 ml of biodiesel and 300 mililiter of glycerol [2]. Appearace of biodiesel is a clear yellow liquid while glycerol at this point is slippery and viscous with dark brown colour. In mixtures of those three liquids with compositions 1:1:1 to 1:1:4, the viscosiy and density gradually decrease as proportion of methanol increase as shown on Figure 3 and Figure 4.

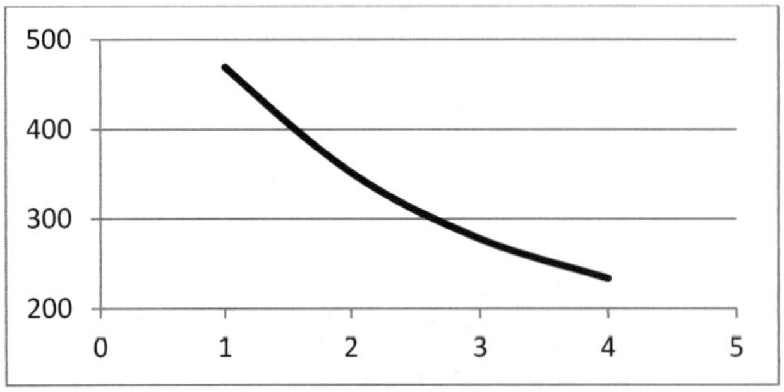

Figure 3. Viscosity of the composition decreasing as methanol percentage increase.

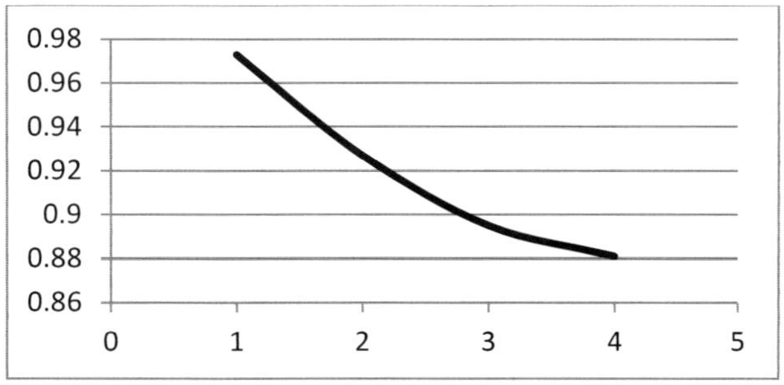

Figure 4. Density of the composition decreasing as methanol percentage increase.

Eventhough waste frying oil was filtered using filter paper and vaccum system to remove solid contents, the colour of biodiesel produced was still dark brown. Impurities in form of tiny particles cannot be removed. Physically, almost all impurities was formed at the bottom layer or glycerol layer. It means that biodiesel produced at the top layer was clear and colourless. After separation, glycerol was filtered one more time [9].

During application of these compositions on boiler burner, viscosity is the major obstacle on this experiment. Normally, using ordinary diesel fuel, the fuel pump works with medium pressure, the colour of flame is blueish red, and the amount of soot is small. However, viscosity of glycerol is very high compared to the other liquids, biodiesel and methanol. Since methanol has the lowest viscosity, proportion of this liquid is the key to have a liquid fuel with reasonable viscosity. Viscous liquid causing fuel pump works harder than usual. Furthermore, the viscous liquid, especially at composition 1:1:1, cannot form droplets easily even the pressure of the fuel pump is high. Consequently, quality of flame created using viscous liquid is unacceptable since the colour of the flame is blackish red with a large amount of soot.

Table 2. Description of composition, fuel pump pressure, colour of flame, and amount of soot.

Composition (v/v)	Fuel Pump Pressure	Colour of Flame	Amount of Soot
1:1:1	Very High	Blackish Red	Large
1:1:2	Very High	Blackish Red	Large
1:1:3	Very High	Blackish Red	Large
1:1:4	High	Blackish Red	Medium

4. Conclusion

Glycerol can be used as additional liquid to create a renewable fuel. The best liquid fuel for boiler burner is a mixture of biodiesel, glycerol, and methanol at proportion of 1:1:4 (v/v) respectively.

References

[1] Hayes K and Pramod K 2007 The Complex Interplay of Palm Oil Fatty Acids on Blood Lipids *European Journal of Lipid Science and Technology* **109** 453
[2] Rubianto L, Sudarminto, Atikah, and Soemarno 2013 Waste Frying Oil as Source of Alternative Energy Research Inventy *International Journal of Engineering and Science* **2**(8) 28-32
[3] Bautista L F, Vicente G, Rodrıguez R, and Pacheco M 2009 Optimisation of FAME Production from Waste Cooking Oil for Biodiesel Use Biomass and Bioenergy **30** 1-11
[4] Hudgens R D, Hercamp R D, Francis J, Nyman D A, Bartoli Y 2007 An Evaluation of Glycerin (Glycerol) as a Heavy Duty Engine Antifreeze/Coolant Base *SAE Technical Paper Series* **1**
[5] Sachs J, Schmidt-Traub G, Kroll C, Lafortune G, and Fuller G 2019 *Sustainable Development*

Report 2019 (New York: Bertelsmann Stiftung and Sustainable Development Solutions Network (SDSN))

[6] Dusenkova I, Malers J, and Berzina L 2015 Rheological Properties of Purified Illite Clays in Glycerol/Water Suspensions *Journal of Physics: Conference Series* **602** 012003

[7] Melero J A, van Grieken R, Morales G, and Paniagua M 2007 Acidic Mesoporous Silica for the Acetylation of Glycerol: Synthesis of Bioadditives to Petrol Fuel *Energy Fuels*

[8] Johnson D T and Taconi K A 2007 The Glycerin Glut: Options for the Value-added Conversion of Crude Glycerol Resulting from Biodiesel Production *Environmental Progress* **26**(4) 338-348

[9] Sims B 2011 Clearing the Way for Byproduct Quality: Why Quality for Glycerin is just as Important for Biodiesel *Biodiesel Magazine* **21**(3) 1782–1791

IOP Conf. Series: Materials Science and Engineering 1073 (2021) 012006 doi:10.1088/1757-899X/1073/1/012006

Biodegradable foam tray based on sago starch with beeswax as coating agent

N Hendrawati*, A A Wibowo, R D Chrisnandari and R Adawiyah

Chemical Engineering Department, State Polytechnic of Malang, Jalan Soekarno Hatta No 9 Malang, Jawa Timur, Indonesia

*nanik.hendrawati@polinema.ac.id

Abstract. Biodegradable foam is an original packaging as a styrofoam alternative. Sago starch is used to produce biodegradable foam because of its huge amount, low-cost, and high biodegradability. Starch as a prospectus foam material is sensitive to water due to its hydrophilic properties. Thus, improvement is necessary to boost its water resistance capability. Coating the foam with a hydrophobic material prevents water contact with the starch. In this work, Beeswax was used as a coating agent due to its hydrophobicity. The purpose of this work to determine the concentration and time of beeswax coating on the mechanical properties of biodegradable foam. Beeswax was varied at 0, 2, 4, 6, 8, and 10% wt concentration with coating duration process for 30, 60, 90, 120, and 150 seconds. Biodegradable foam was produced by baking process at 80°C for 1 hour 20 minutes. The results showed that the beeswax coating process increases the water resistance, although it does not significantly affect the tensile strength. Beeswax addition as much as 4% wt. for 150 seconds had improved the biodegradable foam properties as much as 1.92%, 73% in 28 days, and 0.09 MPa for water absorption capability, biodegradability, and tensile strength, respectively.

1. Introduction

Styrofoam or polystyrene is a type of container with superior properties for packaging such as high and low-temperature resistance, water resistance, lightweight, and durable. The Republic of Indonesia Food and Drug Supervisory Agency (BPOM) stated that styrofoam contains hazardous styrene substance. Moreover, styrofoam has a difficult nature decomposed by microorganisms; thus, the Environmental Protection Agency (EPA) established styrofoam as the fifth largest generated waste in the world during the manufacturing process. It is necessary to replace styrofoam with environmentally friendly foam polymer, such as starch-based biodegradable foam. Starch has a high biodegradability property, inexpensive, and abundant availability, thus it is suitable as raw materials for biodegradable foam instead of Styrofoam [1].

In general, biodegradable foam from natural starch can bind the water, so it is sensitive to a humid environment while being stored and used [2]. Hendrawati et al, study explained the water absorption on biodegradable foam from natural and modified starch as much as ± 38.26% and ± 10%. Biodegradable foam with modified starch is less economical than the natural one because it still requires chemicals and special treatment in the form of alcoholic acid hydrolysis [3]. Coating natural starch with other materials is another way to reduce the water absorption issue on biodegradable foam.

Several researchers had conducted the study regarding the coating method on biodegradable foam. Bergel et al applied chitosan as a coating substance on Thermoplastic Starch Foam (TPS) derived from potato, cassava, and maize starch using the thermo-pressing method [4]. The water absorptions of coated and uncoated TPS were 140% and 280%. The tensile strength of coated TPS derived from maize starch was 1.61 MPa. Moreover, Bergel et al studied TPS production from potato starch with 25%, 4%, 6% w/v polylactic acid as the coating agent [5]. The addition of 6% w/v polylactic acid mostly decreased the water absorption capability as much as four times from 2.24 to 0.56 *g* of water/*g* of solid. The uncoated TPS had 1.332 MPa of tensile strength value. Reis et al implemented beeswax as a coating agent of biodegradable tray derived from cassava starch. The concentration was varied as much as 0; 1; 2; 3 g wax/g solution [6]. The lowest water vapor permeability was 0.2×10^{-11} g/m.s.Pa with 3 g wax / g solution of beeswax. The value was increased without beeswax addition as much as 7.9×10^{-11} with 8.875 MPa of tensile strength value. Based on the description above, research on making biodegradable foam from natural starches becomes interesting by using beeswax as a coating material to increase the water-resistant capability because of its high hydrophobicity.

2. Materials and methods

2.1. Materials

Materials used were include sago starch with brand Sapapua, soy protein isolate (local), carrageenan (local), beeswax and chitosan were bought from the local grocery store in a food-grade label. Magnesium stearate (Macron), glycerol (Merck), acetate acid (Merck), citrate acid (MultiChem), polyvinyl alcohol(Merck), sodium bicarbonate (Pudak Scientific, 99%), ethyl alcohol and tween 80 were purchased in technical grade.

2.2. Biodegradable foam production

Sago starch was initially dried at 80 °C for 24 hours. The drying process removes the water content of the sago starch. Chitosan, 30% weight of starch, was dissolved in of 2% acetate solution until homogenous. Pure protein isolate 11.12% w/w starch, magnesium stearate 5.56% w/w starch, carrageenan 2.08% w/w starch, glycerol 16.67% w/w starch, chitosan solution, $NaHCO_3$:citric acid (1.3:1) of 12% w/w starch dissolved in 100 ml distilled water with rapid agitation speed (500 rpm) until homogeneous then the mixture is left for 15 minutes to maximize the process. Moreover, 36 g of dried sago starch was added to the mixture with slow agitation speed until homogenous and gelatinized at 72 °C until the gel was formed. The mixture was poured into the mold until half the baking pan's volume and baked in an oven at 80 °C for 80 minutes. The biodegradable foam was leftover in ambient temperature overnight. The foam was cut accordingly, coated, and analyzed.

2.3. Coating

The beeswax was weighted as much as 0, 2, 4, 6, 8, and 10 g. Beeswax was melted at 65 °C and mixed in 100 g of hot ethyl alcohol and tween 80 as much as 25% of beeswax weight. The solution was stirred at 200 rpm for 2 minutes. The biodegradable foam was coated by immersion in the beeswax solution for 30, 60, 90, 120, and 150 seconds.

2.4. Water absorption analysis

The biodegradable foam was cut 2.5 x 5 cm and weighted accordingly. The sample was immersed in the water for 1 minute. The sample, then, was wiped from water and weighted. Water absorption analysis was analyzed based on ABNT NBR NM ISO 535, 1999 standard. The equation to calculate the water absorption capability is presented below:

$$\text{Water absorption } (\%) = \frac{W1-W0}{W0} \; x \; 100\% \tag{1}$$

W_0 = Initial sample weight, g
W_1 = Final sample weight, g

IOP Conf. Series: Materials Science and Engineering 1073 (2021) 012006 doi:10.1088/1757-899X/1073/1/012006

2.5. Biodegradability analysis

The foam biodegradability was analyzed by calculating the weight difference of biodegradable foam buried in the soil for 28 days. The observation was conducted every seven days. The sample was firstly cleaned and weighted. The biodegradability was measured with this following equation:

$$Weight\ loss\ (\%) = \frac{W_0 - W_1}{W_0}\ x\ 100\% \qquad (2)$$

W_0 = Initial sample weight, g
W_1 = Final sample weight, g

2.6. Tensile strength analysis

The biodegradable foam was cut accordingly as standardized. The sample was analyzed using the Technical Association of the Pulp and Paper Industry (TAPPI) standard Number T404 to generate the samples' maximum stress. The tensile strength of the sample is calculated with the following equation:

$$\sigma = \frac{Fmaks}{A} \qquad (3)$$

σ = Tensile strength, MPa
F_{max} = Maximum stress, N
A = Surface area, mm^2

3. Results and Discussions

3.1. Water absorption analysis

Water absorption analysis is used to determine biodegradable foam resistance to water by calculating sample mass changes before and after the biodegradable foam is immersed in water [7]. As a styrofoam substitute, the biodegradable foam should have a similar characteristic; thus, water absorption capability in the study referred to ASTM C578 *"Standard Specification for Rigid, Cellular Polystyrene Thermal Insulation."* The water absorption result is presented in Figure 1.

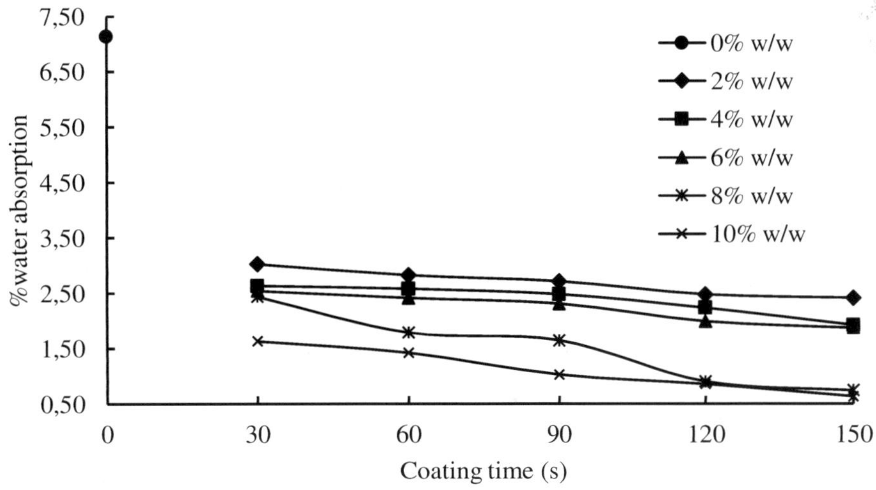

Figure 1. Water absorption capability of biodegradable foam.

Figure 1 shows that the higher the concentration and the longer the coating duration, the lower the water absorption capability. Water absorption value on biodegradable foam with beeswax coating was 0.63% – 3.02%, which is under ASTM C578; less than 4%. The uncoated biodegradable foam had a 7.14%

water absorption capability. This research is in line with Reis et al, who used beeswax as a coating material and applied it to biodegradable trays, It stated that the more beeswax content in the coating the transmission of water vapor permeability of biodegradable trays are getting smaller [6]. It is also following Li et al, who used beeswax solution for wood impregnation and explained that the longer the immersion time, the value of water absorption decreases [8]. It is caused by more beeswax solution trapped in the pores, which increases its hydrophobicity. Beeswax's hydrophobic nature is due to the content of esters, alcohols, alkanes, and a long chain of fatty acids so that it could limit the solubility and diffusion of water through it [9].Beeswax is classified as animal fat [10]. If animal fat is being cooled, the lost heat will slow the molecular movement. Thus, the distance between molecules becomes smaller and attractive forces between molecules called Van der Walls force to occur. The force caused the fatty acids in the fat molecules to be arranged in a row, overlapping, and bond together to form crystals [11]. This procedure is occurred by beeswax as a coating agent.

3.2. Biodegradability

Biodegradability test is used to determine the duration of biodegradable foam degraded in the environment. The standard biodegradability value in the study applies the European standards listed in EN 13432. The standard explains the degradation time of biodegradable polymers such as bioplastic will be degraded by 90% in 12 weeks or 30% in 28 days. The result of biodegradability analysis in this study is presented in Figure 2.

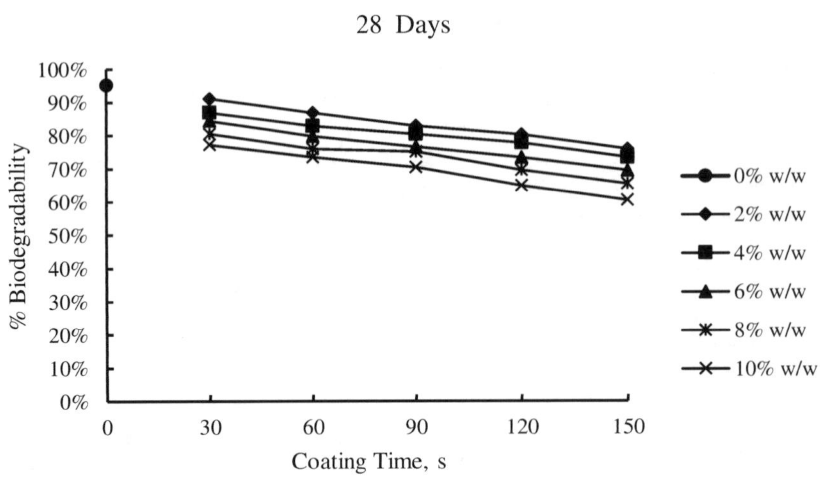

Figure 2. Biodegradability of the sample as coating time function for 28 days.

Figure 2 displays that the lower the concentration and the shorter the biodegradable foam coating time, the more easily degraded the foam were. In 28 days, the degradation of the foam was 95%, 76-91%, 73-81%, 69 -84%, 65-81%, 60-77% for 0, 2%, 4%, 6%, 8%, and 10% addition of coating agent, respectively. It is following Lewkittayakom et al who implemented beeswax as a coating agent for biodegradable plates; the higher the beeswax concentration, the lower the degradable property was [12]. As explained further, the longer the coating time, the hydrophobicity value of biodegradable foam is higher. It means that the water absorbed is less than others. In contrast, microorganisms need enough water for metabolism to degrade biodegradable foam [13]. Svagan et al stated that the moisture content in the biodegradable foam could reduce the strength and soften the cell walls, making it easier for microorganisms to degrade the biodegradable foam [14]. Soil conditions will also affect this analysis, according to Widyati Soils with clay textures are suitable for the growth of worms and soil organisms, and vice versa, sand-textured soils with low water holding capacity are not suitable for the organisms'

IOP Conf. Series: Materials Science and Engineering 1073 (2021) 012006 doi:10.1088/1757-899X/1073/1/012006

growth[15]. Based on Stres et al study, the moisture content of soil regulates the oxygen diffusion in which maximum value occurred at 50%–70% [16]. High water content reduces the rate of decomposition of organic matter due to the low oxygen supply. It is caused by diffusion in water is smaller (10^4 times) than in air [17] . Otherwise, low moisture reduces microbial activity by reducing microbial mobility.

3.3. Tensile strength

The tensile strength test on biodegradable foam is used to determine biodegradable foam's maximum capability when under load. The tensile strength value of biodegradable foam refers to the ASTM C578 standard "Standard Specification for Rigid, Cellular Polystyrene Thermal Insulation," 10-100 psi or 0.06-0.6 MPa. The tensile strength analysis of biodegradable foam with beeswax coating can be seen in Figure 3.

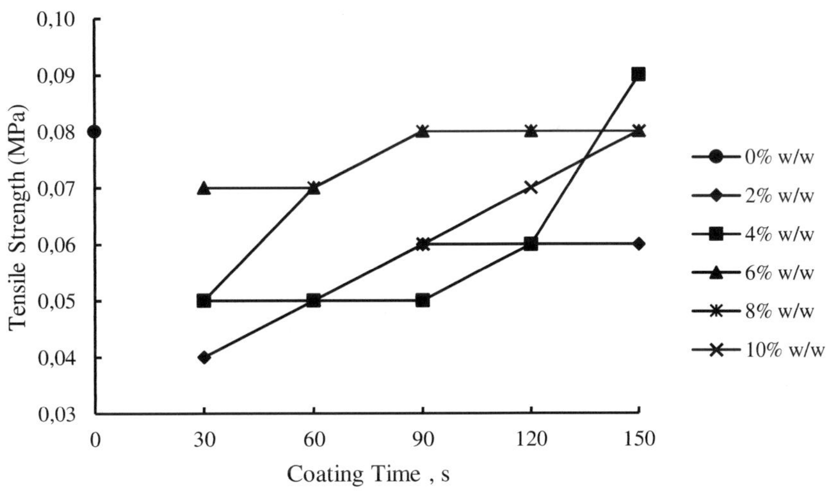

Figure 3. The relationship of coating time and tensile strength of the biodegradable foam.

Figure 3 shows the tensile strength of biodegradable foam with beeswax coatings were ranging from 0.04 MPa to 0.09 MPa. Meanwhile, the tensile strength of biodegradable foam without coating was 0.08 Mpa. The beeswax coating has no significant effect on the tensile strength value because it occurs on the surface, and beeswax is not an ingredient in the formulation [6]. Khwaldia indicated that the coating concentration using paraffin wax could not affect the tensile strength [18]. The tensile strength in Figure 3 shows fluctuate values because the unsteady of the stirring process on biodegradable foam affect the bubble production within the solution. Hou and Wang that the higher the agitation speed, the higher the bubble produced; thus, the porosity of biodegradable foam increases [19]. The increasing value of porosity and gas shows that biodegradable foam can expand excessively. It causes the polymer to be easily damaged or deformed, so the tensile strength of biodegradable foam is lower [20]. The polymer chain breakage is caused by the amount of gas production, which increases the polymer chain's pressure. Thus, the cell walls get thinner, and the lower the tensile strength of biodegradable foam is. Moreover, the study also used NaHCO3 in biodegradable foam production as a chemical blowing agent. Sodium bicarbonate generates carbon dioxide if it reacts with water or reacts with citric acid. The addition of a chemical blowing agent can increase the amount of porosity and pressure formed in the polymer to affect the tensile strength of the biodegradable foam.

4. Conclusions

The higher the concentration and the longer the coating duration, the lower the value of water absorption and biodegradability was. The concentration and time of coating had no significant effect on the tensile strength value. The best biodegradable foam was a coating concentration of 4% beeswax with a coating time of 150 seconds which has a water absorption value of 1.92%, with a value of biodegradability were 73-87% in 28 days., and tensile strength were 0.09 MPa.

Acknowledgments

The author would like to thank the State Polytechnic of Malang (DIPA Fund Number SP DIPA - Politeknik Negeri Malang Nomor SP DIPA – 042.01.2.401004/2020 for the financial support

References

[1] Etikaningrum 2017 *Pengembangan Berbagai Modifikasi Serat Tandan Kosong Sawit Pada Pembuatan Biofoam* (Institut Pertanian Bogor)

[2] Xu Y and Hanna M A 2005 Preparation and properties of biodegradable foams from starch acetate and poly(tetramethylene adipate-co-terephthalate) *Carbohydr. Polym.* **59** 521–9

[3] Hendrawati N, Dewi E N, Santosa S, Kimia J T, Malang P N and No J S 2019 Karakterisasi Biodegradable Foam dari Pati Sagu Termodifikasi dengan Kitosan Sebagai Aditif **3** 47–52

[4] Bergel B F, da Luz L M and Santana R M C 2017 Comparative study of the influence of chitosan as coating of thermoplastic starch foam from potato, cassava and corn starch *Prog. Org. Coatings* **106**

[5] Bergel B F, da Luz L M and Santana R M C 2018 Effect of poly(lactic acid) coating on mechanical and physical properties of thermoplastic starch foams from potato starch *Prog. Org. Coatings* **118** 91–6

[6] Reis M O, Olivato J B, Bilck A P, Zanela J, Grossmann M V E and Yamashita F 2018 Biodegradable trays of thermoplastic starch/poly (lactic acid) coated with beeswax *Ind. Crops Prod.* **112** 481–7

[7] Hendrawati N, Sa'Diyah K, Novika E and Wibowo A A 2020 The effect of polyvinyl alcohol (PVOH) addition on biodegradable foam production from sago starch *AIP Conf. Proc.* **2197**

[8] Properties S 2020 Effect of Beeswax Impregnation on the Dimensional Stability, Surface Properties, and Thermal Characteristics of Wood **15** 2181–94

[9] Kristo E, Biliaderis C G and Zampraka A 2007 Water vapour barrier and tensile properties of composite caseinate-pullulan films: Biopolymer composition effects and impact of beeswax lamination *Food Chem.* **101** 753–64

[10] Sarungallo Z L, Soekarto S T and Budijanto S 2002 KAJIAN PENURUNAN TITIK LELEH LILIN LEBAH (Apis cerana) DALAM PEMBUATAN MARGARIN OLES RENDAH KALORI I) [Study on Decreasing the Melting Point of Beeswax (Apis cerana) in the Production of Low -Calorie Margarine] *Teknol. clan Zndustri Pangan Univ. Negeri Papua* **Xm**

[11] Sinaga, Ricky Alberto, Budijanto Slamet W N *STUDY OF CPO ' S QUALITY AND ANALYSIS OF CHARACTERISTICS OF THE OLEIN AND STEARIN AS A RESULT OF PALM OIL FRACTIONATION Ricky Alberto Sinaga , Slamet Budijanto , Nur Wulandari Department of Food Science and Technology , Faculty of Agricultural Technology*

[12] Iewkittayakorn J, Khunthongkaew P, Wongnoipla Y, Kaewtatip K, Suybangdum P and Sopajarn A 2020 Biodegradable plates made of pineapple leaf pulp with biocoatings to improve water resistance *J. Mater. Res. Technol.* **9** 5056–66

[13] Foam B and Process B 2015 Pengaruh Penambahan Magnesium Stearat dan Jenis Protein Pada Pembuatan Biodegradable Foam Dengan Metode Baking Process *J. Bahan Alam Terbarukan* **4** 34–9

[14] Svagan A J, Berglund L A and Jensen P 2011 Cellulose nanocomposite biopolymer foam-hierarchical structure effects on energy absorption *ACS Appl. Mater. Interfaces* **3** 1411–7

IOP Conf. Series: Materials Science and Engineering 1073 (2021) 012006 doi:10.1088/1757-899X/1073/1/012006

[15] Widyati E 2013 Pentingnya Keragaman Fungsional Organisme Tanah Terhadap Produktivitas Lahan *Tekno Hutan Tanam.* **6** 29–37

[16] Stres B, Danevčič T, Pal L, Fuka M M, Resman L, Leskovec S, Hacin J, Stopar D, Mahne I and Mandic-Mulec I 2008 Influence of temperature and soil water content on bacterial, archaeal and denitrifying microbial communities in drained fen grassland soil microcosms *FEMS Microbiol. Ecol.* **66** 110–22

[17] Yan N, Marschner P, Cao W, Zuo C and Qin W 2015 Influence of salinity and water content on soil microorganisms *Int. Soil Water Conserv. Res.* **3** 316–23

[18] Khwaldia K 2010 Water Vapor Barrier And Mechanical Properties Of Paper-Sodium Caseinate And Paper-Sodium Caseinate-Paraffin Wax Films *J. Food Biochem.* **34** 998–1013

[19] Hou Q and Wang X 2017 The effect of PVA foaming characteristics on foam forming *Cellulose* **24** 4939–48

[20] M.A L and R.F.A Z 2020 *Pembuatan Biodegrdable Foam dari Pati Sagu Alami dan Terasetilasi dengan Penambahan NaHCO3 dan Asam Sitrat sebagai Blowing Agent* (Politeknik Negeri Malang)

IOP Conf. Series: Materials Science and Engineering 1073 (2021) 012007 doi:10.1088/1757-899X/1073/1/012007

The effect of fermentation time and addition of crude cellulase to concentration of bioethanol in bagasse fermentation

S Rulianah*, P Prayitno, A Indiastari and D Fatmawati

Chemical Engineering Department, State Polytechnic of Malang, Malang, Indonesia

*sri.rulianah@polinema.ac.id

Abstract. Bagasse is a solid waste from the sugar cane milling process in the sugar industry. Bagasse contains lignin, cellulose and hemicellulose, which through the fermentation process by *Phanerochaete chrysosporium* can produce crude cellulase, furthermore cellulase and other enzymes can be used in bagasse fermentation to produce bioethanol. The purpose of this study was to determine the effect of fermentation time and the addition of crude cellulase to the yield and concentration of bioethanol produced in bagasse fermentation. The research was carried out with the stages of the process: Preparation of raw materials (bagasse), Preparation of crude cellulase, Fermentation process, Product analysis (Bioethanol). The variables used in the experiment were fermentation time (96, 120, and 144 hours) and the addition of crude cellulase (10%, 20%, 30%, 40%, and 50% (v/v)). Analysis of crude cellulase activity using the DNS method, while the analysis of bioethanol concentration using the chromatography methods. The results showed that the maximum yield was 16.24% and the highest bioethanol concentration of 11.04% was obtained at the time of fermentation of 144 hours and the addition of crude cellulase by 50% (v/v).

1. Introduction

Bagasse is a solid waste from the sugar cane milling process in the sugar industry. Bagasse contains lignin, cellulose and hemicellulose, which through the fermentation process by *Phanerochaete chrysosporium* can produce crude cellulase, furthermore cellulase and other enzymes can be used in bagasse fermentation to produce bioethanol [1–3]. Bagasse (sugarcane bagasse) is included as lignocellulosic waste which can be used as raw material for making biofuels, such as bioethanol. Bagasse is biomass from vegetable. Biomass from vegetable source is renewable and has a great potential for energy generation. The composition of bagasse are cellulose 44.43 %, hemicellulose 22.90 %, and lignin 17.52 % [4]. On the other hand, *Phanerochaete chrysosporium* has the ability to produce cellulase enzymes from substrates containing cellulose and simultaneously produce enzymes that can break down lignin [5,6]. So there is no need for a delignification process to break down the lignin in bagasse [7].

In a study conducted by Rulianah et al [5] which used bagasse as raw material for crude cellulase production by *Phanerochaete chrysosporium* with fermentation times: 9, 11, 13, 15 and 17 days, and the concentration of bagasse added: 5, 6, and 7%. In this study, the highest crude cellulase activity was obtained at the longest fermentation time of 17 days. From the results of this study, further research was carried out for crude cellulase production by extending the fermentation time to 29 days, and the results of crude cellulases were used for bioethanol production using the Simultaneous Sacharification and Fermentation (SSF) process [3]. In this study, the highest fermented bioethanol content was 9.22% [3]. These results are quite good when compared to using crude cellulases from other types of microbes, such as research conducted by Hernawan et al [4], which obtained ethanol levels of 6.83 ± 0.07%, and research [8], using *Pachysolen tannophilus* MTCC 1077, which produced ethanol with a concentration of 9.15 g/L at an incubation time of 72 hours.

The purpose of this study was to determine the effect of fermentation time and addition of crude cellulase produced by *Phanerochaete chrysosporium* on bioethanol levels produced from the fermentation process by Simultaneous Sacharification and Fermentation (SSF) using *Saccharomyces cerevisiae*. This research is a development from previous research [3], by adding CMC to the production medium of crude cellulase from bagasse using *Phanerochaete chrysosporium*, with the hope of increasing the activity of the crude cellulase enzyme produced.

2. Methods and materials

2.1. Materials and equipment

2.1.1. The materials used in this research. Sugarcane baggase was obtained from a local of sugar factory Kebun Agung, Malang, East Java, Indonesia. Potato Dextrose Agar (PDA) Difco ™ , $CaCl_2$, $MgSO_4$. H_2O, CTAB (Cetyltrimetyl Ammonium bromide), glucose, acetic acid, sodium acetate, CMC (Carboxy Methyl Cellulose), and ethanol p.a were purchased from Merck. H_2SO_4 98% was purchased from Smart Lab, ethanol 70%, cotton fatty, pH universal, NPK, spirit, distilled water, and vitamin B1 were purchased from local market. dry *Saccharomyces cerevisiae* (fermipan), *Phanerochaete chrysosporium* was cultured in Bioprocess Laboratory Politeknik Negeri Malang, Nitrogen Limited Media (NLM) consisting of: NPK (nitrogen, phosphorus, potassium) as a fertilizer, $MgSO_4{\cdot}H_2O$, $CaCl_2$, glucose, and molasses from the Kebon Agung sugar factory, Malang, East Java.

2.1.2. The equipment used. Autoclave (Selecta), incubator shakers (New Brunswick Scientific CO., INC), incubator oven (Memmert), analytical balance (Yoke), vibrating screening (Retsch), spektofotometri UV-VIS (Perkin Elmer), furnace (Stuart), sentrifuge (Hitachi), water bath (Memmert), vacuum pump (Krisbow), oven (Memmert), microskop (Olympus), *Gas Chromatography* (GC) type HP 5890, measuring cup, erlenmeyer, measure pipette, pipette-sized (Schott Duran), micropipette (FinnpipetteTM F2), pH meter, RBF 250 mL dan 500 mL (Schott Duran), desiccator (unterdruck geprüft), and Whatman filter paper, and a set of distillation equipment (Preciso).

2.2. Pre-treatment of sugarcane baggase

Baggase was obtained from a local sugar factory Kebun Agung, Malang, East Java, Indonesia, dried up under sunlight, then reduced in size by a grinder to a size of ± 27-48 mesh, and the water level will be analysed.

2.3. Analysis water content of sugarcane bagasse

Weighing bagasse as much as 1 gram, then dry it in the oven for ± 1 hour at a temperature of 100-105°C. After completion of the oven, the bagasse is cooled in a desiccator for ± 30 minutes. Then weighed. This treatment is repeated until a constant mass is obtained. Water content of sugarcane baggase is calculated by the following formula:

$$\text{Water content (\%)} = \frac{\text{initial weight (g)} - \text{final weight (g)}}{\text{initial weight (g)}} \times 100\% \qquad (1)$$

Determination of levels of ADF, NDF, Hemicellulose, lignin and cellulose using the Van Soest method [7].

2.4. Production of crude cellulase
Crude cellulase production is carried out by making NLM consisting of NPK 6 g/L, $MgSO_4.7H_2O$ 0.5 g/L, vitamin B1 0.001 g/L, $CaCl_2$ 0.1 g/L, and glucose 10 g/L, dissolved in a pH buffer of 5 to 1 litter, then 330 mL of NLM are put into each 1 L erlenmeyer, add 7% (w/v) of bagasse, 0.5% (w/v) of CMC, and molasses of 0.4% (v/v). This media is sterilized at 121°C for 30 minutes. The sterilized media was cooled, then added inoculum *Phanerocahete chrysosporium* (10% (v/v) of the total volume). The inoculum contains 8.51 x 106 spores/mL. Then the incubation was carried out for 30 days, at a temperature of 35 °C and a shaker incubator speed of 150 rpm.

After the incubation process is complete, the results are filtered in a vacuum, separated between the supernatant and the biomass. Then the supernatant is analysed by the DNS method to determine the activity of crude cellulose enzymes.

2.5. Bioethanol production using Simultaneous Saccharification and Fermentation (SSF) Process
Ethanol production from bagasse media was carried out by making a Nutrient Medium consisting of 10 g/L (NH_4) $2PO_4$, 0.5 g/L $MgSO_4.7H_2O$, 10 g/L glucose and 10 g/L yeast extract. Then dissolved with aquadest up to 1 L. The production of bioethanol is done using a brown bottle. In each bottle, 8% bagasse of the media volume and 100 ml Nutrient Medium were added, then sterilized at 121 °C for 30 minutes in autoclave. After being sterilized, crude cellulase was added to the media according to the specified variables, namely 10%, 20%, 30%, 40%, and 50% (v/v). Then, added dry *Saccharomyces cerevsiae* (fermipan) directly as much as 2% of the media volume (100 mL). After that, it was incubated (fermented) according to the specified variables, namely 96, 120 and 144 hours with a temperature of 30 °C. After the fermentation process, the fermented products were distilled at 100 °C to separate the liquid and solids, then the distillate results were measured, their mass and volume, and the ethanol content were analysed using Gas Chromatography (GC) type HP 5890.

3. Results and discussions

3.1. The results of the initial and final analysis of the bagasse, before and after fermentation
The results of the initial analysis of bagasse content using the Van Soest method [7] are shown in Table 1. In the table it is known that bagasse used as raw material has hemiseulose, cellulose and lignin levels that are different from previous researchers [9]. This difference occurs due to the levels of bagasse from different areas, different moisture content and different analysis methods, causing the results of different levels of lignin, cellulose and hemicellulose.

Table 1. Chemical compositions of the sugarcane bagasses before and after fermentation.

Component	Composition of bagasse before fermentation (%)	Composition of bagasse after fermentation (%)
Water	10,3	10,22
ADF	48,23	17,46
NDF	77,84	5,94
Hemicellulose	29,56	7,79
Cellulose	32,03	1,51
Lignin	15,67	0,39
Insoluble Ash	0,53	10,22

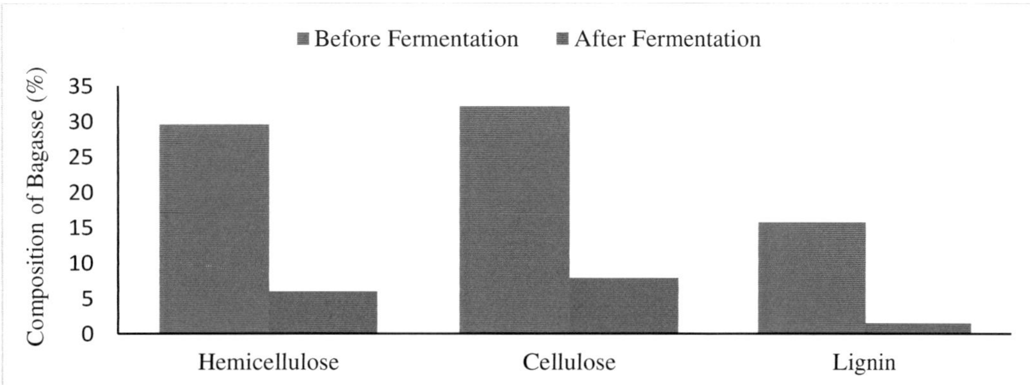

Figure 1. Hemicellulose, cellulose, and lignin contained in bagasse, before and after fermentation.

Based on Figure 1 above, there was a decrease in the levels of lignin, cellulose, and hemicellulose in bagasse after fermentation for 30 days. This was caused by the lignocellulose degradation process carried out by *Phanerochaete chrysosporium*. *P. chrysosporium* produces laccase enzymes, lignin peroxidase (LiP), and manganese peroxidase (MnP) which are used to degrade lignin. LiP activated by H_2O_2 can catalyze non-phenolic and phenolic lignin compounds. Meanwhile, MnP acts as a breaker of phenolic lignin units by oxidizing Mn^{2+} to Mn^{3+} [10,11] . Based on research conducted by Yakin et al [12] using the kakako fruit peel substrate, the enzyme activity of the *Phanerocahaete chrysosporium* fungus LiP and MnP was 0.22 U/mL and 0.09 U/mL, the activity of the enzymes LiP and MnP of *Pleurotus ostreatus* fungi. amounted to 0.13 U/mL and 0.05 U/mL, while the enzyme activity of the *Schizophyllum commune* mold LiP and MnP was 0.16 U/mL and 0.06 U/mL with fermentation time of 7 days. *Phanerochaete chrysosporium* also produces endoglucanase, exogluconase, and β-glycosidase enzymes [13–15]. These enzymes play a role in degrading cellulose into glucose. The endogilanase enzyme randomly attacks the β- (1,4) glycosidic bonds from the amorphous region and randomly cuts the internal bonds of the glycosidic chain. The exoglucanase enzyme plays a role in removing cellobiose from the ends of the free chain. The β-glycosidase enzyme acts to increase the hydrolysis of cellulose by converting cellobiose to glucose [16]. Meanwhile, hemicellulose has a complex oligomeric structure and branching pattern, making it difficult to degrade [17]. *Phanerochaete chrysosporium* mold produces the enzyme xylanase which acts to break down xylan into oligosaccharides. This causes a decrease in hemicellulose levels after fermentation.

In this study, there was a decrease in the levels of the bagasse components used. Decreased levels of bagasse content, namely 78.81% ADF, 77.57% NDF, 79.91% hemicellulose, 75.65% cellulose, and 90.36% lignin. According to Rulianah et al [5] shows that the best fermentation conditions are obtained at the addition of 0.8% CMC for 21 days with a decrease in lignin levels by 69.22% and a decrease in cellulose levels by 75.26%, on the fermentation of patchouli leaves. While the research conducted by Boonyuen et al [18], using the microorganisms *Aspergillus flavus* and *Penecillium citrinum* to degrade bagasse, obtained a decrease in lignin by 9.21%, cellulose by 9.08%, and hemicellulose 21.03% with fermentation time for 90 days. When compared with other studies, the decrease in bagasse content in the study was greater. This shows that using *Phanerochaete chrysosporium* can reduce levels of lignin, cellulose and hemicellulose higher than other types of fungi, such as *Aspergillus flavus*, *Penecillium citrinum*, and *Trichoderma viride*.

IOP Conf. Series: Materials Science and Engineering 1073 (2021) 012007 doi:10.1088/1757-899X/1073/1/012007

3.2. The effect of fermentation time and addition of crude cellulase to concentration of bioethanol in bagasse fermentation

The bioethanol fermentation process is carried out using *Saccaromyces cereviceae*. *Saccaromyces cereviceae* is a type of yeast in yeast that can convert sugar into other products in the form of alcohol (ethanol). The yeast produces invertase and zimase enzymes. Invertase enzyme functions to break down polysaccharides and sucrose which have not undergone the hydrolysis process into glucose (monosaccharide). While the zimase enzyme functions to convert glucose (monosaccharide) into alcohol products (ethanol) [19]. The following is the reaction for the formation of ethanol:

$$C_6H_{12}O_6 \xrightarrow{S.\ cereviceae} 2\,CH_3CH_2OH + 2CO_2 \quad (2)$$

Glucose Ethanol

In this study, ethanol was prepared using the SSF method. The SSF method, which is a process in which cellulose hydrolysis and sugar fermentation are carried out simultaneously or simultaneously in one fermenter vessel [3,4,20–22]. In this study, the SSF method was carried out at 30 °C, 8% (w/v) substrate concentration, 10-50% (v/v) crude cellulase concentration and 2% (w/v) *Saccharomyces cereviceae* concentration. The fermentation process was carried out under anaerobic conditions for 96, 120 and 144 hours.

The results of bagasse fermentation to produce bioethanol using the SSF method, using crude cellulase produced by *Phanerochaete chrysosporium*, and yeast type *Saccaromyces cereviceae* are shown in Figure 2 and 3. This figure shows that the longer the fermentation time is up to 144 hours, the higher the bioethanol level produced, and it will increase the yield of bioethanol against bagasse. The figure also shows that the greater the amount of crude cellulase added up to 50%, it will increase the bioethanol content produced and will increase the yield of bioethanol against bagasse. Because the more amount of crude added will speed up the hydrolysis process, so that more sugar is available to be converted into bioethanol. Cellulase enzymes that hydrolyse cellulose have increased the amount of glucose so that *Saccharomyces cereviceae* will ferment glucose in a larger amount and produce higher levels of bioethanol as a result of its fermentation. However, at certain concentrations the rate of cellulose hydrolysis by enzymes will reach an optimal point so that the enzyme can no longer hydrolyse cellulose beyond its optimal point.

The fermentation time has a direct effect on the ethanol content produced. In this fermentation, fermentation time is closely related to the anaerobic growth of *Saccaromyces cereviceae*. In anaerobic conditions, *Saccharomyces cereviceae* tends to produce metabolite products in the form of enzymes that can catalyze glucose into bioethanol. It can be seen in Figure 2 that the ethanol content from 96 hours to 144 hours fermentation has increased, this is because up to 144 hours, *Saccharomyces cereviceae* is still alive and still produces enzymes that function as biocatalysts in the process of bioethanol formation. Based on the results of the research, the highest ethanol content was obtained at the addition of 50% crude and a fermentation time of 144 hours, which was 11.04% with a yield of 16.24%. According to research conducted by Silva [23], it shows that using a strain of *Kluyveromyces marxianus* as much as 30 FPU and a fermentation time of 72 hours using the SSF method from bagasse yields the highest ethanol concentration of 2.92%.

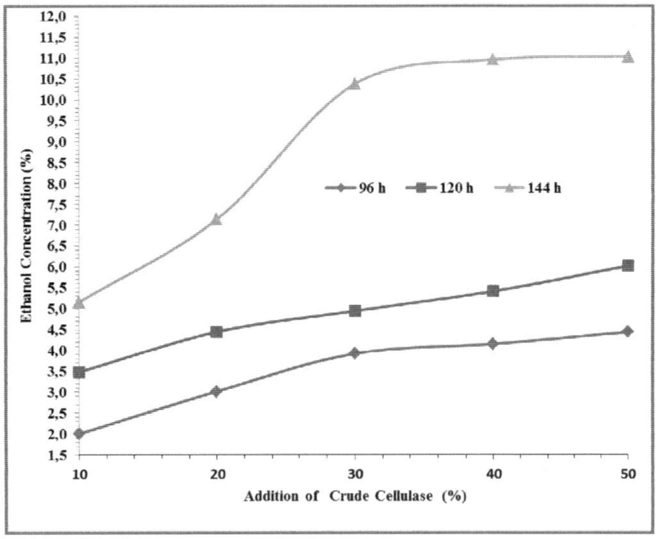

Figure 2. The relationship between the addition of crude cellulase and fermentation time to the ethanol content produced in the SSF method.

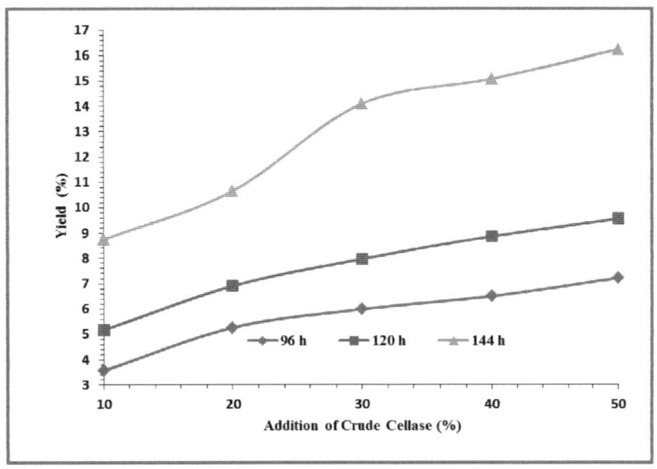

Figure 3. The relationship between the addition of crude cellulase and fermentation time to the yield ethanol.

Meanwhile, research conducted by Oktaviani [24], showed that the addition of 50 FPU commercial cellulase enzyme (Meicellase) and a fermentation time of 72 hours using the SSF method from alkaline delignified bagasse resulted in the highest ethanol concentration of 2.264%. Hashmi et al [25] obtained ethanol levels of 42.24 g/L equivalent to 4.224%. Hernawan et al [4] used the enzyme cellulase and combination of cellulase enzymes and β-glucosidase in the SSF process, resulting in bioethanol levels 5.87 ± 0.78% and 6.83 ± 0.07% respectively. If the results of this study are compared to the four previous studies above, it can be seen that the ethanol content in this study is much higher. This is due to the longer fermentation time and the cellulase enzyme used from the *Phanerochaete chrysosoporium* fungus, which is able to hydrolyse lignocellulose to glucose, which is higher than other fungi.

Meanwhile, bioethanol yield is directly proportional to bioethanol content, the higher the bioethanol content, the higher the yield produced. In this study, obtained higher results than previous studies [3], because in this study the media for crude cellulase production was added with CMC.

4. Conclusions

The results showed, the longer the fermentation time was up to 144 hours, and the higher the addition of crude cellulase from Phanerochaete chrysosporium to 50%, the higher the levels and yield of bioethanol produced. The best operating conditions were obtained when the addition of crude cellulase was 50% and the fermentation time was 144 hours, the ethanol content was 11.04% and the ethanol yield was 16.24%.

Suggestions for further research are to develop crude cellulase production technology that can increase the concentration of crude cellulase (crude cellulase activity) and develop bioethanol production from biomass (lignocellulose) raw materials.

Acknowledgements

The authors gratefully acknowledge the financial support from State Polytechnic of Malang who have provided financial support through the DIPA research fund. Besides that, the researcher thanked the students and staff who collaborated for this research.

Reference

[1] Rulianah S, Sindhuwati C and Prayitno 2019 Produksi Crude Selulase dari Limbah Kayu Mahoni Menggunakan Phanerochaete chrysosporium *J. Tek. Kim. dan Lingkung.* **3** 39

[2] Rulianah S, Irfin Z, Mufid M and Prayitno P 2017 Produksi Crude Selulase dari Bahan Baku Ampas Tebu Menggunakan Kapang Phanerochaete chrysosporium *J. Tek. Kim. dan Lingkung.* **1** 17

[3] Rulianah S, Prayitno, Hendrawati N and Nafisa K N 2020 Production of bioethanol from bagasse with a simultaneous saccarification and fermentation (SSF) process using crude cellulase from Phanerochaete chrysosporium *AIP Conf. Proc.* **2197**

[4] Hernawan H, Maryana R, Pratiwi D, Wahono S K, Darsih C, Hayati S N, Poeloengasih C D, Nisa K, Indrianingsih A W, Prasetyo D J, Jatmiko T H, Kismurtono M and Rosyida V T 2017 Bioethanol production from sugarcane bagasse by simultaneous sacarification and fermentation using Saccharomyces cerevisiae *AIP Conf. Proc.* **1823**

[5] Rulianah S, Profiyanty H S, Maryanty Y and Irawan B 2017 The effect of fermentation time and addition of CMC to decrease of lignin content in fermentation of patchouli leaves using Phanerochaete chrysosporium *Adv. Sci. Lett.* **23** 5666–8

[6] Suharti P H, Rulianah S, Maryanty Y, Irawan B, Chen F and Tsai M J 2017 The degradation of cellulose in the fermentation of patchouli leaves using Phanerochaete chrysosporium *Adv. Sci. Lett.* **23** 5669–71

[7] Rulianah S, Prayitno, Sindhuwati C, Ayu D R A and Sa'diyah K 2020 Penurunan Kadar Lignin pada Fermentasi Limbah Kayu Mahoni Menggunakan Phanerochaete chrysosporium *J. Tek. Kim. dan Lingkung.* **4** 81

[8] Bhatia L and Johri S 2016 Optimization of Simultaneous Saccharification and Fermentation Parameters for Sustainable Ethanol Production from Sugarcane Bagasse by Pachysolen tannophilus MTCC 1077 *Sugar Tech* **18** 457–67

[9] Sutikno, Marniza N S 2015 Pengaruh Perlakuan Awal Basa dan Hidrolisis Asam Terhadap Kadar Gula Reduksi Amaps Tebu *J. Teknol. Ind. Has. Pertan.* **20** 65–72

[10] Thanh Mai Pham L, Eom M H and Kim Y H 2014 Inactivating effect of phenolic unit structures on the biodegradation of lignin by lignin peroxidase from Phanerochaete chrysosporium *Enzyme Microb. Technol.* **61–62** 48–54

[11] Pham L T M and Kim Y H 2014 Accelerated degradation of lignin by lignin peroxidase isozyme H8 (LiPH8) from phanerochaete chrysosporium with engineered 4-O-methyltransferase from clarkia breweri *Enzyme Microb. Technol.* **66** 74–9
[12] Yakin E A, Bachruddin Z, Utomo R and Millati R 2017 Effect of Lignoselulolitic Fungus To Enzimatic Activity, Fiber Frction, and Digestibility on Fermentation Process of Cocoa POD *Bul. Peternak.* **41** 250
[13] Asina F N U, Brzonova I, Kozliak E, Kubátová A and Ji Y 2017 Microbial treatment of industrial lignin: Successes, problems and challenges *Renew. Sustain. Energy Rev.* **77** 1179–205
[14] Pérez J, Muñoz-Dorado J, De La Rubia T and Martínez J 2002 Biodegradation and biological treatments of cellulose, hemicellulose and lignin: An overview *Int. Microbiol.* **5** 53–63
[15] Singh D and Chen S 2008 The white-rot fungus Phanerochaete chrysosporium: Conditions for the production of lignin-degrading enzymes *Appl. Microbiol. Biotechnol.* **81** 399–417
[16] Subsamran K, Mahakhan P, Vichitphan K, Vichitphan S and Sawaengkaew J 2019 Potential use of vetiver grass for cellulolytic enzyme production and bioethanol production *Biocatal. Agric. Biotechnol.* **17** 261–8
[17] Kumneadklang S, Larpkiattaworn S, Niyasom C and O-Thong S 2015 *Bioethanol Production from Oil Palm Frond by Simultaneous Saccharification and Fermentation* vol 79 (Elsevier B.V.)
[18] Boonyuen N, Manoch L, Luangsa-ard J J, Piasai O, Chamswarng C, Chuaseeharonnachai C, Ueapattanakit J, Arnthong J and Sri-indrasutdhi V 2014 Decomposition of sugarcane bagasse with lignocellulose-derived thermotolerant and thermoresistant Penicillia and Aspergilli *Int. Biodeterior. Biodegrad.* **92** 86–100
[19] Novia, Windarti A and Rosmawati 2014 Pembuatan Bioetanol dari Jerami Padi dengan Metode Ozonolisis – Simultaneous Saccharification and Fermentation (SSF) *J. Tek. Kim.* **20** 38–48
[20] Dahnum D, Tasum S O, Triwahyuni E, Nurdin M and Abimanyu H 2015 Comparison of SHF and SSF processes using enzyme and dry yeast for optimization of bioethanol production from empty fruit bunch *Energy Procedia* **68** 107–16
[21] Jayus J, Suwasono S and Wijayanti I 2017 Produksi Bioetanol Secara SHF dan SSF Menggunakan Aspergillus niger, Trichoderma viride DAN New Aule Instant Dry Yeast Pada Media Kulit Ubi Kayu *J. Agroteknologi* **11** 61
[22] de Araujo Guilherme A, Dantas P V F, Padilha C E de A, dos Santos E S and de Macedo G R 2019 Ethanol production from sugarcane bagasse: Use of different fermentation strategies to enhance an environmental-friendly process *J. Environ. Manage.* **234** 44–51
[23] Silva G M, Giordano R L C, Cruz A J G, Ramachandriya K D, Banat I M and Wilkins M R 2015 Ethanol production from sugarcane bagasse using SSF process and thermotolerant yeast *Trans. ASABE* **58** 193–200
[24] Oktaviani M, Fajriutami T and Hermiati E 2016 Produksi Etanol dari Ampas Tebu Terdelignifikasi Alkali melalui Proses Sakarifikasi dan Fermentasi Serentak *Seminar Nasional Inovasi Dan Aplikasi Teknologi Di Industri* pp 45–51
[25] Hashmi M, Shah A A, Hameed A and Ragauskas A J 2017 Enhanced production of bioethanol by fermentation of autohydrolyzed and C4mimOAc-Treated sugarcane bagasse employing various yeast strains *Energies* **10**

IOP Conf. Series: Materials Science and Engineering 1073 (2021) 012008 doi:10.1088/1757-899X/1073/1/012008

Analysis of the condensate phase in offshore pipe gas company

W Zamrudy*, A Rahmatullah, P Prayitno and A A Wibowo

Chemical Engineering Department, State Polytechnic of Malang, Malang, Indonesia

*windimlg@gmail.com

Abstract. Gas transportation is usually done using pipes. As was done in one of the Gas Companies. In gas transportation consists of suppliers (Station A) and receivers (station C). From the supplier to the receiver goes through a very far pipe. Gas that passes through the pipeline has a composition consisting mostly of methane, but it is undeniable that there is still heavy carbon and impurities in the gas. Heavy carbon and various impuritis contained in the gas will disrupt gas transportation in the pipeline because it can cause condensate. Condensate consists of heavy carbon and impuritis which changes phase into liquid. The occurrence of liquid is caused by external and also internal influences, especially on the temperature and pressure that is treated to the gas. Condensate in the pipe will cause obstruction or decrease the volume in the pipe and otherwise it will cause internal corrosion. In the Gas Company there is an offshore pipeline that has the deepest depth of 70 meters. This elevation allows condensate inundation to occur in the offshore pipeline. The purpose of this study was to determine the effect of operating conditions on the occurrence of the existing condensate phase in the offshore pipeline between B Station B and C Station, which is 105 km away. The method used is the collection of gas composition data through gas chromatography analysis at station A, piping specifications, operating conditions at station B, and Hysys Aspen Simulation. It can be concluded that the offshore pipeline between B station and C station formed condensate (liquid phase) on 2nd day. The operating pattern of the envelope phase shows that in this condition natural gas condensate is formed because it is in the gas-liquid phase. If the methane composition and sea temperature are getting smaller, the possibility of natural gas condensate formation will be even greater.

1. Introduction

In the process of natural gas transmission, onshore and offshore pipes are often used, in addition to still using tankers or tank cars. Offshore or offshore pipelines are used as a means of transmitting natural gas through the sea. One example of an offshore pipeline that is used is the gas transmission pipeline network between Station B - station C. This pipeline network is parallel to the Station A – D Station pipeline network [1]. This pipeline network is known as the South Sumatra - West Java (South Sumatra West Java - SSWJ) transmission line with the aim of increasing the supply of natural gas for industries in Java. In addition, the construction of an offshore pipeline is used to connect gas from the supply point to gas customers or consumers and land receiving facilities [2]. The use of offshore pipelines is a more effective way of transmitting gaseous fluids at the point of supply in terms of capacity and distance traveled. Several technical obstacles often occur in the process of natural gas transmission using offshore pipelines. Sea temperature, operational pressure and elevation in pipelines are technical constraints in the gas transmission process with offshore pipelines. Deep sea temperatures and pressures that tend to be lower often produce condensate in natural gas. The high temperature fluid flowing in the pipe will come into contact with the cross section pipe which has a low temperature and will convert some of the fluid into gas and liquid [3].

The pressure and temperature factors take into account the binding forces between the gas molecules. In addition, the gas fluid through some local bending will collide with a cross section that is formed like an elbow. This resulted in the formation of a liquid phase at several elevations in the offshore pipe [4].

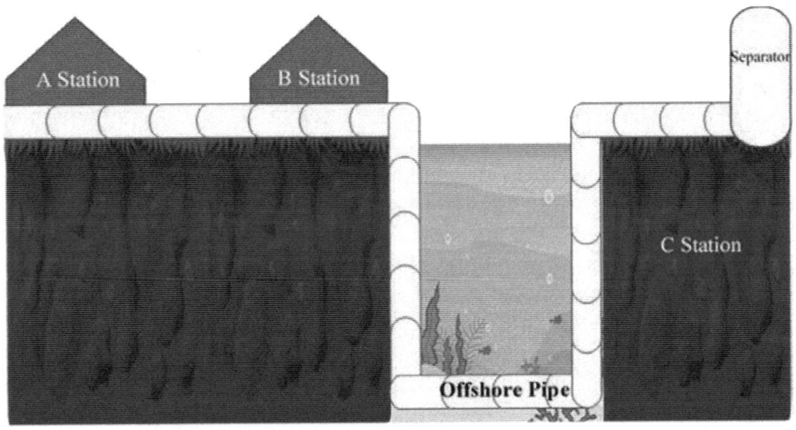

Figure 1. Illustration of offshore pipe (A Station, B Station, and C Station).

On 4th day there is condensate at station C which exceeds the sumptank level. There is concern that this condition can affect the flow of gas in the offshore pipeline. The presence of condensate is also detrimental to the company due to the obstruction of gas distribution in the pipeline. Condensate in natural gas can damage piping installations because the content in this condensate is in the form of heavy carbon and gas impurities. Damage that can occur such as corrosion of natural gas distribution pipes. There are two offshore pipes owned by National Gas Company. The first one stretches between station B - station C and the second stretches between Station B - Station D. The offshore pipe has a pipe length of 105 km and a deepest of 70 meters. The offshore piping system does not have a tool to detect condensate. Based on the description above, the purpose of this study is to determine the effect of operating conditions on the occurrence of the existing condensate phase in the offshore pipeline network between B Station and C Station which is 105 km away.

1.1. Offshore pipe specifications and environmental conditions

1.1.1. Offshore pipe specifications. Inline Inspection: in the form

- internal diameter : 30.5 inches
- out diameter : 32 inch
- pipe length : 101 km
- wall thickness : 19.05 mm
- Roughness : 0.0005 inch
- nominal diameter : 32 inch
- ambiance temperature : 68 ° F
- pipe material : carbon steel
- std (standard pipe) : API 5L

1.1.2. Environmental conditions

- Sea temperature : 39.3 ° C

1.2. Natural gas

Natural gas in Indonesia has a fairly dominant role after the role of oil as the main energy source has begun to be reduced. Moreover, with the commitment given by the government in the Clean Development Mechanism to the Kyoto Protocol, natural gas has begun to be chosen because of its lower pollution levels.

The Potential of Natural Gas in Indonesia apart from petroleum, Indonesia has quite large natural gas reserves, namely 170 TSCF and annual production of 2.87 TSCF, with this composition Indonesia has a reserve to production (R / P) of 59 years. Natural gas also has a stable price because it is far from political content, unlike petroleum. The products of natural gas used are LPG (Liquid Petroleum Gas), CNG (Compressed Natural Gas), LNG (Liquid Natural Gas) and Coal Bed Methane (CBM) which are non-conventional sources that are being developed in Indonesia (Kementerian ESDM, 2018- 2027).

1.3. Composition of natural gas

Natural gas or natural gas also has a chemical composition like crude oil, only more simply, consisting of a mixture of compounds whose main elements consist of hydrogen atoms (H) and carbon atoms (C) or often referred to as hydrocarbon compounds, starting from C1 (methane) to C4 (Butane) sometimes there is also C5 + (pentane and heavier) which is already present in liquid form as condensate. From this it can be concluded that the composition of natural gas in terms of its carbon molecular compounds only consists of C1 to C4. So the constituent of natural gas only consists of CH_4 (methane), C_2H_6 (ethane), C_3H_8 (propane), and C_4H_{10} (butane) [5]. So compared to petroleum, in natural gas the number of carbon atoms is shorter. In general, natural gas is composed of the largest component of methane so that natural gas that has met the specifications for selling gas has properties close to methane properties.

Natural gas [6] can occur alone or exist together with crude oil. In addition, other compounds that are often present with it are carbonic acid gas (carbon dioxide or CO_2), helium gas (He), mercaphthan (RSH) and water vapor (H_2O) and metals / metals. The heavy metals present are Vanadium (V) and mercury or mercury (Hg). The content of other compounds in natural gas is undesirable, because these compounds are impurities or are often called impurities, which can disturb the next gas process [7].

1.4. Condensate

Natural gas condensate is a simple condensate that sometimes contains natural gasoline because it contains hydrocarbons which have the same boiling point as gasoline. Natural gas condensate usually consists of hydrocarbons ranging from C4 to C15 which mostly contain 25-95% by weight of organic compounds C6, C7 and C8. In natural gas the condensate consists of hexane, cyclohexane and benzene as C6 hydrocarbons; heptane and toluene as C7 hydrocarbons while octane, ethyl benzene, and xylene are considered C8 hydrocarbons [8]. Condensate is formed by condensation of steam or gas in the form of liquid hydrocarbons which will be separated from natural gas due to changes in temperature and pressure when it reaches the well for separation on the surface. The condensed liquid usually returns to the boiler upstream but if there is natural gas condensate present in the pipe it will be separated and collected through natural gas separation in the slug catcher, scrubber outlet and delivery scrubber to get natural gas free from condensate, liquids and particles. impurity.

This condensate contains hydrocarbon components of important and valuable raw material fuels. Natural gas or natural gas can occur alone or exist together with crude oil. In addition, other compounds that are often present with it are carbonic acid gas (carbon dioxide or CO_2), helium gas (He), mercaphthan (RSH) and water vapor (H_2O) and metals / metals. The heavy metals present are Vanadium (V) and mercury or mercury (Hg). The content of other compounds in natural gas is undesirable, because these compounds are impurities or are often called impurities, which can disturb them. the next gas process [9].

1.5. Gas distribution system through piping

The distribution or distribution of gas through pipelines must have adequate infrastructure in the construction or design of the pipe. Piping infrastructure has begun to develop in Indonesia due to its effectiveness and efficiency in all sectors. The piping industry can be divided into two broad areas, namely transmission and distribution. Transmission pipelines are used to transport large amounts of natural gas over long distances and at high pressure (300-1400 psig). Distribution pipes are used to deliver gas from the point of supply to consumers at a lower pressure than transmission pipes and usually take supply from transmission pipes [1].

The transmission pipeline network is a pipe network that has a large diameter and high pressure, so it has a large gas distribution capacity. The pressure in this transmission pipeline can be anywhere from 200 psig to 1500 psig. This pressure is used to provide potential energy to the gas so that it can flow in the pipe. The transmission pipe consists of at least 3 main components, namely the transmission pipe itself, the compressor and the offtake station.

The distribution of gas through distribution pipes is the last stage before the gas enters the customer's installation. The distribution pipe gets gas from the transmission pipe through the delivery point which is called the city gate. At this city gate the gas from the transmission pipe is given an odorant (odor) a kind of mercaptant compound, as an anticipation that the gas can be smelled when a gas leak occurs in the distribution pipeline network. Then the pressure regulation process is also carried out, so that the pressure is in accordance with the pressure regime in the distribution pipe which is generally lower than the transmission pipe.

2. Research methodology

The methodology in this research is as follows:

2.1. Gas chromatography data collection at a station

The composition of natural gas at A station is recorded including methane, ethane, propane, i-butane, n-butane, i-pentane, n-pentane, n-hexane, n-heptane, n-octane, n-nonane, nitrogen, carbon dioxide, water, and sulfur.

2.2. Collecting data at B station

- The operating conditions at B station are recorded including pressure, temperature and flowrate.
- Recording is done every hour and every day so that you get the operating conditions before autopressure, during autopressure, and after autopressure.

2.3. Retrieval of offshore pipe specification data.

The data for the B Station - C Station offshore pipe specifications were taken in the form of internal diameter, out diameter, pipe length, wall thickness, roughness, nominal diameter, ambiance temperature, pipe material, std (pipe standard).

2.4. Observation at C station.

- The operating conditions at C station were observed including pressure and temperature.
- Observations were taken 3 different points each day which represented the conditions before autopressure, during autopressure, and after autopressure.

2.5. Simulation using Hysys software.

- Field data that has been taken on gas chromatgraphy from A Station and operating conditions from B Station are processed as input data used in simulations with Hysys software.
- Three points of operating conditions (before auto pressure, during auto pressure, and after autopressure) that occur every day starting on 1st day 1 to 4th day were used as experimental variables.
- The design for operation on the Hysys software is made according to field conditions, starting from the downstream of B Station to the upstream of C Station.
- The simulation is run at steady state

2.6. Making phase envelope curve using Microsoft Excel

- Bubble points and dew points obtained through simulation results from Hysys software, graphed using Microsoft Excel. This graph from Microsoft Excel is called a phase envelope curve.
- Then the operating conditions at the three station points are combined in a phase envelope curve as material for analysis.

2.7. Graphs formed from data input in excel are analyzed every day from 1st day to 4th day.

3. Results and discussion

Based on the analysis of the data on operating conditions and gas composition on 1st to 4th day as well as data on the operating conditions performed by autopressure during the natural gas transmission process, it is seen that condensate is formed in the offshore pipeline between A station and C station. It can be seen from the phase envelope graph in Figure 4, that the operating condition line has entered the two-phase part of the curve. The phase envelope is a relationship between temperature and pressure which shows the conditions from the same side between the various phases of a chemical compound, a mixture of compounds, and a solution [10]. From the phase envelope curve, a line of operating conditions is inputted, the relationship between temperature and pressure on a daily basis, starting from the A Station input to the C Station output. Creating a phase envelope requires bubble point and dew point data[10]. Bubble point is a temperature that separates the liquid phase and partially the two phases (gas-liquid phase). While the dew point is a temperature that separates the gas phase and partially the two phases (liquid-gas phase) [10].

Table 1. Operating Conditions Of National Gas Company at B station for 1st to 4th day.

Operating Condition	units	B Station	Inlet Offshore Pipe	Outlet Offshore Pipe
1stDay				
Before Autopressure	T (°F)	81,806	81,734	68,072
	P (Psig)	794,662	794,662	755,213
AutoPressure	T (°F)	72,302	58,478	67,964
(set Pressure :400 psig)	P (Psig)	861,236	645,274	641,213
After *Autopressure*	T (°F)	57,002	27,833	66,812
	P (Psig)	840,205	414,374	363,03
2ndDay				
Before Autopressure	T (°F)	156,38	145,634	76,748
	P (Psig)	807,8617	584,358	473,984
AutoPressure	T (°F)	76,208	65,318	67,964
(set Pressure :500 psig)	P (Psig)	473,984	680,083	671,961
After *Autopressure*	T (°F)	63,194	42,5624	66,812
	P (Psig)	838,03	514,74	304
3rdDay				
Before Autopressure	T (°F)	73,292	63,158	67,334
	P (Psig)	823,526	664,854	585,373
AutoPressure	T (°F)	77,108	68,72	67,982
(set Pressure :500 psig)	P (Psig)	834,114	699,663	695,602
After Autopressure	T (°F)	63,104	41,1638	67,928
	P (Psig)	843,831	508,648	481,816
4thDay				
	T (°F)	68,702	54,5	66,65
	P (Psig)	809,167	592,6253	505,4574

Table 2. Composition of Gas - Condensate at A Station.

No	Day Components	Symbol	1st	2nd	3rd	4th
			Mole%			
1	Methane	C_1	84.10	82.10	83.76	84.03
2	Ethane	C_2	5.82	5.63	5.77	5.97
3	Propane	C_3	1.87	3.24	2.00	1.64
4	i-Butane	iC_4	0.33	0.60	0.38	0.31
5	n-Butane	nC_4	0.45	0.78	0.50	0.43
6	i-Pentane	iC_5	0.17	0.30	0.19	0.16
7	n-Pentane	nC_5	0.11	0.20	0.13	0.11
8	n-Hexane	nC_6	0.13	0.24	0.15	0.12
9	n-Heptane	nC_7	0.07	0.12	0.07	0.06
10	n-Octane	nC_8	0.02	0.04	0.02	0.02
11	n-Nonane	nC_9	-	0.00	0.00	0.00
12	Nitrogen	N_2	1.34	1.27	1.33	1.33
13	Carbon Dioxide	CO_2	5.57	5.49	5.69	5.80
14	Water	H_2O	-	0	0	0
15	Sulphur	H_2S	-	0	0	0

The autopressure condition is carried out to reduce the fluid pressure at B station outlet before transmission through the offshore pipe to the C station. This operating condition was carried out because C station had fulfilled customer needs and with the autopressure that occurred, several gas flows would be channeled directly to D station, which had a shortage of gas supply. With a change in the operating pattern that occurred for several days, on 4th Day there was an increase in the level of condensate formed at the Bojonegara offtake Station and closed the shut down valve. The parameters used to determine the formation of condensate in the offshore pipeline include natural gas composition, operating conditions during autopressure, operating conditions before and after autopressure. The outlet temperature of the offshore pipe will increase after autopressure occurs due to the increase in operating pressure.

3.1. Phase diagram

According to thermodynamic definition phase diagram (phase envelope) is a graph showing the pressure at which transition of different phases from a compound, respect to temperature [10,11]. Here an example of phase envelope of a compound and its region phases.

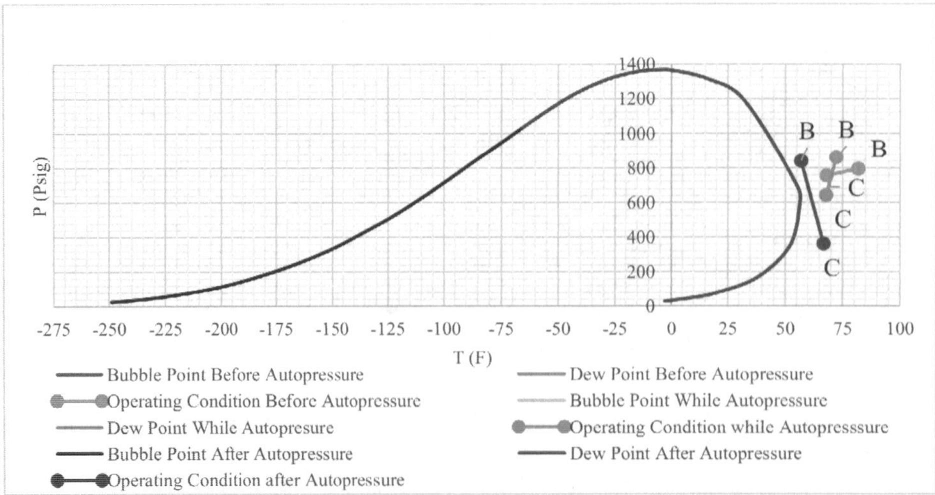

Figure 2. Phase envelope of natural gas on 1st day.

On 1st day, there is an autopressure mode in the natural gas transmission process. Figure 2 shows the characteristics of an offshore gas outlet pipe in the form of a phase envelope diagram. The operating temperature does not exceed the dew point formed and is still in the gas phase part of the envelope phase. This shows that at the operating temperature before autopressure, during autopressure, and after autopressure, two phases are not formed and no condensation occurs in the offshore pipe. The dew point formed was 56.27 °F with a pressure of 574.48 Psig. Dew point data is obtained from envelope utility in Hysys software. The operating conditions after autopressure on day 1 showed that the temperature and pressure were close to the dew point, namely at a temperature of 57 °F and a pressure of 840.21 Psig. This figure is obtained from the simulation results that have been running on the Hysys software in accordance with the specified operating conditions. If the pressure is lowered again, the operating conditions will reach their dew point. After passing through the offshore pipeline, natural gas will go through the PCV to reduce the pressure according to customer distribution needs.

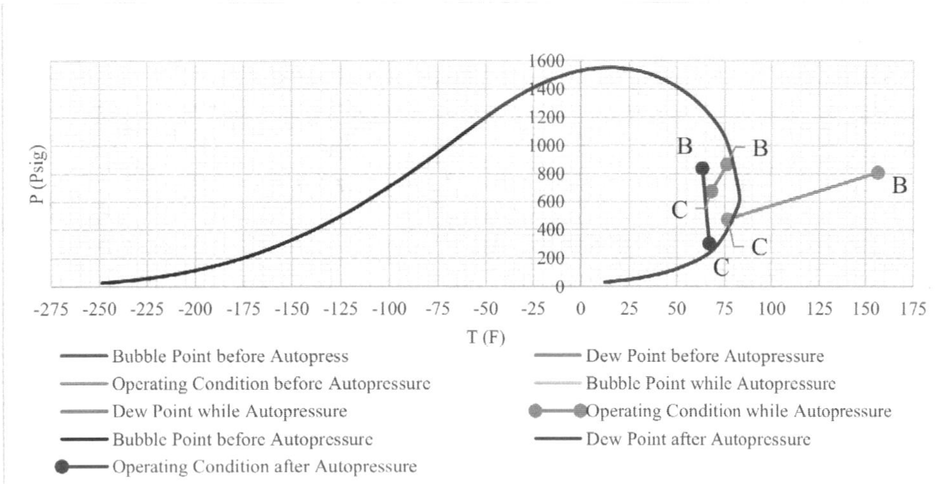

Figure 3. Phase envelope of natural gas on 2nd day.

On 2nd day there is an autopressure mode in the natural gas transmission process. Figure 3. shows the characteristics of an offshore gas outlet pipe in the form of a phase envelope diagram. The operating temperature does not exceed the dew point formed and is still in the gas phase part of the envelope phase. This shows that at the operating temperature before autopressure, during autopressure, and after autopressure two phases are formed and condensation occurs in the offshore pipe. The dew point curve formed was 82.79 ° F with a pressure of 675.44 Psig. Dew point data is obtained from envelope utility in Hysys software.

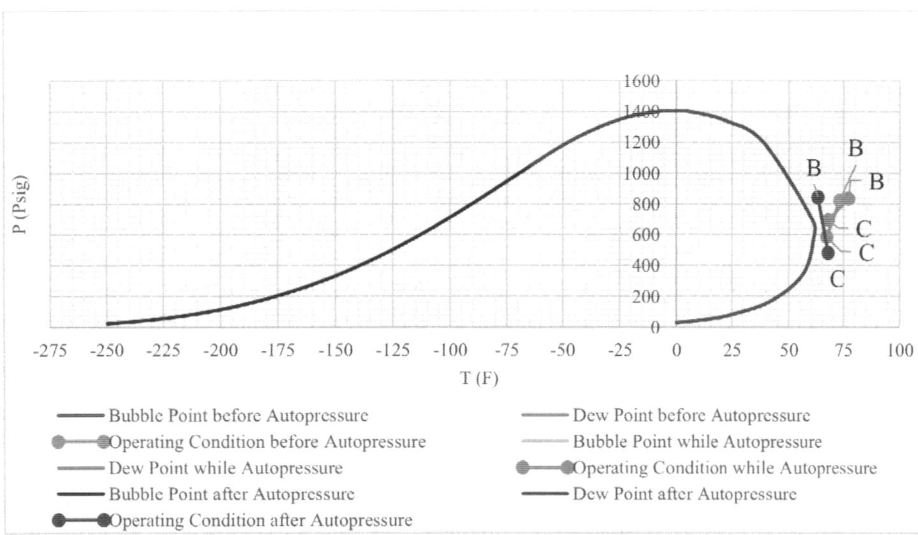

Figure 4. Phase envelope of natural gas on 3rd day.

IOP Conf. Series: Materials Science and Engineering 1073 (2021) 012008 doi:10.1088/1757-899X/1073/1/012008

On 3rd days there is an autopress mode in the natural gas transmission process. Figure 4 shows the characteristics of an offshore gas outlet pipe in the form of a phase envelope diagram. The operating temperature does not exceed the dew point formed and is still in the gas phase part of the envelope phase. This shows that at the operating temperature before autopressure, during autopressure, and after autopressure, two phases are not formed and condensation occurs in the offshore pipe. The dew point curve formed was 61.34 °F with a pressure of 592.07 Psig. Dew point data is obtained from envelope utility in Hysys software.

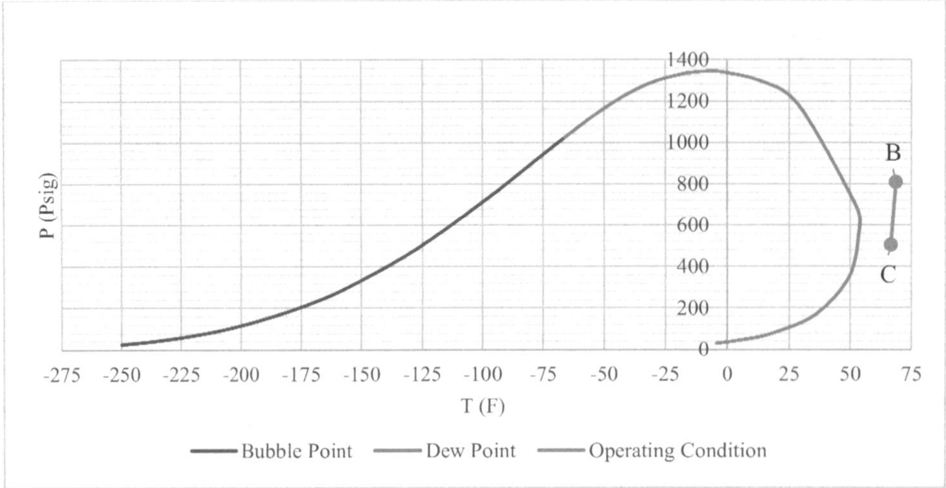

Figure 5. Phase envelope of natural gas on 4th day.

On 4th day there is an autopressure mode in the natural gas transmission process. Figure 5 shows the characteristics of an offshore gas outlet pipe in the form of a phase envelope diagram. The operating temperature does not exceed the dew point formed and is still in the gas phase part of the envelope phase. This shows that at the operating temperature before autopressure, during autopressure, and after autopressure, two phases are not formed and condensation occurs in the offshore pipe. The dew point curve formed was 53.55 ° F with a pressure of 565.15 Psig. Dew point data is obtained from envelope utility in Hysys software.

The condensate formed in the offshore pipeline occurs due to internal and external factors. Internal factors are the composition of natural gas and operating conditions such as pressure, temperature and flowrate. If the composition of methane gas is getting smaller, then the possibility of forming natural gas condensate will increase. Meanwhile, external factors are sea elevation and temperature. With the deepest offshore pipe depth at 70 meters, the hydrocarbons in natural gas will experience a phase change. In this case, the factor that influences the formation of condensate is the composition of natural gas which is not good because the composition of methane is low so that the heavy carbon composition increases. The increasing heavy carbon composition shows that the dew point on the phase envelope graph is getting to the right.

Operating conditions such as temperature and pressure must be maintained so that condensate does not form in offshore pipes. The composition of natural gas must be treated to clean natural gas from impurities. Methods that can be done in cleaning natural gas from impurities and carbon by weight can be done by:

3.1.1. Doing pigging. Pigging aims to clean the pipes from dirt or condensate trapped in the pipes during the distribution process. Pigging can push impurities such as condensate and scale in the pipe. The edge is done by inserting the pig in the pig launcher which will be channeled using natural gas from the pipe to the pig receiver.

3.1.2. The presence of CO_2 removal. CO_2 is a gas impurities that must be removed because this content can cause corrosion in pipes. CO_2 removal can be done in 2 ways, namely using absorption and using a membrane.

3.1.3. Optimization of separator / scrubber performance. The separator / scrubber is used to separate the gas from the liquid contained in it by separating it based on density.

4. Conclusion

The offshore pipe between B station and C station formed condensate (liquid phase) on day 2. The operating pattern of the envelope phase shows that in this condition natural gas condensate is formed because it is in the gas-liquid phase. If the methane composition and sea temperature are getting smaller, the possibility of natural gas condensate formation will be even greater. The composition of natural gas must be treated, such as cleaning through a separator or by CO_2 removal. Another method that can be done is pigging the offshore pipe. Pigging aims to clean the pipes from dirt or condensate trapped in the pipes during the distribution process.

Acknowledgements

The authors wish to thank to Research and Community Service State Polytechnic of Malang who has contributed funding and supporting facilities in this research.

References

[1] Ardiansyah 2008 *Perancangan Sistem Perpipaan Transmisi Gas dari Muara Bekasi ke Muara Tawar Melalui Jalur Lepas Pantai.* (Universitas Indonesia. Jakarta.)

[2] Ansory S and Fathaddin M T 2016 Optimasi Trnasportasi Gas Alam dari Lepas Pantai ke Fasilitas Penerima Darat *Semin. Nas. Cendekiawan. Univ. Trisakti.*

[3] Rachmat B, Purwanto C and Raharjo P 2011 Kajian Identifikasi Infrastruktur Jaringan Pipa Migas Bawah Laut Di Perairan Sebelah Utara Provinsi Banten *J. Tek. Geol. Kelaut.* . **9(2)** 79–96

[4] Supahar 2005 Kajian Numerik Sifat Densitas Sistem Hidrokarbon Gas Kondensat Berbasis Komposisi Menggunakan Persamaan Soave Redlich-Kwong". *Pros. Semin. Nas. Penelitian, Pendidik. Penerapan MIPA. Univ. Negeri Yogyakarta.* 1–11

[5] Sujan S M A, Jamal M S, Hossain M, Khanam M and Ismail M 2015 Analysis of gas condensate and its different fractions of Bibiyana gas field to produce valuable products **50** 59–64

[6] Medeiros L De, Arinelli L D O, Teixeira A M, Araújo D Q F, Química E De and Janeiro R De 2019 Supersonic separator for cleaner offshore processing of natural gas with high carbon dioxide content : Environmental and economic assessments **233** 510–21

[7] Chotler Desemsi P 2011 *Modifikasi Proses Pemisahan CO2 Kadar Tinggi dari Gas Alam* (Universitas Indonesia)

[8] Goverment of Canada 2016 Risk Management Approach for Natural Gas Condensates *Enviroment, Clim. Chang. Heal. Canada*

[9] Kemendikbud 2015 Gas Proccesing. Jakarta

[10] Soetikno D, Kusdiantara R, Puspita D, Sidarto K A, Siagian U W R, Soewono E and Gunawan A Y Critical Point Analysis of Phase Envelope Diagram **5** 1–4

[11] Davarikhah Q 2020 Prediction of a wellhead separator efficiency and risk assessment in a gas condensate reservoir

IOP Conf. Series: Materials Science and Engineering 1073 (2021) 012009 doi:10.1088/1757-899X/1073/1/012009

Literature study of production dry cellulase from *Trichoderma reesei*, *Aspergillus niger*, and *Bacillus subtilis*

Y Maryanty*, D R Wulan, M D Hidayati, D A Rizal and F Aliffandri

Chemical Engineering, State Polytechnic Malang, Malang, Indonesia

*yanty.maryanty@polinema.ac.id

Abstract. Cellulase is an enzyme that can degrade cellulose through a catalytic process. The enzyme that works synergistically to release sugar (glucose). This review aims to determine cellulase enzymes' activation from three different microorganisms, Aspergillus niger, Bacillus subtilis, and Trichoderma reesei. Enzymes' activation using various kinds of lignocellulosic substrates and the effect of adding carrier agents on the freeze-drying process. The cellulase enzyme from Trichoderma reesei has an activity of 3.4 IU / mL. This enzyme is produced by pre-treatment 1% H2O2 in lignocellulosic media. Aspergillus niger has a cellulase enzyme activity of 0.229 IU / mL using a sugarcane bagasse substrate that has been pre-treated by shiitake mushrooms. The cellulase enzyme from Bacillus subtilis has an activity of 0.907 IU / mL, with the addition of CMC levels of 5% as an inducer. Cellulase enzymes in liquid form are susceptible to denaturation during storage. The production of cellulase enzymes in solid form is expected to maintain the stability of the resulting cellulase enzyme activity longer. The addition of non-reducing sugar as a carrier agent in the freeze-drying process is reported to preserve the biological activity. The addition of sucrose and trehalose, each with a concentration of 300 mM, maintained the amylase enzyme activity up to 90%. The addition of sucrose with an optimum concentration of 60 mM protects both total protein stability and good lysozyme activity. It is suspected that a carrier agent with an optimal concentration can maintain protein stability in cellulase enzyme activity in this review.

1. Introduction

Cellulase is a complex enzyme that cuts down the cellulose chain into glucose. At least three groups of enzymes are involved in the enzymatic hydrolysis process on cellulose. The enzymes are endoglucanase, exoglucanase or cellobiohydrolase, and β-glucosidase [1]. Microorganisms produce cellulase during its growth in cellulosic material. Although many organisms are capable of degraded cellulose, only a few of them produce large amounts of bioactive cell-free compounds that can fully hydrolyze cellulose crystals *in vitro* [2].

Type Trichoderma Fungi Reesei can produce endo-ß 1.4-glucanase and Ekso-ß-1.4-glucanase up to 80% [3]. Still, the ß-Glucosidase is low, while the Aspergillus Niger type fungi can produce higher glucosidase when compared to Endo-ß-1.4-glucanase and Ekso-ß-1.4-glucanase. But according to Menendez *et al.* [4], the ability of cellulolytic bacteria has been widely utilized for a wide range of industrial purposes. Because the cost of production is cheap, short production time, producing complex multienzyme, and tend to be stable in extreme conditions. The difference in the ability to produce cellulase enzymes from these three microbes can have compared through the value of enzyme activity.

IOP Conf. Series: Materials Science and Engineering 1073 (2021) 012009 doi:10.1088/1757-899X/1073/1/012009

In general, enzymes are fluid-shaped and have a weakness in which enzymes become inactive due to changes in structures caused by denaturation. Denaturation has usually caused by external influences (*stresses*) such as heat, pH change, and chemical interactions. This *Stresses* can have avoided if the proteins have converted to dry form in the presence of stabilizing substances that prevent protein denaturation and help preserve protein. *Freeze Drying* can minimize thermal *stress* and reduce *stress* due to dehydration, making it an extraordinary tool to preserve enzymes. Therefore, it is necessary to produce enzymes in the form of solids, such as powders and granules. However, the step of the enzyme drying process often causes Enzim to lose its activity, and the end product is still vulnerable to inactivation. The proper design of the drying process should ensure the value of high enzyme activity [5].

The principle of this *freeze-drying* method is to freeze the sample rapidly with a vacuum condition to avoid forming crystals – the ice crystals that can damage the sample. Still, this process also often damages the structure of proteins that will eventually cause the denaturation and aggregation of proteins when reconstituted in water [6].

The protein conformation has essential to protection. The lyophilization is processing to maintain pharmacological effects and reduce the risk of product immunogenicity. Sucrose has the ability as stabilizers, can protect proteins during the lyophilization process and subsequent storage processes. The interaction between the hydroxyl groups of non-reducing sugars with protein molecules in powder form can prevent the release of hydrophobic sites, which can then lead to aggregation [6]. The interaction between the hydroxyl group of non-reducing sugar with the protein molecule of the powder pad can prevent the discharge of hydrophobic sites, which can then lead to aggregation [6].

In many cases, a disaccharide or even a mixture of two different disaccharides. They seem most effective in stabilizing enzymes during the freeze-drying procedure. These cases include Chymotrypsinogen protection in the presence of 300 MM sucrose [7], complete protection of four restriction enzymes by 15% trehalose [8] and stabilization of recombinant human serum albumin [9]. The type and concentration of disaccharides influence protein stabilization during the Liofilisation [10]. If the freeze-Drying operation has appropriately done, the process will protect most of the initial biological activity of the substance in a dry state [11].

Study of this literature aims to compare the activity of the enzyme cellulase produced from *Trichoderma reesei*, *Aspergillus niger*, and *Bacillus subtilis* as well as knowing the influence of *carrier agent* (sucrose) in the process of *freeze-drying* of the activity of cellulase Enzyme.

2. Methods

This literature study method by conducting a study of experimental methods conducted by previous researchers related to the topic of research is qualitative descriptive and identified in each journal. The information to analyze the ability of *Trichoderma reesei*, *Aspergillus niger,* and *Bacillus subtilis* in the production of cellulase enzymes. The *Freeze Drying* method has used to acquire enzymes in the form of *dry* cellulose (powder). The addition of sucrose as a *carrier agent has* expected to maintain a stable enzyme that has remained a steady *freeze-drying* process. The methodology below is a result of studies and comparison of several previous journals that will be used as a reference when researching in the laboratory.

2.1. Pra-treatment media

Media that contains a mixture of wood powder sengon, bran, and lime are placed in a tray to be in the count to reduce the size of the media. The media that has minced has dried under the heat of the Sun for a full-day. After experiencing the drying process is expected Moisture content in the media can be less and the media has sifted on 80 mesh to expand the surface area, the enzyme has a substantial enough chance to degrades the cellulose in the media.

IOP Conf. Series: Materials Science and Engineering 1073 (2021) 012009 doi:10.1088/1757-899X/1073/1/012009

2.2. Analysis media
Media character analysis (NDF degradation analysis, ADF, cellulose, hemicelluloses, and lignin). Dry solids (residue) can have used to determine the content and degradation of NDF, ADF, cellulose, and hemicellulose using *Van Soest* method.

2.3. Inoculum production
Inoculum produces by taking spores from *the microbe (aspergillus niger, Trichoderma reesei,* and *Bacillus subtilis).* They have then resuspended in 8 mL of sterile aquadest. Suspense has taken each containing 13 mL of Liquid sterile media. Subsequently, the medium has incubated in a shaker for 49 hours.

2.4. Media preparation
Pleurotus ostreatus Media in the form of a mixture is taken as much as 1.8 grams and added nutrients in the way of BHM (*Basal Hard Medium*) to 18 mL. Subsequently, the media was given an inoculum suspension for the samples. To be tested and blank without being given inoculum microorganisms.

2.5. Production of crude cellulase
Wahyuningtyas Research [3] to produce cellulase started by mixing 5 grams of straw powder with 25 mL of nutrients into the Erlenmeyer 250 mL, then covered with sterile cotton and coated with aluminum foil, paper, and tied with yarn. A mixture of substrates and media has then sterilized using the autoclave at 121°C temperature for 15 minutes. The media is then cooled down to room temperature before the aseptic inoculation process has been performed. Spores grow in a single reaction tube. The spore has resuspended into 1ml of 0.1% *tween* 80 and then inoculated aseptically into the media. The results have subsequently incubated for 4.6 and 8 days at 27 °C, 30 °C, and 35 °C with pH variations in liquid media of 4.5 and 6, respectively.

2.6. Glucose standard curve
In the Julaeha research [12] has taken the methodology on the creation of the standard curve of glucose that is utilizing a standard stock glucose curve is made using glucose concentrations 0.25; 0.5; 0.75; 1; 1.25; 1.5 ppm. The standard curve produces by glucose solution. The solution has dissolved as much as 250 mg of glucose with aquadest up to a volume of 50 mL. The reducing sugar test has done by taking 1 mL of the sample into the tube, then added 1 ml of the DNS reagent and 2 ml of the aquadest, attached to each of the test tubes using a pipette. The Reaction tube is heated inside the *water bath* for 5 minutes to react to glucose and DNS. The absorption of each solution has measured at 540 nm [13]. Its absorbable absorption rate is linked to the standard curve to determine the concentration of glucose in the sample.

2.7. The activity test of cellulase enzyme
According to Rosyda Study [14], determining enzyme activity was tested using the CMCase method in units of *International units* (I.U.) with the reagent *dinitro-salicylic Acid* (DNS) [15]. Cellulase activity has determined by utilizing each liquid enzyme. The production taken an of 1.5 mL to has inserted into the *microtube. Then* The sample is centrifuged at a speed of 10000 rpm for 5 minutes at a temperature of 4 °C. The supernatant has taken as much as 0.5 mL. And inserted into the closed reaction tube and added 0.5 mL of the 1% CMC substrate was then homogenized and incubated for 30 minutes at a temperature of 50 °C. After that, it is added 3 mL of DNS and heated for 5 minutes on boiling water. Each solution in the tubes measured its absorption with a UV-Vis spectrophotometer at a wavelength of 550 nm. The calculation to get activity volume enzyme CMC-ASE has based on one μ-mol glucose = 0.18 mg, and 1 unit of CMCase activity is one μ-mol glucose produced per minute.

2.8. Drying enzyme process using freeze-drying method
Shariat Research [11] uses the enzyme lactoperoxidase as a sample that will be freeze-drying. The first step is 15 mL of the enzyme solution that has mixed with a variation of the non-reducing sugar

concentration is inserted into the freeze-Drying tube and frozen at a temperature of -70°C for 24 hours. The frozen samples have transferred to the vacuum rubber at the freeze Dryer. The temperature has set at a temperature of -80 °C. A *vacuum* pressure of 0.001 bar has awaited until the sublimation process in *freeze-drying* was completed with the marked formation of solid Powder in the resulting sample.

3. Results and discussion

3.1. Fibre component values

The analysis of the media content using the Van Soest method has done by determining the rate of NDF, ADF, hemicellulose, cellulose, and lignin.

Table 1. Water content analysis, ADF, NDF, cellulose, and lignin.

Analysis	Baglog Age		
	0 months	Two months	Four-months
Water content average (%)	8.59 ± 0.26	5.62 ± 0.58	3.98 ± 0.27
ADF average	62.14 ± 15.36	70.05 ± 3.71	88.04 ± 1.29
NDF average	69.34± 11.63	74.89 ± 0.19	97.93 ± 0.18
Cellulose average (%)	46.41 ± 13.82	64.08 ± 4.21	85.25 ± 1.51
Lignin average (%)	15.59 ± 1.59	5.52 ± 0.03	2.67 ± 0.12

*each data was report as average ± standard deviation obtained from two times measurement.

The results showed that *Pleurotus ostreatus* media analysis containing water content decreased as the age while percentage of cellulose increased. According to Setyowati [16] that the content of cellulose in the mixed Media of sengon powder and bran contains a high content of cellulose, as is the case in Bran as a diverse medium that has bacteria *acetobacter xylinium* which plays a role in converting carbohydrate waste in the media into cellulose.

ADF (*acid Detergent Fiber*) is a component of cell Wall that is soluble in acidic detergents. The ADF has used as a preparatory step to determinate lignin. That hemicellulose can have estimated from the difference in cell wall structures [17] NDF (*Neutral Detergent Fiber*) is a component of the cell wall that is soluble in neutral detergent [17]. From the data above, obtained NDF rate at the age of 0 and 2 months. The data decreased at a period of 4 months. The rate is rising. It is in line with Zenebe Research 2016 [18] stating that any *Pleurotus ostreatus* media treatment has followed by a decreased percentage of NDF, ADF, cellulose, and lignin content with the same pattern.

Table 2. Changes in fiber content (NDF, ADF, cellulose, lignin) on the various moisture content of substrates and inoculum doses of *Pleurotus ostreatus*.

Treatment	Response variables			
	NDF	ADF	Cellulose	Lignin
Substrate Moisture				
62.50%	-8.99^{a}	-7.74^{a}	-9.18^{a}	-13.78^{a}
70.00%	-13.19^{b}	-11.75^{b}	-14.65^{b}	-15.76^{a}
77.50%	-27.13^{c}	-15.45^{c}	-19.33^{c}	-21.73^{b}
85.00%	-26.64^{c}	-15.88^{c}	-19.88^{c}	-21.98^{b}
Inoculum Dose				
10g	-15.58^{a}	-7.95^{a}	-9.5^{a}	-14.98^{a}
15g	-19.18^{bc}	-12.73^{b}	-15.66^{b}	-17.4^{b}
20g	20.24^{c}	-14.71^{c}	-18.68^{c}	-19.97^{c}
25g	-21.15^{c}	-15.42^{c}	-19.2^{c}	-20.97^{c}

At each dosing treatment inoculum, *Pleurotus ostreatus* has been able to lower the content of all Fiber Components (NDF, ADF, cellulose, and lignin) so that the content of each component decreases (The percentage of decline becomes higher).

3.2. Cellulase enzyme activity

The best result of the literature study, the highest value of cellulase activity gained from the *Trichoderma reesei*, amounted to 3.4 U/mL. The result has shown in research conducted by Zhao [19] at Optimum conditions incubation time for 139 hours using a lignocellulose substrate of 1% residual furfural, which has a content of cellulose 38.06% and lignin 40.23%. It's quite a difference for *Aspergillus niger* 0.229 U/mL, 216 hours [14]. *Bacillus subtilis* 0.907 U/mL with 2% bagas sugarcane, 72 hours, and using CMC 5% [20].

Table 3. Study of the optimal activity literature of cellulase enzyme.

Microbes	Source	Substrate	Operating conditions	Enzyme activity of cellulase(U/ml)
Trichoderma reesei	Zhao (2018)	Residu Furfural 1%	50 °C	3.4
Aspergillus niger	Rosyida (2018)	Sugar Cane Bagas 5%	50 °C	0.229
Bacillus subtilis	Yandri (2012)	CMC 5%	55 °C	0.907

Based on the studies of the above literature compared to the variables we will use in the media waste baglog planting oyster mushroom with the utilization of *Pleurotus ostreatus*. The value of the enzyme activity generated in the research of Zhao [19] by using furfural waste Media 1% with *pre-treatment* using H_2O_2 1% can produce enzyme Activity of 3.4 U/mL. It has also has been influenced by incubation conditions of 50 °C and pH 6. The content of furfural waste that does not contain hemicelluloses. On Rosyida [14] using the media bagas sugarcane 5% and has cellulose content of 52.42% therein. The media has also undergone a delignification process assisted by shiitake mushroom. Other conditions that affect also use an optimal temperature of 50 °C. Next, the activity assay of cellulase enzyme Using *Bacillus subtilis* has the most substantial activity value produced at 0.907 U/mL [20]. It has been influenced by the absence of the addition of lignocellulose substrate instead of adding CMC Nutrients of 5 grams and contains cellulose content of 73.4%. Besides, the incubation time also affects Yandri Research [20], which uses the condition for 72 hours at 40 °C temperature, which can obtain optimum health to determine the activity of the cellulase enzyme. From the data collected on some sources in this literature, the study can be a reference for using optimum operating conditions on each cellulase-producing microorganisms.

3.3. Freeze drying and carrier agent

In Gianini research [21] compares the resulting proteolytic activity using a variety of Drying methods such as *solar drying, cabinet drying, vacuum drying,* and *freeze-drying*. Operating conditions in various types of drying methods have their variable differences. The best lyophilization is 4 °c for three months [22]. The stability of the protein in the same period 0,1, and 2 are 42,40 and 41 %, respectively. The lysozyme activity of 1140 U/mg, 1140 U/mg, and 950 U/mg were the most stable results when compared with other variables. Besides, the test for the water content lost during lyophilization or freeze-drying was the highest at 76.12%, with the addition of the concentration of the composition of the stabilizer or carrier agent CMC 0.3% and 25% sucrose conducted by Cahyadi et al. [23].

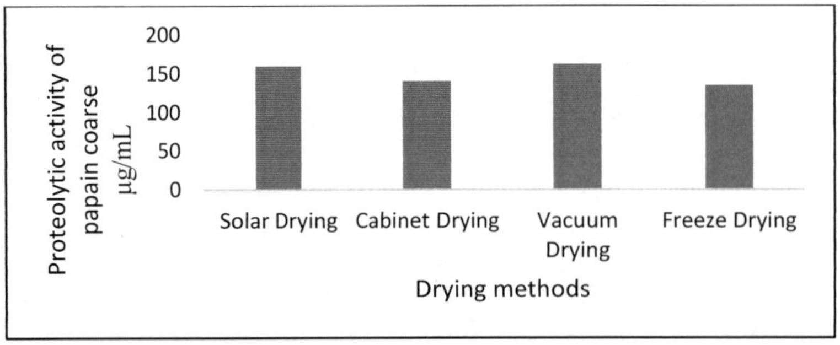

Figure 1. Proteolytic activity of papain coarse with various drying methods.

Research by De Jesus and Michael Fillio [24] shows that the drying process using freeze-drying can maintain better biological activity than Permata's study [25]. In the freezing stage, the enzyme has carried out using liquid nitrogen so that it was able to reduce water activity to 0.2 for 4 hours, and the results we're able to maintain enzyme activity by 98% after 16 hours of the process [24]. So, the fast cooling process, of course, also affects the freeze-drying process and the resulting activity because the quicker the water activity decreases, the higher the enzyme activity is maintained in the freeze-drying process.

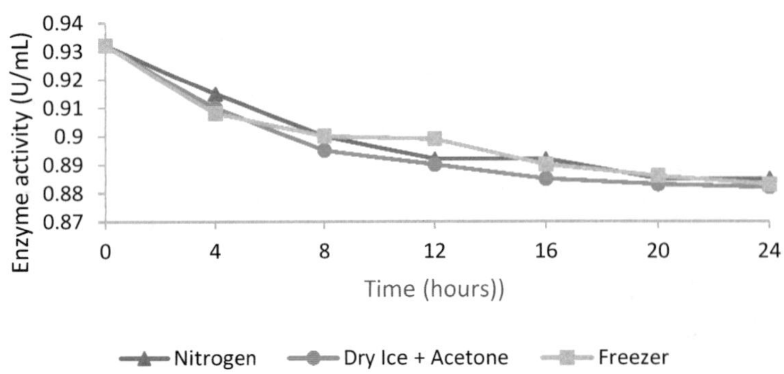

Figure 2. Influence of drying time of the enzyme activity after the primary Cooling is carried out using: (Δ) liquid nitrogen; (○) Dry Ice and acetone; (□) Conventional freezer. (Source: De Jesus and Michael Fillio, 2014).

Table 4. A *freeze drying* literature study and *carrier agent.*

Test	Save time and the concentration of sucrose	Activity value (mg/mL)	
		At temperature 4°C	At temperature Spaces
Kadar Total	0 months (60 MM)	42	42
Protein Serum	1 months (60 mM)	40	37
Otologus	3 months (60 MM)	41	35
Test	Save time and the concentration of sucrose	Activity value (U/mg)	
		At temperature 4°C	At temperature Spaces
Activities	0 months (60 MM)	1140	1140
Lisozim	1 months (60 mM)	1140	1000
Serum	3 months (60 MM)	950	690
Otologus			

Based on some literature, adding a carrier agent in the form of sucrose with various concentrations with storage at room temperature, there are differences in the operating conditions of the process. The addition of a stabilizer of sucrose (non-reducing saccharide) is a stabilizer that can protect protein and can replace the hydrogen bonds that occur between proteins (enzymes) and water. The OH- group present in sucrose can bind to proteins and protect chemical and physical degradation in aqueous solutions during the freeze-drying process and the storage process. At a later stage, the addition of a carrier agent can improve the performance of reducing water content during freeze-drying. Enzyme or protein storage at lower temperatures shows a very significant effect and can maintain enzyme activity longer than storage using room temperature or higher. High temperatures accelerate the physical aggregation of proteins in a dense and denatured state. Increasing the temperature affects protein indirectly. Heat can break the hydrogen bonds of proteins. Storage temperature has an essential factor. Temperatur influence the denaturation of the enzyme, a decrease in the stability of the enzyme activity or protein content. In the research of Shariat (2015) also confirmed that the use of optimum sucrose concentration and storage at low temperatures was able to maintain the enzyme activity produced by 90% after 40 days.

4. Conclusion

From the results of the literature study that we have compared, it can have to be concluded that *Trichoderma reesei, Aspergillus niger*, and *Bacillus subtilis* can produce cellulase enzymes with varying capabilities. The difference in ability can be seen from the activity value of the resulting cellulase enzyme and factor the factors that affect it involved. Substrate selection and optimal operating conditions can produce cellulase enzymes with high enzyme activity values—the use of carrier agents to maintain enzyme activity at an optimum concentration during the freeze-drying process.

Acknowledgments

This work was supported by a grant from research project for industrial collaboration, Malang State Polytechnic DIPA in 2020.

References

[1] Hermiati E 2010 *Utilizing ligneselulosa biomass of sugarcane for the production of bioethanol* (Bogor: Faculty of Agricultural Technology, IPB)

[2] Bai S, Kumar M R, Kumar D M, Balashanmugam P, Kumaran M B and Kalaichelvan P T 2012 Cellulase production by Bacillus subtilis isolated from cow dung *Arch. Appl. Sci. Res.* **4**(1) 269-279

[3] Wahyuningtyas P, Argo B D and Nugroho W A 2013 The study of the enzyme-making cellulase of Microfungi Trichoderma reesei with a rice straw substrate as a catalyst of enzymatic hydrolysis in bioethanol production *Journal of Tropical Commodity bioprocesses* **1**(1)

[4] Menendez E, Garcia-Fraile P and Rivas R 2015 Biotechnological applications of bacterial cellulases *AIMS Bioengineering* **2**(3) 163

[5] Mujumdar A S (Ed.) 2007 *Handbook of industrial drying* (New York: Dekker) p 948

[6] Melpin R, Maksum I P, Subroto T, Enus S and Soemitro S 2015 *Stabilization of lysozyme activity in dry frozen powder preparations of Otologus serum using sucrose lioprotectant* (Thesis, Bandung: University of Padjadjaran)

[7] Allison S D, Dong A and Carpenter J F 1996 Counteracting effects of thiocyanate and sucrose on chymotrypsinogen secondary structure and aggregation during freezing, drying, and rehydration *Biophysical journal* **71**(4) 2022-2032

[8] Colaco C, Sen S, Thangavelu M, Pinder S and Roser B 1992 Extraordinary stability of enzymes dried in trehalose: simplified molecular biology *Bio/technology* **10**(9) 1007-1011

[9] Han Y, Jin B S, Lee S B, Sohn Y, Joung J W and Lee J H 2007 Effects of sugar additives on protein stability of recombinant human serum albumin during lyophilization and storage *Archives of pharmacal research* **30**(9) 1124

[10] Prestrelski S J, Tedeschi N, Arakawa T and Carpenter J F 1993 Dehydration-induced conformational transitions in proteins and their inhibition by stabilizers *Biophysical journal* **65**(2) 661-671
[11] Shariat S S, Jafari N, Tavakoli N and Najafi R B 2015 Protection of lactoperoxidase activity with sugars during lyophilization and evaluation of its antibacterial properties *Research in pharmaceutical sciences* **10**(2) 152
[12] Julaeha E, Rustiyaty S, Fajri N N, Ramdani F and Tantra R 2016 Pemanfaatan Tepung Gadung (Dioscorea Hispida Dennst.) Pada Produksi Amilase Menggunakan Bacillus sp. *EDUFORTECH* **1**(1)
[13] Ceirwyn J S 1995 *Analytical Chemistry of Foods* (London: Blackie Academic and Professional) pp 71-135
[14] Rosyida V T, Hayati S N, Indrianingsih A W, Maryana R, Purwesti Y A and Ayesda C S 2018 Enzim Selulase Kasar Aspergillus niger FNCC 6018 untuk Produksi Bioetanol melalui Proses Sakarifikasi dan Fermentasi Serentak *Jurnal Mikologi Indonesia* **2**(2) 77-89
[15] Miller G L 1959 Use of dinitrosalicylic acid reagent for determination of reducing sugar *Analytical chemistry* **31**(3) 426-428
[16] Reyeki S and Asngad A 2013 *Pemanfaatan Serbuk Gergaji Kayu Sengon (Albizia falcataria) dan Bekatul Sebagai Media Tanam Budidaya Jamur Tiram Putih (Pleurotus ostreatus) dengan Penambahan Serbuk Sabut Kelapa (Cocos nucifera)* (Surakarta: Universitas Muhammadiyah Surakarta)
[17] Rulianah S, Gunawan P, Hendrawati N and Nafisa K N 2020 Production of bioethanol from bagasse with a simultaneous saccarification and fermentation (SSF) process using crude cellulase from Phanerochaete chrysosporium *AIP Conference Proceedings* **2197**(1) 030007
[18] Girmay Z, Gorems W, Birhanu G and Zewdie S 2016 Growth and yield performance of Pleurotus ostreatus (Jacq. Fr.) Kumm (oyster mushroom) on different substrates *Amb Express* **6**(1) 87
[19] Zhao C H, Liu X, Zhan T and He J 2018 Production of cellulase by Trichoderma reesei from pretreated straw and furfural residues *RSC advances* **8**(63) 36233-36238
[20] Yandri Herasari D, Suhartati T and Hadi S 2009 The effect of chemical modification on the thermal stability of protease from local isolate bacteria, Bacillus subtilis ITBCCB148 *Nat Sci* **7** 68-75
[21] Apati G P, Furlan S A and Laurindo J B 2010 Drying and rehydration of oyster mushroom *Brazilian Archives of Biology and Technology* **53**(4) 945-952
[22] Jalali M, Abedi D, Varshosaz J, Najjarzadeh M, Mirlohi M and Tavakoli N 2012 Stability evaluation of freeze-dried Lactobacillus paracasei subsp. tolerance and Lactobacillus delbrueckii subsp. bulgaricus in oral capsules *Research in pharmaceutical sciences* **7**(1) 31
[23] Cahyadi W and Widiantara T 2018 Penambahan konsentrasi bahan penstabil dan sukrosa terhadap karakteristik sorbet murbei hitam *Pasundan Food Technology Journal (PFTJ)* **4**(3) 218-224
[24] De Jesus S S and Maciel Filho R 2014 Drying of α-amylase by spray drying and freeze-drying-a comparative study *Brazilian Journal of Chemical Engineering* **31**(3) 625-631
[25] Permata D A, Ikhwan H and Aisman A 2016 Aktivitas proteolitik papain kasar getah buah pepaya dengan berbagai metode pengeringan *Jurnal Teknologi Pertanian Andalas* **20**(2) 58-64

IOP Conf. Series: Materials Science and Engineering 1073 (2021) 012010 doi:10.1088/1757-899X/1073/1/012010

The study of xylanase immobilize enzyme using sol-gel method

S Hadiantoro, Y Maryanty*, D R Wulan, S A Putri, E M Putra and N S Achmadin

Chemical Engineering Department, State Polytechnic of Malang, Jalan Soekarno-Hatta No. 9 Malang, Indonesia

*yanty.maryanty@polinema.ac.id

Abstract. Xylanase has many benefits, but many obstacles to produce it. Common microorganisms produce fungi and bacteria such as Aspergillus niger, Trichoderma viride, and Bacillus subtilis. Various methods can do enzyme immobilization; one of them is entrapment in the form of sol-gel. This study discusses the xylanase activity. The enzymes have produced from Aspergillus niger, Trichoderma viride, and Bacillus subtilis. After that, the xylanase enzyme has immobilization using the sol-gel method. This study is doing by lab work and study literature. Lab work is for knowing fiber amount by van Soest method with amount of % hemyselulosa 4,8% 7,33%, 9,1% for media 0 months, two months, and four-month. The study literature doing with compare xylanase amount by Aspergillus niger, Trichoderma viride, and Bacillus subtilis the highest enzyme activity by Trichoderma viride with optimum condition temperature 25°C, pH five and incubation time 36 hours with enzyme activity 26,67 U/mL. The immobilization obtained xylanase activity of 28.88 U / mL. The matrix composition of TEOS, 5% xylan, and 5% calcium alginate during fermentation for 168 hours. This value is higher than the fermentation process without immobilization carried out for 72 hours with an activity value of 22.04 U / mL.

1. Introduction

Second of the most significant plant constituent components, in addition to cellulose, is xylan, which covers cell wall material of annual plants accounts for 30, 15–30% of hardwoods and 7–10% of softwoods. Xylan is a polymer from xylose, which has founded in hemicellulose. Hemicellulose, lignin, and cellulose are components that make up the wooden structure.

Xylanase (EC 3.2.1.8) is an enzyme that plays a role in the hydrolysis of xylan (hemicellulose) into xylooligosaccharides and xylose. The application of this enzyme is quite widespread, among them are used in the pulp industry for the eco-friendly process of deinking and bleaching [1]; in livestock feed industry; in food and beverages industry for dough handling improvement, for the extraction of coffee, plant oils, for clarification of fruit juices [2] and starch assisted in the pre-treatment of biomass in the bioenergy industry [3].

Xylanase has various benefits, but many obstacles have been an encounter in the production process. One of them is the lack of availability of superior xylanase-producing microbial cultures. The use of microbes as enzyme-producing agents has several advantages, including high speed of microbial growth so that it can be produced in a short time, easy to control, and relatively cheap production costs [4]. Xylanase can have been delivering from bacteria, fungi, and yeast, including Aspergillus, Trichoderma, Streptomycetes, Phanerochaetes, Chytridiomycetes, Ruminococci, Fibrobacteres, Clostridia, and Bacilli

[3,5–8]. Xylanase producing organisms require xylan as a carbon source. Because xylan has a relatively high price, the use of pure compounds directly in the production medium is quite expensive.

Xylan is an essential component in the agricultural biomass that contains hemicellulose on the cell wall of the plant. Xylan components are also abundant in agrarian waste such as wheat bran 12.3%, Bagas sugarcane 9.6% and rice husk 12.1% [3]. Based on the xylan content in the trunk is quite high, then in this research used wood powder from Sengon tree as a source of carbon. Wood powder from the Sengon tree that has utilized as media grows oyster mushroom contains an amount of cellulose about 45.42%, and the rest is hemicellulose 21%, lignin 26.50% and ash 7.08% [9]. Hemicellulose contained in wood powders can produce xylan after the process of de lignification by white weathered mushrooms.

Enzyme immobilization is a process whereby the movement of enzyme molecules in the chamber where the reaction has held in such a way that the enzyme system forms active and insoluble water [10]. Enzyme immobilization synthesis with various methods. The selection of enzyme immobilization methods depends on how technology affects the activity of enzyme catalysts. There are several methods of immobilization of enzymes with multiple types of support, namely: entrapment, carrier-binding, adsorption, and cross-linking. Among these three methods, one new approach to development in the mobilization of enzymes is the method of encapsulation/entrapping. The encapsulation method is the packing of a protein molecule in the sol-gel matrix that is not covalent, but the particle has physically trapped in the tissue of the gel that has grown around it.

2. Material and methods

This literary study was the information to analyze the ability of *Trichoderma reesei*, *Aspergillus niger*, and *Bacillus subtilis* in the production of cellulase enzymes. The Freeze-Drying method use to acquire enzymes in the form of dry cellulose (powder). The addition of sucrose as a *carrier agent has* expected to maintain a stable enzyme that has remained a steady freeze-drying process. The methodology below is a result of studies and comparison of several previous journals that will be used as a reference when researching in the laboratory.

2.1. *Pra-treatment media*

Media that contains a mixture of wood powder sengon, Bran and lime are placed in a tray to be in the count to reduce the size of the media. The media that has minced has dried under the heat of the Sun for a full day. After experiencing the drying process is expected Moisture content in the media can be less and the media has sifted on 80 mesh to expand the surface area, the enzyme has a substantial enough chance to degrades the cellulose in the media.

2.2. *Media character analysis (NDF degradation analysis, ADF, cellulose, hemicelluloses, and lignin)*

Dry solids (residue) can have used to determine the content and degradation of NDF, ADF, cellulose, and hemicellulose using *Van Soest* method.

2.3. *Media preparation*

Pleurotus ostreatus media in the form of a mixture is taken as much as 1.8 grams and added nutrients in the way of BHM (Basal Hard Medium) to 18 ml. Subsequently, the media was given an inoculum suspension for the samples. To be tested and 1 blank without being given inoculum microorganisms.

2.4. *Enzyme production*

Three 250 ml Erlenmeyer has prepared, each containing 150 ml of liquid media, then sterilized in an autoclave at a temperature of 121 °C with a pressure of 15psi for 15 minutes. After cooling, add 15 ml of inoculum aseptically. Furthermore, the media has incubated in a shaker. The speed of 150 rpm at room temperature for 98 hours, after which the enzyme isolation has to carry out. The enzyme filtrate was centrifuged at 2000 rpm for 20 minutes at 4 °C. The supernatant obtained was a crude extract of the xylanase enzyme. The xylanase enzyme crude extract has then tested for enzyme activity.

2.5. *Immobilized with sol-gel method*
In general, the sol-gel method process can be summarize as follow: preparation precursors solution, hydrolysis and partial condensation of alkoxides forming “sol”, polycondensation of hydrolyzed precursors forming “gel”, drying (solvent evaporation) forming a dense “xerogel” via collapse of the porous network [11].

3. Results and discussion
Based on several journals used in analyzing differences in xylanase enzyme activity against Aspergillus niger, Trichoderma viride, and Bacillus subtilis with various substrates from various kinds of literature. Knowing how the immobilization process with the sol-gel method from some journals.

3.1. *Trichoderma viride*

Table 1. Literature study of the activity of the enzyme xylanases from *Trichoderma viride*.

Media	Temperature (°C)	pH	Incubation time (Hours)	Enzyme (U/ml) activity	Source
Corn Klobot	25	5	60	14,53	[12]
Corn Klobot	30	5	36	17,01	[13]
CORN COB	25	5	36	26,67	[10]
Corn Klobbot	40	5	60	2,97	[14]

Research conducted *Trichoderma viride* [12]. Cornhusk used as a substrate with *Trichoderma viride* inoculum. Incubation time 60 hours at room temperature ± 25oC. Immobilization has carried out with a bentonite matrix with a shaking time of 1, 2, 3, 4, and 5 hours. The optimum time for shaking was 3 hours, with the enzyme activity of 10.245 units/ml. A study related to xylanase immobilization using a Ca-Alginate-chitosan. Matrix capable of producing enzymes with an activity value of 17.008 U/ml, with optimum conditions at pH 5, temperature 30 °C, and an incubation time of 36 hours [13].

Trichoderma viride can produce endo-1.4-β-xylanase enzymes that can degrade xylan [10]. Corn Cob can have utilized as the media production of the enzyme xylanases from *Trichoderma viride*. The increase xylanases enzyme activity of *Trichoderma viride* needs to has done optimizing by combining several variables that can affect the activity of enzymes such as temperature, pH, the concentration of enzymes, substrate concentration, activator, and inhibitors. This study aims to determine the effect of temperature and pH on the xylanase enzyme activity of *Trichoderma viride*. Growing media in the form of corn cobs. The media contains lignocellulose. The highest xylanase enzyme activity has obtained at a temperature of 25 °C at pH 5 and an incubation time of 36 hours. The value of xylanase enzyme activity was 26.67 units.

At the influence of temperature, enzyme activity at 60 °C storage has the highest activity. Based on the graph produced at 60 °C is the optimum temperature. At a temperature of 70 °C, the enzyme activity is the lowest because it has undergone denaturation. Denaturation can cause conformational changes in the enzyme. The amount of substrate can have bounded by the active site of the enzyme decreases, and the enzyme activity decreases. Enzyme activity at 40oC at storage time of 15-25 hours has lower activity when compared to storage at 30 °C. This condition can have cause by the presence of the protease enzyme, which has an optimum temperature of 35°C. In this condition, the xylanase enzyme has degraded more. The stability of the xylanase enzyme can have seen from the activity of the remaining enzymes. The enzyme is said to be stable when the remaining enzyme activity is more than 50% of the initial enzyme activity. The activity of the residual enzymes decreases with the more extended the xylanase is stored. The highest residual enzyme activity is at a storage temperature of 60°C.

Meanwhile, the lowest residual enzyme activity has stored at 70 °C. In general, and xylanase is still stable up to 25 hours of storage. However, for storage of 40 °C, xylanase is stable for up to 20 hours, with the residual enzyme activity of 56.95% (2.973 units). Permanent xylanase enzyme storage at 70 °C, for up to 15 hours with residual enzyme activity 54.88% (2.865 units).

From several studies discuss the incubation time. The xylanase enzyme activity was 26.67 U / ml with an incubation time of 36 hours, 14.528 U / ml with an incubation time of 60 hours [12]. This difference in incubation time affects the value of enzyme activity. The longer the incubation time, the lower the enzyme activity value. From the data, conducted with an incubation time of 36 hours when adjusted for the variables that future researcher can do at 40, 45, 50, 55, and 60 minutes, it is very different. The different incubation times will produce different values so that in the experiment, and the incubation time will have extended so that future researcher can find out the amount of the enzyme activity. From 0 hours, the value of enzyme activity will be seen whether the amount increases with increasing incubation time.

3.2. *Aspergillus niger*

Fungi growth is very related to breeding that affects the increase in the number or volume of cells [4]. The reproduction of fungi is asexual in general is through the formation of spores, to know the growth can have done the calculation of spore density. Spore density is measured using hemacytometer based on the way, which is done every consecutive day until the densest spores to has used as a working culture. The relationship between the number of spores *Aspergillus niger* with the time of spore calculation has shown in Figure 1, seen increasing the number of spores a. Niger until the sixth day. Based on Figure 1, the number of spores *A. Niger* increased from the first day, i.e., 1, $12x10^7$ cells/ML until the second day entered the exponential phase of the cell multiplication phase.

Moreover, reached the peak of stationary on the third day, with the number of spores from the highest *A. Niger* that is 1, $65x10^8$ cells after the fourth day the number of spores A. Niger starts to decrease but relatively still stable until the sixth day with the number of spores 1, $46x10^8$ cells/ml. On the fourth day, it begins the death phase. Thus, to obtain the optimum production of the enzyme is done harvesting at incubation time 2-3 days because, in that phase, the number of Spore *A. Niger* is at the peak point. With some of the spores, *Aspergillus niger* has expected its enzymatic production is also high.

Table 2. Literature Study of xylanase enzyme activity of *Aspergillus niger*.

Media	Temperature (°C)	pH	Incubation Time (Hours)	Enzyme (U/ml) activity	Source
Bran	30	6	120	1.4	[15]
Rice straw	37	6	56	0,055	[4]

The fermentation process was carried out in a Shaker incubator at 37 °C, stirring at 200 rpm and pH 6. The reduction test of sugar content used the DNS method to obtain enzyme activity. The results showed that the addition of molasses to rice straw media could increase the growth of *A. niger* but did not significantly increase the xylanase enzyme activity and required a longer incubation time. The most optimal concentration of molasses for xylanase enzyme production is 1%, with the highest enzyme activity of 0.055 U / ml and an incubation time of 56 hours.

Based on the research in Table 2, the optimum condition gained; differently 56 hours, and 120 hours for the optimum condition incubation time. At the optimum state of pH obtained at pH 6. The optimum temperature varies from 37 °C [4] and 30°C [15]. The difference quite far lies in the value of enzyme activity of 0.055 U/ml and 1.4 U/ml.

The difference between the two studies affects the value of its enzymatic activity. The longer the incubation time than the smaller the value of the enzymatic activity. However, if the incubation time equation does further research, then the value of the resulting enzyme activity will not differ considerably. It has expected to conduct experiments using the conditions made by Kanimozhi and Nagalakshmi [15] at pH 6; temperature 30 °C and the incubation time 120 with the test data that future research can have obtained optimum condition. The variables we use are in the range of pH 5, 6, 7, 8, 9; Temperatures (40, 45, 50, 55, 60) °C and incubation times 40, 45, 50, 55, and 60 minutes. If based on Wahyu et al. [4] research, future researcher need to adjust from the incubation time and the temperature.

3.3. *Bacillus subtilis*

Based on research conducted by Erika et al. [16] in the production of xylanases using the Inoculum *Bacillus subtilis* which is grown in 50 ml of sterile media with the composition: xylan beechwood 0.5%; *Yeast extract* 0.35%; Tryptophan 0.35%; NaCl 0.2%; KH_2PO_4 0.245%; $MgSO_4$ 0.035%; $(NH_4)_2SO_4$ 0.175%) and Corn Cob. The determination of xylanases activity has done by the method of 3.5-dinitrosalicylic acid (DNS). The enzyme activity is measured based on the absorption rate of the sample solution at 575 nm wavelength. As a blank, the addition of enzymes into the reaction tube has performed after the incubation process. One unit of xylanase activity (U/ml) has been defining as the number of enzymes that release the µmol of xylose per minute. The value of Xylanase activity generated is 0.2 U/ml.

One unit of xylanase activity is the number of enzymes needed to break the xylan into 1 µmol of reducing sugar per minute in test conditions. This enzyme, when manufactured using *xylan broth* Media in temperature conditions of 30°C, 150 rpm with starter containing total bacteria (2, 43, 3) x 10^8 CFU/ML or OD600 0.715 as much as 10% takes two days. The resulting xylanases enzyme works optimum at PH 8 and temperature 70 °C with the activity value of 5.17 U/ml.

Based on some studies conducted by Erika et al. [16] and Ardiansyah et al. [17]. There are different outcomes due to differing work on each researcher, and the substrate they use the optimum condition of the enzyme obtained 12 hours, while in the research of Fawzya et al. [18] received 24 hours. pH 8 is the optimum pH achieved by Fawzya et al. [19] and Ardiansyah et al. [17], while Erika gets pH six as the optimum pH of the enzyme. In the optimum temperature test, the enzyme got the result that the temperature 40°c is the optimum temperature of the enzyme [16,17], while Fawzya et al. [19] got at 70°C as the optimum temperature. Highest activity value obtained with activity value 6.088 U/ml with optimum conditions incubation time 12 hours pH eight and temperature 40°C [17].

In this experiment the range of pH 5, 6, 7, 8, 9, temperature (40, 45, 50, 55, 60) °C and incubation time 40, 45, 50, 55, and 60 minutes. When it has based on the optimum condition of Ardiansyah et al. [17]. At the incubation time, 12 Hours, PH 8, and temperature 40oC are quite the same. The optimum condition of the enzyme that we do can get optimum value because the type of media used is different from the content of the xylan in the media is also different so that each media also has a specific optimum condition that results in the difference in the value of the enzyme activity itself.

Table 3. A freeze-drying Literature study and carrier agent.

Media	Temperature (°C)	pH	Incubation time (Hours)	Enzyme (U/ml) activity	Source
CORN COB	40	6	12	0,2	[16]
Xylan broth	70	8	48	5,17	[18]
Rice husk	40	8	12	6,088	[17]

3.4. *Immobilization techniques*

The activity of xylanase of 28.88 U/ML received from a culture that has mobilized in *sol-gel* with a matrix composition of TEOS, 5% Xylan, and 5% calcium alginate at the time of fermentation for 168 hours [20]. The value is higher than that of the fermentation process, with no immobilization conducted for 72 hours with activity value 22,04 U/ML (Figure 1).

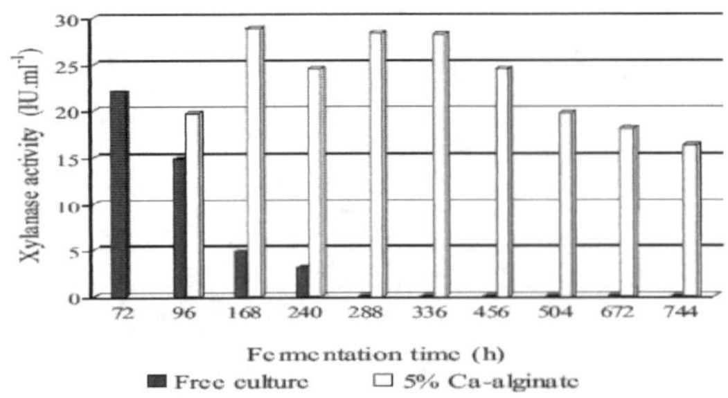

Figure 1. Production of free xylanase and the mobilization of the culture *aspergillus* on hybrid matrices with TEOS composition, 5% xylan, and 5% Calcium alginate [20].

Calcium alginate provides a positive effect on the production of the immobilization of xylanase. Mobilized cultures can maintain biosynthetic capabilities over a long period during the fermentation process. The *sol-gel* matrix used in research studies provides high porosity for the diffusion of nutrients and better products. However, a hybrid matrix determination analysis is composed of TEOS, 5% Xylan, and 5% calcium alginate as a suitable ingredient as a *sol-gel* immobilization. The mobilized culture can increase the production capacity of xylanases. Furthermore, maintain biosynthetic capability during the development process versus free culture. The use of calcium alginate as an organic component in the matrix composition also solves the problem of diffusion in the Support section.

Immobilized by using *the Sol-gel* method as an increase in enzyme activity, the research enzyme that has immobilized with *Sol-gel* matrix containing 10% PEO get the value of activity 1297.44 U/ml [20]. The amount of the activity is higher than the enzyme value that has not been immobilizing with the activity value of 542.76 U/ml. From these results, the mobilized culture can maintain the biosynthetic capability for a long time during the fermentation process. The use of calcium alginate in the hybrid composition of *the sol-gel* matrix proves that calcium alginate gives a better effect on nutrient absorption, which causes the increased value of enzyme activity and reduces fermentation time.

The main goal in the experiments conducted by Peralta-Pérez et al. [21] was to immobilize xylanases from *Aspergillus niger* in the *sol-gel* matrix. The decreased activity caused by the gel pores formed. Optimization-based of immobilization techniques and the subsequent work that is to increase the value of the activity of the mobilization enzyme. Although there is a decrease in the amount of enzyme activity, it can has used periodically.

Research conducted by Pal and Khanum [22] shows that the immobilized enzyme has increased activity value from 7092 U/ml to 8000 U/ml with Optimum pH condition of 5 to 5.5 and the temperature 40 to 45°C. Immobilized enzymes can also be used repeatedly five times and can withstand a total of 85% of the first activity value. The Immobilization method can also address the problem of decreasing xylan loss.

Enzyme immobilization Process using the Sol-gel method when comparing between various journals can be very different. So, if to immobilize using the Sol-gel method, it is good to do the research first from previous research. Which sol-gel method fits and corresponds to the research time.

An increase in the value of the activity of free enzymes with enzymes held immobilization, i.e., from 3333.33 U/ml to 5000 U/ml. In this study, pH stability and temperature could increase as well as the immobilized enzyme can have used up to 10 times with the remaining 60% of the initial activity. The maximum activity of free xylanases and xylanase mobilization is at pH 7. However, the activity of the generally mobilized enzymes is higher when compared to free enzymes. The pH stability indicates that the immobilized enzyme shows excellent stability when compared to free enzymes [23].

The process of enzyme immobilization is an increased process of enzyme activity where the value of enzyme activity has expected to stabilize within a given time. However, this process requires additional costs because the enzyme immobilization requires a matrix. Many types of matrices also affect the stability of the enzyme activity itself. In the experiments, we will do that is using hydrogel.

The enzymes that have produced will be adding to the hydrogel. In hopes of increasing the time, the enzymes in the enzyme hydrogel will have trapped in the hydrogel.

4. Conclusion

The best microorganisms capable of producing the activity value of the highest xylanase enzyme are from *Trichoderma viride* with corn cob media with the optimum condition at 25°c, pH 5, and incubation time 36 hours with activity value of 26.67 u/ml. An enzyme that has immobilized with a *sol-gel* matrix containing 10% PEO gets the amount of 1297.44 u/ml activity. The value of the activity is higher than the enzyme value that has not been immobilizing with an activity value of 542.76 u/ml. From these results, the mobilized culture can maintain the biosynthetic capability for a long time during the fermentation process. Although there is a decrease in the value of enzyme activity, it has advantages that have use periodically.

Acknowledgments

The authors would like to acknowledge The Ministry of Education and Culture of Republic Indonesia for financial research support through "Penelitian DIPA Riset Terapan 2020" State Polytechnic of Malang.

References

[1] Walia A, Guleria S, Mehta P and Chauhan A 2017 Microbial xylanases and their industrial application in pulp and paper biobleaching : a review *3 Biotech* **7** 1–12

[2] Ghimire S, Shakep B, Sailesh P, Bimal T and Bhupal G S 2016 Isolation and Screening of Potential Cellulolytic and Xylanolytic Bacteria from Soil Sample For Degradation of Lignocellulosic Biomass *The Journal of Tropical Life Science* **6**(3) 165 – 169

[3] Richana N, Lestina P and Tedja Irawadi T 2014 Karakterisasi Lignoselulosa dari Limbah Tanaman Pangan dan Pemanfaatannya untuk Pertumbuhan Bakteri RXA III-5 Penghasil Xilanase *J. Tanam. Pangan* **23**

[4] Wahyu N U R, Pangesti I and Pangastuti A 2012 Pengaruh penambahan molase pada produksi enzim xilanase oleh fungi Aspergillus niger dengan substrat jerami padi *Bioteknologi* **9** 41–8

[5] Houfani A A, Větrovský T, Navarrete O U, Štursová M, Tláskal V, Beiko R G ... and Jorquera M A 2019 Cellulase– Hemicellulase Activities and Bacterial Community Composition of Different Soils from Algerian Ecosystems *Microbial ecology* **77**(3) 713-725

[6] Sunna A and Antranikian G 1997 Xylanolytic enzymes from fungi and bacteria *Crit. Rev. Biotechnol.* **17** 39–67

[7] Bala A and Bijender S 2018 Cellulolytic and xylanolytic enzymes of thermophiles for the production of renewable biofuels *Renewable Energy* 1-61

[8] Bischof R H, Jonas R and Bernhard S 2016 Cellulases and beyond: the first 70 years of the enzyme producer *Trichoderma reesei. Microbial Cell Factories* **15** 106

[9] Hartati S, Sudarmonowati E, Fatriasari W, Hermiati E, Dwianto W, Kaida R, Baba K and Hayashi T 2010 Wood Characteristic of Superior Sengon Collection and Prospect of Wood Properties Improvement through Genetic Engineering *Wood Res. J.* **1** 103–5

[10] Sutrisno S, Roosdiana A, Prasetyawan S and Sari I P 2017 Determination of Immobilization Optimum Conditions of Trichoderma viride's Xilanase on Acid-Activated Zeolite Matrix *JKPK (Jurnal Kim. dan Pendidik. Kim.* **2** 97

[11] Esposito S 2019 "Traditional" Sol-Gel Chemistry as a Powerful Tool for the Preparation of Supported Metal and Metal Oxide Catalysts *Materials (Basel).* **12** 1–25

[12] Mardiana, Sutrisno and Mahdi C 2013 Optimasi Amobilisasi Xilanase dari Trichoderma viride Menggunakan Matriks Bentonit *Kim. Student J.* **2** 435–40

[13] Mesla W, Mahdi C, Kimia J, Matematika F, Alam P and Brawijaya U 2014 Optimasi Amobilisasi Xilanase dari Trichoderma viride Menggunakan Matriks Ca-Alginat-Kitosan *Kim. Student J.* **2** 428–34

[14] Sukmana M E, Sutrisno and Roosdiana A 2014 Pengaruh Suhu dan Lama Penyimpanan Terhadap Kestabilan Enzim Xilanase Dari Trichoderma viride *Kim. Student J.* **2** 340–4

[15] Kanimozhi K and Nagalakshmi P K 2014 Original Research Article Xylanase production from Aspergillus niger by Solid State Fermentation using Agricultural waste as substrate *Int. J. Curr. Microbiol. App. Sci.* **3** 437–46

[16] Erika E, Agustrina R, Sumardi S and Mulyono M 2016 OPTIMASI MEDIA PRODUKSI ENZIM XILANASE DARI Bacillus sp. (*Medium Optimization of Xylanase Production from Bacillus sp.*) *J. SELULOSA* **6** 19–26

[17] Ardiansyah Y T, Mulyani N S and Sarjono P R 2014 Isolasi dan Karakterisasi Enzim Xilanase dari Bacillus Subtilis pada Media Nutrient Broth dengan Penambahan Xilan Hasil Isolasi Jerami Padi *J. Kim. Sains dan Apl.* **17** 95–9

[18] Fawzya Y N, Prima R E, Mangunwardoyo W, Munifah I and Patantis G 2013 Produksi dan Karakterisasi Xilanase dari Isolat Bakteri M-13.2a Asal Air Laut Manado *J. Pascapanen dan Bioteknol. Kelaut. dan Perikan.* **8** 64

[19] Fawzya Y N, Prima R E, Mangunwardoyo W, Munifah I and Patantis G 2013 Produksi dan Karakterisasi Xilanase dari Isolat Bakteri M-13.2a Asal Air Laut Manado *J. Pascapanen dan Bioteknol. Kelaut. dan Perikan.* **8** 55

[20] Evstatieva Y, Yordanova M, Chernev G, Ruseva Y and Nikolova D 2014 Sol-Gel immobilization as a suitable technique for enhancement of A-Amylase activity of aspergillus oryzae PP *Biotechnol. Biotechnol. Equip.* **28** 728–32

[21] Peralta-Pérez M R, Martínez-Trujillo M A, Nevárez-Moorillón G V, Pérez-Bedolla R and García-Rivero M 2011 Immobilization of Aspergillus niger sp. in sol gel and its potential for production of xylanases *J. Sol-Gel Sci. Technol.* **57** 6–11

[22] Pal A and Khanum F 2011 Covalent immobilization of xylanase on glutaraldehyde activated alginate beads using response surface methodology: Characterization of immobilized enzyme *Process Biochem.* **46** 1315–22

[23] Nagar S, Mittal A, Kumar D, Kumar L and Gupta V K 2012 Immobilization of xylanase on glutaraldehyde activated aluminum oxide pellets for increasing digestibility of poultry feed *Process Biochem.* **47** 1402–10

IOP Conf. Series: Materials Science and Engineering 1073 (2021) 012011 doi:10.1088/1757-899X/1073/1/012011

Hydrothermal synthesis and characterization of nano-particles γ-Al_2O_3

H P Buwono*, S Adiwidodo, H Wicaksono and H I Firmansyah

Mechanical Engineering Department, Politeknik Negeri Malang, Jl. Soekarno Hatta No. 9, Jatimulyo, Kec. Lowokwaru, Kota Malang, Jawa Timur 65141, Indonesia

*haris.puspito@polinema.ac.id

Abstract. The study of hydrothermal treatment on the phase formation and the crystallites of gamma-alumina (γ-Al_2O_3) was performed. Different reaction time and temperature of hydrothermal treatment were investigated. The preparation of γ-Al_2O_3 includes precipitation of aluminum nitrate nonahydrate by ammonia, aging, washing, transferred the gel into the hydrothermal reactor, drying, and calcination at 600 °C. The precipitation was stopped at pH equals 7.5. The reaction times of hydrothermal treatment were 1 – 2 hours, and the temperatures were 140 – 200 °C. X-Ray Diffraction analysis was used to investigate the phase formation and the particle size. The results show all the samples represented pure nano γ-Al_2O_3. The examination of hydrothermal treatment of γ-Al_2O_3 reveals as the reaction time longer, the crystallite grows bigger, as well as the higher temperature applied. The smallest γ-Al_2O_3 obtained based on the Scherrer equation is 4 nm and the biggest is 6.5 nm. Interestingly, the peaks of the γ-Al_2O_3 prepared by hydrothermal treatment at the longest time and the highest temperature have shifted into a higher diffraction angle. In the case of this sample, the peaks shifted into a higher diffraction angle. That is means there is a change in the lattice structure of γ-Al_2O_3 probably was affected by the lattice of hydroxyl content.

1. Introduction

Alumina, Al_2O_3, is a ceramic material used in across-the-board applications for adsorbents, abrasives, catalysts, and catalyst supports [1–7]. The extensive use of Al_2O_3 in heterogeneous catalysis has ascended since its high chemical and thermal stability as well as its desirable textural properties such as porosity, surface area, and acidity/basicity [3,4,8]. Alumina has many phases depend on the condition. The phase transformation of alumina is closely related to temperature [9–11]. The most stable phase is alpha-alumina (α-Al_2O_3) present at a temperature higher than 1100 °C. The highest specific surface area is gamma-alumina (γ-Al_2O_3) present at 400 – 600 °C. So that γ-Al_2O_3 is a metastable phase among alumina phases. Gamma-alumina is also frequently used as a support of heterogenous catalysts due to its high specific surface area, high thermal stability, and desirable acidity/basicity. It is also used widely as catalysts and adsorbents for chemical processes [2,12–19]. Commercially γ-Al_2O_3 has been prepared through precipitation and sol-gel route [1,2,8,9].

Nowadays the attention has shifted towards nanostructured materials since changing the morphology at the nanoscale can probably lead to materials with different physical and chemical properties, regardless of the widespread use of bulk γ-Al_2O_3 [2,9]. By controlling the morphology, it can lead to selective contact of support and metal so that different metal support interactions are observed [20–24].

These specific interactions can offer stability to metal nanoparticles. The surface properties of the nanoparticles also play an important role in their catalytic properties [19,25,26]. For instance, the surface area of TiO was closely related to a hydrothermal parameter such as filling factor, temperature, stirring, raw materials specific surface area, and reaction time [27]. The hydrothermal treatment also affected the total acidity, acidity distribution on γ-Al_2O_3. However the textural properties of the γ-Al_2O_3 like specific surface area prepared by hydrothermal treatment almost the same to those samples prepared by precipitation, as well as the pore volume of samples [28]. The hydrothermal treatment also affected the binding energy of Pt on γ-Al_2O_3 so that it increases the selectivity of propane dehydrogenation and its turn over frequency (TOF) [25].

To prepare nanoparticle different structure-direct-agent alike surfactant and others were used in the synthesis using either room temperature such as precipitation or hydrothermal treatment method under high temperature over 100 ^{0}C and high pressure by autoclave reactors [1,15,29]. In this present work, we prepared nanoparticle of γ-alumina by hydrothermal treatment compared with precipitation method. The samples crystallized well and the growth of diameter crystallize size was observed at hydrothermal treatment. Then we investigated the effect of crystallization temperature and reaction time under hydrothermal treatment. The prepared γ-alumina was characterized by X-Ray Diffraction (XRD) analysis.

2. Experimental

2.1. Preparation

Gamma-alumina (γ-Al_2O_3) was prepared from aluminum nitrate nano hydrate (Merck, 98.5% in wt.). All the chemicals were used as received without any further purification. Initially, aluminum nitrate nano hydrate was dissolved in water under vigorous stirring. The molar ratio of water to aluminum nitrate nano hydrate was equal to 1. An aqueous ammonia solution (Merck, 25% in wt.) was added dropwise until the pH of the supernatant was 7.5. After aging for 1 hour, the slurry was transferred into a hydrothermal reactor and it was treated at 140 – 200 ^{o}C for 1 – 2 hours. The white gel thus obtained was recovered by filtration and washed with deionized water. After drying overnight at 110 °C, the solid thus obtained was calcined in air at 600 °C for 6 h. As a reference, another γ-Al_2O_3 was prepared by precipitation method in the same way as the above without hydrothermal treatment.

2.2. Characterization

Powder X-Ray Diffraction (XRD) analysis was performed using monochromatic Cu Kα radiation (PanAnalytical). The diameter crystallite sizes were calculated by the Scherrer equation.

3. Results and discussion

3.1. Peak intensity

Figure 1 shows the XRD pattern of γ-Al_2O_3 synthesized by precipitation (1a) and hydrothermal treatment (1b-d) after calcination at 600 ^{o}C, for 6 h. Three intense peaks were identified. According to US Patent US Patent No. US 2009/0104.108 1A, three peaks of typical γ-Al_2O_3 appeared at 38°, 45°, and 67° after calcined at 600 °C for 6 h [30]. According to Rahmanpour, the peaks of calcined g-alumina at 550 ^{o}C for 4 h appeared at 37.8^o, 45.7 ^{o}, and 66.9 ^{o} [2]. Moreover, the pattern was compared with Joint Committee on Powder Diffraction Standards (JCPDS) 29-0063 as supplied in the supporting information (see Figure S1) and it also showed three intense peaks at 38°, 45°, and 67° assigned to (3 1 1), (4 0 0), and (4 4 0) consecutively with the lattice is face-centered cubic (fcc). Therefore, the prepared material represents the single phase of γ-Al_2O_3 so that it was synthesized successfully via both precipitation and hydrothermal treatment.

Figure 1 also shows the effect of hydrothermal treatment on the diffraction peaks intensity. All γ-Al_2O_3 prepared by hydrothermal treatment (Figure 1 b-d) shows the diffraction peaks intensity higher than that prepared by precipitation. It also shows under hydrothermal treatment that as the temperature increase, the diffraction peaks intensity increases as well as the reaction time. That figure shows that under hydrothermal treatment, the particles were recrystallized and it grew into bigger particles. It is also clear that the γ-Al_2O_3 under hydrothermal treatment has good crystallinity compared with that prepared by precipitation.

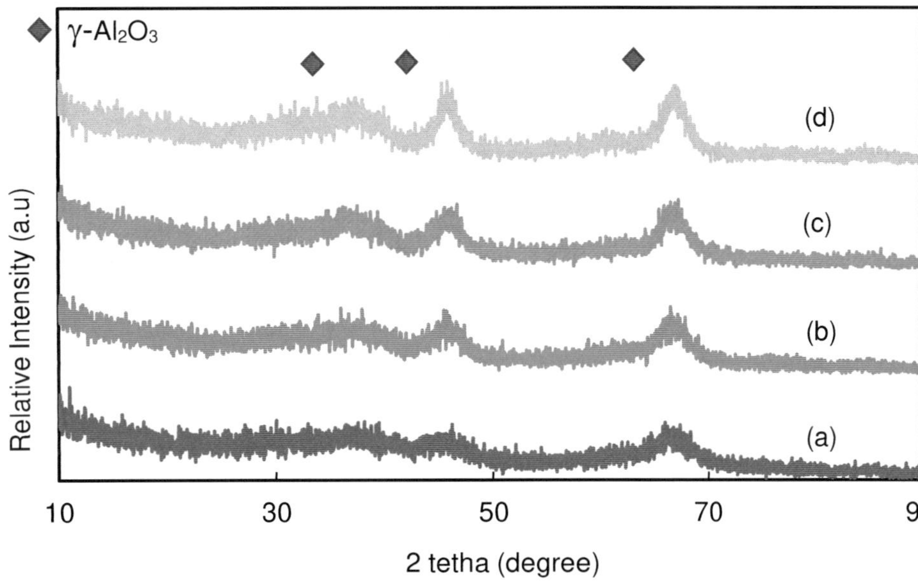

Figure 1. X-Ray Diffraction of γ-Al_2O_3 after calcination at 600 °C for 6 h for precipitation (a) and hydrothermal (b-d) at different conditions b) 140 °C for 1 h, c) 140 °C for 2 h, and d) 200 °C for 2 h.

3.2. Average crystallite size

The boarding of the peaks occurs in the g-alumina prepared by precipitation as shown in Figure 1 (a). The hydrothermal treatment at 140 °C – 200 °C for 1 – 2 h leads to more matching of the (d) spacing and increase the intensity. That is means the crystallites grow bigger than that prepared by the precipitation method. To confirm the size of the diameter crystallite, Scherrer calculation was performed. The Scherrer equation was shown in equation 1 where D is the diameter crystallite, k is a dimensionless shape factor, λ is the X-ray wavelength, β is the line broadening at half the maximum intensity (FWHM), and θ is the Bragg angle.

$$D = \frac{k\ \lambda}{\beta \cos\theta} \quad (1)$$

Table 1. Diameter crystallite sizes of γ-Al_2O_3 after calcination at 600 °C for 6 h for precipitation (a) and hydrothermal (b-d) at different conditions b) 140 °C for 1 h, c) 140 °C for 2 h, and d) 200 °C for 2 h.

Identity	Sample	D (nm)
a	γ-Al_2O_3 - Precipitation	3.73
b	γ -Al_2O_3 - HT 140 °C, 1h	4.08
c	γ -Al_2O_3 - HT 140 °C, 2h	4.59
d	γ -Al_2O_3 - HT 200 °C, 2h	6.53

Table 1 shows the diameter of the crystallite size. It shows the crystallite sizes of samples prepared by hydrothermal treatment growth bigger than that prepared by precipitation. During hydrothermal treatment, the crystallite restructure. Jun-Cheng et. al showed the hydrothermal treatment of γ-Al_2O_3 for prolongation of the reaction time led to a decrease in the specific surface area which closely related to the overgrowth of the surface crystallites [28]. Based on the crystallite size, it could be inferred that the bigger crystallite size, the surface area probably also decreased. The specific surface area of samples at moderate hydrothermal conditions (140 °C, 1 – 2 h) is quite the same. However, the specific surface area at a high temperature (200 °C), even for moderate reaction time, is decreased.

3.3. Shifted peaks at a higher diffraction angle

Figure 2 shows the XRD of samples at 55 – 75 degrees. It clearly shows that at the moderate hydrothermal condition (140 °C, 1-2 h) the peaks quite the same with the sample prepared by precipitation. However, at the high temperature (200 °C) the peaks shifted into a higher diffraction angle. Compared with samples prepared by precipitation, the shifted peaks at hydrothermal treatment were observed for ZnS, $ZnAl_2O_4$, and $BaTiO_3$ [31–33]. Hayashi and Ebina showed that as the hydrothermal temperature increase, the peaks intensity of diffraction pattern increase and shifted, and the lattice hydroxyl content decreased thus, tetragonality might be closely related to the lattice hydroxyl content along with to the particle size [33]. A similar state probably applied to this sample that at the high-temperature hydrothermal treatment the peaks intensity increased and shifted so that the structure was affected by the lattice of hydroxyl content.

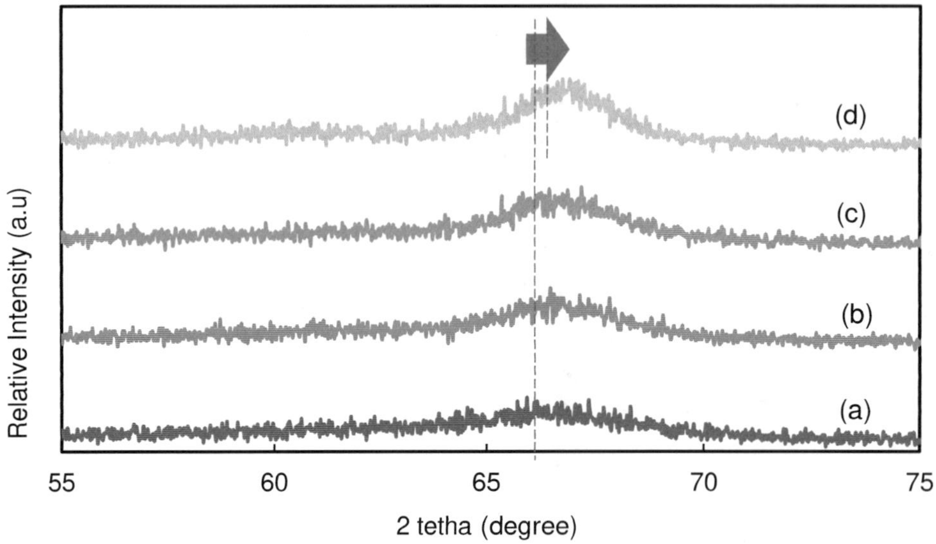

Figure 2. X-Ray Diffraction of γ-Al_2O_3 after calcination at 600 °C for 6 h for precipitation (a) and hydrothermal (b-d) at different conditions b) 140 °C for 1 h, c) 140 °C for 2 h, and d) 200 °C for 2 h.

4. Conclusion

Nanoparticles of γ-Al_2O_3 were successfully synthesized through hydrothermal treatment. The crystallites size diameter of γ-Al_2O_3 prepared by a hydrothermal treatment was bigger than those prepared by precipitation. They are 4 – 6.5 nm, in contrast to those prepared by precipitation (3.7 nm). The prolonged hydrothermal treatment time, as well as the higher temperature, the crystallites grow bigger. The shifted peaks into higher diffraction angle of γ-Al_2O_3 were observed on the 200 °C and 2 h of hydrothermal treatment indicating the crystal lattices were affected by the lattice of hydroxyl content.

5. Supporting information

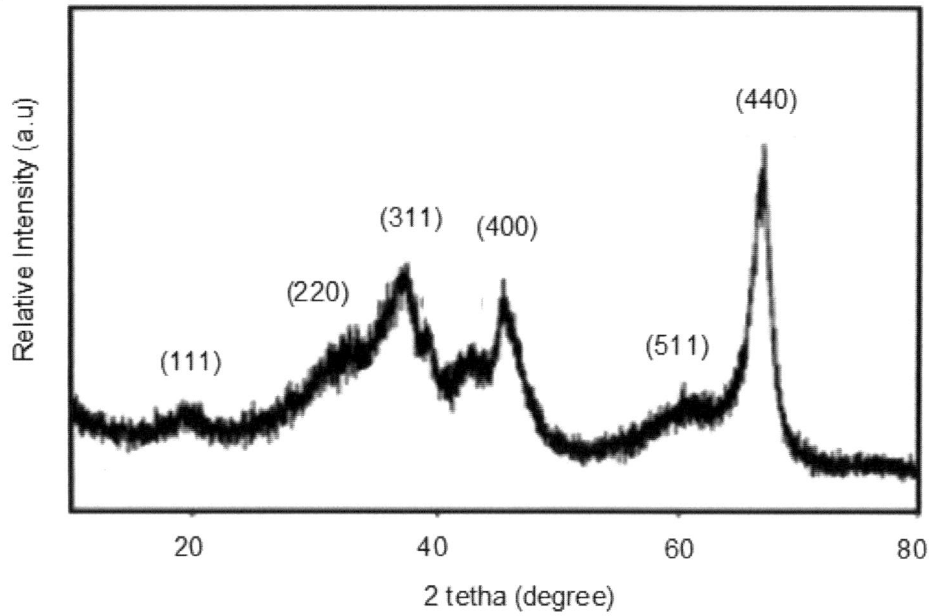

Figure 3. X-Ray Diffraction (XRD) patterns of gamma-alumina (γ-Al_2O_3) of Joint Committee on Powder Diffraction Standards (JCPDS) 29-0063.

Acknowledgments
The author would like to thank you to UPT P2M Politeknik Negeri Malang DIPA No. SP DIPA 023.18.2.677606/2020 with contract no. 5272/PL2.1/HK/2020 for financial support.

References

[1] Bell T E, González-Carballo J M, Tooze R P and Torrente-Murciano L 2017 *RSC advances* **7**(36) 22369-22377
[2] Rahmanpour O, Shariati A and Nikou M R K 2012 *Int J Chem Eng Appl.* 125–8
[3] Yao H C, Japar S and Shelef M 1977 *J Catal* **50**(3) 407–18
[4] Burch R and Loader P K 1996 *Appl Catal A Gen.* **143**(2) 317–35
[5] Weckhuysen B M, Verberckmoes A A, Debaere J, Ooms K, Langhans I and Schoonheydt R A 2000 *J Mol Catal A Chem.* **151**(1–2) 115–31
[6] Shi H, Chen J, Yang Y and Tian S 2014 *Fuel Process Technol.* **118** 161–70
[7] Bautista F M, Campelo J M, Garcia A, Luna D, Marinas J M and Romero A A et al. 1994 *J Catal.* **145** 107–25
[8] Keshavarz A R, Rezaei M and Yaripour F 2011 *J Nat Gas Chem.* **20**(3) 334–8
[9] Pimsorn P, Tipakonthitikul R, Niyompan A, Kanchanarat N and Tunkasiri T 2008 *Adv Mater Res.* **55–57** 653–6
[10] Paranjpe K Y 2017 *Pharma Innov.* **6**(11) 236–8
[11] Khairani R, Setiawan H and Sunendar B 2016 *J Keramik dan Gelas Indones.* **25**(2) 89–95
[12] Chen Y-W and Wang W-J 2000 *React Kinet Catal Lett.* **71**(1) 41–6
[13] Buwono H P, Minami S, Uemura K and Machida M 2015 *Ind Eng Chem Res.* **54**(29) 7233–40
[14] Park J, Sato H, Shibata K, Shao S, Nishihama S and Yoshizuka K 2012 *J Ion Exch.* **23**(3) 82–7
[15] Ali S, Abbas Y, Zuhra Z and Butler I S 2019 *Nanoscale Adv.* **1**(1) 213–8
[16] Bahrin D, Subagjo S and Susanto H 2016 *Bull Chem React Eng Catal.* **11**(1) 93–9

[17] Liu Y, Wang X, Tian F and Jia C 2004 *Cuihua Xuebao.* **25**(9) 721–6
[18] Ibrahim M M 2019 *J Environ Chem Eng.* **7**(1) 102848
[19] Zhang H, Ruan Y, Feng Y, Su M, Diao Z and Chen D 2019 *J Colloid Interface Sci.* **536** 180-8
[20] Srifa A, Viriya-empikul N, Assabumrungrat S, Faungnawakij K, Stakheev A Y and Kustov L M 2016 *J Catal.* **5**(1) 3693–705
[21] Lee J, Burt S P, Carrero C A, Alba-Rubio A C, Ro I and O'Neill B J 2015 *J Catal.* **330** 19–27
[22] Jun J W, Suh Y W, Suh D J and Lee Y K 2018 *Catal Today.* **302**(September 2016) 108–14
[23] Arun N, Sharma R V and Dalai A K 2015 *Renew Sustain Energy Rev.* **48** 240–55
[24] Stakheev A Y and Kustov L M 1999 *Appl Catal A Gen.* **188**(1) 3–35
[25] Mironenko R M, Belskaya O B, Talsi V P, Gulyaeva T I, Kazakov M O and Nizovskii A I 2014 *Applied Catal A: Gen.* **469** 472–82
[26] Li J, Xiang L, Feng X and Wang Z 2008 **254** 2077–80
[27] Safaei M, Sarraf-Mamoory R, Rashidzadeh M and Manteghian M 2010 *Physica Status Solidi C* **7**(11–12) 2727–30
[28] Jun-Cheng L, Lan X, Feng X, Zhan-Wen W and Fei W 2006 *Appl Surf Sci.* **253**(2) 766–70
[29] Mahajan S 2001 Hydrothermal synthesis reactor *Encycl Mater Sci Technol.* 3989–92
[30] Jun K-W, Lee Y-J, Kim S-M and Jeon Y K 2009 Patent Application Publication Pub. No.: US 2009 / 0186126A1.1(19) 2–5
[31] Tuan N T, Trung-Kien N D, Huy P T and Tung N H 2011 *e-Journal Surf Sci Nanotechnol.* **9**(March 2015) 521–5
[32] Kumari P, Dwivedi Y and Bahadur A 2018 *Optik (Stuttg).* **154**(October) 126–32
[33] Hayashi H and Ebina T 2018 *J Ceram Soc Japan.* **126**(3) 214–20

IOP Conf. Series: Materials Science and Engineering 1073 (2021) 012012 doi:10.1088/1757-899X/1073/1/012012

Syngas production process from catalytic cracking of glycerol based on modified zeolite catalyst Ni/ZSM-5

H Dewajani*, D Moentamaria, A R Hakim, M A I Iswara, I Z Mahsunah and A F Rahima

Chemical Engineering Program, The State Polytechnic of Malang, Jalan Sukarno Hatta 9, Malang, 65141, Indonesia

*henyhp@gmail.com

Abstract. Increased biodiesel production has an effect on increasing the availability of glycerol, where glycerol is a by-product of biodiesel production. So far, the use of glyserol is less than optimal where only a small portion is used as an additive in the food industry. So it needs efforts to utilize glycerol becoming a more valuable product. Based on its characteristics, glycerol is potentially converted with catalytic cracking process produced gas that are enrich in H_2, CO which is known as synthesis gas (Syngas). Syngas is the main ingredient in the petrochemical industry, besides syngas can also be used as an alternative fuel do to its heating value. The main objective of this research is to study the conversion of glycerol to syngas with the catalytic cracking process, the effect of the ratio of catalyst to glycerol and reaction temperature on product yield. This research was carried out in a semi batch reactor filled with Ni/ZSM-5 catalyst which was flowed by Nitrogen gas at 350, 400, 450 and 500°C reaction temperature variations with catalyst ratio of 0, 3, 4 and 5% (w/w). Gas products were analyzed using Gas Chromatography (GC) to determine the composition of the constituent gases which are CO, CO_2, H_2 and the catalyst characterization was carried out by the BET, XRD and AAS methods. The highest yield was obtained at a proocess temperature of 500 °C with catalyst ratio of 3 wt % which produced 49.45% of Syngas with the hydrogen selectivity of 20.94%.

1. Introduction

Biodiesel is one of the alternative fuels to replace biofuels which are currently being produced. In the process of production biodiesel a by-product is produced in the form of glycerol as much as 10% [1]. Government policies related to renewable energy that encourage increased biodiesel production have an impact on the abundance of glycerol produced. Currently, glycerol is widely used in the production of cosmetic, medicine and even as a flavor enhancer in food. Glycerol has not been used optimally compared to its availability, therefore it is necessary to make some efforts to utilize it so that glycerol does not only become waste from the biodiesel production process. Glycerol can be used to produce syngas which in the process does not need to be reacted with other chemicals.

Syngas or commonly called synthesis gas is a gas mixture with the main components are hydrogen gas (H_2) and monoxide gas (CO), while the side components are carbon monoxide gas (CO_2) and methane gas (CH_4). Syngas can be used as fuel for internal combustion engines and diesel combustion engines. In addition, syngas is also used as raw material for producing ethanol through the Fischer-Tropsch process and urea production process [2].

In the syngas production process, glycerol can be converted through several process using the utilization of a catalyst, for example the reforming and catalytic cracking process. About 95% of hydrogen is still produced from fossil fuel sources through the pyrolysis or steam reforming process, and only about 5% is produced through renewable sources, including through the electrolysis of water [2]. However, using some of these process still requires a lot of heat. In addition to the above processes, syngas can be produced by catalytic cracking of glycerol which does not require high temperatures.

Several previous studies on the conversion of glycerol to syngas have been carried out, such as an experiment by Shahirah et al. [3]. In this study, syngas was produced from cracking of glycerol using a praseodymium-Ni/Al_2O_3 catalyst. The research includes catalyst synthesis and catalyst characteristics. The reaction was carried out using a stainless steel fixed bed reactor with operating conditions 973 K – 1073 K and at atmospheric pressure. The results of this study obtained a hydrogen yield of 29.04% [3]. In addition, other research by Harun et al. [4] has converted glycerol to hydrogen using the dry reforming method using an Ag-Ni/Al_2O_3 catalyst. Research includes catalyst preparation, glycerol dry reforming, and product analysis. $Ni(NO_3)_2.6H_2O$ solution and $AgNO_3$ solution were used to impregnate Al_2O_3. The reaction was carried out using a fixed bed reactor with operating conditions 873 K – 1173 K. The results of this study obtained a glycerol conversion 40.70% and H_2 yield 32% [4].

In this study, the conversion of glycerol to syngas was carried out using a catalytic cracking process where this process does not require a high temperature of around 300ºC-500ºC compared to the steam reforming process which uses temperatures above 600ºC. While the catalyst was modified using nickel metal impregnation so that it became Ni/ZSM-5 to increase the active site of the catalyst. Nickel is a transition metal that has one empty d orbital, so it can bind the split carbon atoms of glycerol [5].

2. Materials and methods

This experiment consits of preparation catalyst and catalytic cracking reaction. In catalyst preparation, ZSM-5 catalyst were prepared by wet impregnation. ZSM-5 mixture with an aqueous solution of $Ni(NO_3)_2$.$6H_2O$ stirred at 80-85°C for 3 h and dried at 120°C for 6 h. The prepared catalysts were calcination at 10°C/min to 600°C for 2 h in the air atmosphere. For the reduction process, the catalysts were reduced with flowing 5 ml/min hydrogen at 600°C for 2 h, and get the Ni/ZSM-5 catalyst. Prior used to the catalytic cracking, the catalysts is mixed with kaolin (5 wt. %) in order to more easily when formed into pellets. The explanation of the catalyst preparation used to experimental has been presented in previous paper [5].

Catalytic cracking gliserol perfomed in a fixed bed reactor as in figure 1 and heated at the temperature reaction of 350-500°C. The catalyst is placed in the reactor which contacted with glycerol vapor. Product were analyzed using GC-MS and the catalyst were characteristic using BET, XRD and AAS.

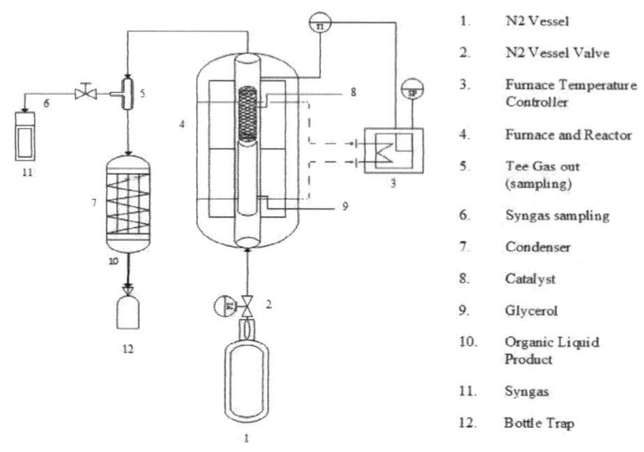

Figure 1. Equipment scheme.

3. Results and discussion

3.1. Characteristic catalyst

3.1.1. BET characteristic catalyst. One of method to estimate the type of pores obtained from the impregnation process of Ni metal on the catalyst is to analyze the absorption pattern of nitorgen (N_2). The N_2 absorption pattern of the ZSM-5 and Ni/ZSM-5 catalyst is shown in figure 2.

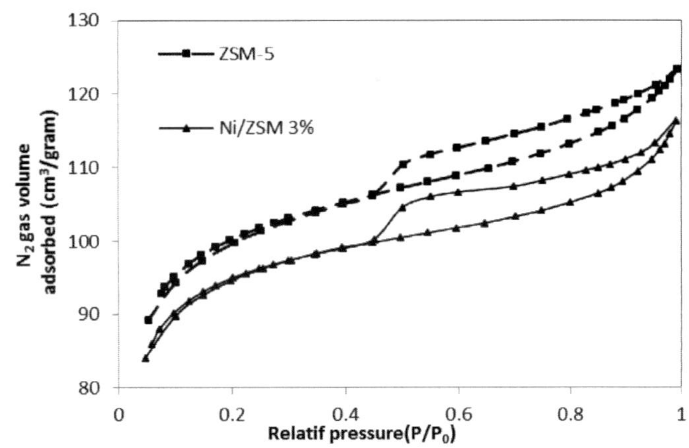

Figure 2. Adsorption and desorption isothermal curve from HZSM-5 and Ni/HZSM-5 catalysts.

Nitrogen adsorption is physical method that use in Isoterm Brunauer, Emmett and Teller (BET) method to determine total surface area and pore structure of a solid [6]. At figure 2, Nitrogen adsorption isotherm all of ZSM-5 sample show a similar pattern in which there is a rapid rise at a relatively low pressure (P/P_0), than increase slowly in the middle and again rapidly as P/P_0 approaches one. The gas absorption pattern at figure 2 clarifies that impregnation Ni metal effect to the catalyst characteristic result. Heating and stirring at impregnation process will caused expansion of space between sheets (space between basalts) in ZSM-5 catalyst and formation of a good pore system. Introduction nikel metal to the pore catalyst caused spesific surface area Ni/ZSM-5 decrease with increasing nikel loaded. From BET result spesific surface area ZSM-5 is 353 m^2/g and compare with microcrystalline synthesis ZSM-5 (MZ) result from sodium silikat, alumunium nitrat, and tetrapropil ammonium hidroksida at Saxena and Vismanadham [7] experiment that has surface area 342.4 m^2/g. Surface area Ni/ZSM-5 decrease to 295 m^2/g at 3% Ni metal impregnation. This result according with Chen et al. [8] experiment that ZSM-5 catalyst have surface area between 300-310 m^2/g and after loaded nikel metal surface area decrease to 230-250 m^2/g. The BET analyzed result from the catalysts is presented in table 1.

Table 1. ZSM-5 catalyst characteristic and Ni/ZSM-5 impregnation result.

Catalyst Characteristic	ZSM-5	3%Ni/ZSM-5
BET surface area (m^2/g)	353	295.3
Pore size distribution BJH :		
Total pore volume (cm^3/g)	0.0408	0.0357
Pore diameter (nm)	3.0598	3.0246
Average diameter (nm)	2.4488	2.4388

Reduction in spesific surface area is proportional to the decrease pore volume. The largest pore volume 0.0408 cm^3/g was obtained on the initial catalyst before nickel metal impregnation. The pore volume decreased with the increase of % Ni metal to 0.0357 in the addition of 3% Ni metal.

In order to determine the effectiveness of the Ni metal loading process in the ZSM-5 catalyst, an AAS (Atomic Absorption Spectrophotometry) analysis was performed. From the results of AAS analysis, it is known that the Ni metal content in the catalyst is 2.88 wt. %.

3.1.2. XRD characteristic catalyst. From the XRD analysis result which presented in figure 3, it can be seen that the ZSM-5 catalyst has an orthorombic crystallinity. For comparison, figure 3 (c) is the XRD results of the ZSM-5 simulation conducted by Treacy and Higgins [9]. The main peaks of the ZSM-5 are expressed in 2θ at 7.9; 8.8; 23.1; 23.9; and 24.4° which corresponds to the characteristics of ZSM-5 with specific peaks of 2θ = 7-9° and 23-25°. The results were similar to the XRD HZSM-5 catalyst synthesized by Vichaphund et al. [10] from Sodium metasilicate pentahydrate, Sodium aluminate, and Tetrapropilammonium. In the synthesis of ZSM-5 catalyst by Chen et al. [8] obtained XRD patterns with a specific peak of 2θ in the range of 6-9° and 22.5-25.0°.

In an experiment conducted by Emori et al. [11], the XRD analysis results of the ZSM-5 catalyst showed the MFI structure with the characteristics of 2θ = 8.04; 8.89; 23.19; 23.32; and 23.78°. The results are similar to the experiments conducted by Zhao et al. [12] where the XRD analysis results of ZSM-5 showed the diffraction peak characteristics of the MFI structure at 2θ = 8.04; 8.89; 23.19; 23.32; and 23.78°.

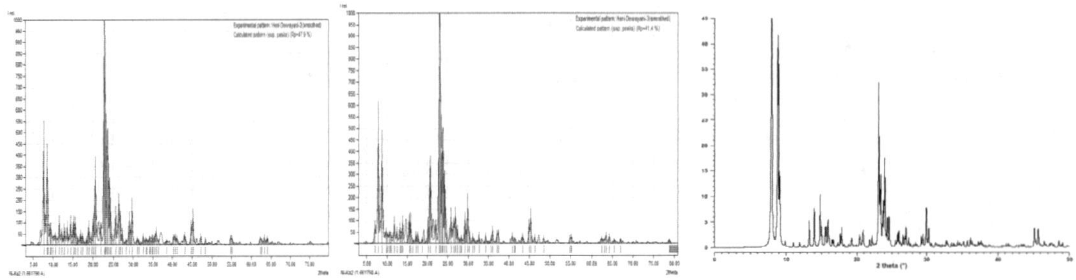

(a) (b) (c)

Figure 3. XRD pattern results (a)ZSM-5; (b)3%Ni/ZSM-5; and (c) ZSM-5 simulation [9].

The diffraction intensity of the ZSM-5 support was seen decrease with the increase in the amount of nickel charged, the peak of the diffraction was very clear showing that the crystallographic structure did not change much after the nickel loading and calcination process. This shows that the addition of Ni metal does not change the structure of ZSM-5. In the calculation of the relative crystallinity calculated based on the number of peak intensities of 2θ, the Ni/ZSM-5 catalyst shows a decrease in crystallinity. This is due to the nickel metal dispersion on the surface of the ZSM-5 support causes a decrease in the degree of crystallinity (%). The crystallinity of the ZSM-5 catalyst from the experiments of Chen et al. [8] showed almost the same results 40.2%. The decrease in the crystallinity of ZSM-5 in Ni impregnation was due to metal dispersion on the surface of the catalyst.

3.2. Effect of reaction temperature on syngas yield

In syngas production process there are several parameters that affect the product, on of them is the reaction temperature. The effect of reaction temperature can be seen in figure 4.

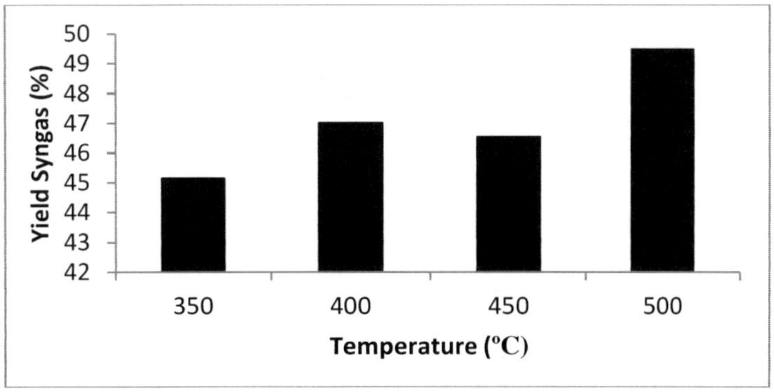

Figure 4. Correlation between reaction temperature and syngas yield.

In the figure 4, it can be seen that the higher the temperature increase syngas yield. This is due to the glycerol decomposition reaction is an endothermic reaction. In theory, glycerol decomposition is an endothermic reaction that requires heat during the reaction [13]. So that the high temperature used will increase the glycerol decomposition reaction and produce greater syngas products.

3.3. Effect of % catalyst on syngas yield

The number of catalyst used can affecct the product of the syngas production process. The effect of % catalyst can be seen in Table 2.

Table 2. Effect of % catalyst on syngas yield.

Temperature (°C)	% catalyst	Hydrogen yield *(%)*
500	0	9,07
500	3	20,94
500	4	17,44
500	5	12,84

Table 2 shows that more catalysts used in the glycerol conversion process, can increase hydrogen yield product and the highest hydrogen yield produced with a catalyst of 3%. This is in accordance with the theory that the use of a catalyst can expand the catalyst contact surface, causing collisions between molecules increase so that the reaction rate increases as well and the resulting product is even greater [14]. The use of a catalyst will reduce the activation energy and speed up the reaction rate by providing an alternative pathway that is easier to produce product.

4. Conclusion

From the experiments that have been carried out can be concluded that glycerol has potentially to be converted to a syngas through a catalyzed cracking reaction process using modified zeolite formed Ni/ ZSM-5 with the experiment results as follow:

- The temperature of the catalytic cracking reaction affects the syngas yield (%) where the higher the temperature, the syngas yield tends to increase.
- The ratio of catalyst to glycerol as a raw material has an effect on syngas yield. However, it did not produce a linear trend, where at the ratio of 3% syngas yield tend to increase but with the addition of 4 and 5% there was a decrease in yield.
- The highest yield was obtained at a proocess temperature of 500°C with catalyst ratio of 3% which produced 49.45% of Syngas with the hydrogen selectivity of 20.94%.

Acknowledgments
The authors would like to acknowledge The Ministry of Research Technology and Higher Education of the Republic of Indonesia for financial support of this work through "Penelitian DIPA Swadana Inovatif 2020" from Politeknik Negeri Malang.

References

[1] Aziz I, Nurbayati S and Luthfiana F 2008 *Jurnal Kimia Valensi* **1** 157-162
[2] Salimafrozi A and Sudaryanto S 2013 *Jurnal Sains Materi Indonesia* **15** 1-6
[3] Shahirah M N N, Ayodele B V, Gimbun J, Lam S S and Cheng C K 2017 *Apllied Thermal Engineering* **112** 871-880
[4] Harun N, Abidin S Z, Osazua O U, Yap Y H T and Azizan M T 2018 *International Journal of Hydrogen Energy* **1** 213-225
[5] Dewajani H, Rochmadi, Budiman A and Purwono S 2016 *AIP Conference Proceedings* **1755** 050002
[6] Haber J, Block J H and Delmon B 1995 *Pure. Appl. Chem.* **67** 1257-1306
[7] Saxena S and Vismanadham N 2014 *Fuel Process Technol* **199** 158-165
[8] Chen H, Wang Q, Zhang X and Wang L 2014 *Appl. Catal. B-Environ* **166-167** 327-334
[9] Treacy M M J and Higgins J B 2001 *Collection of Simulated XRD Powder Patterns for Zoelites* (Amsterdam: Elsevier)
[10] Vichaphund S, Aht-ong D, Sricharoenchaikul and Atong D 2014 *Renew. Energy* **65** 70-77
[11] Emori E Y, Hirashima F H, Zandonai C H, Ortiz-Bravo C A, Machado N R C F and Scaliante M H N O 2016 *Catal Today* **279** 1-9
[12] Zhao X, Wei L, Cheng S, Huang Y, Yu Y and Julson J 2015 *Fuel Process Technol* **139** 117-126
[13] Arif N N M, Vo D V N, Azizan M T and Abidin S Z 2016 *Bulletin of Chemical Reaction Engineering & Chemical Industry* **11** 200-209
[14] Purnami, Wardana I N G and Veronika K 2015 *Jurnal Rekayasa Mesin* **6** 51-59

D-limonene from orange (Citrus Maxima) peel extraction as destructive agent of styrofoam waste

H Hardjono, E N Dewi*, C E Lusiani, I Febriansyah and R I Bachtiar

Department of Chemical Engineering, Politeknik Negeri Malang, Soekarno-Hatta Street No. 9, Malang 65141, Indonesia

*ernianovika@polinema.ac.id

Abstract. Styrofoam is widely used for food packaging because it has good weight, stability, and stiffness but it cannot be decomposed naturally by microorganisms. D-Limonene, one of the monoterpene derivatives, is an environmentally friendly styrofoam solvent. This monoterpene is the main component of essential oil that can be extracted from orange peel. The aim of this study is to destruct styrofoam wastes (foods packaging, electronics packaging, foam boards and cup noodles) using d-limonene from orange (Citrus Maxima) peel extraction by water distillation process. The orange peel after size reduction is extracted by water distillation process at 94°C for 7 hours to produce essential oils. It was used to destruct styrofoam waste at the different volume ratio (1:0:0 ; 1:0:3 ; 1:1:2 ; 1:2:1) of the essential oil mixture (essential oil : ethanol: water). The result of this study shows that styrofoam wastes can be destructed using d-limonene from orange (Citrus Maxima) peel extraction. D-limonene of essential oil from orange (Citrus Maxima) peel can be used as destructive agent of styrofoam waste. The fastest destruction time of styrofoam waste occurs at electronic packaging with essential oil mixture by volume ratio of 1:1:2.

1. Introduction

Styrofoam or EPS (*Expanded Polystyrene*) is a synthetic polymer consisting of 5% polystyrene (PS) and 95% air. Because of this composition, styrofoam has good weight, stability, and stiffness. However, styrofoam cannot be decomposed naturally by microorganisms then it is polluting the environment. Several methods were developed to overcome the environmental problems caused by styrofoam waste. Liu et al studied EPS recycle by breaking down PS into its monomers by pyrolysis methods at a temperature of 450-700 °C [1]. Recycle EPS using the pyrolysis method requires high energy, so it is less economical to process styrofoam waste. Conventional methods of recycling EPS have the disadvantages of molecular degradation caused by oxidation and contamination by other materials, resulting in a reduction of the quality of the recycled polymer. A method of destruction using a natural solvent called d-limonene is promising for improving the quality of recycled PS [2].

Limonene is one of the most common essential oil constituents of aromatic plants. D-limonene is the main compound in the essential oils of the peels of Citrus sp. Limonene is widely used as a flavor and fragrance additive in consumer products such as perfume, beverages, detergents, and soaps [3]. D-limonene has been clinically used to dissolve cholesterol-containing gallstones. It has also been used for the relief of heartburn, because of its gastric acid neutralizing effect and improvement of peristalsis. D-limonene has well-established chemo-preventive activity against many types of cancers [4]. D-limonene is the main component of essential oils from orange peels with 82-99 % content [5–8]. The amount of

IOP Conf. Series: Materials Science and Engineering 1073 (2021) 012013 doi:10.1088/1757-899X/1073/1/012013

limonene in orange peels has not been utilized properly. Orange fruits are only used for consumption as a source of vitamin C while the peels and up in the trash. The peels of orange (Citrus Maxima) which are mostly produced from Nambangan, Madiun is one of them that has not been widely used. This type of orange has a unique characteristic, has a large size, fresh taste, and long shelf life which can reach 4 months [9]. The long shelf life of orange (Citrus Maxima) peels is a supporting factor in the development of research to increase its economic value.

The high limonene content of orange (Citrus Maxima) peels essential oil (77.7 – 95.4%) has the potential to be converted into more useful products [10,11]. One of the uses of limonene as a destructive agent of styrofoam waste.

Several methods have been developed to produce essential oils from orange peels ranging from simple technology to the latest technology. Kurniawan et al compared distillation, pressing, and leaching methods to produce essential oil of orange peel with the best quality and highest yield. From the result of this research, the best quality of citrus peels essential oil obtained by the distillation method [12]. Giwa et al using steam distillation, water distillation, and solvent extraction to produced essential oil from orange peels. From the results obtained, it was discovered that the orange peels could give the maximum yields of essential oil 3.47% when water distillation was employed [13].

Based on the above description, it is necessary to develop research on the use of d-limonene from orange (Citrus Maxima) peels as a styrofoam destruction agent. This study aims to destruct styrofoam waste using d-limonene from orange (Citrus Maxima) peel extraction by water distillation process. Styrofoam wastes from foods packaging, electronics packaging, foam boards, and cup noodles were destructed at the different volume ratios (1:0:3; 1:1:2; 1:2:1) of the essential oil mixture (essential oil : ethanol: water). The destruction time is measured to obtain the fastest destruction time and the best composition of the destructive agent.

2. Materials and methods

D-limonene, the highest essential oil component from orange peel, was extracted by water distillation process. The essential oils were used to destruct styrofoam wastes from foods packaging, electronics packaging, foam boards, and cup noodles in certain compositions of the essential oil mixture with ethanol and water. The essential oil was analyzed by GCMS (Gas Chromatography-Mass Spectrometry) analysis to determine the essential oil content. Density and viscosity of the essential oil were also analyzed. The essential oil mixture (Essential Oil : Ethanol : Water) was used as destructive agent for styrofoam wastes. The destruction time for each type of styrofoam waste was determined at various volume ratios of the essential oil mixture.

The experiment can be divided into four steps, i.e.:

- Pretreatment of orange (Citrus Maxima) peel
- Extraction of essential oil by water distillation process
- Destruction of styrofoam wastes
- Analysis

2.1. Materials

Materials used in this research included orange (Citrus Maxima) peel that collected from local market, aquadest, ethanol with the composition of 96%, styrofoam wastes (foods packaging, electronics packaging, foam boards, and cup noodles) that collected from household waste.

2.2. Methods

2.2.1. Pretreatment of orange (Citrus Maxima) peel. The collected orange peel was separated from the impurities manually then it was processed by size reduction using a blender.

2.2.2. Extraction of essential oil by water distillation process. After size reduction, 220 grams of the orange peel were put into the erlenmeyer and added 220 ml of aquadest. Then, it was processed by distillation to extract the essential oil at temperature of 94°C for 7 hours. After the distillation process, the distillate containing the water-essential oil mixture (hydrosol) was separated using a separating funnel. The water distillation apparatus can be seen in Figure 1.

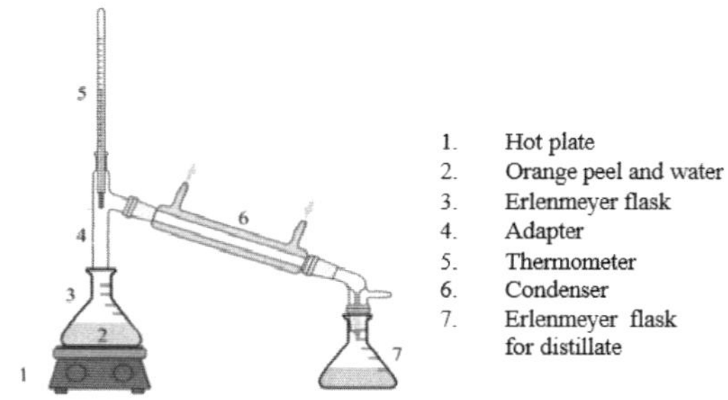

Figure 1. The water distillation apparatus.

2.2.3. Destruction of styrofoam wastes. The types of styrofoam waste used in this study were foods packaging, electronics packaging, foam boards, and cup noodles. The wastes cut into uniform sizes (1 x 1 x 0.25 cm) then put it into a bottle sample containing the essential mixture as destructive agent. Styrofoam wastes were destructed using essential oil mixture with ethanol and water at various volume ratios, i.e. 1:0:0; 1:0:3; 1:1:2; 1:2:1. The destruction time was measured from the beginning of mixing until the styrofoam was completely destructed then it was determined the best composition of the destructive agent.

2.2.4. Analysis. The essential oil extracted from orange (Citrus Maxima) peel was analyzed using GCMS analysis to determine the content of the essential oil. The density of essential oil from orange peels was measured by pycnometer and the viscosity of essential oils was measured by analyzing the intrinsic viscosity using the Ostwald viscometer instrument.

3. Results and discussion

3.1. The components of essential oil from orange (Citrus Maxima) peel extraction

The essential oil product from orange (Citrus Maxima) peel extraction using water distillation process was analyzed by GCMS to determine the constituent components. The results of GCMS analysis as presented in Table 1 show that d-limonene is the highest essential oil component with the compositon of 90.43% w/w. It is consistent with the result of other studies that analyzed the essential oil content of various varieties [5-7]. The highest essential oil component from orange peel extraction is d-limonene. Therefore, the essential oil is also known as limonene oil [8].

Table 1. The results of GCMS analysis of essential oil from orange (*Citrus Maxima*) peel extraction.

Peak	Component	Content (%w/w)	Peak	Component	Content (%w/w)
1	*α-Pinene*	1.12%	6	*Sabinene*	0.48%
2	*Sabinene*	0.31%	7	*α-Terpineol*	0.23%
3	*Myrcene*	4.96%	8	*Geranyl acetate*	0.42%
4	*α-Phellandrene*	0.58%	9	*β-Caryophyllene*	0.19%
5	*Limonene*	90.43%	10	*Germacrene-D*	1.28%

Monoterpene derivatives, such as d-limonene, isoamyl acetate, benzyl propionate, and ethyl butyrate, are good solvents for styrofoam. D-limonene, which is a natural vegetable oil extracted from the orange peel, is the best destructive agent for styrofoam because of its high solubility, stable supply, and fragrance [2]. The results of the previous studies showed that the essential oil could destruct styrofoam [13,14].

The physical properties of the essential oil from orange (Citrus Maxima) peel extraction were also determined the density and viscosity. Based on the measurement, the density of the essential oil is 0.8494 g/mL and the viscosity is 8.636 x10^{-3} g/cm.s.

3.2. Destruction of styrofoam wastes

Essential oil product by water distillation process was used as destructive agent for styrofoam waste. It was mixed with ethanol and water with various volume ratios, i.e. 1:0:0; 1:0:3; 1:1:2; 1:2:1. Ethanol was added to the mixture to facilitate the penetration of essential oil particles into the styrofoam waste. The mixture of essential oil, ethanol, and water was used to destruct styrofoam wastes from foods packaging, electronics packaging, foam boards, and cup noodles. The destruction time of styrofoam wastes at various volume ratios of the mixture are presented in Figure 2.

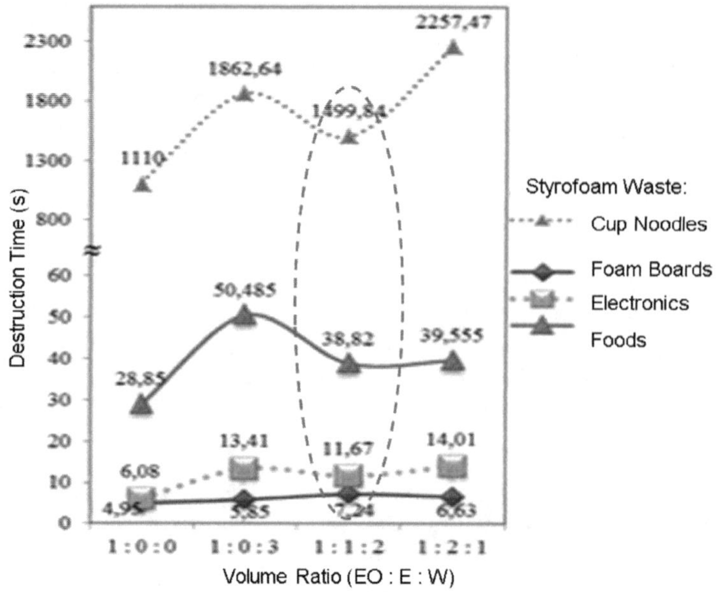

Figure 2. The destruction time at various volume ratios with the various styrofoam waste types.

Figure 2 shows that the destruction time tends to decrease at the composition of 25% v/v ethanol addition (at volume ratio of 1:1:2) but increase at the composition of 50% v/v ethanol addition (at volume ratio of 1:2:1). It is consistent with the research of Noguchi et al who reported that ethanol is a poor destructive agent for styrofoam. The addition of 3% by weight of pure limonene decreased the viscosity of limonene however maximized the penetration of the essential oil mixture to styrofoam and accelerated the destruction time [2]. The fastest destruction time occurs at the essential oil mixture by the volume ratio of 1:1:2 (Essential Oil: Ethanol: Water). The best composition mixture is determined on the composition with the minimum amount of essential oil because the price of essential oil is high. The composition of 1:0:0 is only for comparison however it is not recommended because the pure essential oil is expensive.

3.3. Effect of styrofoam waste types on the destruction percentage

Destruction percentage of styrofoam waste can be used to determine the amount of the destructed styrofoam waste at the various styrofoam waste types with the same volume ratio of the essential oil mixture. The destruction percentage was calculated by the mass ratio of the destructed styrofoam to the destructive agent. The results of this calculation be seen in Table 2.

Table 2. Destruction percentage of styrofoam waste.

Volume Ratio	Styrofoam Waste Types	Destruction Percentage (%)
1:1:2	Foods Packaging	8.48
	Electronics Packaging	9.37
	Foam Boards	6.95
	Cup Noodles	7.20

At the same volume ratio of essential oil mixture, Table 2 shows that the lowest percentage of destruction occurs at cup noodles with the percentage of 7.2 %. The electronics packaging result the highest percentage of destruction with the percentage of 9.37%. It is caused by the density of the polymer in the styrofoam varies depending on the types of styrofoam.

4. Conclusion

Styrofoam wastes can be destructed using d-limonene from orange (Citrus Maxima) peel extraction by water distillation process. D-limonene of the essential oil (90.43% w/w) can be used as destructive agent of styrofoam wastes from foods packaging, electronics packaging, foam boards, and cup noodles. The fastest destruction time of styrofoam waste occurs in essential oil mixture by volume ratio of 1:1:2. The highest destruction percentage (9.37%) of styrofoam waste occurs at electronics packaging.

Acknowledgments

The authors would like to acknowledge The Ministry of Research Technology and Higher Education of the Republic of Indonesia for financial support of this work through "Penelitian DIPA Reguler Kompetisi 2020" from Politeknik Negeri Malang.

References

[1] Liu Y, Qian J and Wang J 2000 Pyrolysis of polystyrene waste in a fluidized-bed reactor to obtain styrene monomer and gasoline fraction *Fuel Processing Technology* **63**(1) 45-55

[2] Noguchi T, Miyashita M, Inagaki Y and Watanabe H 1998 A new recycling system for expanded polystyrene using a natural solvent *A new recycling technique. Packaging Technology and Science: An International Journal* **11**(1) 19-27

[3] Erasto P and Viljoen A M 2008 Limonene-A review: Biosynthetic, ecological and pharmacological relevance *Natural Product Communications* **3**(7) 1934578X0800300728

[4] Sun J 2007 D-limonene: Safety and clinical applications *Altern. Med. Rev.* **12** 259–264

[5] Lestari A and Arreneuz S 2014 Uji Bioaktivasi Minyak Atsiri Kulit Buah Jeruk Pontianak (Citrus nobilis Lor) terhadap Rayap Tanah (Coptotermes curvignathus sp) *Jurnal Kimia Khatulistiwa* **3** 38–43

[6] Muhtadin A F, Wijaya R, Prihatini P and Mahfud M 2013 Pengambilan Minyak Atsiri dari Kulit Jeruk Segar dan Kering dengan Menggunakan Metode Steam Distillation *Jurnal Teknik ITS* **2**(1) F98-F101

[7] Nata I F, Ma'rifah Y N and Herlina H 2014 Minyak Kulit Jeruk Pakis Sebagai Essential Oil Dalam Pembuatan Sabun: Ekstraksi Dan Karakterisasi *Konversi* **3**(2) 67-73

[8] Hidayati 2012 Distilasi Minyak Atsiri dari Kulit Jeruk Pontianak dan Pemanfaatannya dalam Pembuatan Sabun Aromaterapi *Biopropal Industri* **3** 39–49

[9] Rahayu A 2017 Karakter morfologi dan kimia kultivar pamelo (Citrus maxima (Burm.) Merr.) berbiji dan tanpa biji *Jurnal Agronomi Indonesia (Indonesian Journal of Agronomy)* **40** 48–55

[10] Thavanapong N, Wetwitayaklung P and Charoenteeraboon J 2010 Comparison of essential oils compositions of Citrus maxima Merr. peel obtained by cold press and vacuum stream distillation methods and of its peel and flower extract obtained by supercritical carbon dioxide extraction method and their antimicrobial activity *Journal of Essential Oil Research* **22**(1) 71-77

[11] Ndiaye E H B, Gueye M T, Ndiaye I, Diop S M, Diop M B, Thiam A and Lognay G 2017 Chemical composition of distilled essential oils and hydrosols of four Senegalese citrus and enantiomeric characterization of chiral compounds *Journal of Essential Oil Bearing Plants* **20**(3) 820-834

[12] Kurniawan A, Kurniawan C and Indraswati N 2017 Ekstraksi Minyak Kulit Jeruk Dengan Metode Distilasi, Pengepresan dan Leaching *Widya Teknik* **7**(1) 15-24

[13] Fitrianti A E, Dheafithraza Y, Handayani N, Afifah N N and Mariyam S 2016 Penentuan Kadar Minyak Atsiri Kulit Jeruk Sunkist (Citrus sinensis L. Osbeck) sebagai Alternatif Peluruh Sterofoam Alami *Indones. J. Pharm. Sci. Technol.* **3** 47-52

[14] Pratama S H A and Ariyanto E 2019 Ekstraksi minyak atsiri dari kulit jeruk sebagai bahan peluruhan styrofoam *Prosiding Applicable Innovation of Engineering and Science Research* **2019** 1016-1019

IOP Conf. Series: Materials Science and Engineering 1073 (2021) 012014 doi:10.1088/1757-899X/1073/1/012014

The effect of tranducer distance on the concrete crack depth measurement using PUNDIT

S Suhariyanto*, D P Arystianto and B A Raharjo

Civil Engineering Department, State Polytechnic of Malang, Indonesia

*suhariyanto.polinema@gmail.com

Abstract. The depth of crack in concrete structural elements can be measured using ultrasonic waves. One tool that can be used is PUNDIT. The research was conducted by measuring the depth of cracks on the concrete beam test object with a variation of the PUNDIT transducer distance. The purpose of this research is to determine the effect of the distance of the transducer (b) on the relative error (% error) measured in the cracked concrete depth. The results showed that the results of concrete crack depth measurements with the smallest relative error (% error) is obtained by adjusting the transducer distance to about 2/3 or 0.67 the thickness of the test object (h). At the distance of the transducer less than 0.67 or more than 0.67 the thickness of the test object, the relative error (% error) will be greater.

1. Introduction

The use of the PUNDIT (Portable Ultrasonic Non-destructive Digital Indicating Tester) as a concrete NDT (Non Destructive Test) tool in recent years has been growing. The results of testing using this tool are strongly influenced by many factors, so a lot of research is needed in order to get the best practice of using PUNDIT as an NDT tool on concrete structural elements in particular.

The depth of cracks in concrete structures can be measured using PUNDIT, whose working principle is based on measuring the travel time of ultrasonic waves that propagate in the concrete structure.

Accuracy, care and persistence are required in determining the depth of the crack, considering that the crack depth determines the handling that will be carried out on these structural elements.

The accuracy of measuring the crack depth using PUNDIT is influenced by the determination of the transducer distance.

Based on this, it is necessary to conduct research to determine the effect of transducer distance on the measurement of the depth of cracks using PUNDIT.

2. Literature review

2.1. Crack on concrete structures

Crack is a state of breaking or splitting of a structure without collapse. With cracks, it can cause the entry of hazardous chemicals, pollution, chloride, etc. into the structure.

Cracks in concrete can be divided into structural and nonstructural cracks. Structural cracks occur due to errors in design or loads that exceed the capacity, while non-structural cracks occur mostly due to physical chemical processes in concrete in the early phase of concrete life. In general, non-structural cracks do not directly result in the weakening of the building structure.

In addition there are cracks called micro cracks, these cracks occur in the concrete hardening phase where these types of cracks will be difficult to detect because they are too small.

The width-narrow, length-shortness and depth of the cracks are influenced by many factors, including the cause of cracks, age of cracks and so on. Cracks that appear in structural elements such as beams, columns or slabs, must be detected immediately. Usually, handling that will be carried out is preceded by an investigation of the crack itself, namely the depth, width and length of the crack. Measuring the width and length of the crack is very easy to do because it is visually very clear, but the measurement of the depth of the crack is very difficult because it is not visible, plus if the cracks that occur are very small or termed micro cracks [1].

3. Non destructive test on concrete structures

Sulistyani R, Dyah et al. [2], stated that Non Destructive Test on concrete structures can be carried out using PUNDIT, which in principle works using the ultrasonic method. The ultrasonic method is a method that is widely used. The ultrasonic methods are used to detect/evaluate defects, measure dimensions, and characterize materials. The ultrasonic method system generally consists of a pulser/receiver, transducer and display equipment. Pulser / receiver is electronic equipment that can generate high voltage electrical pulses. Controlled by the pulser, the transducer generates high frequency ultrasonic energy. Sound energy is transferred and propagates through the material in the form of waves. When there is a discontinuity (such as a crack) along the wave path, some of the energy will be reflected back from the defect surface. A discontinuity will reflect the applied ultrasonic waves, so that the reflected wave recording device records the pulses from the reflected waves that are not from the surface of the test object. The reflected wave signal is then converted into an electrical signal by the transducer and displayed on the screen. The signal travel time can be converted directly to the distance the signal travels. From the signal, information about the location of discontinuities, size, orientation and other properties can be obtained. Ultrasonic waves are sound waves that have a frequency above the limit of human hearing. Based on the frequency, sounds can be distinguished:

- Infrasonic: $f < 16$ Hz
- Audible sonic: $16 \text{ Hz} < f < 20$ KHz
- Ultrasonic: $f > 20$ KHz
- Hypersonic: $f > 1$ GHz

In the ultrasonic method, the frequency regions suitable for testing / examining a type of material are as follows:

- Aluminum, ceramics, steel: 2 - 10 MHz
- Cast iron: 0.5 - 2 Hz
- Concrete: $f < 0.5$ MHz

Some of the advantages of ultrasonic methods are as follows:

- Sensitive to surface and subsurface discontinuities.
- The depth of penetration for defect detection or measurement is outperformed by other NDT methods.
- Only one-sided access is required when using pulse-echo technique.
- Highly accurate in determining reflector position and estimating the size and shape of defects.
- The preparation of objects is minimal.
- Electronic equipment produces instant results.
- Detailed drawings can be created by an automatic system.
- Can be used to measure thickness, apart from flaw detection.

IOP Conf. Series: Materials Science and Engineering 1073 (2021) 012014 doi:10.1088/1757-899X/1073/1/012014

Like all other NDT methods, the ultrasonic method also has limitations, namely:

- The surface of the test specimen must be able to transmit ultrasound.
- The skills and knowledge are more extensive than some of the other methods.
- Generally, a coupling medium is needed to help transfer sound energy into the object.
- Materials that are coarse, abnormally shaped, very small, very thin or inhomogeneous will be difficult to inspect.
- Pig iron and other coarse grained materials are difficult to inspect due to low sound transmission and lots of noise.
- Linear defects whose orientation is parallel to the direction of the sound may be difficult to detect.

3.1. Measurement of the depth of concrete cracks using pundit

PUNDIT (Portable Ultrasonic Non-destructive Digital Indicating Tester) as an NDT (Non Destructive Test) tool or many also call it the UPV Test, works based on measuring the travel time of ultrasonic waves that propagate in the concrete structure.

The ultrasonic waves are transmitted from the transducer transmitter which is placed on the concrete surface through the concrete material to the transducer receiver and the wave travel time is measured by the PUNDIT (Portable Unit Non Destructive Indicator Tester) Read-Out unit in m / second.

The equipment used for the concrete crack test with the Ultrasonic Pulse Velocity Test consists of:

- One Read-out Unit PUNDIT (Portable Unit Non Destructive Indicator Tester)
- Two 54 kHz Transducers (one each as a transmitter and receiver).
- One Calibration Bar and cables and connectors

One type of PUNDIT can be seen in the following figure:

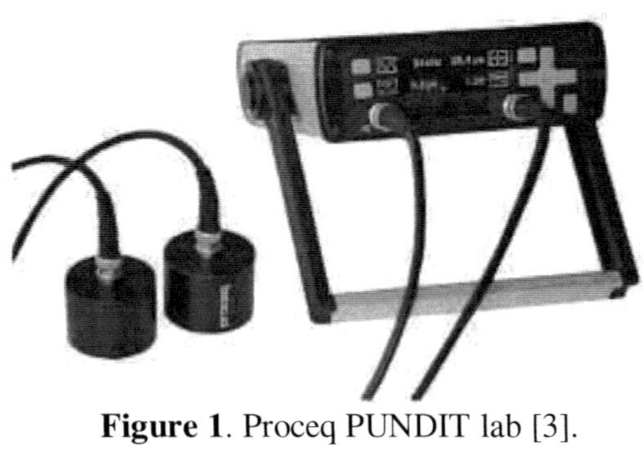

Figure 1. Proceq PUNDIT lab [3].

The method used to measure the depth of cracks using PUNDIT is the Indirect Method. This method is used to measure the time the waves propagate from the transmitter to the receiver in one surface area, where when crossing the crack line there is a time jump.

To determine the depth of the cracks, 2 (two) measurements of wave propagation were carried out. The first is that the transmitter and receiver are placed opposite in one surface area with the same distance from the surface crack line, namely at a distance of X1, and then at a distance of X2. Measurement illustration as in the following figure 2:

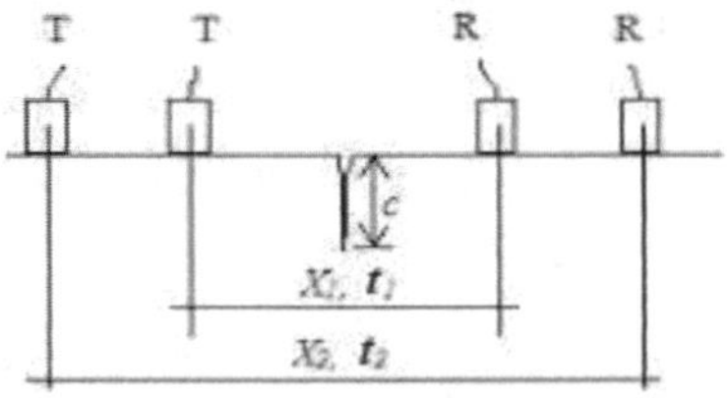

Figure 2. Transducer distance (X1, X2), travel time (T1, T2) and crack depth (c) parameters [3].

Then the crack depth can be calculated by the following equation:

$$c = \frac{1}{2}\sqrt{\frac{X_2^2 T_1^2 - X_1^2 T_2^2}{T_2 - T_1}}$$

With:
X1 = distance between transducers (transmitter and receiver) on the first measurement
X2 = distance between transducers (transmitter and receiver) in the second measurement
t1 = time that the wave propagates from the transmitter to the receiver at the first observation
t2 = time that the wave propagates from transmitter to receiver in the second observation

If in the first measurement the distance between the cracked position and the transmitter is b, and the distance between the receiver and the cracked position is also b in the opposite direction, then X1 = 2b. Furthermore, if in the second measurement the distance between the cracked position and the transmitter is 2b, and the distance between the receiver and the cracked position is also 2b in the opposite direction, then X2 = 4b. Then the measurement illustration can be described as:

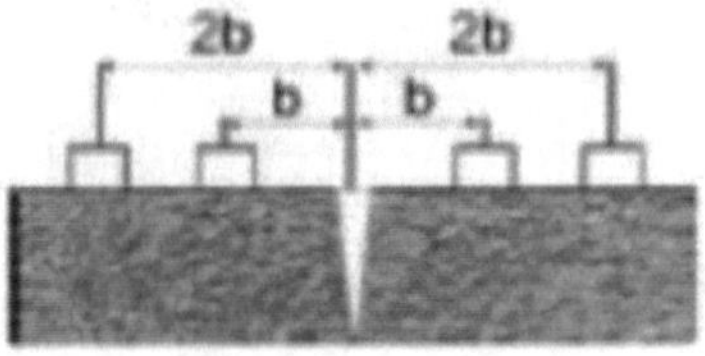

Figure 3. Parameters of Transducer Distance (b), Travel Time (T1, T2) and Crack Depth (c) [3].

Then the crack depth can also be calculated with the following formula:

$$c = b\sqrt{\frac{4{t_1}^2 - {t_2}^2}{{t_2}^2 - {t_1}^2}}$$

The procedure for measuring the depth of concrete cracks using PUNDIT in outline show in figure 4.

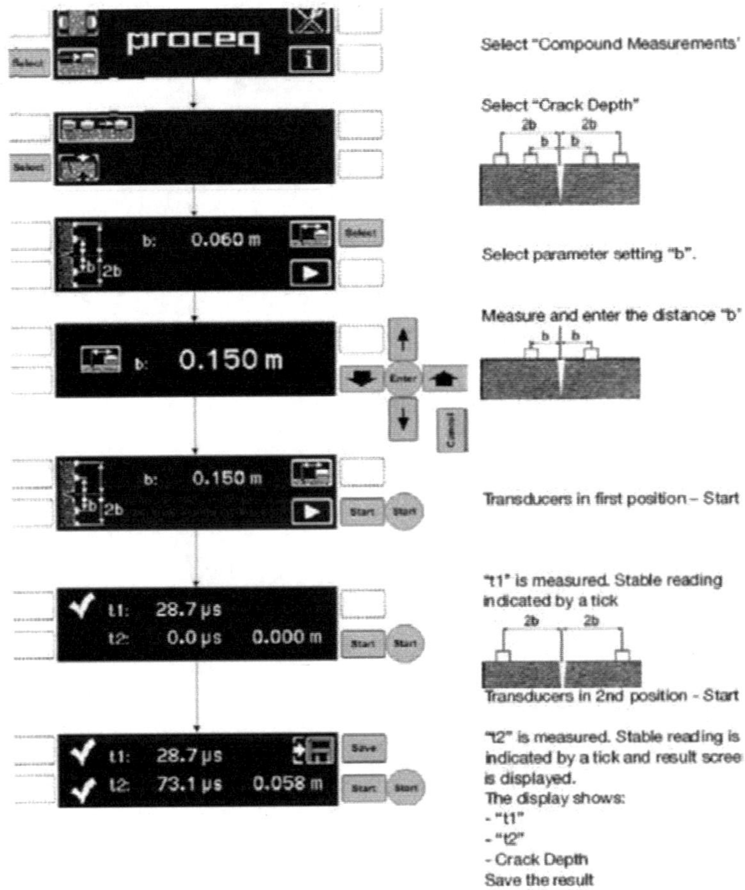

Figure 4. Procedure of crack depth measurement using proceq PUNDIT lab [3].

4. Methodology

The research of the effect of the distance of the transducer on the measurement of the depth of concrete cracks using PUNDIT was carried out on 2 specimens as follows:

- The test object no 1 is a concrete beam (K350) with a length of 61cm, width 15cm and a depth of 13,5 cm and a crack depth of 3.25 cm and 7.5 cm in the mid span of beam. The crack depth measurements of the test object with 5 cm, 7.5 cm and 10 cm transducer distance variations.
- The test object no 2 is a concrete beam (K350) with a length of 75 cm, width 15 cm and height 15 cm and a crack depth of 8 cm in the mid span of beam. The crack depth measurement of the test object with 5 cm, 10cm and 15 cm transducer distance variations.

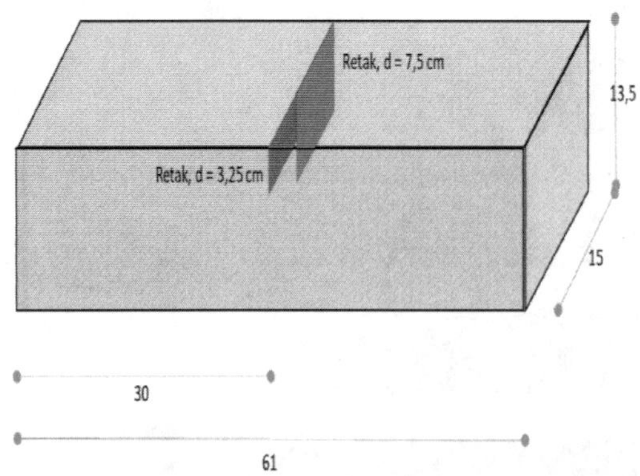

Figure 5. Skecth of test object no 1.

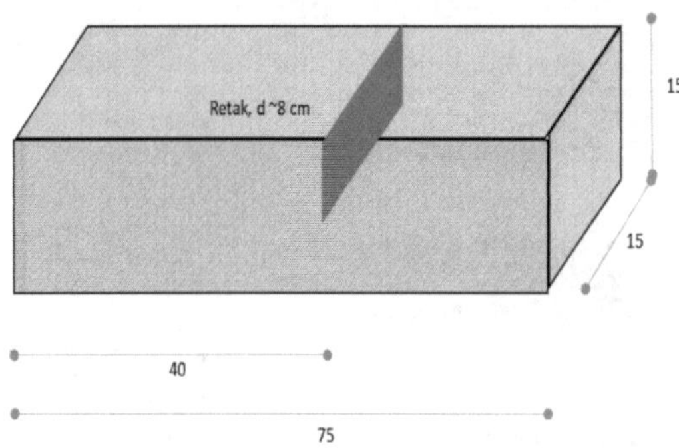

Figure 6. Skecth of test object no 2.

The picture of test object for this research are as follows figure 7:

Figure 7. Photo of object test.

The test is carried out by two people, one person tests the specimens using the PUNDIT tool while one person record the testing results. Photographs of test equipment and test implementation are as follows in figure 8 and 9:

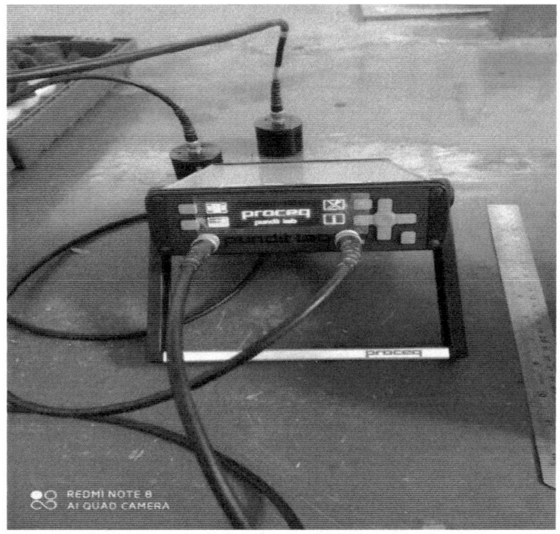

Figure 8. Photo of PUNDIT equipment.

Figure 9. Photo of crack depth measurement.

5. Results and discussion

Measurement of crack depth with a variation of the transducer distance was carried out on concrete test object that were 28 days old with an estimated compressive strength of about 350 kg/cm^2 (direct speed measurement using PUNDIT around 3926 m/s).

The plotting graph of crack depth measurement results with variations in transducer distance is as follows figure 10,11 and 12:

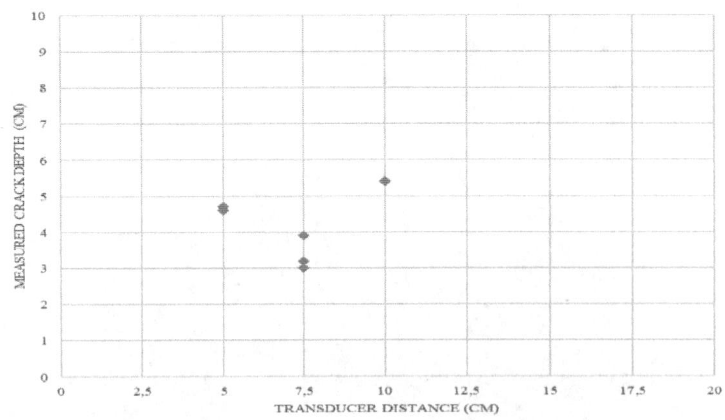

Figure 10. Plotting the measurement of crack depth with a variation of the transducer distance in Test Object No 1 da = 3.25 cm.

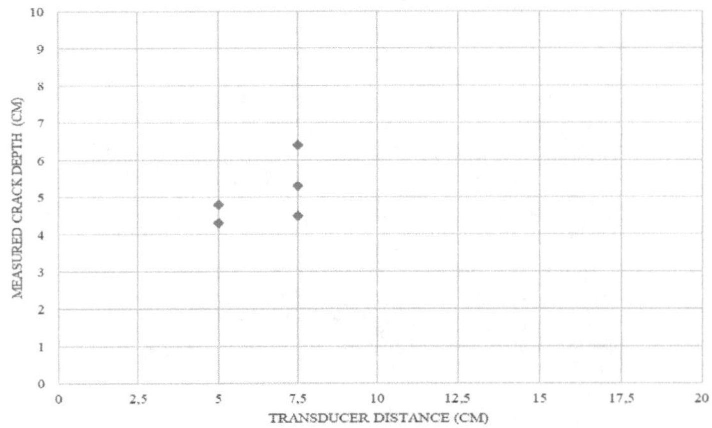

Figure 11. Plotting the measurement of crack depth with a variation of the transducer distance in Test Object No 1 da = 7,5 cm.

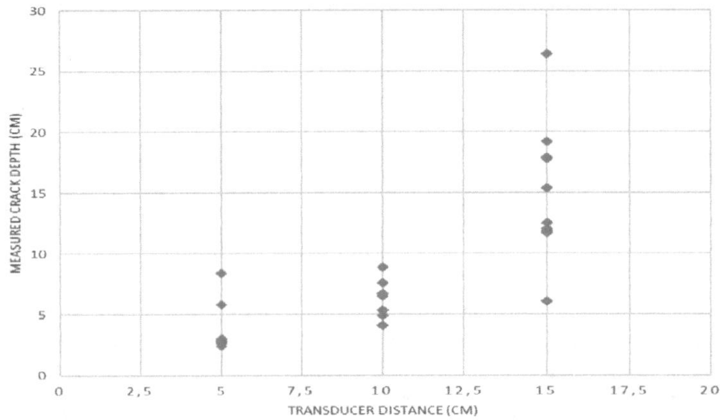

Figure 12. Plotting the measurement of crack depth with a variation of the transducer distance in Test Object No 2 da = 8 cm.

IOP Conf. Series: Materials Science and Engineering 1073 (2021) 012014 doi:10.1088/1757-899X/1073/1/012014

The relative measurement error (% error) of each test can be seen in the following table 1:

Table 1. The relative measurement error (% Error) in test object no 1 da = 3,25 cm.

Transducer Distance	Crack Depth		% Error		
	Measured	Actual			
b (cm)	dt(cm)	da(cm)			
	5,00	3,25	53,85	min	41,54
5	4,60	3,25	41,54	max	53,85
	4,70	3,25	44,62	average	46,67
	3,00	3,25	7,69	min	1,54
7,5	3,20	3,25	1,54	max	20,00
	3,90	3,25	20,00	average	9,74
	5,40	3,25	66,15	min	66,15
10	NA	3,25	NA	max	NA
	NA	3,25	NA	average	NA

Based on the Table 1, it is known that in the testing of Test Object No 1 da = 3.25 cm:

- At 5 cm transducer distance, the relative error (% error) of concrete crack depth measurements is between 41.54% to 53.85% and the average% error of 46.67%
- At 7,5 cm transducer distance, the relative error (% error) of concrete crack depth measurements is between 1.54% to 20% and an average% error of 9.74%
- At the transducer distance of 10 cm, the relative error (% error) cannot be calculated, because the depth of the crack is unreadable.

Table 2. The Relative Measurement Error (% Error) in Test Object No 1 da = 7,5 cm.

Transducer Distance	Crack Depth		% Error		
	Measured	Actual			
b (cm)	dt(cm)	da(cm)			
	3,90	7,50	48,00	min	36,00
5	4,80	7,50	36,00	max	48,00
	4,30	7,50	42,67	average	42,22
	6,40	7,50	14,67	min	14,67
7,5	5,30	7,50	29,33	max	40,00
	4,50	7,50	40,00	average	28,00
	NA	7,50	NA	min	NA
10	NA	7,50	NA	max	NA
	NA	7,50	NA	average	NA

Based on the Table 2 it is known that in the testing of Test Object No 1 da = 7.5 cm:

- At 5 cm transducer distance, the relative error (% error) of concrete crack depth measurements is between 36% to 48% and the average% error of 42.22%
- At 7,5 cm transducer distance, the relative error (% error) of concrete crack depth measurements is between 14,67% to 40% and an average% error of 28%
- At the transducer distance of 10 cm, the relative error (% error) cannot be calculated, because the depth of the crack is unreadable.

Table 3. The relative measurement error (% Error) in test object no 2 da = 8 cm.

Transducer Distance	Crack Depth		% Error		
	Measured	Actual			
b (cm)	dt(cm)	da(cm)			
	2,80	8,00	65,00	min	5,00
5	5,80	8,00	27,50	max	70,00
	2,80	8,00	65,00	average	55,63
	3,00	8,00	62,50		
	8,40	8,00	5,00		
5	2,80	8,00	65,00		
	2,80	8,00	65,00		
	2,60	8,00	67,50		
5	2,40	8,00	70,00		
	2,90	8,00	63,75		
	5,30	8,00	33,75	min	5,00
10	8,90	8,00	11,25	max	48,75
	6,50	8,00	18,75	average	21,50
	6,70	8,00	16,25		
10	6,72	8,00	16,00		
	4,10	8,00	48,75		
	7,60	8,00	5,00		
10	4,90	8,00	38,75		
	7,60	8,00	5,00		
	17,80	8,00	122,50	min	23,75
15	26,40	8,00	230,00	max	230,00
	17,90	8,00	123,75	average	93,25
	15,40	8,00	92,50		
	19,20	8,00	140,00		
15	12,50	8,00	56,25		
	12,00	8,00	50,00		
	6,10	8,00	23,75		
15	11,80	8,00	47,50		
	11,70	8,00	46,25		

Based on the Table 2, it is known that in the testing of Test Object No 2 da = 8 cm:

- At 5 cm transducer distance, the relative error (% error) of concrete crack depth measurements is between 5% to 70% and the average% error of 55,63%
- At 10 cm transducer distance, the relative error (% error) of concrete crack depth measurements is between 5% to 48,75% and an average% error of 21.5%
- At 15 cm transducer distance, the relative error (% error) of concrete crack depth measurements is between 23,75% to 230% and an average% error of 93,25%

Based on measurements of the depth of concrete cracks in test object No 1 and 2 known:

- For specimen No. 1 with a thickness of 13.5 cm, the smallest relative error of measurement occurs at the transducer distance of 7.5 cm or at the distance of the transducer (b) ~ 0.56 thickness of the test object.
- For specimen No. 2 with a thickness of 15 cm, the smallest relative error of the test occurs at the transducer distance of 10 cm or at the distance of the transducer (b) ~ 0.67 thick of the test object.

The relationship between the transducer distance and the relative error (% error) of the concrete crack depth measurement can be seen in Figure 13.

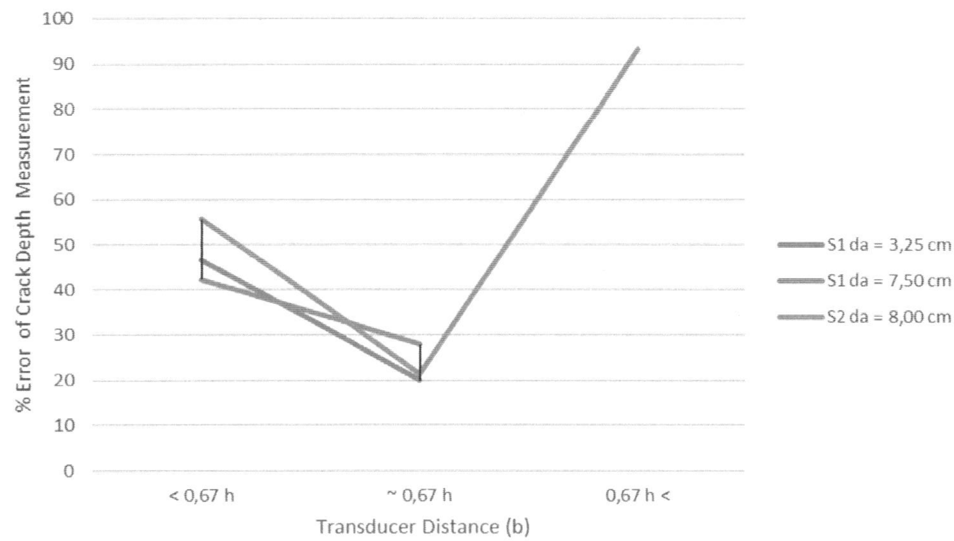

Figure 13. Relationship between transducer distance and relative error (% error) concrete crack depth measurement.

Based on Figure 13 Relationship between Transduser Distance and Ralative Error (%Error) Concrete Crack Depth Measurement known;

- The smallest relative error (% error) of measurement of concrete crack depth occurs at transducer distance (b) around 0.67 the test object thickness.
- At the distance of the transducer (b) less than 0.67 or more than 0.67 the thickness of the test object, the relative error (% error) will be greater
- To get the results of the measurement of concrete crack depth with the smallest relative error (% error), it can be done by adjusting the distance of the transducer around 2/3 or 0.67 the height of the test object.

6. Conclusion

Based on research that has been done can be concluded:

- The smallest relative error (% error) of measurement of concrete crack depth occurs at transducer distance (b) around 0.67 the test object thickness (h)
- At the distance of the transducer (b) less than 0.67 or more than 0.67 the thickness of the test object (h), the relative error (% error) will be greater
- To get the results of the measurement of concrete crack depth with the smallest relative error (% error), it can be done by adjusting the distance of the transducer (b) around 2/3 or 0.67 the thickness of the test object (h)
- To test the validation of the results of this research it is necessary to conduct a similar research by increasing the number of test object and repetition of testing. This research also needs to be developed for other concrete structural elements such as plates and columns.

Acknowledgments

The author would like to acknowledge the support provided by the DIPA P2M State Polytechnic of Malang

References

[1] Linggasari D 2019 Estimating the Depth of Cracks in Concrete Using Ultrasonic *Waves Journal of Muara Sains* **3**(1) 142-152

[2] Sulistyani R, Dyah and Sumaryanto 2010 Detection of the Depth of Concrete Cracks Using the Ultrasonic Method *Proceedings of PPI-PDIPTN* 51-55
[3] Operating Instructions Pundit Lab / Pundit Lab+ Ultrasonic Instrument 2017 Proceq

Usability measurement for *Sistem Proyek Akhir Manajemen Rekayasa Konstruksi* (SIPROK-MRK) measured with the system usability scale

W Wahiddin[1], A R Syulistyo[2,*] and M N Shoumi[1,2]

[1] Department of Civil Engineering, State Polytechnic of Malang, Indonesia
[2] Department of Information Technology, State Polytechnic of Malang, Jl. Soekarno-Hatta No. 9, Malang 65141, East Java, Indonesia

*arie.rachmad.s@polinema.ac.id

Abstract. During the corona virus pandemic, the thesis examination in the Construction Engineering Management study program was conducted online. This is in accordance with the rules of the director at State polytechnic of Malang. Based on this background, a thesis examination system was developed to enable the tests to be conducted online. The main features of this application are the guidance assessment, exam evaluation and revision can be done online. To evaluate the system that has been developed, the usability level of the application is measured using the System Usability Scale. System usability scale is a technique used to measure usability score, system quality, the information presented and the system interface. Based on the questionnaire filled by user the usability score of SIPROK-MRK was 0,82.

1. Introduction

The Covid-19 pandemic has a very wide impact throughout Indonesia, in the fields of economy, tourism, and education [1]. One of the effects of covid-19 in the education sector is that the learning process is required to take place online, to prevent the spread of the corona virus [2]. The online teaching and learning process is also implemented by the State Polytechnic of Malang, where theoretical and practical lecture activities, final assignment guidance, and final assignment examinations are carried out online, without face to face meeting. In order to simplify the systematics of the final assignment examinations for students, a system is needed to be able to assist the series of the final project examination process. Construction Engineering Management which is one of the study programs in the Department of Civil Engineering, State Polytechnic of Malang, began to develop an information system to assist the implementation of online final exams, which is named *Sistem Proyek Akhir Manajemen Rekayasa Konstruksi* (SIPROK-MRK).

Currently there are three parties that can use SIPROK-MRK, there are the final project examination committee, lecturers, and students. Through SIPROK-MRK, the committee can arrange exam schedules and manage data, such as lecturer and student guidance data. Lecturers can see the exam schedule and give grades to the students they supervised and the students they tested. Meanwhile, students can see the exam schedule and download the revised list provided by the examining lecturer. SIPROK-MRK has been implemented and used for the final oral examination in the even semester of the 2019/2020 school year which consists of 5 stages. To find out whether this information system

IOP Conf. Series: Materials Science and Engineering 1073 (2021) 012015 doi:10.1088/1757-899X/1073/1/012015

really facilitates the online exam process, as well as to find out the level of its usefulness for the organizing committee, lecturers, and students, a method of calculating the level of usefulness of this system is needed. Therefore, a measurement of the usability level of SIPROKMRK was carried out using the System Usability Scale.

Measurements are carried out on four components, there are the usability of the entire system, the quality of information system, the quality of information conveyed to users, and the interface quality of this information system [3]. Usability ratings measurement of this information system is useful as an evaluation material in the process of further system development, and of course to determine the usefulness of this information system [4].

2. Background Study

2.1. System Usability Scale (SUS)

This research uses SUS with the Post-Study System Usability Questionnaires (PSSUQ) questionnaire model [5]. The System Usability Scale (SUS) provides a "quick and dirty", reliable tool for measuring the usability [6,7]. PSSUQ consists of a 16 items question with 7 response options for respondents; from Strongly disagree to Strongly agree. The list of question is as follow [8].

Table 1. PSSUQ with 16 items question.

1.	Overall, I am satisfied with how easy it is to use this system.
2.	It was simple to use this system.
3.	I was able to complete the tasks and scenarios quickly using this system.
4.	I felt comfortable using this system.
5.	It was easy to learn to use this system.
6.	I believe I could become productive quickly using this system.
7.	The system gave error messages that clearly told me how to fix problems.
8.	Whenever I made a mistake using the system, I could recover easily and quickly.
9.	The information (such as online help, on-screen messages, and other documentation) provided with this system was clear.
10.	It was easy to find the information I needed.
11.	The information was effective in helping me complete the tasks and scenarios.
12.	The organization of information on the system screens was clear.
13.	The interface of this system was pleasant.
14.	I liked using the interface of this system.
15.	This system has all the functions and capabilities I expect it to have.
16.	Overall, I am satisfied with this system.

The formula to calculate the score of SUS is as follow:

$$Score = \frac{total\ score\ of\ respondent}{maximum\ score} \qquad (1)$$

3. Research methodology

This research uses the prototype method, because the software development time is short, which is 1 month [9]. So that by using prototype, the software development team can receive input from users quickly in an adaptive manner so that it can be changed according to user needs. The prototype model stage can be seen on Figure 1, user can evaluate the software and give suggestion for the development team as many as possible [10].

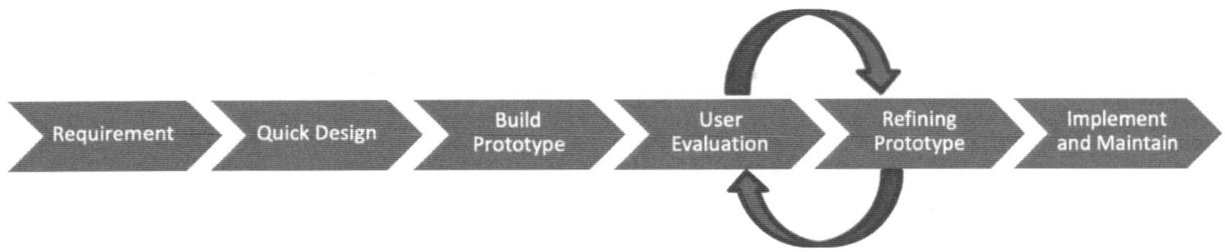

Figure 1. Prototype model stage.

3.1. Information system development

SIPROK-MRK is a web-based information system developed using the Laravel framework and MySQL database. There are several main features that can be accessed by lecturers, students, and final project examination committee, there are:

- Schedule page, which contains a list of the final oral exam schedule. The information displayed is student's name, the final project title, the exam room, the academic year, the date and time of the exam, and the examiner's name. This page can be accessed by lecturers, students, and the organizing committee.
- Revision page, which contains revision notes provided by examiners. Revision notes are provided in PDF format and can be downloaded by students. This page can be accessed by lecturers, students, and the organizing committee.
- The assessment page, which is used by the lecturer to provide an assessment of the students he supervises, the students being tested, and to provide revisions to the students being tested. This page can be accessed by lecturers.
- Admin page, which is used to manage all data. This page can be accessed by the admin and the organizing committee.

4. Results and discussion

4.1. Implementation of SIPROK-MRK

Some examples of the interface from SIPROK-MRK can be seen in Figure 2.

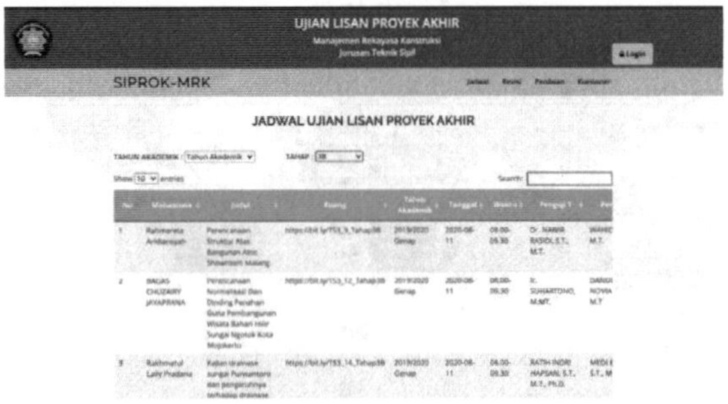

(a) Schedule page

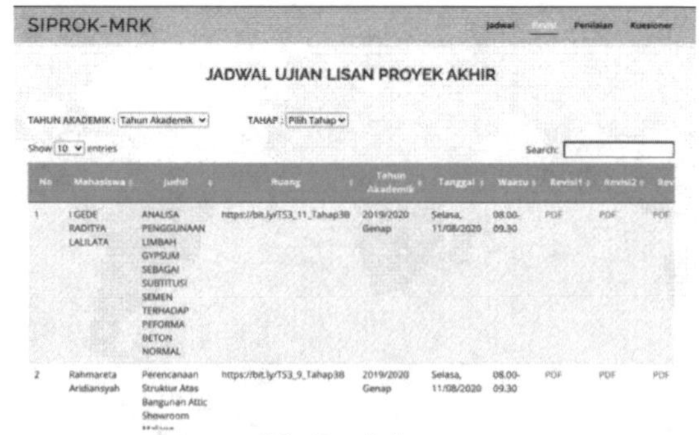

(b) Revision page

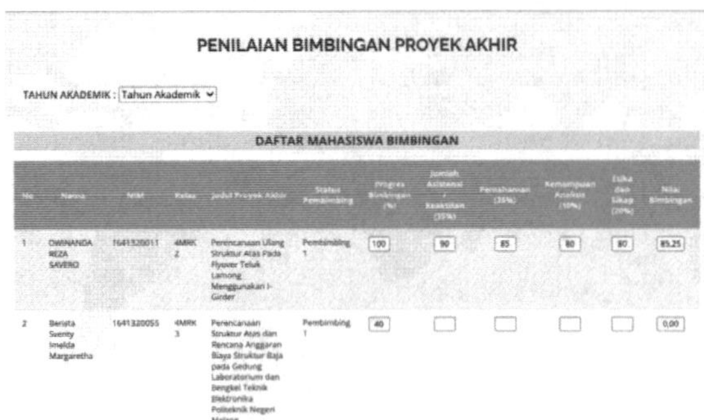

(c) Grading page for tutoring students

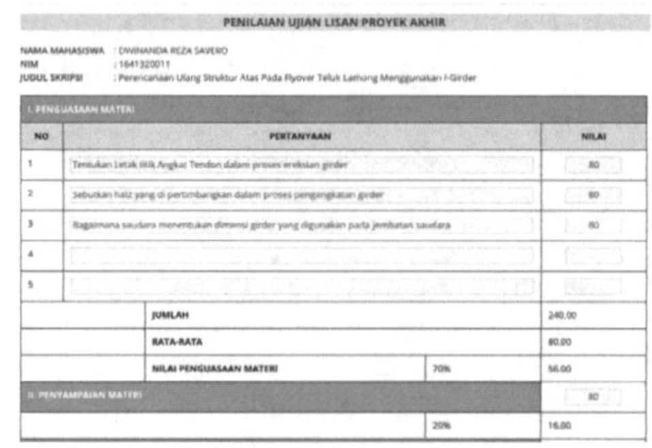

(d) Assessment page for the student being tested

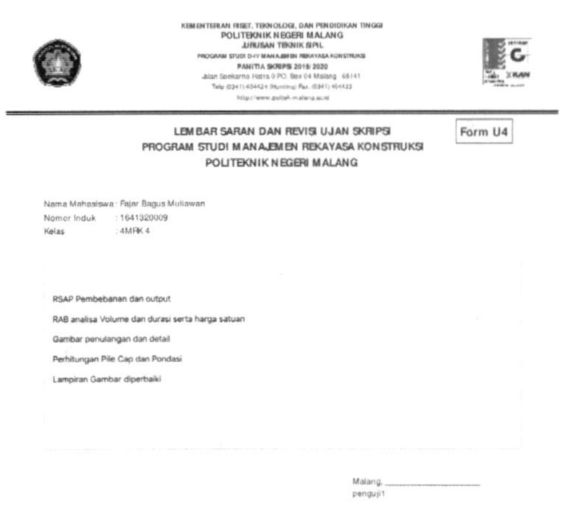

(e) Form U4, contains revision

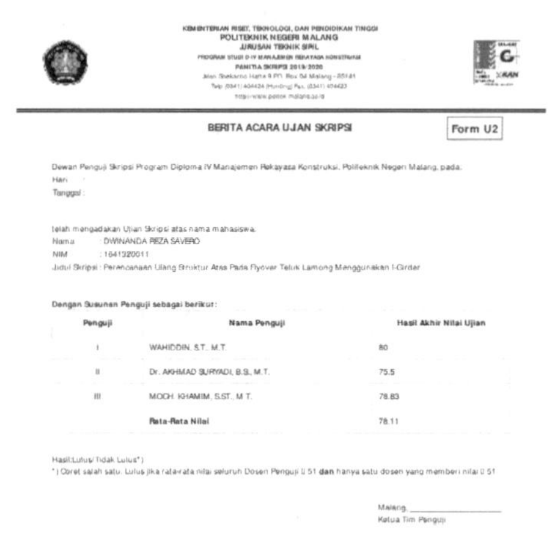

(f) Form U2, contains detailed information about the exam

Figure 2. Examples of the interface in SIPROK-MRK.

The main page of SIPROK-MRK is the exam schedule page as can be seen in Figure 2 (a), where this page can be accessed by students and lecturers. On this page, users can filter by academic year, exam stage, and search based on desired keywords. On the revision page (Figure 2 (b)), students can download and print Form U4 (Figure 2 (e)) which is a revision note given by three examiners. On the assessment page, there are two types of assessments, assessments for guidance students (Figure 2(c)) and assessments for students being tested (Figure 2 (d)). This assessment page can only be accessed by lecturers by first logging in using the employee identification number. While the U2 form in Figure 2 (e) is an examination report containing the scores of the three examiners, this form can only be obtained by the first examiner.

4.2. Usability measurement of information system

The questionnaire for SIPROK-MRK can be accessed publicly by opening the questionnaire menu on the SIPROK-MRK page. The questionnaire consists of 16 questions that are used to measure the usability of the system that has been developed. And there are additional questions as system development suggestion for the next semester. Based on the recapitulation of the questionnaire, there were 47 users who filled out the SIPROK-MRK, consisting of 34 students and 13 lecturers. The questionnaire is used to measure 4 things, such as:

- Overall usability score was 0,82 which obtained from the recapitulation of questions 1-16
- Quality System (SysQual) score was 0,86 which obtained from the recapitulation of questions 1-6
- Information Quality (InfoQual) score was 0,80 which obtained from the recapitulation of questions 7-12
- Interface Quality (IntQual) score was 0,77 which obtained from the recapitulation of questions 13-15

Based on the result of the usability measurement interface quality should give more attention because the score lowest compare to other scores. Based on the analysis, this can be caused because the interface that displays the table has to be scrolled horizontally to be able to see information on all its columns.

5. Conclusion

Sistem Proyek Akhir Manajemen Rekayasa Konstruksi (SIPROK-MRK) has been used to carry out the thesis examination in the Civil Engineering Department, State Polytechnic of Malang. Based on the questionnaire results of the overall usability score was good with the results of 0.82. However, there are many suggestions that can be used to improve SIPROK-MRK so that it can increase the usability score. Especially on the quality of the interface which should be improve immediately so the users fell comfort when using SIPROK-MRK.

References

[1] Djalante R, Lassa J, Setiamarga D, Mahfud C, Sudjatma A, Indrawan M, Haryanto B, Sinapoy M S, Rafliana I, Djalante S and Gunawan L A 2020 Review and analysis of current responses to COVID-19 in Indonesia: Period of January to March 2020 *Progress in Disaster Science* **6** 100091

[2] Mulyanti B, Purnama W and Pawinanto R E 2020 Distance Learning in Vocational High Schools during the COVID-19 Pandemic in West Java Province, Indonesia *Indonesian Journal of Science & Technology* **5**(2) pp 271-282

[3] Lewis J R 2018 The System Usability Scale: Past, Present, and Future *International Journal of Human-Computer Interaction* **34**(7) pp 577-590

[4] McLellan S, Muddimer A and Peres S C 2012 The Effect of Experience on System Usability Scale Ratings *Journal of Usability Studies* **7**(2) pp 56–67

[5] Bangor A, Kortum P T and Miller J T 2008 An Empirical Evaluation of the System Usability Scale *International Journal of Human–Computer Interaction* **24**(6) pp 574-594

[6] Borsci S, Federici S, Bacci S, Gnaldi M and Bartolucci F 2015 Assessing User Satisfaction in the Era of User Experience: Comparison of the SUS, UMUX, and UMUX-LITE as a Function of Product Experience *International journal of human-computer interaction* **31**(8) pp 484-495

[7] Sauro J 2015 SUPR-Q: A Comprehensive Measure of the Quality of the Website User Experience *Journal of Usability Studies* **10**(2) pp 68-86
[8] Klug B 2017 An Overview of the System Usability Scale in Library Website and System Usability Testing *Journal Of Library User Experience* **1**(6)
[9] Susanto A and Meiryani 2019 System Development Method with The Prototype Method *International Journal of Scientific & Technology Research* **8**(07) pp 141-144
[10] Al-Husseini K and Obaid A 2018 Usage Of Prototyping in Software Testing *Multi-Knowledge Electronic Comprehensive Journal For Education And Science Publications (MECSJ)* **14** pp 1-10

IOP Conf. Series: Materials Science and Engineering 1073 (2021) 012016 doi:10.1088/1757-899X/1073/1/012016

Potential study of motorcycle parking on State Polytechnic of Malang

B Burhamtoro*, N D Raharjo, and M Marjono

Civil Engineering, State Polytechnic of Malang, Indonesia

*burhamtoro@polinema.ac.id

Abstract. Parking is a vehicle that stops or does not move for a while and is abandoned by the driver. requires infrastructure for parking purposes while the available land is increasingly limited. The method of field research (Field Research Method) is direct observation at the location to see the physical condition of the movement of vehicles entering and leaving the location during rush hour. Meanwhile, research on the area of parking was carried out by means of digital map measurement (Digital Map Measurement Research Method), namely by measuring based on measurement results from digital maps. The number of motorbikes that entered was 5716 vehicles, while 1373 vehicles exiting with a volume of 4546 vehicles were parked. The parking area for motorbikes is 7,480.44 m2 with a parking area of 5,217.67 m2 and a circulation area of 2,262.77 m2.

1. Introduction

Parking according to Tamin [1] can be interpreted as a place to stop a vehicle for a while, meanwhileAccording to the Law on Traffic and Road Transportation Parking is a vehicle that stops or does not move for a while and is abandoned by the driver [2]. Meanwhile, according to Nawawi, et al. [3] stated that parking is a temporary immovable condition of a vehicle because it is abandoned by the driver. According to Gusnadi [4], it is legally prohibited to park in the middle of the highway, but parking on the side of the road is generally allowed. Included in the definition of parking is any vehicle that stops at certain places, whether declared with traffic signs or not, and not solely for the purpose of raising and or unloading goods or people [5]. From the above understanding, it can be concluded that parking is a condition in which the vehicle does not move or stops for a certain duration of time.

Parking facilities are locations that are determined as non-temporary stopping places for carrying out activities at a certain time. Where is the parking lot on the street, (on street parking) and off street parking facilities [6]. The parking problem is a problem that is often encountered in the transportation system, in an effort to deal with this problem, it is necessary to provide sufficient parking space, where the need for parking space (Supply) and vehicles to be parked (supply) must be balanced and adjusted to parking characteristics [7].

The parking space unit (SRP) is a unit of measure needed to park a vehicle according to various forms of provision. The amount of parking space is influenced by the dimensions of the vehicle, vehicle free space and the width of the vehicle door openings [8].

IOP Conf. Series: Materials Science and Engineering 1073 (2021) 012016 doi:10.1088/1757-899X/1073/1/012016

Table 1. Determination of parking space units.

Transportation type	SRP (m2)
1. a. Passenger cars for goals. I	2.30 x 5.00
b. Passenger cars for goals. II	2.50 x 5.00
c. Passenger cars for goals. III	3.00 x 5.00
2. Bus / truck	3.40 x 12.50
3. Motorcycle	0.75 x 2.00

Source: Directorate General of Land Transportation [6].

Parking volume is the number of vehicles parked per time period in minutes or hours [9], is the number of vehicles parked in an area at a certain time.

$$\text{Parking volume} = Ei - Ex + X$$

Information:
Ei = Entry (vehicles that enter the parking location)
Ex = Extry (vehicles that leave the parking location)
X = Number of existing vehicles

Volume This can be used as a measure of parking space requirements at the research location [10].

Parking duration is the length of parking time used by each vehicle to stop at a parking space. The average length of parking is expressed in hours / vehicle.
From the results of the calculation of the average length of parking time, it will be known the time taken by parkers to park the vehicle on the parking lot [11].

$$\mathbf{D = \text{time out} - \text{time of entry}}$$

Information:
D = parking time / duration (hour / vehicle)

2. Methodology
This research was conducted to determine the number of motorbikes parked during peak hours by counting the number of vehicles entering and exiting the research location during peak hours. To get the number of vehicles parked by mfield research method (Field Research Method), namely direct observation at the location to see the physical condition of the movement of vehicles entering and leaving the location during rush hour. Meanwhile, research on the area of parking was carried out by means of digital map measurement (Digital Map Measurement Research Method), namely by measuring based on measurement results from digital maps.

3. Discussion
The parking problem is very important to be studied more deeply, because almost all outdoor activities require a parking space. The required parking space must be available adequately. Because with the greater the volume of activity that either leaves or goes to the center of activity, the greater the need for parking space.

Observations for all gates (6 locations) were carried out simultaneously on 27 August 2019 and 28 August 2019 in the morning hours (06: 00-09: 00) and during the day (11: 00-14: 00), by recording the number of motorbikes and cars, both entering and exiting, with an observation time interval of 10 minutes.

The recapitulation of the number of incoming and outgoing vehicles is visible on 27 August 2019 all gates are:

- Motorbikes enter 5566 vehicles, exit 1020 vehicles
- Motorbikes parked 4546 vehicles

While the Recapitulation of the Number of Motorcycles Entering and Exiting on August 28, 2019 for all the gates are:

- Motorbikes enter 5716 vehicles, exit 1373 vehicles
- Motorbikes parked 4373 vehicles

From the results of these calculations it can be obtained data that the number of parking vehicles can be used as a basis in calculating current predictions to prepare motorcycle parking facilities with the largest value of 4546 vehicles.

Table 2. Recapitulation of the number of vehicles in-out the morning session on Tuesday, 27 August 2019.

City : Malang
Location : Politeknik Negeri Malang
Surveyor :
Day / Date : Tuesday, 27 August 2019
Session : Morning

Time	In				Out			
	Motorbikes	Car	Exectra	Amount	Motorbikes	Car	Exectra	Amount
06.00 - 06.10	85	3	0	88	1	0	0	1
06.10 - 06.20	140	6	1	147	16	2	0	18
06.20 - 06.30	225	5	1	231	6	1	0	7
06.30 - 06.40	441	16	0	457	21	6	0	27
06.40 - 06.50	864	23	2	889	45	2	0	47
06.50 - 07.00	1222	29	0	1251	57	4	0	61
07.00 - 07.10	775	32	1	808	47	11	0	55
07.10 - 07.20	392	28	1	421	40	6	1	47
07.20 - 07.30	233	21	0	254	76	6	0	82
07.30 - 07.40	203	19	0	222	72	14	0	86
07.40 - 07.50	165	15	1	181	62	9	0	71
07.50 - 08.00	138	18	0	156	42	17	0	59
08.00 - 08.10	82	18	0	100	71	10	0	81
08.10 - 08.20	114	17	0	131	82	8	0	90
08.20 - 08.30	104	12	0	116	84	4	0	88
08.30 - 08.40	130	13	2	145	91	9	0	100
08.40 - 08.50	117	16	1	134	113	13	0	126
08.50 - 09.00	136	11	0	147	94	8	0	102
Total	5566	302	10	5878	1020	130	1	1151

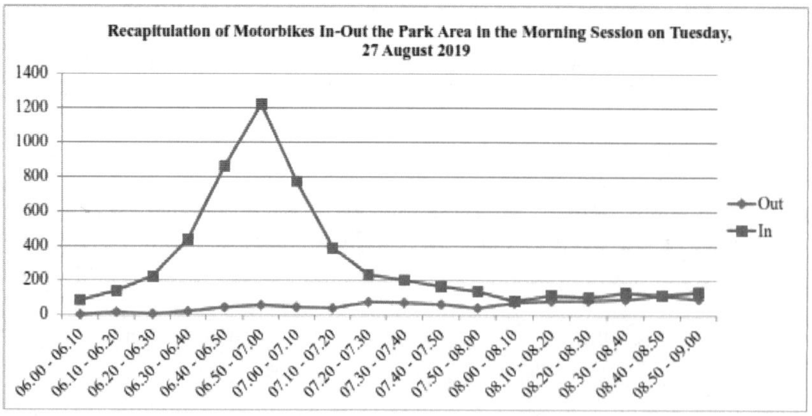

Figure 1. Recapitulation of motorbikes in-out the park area in the morning session on Tuesday, 27 August 2019.

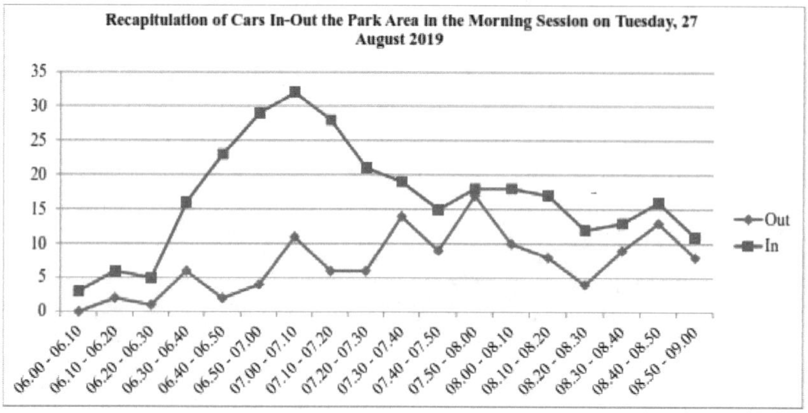

Figure 2. Recapitulation of cars in-out the park area in the morning session on Tuesday, 27 August 2019.

Table 3. Recapitulation of the number of vehicles entering and exiting the morning session on Wednesday, 28 August 2019.

City : Malang
Location : Politeknik Negeri Malang
Surveyor :
Day / Date : Wednesday, 28 August 2019
Session : Morning

Waktu	In				Out			
	Motorbikes	Car	Exectra	Amount	Motorbikes	Car	Exectra	Amount
06.00 - 06.10	49	6	28	83	12	5	0	17
06.10 - 06.20	75	4	54	133	21	5	0	26
06.20 - 06.30	143	6	131	280	36	3	0	39
06.30 - 06.40	226	5	273	504	127	9	1	137
06.40 - 06.50	516	14	468	998	82	9	0	91
06.50 - 07.00	619	16	448	1083	91	15	0	106
07.00 - 07.10	514	20	412	946	110	16	0	126

Table 3. Cont.

07.1 0 - 07.20	225	17	236	478	70	16	1	87
07.20 - 07.30	141	18	115	274	48	13	0	61
07.30 - 07.40	100	10	91	201	41	13	0	54
07.40 - 07.50	72	9	95	176	55	9	0	64
07.50 - 08.00	54	4	66	124	41	21	0	62
08.00 - 08.10	57	5	57	119	92	9	2	103
08.10 - 08.20	55	9	56	120	84	13	0	97
08.20 - 08.30	60	7	65	132	77	7	0	84
08.30 - 08.40	104	13	67	184	133	13	0	146
08.40 - 08.50	161	11	52	224	142	1 0	0	152
08.50 - 09.00	62	7	62	131	111	5	1	1 17
Total	3233	181	2776	6190	1373	191	5	1569

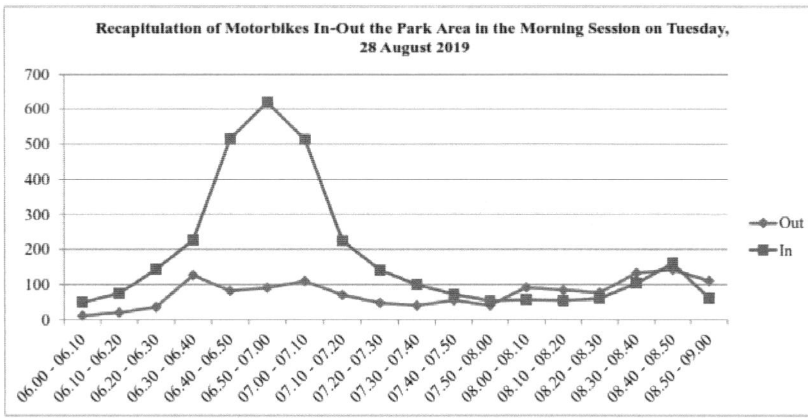

Figure 3. Recapitulation of motorbikes in-out the park area in the morning session on Tuesday, 28 August 2019.

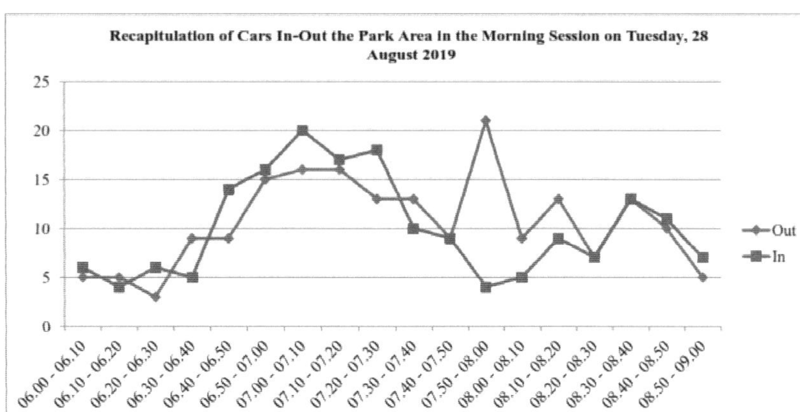

Figure 4. Recapitulation of cars in-out the park area in the morning session on Tuesday, 28 August 2019.

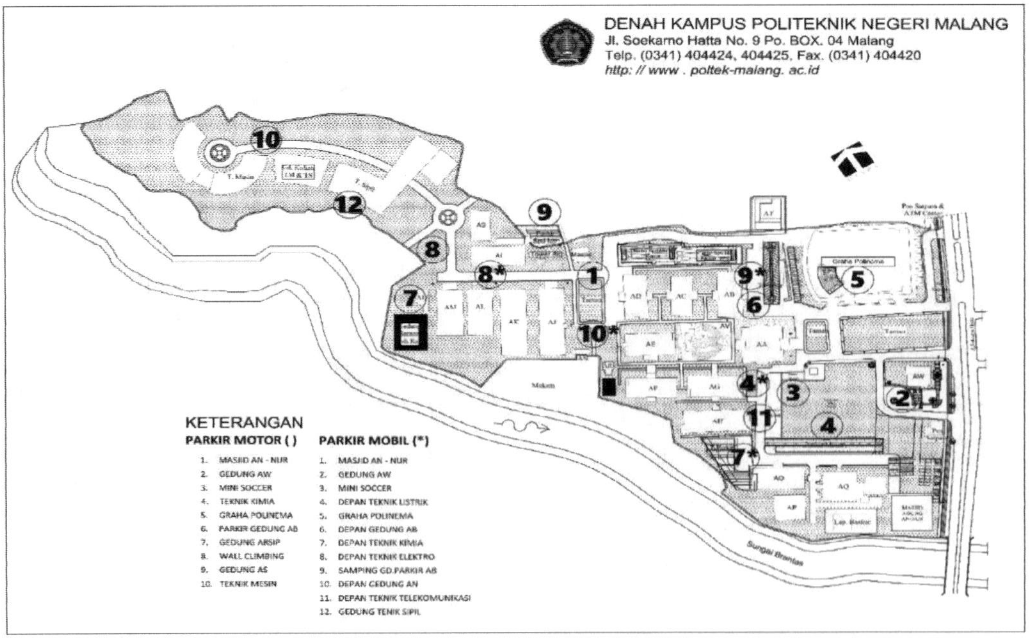

Figure 5. Master plan of State Polytechnic of Malang.

From the Study of the Situation Map Open land and part of the closed land (parking building), the area of land used for motorcycle parking is 7,480.44 m2, while for cars it is 3,659.35 m2 or all of 11,139.79 m2, while the total area of Polinema Campus is 13.8 Ha or 130,800 m2 means that almost 10% of the available land is used for parking needs.

4. Conclusion

The pattern of vehicle movement both in and out gives an overview of activity activities in Polinema which are more dominant in the morning with specific patterns giving a certain value. Meanwhile, during the day, it provides an overview of the activity transition, which value is lower than that in the morning when viewed by the number of vehicles entering or leaving. Collecting data using the field research method (Field Research Method) obtained by motorbikes the number of vehicles that entered was 5716 vehicles while those that left were 1373 vehicles. Accumulated parking is the difference between incoming and outgoing vehicles, the largest accumulated parking for motorbikes when viewed from the difference between incoming and outgoing vehicles during the observation that occurred in the morning of 4546 vehicles. Research on the area of parking was carried out by measuring the digital map (Digital Map Measurement Research Method), namely by measuring based on the measurement results from thc digital map, while the measurement of the parking area was carried out directly in the field. The parking area for motorbikes is 7,480.44 m2 with a parking area of 5,217.67 m2 and a circulation area of 2,262.77 m2 or 34% of the motorbike parking area. The total parking area is 11,140 m2, if calculated with the land area of Malang State Polytechnic Campus 13.8 Ha (138,000 m2) then 8.07% is used for Parking area.

References

[1] Tamin O Z 2000 *Perencanaan & Pemodelan Transportasi* (Institut Teknologo Bandung, Bandung)

[2] Undang-Undang Republik Indonesia Nomor 22 Tahun 2009. Lalu Lintas dan Angkutan Jalan. Jakarta

[3] Nawawi S N S, Sulistyorini R and Martono Y 2015 Studi Optimalisasi Perparkiran dan Pedestrian di Fakultas Teknik Jurusan Sipil Universitas Lampung *JRSDD* **3**(1) pp 71-80 (ISSN: 2303-0011) (Lampung: Jurusan Teknik Sipil Universitas Lampung)

[4] Gusnadi I 2012 Program Kebijakan Perparkiran Kota *Jurnal Teknik Sipil* **3**(1)

[5] Wibowo F M and Syifa F T 2020 Design and Build a Vehicle Positioning System in Parking Areas Using Bluetooth Technology *Jurnal RESTI (Rekayasa Sistem Dan Teknologi Informasi)* **4**(1) pp 77-82

[6] Departemen Perhubungan 1996 *Keputusan Direktur Jenderal Perhubungan Darat No. 272/HK.105/DJRD/96 ten- tang Pedoman Teknis Penyelenggara- an Fasilitas Parkir* Jakarta

[7] Burhamtoro 2019 Laporan Penelitian *Model Kebutuhan Lahan Parkir Di Lingkungan Kampus (Studi Kasus Kampus Politeknik Negeri Malang)* (Politeknik Negeri Malang)

[8] Departemen Perhubungan 1994 *Keputu san Menteri Perhubungan No. KM. 4 Tahun 1994 tentang Tata Cara ParkirKendaraan Motor di Jalan* (Jakarta)

[9] Kasuma I N 2011 Analisis Kelayakan Finansial Rencana Pembangunan Gedung Parkir Bertingkat Di Pasar Lokitasari (Doctoral dissertation, Thesis). Available at: http://www. pps. unud. ac. id/thesis/pdf_theis/unud-316-442774867-tesis% 202. pdf)

[10] Wikrama A A J 2010 Analisis karakteristik dan kebutuhan parkir di pasar Kreneng *Jurnal Ilmiah Teknik Sipil*

[11] Hoobs F D 1995 *Traffic Planning and Engineering, Second Edition* (Yogyakarta: Gajah Mada Press)

IOP Conf. Series: Materials Science and Engineering 1073 (2021) 012017 doi:10.1088/1757-899X/1073/1/012017

Analysis of the needs road facilities in Jalan Soekarno Hatta Malang City

D Ratnaningsih*, W Wahiddin, U Subagyo and R Sasongko

Civil Engineering, Politeknik Negeri Malang, Malang, Indonesia

*dwi.ratnaningsih@polinema.ac.id

Abstract. Pedestrians are an element of urban transportation. The Pedestrian movement includes down the road, crossing both on the road and at the intersection. The existence of adequate pedestrian facilities is needed, including crossing facilities. The number of pedestrians on the Soekarno Hatta road is very high while gap vehicles are short so that it is difficult for pedestrians to cross. The purpose of this study was to determine the need for crossing facilities on the Soekarno Hatta street in Malang. Primary data collection was carried out on March 9 -11, 2020. Analysis results obtained by the number of pedestrians is 322 people and volume traffic 4867,1 pcu/hour, PV2 = 6,3. 109 so that appropriate pedestrian facilities are pedestrian platform.

1. Introduction

Roads parts of the road, including complementary buildings and equipment intended for public traffic, which are on the ground level, above the ground level, below the ground and /or water level, and above the water level, except for rails and cable roads. A pedestrian is any person who walks in a road traffic space [1].

Pedestrians are an important form of transportation in urban areas. These pedestrians are in a weak position when mixed with vehicles, especially for pedestrians crossing the road, indirectly this activity will lead to sharp conflicts in traffic, namely the occurrence of traffic delays and even accidents. So that infrastructure for pedestrians that meets security, safety, comfort, and able to minimize delays or traffic accidents when crossing needed.

The movement of pedestrians includes walking along roads, crossing roads and intersections. Facilities for pedestrians an example important transportation infrastructure for users. The infrastructure needed to support transportation facilities for pedestrians is to build a complementary building for pedestrian comfort, smoothness, and safety.

Pedestrians is one element of road users. Most pedestrians and pedestrians are located in shopping centers, educational centers, hospitals, terminals, stations, and others. Thus, this location requires the provision of adequate, safe, and comfortable public facilities for road crossers. These facilities include zebra crossings, pelicans, control lights, and pedestrian bridges

Soekarno Hatta street, especially in front of State Polytechnic of Malang Campus, is often a crossing point for pedestrians, especially Student State Polytechnic of Malang because the majority of students live across the campus.

Figure 1. Pedestrians in front of State Polytechnic of Malang Campus.

Figure 2. The wader on Soekarno Hatta street.

The less of pedestrian facilities in the area makes it difficult for pedestrians to cross, while the heavy volume of vehicles also increases the risk of pedestrians. The purpose of this research is to determine the proper facilities for pedestrians so that they can cross comfortably and safely.

According to UU no. 22 of 2009, traffic is the movement of vehicles and people in the road traffic space [1]. Road traffic space is infrastructure designated for moving vehicles, people and /or goods in the form of roads and supporting facilities. Roads are all parts of the road, including complementary buildings and equipment intended for public traffic on the ground surface, above the surface of the ground, below the surface of the land and /or water, and above the water surface, except for rails and cable roads. A pedestrian is any person who walks in a road traffic space

According to the Minister of Public Works Regulation No. 03 / PRT / M / 2014 A pedestrian is anyone who walks in the road traffic space. Pedestrian networks are pedestrian segments, either integrated or separate from roads, which are designated for pedestrian infrastructure and facilities as well as connecting activity centers and/or modal change facilities [2]. Pedestrian network infrastructure and facilities are the facilities provided along the pedestrian network to ensure the safety and comfort of pedestrians.

According to pedestrian facilities guidelines the completeness of pedestrian facilities includes pedestrian paths (sidewalks) and wading which consists of level waders and waders are not level [2].

2. Pedestrian crossings

There are two types of pedestrian crossings, as follows:

2.1. *Level crossing*

The criteria for selecting a level crossing are as follows:

- Based on the empirical formula (PV2), where P is the flow of pedestrians crossing the road as long as 100 meters per hour (pedestrian/hour) and V is the hourly flow of vehicles in two directions (vehicle/hour).
- P and V are the average flows of pedestrians and vehicles during peak hours, with initial recommendations as shown in the table below:

Table 1. Selection of based on PV^2 pedestrian facilities [3].

PV^2	P	V	Recommendation
> 10^2	50 -1100	300 - 500	Zebra crossing
> 2 x 10^8	50 - 1100	400 - 750	Zebra cross waiting stalls
> 10^8	50 - 1100	> 500	Pelican Cross
> 10^8	>1100	> 300	Pelican Cross
> 2 x 10^8	50 - 1100	> 750	Pelican Cross waiting for stalls
> 2 x 10^8	>1100	> 400	Pelican Cross waiting for stalls

Where:
P = 100 pedestrian crossing traffic flow, expressed in people / hour
V = Two-way traffic flow per hour, expressed as vehicles/ hour

Level crossing can be applied to intersections and road sections. A level crossing can be:

2.1.1. *Zebra crossing*

- Installed at the foot of the intersection without or with traffic signaling devices or on roads.
- If the intersection is regulated by traffic control lights, the provision of crossing time for pedestrians becomes one unit with the crossing traffic control lights.
- If the intersection is not regulated by traffic control lights, then the motor vehicle speed limit criterion is <40 km/ hour.
- The zebra crossing refers to the Road Marking Implementation Guidelines.

2.1.2. *Pelican crossings*

- Installed on the road, at least 300 meters from the intersection, or
- On roads with the average operating speed of vehicle traffic> 40 km / hour.

2.1.3. Pedestrian platform. The pedestrian platform is a pedestrian path in the form of level crossing facilities whose surface is higher than the road surface.

2.2. *Non level crossing*

Non-level crossing is used when:

- The level crossing facility has disrupted the existing traffic flow.
- The frequency of accidents involving pedestrians is quite high.
- On roads with a design speed of 70 km / hour.
- In a strategic area, but does not allow road crossers to cross the road other than at non-level crossings.

Some of the provisions that must be considered in planning the plot of the crossing facilities:

- non-level crossings should be easily accessible to the disabled, for example by adding RAM or an elevator.
- The crossing facilities must be equipped with good lighting which can increase safety for pedestrians.

- The location and building must pay attention to the aesthetic value and the needs of pedestrians.

The selection criteria for an off-level crossing are shown in the table below.

Table 2. Criteria for determination of non-level crossing facilities [2].

P (people/hour)	V (Vehicle/hour)	PV^2	Recommendation
> 1100	> 750	$> 2 \times 10^8 10^8$	The crossing is not level

3. Headway

According to the Highway Capacity Manual the headway is the time between (in seconds) between two consecutive vehicles as they pass through a point in the road section measured by the common features of both vehicles (for example front axle or front bumper) [4]. According to Tentero Rianti, critical Gap and Gap are Important factors to consider pedestrians who will cross the road are the availability of gaps/gaps or time/ distance between vehicles on the main traffic flow which is sufficient to join and cross and cross into the traffic flow [5,6].

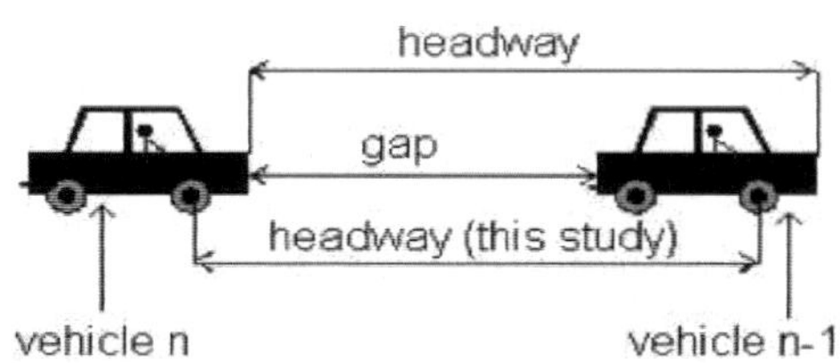

Figure 3. Headway and gap.

4. Methodology

The research location is located on Soekarno Hatta street, Malang City.

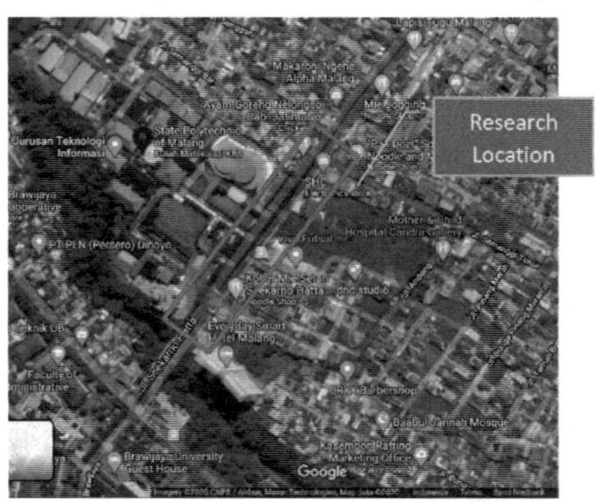

Figure 4. Research location (www. google map.com).

The data required includes:

4.1. Primary data

- Geometric path data
- Speed data
- Pedestrian data
- Wader data

IOP Conf. Series: Materials Science and Engineering 1073 (2021) 012017 doi:10.1088/1757-899X/1073/1/012017

- Vehicle gap data

4.2. Secondary data

The secondary data required are:

- Location map
- Total population of Malang city

Data collection is carried out on working days from 06.00-17.00 WIB

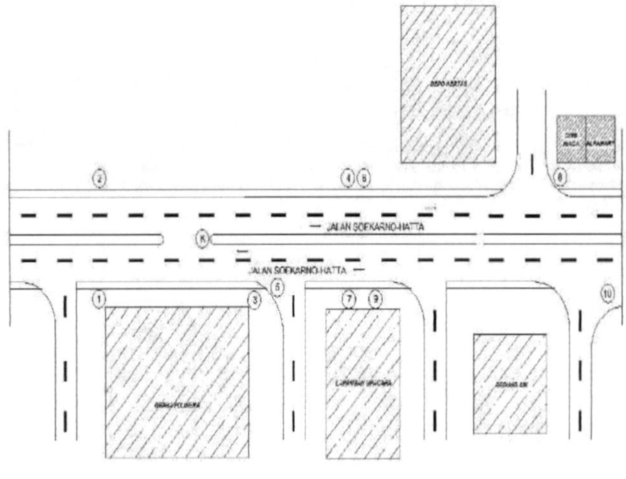

Figure 5. Placement of surveyors.

5. Discussion

Geometric road:

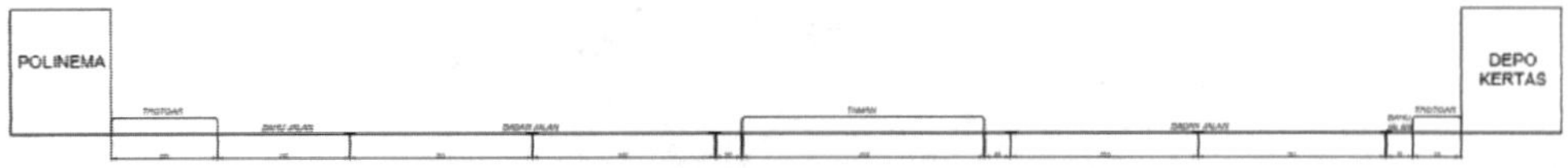

Figure 6. Geometric road Jalan Soekarno Hatta.

Geometric data is obtained from a direct survey of geometric road conditions. The geometric data for Jalan Soekarno Hatta Malang are as follows:

5.1. Geometric conditions and road facilities

- Type of road: 4/2 D
- Width per lane: 3.5 m
- Sidewalk width: 2.00 m
- Median: 4.5m

5.2. Traffic

The composition of traffic passing through Soekarno Hatta road is as follows:

- Motorcycles (MC), i.e. two or three-wheel vehicles.

- Light vehicles (LV), which are four-wheeled motorized vehicles with two-axle distances of 2.0 - 3.0 m (including passenger vehicles, pick-ups and small trucks).
- Heavy vehicles (HV), namely motorized vehicles with two axles distances of more than 3.50 m. Usually more than four wheels including buses, 2 axles, and 3 axles trucks).

Table 3. Traffic volume.

Time Period	pcu						Total
	Direction 1 (N-S)			Direction 1 (S-N)			
	MC	LV	HV	MC	LV	HV	pcu/hour
06.00-07.00	1278,25	1090	6	1042,25	847	10,8	4274,3
06.15-07.15	1467,25	1062	2,4	1268,25	919	13,2	4732,1
06.30-07.30	1523,5	974	2,4	1395,25	936	18	4849,15
06.45-07.45	1583	936	2,4	1446,5	874	25,2	4867,1
07.00-08.00	1577	889	0	1366,25	916	26,4	4774,65
07.15-08.15	1574,5	926	3,6	1236,25	950	33,6	4723,95
07.30-08.30	1514,5	962	7,2	1180	993	40,8	4697,5
07.45-08.45	1443	994	8,4	1083	1012	40,8	4581,2
08.00-09.00	1395,25	1027	9,6	1025,5	995	45,6	4497,95

The peak hour traffic volume occurred at 06.45-07.45 with the number of vehicles 4867.1 pcu / hour.

5.3. Number of pedestrians

Based on the survey results, it was found that the number of pedestrians during peak hours was 322 people.

Table 4. Number of pedestrians.

Time	Side Friction				Total
	Pedestrian	Stop/Parking	In/Out	UM	
06.00-07.00	233	79	2732	49	3093
06.15-07.15	291	77	3850	59	4277
06.30-07.30	322	83	4384	66	4855
06.45-07.45	280	93	4476	87	4936
07.00-08.00	249	105	3849	93	4296
07.15-08.15	197	103	3303	77	3680
07.30-08.30	178	114	3090	66	3448
07.45-08.45	201	130	3088	49	3468
08.00-09.00	241	134	2995	32	3402

Take advantage of the time between vehicles. If the time between vehicles is long, the pedestrian can cross the road but if the time between vehicles is small, the pedestrian cannot cross the road.

Based on the results of the analysis, it is obtained that the average gap on the Soekarno Hatta road can be seen in Table 5.

Table 5. Gap average on Soekarno Hatta street.

No	Transp. Type	Direction of Movement	Gap Aver. (sec)
1	LV	N-S	3,35
		S-N	2,71
2	MC	N-S	0,52
		S-N	0,39

Gap average on Jalan Soekarno Hatta was very short. That is 0.39 seconds. Based on table 3 it is included in the high density category so that the wading will be difficult to cross.

Table 6. Flow determination criteria.

No.	Category	Time Headway
1.	High Density	< 2,5 second
2.	Density	2,5 - 9 second
3.	Low Density	> 9 second

5.4. Vehicle speed

The Speed of the vehicle in question is the instantaneous velocity of vehicles that will interact with pedestrians. When there are road crossers, it is hoped that the vehicle will provide space for the pedestrian to cross the road by reducing its speed. The vehicles being measured are also differentiated for each type of vehicle into groups of light vehicles, heavy vehicles, and motorbikes. Based on the analysis results obtained the average speed

Figure 7. Average speed.

5.5. Selection of crossing facilities

From the side obstacle survey data, the data on the number of pedestrian crossings at peak hours is 322 people/hour (P). And the traffic volume from two directions is 4867,1 pcu/hour (V).
Information:
p: Number of Pedestrians (person/hour)
v: Amount of Two-Way Traffic (vehicle / hour)

So that 322/4867,12 = 6,3 .10 9 and PV2> 2 x 108 then the appropriate pedestrian facilities are the Pedestrian Platform. The surface is made in contrast so that is clearly visible to the driver and is equipped with markings near the ramps so that the driver can see the upper edge of the pedestrian platform, a zigzag mark / line of white paint that can sparkle/reflect light and must be mounted across the full width of the approaching ram.

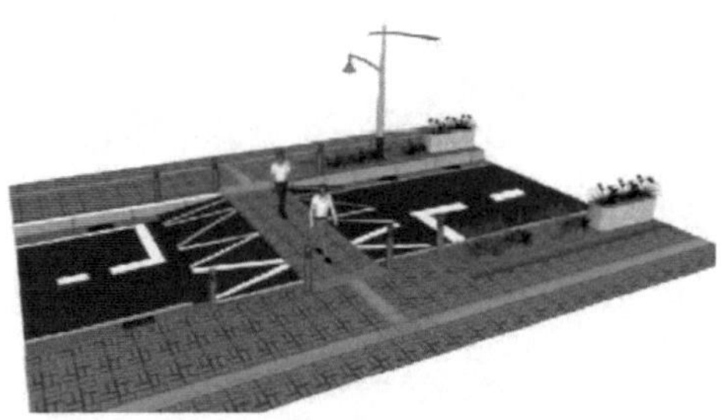

Figure 8. Pedestrian platform on Jalan Soekarno Hatta.

6. Conclusion

- Total peak hour traffic volume is 4867.1 pcu/hour

- The number of pedestrians during peak hours 322 people
- The right crossing facility for Soekarno Hatta Street is the Pedestrian Platform

References

[1] UU No 22 of 2009 *concerning Road Traffic and Transportation* (Jakarta)
[2] Minister of Public Works Regulation 2014 *Guidelines, Provision, and Utilization of Pedestrian Network Infrastructure and Facilities in Urban Areas* No.03 / PRT.M / 2014
[3] Ministry of Public Works and Public Housing 2018 *Pedestrian Facility Technical Planning* No 02/SE/M/2108
[4] Transportation Research Board 2000 *Highway Capacity Manual*
[5] Tentero R 2015 The Requirement of Road Crossing Facilities Based on Critical Gap at The Wolter Mongisidi Road Freshmart Bahu Mall Manado *Jurnal Sipil* **3** 8
[6] Ikbal M and Mashuri M 2011 Study of Pedestrian Characteristics and Selection of Types of Pedestrian Crossing Facilities in the City of Palu (Case Study: Jl. Emmi Saelan Front Taruta Mall, Palu City) *Unpublished Journal. Palu: Civil Engineering Department of Tadulako University*

IOP Conf. Series: Materials Science and Engineering 1073 (2021) 012018 doi:10.1088/1757-899X/1073/1/012018

Performance evaluation of soil water content observation from satellite images in sediment disaster-prone area

R I Hapsari[1,*], R I Putri[2] and G Aponno[1]

[1] Department of Civil Engineering, State Polytechnic of Malang, Malang, Indonesia
[2] Department of Electronic Engineering, State Polytechnic of Malang, Malang, Indonesia

*ratih@polinema.ac.id

Abstract. Soil water content observations from remote sensing are growing for landslide disaster management due to the potential of wide coverage and fine resolution. Characteristic evaluation of various soil water content products is necessary for improving the accuracy. This study assesses soil water content products from Soil Moisture Active Passive from Global Modelling and Assimilation NASA, Soil Moisture and Ocean Salinity from ESA, and Soil Water Index (SWI) from Copernicus Global Land Service. Results are compared with soil test in terms of spatial and temporal distribution, time series, and quantitative performance indices. The study is conducted in Brantas River Basin Indonesia, where the upstream is vulnerable for landslide disaster. The investigation has shown that with correlation coefficient 0.81, SWI is suitable for further application in the study area. This finding has important implications for modelling the landslide hazard using quantitative soil water content product from satellite imagery.

1. Introduction

Landslide cases are highly triggered by rainfall. The rainfall intensity and the runoff influence the soil water content. Landslide in hilly regions around the world are often affected by high soil water content. Therefore, landslide early warning system should be more effective by tracking changes in the condition of the soil water content [1,2]. A variety of methods have been used, i.e. field sensor, direct test, remote sensors, and even hydrological or land surface model assimilation to estimate moisture [3-5]. Each has its own advantages and disadvantages. Every remote sensing imagery has geometrical distortions and some other errors due to atmospheric effects. However, it facilitates wide and long data coverage.

Soil water content observation from satellites has recently been advanced and widely used. Because most satellite observed data require similar long processed before being used, it is preferable to use higher level for operation purpose. The Soil Moisture Active Passive (SMAP) from Global Modeling and Assimilation NASA provides Level 4 Surface and Root Zone Soil Moisture Product [6]. The Soil Moisture and Ocean Salinity (SMOS) mission, which is one of ESA missions, observes variability in soil moisture globally in Level 2 and 3 product [7]. It also provides global Level 3 data for limited access and Level 4 data for European area. Soil Water Index (SWI) and Surface Soil Moisture from Copernicus Global Land Service is available at global scale for direct utilization as it has been verified with global Land Surface Models output and with in-situ measurements [8]. Several attempts have been made to make inter-comparison of soil water content estimates from satellite [9]. The accuracy is evaluated in terms of data availability, spatial and temporal resolution, spatial and temporal variations, error, RMSD, MSE, bias, and correlation. However, few of them discuss the comparison with ground-truth data.

Brantas River in Indonesia is Java Island's second largest river. The upstream of this river is particularly vulnerable to landslide disasters because of its morphology and anthropogenic activities [10]. Until recently, there has been little attention paid into the landslide mitigation specifically in this area. Most studies in upper Brantas have only focused on soil water content ground measurement, erosion and sedimentation, and landslide disaster management efforts [10,11].

The study will review the benefit of soil water content measurement from SMOS, SMAP, and SWI imagery products. The major objective is to propose the recommended satellite image for operational purpose in upper Brantas Basin. Both qualitative and quantitative measures are used in investigating the performance of these images. By comparing the remote sensing data with the results of soil test, this study also attempts to validate the satellite imagery. The study offers some important insights into the use of continuous wide coverage of soil water content information for landslide disaster prediction.

2. Methods

2.1. Study area

To examine the remote sensing data, the catchment area of upper Brantas River Basin (Figure 1) is selected as a study area (112.54 km2). The area is covered by forest in the upstream and dense residential area in the downstream. With relatively high mean annual rainfall of 2000 mm/year, 85% of the rain falls in the rainy season. Due to rapid land use change and prolonged precipitation in a common occurrence on the sloping downhill in the upstream (10%-25%), this catchment is vulnerable for landslide. The upstream is formed mainly Andosols soil type. At least seven landslides in 2018 caused 28 deaths [10].

2.2. Soil water content satellite products

Different satellite mission have measured soil water content in different way and level. Recently, higher level of satellite soil water content measurement have been published. In Figure 2, we show the comparison of these data. The spatial data for this study are taken from one week in the rainy season of 2020 covering whole Brantas catchment area. A short sample is chosen because of the expected difficulty of obtaining the continuous data. This data are selected because during this period, there was observation campaign of soil testing. This period is chosen for its representativeness of rainy and not rainy days. In Figure 2 the availability of data during observation campaign is shown.

2.3. Comparison of the observation

Investigating the accuracy and availability is a continuing concern within the utilization of soil water content product form satellite. For this study, ground truth data from previous research is used to explore the advantage and drawback of each observation. Furthermore, the comparison of each satellite data in terms of spatial and temporal availability, spatial distribution, quantitative value.

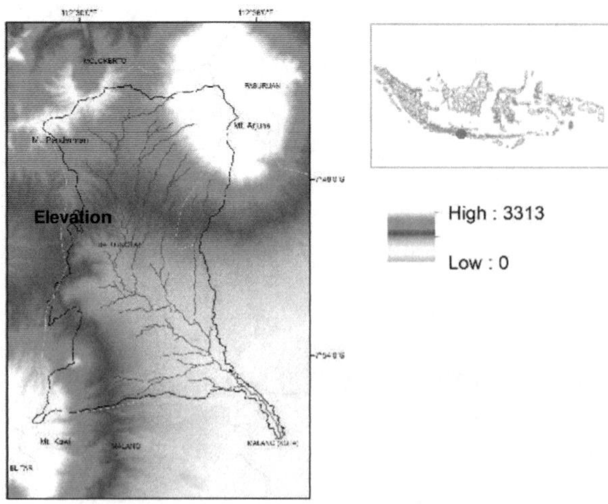

Figure 1. Upper Brantas River Basin.

Table 1. Comparison of satellite products.

	SMAP	SMOS L2	SWI
Provider	NASA, JPL CIT	ESA	Copernicus GLS
Level	4	2	2
Spatial resolution	10 km	15 km	11 km
Temporal resolution	3 hours	1 day	1 day
Type	Active passive	Passive	Active
Reference	Kimball et al., [6]	Bengoa et al., [7]	Bauer-Marschallinger et al., [8]
Temporal coverage	31 Mar 2015 - date	6 Jun 2010 - date	20 Feb 2007 - date
Spatial coverage	Global	Global	Global
Unit	%	%	%
Web	https://nsidc.org/data/SPL4SMGP	https://smos-diss.eo.esa.int/socat/SMOS_Open/	https://land.copernicus.eu/global/products/swi

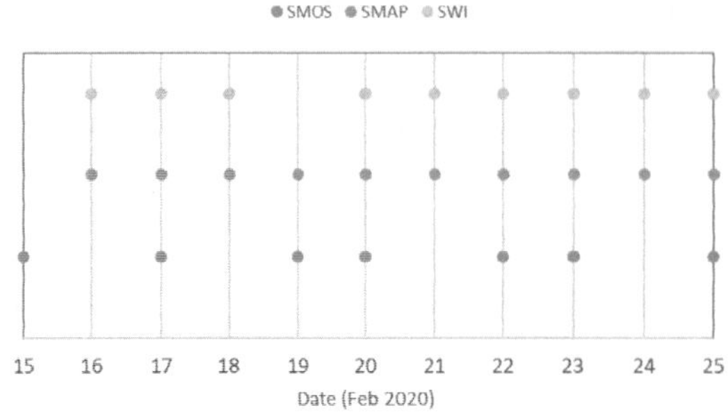

Figure 2. Data availability during observation campaign.

The performance indices i.e. mean absolute error, root mean square error, correlation coefficient, percent bias. In order to examine the trend significance, Mann Kendall analysis is used. This assessment

IOP Conf. Series: Materials Science and Engineering 1073 (2021) 012018 doi:10.1088/1757-899X/1073/1/012018

provides an important contribution to advance the understanding of the available soil water content observation over East Java area.

3. Results and discussion

3.1. Spatial and temporal distribution

The maps in Figure 3 presents the example of soil water content spatial distribution in Brantas River Basin from 2 observation time. From the map we can see that the SMOS data has some blank grids and the spatial resolution is rather rough. Regarding SMAP and SWI, both has similar finer spatial resolution. However, quantitative value of SMAP tends to be lower than SWI. In terms of temporal availability, from Figure, it is clear that SMAP is relatively continuous. Though SWI is missed in 1 day of observation, overall, this product is quite reliable. SMOS product is rather incomplete, which requires preprocessing as consequences. SMOS is found also to have high spatial variation of moisture value.

3.2. Soil water content quantitative estimate

The consistency of the observation to provide the accurate estimation of soil water content is evaluated by showing the time series chart. Figure 4 shows the temporal variation of soil water content observation spatially averaged over the upper catchment, middle catchment, and lower catchment. Interpolation is done to construct data points between the range of known values. It is apparent from the figure that the quantitative amount of soil water content from SWI is close to the true value. However, in terms of tendency, SMOS data provides a better trend visually compared to the true value.

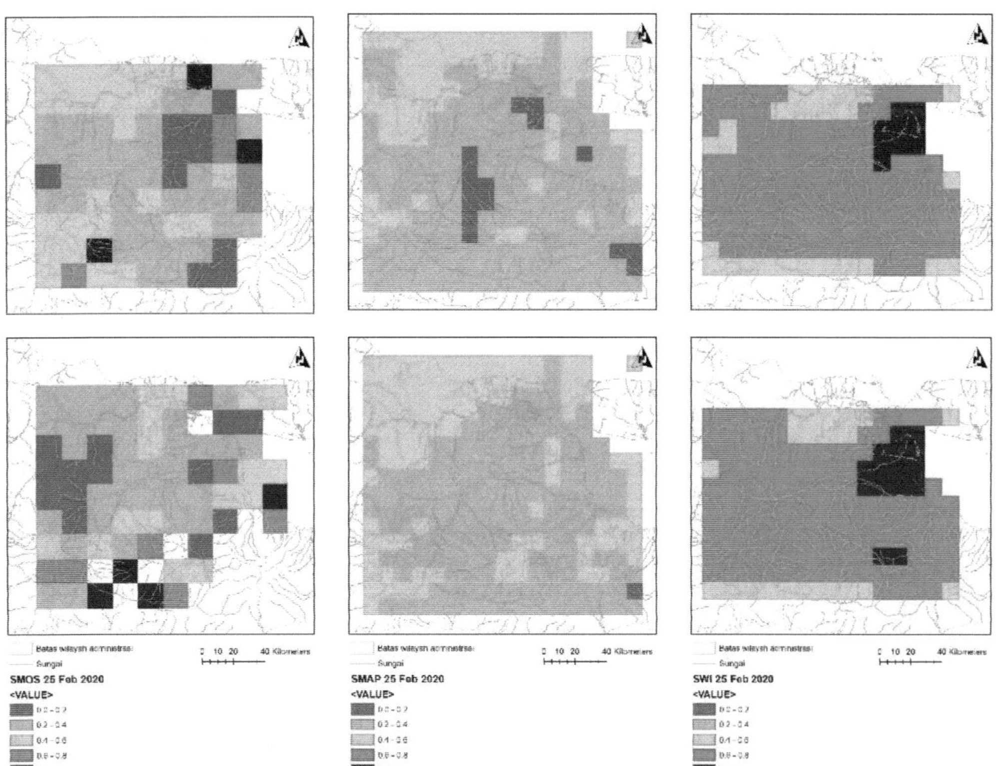

Figure 3. Soil water content spatial distribution in whole Brantas River Basin.

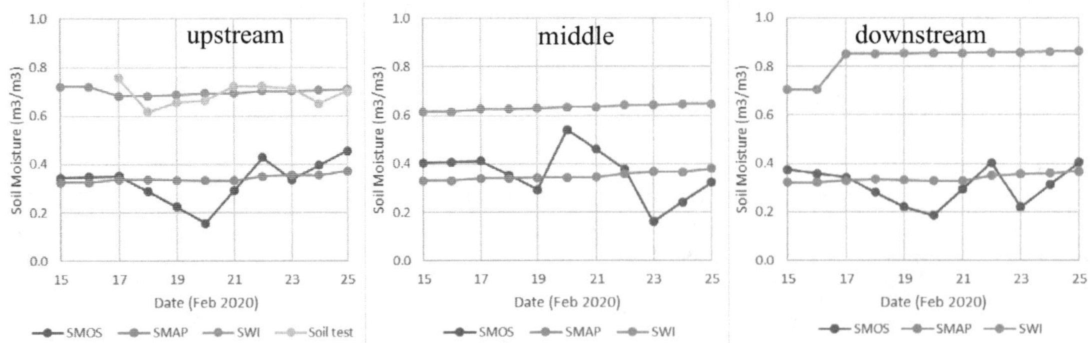

Figure 4. Time series of soil water content observation in upstream, middle, and downstream.

All the observations indicate positive trend with true value with Z value ranges from 0.78 to 3.10. SWI is observed to have similar trend of the water content with true value indicated by Mann Kendal test that shows Z test and significance value which is close (Table 1). Contrary, the study does not find a similar trend of SMOS, though visually it shows a similarity with true value. SMOS has highly positive and significant tendency. It seems possible that this result is due to high variance of SMOS measurement, compared to other three measurements.

From the previous discussion, the analysis reveals the advantage of SWI in terms of qualitative assessment. It is now necessary to examine the quantitative estimates of soil moisture. The statistic of all data is summarized as box plot in Figure 5. Maximum, minimum, mean, standard deviation, 25% percentile, median, and 75% are set out. SMOS is obviously highly variable. Many grids are observed.

Table 2. Indices comparison of satellite products.

	SMOS	SMAP	SWI	Soil Test
Mann-Kendall Z	1.65	3.1	1.45	0.78
Trend significant at a	0.1	0.01	>0.1	>0.1
MAE	0.40	0.31	0.11	
RMSE	0.43	0.32	0.12	
P-Bias	-57.69%	-48.18%	17.41%	
r	0.46	0.66	0.81	

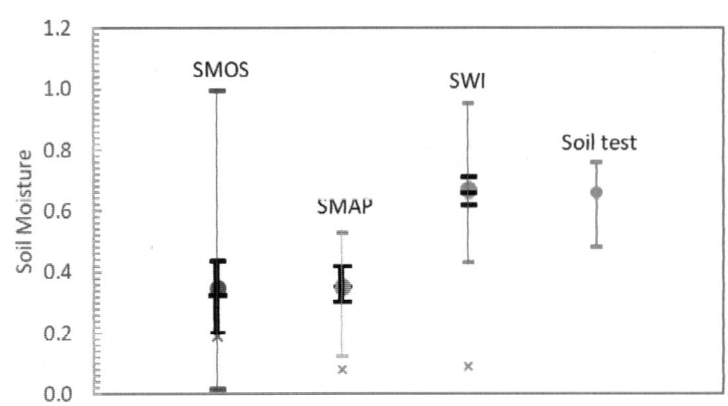

Figure 5. Distribution of the data indicating data range, 25, 50, and 75 percentiles (black markers), mean, and standard deviation.

Similar values views at lower parts of the span, but in other grids values are more variable. Meanwhile, SMAP has low variation indicated by small span. Although the range of data is a little higher, there appears to be some agreement between that SWI central tendency and the true value. In addition, the standard deviation is small, indicating that the values tend to be near to mean or expected value.

3.3. Performance evaluation

The performance indices show a good correspondence between the estimated soil water content from SWI and observation. There is positive strong direction of a linear relationship between SWI and true value indicated by correlation coefficient of 0.81. Although SMOS and SMAP have shown positive correlation coefficient too, the values suggest a weak relationship exists. MAE, RMSE, and P-bias indices reflect low accuracy of estimated soil moisture from SMOS and SMAP for direct application in upper Brantas River Basin. On average, SWI is shown to have good performance indicated by MAE, RMSE, and P-bias which are outperform others. The better performance in SWI could be attributed to data fusion approach that is pre-processed to the data from sensors.

The discrepancy of performance of each satellite observation is associated with the characteristic of the sensors and algorithm. As given in Figure 2, densely populated are tends to have high moisture value and vegetated area tends to have low moisture value. This study supports the idea of Wang et al., that vegetation characteristics are influencing the soil water content [4]. Interestingly, very high soil water content is found in similar spots from SMOS observation, which is at middle south because there is large reservoir at this area. Low moisture is found in some area that does not have water body, such as mountain top. These finding is in agreement with Su et al., which states that SMOS is highly affected by water bodies [9]. SMOS can capture the variation of water content as actual value. However, as with common remote sensing data, the data therefore need to be processed by renormalization [9].

4. Concluding remarks

In this research soil water content products from Soil Moisture Active Passive from Global Modelling and Assimilation NASA (SMAP), Soil Moisture and Ocean Salinity from ESA (SMOS), and Soil Water Index (SWI) from Copernicus Global Land Service are compared with soil test in terms of qualitative and quantitative assessments. Regarding the temporal availability, SMAP product is relatively continuous. Both SMAP and SWI provides good availability spatially. SWI is observed to have similar trend with true moisture value indicated by Mann Kendal Z value. SWI is suitable for further application in the study area with 0.81 correlation coefficient. Although SMOS and SMAP have shown positive correlation coefficient too, i.e. 0.46 and 0.66 respectively, the values show a weak relationship exists. RMSE value of 0.43 and 0.32 for SMOS and SMAP respectively suggests bias correction procedure for their application. This finding may contribute to the modeling of landslide phenomenon in slope hill by introducing quantitative soil water content product from satellite imagery to hydrological model and land surface model.

References

[1] Yuliza W, Habil H, Munir M M, Irsyam I, Abdullah M and Khairurrijal 2015 Study of soil moisture sensor for landslide early warning system: Experiment in laboratory scale *Journal of Physics: Conference Series* **739** 012034

[2] Kim K, Han D, Kim D, Wang W, Jung J, Kim J and Kim H S 2019 Combination of Structural Measures for Flood Prevention in Anyangcheon River Basin, South Korea *Water* **11** 2268

[3] Segoni S, Rosi A, Lagomarsino D, Fanti R and Casagli N 2018 Brief communication: Using averaged soil moisture estimates to improve the performances of a regional-scale landslide early warning system *Natural Hazards and Earth System Sciences* **18**(3) 807-812

[4] Wang J R 1985 Effect of vegetation on soil moisture sensing observed from orbiting microwave radiometers *Remote Sensing of Environment* **17**(2) 141-151

[5] Famiglietti J S, Ryu D, Berg A A, Rodell M and Jackson T J 2008 Field observations of soil moisture variability across scales *Water Resources Research* **44**(1) 1-6

[6] Kimball J S, Jones L A, Glassy J, Stavros E N, Madani N, Reichle R H, Jackson T and Colliander A 2016 *Soil Moisture Active Passive Mission L4_C Data Product Assessment* (Greenbelt: NASA Goddard Space Flight Center)

[7] Al Bitar A, Mialon A, Kerr Y H, Cabot F, Richaume P, Jacquette E, Quesney A, Mahmoodi A, Tarot s, Parrens M, Pellarin T, Rodriguez-Fernandez N, Wigneron J and Al-Yaari A 2017 The global SMOS Level 3 daily soil moisture and brightness temperature maps *Earth Syst. Sci. Data* **9**(1) 293-315

[8] Sánchez-Zapero J 2018 *Copernicus Global Land Operations "Vegetation and Energy"* (Ebro River Delta: Copernicus europe's eyes on earth)

[9] Su C H, Ryu D, Young R I, Western A W, amd Wagner W 2013 Inter-comparison of microwave satellite soil moisture retrievals over the Murrumbidgee Basin, southeast Australia *Remote Sensing of Environment* **134** 1-11

[10] Anggraeni F A K, Minarto E and Realita A 2017 Application of Resistivity Method Wenner-Schlumberger Array to Evaluate Brantas River Embankment as Avoid Landslide *IPTEK Journal of Proceedings Series* **3**(2) 1-4

[11] Uchida T, Takahashi F, Onda Y, Sisinggih D, Kato H, Noro T and Osanai N 2009 Estimating soil erosion rate and sediment sources using radionuclide Pb-210ex in upper Brantas river basin in Indonesia *Journal of Japan Society of Hydrology and Water Resources* **22**(3) 188-197

IOP Conf. Series: Materials Science and Engineering 1073 (2021) 012019 doi:10.1088/1757-899X/1073/1/012019

Flood analysis of Buntung River and the structural measures in Sidoarjo Regency

R A Sulhan*, R I Hapsari, M Efendi and M S Safitri

Department of Civil Engineering, State Polytechnic of Malang, Malang, Indonesia

*adipatijhon@gmail.com

Abstract. Sidoarjo Regency as one of the downstream areas of the Brantas River Basin is prone to flooding. This study aims to simulate the flood discharge and inundation in Buntung River, a river passes through Sidoarjo City area with several time return periods and to discuss its physical flood mitigation. Data needed in this study are Rainfall Data for 3 Rain Stations, topography map, cross section of river stream, and land use map. Based on these data, a flood flow analysis plan is then performed using the Nakayasu synthetic hydrographs method. The inundation is simulated using HEC-RAS with 10, 25, and 50 years return periods. Based on the calculation, Buntung River flood discharge with return period of 10, 25, and 50 years return periods is 151,342, 196,923, and 233,981 m3/s respectively. The inundation extend of 25 years return period is reaching 0,53 m height above the bank along Sta. P0 to P1. The study reveals that the existing cross section cannot accommodate the design flood for at least 25 years return period. Therefore, normalization is essential for river improvement. The channel is design with river cross-sectional shape of double trapezoidal. The size of the earth dike is 0.60 m above the flood water level with 3.0 m width and 1:2 channel slope. The designed river channel improvement increase the channel depth from -3,56 m to -4,07 m in Sta. P1. The normalized channel could accommodate the design flood and would be beneficial for increasing the urban area safety from flood disaster.

1. Introduction

Rapid development, particularly in metropolitan city, increases the risks of flooding. Some of the causes of flood are decreasing of the soil capability to infiltrate the water due to impermeable surface and sedimentation that reduces the channel capacity. Furthermore, prolonged rain is unfortunate especially for flat river in relatively plane catchment. For flood countermeasures, combination of structural and non-structural efforts must be carried out [1]. Structural measure consists of construction of hydraulic structures to increase carrying capacity of the river, such as levee, river normalization, and shortcut. Non-structural, such as flood simulation by various hydrology and hydraulic models, forecasting and warning as well as awareness awareness and law enactment can be more effective to minimize the flood damage.

Sidoarjo Regency in Indonesia is located in the downstream areas of the Brantas River Basin. Sidoarjo are is a buffer of Surabaya metropolitan city. Tambak Oso Village, Waru District, Segoro Tambak Village, and Sedati District were reported to experience inundation. Afvour Buntung is vulnerable from flooding in these area. The river morphology is meandered, eroded in some places, and shallow due to sedimentation. These condition would have a major impact on the occurrence of flood.

IOP Conf. Series: Materials Science and Engineering 1073 (2021) 012019 doi:10.1088/1757-899X/1073/1/012019

There has been several researches regarding flood disaster mitigation combining inundation simulation combined with the hydraulic structure [2,3]. In Buntung River, Ikhsan has studied the drainage system and simulate the flood using 25 years return period and propose construction of levee [4]. However, it did not use hydraulic simulation model. Azzahra studied the stream pattern and sedimentation of Buntung River [5]. It was mentioned that tides affect the flood highly in the downstream of Buntung River.

In this study, the flood and overtopping simulation in Buntung River using several return periods is carried out. As non-structural countermeasures. By understanding the river overflowing in particular rainfall amount, early warning can be designed carefully. As for physical countermeasures, levee and normalization are studied with 25 years return period in order to prevent the flood plain river being inundated.

2. Methods

The study area is Buntung River with length, L of 4 km that passes through Sidoarjo Regency and Surabaya City. The catchment area, A is 15864.93 Figure 1 shows the map, photo, and catchment area of Buntung River. The required data are daily rainfall data of 2009-2018 from Ketegan, Bono, and Wonorejo Stations, flood report, catchment map, river cross sections, and tides data at the river estuary. Rainfall data are pre-processed using double-mass curve method and arithmetic means to obtain areal rainfall [6]. Design rainfall (mm/day) with return period of 10, 25, and 50 years are analyzed by using Log-Pearson Type III distributions [7]:

$$\log X_T = \overline{\log X} + K.s$$

Where $\overline{\log X}$ is average logarithmic of daily rainfall (mm/day), K is constant, and s is standard deviation (mm/day). Flood hydrograph is analyzed by Nakayasu synthetic hydrograph method [8]:

$$Tg = 0{,}4 + 0{,}058 \times L$$
$$Tr = (1 - 0{,}5) \times Tg$$
$$Tp = Tg + 0{,}8 \times Tr$$
$$T_{0,3} = \alpha \times Tg$$
$$Q_p = \frac{CA.R_o}{3{,}6\,.\,(0{,}3T_p + T_{0,3})}$$
$$Q_t = Q_p\left(\frac{t}{T_p}\right)^{2{,}4}$$
$$Q_r = Q_P\,.\;0{,}3^{(t-T_p)/T_{0,3}}$$
$$Q_t = Q_p\,.\;0{,}3^{[(t-T_p)+(0{,}5\,.\,T_{0,3})]/(1{,}5T_{0,3})} \text{ and } Q_t = Q_p\,.\;0{,}3^{[(t-T_p)+(1{,}5T_{0,3})]/(2\,.\,T_{0,3})}$$

Basin parameters, T, and Q are depicted in Figure 1 and Figure 2. Flood hydrograph is calculated by make a superposition from unit hydrograph from hourly rainfall distribution. Rt:

$$Rt = \frac{R_{24}}{5} x \left(\frac{5}{t}\right)^{2/3}$$

Where R24 is designed rainfall (mm/day), t is time step (h).

In order to simulate the overflowing, HEC-RAS 5.0.1 1D hydraulic model is applied. By using this model, it is easier and timely to estimate water level profile, water flow velocity, and Froude number. Afterward, the intended water level during rainfall is determined in particular return period, i.e. 25 years. According to this scenario, the river capacity is design to meet the specific discharge. Menu of Channel Modification in HEC-RAS is applied to simulate the best scenario by considering existing cross section as well as local condition [9].

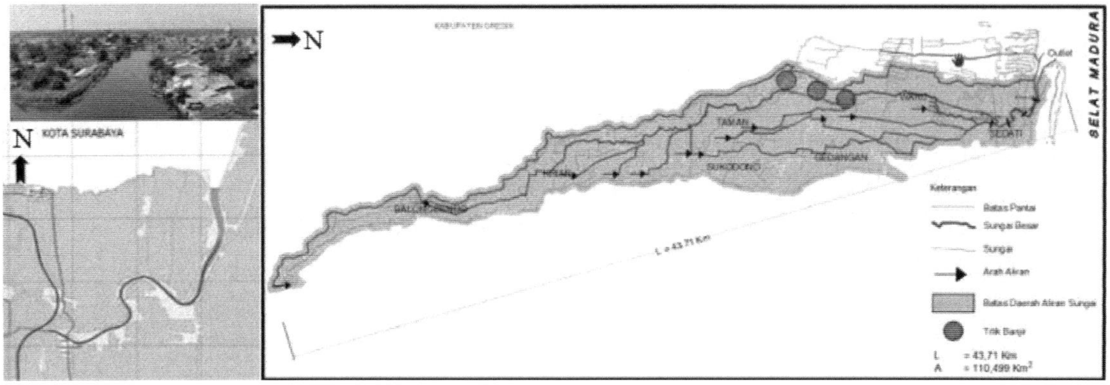

Figure 1. Map, photo, and catchment area of Buntung River (Source: BBWS Brantas).

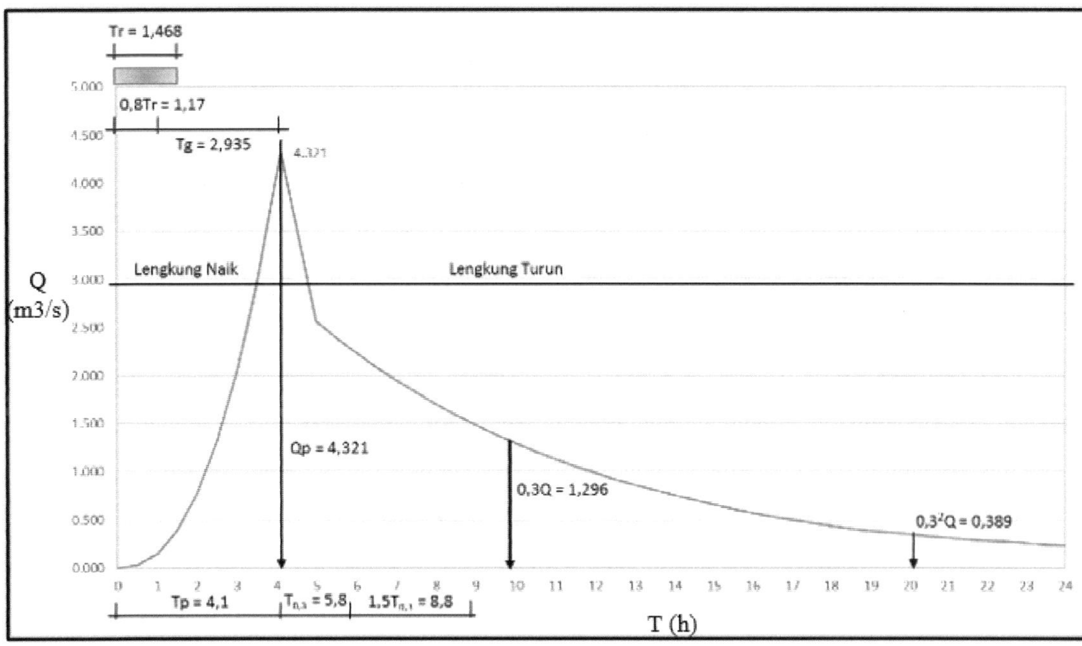

Figure 2. Unit hydrograph elements.

3. Results and discussion

3.1. Rainfall-runoff transformation

Analysis of hydrology shows that the design rainfall for 10, 25, and 50 years return period are 141.356 mm/day, 158.975 mm/day, and 177.520 mm/day respectively. The rainfall intensity hourly distribution is given in Figure 3. These data are introduced to the Nakayasu model and the results are shown in runoff as illustrated in Figure 4. Due to lack of flood observation data, validation of flood simulation is conducted by comparing the analysis with the reference from past study, i.e. simulation carried out by Ikhsan [4] which is shown by black curve in Figure 4. It can be concluded that the simulation could provide the plausible result despite slightly lower value of the calculation.

IOP Conf. Series: Materials Science and Engineering 1073 (2021) 012019 doi:10.1088/1757-899X/1073/1/012019

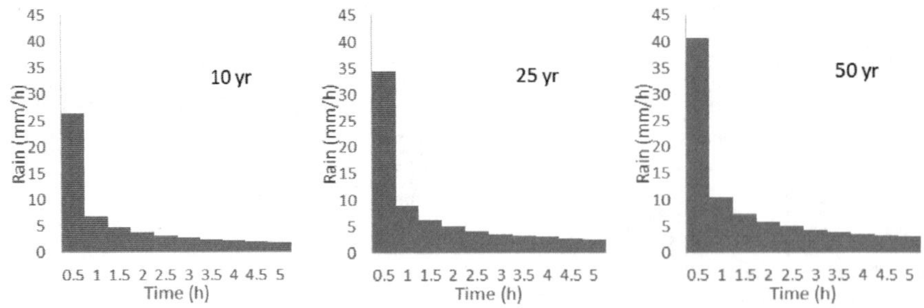

Figure 3. Rainfall intensity hourly distribution.

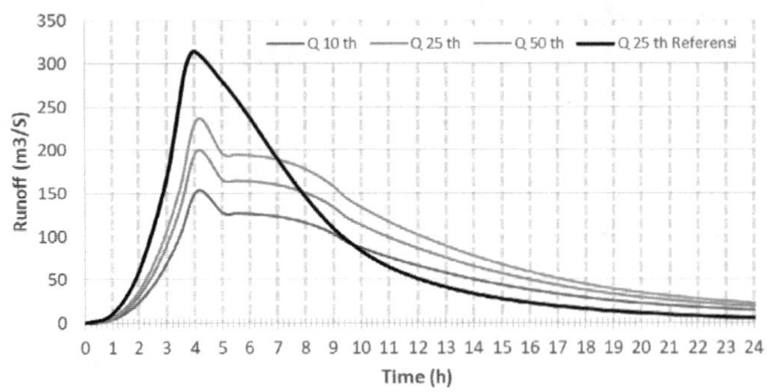

Figure 4. Flood hydrograph as upstream boundary condition.

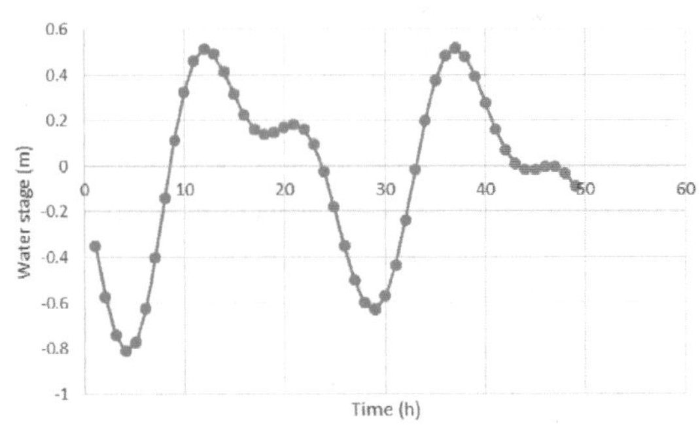

Figure 5. Tidal stage as downstream boundary condition (right).

3.2. Simulation of inundation from river overtopping

Runoff from Nakayasu becomes an upstream (P81) boundary condition of hydraulic model/HEC-RAS. As for the downstream (P0), the boundary condition is set as stage/flow hydrograph which is obtained from tides data. Friction coefficient or Manning's n values is 0,035 representing rough natural stream. Contraction and expansion coefficients are 0.1 and 0.3 respectively. The results of the simulation is shown in Figure 6 for upstream (P80), middle (P30), and downstream (P0).

For model validation, comparison with the inundation occurrence is presented. According to the flood report, on 22 February 2017 and 15 December 2018, flood occurred at Kedungrejo and Ketegan area respectively. The recorded rainfall was up to 21 mm/day and 110 mm/day respectively. The inundation depth in Buntung River vicinity ranges from 0.15 m to 0.5 m. This report conform to the simulation from HEC-RAS, though the simulation shows little higher value. This may be attributable to

IOP Conf. Series: Materials Science and Engineering 1073 (2021) 012019 doi:10.1088/1757-899X/1073/1/012019

the location of flood survey report that are located higher than the simulated points of simulation. Nevertheless, the simulation would result in more safe countermeasures.

3.3. *River normalization as structural measure*

In order to overcome the flooding, channel cross section normalization is designed. The proposed dimensions of the river is a double trapezoidal cross section. By trial and error of new cross section in HEC-RAS Channel Modification, the ideal cross section to alleviate flood can be obtained. Trial is conducted to reduce the water level at least 0.88m, 0.81m, and 0.53m at downstream, middle stream, and upstream respectively. Figure 7 shows the result of channel modification. After being trialed, the channel with width of 21 m, bank slope of 1:2, and dyke height of 0.6 m is sufficient to convey the design flood with 25 years return period.

This study has shown the effect of normalization and dyke structure to reduce the occurrence of river overflowing in 25 years return period rainfall. However, there is still possibility that Buntung River experiences inundation in the upcoming years when the rain occurs with probability of less than 4%. Therefore, it is necessary to evaluate the robustness of these non-structural and structural countermeasures by studying the benefit and cost of the measures. Intangible benefit, such as the flood damage reduction as well as public willingness to pay the structural measure would be beneficial for operational purposes [10,11].

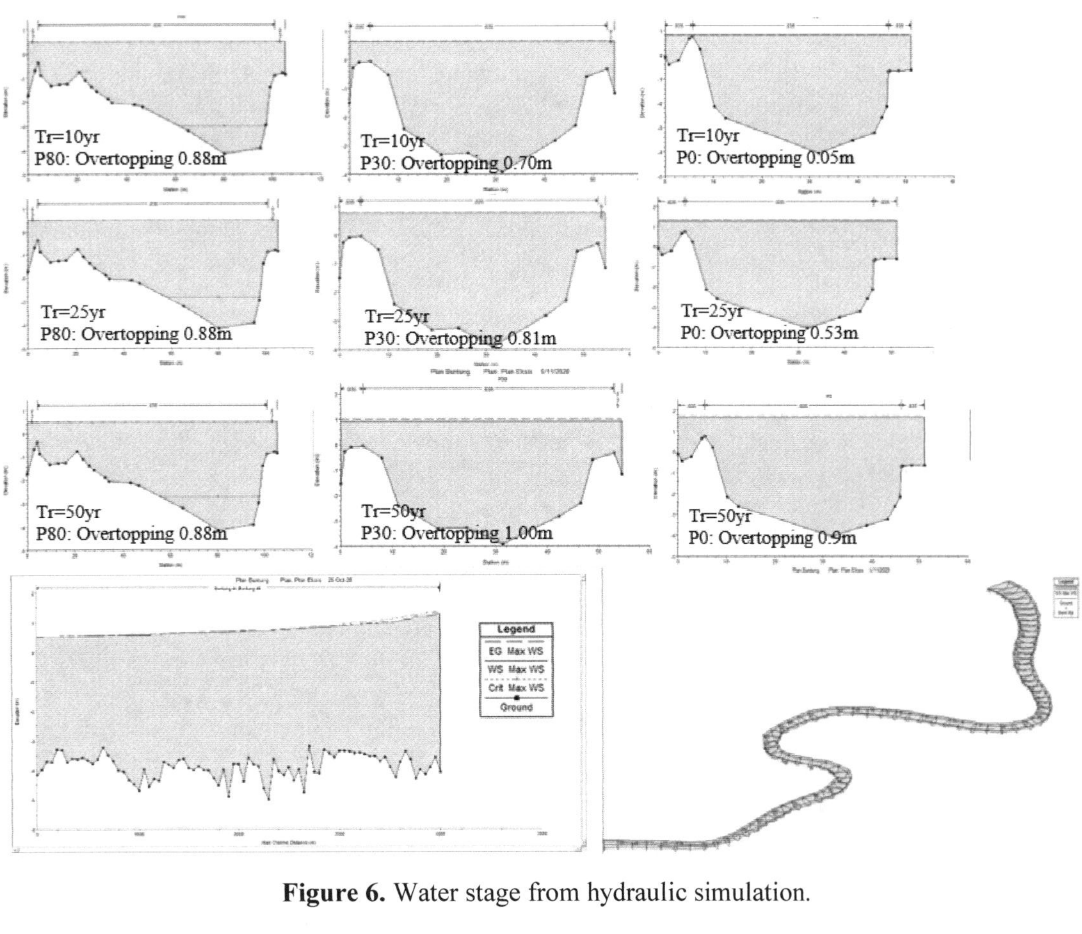

Figure 6. Water stage from hydraulic simulation.

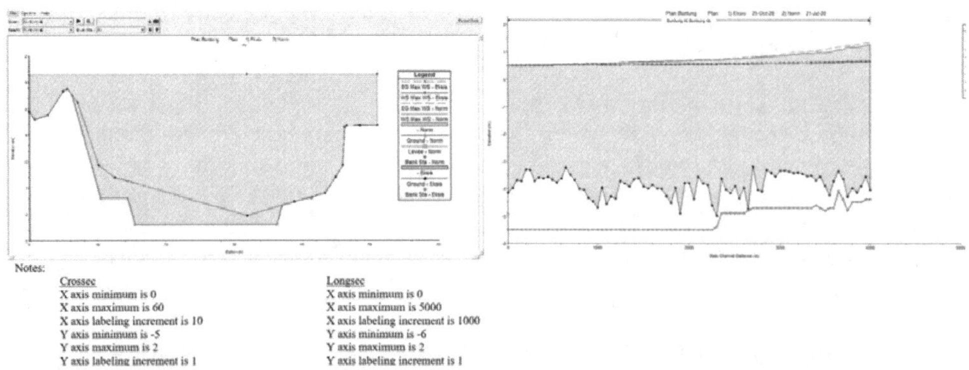

Figure 7. Channel modification and water profile for cross section (left) and long section (right) at downstream (P0).

4. Conclusions

Simulation of rainfall-runoff and open channel hydraulic of Buntung River Sidoarjo has been presented. Design flood with return period of 25 years is 196.923 m^3/s. It is found that existing channel capacity cannot accommodate the runoff that causes inundation in surrounding area. Validation with flood report and survey shows that the simulation shows a reliable results. As for the structural measures, river normalization and levee construction is proposed. The channel with width of 21 m, bank slope of 1:2, and dyke height of 0.6 m is sufficient to convey the design flood with 25 years return period. For further study, it is recommended to simulate the submerged area due to inundation as the flood plain is relatively flat. Validation with flood depth from water level recorder or direct survey is necessary to gain the confidence of the modeling. Studying the feasibility of the physical countermeasures by including the intangible cost/benefit of flood damage is recommended to provide integrated flood alleviation in the study site.

References

[1] Kim K, Han D, Kim D, Wang W, Jung J, Kim J and Kim H S 2019 Combination of Structural Measures for Flood Prevention in Anyangcheon River Basin, South Korea *Water* **11**(11) 2268

[2] Ferrari A, Dazzi S, Vacondio R and Mignosa P 2020 Enhancing the resilience to flooding induced by levee breaches in lowland areas: a methodology based on numerical modelling *Natural Hazards & Earth System Sciences* **20**(1) 59–72

[3] Mazzoleni M, Dottori F, Brandimarte L, Tekle S and Martina M L 2017 Effects of levee cover strength on flood mapping in the case of levee breach due to overtopping *Hydrological Sciences Journal* **62**(6) 892-910

[4] Rohmaniyah M and Rohmadiani L D 2020 Bentuk dan tingkat partisipasi masyarakat terhadap banjir Sungai Buntung di Kecamatan Waru *Waktu* **18**(01) 15-25

[5] Azzahra A D 2015 *Analisa pola arus dan sedimentasi di daerah muara Sungai Kali Buntung, Tambakoso, Surabaya* (Surabaya: Institut Teknologi Sepuluh Nopember)

[6] Soemarto C D 1987 *Hidrologi Teknik* (Surabaya: Usaha Nasional)

[7] Soewarno 1995 *Hidrologi* (Bandung: Nova)

[8] Limantara L M 2019 *Rekayasa Hidrologi* (Yogyakarta: Andi)

[9] Brunner G W 2002 Hec-ras (river analysis system) *North American Water and Environment Congress & Destructive Water* 3782-3787

[10] Wurbs R A 1983 Economic feasibility of flood control improvements *Journal of Water Resources Planning and Management* **109** 29-47

[11] Thunberg T and Shabman L 1992 *Determinants of landowner's willingness to pay for flood hazard reduction* Journal of the American Water Resources Association **27** 657-665

IOP Conf. Series: Materials Science and Engineering 1073 (2021) 012020 doi:10.1088/1757-899X/1073/1/012020

Optimization of heavy equipment for earthwork in the construction of Mainroad Section X of Pandaan-Malang Toll Road

S S Riskijah*, B A Pradipta and D Lidyaningtyas

Civil Engineering Department, State Polytechnic of Malang, Malang, Indonesia

*ririssafiatus@gmail.com

Abstract. The construction of Mainroad Section X of Pandaan-Malang Toll Road requires of heavy equipment for excavation and embankment works. The type and number of the heavy equipment can be varied depending upon the construction volume, equipment capacity, and the budget. Therefore, the combination of heavy equipment must be chosen carefully to minimize the cost. This study aims to determine the optimum number of heavy equipment combinations with constraints of work volume, schedule, and resources availability to minimize the construction cost. The data needed is site lay-out plan, cross section of the road, project time schedule, specifications of the heavy equipment, and the resources unit price. The method for optimization is Linear Programming using Simplex Method. The results shows that the optimum combination of heavy equipment for excavation work are 14 units of bulldozer type Zoomlion ZD160-3, 5 unit excavator type Komatsu PC200-8MO, and 19 units dump trucks type Mitsubishi Fuso-FN627 at Rp 14.839.362.044,00. As for the embankment work, the optimum equipment are 5 units excavators type Komatsu PC200-8MO, 19 units of dump truck type Mitsubishi Fuso-FN627, 19 units of motor grader type Caterpillar 120K, 18 units of sheep foot roller type Ammann ASC-200, and 23 units of water tank truck type Mitsubishi-125HD at Rp 22.979.058.760,00.

1. Introduction

The Toll Road Construction Project of Pandaan-Malang region is the extension of the Trans Java Toll Road Project which connect Surabaya - Pandaan - Malang. Considering its huge volume of workloads, specific methods, and large cost, the Pandaan-Malang Toll Road construction project requires an effective method of implementation. The mainroad's earthwork uses heavy equipment to ease and improve the work, hence it must be managed accordingly to avoid its idleness, which could increase the equipment cost. By taking the work and length of this project into consideration, this study only observes the mainroad section X of Pandaan - Malang Toll Road.

This study aims to determine the optimal combination of heavy equipments usage with minimum cost for mainroad's earthwork. Optimization problems can be solved by using the linear programming method which is an optimization technique and a linear constraint system that aims to maximize profits or minimize costs, restricted by certain limitations [1]. Optimization problems using linear programming sometimes involve many constraints and variables, thus a series of mathematical procedures is needed to find solutions to more complex problems. The procedure that can be used is the simplex method [2].

IOP Conf. Series: Materials Science and Engineering 1073 (2021) 012020 doi:10.1088/1757-899X/1073/1/012020

Similar research has been conducted by several people, namely: Yadam [3] who found optimum cost Suzuki Negara Stock Yard Development Project, Bali of Rp. 560,960,000; Maddeppungeng et al. [4] who found optimum cost Port Development Projects of PT. Cemindo Gemilang of Rp. 10.395.810.000; Wibowo and Zacoeb [5] who found optimum cost Villa Bukit Tidar Malang Housing Project of transportation cost of Rp.248.296.400, and embankment cost of Rp.111.571.400; and Suhendra et al. [6] who found optimum cost Providing Selected Landfills for Improving Network Access of Clean Water Pipeline Network in West Tanjung Jabung Regency, Booster Senyerang reduced cost of Rp.6.455.000.

The most crucial step in linear programming is to formulate a linear program. This step includes identifying goals and boundaries. Some elements commonly used are the formulation of decision variables, objective functions, constraint functions, and variable constraints [1].

The standard linear programming problems are as follows [7]: Maximize or minimize

$$Z = c_1 \times 1 + c_2 \times 2 + c_3 \times 3 + \cdots + c_j \times j \tag{1}$$

Limiting resources (constraints):

$$Za_{11} \times 1 + a_{12} \times 2 + a_{13} \times 3 + \cdots + a_{1j} \times j \leq b_1 \tag{2}$$

$$a_{21} \times 1 + a_{22} \times 2 + a_{23} \times 3 + \cdots + a_{2j} \times j \leq b_2$$

$$\ldots$$

$$a_{j1} \times 1 + a_{j2} \times 2 + a_{j3} \times 3 + \cdots + a_{jn} \times j \leq b_i$$

Where c_j = cost per unit for each x_j; x_j = the jth decision variable; a_{ji} = the i resource requirements for each x_j; b_i = number of available resources; j = number of decision variables, starting from 1,2,3…j; and i = number of resources usage, starting from 1,2,3…i.

In the use of heavy equipment, it is necessary to pay attention to the type of heavy equipment used, knowledge of the capacity and capability of heavy equipment in order to meet the requirements of use, namely not causing a waste of labor, capital, and meeting safety needs [8]. Equipment cost per unit is also determined by the productivity. To calculate the productivity of each heavy equipment used [9]:

a. Excavator productivity:

$$Q = \frac{V \times F_a \times F_b \times 60}{T_{s1} \times F_y} \tag{3}$$

Where V = bucket capacity; F_a = the tool's efficiency factor; F_b = bucket factor; T_s = cycle time; F_y = conversion factor; and 60 is hour to minute conversion.

b. Dump truck productivity:

$$Q = \frac{V \times F_a \times 60}{\mathrm{D} \times T_S} \tag{4}$$

Where V = body capacity; F_a = the tool's efficiency factor; F_k = material construction factor; D = material mass (loose); T_s = cycle time; and 60 is hour to minute conversion.

c. Bulldozer productivity; stripping work:

$$Q = \frac{q \times F_b \times F_m \times F_a \times 60}{Ts} \tag{5}$$

and leveling work:

$$Q = \frac{l \times \{n(L - Lo) + Lo\} \times F_b \times F_m \times F_a \times 60}{N \times n \times T_s} \tag{6}$$

Where q = stripping capacity; F_b = blade factor; F_m = blade inclination factor; F_a = the tool's efficiency factor; l = stripping distance; n = number of paths; Lo = overlap width; N = number of stripping paths; T_s = cycle time; and 60 is hour to minute conversion.

d. Sheep foot roller productivity:

$$Q_p = W_e \times V \times H \times 1000 \times \frac{E}{N} \quad (7)$$

Where Q_p = sheep foot roller productivity; W_e = effective compaction width; V = average tool's velocity; H = compaction thickness; E = tool's efficiency factor; N = number of paths; and 1000 is km to m conversion.

e. Motor grader productivity; leveling work:

$$Q = \frac{L_h \times \{n(b - b_0) + b_0\} \times F_a \times 60}{N \times n \times T_s} \quad (8)$$

and leveling solid overlay:

$$Q = \frac{L_h \times \{n(b - b_0) + b_0\} \times F_a \times 60 \times t}{N \times n \times T_s \times F_k} \quad (9)$$

Where L_h = the stretch length; n = number of paths; b_0 = overlap width; F_a = work efficiency factor; N = number of stripping per track; T_s = cycle time; t = solid overlay thickness; F_k = material construction factor; and 60 is hour to minute conversion.

f. The water tank truck productivity:

$$Q = \frac{P_a \times F_a \times 60}{W_c \times 1000} \quad (10)$$

Where P_a = water pump capacity; F_a = tool's efficiency factor; W_c = water needs per m3 solid material; 60 is hour to minute conversion; and 1000 is km to m conversion.

The calculation of the rental and operational costs of each heavy equipment per hour is as follows [9];

a. Definitive cost:

$$G = (E + F) = \frac{(B - C) \times D}{W} + \frac{Ins \times B}{W} = \frac{(B - C) \times D + (Ins \times D)}{W} \quad (11)$$

Where G = hourly estimated cost (Rp); E = capital return cost; F = insurance, taxes, and other cost/year; B = local cost of equipments; C = tool's residuel value; D = installment/return factor; W = number of equipments working hour/year, Ins = Insurance (%).

b. Operational cost:

$$P = \mathrm{H + I + J + K + L + M} \quad (12)$$

Where H = fuel usage (liter)/hour; I = lubricant usage (liter)/hour; J = workshop maintenance cost/hour; K = service cost/hour including parts replacement; L = operator/driver's hourly wage/hour; M = assistant operator/driver's hourly wage/hour.

One of the computer applications that can easily, quickly and accurately solve linear programs is LINDO (Linear Interactive Discrete Optimizer) software. The main working principle of the LINDO program is entering formulas, solving them and estimating the correctness and appropriateness of formulas based on their solutions [10].

2. Methods
The optimization of heavy equipment usage for main road section X of Pandaan-Malang Toll Road consist of several stages. The first stage is collecting secondary data such as site plan drawings, cross section drawings, specifications and types of heavy equipment as well as its pricelist, and the project master schedule. The next stage are identifying the type of earthworks; which consists of excavation and embankment work, then calculating the volume based on cross section data, determining the method to carry out excavation and embankment work, determining the type of heavy equipment that can be used according to work conditions and calculate its productivity, determining the combination of heavy equipment for existing conditions and alternative combinations [11], then calculating its cost, respectively. The above data are used to determine the optimal heavy equipment usage for earthworks using linear programming with simplex method in the form of decision variables, objectives, and constraints as well as transforming those variables into objective and constraints functions. In order to minimize the heavy equipment cost, the objective and constraints function are transformed into a standard model according to the provisions of LINDO 6.1 to yield the optimized solution.

3. Results and discussion

3.1. Work volume calculation

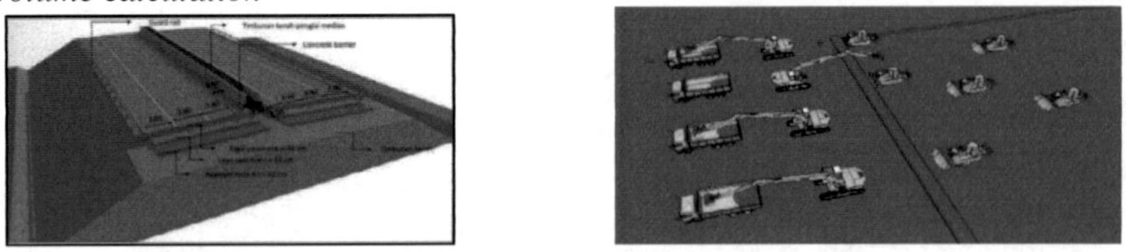

(a) Detailed Structure (b) Movement of Heavy Equipment

Figure 1. Detailed structure of mainroad works and movement of heavy equipment.

The excavation volume defines the required amount of peat soil to be disposed from the former paddy fields, ranging from STA 35+375 – STA 38+488 as deep as 30 cm. The calculation was based on the cross-section data, which resulting excavation volume of 158.355 m3 and soil embankment volume of 696.776 m3. Figure 1 below shows the detailed structure of main road works.

3.2. Excavation work method
The excavation work aims to disposing peat soil from the embankment location to disposal area, by stripping the surface soil on former paddy fields. The heavy equipment used are bulldozer, excavator, and dump truck. The movement flow of each heavy equipment used in excavation work is illustrated in Figure 1(b). As the bulldozer clears the location, it moves forward; pushing the excavated ground and cleans the location towards the excavator's rotation area, in back and forth motion. The excavator only rotates 180°, processing the loaded excavation material into a dump truck. The dump truck queues the loading process at the site, process the loading from excavator, transports the excavation materials to quarry location, disposes the material at the quarry location, and travels back to the project site, respectively.

3.3. Embankment work method
The embankment work which carried out through the STA. 35+375 – STA 38+488, aims to achieve the desired elevation by using materials from the borrow area. The heavy equipment used for this work are excavator, dump truck, motor grader as the landfill spreader tool, sheep foot roller as the soil compactor, and water tank truck as watering tool. The work includes soil extraction from the quarry using excavator which then transported to the project site by dump truck, laying each layer of materials using motor grader, then levelled by the bulldozer. The sheep foot rollers and water tanks then work simultaneously to achieve the desired solid embankment as planned.

The movement flow of each heavy equipment used in embankment work is: the excavator processes the loading excavated soil into a dump truck, thus it only rotates 180°. The dump truck queues the loading process at the quarry, process the loading from excavator, transports the materials to project site, dumps the materials at the project site, respectively in back and forth motion. The motor grader moves forward to level the soil, also in back and forth motion. The sheep foot roller and water tanks simultaneously compact and watering the soils in forward and backward motions.

3.4. Selection of heavy equipment

The selected heavy equipment used in the mainroad earthwork based on Ministry of Public Works [11], are EXT1 (Excavator - Caterpillar 320D2), EXT2 (Excavator - Komatsu PC 200-8MO), DT1 (Dump Truck - Hino FM 260JM), DT2 (Dump Truck - Mitsubishi Fuso FN 627), BD1 (Bulldozer - Zoomlion Zd 160-3), BD2 (Bulldozer - Komatsu D85E SS-2 Angle Dozer), SFR1 (Sheep Foot Roller - Ammann ASC-200), MG1 (Motor Grader - Caterpillar 120 K), MG2 (Motor Grader - Mitsubishi MG330), WTT1 (Water Tank Truck - Mitsubishi 125HD).

3.5. Calculation of productivity and equipment cost

Table 1. Heavy equipment productivity and rental and operating cost.

Excavation (m3/hour)		Embankment (m3/hour)		Operational Rental Price / hour	Mobilization Fee
BD1	65.652			Rp 696,608	Rp 3,775,000
BD2	64.740			Rp 729,112	Rp 3,775,000
EXT1	160.551	EXT1	160.551	Rp 844,647	Rp 5,285,000
EXT2	147.266	EXT2	147.266	Rp 625,052	Rp 5,285,000
DT1	20.659	DT1	13.616	Rp 863,130	Rp 566,250
DT2	20.659	DT2	13.616	Rp 814,435	Rp 566,250
		MG1	27.321	Rp 811,686	Rp 3,775,000
		MG2	34.471	Rp 837,075	Rp 3,775,000
		SFR1	104.197	Rp 782,856	Rp 3,775,000
		WTT1	71.143	Rp 439,074	Rp 566,250

The heavy equipment productivity presented at Table 1, are calculated using Equation 3-10. Since each type of equipment used has rental, operational, and mobilization cost, the total equipment cost at Table 1 are calculated using Equation 11-12. Calculation of wages and equipment rental costs based on AHSPK [12].

3.6. Heavy equipment alternative combination

Table 2. Alternative combinations of heavy equipments.

Alternative								Work
1	2	3	4	5	6	7	8	
DB1	BD1	BD1	BD1	BD2	BD2	BD2	BD2	Excavation
EXT1	EXT1	EXT2	EXT2	EXT1	EXT1	EXT2	EXT2	
DT1	DT2	DT1	DT2	DT1	DT2	DT1	DT2	
EXT1	EXT1	EXT2	EXT2	EXT1	EXT1	EXT2	EXT2	Embankment
DT1	DT2	DT1	DT2	DT1	DT2	DT1	DT2	
MG1	MG1	MG1	MG1	MG2	MG2	MG2	MG2	
SFR1	SFR1	SFR1	SFR1	SFR1	SFR1	SFR1	SFR1	
WTT1	WTT1	WTT1	WTT1	WTT1	WTT1	WTT1	WTT1	

The alternative combination of heavy equipment are selected based on available types and specifications [11], which then optimized accordingly. Determining the combination of heavy equipment are the work volume, limited project site area, and each of equipment's dependency. The existing combination is described as alternative 1 in Table 2, followed by alternative combinations 2-8.

3.7. Optimization

Aiming to get the optimum number of each heavy equipment for the mainroad earthwork as well as its minimum cost, the optimization of excavation and embankment work are carried out separately by considering the differences in working conditions and characteristics.

3.7.1. Optimization for excavation work (Alternative 1). The three pre-determined main factors are: Decision variables: X_1 = Number of DB1 units; X_2 = Number of EXT1 units; and X_3 = Number of DT1 units; Objective function: referring the Equation 1 and Table 2, Z min = 696,608X1 + 844,647X2 + 863,130X3; and Constraints function: Excavation volume constraints: 65.652 $X_1 \geq$ 306.269; 160.551 X2 $\geq$ 306.269; and 20.659 X3 $\geq$ 306.269; Area constraints: 383X1 + 1327X2 + 636X3 $\leq$ 281250; 6.630X1 $\leq$ 90; 18.760X2 $\leq$ 90; and 4.955X3 $\leq$ 90; Equipment dependency constraints: $10X_1 + 4X_2 + 31X_3 \geq$ 158353,4; and Non-negativity constraints: X_1, X_2, $X_3 \geq 0$. The optimization for excavation work yields BD1 = 14 units, EXT1 = 5 units, and DT1 = 19 units as the number of heavy equipment per hour with minimum cost of Rp. 30,375,207.

3.7.2. Optimization for embankment work (Alternative 1)

Table 3. Recapitulation of total heavy equipment cost for excavation work and embankment work.

Excavation Work of Mainroad

Alt	Volume (m^3)	Hour	Optimum Number of Unit/Hour			Productivity (m^3/Hour)			Rental Cost/Hour (Rp)	Mobilization Cost (Rp)	Total Rental Cost (Rp)	Total Cost (Rp)
			BD	EXC	DT	BD	EXC	DT				
1	158.355	517	14	5	19	65,65	160,55	20,66	Rp 30.375.207	Rp 180.067.500	Rp 15.705.362.651	Rp 15.885.430.151
2	158.355	517	14	5	19	65,65	160,55	20,66	Rp 29.450.016	Rp 180.067.500	Rp 15.226.996.883	Rp 15.407.064.383
3	158.355	517	14	5	19	65,65	147,27	20,66	Rp 29.277.233	Rp 180.067.500	Rp 15.137.660.312	Rp 15.317.727.812
4	158.355	517	14	5	19	65,65	147,27	20,66	Rp 28.352.042	Rp 180.067.500	Rp 14.659.294.544	Rp 14.839.362.044
5	158.355	517	14	5	19	64,74	160,55	20,66	Rp 30.830.272	Rp 180.067.500	Rp 15.940.652.107	Rp 16.120.719.607
6	158.355	517	14	5	19	64,74	160,55	20,66	Rp 29.905.081	Rp 180.067.500	Rp 15.462.286.339	Rp 15.642.353.839
7	158.355	517	14	5	19	64,74	147,27	20,66	Rp 29.732.298	Rp 180.067.500	Rp 15.372.949.767	Rp 15.553.017.267
8	158.355	517	14	5	19	64,74	147,27	20,66	Rp 28.807.108	Rp 180.067.500	Rp 14.894.584.000	Rp 15.074.651.500

Embankment Work of Mainroad

Alt	Volume (m^3)	Hour	Optimum Number of Unit/Hour					Heavy Equipment Productivity (m3/Hour)					Rental Cost/Hour (Rp)	Mobilization Cost	Total Rental Cost
			EXC	DT	MG	SFR	WTT	EXC	DT	MG	SFR	WTT			
1	696.776	399	5	19	19	18	23	160,55	13,62	27,32	104,20	71,14	Rp 58.682.363	Rp 379.765.000	Rp 23.406.260.586
2	696.776	399	5	19	19	18	23	160,55	13,62	27,32	104,20	71,14	Rp 57.757.172	Rp 379.765.000	Rp 23.037.235.565
3	696.776	399	5	19	19	18	23	147,27	13,62	27,32	104,20	71,14	Rp 57.584.389	Rp 379.765.000	Rp 22.968.318.781
4	696.776	399	5	19	19	18	23	147,27	13,62	27,32	104,20	71,14	Rp 56.659.198	Rp 379.765.000	Rp 22.599.293.760
5	696.776	399	5	19	19	18	23	160,55	13,62	34,47	104,20	71,14	Rp 59.164.754	Rp 379.765.000	Rp 23.598.669.064
6	696.776	399	5	19	19	18	23	160,55	13,62	34,47	104,20	71,14	Rp 58.239.563	Rp 379.765.000	Rp 23.229.644.043
7	696.776	399	5	19	19	18	23	147,27	13,62	34,47	104,20	71,14	Rp 58.066.781	Rp 379.765.000	Rp 23.160.727.259
8	696.776	399	5	19	19	18	23	147,27	13,62	34,47	104,20	71,14	Rp 57.141.590	Rp 379.765.000	Rp 22.791.702.238

This following model for alternative 1 is calculated on LINDO: Min = 844647X1+ 863130X2 + 811686X3 + 969608X4 + 439074X5; 160.551 X_1 >= 1746,902; 13.616 X_2 >= 1746,902; 27.321 X_3 >= 1746,902; 104.197 X_4 >= 1746,902; 71.143 X_5 >= 1746,902; 1327X1 + 295X2 + 348X3 + 289X4 + 117X5 <= 281250; 18.760X1 <= 90; 4.955X2 <= 90; 4.895X3 <= 90; 5.100X4 <= 90; 4.020X5 <= 90; $4X_1 + 47X_2 + 24X_3 + 6X_4 + 9X_5$ >= 0; X1, X2, X3, X4, X5 >=0. The optimization for embankment work yields EXT1 = 5 units, DT1 = 19 units, MG1 = 19 units, SFR1 = 18 units, and WTT1 = 23 units, as the number of heavy equipment per hour with minimum cost of Rp. 58,682,263. Following the same calculation for other alternative combinations, each optimum cost per hour for each combination. The total cost of each Excavation and embankment work equipment can be calculated by multiplying the duration of the work per hour with operational costs and additional mobilization costs, hence the results are presented in Table 3. Based on Table 3, it is known that alternative 4 (gray block) is the best alternative for excavation work with a minimum cost of Rp. 14,839,362,044, - and alternative 4 (gray block) is also the best alternative with a minimum cost of Rp. 22,979,058,760.

4. Conclusions

Based on the results of the calculation using LINDO, the study findings can be concluded as follows:

- The optimum alternative for excavation work is alternative 4 with total cost of Rp. 14,839,362,044, - following the combination of 14 units Bulldozer Type Zoomlion ZD 160-3 Type, 5 units Excavator Type Komatsu PC 200-8MO, and 19 units Dump Trucks Type Mitsubishi Fuso FN 627.
- The optimum alternative for embankment work is alternative 4 with total cost of Rp. 22,979,058,760, following the combination of 5 units Excavators Type Komatsu PC 200-8MO, 19 units Dump Trucks Type Mitsubishi Fuso FN 627, 19 units Motor Graders Type Caterpillar 120 K, 18 units Sheep Foot Rollers Type Ammann ASC-200 Type, and 23 Units Water Tank Trucks Type Mitsubishi 125HD.

It is expected that the further research can present more varied alternative combinations of heavy equipment according to the types of equipment available in the catalogue to improve optimization results. Furthermore, we need to pay more attention to the limitations on the field in order to formulate the overall constraints. The contractor is advised to use combination IV as a combination of heavy equipment used for excavation and landfill in the project, which described in the conclusion so that the contractor can minimize heavy equipment usage cost.

References

[1] Rafflesia U and Widodo F H 2014 *Pemrograman Linier* (Bengkulu: Faculty of Agriculture UNIB)

[2] Notoprasetio D D 2017 *Optimasi Biaya Penggunaan Alat Berat Pada Proyek Pembangunan Underpass Mayjen Sungkono Surabaya* (Surabaya: Institut Teknologi Sepuluh Nopember)

[3] Yadam R W 2015 *Optimalisasi penggunaan alat berat pada pekerjaan galian tanah (Studi kasus: Proyek pembangunan stock yard Suzuki Negara, Jln. Denpasar-Gilimanuk, Kec. Melaya, Kab. Jembrana, Bali)* (Denpasar: Universitas Udayana)

[4] Maddeppungeng A, Suryani I and Febriana D 2018 Optimalisasi komposisi alat berat pada proyek pembangunan Pelabuhan PT. Cemindo gemilang *Konstruksia* **9**(1) 59-67

[5] Wibowo S A and Zacoeb A 2018 Kajian Biaya Penggunaan Alat Berat Untuk Pekerjaan Pengangkutan Dan Penimbunan Pada Proyek Perumahan Villa Bukit Tidar Malang Dengan Program Linier Metode Simpleks *Jurnal Mahasiswa Jurusan Teknik Sipil* **1**(1) 001

[6] Suhendra S, Dwiretnani A and Endika E 2019 Optimasi Alat Berat pada Pemindahan Tanah Mekanis (Studi Kasus Penyediaan Timbunan Tanah Pilihan Pada Pekerjaan Peningkatan Akses Jaringan Pipa Air Bersih Kabupaten Tanjung Jabung Barat, Booster Senyerang) *Jurnal Talenta Sipil* **2**(1) 8-15

[7] Anggoro B S, Rosida R M, Mentari A M, Novitasari C D and Yulista I 2019 Profit Optimization Using Simplex Methods on Home Industry Bintang Bakery in Sukarame Bandar Lampung *Journal of Physics: Conference Series* **1155** 012010

[8] Kurniawan R 2018 Optimalisasi biaya alat berat pada pekerjaan cut and fill pembangunan jalan alternatif Srigonco-Jls Kabupaten Malang (Malang: University of Muhammadiyah Malang)

[9] Regulation of the Minister of Public Works and Public Housing No 28 / PRT / M / 2016 concerning Analysis of Unit Prices for Public Works Works

[10] Rosiyanti H 2016 Penggunaan software lindo dengan metode pembelajaran penemuan terbimbing untuk meningkatkan motivasi belajar mahasiswa matematika angkatan 2013 pada matakuliah program linier *FIBONACCI: Jurnal Pendidikan Matematika dan Matematika* **2**(2) 19-27

[11] Ministry of Public Works 2013 *Heavy Equipment Catalog* (Ministry of Public Works)

[12] Analisis Harga Satuan Pokok Kegiatan (AHSPK) 2019 *Analisis Harga Satuan Pokok Kegiatan (AHSPK) Pemerintah Kota Malang* (Malang)

The development of interlocking models of lightweight bricks and shear failure assessment of its wall structures

T Rochman*, N Rasidi and Y Aditya

Civil Engineering Department, State Polytechnic of Malang, Malang, Indonesia

*taufiq.rochman@polinema.ac.id

Abstract. In this paper the development of interlocking models of lightweight bricks is proposed and its shear failure assessment in wall structures has been studied. Walls are a brittle infill structures and quiet vulnerable to collapse compared among all building components when a strong earthquake occurs. An interlocking system approach is such a simple method that is proposed to make sure the wall behave keep strong and stiff during seismic load. This research was performed to improve the shear capability and modelling of lightweight brick as well, which is interlocked each other. The data from the specific gravity testing results and its compressive strength then inputted as one of the modelling parameters. An experiment was carried out utilizing the Taguchi method to find out the optimum mixture used in shear testing in medium scale of walls. The compressive strength and shear strength testing methods follow the Indonesian code of SNI 03-1974-1990 and ASTM E2126-2011. The cube specimens of 15x15x15cm were used for compressive experiments, while a set of wall structures of 100x100 cm that was surrounded by reinforced concrete practical frame of columns and beams of 13x13 cm was used as shear test specimens. From shear testing results that using cyclic loads, it was found that the system can resist a maximum load of 45 kN by a lateral deflection of 45 mm. The results also obtained elastic stiffness of 2.55 kN / mm, the yielding behaviour occurs at the load, lateral deflection, ductility factor $\Delta f/\Delta y$ and shear strength of 38.25 kN, 14,961 mm, 2.61 and 35,714 kN / mm respectively, which means that each width of 1 mm wall can be resist a shear load of 35,714 kN. It can be simply concluded that in cyclic shear loads, the more the wall is getting damaged or cracked, the structural stiffness will decreased as shown in a hysteresis behaviour.

1. Introduction

Masonry walls are a brittle structure and one of the most vulnerable to collapse of the entire building components during a strong earthquake shaking. A simple method that makes wall behaves strongly and stiffly during earthquake is to make the wall behave in the same way as a box, together with the roof at the top and with the foundation at the bottom. In a construction, aspects are required to ensure the action of this box. One of them is the connection between the walls must be good. This aspect can be achieved by ensuring the masonry is interconnected [1].

Ramadhan has conducted research on *interlocking masonry bricks* , from this research, the results of static testing of the *interlocking masonry brick* wall portals showed that the specimen was able to receive a pushover load of 63931.2 N which resulted in an ultimate displacement of 35 mm with a crack thickness of 4 mm. The wall shear strength obtained is 21.31 N / mm. Based on the energy equivalent elastic plastic (EEEP) curve analysis, the wall elastic stiffness is 5905.88 N / mm, and the wall yield load is 57237.55 N with a yield displacement of 9.69 mm [2]. Based on the description above, it is necessary to conduct research to improve the interlocking lightweight brick model and test the shear strength of the walls made of interlocking lightweight bricks.

2. Materials

2.1. Shear failure

First failure type occurs at low normal stress and usually named by shear failure. Cracks often propagate follow stairs mode. Many researchers state that the failure occurs due to the damage of bond at bed joint or connection [3]. Several shear failure issues can be presented below.

2.1.1. Stepped cracks shear failure. Stepped cracks shear failure is shear damage along mortar connection in the wall in brick's side and little cracks appear in the brick's part (Figure 1).

2.1.2. Horizontal sliding shear failure. Horizontal sliding shear failure is shear damage along horizontal wall near or right on half infill wall panel (Figure 2) [3].

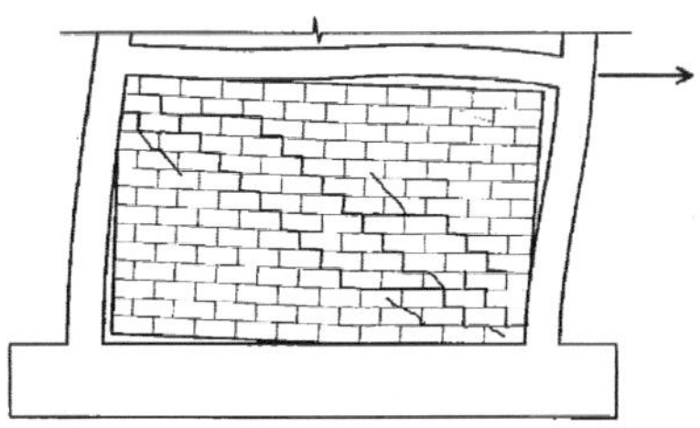

Figure 1. Stepped cracks shear failure pattern.

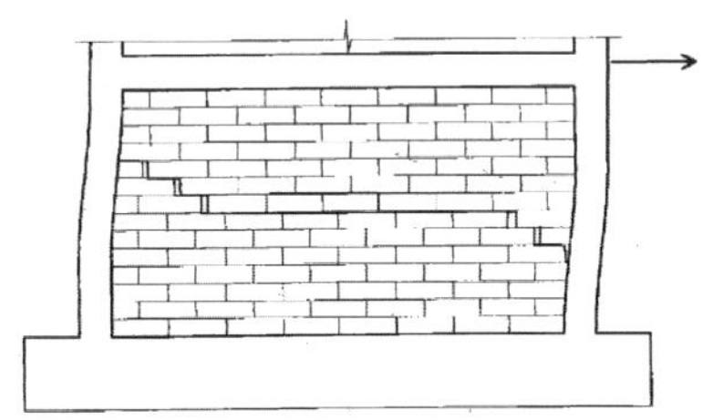

Figure 2. Horizontal sliding shear failure pattern.

2.2. Bricks stress failure

For moderate to high values of normal stresses, tensile failure of the brick generally occurs. These failure modes are called tensile failure cracks. The shear strength of the mortar joint increases due to the effect of normal compressive stress. Therefore, cracks appear in the brick instead of in the joint, as a result of the tensile stress caused by the compressive stress state. As observed in the **Figure 3**, the cracks follow the head joints and pass through the bricks with a tendency that depends on the orientation of the main stresses on the bricks.

When panels are horizontally reinforced or when the condition of the panel-frame joint is improved, as occurs in frame pairs, cracks are usually small and distributed in wide zones along the diagonal. In other cases, the damage is concentrated in one or two large cracks.

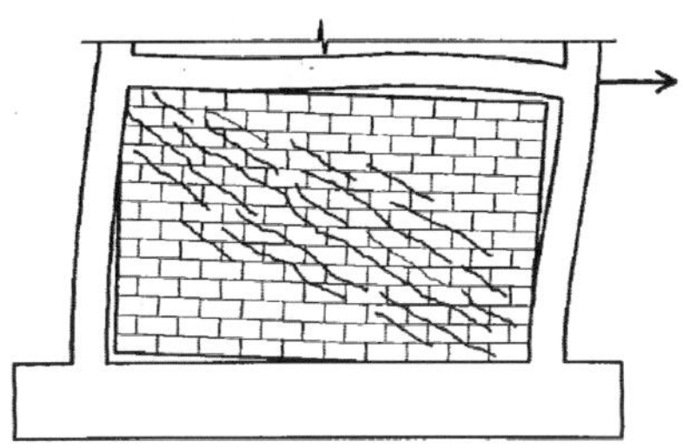

Figure 3. Diagonal cracks due to the bricks tensile stress failures.

Mahdizadeh suggests that damage to masonry walls is usually characterized by cracks [4]. The crack category based on its dimensions for the pair wall structure can be divided into 2 categories. The first category has no effect on the structure except the aesthetics of the building, these crack sizes range from

IOP Conf. Series: Materials Science and Engineering 1073 (2021) 012021 doi:10.1088/1757-899X/1073/1/012021

0.1 mm to 2.0 mm. The second category of cracks that will affect the use and service of the building needs to be repaired if necessary because it can lead to building collapse. The dimensions of the cracks start from 2.0 mm until the most dangerous is 25 mm.

3. Methodology

3.1. Foamed bricks

In this study, several experiments on the mixture composition of lightweight bricks along with compressive testing were carried out , developing a new model of interlocking lightweight bricks, apply the interlocking lightweight brick model as a masonry wall, and testing the shear strength of the masonry wall with the cyclic load. The sample in this study were 16 cube specimens with dimensions of 15x15x15 cm and 10 units of interlocking lightweight bricks. Materials and equipment that will be used for the manufacture of lightweight bricks are water, white cement, fiberglass, and foam agent.

3.2. Wall infill reinforced concrete frame

Whereas for the manufacture of masonry reinforced concrete frame in the wall shear test is 4D10 mm for the main reinforcement and Ø6 mm for stirrups with 200 mm spacing between stirrups, the composition of the portal concrete mixture is 1 cement: 2 sand: 3 gravel. The equipment for testing the compressive strength is the compaction testing machine, while the test equipment for testing the shear strength is the loading frame, dial gauge, hydraulic jack and hydraulic jack pump. The following are 16 plans of composition variations for the trial mix obtained from the Taguchi method.

Table 1. Composition of trial mix of cube specimens.

Run	Cement (gram)	Fiberglass (gram)	Paste Water (ml)	Water for foaming (ml)	Foam Agent (ml)
1	2500	25	416.7	833.3	41.7
2	2500	37.5	416.7	833.3	55.6
3	2500	50	416.7	833.3	66.7
4	2500	62.5	416.7	833.3	83.3
5	2500	50	458.3	916.7	45.8
6	2500	62.5	458.3	916.7	61.1
7	2500	25	458.3	916.7	73.3
8	2500	37.5	458.3	916.7	91.7
9	2500	62.5	500.0	1000.0	50.0
10	2500	25	500.0	1000.0	66.7
11	2500	37.5	500.0	1000.0	80.0
12	2500	50	500.0	1000.0	100.0
13	2500	37.5	541.7	1083.3	54.2
14	2500	50	541.7	1083.3	72.2
15	2500	62.5	541.7	1083.3	86.7
16	2500	25	541.7	1083.3	108.3

The procedure for making lightweight bricks is putting water and cement in a row into the bucket and stirring with an electric drill, making foam using a foam generator, putting the foam into the bucket, put the fiberglass, and then stirring until well mixed, check the wet density of the lightweight brick mixture, then put the lightweight brick mixture into the mold. Remove the lightweight bricks from the mold then put it in the curing tank. After 28 days, the compressive strength test is carried out based on SNI 03-1974-1990. The compressive strength of concrete load is the amount of load per unit area, which causes the concrete specimen to collapse when loaded with a certain compressive force, which is generated by a compression machine [5]. Specific strength is the ratio of strength to density of material. This means that the material has a light weight, but it also has high strength [6].

3.3. Setup of wall-RC frame test

After the curing has completed, then a shear testing was performed to measure its strength of the lightweight masonry wall in relation to the dimensions (1.0 x 1.0 x 0.125) m which is placed above the tie beam according to the setting layout to make sure that there is no lifted tie beam occurs during the test. The process of making the specimen wall done by assembling the beam anchored to the loading frame, then pouring the tie beam with a size of 0.13 m x 0.13 m. First, the installation of interlocking lightweight brick walls is carried out by pouring the interlocking lightweight brick with the tie beam. Then arrange interlocking lightweight brick without using mortar. Then, a 0.13 m x 0.13 m column, and then the entire surrounding RC frame of 0.13 m x 0.13 m poured simultaneously, when all the bricks had been installed. The shear test for the masonry wall specimen then was performed after 7 days.

The shear strength test for interlocking lightweight brick masonry walls with cyclic loads refers to ASTM E2126-11. The setting of the shear strength test with cyclic loads is done by installing a hydraulic jack on the upper corner on the surrounding RC frame, placed parallel to the in-plane wall and repeated cyclic loading in both directions of the wall, from the left loads and vice versa. While the front side of the hydraulic jack is installed with a dial gauge to measure the lateral displacement that occurs due to cyclic loads. The load is given alternately on the left and right of the wall specimen and added gradually with the same increment interval of 5 kN, 10 kN, 15 kN and so on until the wall collapses. Each load increase for each bar is recorded by the dial load, dial indicator, observing the crack pattern that occurs and marked it directly on the specimen wall as the loading increased and the cracks were propagated. The test setup of the hydraulic jack can be seen in **Figure 4**.

Figure 4. The test setup for the shear strength of the interlocking lightweight masonry brick.

The first observation is made after the loading of the first cycle from the left to the right, after that the second observation is made in the first cycle from the right to the left, and so on until the last cycle. The following is the modeling and results of interlocking lightweight brick casting (Figure 5 and Figure 6).

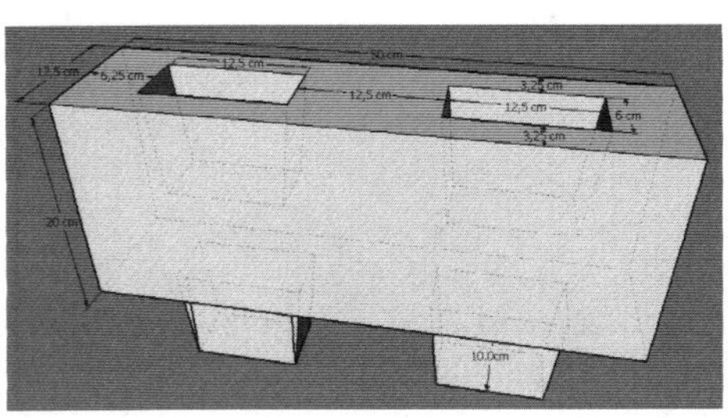

Figure 5. a-3D shape model, single interlocking brick.

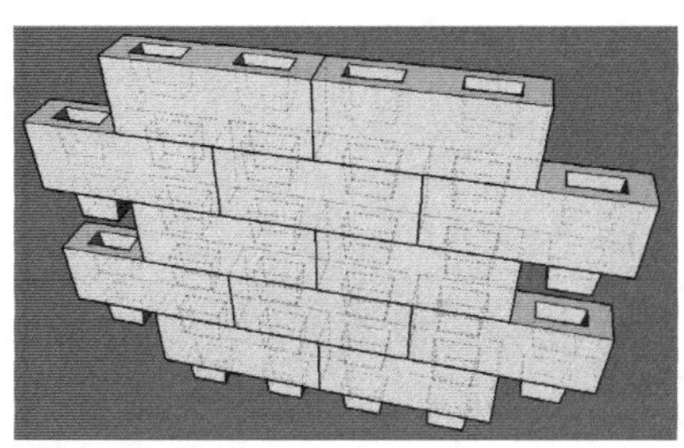

Figure 6. Interlocking system of lightweight brick masonry wall structures.

Figure 7 and **Figure 8** show respectively, the interlocking lightweight single brick and masonry bricks wall that its best composition based on strength has been determined using the *Taguchi* method after the compressive test results has finished.

Figure 7. Pouring result, interlocking lightweight bricks.

Figure 8. Installation result, interlocking lightweight brick masonry wall.

4. Results and discussion

Before the implementation of wall shear testing, a preliminary test is carried out on the trial mix of cube specimens that optimum specific strength will be selected among results and will be used as an interlocking lightweight brick from the selected optimum mix composition. The trial mix test covered density compressive strength testing. The test specimen is cube 15 cm x 15 cm x 15 cm. The following of regarding results of the trial mix cube specimens were depicted in Table 2.

Table 2. The results of the *trial mix* density test.

Run	Mixture variations			Weight (kg) (A)	Volume (m^3) (B)	Density (kg / m^3) (C)=(A/B)
	w/c ratio	*Foam/water* ratio	*Fiberglass content*			
1	0.50	1:20	1%	3	0.003375	888.93
2	0.50	1:15	1.5%	2.85	0.003375	844.44
3	0.50	1: 12,5	2%	2,6	0.003375	770.40
4	0.50	1:10	2.5%	3.07	0.003375	909.67
5	0.55	1:20	2%	2.7	0.003375	800.00
6	0.55	1:15	2.5%	2.45	0.003375	725.93
7	0.55	1: 12,5	1%	3	0.003398	883.00
8	0.55	1:10	1.5%	3,3	0.003353	984.34
9	0.60	1:20	2.5%	2.45	0.003353	730.80
10	0.60	1:15	1%	3,1	0.003375	918.52
11	0.60	1: 12,5	1.5%	2.35	0.003398	691.69
12	0.60	1:10	2%	3.5	0.003375	1037.0
13	0.65	1:20	1.5%	1.8	0.003375	533.33
14	0.65	1:15	2%	2	0.003375	592.59
15	0.65	1: 12,5	2.5%	2.5	0.003375	740.77
16	0.65	1:10	1%	3.12	0.003375	924.44

From the Table 2, it can be noted that the smallest density is 533.33 kg/m^3 and the biggest is 1037.03 kg/m^3. All specimens can float in the water due to their low density compared to water. Then, the compressive test was performed to all specimens and hence, the results of the resisted compressive strength can be shown in Table 3.

Table 3. The trial mix compressive strength results of the cubic specimen.

Run	Mixture variations			Compressive Strength
	w/c ratio	***Foam/water* ratio**	***Fiberglass content***	**(kg / cm^2)**
1	0.50	1:20	1%	27.66
2	0.50	1:15	1.5%	29.98
3	0.50	1: 12,5	2%	27.84
4	0.50	1:10	2.5%	38.20
5	0.55	1:20	2%	25.70
6	0.55	1:15	2.5%	23.56
7	0.55	1: 12,5	1%	32.03
8	0.55	1:10	1.5%	40.59
9	0.60	1:20	2.5%	21.56
10	0.60	1:15	1%	47.11
11	0.60	1: 12,5	1.5%	23.56
12	0.60	1:10	2%	72.16
13	0.65	1:20	1.5%	12.85
14	0.65	1:15	2%	17.13
15	0.65	1: 12,5	2.5%	21.42
16	0.65	1:10	1%	59.42

From the Table 3, it can be summarized that the largest compressive strength is 72.16 kg / cm^2 and the smallest is 12.85 kg / cm^2. Based on SNI 03-0349-1989 and the results of the trial mix compressive strength test, the optimal cube specimen is included in category II [7]. The following Figure 9 represent the relationship between lightweight bricks specimen of compressive strength and its density.

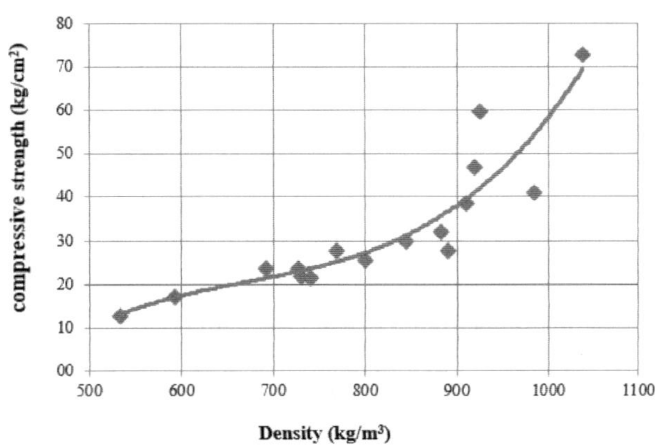

Figure 9. The relationship between trial mix compressive strength and its density.

From **Figure 9** above, it can be pointed out that the increase in density is slightly proportional to the value of compressive strength. It exhibits that the larger the density of lightweight bricks, the larger the compressive strength value will. The following is the trial mix specific strength was shown in Table 4.

Based on Table 4, it can be denoted that RUN-12 has the largest specific strength, which means that it has a quite small weight but has the largest strength. Shear testing of an interlocking lightweight brick wall system that surrounded by reinforced concrete frame then carried out when the brick's age reach 28 days.

Table 4. Specific Strength of the trial mix.

RUN	Density (kg / m$^{3)}$ (A)	Compressive Strength (N / m^2) (B)	Specific Strength (Nm / kg) (C) = (B / A)
1	888.93	2712136.42	3051.02
2	844.44	2940234.05	3481.86
3	770.40	2730338.68	3544.03
4	909.67	3780468.94	4155.87
5	800.00	2520200.61	3150.25
6	725.93	2310183.89	3182.40
7	883.00	3150 250.77	3567.66
8	984.34	4017098.29	4081.01
9	730.80	2114262,26	2893.09
10	918.52	4620367.79	5030.24
11	691.69	2310183.89	3339.94
12	1037.04	7140568.40	6885.55
13	533.33	1260 100.31	2362.69
14	592.59	1680133.74	2835.23
15	740.77	2100260.52	2835.23
16	924.44	5880468.10	6361.08

Cyclic loading is applied to the joints of beam-column of the wall specimen. A certain cyclic loading pattern applied to the loading jack actuator from left to right in repeated cyclic stages until the system was collapsed or reach cracks width of 4 mm, then was continued until the ultimate load and then the load was decreased (unloading), and repeated based on the cyclic loading pattern. The following *hysteresis loop* curve then obtained from the shear test results using cyclic loads such as depicted in **Figure 10**.

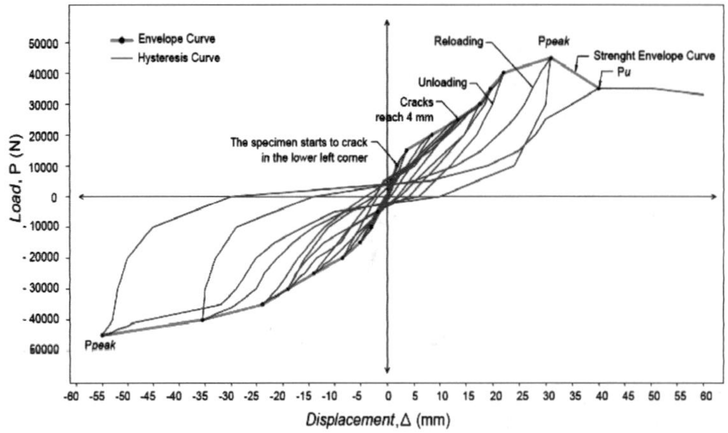

Figure 10. Hysteresis curve and envelope curves of wall structures.

From the results above in Figure 10, it can be concluded that the more the load increases, the more the lateral displacement will occurs, and a maximum load of 45000 N can be achieved by the masonry wall specimen. The test then stopped whenever the wall specimen is cracked by width of 4 mm [8]. At stopped loading, a load of 25000 N can be resisted by the wall and a lateral displacement of 13.44 mm was obtained. Then the repeated loading is carried out until the wall reach its maximum load and maximum lateral displacement.

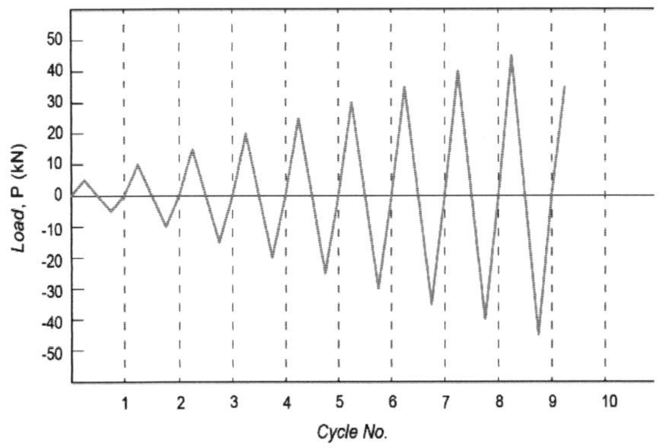

Figure 11. The pattern of cyclic loads applied.

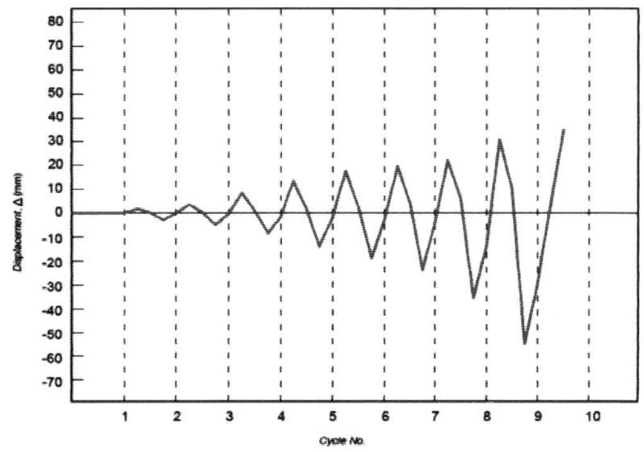

Figure 12. Lateral displacement pattern to cycle loads.

From **Figure 12** above, it seems that the lateral displacement at each cycle does not always return to 0 mm. This is behaving so due to decreasing of its stiffness as the cyclic load repeated and as the loading was increased, hence that the returning ability of the wall to its original position was decreased. Therefore, when the load is beyond than the elastic stiffness of 0.4P *peak* or 18000 N, obviously noted that the lateral displacement at zero load on the next cycle does not return to zero displacement.

Based on ASTM E 2126-2011 the data from the results of this test, namely the loads-displacement data, the values of elastic stiffness, shear strength, yield load, yield displacement, ductility, shear modulus and EEEP curve can be obtained respectively. The following are the results of the calculations obtained by the formula for the ASTM E 2126-2011 [9].

a) Elastic Stiffness

$0.4\,P_{peak} = 0{,}4 \times 45000\ \text{N} = 18000\ \text{N}$

$\Delta_e =$ Deviation at $0.4\,P_{peak} = 7.04$ mm

Elastic Stiffness $(K_e) = \frac{0{,}4\,P_{peak}}{\Delta e} = \frac{18000}{7{,}04} = 2556.82\ \text{N / mm} = 2556818.18\ \text{N / m}$

b) Shear Strength

Shear Strength $(\,v_{peak}) = \frac{P_{peak}}{L} = \frac{45000\ N}{1260\ m} = 35714{,}286\ \text{N / m}$

c) Yield Load

A (curve area from *autocad* application calculation) = 1209,525 Nm

$\Delta_u = 39.1\ \text{mm} = 0.0391\ \text{m}$

$$P_{yield} = \left((\Delta_u - \sqrt{\Delta_u{}^2 - \frac{2A}{K_e}} \right) = 38252.49\ \text{N}$$

d) Yield displacement

$$(\Delta_{yield}) = \frac{P_{yield}}{K_e} = \frac{38252.49\ \text{N}}{2556818.18\ \text{N/m}} = 0.01496\ \text{m}$$

e) Cyclic Ductility Ratio

$$(\,D\,) = \frac{\Delta_{ultimit}}{\Delta_{yield}} = \frac{0{,}0391}{0{,}01496} = 2.61$$

From the results of the analysis above, these values can be described as the *equivalent energy elastic plastic* curve as depicted in **Figure 13** below:

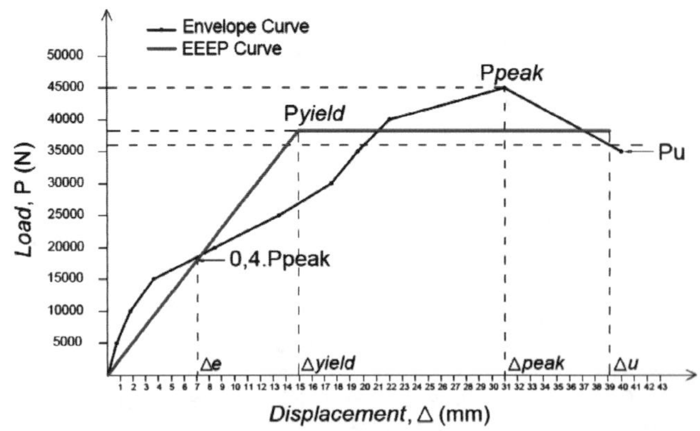

Figure 13. EEEP curve of interlocking lightweight brick masonry wall.

The following is a FE analysis and modeling that show that axial compressive stress Sx of 0.648 MPa occurs in lightweight brick due to the loading tip position of surrounding RC frame, such as depicted in **Figure 14**.

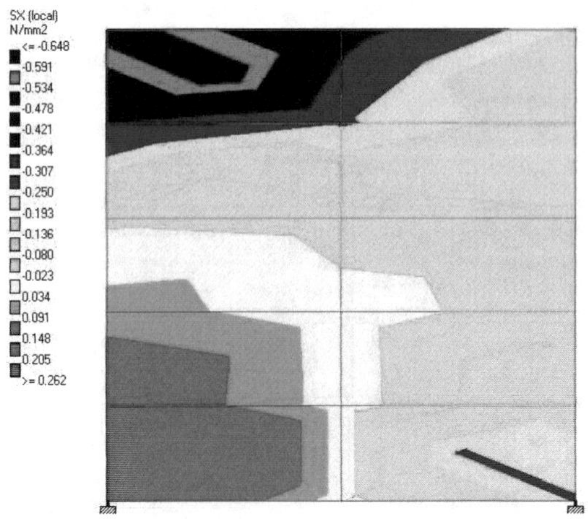

Figure 14. Maximum compressive stress Sx that occurs in interlocking lightweight masonry wall.

The cracked pattern of interlocking lightweight brick masonry wall was indicated after a shear test was carried out at maximum load and with maximum displacement. The crack pattern was observed after loading in each cycle and the crack pattern that occurred then was drawn on the masonry wall using a marker.

The first crack pattern occurs at the load of 10 kN to the right with a displacement of 1.8 mm which results in hair cracks in the lower left corner. At the first crack occurs, the wall stiffness still at 5555.56 N / mm. Then a hair crack in the lower right corner also occur at the increased load of 15 kN and a displacement of 3.62 mm with a wall stiffness of 4143.65 N / mm. Accordingly, at a load of 25 kN to the right, the crack width occur at 4 mm and the interlocking joints between the 3rd and 4th layer bricks is broken and its resulting in 5 mm gap joints with a stiffness of 1860.12 N / mm. At a load of 30kN with a displacement of 17.56 mm the gap width of the *interlocking joints* increases to 8 mm and the wall stiffness can be measured as 1708.43 N / mm.

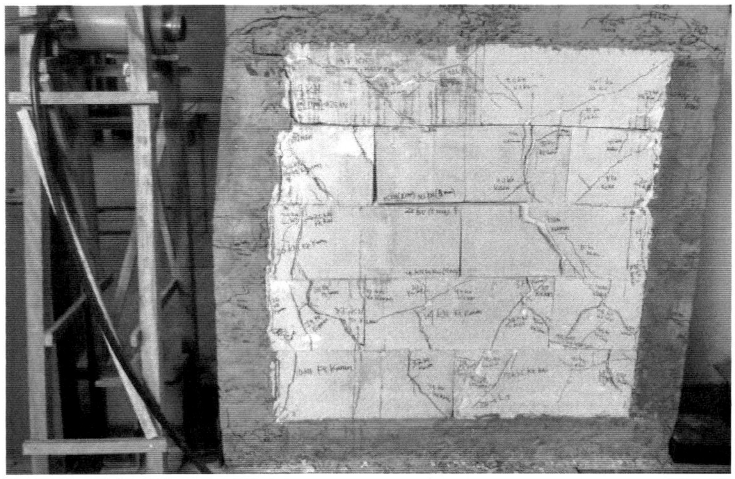

Figure 15. Crack pattern at ultimate load of interlocking lightweight brick masonry wall.

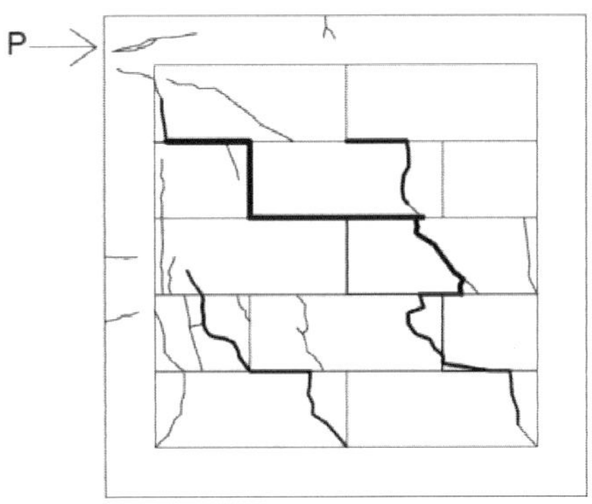

Figure 16. Maximum load crack pattern of interlocking lightweight masonry walls at left to the right direction of cyclic loading.

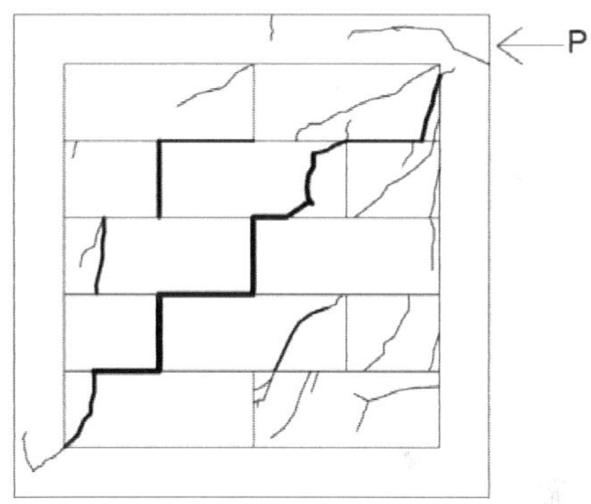

Figure 17. Maximum load crack pattern of interlocking lightweight masonry walls at right to the left direction of cyclic loading.

From shear test results, it can be concluded that the more the wall is damaged or cracked, the stiffness of the structure will obviously decreased, and this was already occurs the first crack appeared, where the wall still had a stiffness of 5555.56 N / mm, but when the crack width reached 4 mm, the wall stiffness decreased by 66.52% to 1860.12 N / mm. Also, as the load get larger of 30 kN, its crack width become 8 mm and consequently, the wall stiffness decreased to 1708.43 N / mm. This was also stated by Li that before cracking, the stiffness would be constant as elastic behavior, such as the reduction of stiffness when loading applied has not occurred, but if the concrete has cracked due to external loads, the stiffness of the structure will also obviously decreased with the addition of crack width [10] .

5. Conclusion

- From the of specific gravity results of the trial mix, it was found that the larger both of the water-cement ratio and the fiberglass content, the smaller the density, but the larger foam agent concentration to the water, the larger the density of the specimen will be found. From the compressive strength results, it was found that the larger both of the water-cement ratio and the fiberglass content, the smaller the compressive strength will be, and contrary with that, the larger the foam agent concentration to the water, the larger the compressive strength value is obtained.

- From the results of the shear strength test of interlocking lightweight brick masonry wall (surrounding by RC frame) with a density of 782 kg / m^3 and a compressive strength of 31.7 kg / cm^2, the shear strength of interlocking masonry wall structures is 35.71 kN / mm, by means that each width of 1 mm wall can resist loads of 35,71 kN. The elastic stiffness is 2.55 kN / mm, the yield displacement is 14.96 mm, its 'yield' load was around 38.25 kN and its ductility was 2.61. From shear test of wall system, it can be concluded that obviously the more the wall is damaged or cracked, the less the structural stiffness will be.
- A new development of an improved interlocking bricks model is proposed, and resulted in low density as a lightweight material, and its wall structures can overcome shear cyclic loading very well and results in a good hysteresis curve which means have a good ductility as well in the seismic resisted structures.

References

[1] Murty C V R 2004 Learning earthquake design and construction *Resonance* **10**(3) 92-95
[2] Karolina R, Handana M A P and Ramadhan P 2020 Shear behavior of interlocking masonry brick wall portal with the substitution of Mount Sinabung volcanic ash against pushover static load *IOP Conference Series: Materials Science and Engineering* **801** 012024
[3] Crisafulli F J 1997 *Seismic behaviour of reinforced concrete structures with masonry infills* (New Zealand: University of Canburry)
[4] Mahdizadeh S F 2008 A Comprehensive Approach Towards The Classification of Cracks in Un-Reinforced Masonry Building *IUST International Journal of Engineering Science, Architect Engineering* **19**(6) 41- 45
[5] The National Standardization Agency for Indonesia 1990 *SNI 03-1974-1990 Testing Methods of Concrete Compressive Strength* Stand Body. Nas. Indonesia
[6] Peirson B 2005 Comparison of specific properties of engineering materials *Laboratory Module* **5**
[7] National B S 1989 *SNI 03-0349-1989* Bata Bet. for wall mates
[8] Prayuda H 2016 Gaya Lateral In-Plane Struktur Portal Dinding Pasangan Bata ½ Batu Melalui Analisis Numerik *Semesta Teknika* **18**(2) 130-139
[9] ASTM E2126 2011 Standard test methods for cyclic (reversed) load test for shear resistance of vertical elements of the lateral force resisting systems for buildings *ASTM E2126 ASTM International West Conshohocken, PA*
[10] Li Y, Walraven J C, Han N, Braam C R, Hoogenboom P C J and Houben I L 2010 *Predicting of the Stiffness of Cracked Reinforced Concrete Structures* (Delf: Delft University of Technology)

IOP Conf. Series: Materials Science and Engineering 1073 (2021) 012022 doi:10.1088/1757-899X/1073/1/012022

Optimization of printed circuit board components placement

R E Purwanto[1,*], Sulistyono[1] and E Mandayatma[2]

[1] Mechanical Engineering Department, State Polytechnic of Malang, Indonesia
[2] Electrical Engineering Department, State Polytechnic of Malang, Indonesia

*r.edypurwanto@polinema.ac.id

Abstract. In order to optimize the throughput rate of these automatic systems, the time taken for the pick and place electronic components for each board has to be minimized. This study focused on optimization problems faced in automated assembly of electronic component in Printed Circuit Board (PCB). Components are placed on PCB where the process of pick-up and placement occurs starting from the machines to pick up the components from the feeder magazine. The number of components to be picked and placed maximum four components, depending on its contribution to minimize tour distance. Increased complexity of PCB assemblies, components and shorter product obsolescence timescales implies machines utilization must be maximized in order to achieve cost efficiencies. The difference in size and shape of components is handled by the head, which brings the optimization problem closer to real machine situation. In hope that the electro-pneumatic arm design can speed up the working time of installing components on the PCB and can add new designs to be registered for patents.

1. Introduction

Mechanization is the process of replacing and using various machines, as well as various technical means that are intended to replace human or animal labor. Pneumatic is a drive system that uses air pressure as its driving force. The advantage of the pneumatic system is that it is environmentally friendly/clean (if there is a leak in the piping system). Electro-pneumatics basically consists of two pneumatic and electronic control systems, control using a power source from the air, and an electric power source with a relatively small voltage and power capacity.

PCB is an object that is shaped like a thin board used to compose a series of electronic components. PCB stands for Printed Circuit Board which in Indonesian is often translated as Printed Circuit Board. PCBs can be found in almost all electronic equipment such as cell phones, televisions, cars, motorbikes, and so on. There are many things a designer must pay attention to in order to design a printed circuit board that can function according to the specified specifications.

2. Previous research

Today's industrial sector requires equipment that can replace human work, which can work automatically to shorten production times, reduce production costs, and eliminate routine work. Automatic transfer of material uses PLC as an electro-pneumatic control actuator and is operated automatically [1].

Electro-pneumatic installation made according to the cascade method. A cascade is a method used to construct pneumatic installations. To take advantage of the cascade method, it is necessary to replace pneumatic components in the input and power signal control systems, with electrical components that have the same function in the installation, thus enabling the use of the Cascade Method in making electro-pneumatic installations [2].

Design of Tofu Printing Machine and Tofu Sticks Using an Electropneumatic System, initially tofu molds in the traditional way, namely by loading with stones which take a long time, require large space, require more employees and have a high risk of employee accidents so that producers know piety and tofu sticks have low effectiveness and productivity. From the above problems, a tofu printing machine and tofu sticks are needed, where to use an electro-pneumatic system [3].

Based on these predecessor researchers, using electro-pneumatic technology can help solve the installation of electronic components quickly and accurately and of good quality.

2.1. PCB (Printed Circuit Board)

PCB is a board that connects electronic components conductively to the track, from copper sheets laminated to a non-conductive substrate. PCB can be in the form of 1 layer, 2 layers, or many layers (multilayer). PCBs can be found in almost all electronic equipment such as cellphones, televisions, cars, motorbikes, and others. There are many things that must be considered by a designer to be able to design a printed circuit board that can function according to the specified specifications.

The following are some of the points and terms used, including the copper thickness, trace width, component footprints, board thickness and layers, path, and component spacing. trace clearance and creepage), via, solder mask and silkscreen.

2.2. Pneumatic

Pneumatics is a theoretical branch of flow or fluid mechanics and includes not only the research of air flows through a duct system, which consists of pipes, hoses, devices, and so on, but also the action and use of compressed air. Compressed air is air taken from environmental air which is then blown forcibly into a relatively small space.

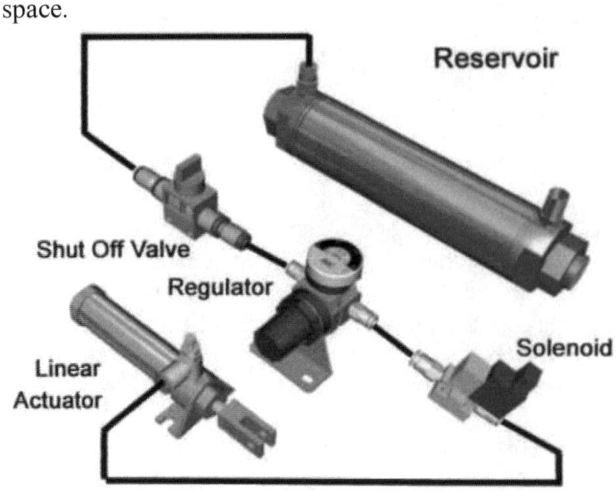

Figure 1. Pneumatic circuit.

2.3. Electro-pneumatic

Electro-pneumatic is the development of pneumatics, where the principle of action is to choose pneumatic energy as the working medium (driving force), while the controlled media uses electric or electronic signals. The electric signal flows to the coil which is attached to the pneumatic valve by activating a switch, sensor or limit switch that functions as a connector or signal breaker. The signal sent to the coil will generate an electromagnetic field and will activate the directional valve as the final element in the pneumatic work circuit. Meanwhile, the pneumatic working media will activate or move the pneumatic working elements such as pneumatic motors or cylinders that will run the system.

IOP Conf. Series: Materials Science and Engineering 1073 (2021) 012022 doi:10.1088/1757-899X/1073/1/012022

2.3.1. Electro-pneumatic system

a. *Electrical signal.* The basic component of an electrical signal is using 24 Volt DC electricity. The simple circuit of the electrical circuit consists of the DC source voltage, the load, and the wiring system.

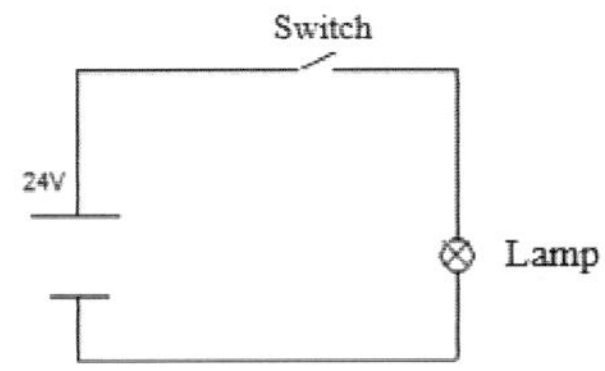

Figure 2. Simple DC electrical circuit.

b. *Switch.* A switch is a component in a circuit that functions to disconnect or connect the current to the load. There are two types of switches, namely, push-button switches and mechanical switches.

c. *Limit switch.* Mechanical limit switches can be set in a certain position or condition. When the workpiece touches the limit switch, it will issue a signal to control a system. This limit switch is usually used to disconnect or connect the current flow.

d. *Relay.* Relay is a component for line connectors and signal controllers, whose energy requirements are relatively small. This relay usually functions with an electromagnet that is generated from the coil. Initially, this relay was used in telecommunication equipment which functioned as a signal amplifier. But now it is common to find in control devices, both on machinery and others.

3. Design the electro-pneumatic arm

The purpose of this research is to optimize the placement time of components on the automatic light fitting PCB board. Automatic light fittings are lamps that are attached to these fittings which can turn on and off independently without an operator according to the user's wishes. Program settings by users using a mobile device. The main parameter that determines the success of the device is the timing accuracy and timing of turning off the lights. The timing of on and off according to the user's wishes is the design of the Automatic Light Fitting system. In general, automatic light fittings consist of 4 (four) main components, namely: (a) an electronic switch; (b) Power supply (power supply); (c) Infrared signal receiver; (d) The main program microcontroller controlling the system.

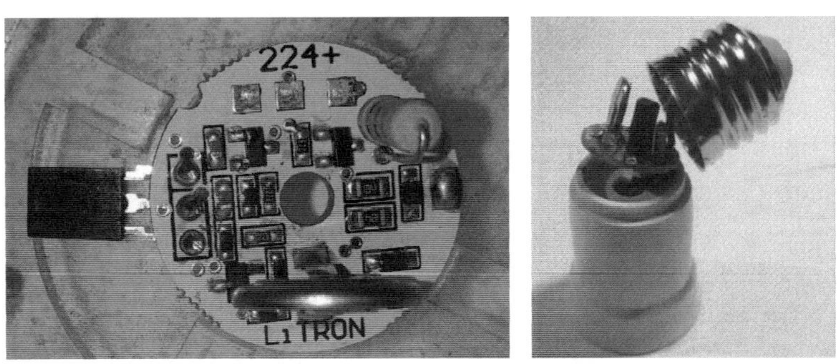

Figure 3. Automatic light fittings.

To optimize the placement of components on the automatic lamp fitting PCB, it is necessary to make an electro-pneumatic sleeve design. The tools and materials used are microcontrollers, aluminum, pneumatic piston, stepper motor, vacuum rubber, and vacuum pump. The working principle of the electro-pneumatic arm is that the proximity sensor is adjusted to locate the component to be taken, after which the arm moves closer to and until it touches the component. Then the vacuum pump is given directions by the microcontroller to suck the components that will be placed on the PCB. After the component is carried away, the robot arm directs and places it according to the coordinates of the component's placement.

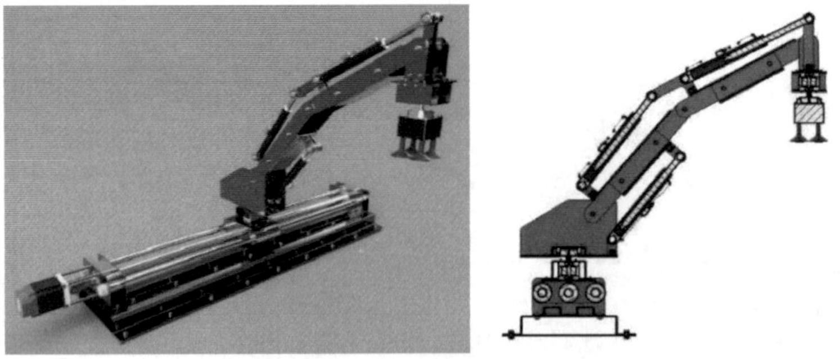

Figure 4. Design electro-pneumatic arm.

4. Conclusion

In hope that the electro-pneumatic arm design can speed up the working time of installing components on the PCB and can add new designs to be registered for patents.

References

[1] Kurniawan R 2008 Rekayasa Rancang Bangun Sistem Perpindahan Material Otomatis dengan Sistem Elektro-Pneumatik *J. Ilm. Tek. mesin CAKRAM* **2** 42–7

[2] Hatuwe A N 2014 Perancangan Instalasi Kontrol Gerak Silinder Elektropneumatik Berdasarkan Prinsip Kerja Metode Cascade *ARIKA* **8** 53–63

[3] Wicaksana D, Poernomo H and Bisono F 2018 Rancang Bangun Mesin Pencetak Tahu Takwa Dan Stik Tahu Menggunakan Sistem Elektropneumatik *Proceedings Conference on Design Manufacture Engineering and its Application* vol 1 pp 212–8

IOP Conf. Series: Materials Science and Engineering 1073 (2021) 012023 doi:10.1088/1757-899X/1073/1/012023

Energy harvesting using microstrip rectenna circular patch on GSM 1800MHz frequency

A Aisah[1,*], A A A Rahman[1], A Hariyadi[1], M Junus[1], S W Yoga[1] and A Rafli[2]

[1] Telecommunication Program Study, Department of Electrical Engineering, State Polytechnic of Malang, Malang Indonesia
[2] Department of Electrical Engineering, Brawijaya University of Malang, Malang Indonesia

*aisahzahra@gmail.com

Abstract. This research discusses the design of rectenna with microstrip antenna with frequency BTS work frequency in Malang city by using a rectifier to convert electromagnetic power into DC output on GSM1800 working frequency and use the energy storage design to be stored. Harvesting this energy is alternative energy conversion. The produced electrical power can be used by low-powered-devices. Since the obtained power is small, additional circuits will be used so that the obtained power can be stored in energy storage. The power obtained from this research is around 1 to 16 mV. It is obtained from BTS around the Malang city.

1. Introduction

The main energy source used by humans as fuel is a non-renewable natural resource, which will gradually decrease in number and will run out. Therefore, there are a lot of researches to overcome this problem such as using alternative energy derived from renewable natural resources [1,2].

Radio frequency is one of the energy sources that has big potential to be used as an alternative energy due to its large number around the community. The usable radio frequency range is from 3 Hz to 300 GHz, which includes TV frequency, GSM, CDMA, 3G, Wi-Fi, and so on [3,4].

In the previous research, Daniel W. Harrist used a quarter-wave antenna designed in such a way as to produce quarter waves that operate at a frequency of 915 MHz. This antenna generates around 5-6 mV DC voltage per second. In another previous research, Rawan Nugraha made rectenna, complete with rectifier, DC amplifier, and antenna. It is capable of working from 900 MHz - 5 GHz and its maximum yield is 790.75 mV. In this research, the obtained power will be increased to 2 V so it can be stored in an assembled battery. As well in another previous research, Aisah designed rectenna of the GSM frequency band 900 MHz for electromagnetic energy harvesting that the resulting output DC voltage is 6,7 mV [5,6].

In this research, a circular microstrip antenna with a rectifier and DC amplifier was made, with an antenna capable of working on GSM frequencies of 1800 MHz. Hopefully, the design of this circular Rectenna microstrip can utilize electromagnetic wave radiation from the Base Transceiver Station (BTS) of the Global System for Mobile communications (GSM) as a new source of voltage [7].

1.1. Microstrip antenna

The microstrip antenna is one of the most popular antennas at the moment. This is due to the fact that microstrip antennas are very suitable for telecommunication devices that currently pay attention to form and size [7,8].

Based on the origin of the word, microstrip consists of two words, namely micro (very small/thin) and strips (blades/pieces). The microstrip antenna is generally divided into three parts, namely patch, substrate dielectric, and ground plane [9].

1.1.1. Patch. In general, patches are made from conductors such as copper or gold which have various forms. The shape of the patch can vary. It can be a circle, square, rectangle, triangle, or annular ring. A patch serves to radiate electro magnetic waves into the air. Patches and feed channels are usually located above the substrate. Patch thickness is made very thin ($t << \lambda 0$; t = patch thickness) [9].

1.1.2. Substrate dielectric. This substrate is made of dielectric materials. Its height (h) is usually between $0.002\lambda 0$ - $0.005\lambda 0$. It serves as a GEM channel media from castings. Its thickness influence the bandwidth on antenna parameters. The thicker it is, the higher the antenna's bandwidth [10-12].

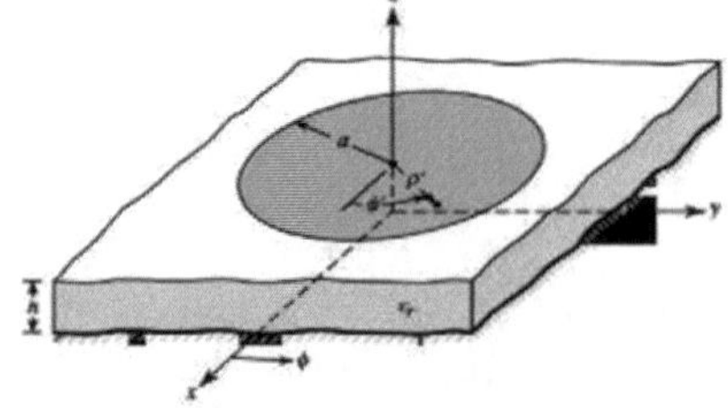

Figure 1. Design of antenna microstrip circular patch.

Modes that can support circular patch antennas can be seen by interpreting the patch, ground plane and material between the two-cavity circles. The value of a small substrate ($h << \lambda$) is TMz where z is taken perpendicular to the patch. Circular ribbons there is only one level of freedom to control (patch radius). Doing this does not change the order of modes but it changes the absolute value of each resonant frequency. Figure 1 shows an image of a microstrip antenna [3,4,9].

1.2. Rectifier

A rectifier is used to change AC (Alternating Current) signal voltage into DC (Direct Current) voltage. The main components in the wave rectifier are voltage drop (transformer), wave rectifier (diode) and filter (capacitor). A full-wave rectifier uses 4 diodes that are installed in bridge shape [2,13].

1.3. Rectenna

A rectenna is a rectifier antenna - a special type of antenna used to convert electromagnetic energy into direct current (DC) electricity. They are used in wireless power transmission systems that transmit power through radio waves. A simple rectenna element consists of a dipole antenna with an RF diode connected across a dipole element [1,13].

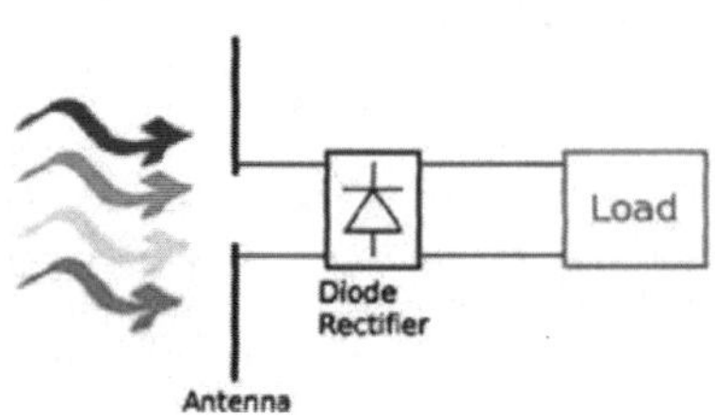

Figure 2. Block diagram of rectenna.

The diode rectifies the AC current induced in the antenna by microwaves to produce DC power, which gives the load strength connected throughout the diode. Schottky diodes are usually used because they have the lowest voltage drop and the highest processing speed, therefore they have the lowest power due to conduction and Rectenna switching designed for energy harvesting applications in GSM bands, UMTS / WCDMA bands, the latest ISM and LTE band [13]. Figure 2 shows the block diagram of how the rectenna work.

1.4. Energy harvesting

Energy Harvesting is a process where energy is captured and converted directly into electricity to be used by small and medium devices such as autonomous wireless sensor nodes, consumer electronics, and vehicles. These energies come from external/unconventional sources such as radio waves (RF energy), heat energy, natural energy, solar energy, and other sources. Other than directly used in small to medium devices, the converted energy might be stored in durable storage cells such as capacitors, super capacitors, or micro-energy cells (MEC), which are in the forms of lithium solid-state batteries. Its system generally includes circuits whose function are to regulate power, protect storage devices and other functions [1-4].

2. Methods

Research methods include the stages and designing of the research starting from the beginning to the implementation of the system. The research phase was prepared with the intention that the research can be carried out in detail. The specific antenna and rectifier configuration is obtained through the process of calculating and designing antenna and rectifier specifications according to the desired specifications. Table 1 shows the desired antenna specifications.

Table 1. Antenna specification.

No	Parameter	Specification
1	Frequency	1800 MHz
2	VSWR	< 2
3	Return Loss	< -10 dB
4	Impedance	50 ohm

After determining the desired specifications, the antenna dimensions are calculated. The calculation results are inputted into CST software to do simulation. The simulation results are shown in Figure 3.

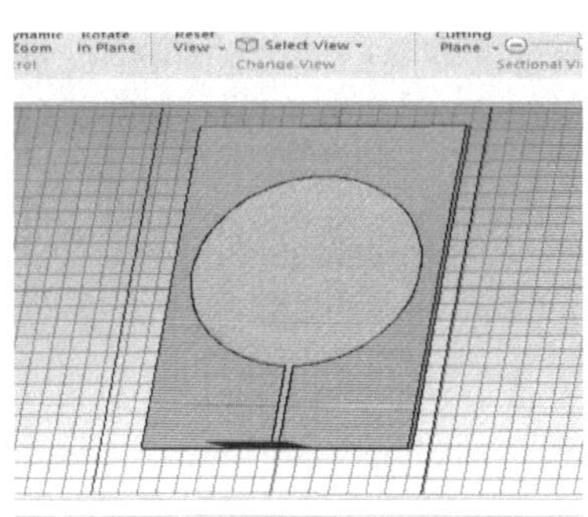

Figure 3. Antenna simulation.

The rectifier is designed using a multimode. It is comprised of 4 Schottky diodes and 100nF capacitors, which are arranged as seen in Figure 4. Microstrip antennas are printed on FR-4 type and patch using copper, whose shape and dimensions can be seen in Figure 5 and Table 2.

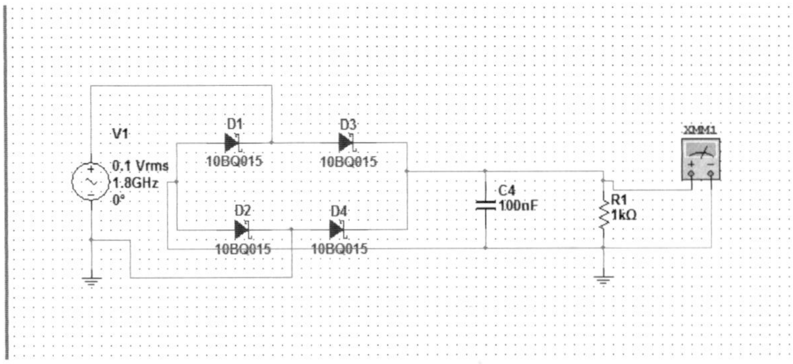

Figure 4. Simulation of rectifier.

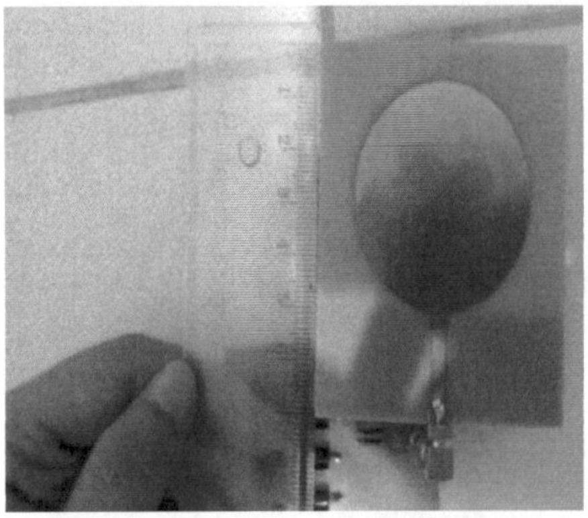

Figure 5. Fabrication of antenna.

Table 2. Fabricated microstrip antenna dimensions.

Variable	Dimension (mm)
Diameter	43
Length patch	19,98
Width patch	2,87
Width substrate	60
Length substrate	71
Width ground	60
Length ground	20

The fabricated antennas and rectifiers were then connected as can be seen in the block diagram system in Figure 6.

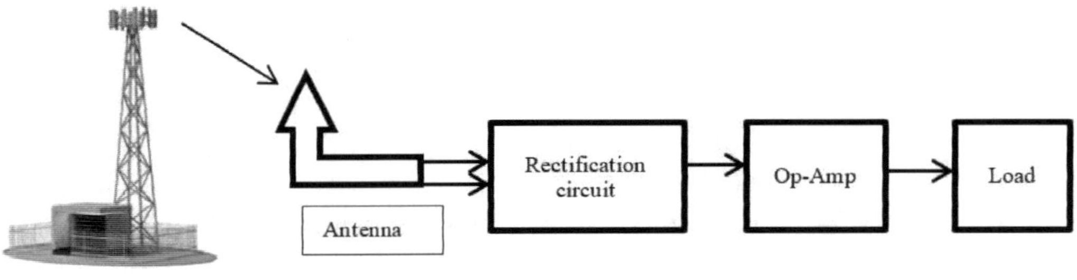

Figure 6. Prototype design block diagram.

3. Results and discussion

Tests on antenna include testing antenna parameters on simulated-antenna and fabricated-antenna. The tested parameters were Return Loss, VSWR, Radiation and Gain Pattern. Return loss is one of the parameters used to find out how much power is lost in the load and return as a reflection. This parameter can also determine to match between the antenna and transmitter.

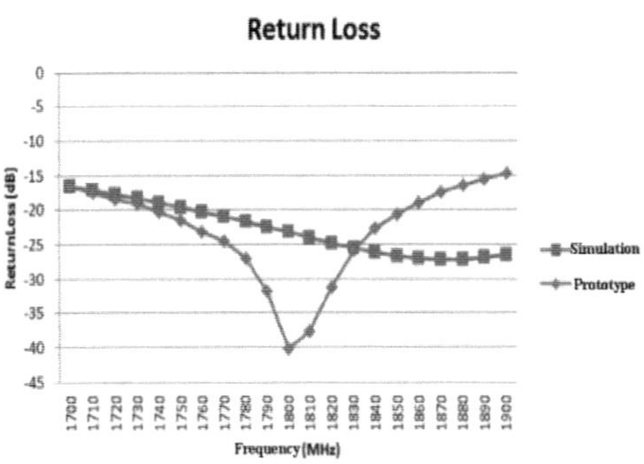

(i) Graph of changes in return loss value

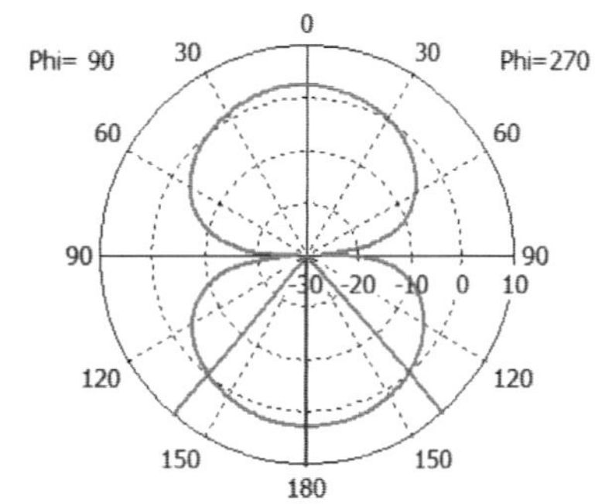

(ii) The direction of the designed antenna beam

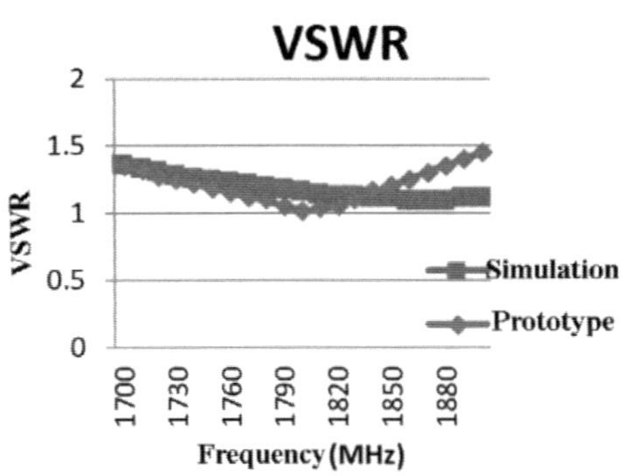

(iii) Graph of changes in VSWR value

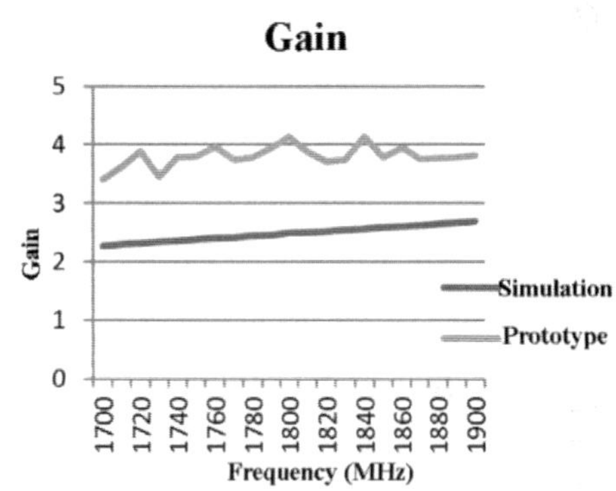

(iv) Graph of gain changes

Figure 7. Graph of parameter value antenna.

The Return Loss value from the simulation results of the microstrip circular patch antenna that works on 1800 MHz frequency can be seen in Figure 7 (i). The Return Loss graph shows the return loss value from the initial frequency of 1700 Hz to 1900 Hz. Since the Return Loss in the frequency range of 1700 MHz up to 1900 MHz is below -10 dB, the antenna can work and the frequency is small during this frequency range. In the GSM frequency Up Link of 1710 MHz-1785 MHz and Down Link frequency of 1805 MHz-1880 MHz, the value of small return loss is around -16 dB to -26.6 dB during simulation while for during test on prototype, it ranges from -14.464 dB to -40.23 dB with it being -23dB at the middle frequency of 1800 MHz.

To ensure that the designed antenna has good capabilities, its radiation pattern and polarization for 1800 MHz frequency, which can be seen in Figure 7 (ii) are checked. Based on Figure 7 (ii), the maximum value of the radiation pattern is 60^0 to -60^0. Based on this data, it can be said that this antenna is bidirectional with its circular polarization following the patch form of this antenna.

The VSWR test results on simulated-antenna and fabricated-antenna are shown in Figure 7 (iii). The simulation uses the initial frequency of 1700 MHz and the final frequency of 1900 MHz. VSWR is obtained in the frequency range of 1700 MHz up to 1900 MHz below 2. This shows that the antenna can still work and the small frequency is given during that frequency range. In GSM frequency Up Link of 1710 MHz-1785 MHz and Down Link frequency of 1805 MHz-1880 MHz, the VSWR value is good, which ranges from 1.3 to 1.09 during the simulation and ranges from 1.45 to 1.05 during the test on the prototype with it being 1.01 at the middle frequency of 1800 MHz VSWR.

Based on Figure 7 (iv), it can be seen that the simulation results of square patch microstrip antennas have increasing gain during the frequency range of 1700-1900 MHz. Meanwhile, the fabricated antenna's gain result is unstable but still has a good average value. This happened because when measuring unstable gain the measuring instrument for the gain itself.

From several rectenna tests without using op-Amp in the morning, afternoon and evening times, the power output samples were obtained as shown in Table 3.

Table 3. Testing results in several areas of Malang.

Areas	Period 06.00-08.00		Period 11.00-13.00		Period 19.00-20.00	
	Minimum voltage (mV)	Maximum voltage (mV)	Minimum voltage (mV)	Maximum voltage (mV)	Minimum voltage (mV)	Maximum voltage (mV)
Soekarno Hatta Rd.	1	6	1	5	1	5
Ikan Trombro Rd.	1	7	1	5	1	8
Sudimoro Rd.	1	15	1	12	1	16
Simpang Candi Panggung Rd.	1	5	1	4	1	4
Mayjen Panjaitan Rd.	1	4	1	2	1	5
Bogor Rd.	1	3	1	3	1	5
Bandung Rd.	1	3	1	2	1	5
Besar Ijen Rd.	1	5	1	3	1	4
Bendungan Sutami Rd.	1	4	1	2	1	3
Veteran Rd.	1	3	1	2	1	3

The analysis of the tests of the output power produced by the rectenna is influenced by the number of BTS and the number of mobile users at the same frequency as the antenna working frequency. From the testing results, the lowest obtained power from all test areas at several different times is 1mV and the highest obtained power is 16mV, which is obtained at 19.00-20.00 at Sudimoro street area.

4. Conclusion

From the test results and discussion, it can be concluded that planning and making renewable energy systems using GSM 1800 MHz frequency-based microstrip antennas can be implemented and realized.

The value of obtained return loss parameters from the fabrication at 1800 MHz frequency is -40.23 dB, which is higher than the simulation results whose value is only -23.2 dB at the same frequency. Antenna's VSWR values in the frequency range of Uplink and Downlink are good, ranging from 1.3 to 1.09 during the simulation and from 1.45 to 1.05 during test on the prototype with it being 1.01 at the middle frequency of 1800 MHz. In the gain value test result, the gain value increases during the simulation and is unstable but still has good average value during the test with the fabricated antenna.

This antenna is bidirectional with circular polarization following the patch form. The test results of the rectenna's produced voltage range from the lowest, 1mV, and the highest, 16 mV. The highest value is obtained during measurements at night and the lowest is obtained during the day. Power acquisition is influenced by the number of users using the same frequency as the rectenna's working frequency, which is GSM 1800MHz frequency. To turn on load or LED, the power produced by the rectenna is added to the power supply of the Operational Amplifier. This research can be developed with other rectifier models to find better energy results. The used antenna material can be replaced with another substrate to get different and better energy absorption. Other tests can be carried out with different frequencies according to available frequencies around the environment. The prospect of the development of research results and application prospects of further studies into the next can be used range frequency from 2G up to 5G, which is frequency used by Base Station in terrestrial spread in Indonesia. That can be used for alternative energy in the next future because of the increasing number of the base station.

Acknowledgments
Authors acknowledged the support of the Malang State Polytechnic for giving the opportunity to attend and present at this conference.

References

[1] Zhang J 2013 Rectennas for RF wireless energy harvesting
[2] Posma S N and Setijadi E 2011 Pemanenan Energi RF 900 MHz menggunakan Rectenna untuk Perangkat Mobile *Semin. Nas. Pascasarj. XI–ITS, Surabaya* **27**
[3] Yuwono R and Wijaya O P Aisah. 2015. Design Microstrip Antenna 900 MHz with Customized Flower Shape Patch *ARPN J. Eng. Appl. Sci.* **10** 9320–4
[4] Harrist D W 2004 Wireless battery charging system using radio frequency energy harvesting
[5] Mattulada A R R, Zulfi Z and Wahyu Y 2018 Perbandingan Daya Keluaran Rectenna Singleband Dan Multiband Pada Rf Energy Harvesting 900–2400 Mhz *eProceedings Eng.* **5**
[6] Aisah R Y and Suryanto F A 2019 Rectenna Design of GSM Frequency Band 900 MHz for Electromagnetic Energy Harvesting *J. Commun.* **14**
[7] Sotyohadi S, Pramono S H and Sarosa M 2016 Design and Fabrication of "Ha ()" Shape-Slot Microstrip Antenna for WLAN 2.4 GHz *Proceedings of Second International Conference on Electrical Systems, Technology and Information 2015 (ICESTI 2015)* (Springer) pp 383–91
[8] Balanis C A 2005 Antenna theory analysis and design, A JOHN WILEY & SONS *Inc., Publ.*
[9] Nakar P S 2004 Design of a compact microstrip patch antenna for use in wireless/cellular devices
[10] Kraus D J 1988 *Antennas*, Tata Mcgraw-Hill, Inc.
[11] Johnson R C and Jasik H 1984 Antenna engineering handbook *mgh*
[12] Stutzman W L and Thiele G A 2012 *Antenna theory and design* (John Wiley & Sons)
[13] Deshmukh U 2017 *Rectenna (Rectifying Antenna) for wireless power transmission*, Linkedin

IOP Conf. Series: Materials Science and Engineering 1073 (2021) 012024 doi:10.1088/1757-899X/1073/1/012024

Design and analysis of a portable web-based monitoring system for electrical parameters measurement

M N Hidayat*, F Ronilaya and I H Eryk

Electrical Engineering Department, State Polytechnic of Malang, Indonesia

*moh.noor@polinema.ac.id

Abstract. Currently, to monitor the parameters of an electrical system, including current, voltage, power, and frequency can be done remotely by using either computers, laptops, or android mobile phones, as long as those gadgets are connected to the internet. Generally, the equipment of this monitoring system is installed in a module at a fixed position. As a measurement tool, it is required that the measurement values are valid. For this reason, to ensure that the measurements shown by the remote monitoring system that has been installed are valid, another monitoring system is needed as a comparison. Based on the problems mentioned above, this research aims to design and analyse a portable web-based monitoring system for electrical parameters measurement. This equipment has a web-based monitoring system, which can remotely monitor the measurement of electrical parameters through the web. In addition, the equipment is light and portable, which can be easily moved from one place to another. Due to its advantages, this equipment can be used as a comparative tool to get valid values for electrical parameters measurement.

1. Introduction

The need for electrical energy is increasing from time to time in line with the increase in population and demand for the availability of electrical energy itself. At the same time, fossil energy sources, as the main energy source in generating electrical energy, are decreasing and becoming increasingly rare in the future. To deal with these problems, the use of renewable energy sources is needed, including wind energy and solar energy.

The obstacles faced in developing renewable energy sources, especially large-scale power plants, are location and cost. So, one way to deal with this problem is to develop small-scale generators, which are relatively easy and cheaper to install. Malang State Polytechnic has a prototype of a Wind Power Plant and a Solar Power Plant, which are located near the Renewable Energy laboratory.

Monitoring of electrical parameters on the prototype generation system can be done remotely because it is equipped with a remote monitoring system using a web-based Arduino. Monitoring can be done using a computer or a mobile phone, as long as it is connected to the internet by accessing the monitoring page.

To ensure that the measurements shown by the remote monitoring system are valid, another monitoring system is needed as a comparison. So, the idea emerged to create a web-based Portable Remote Monitoring System, where monitoring of the Wind Power Plant (WPP) prototype above can be done remotely, either via the web or via an Android phone. With this system, the supervision and system regulation will be more efficient and optimal. In addition, this system is also portable, where monitoring equipment can be brought and moved at any time. So that this tool can also function as a comparison of the measurement results of the monitoring tools installed on the WPP panel.

2. Wind power plant

2.1. Work principle of a wind power plant
The Wind Power Plant converts wind energy into electrical energy by using a wind turbine or windmill. The way it works is quite simple, the wind energy that rotates the wind turbine is continued to rotate the rotor on the generator, so that it will produce electrical energy [1-3]. Electrical energy is usually stored in batteries before it can be utilized. According to the arrangement and function of several important components in the power generator turbine, their respective duties and functions can be described.

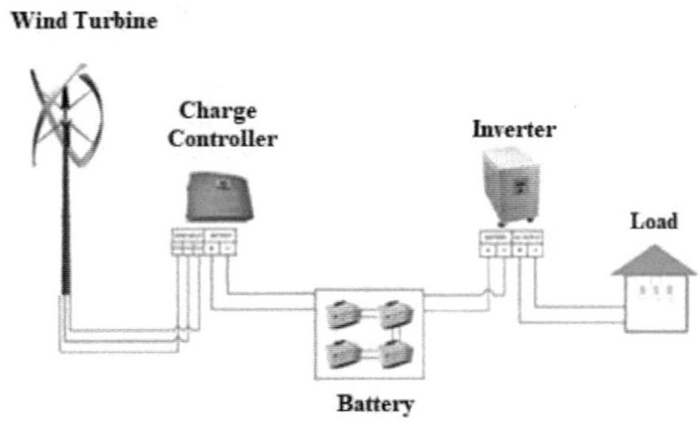

Figure 1. Wind power plant scheme.

Figure 1 shows the wind power generation scheme. The main components in this generation system can be described as follows [3]:

2.1.1. Wind turbine. Wind turbine takes wind energy through two or three propellers which are designed like airplane wings. To get a wind speed that is high enough, constant, and not too much turbulence, a wind turbine is installed on top of a tower. When the wind blows through the blades, there is low pressure air at the bottom of the blades. Low air pressure will pull the propeller into the area. The force that is generated is called the lift. The amount of lift is usually stronger than the pull force. The combination of lift and pull causes the rotor to rotate like a propeller and turn the generator. The components of a wind turbine can consist of:

- Blades, most wind turbines have 2 or 3 fan blades. The wind that blows causes the turbine to spin.
- The rotor, the shaft of the fan blade rotates according to the rotation of the fan blade.
- Pitch, the angle of the fan blade can be adjusted according to the desired rotor speed, depending on wind conditions: low or high.

2.1.2. Controller / charge controller. This section functions to regulate the charging of electric current into the battery so as not to damage the battery due to excessive charging (over charging).

2.1.3. Battery. This section will store the electric current generated by the electric generator so that it can be used at any time. The charge controller controls the storage of electric power in the battery.

2.1.4. Inverters. This section functions to convert the DC voltage from the battery to AC mains voltage (110V or 220V) for equipment with the appropriate voltage rating.

2.1.5. Load. The load is an electric load with a voltage rating that matches the inverter output (110V or 220V)

2.2. Internet of things

In 1999, Kevin Aston has introduced The Internet of Things and it has continued to develop. This term has an aim to expand the benefit of internet which is connected continuously. For example, it can be utilized for data sharing, remote control, etc., including objects in the real world. Almost all human needs, nowadays require the existence of internet connection, such as food, electronic equipment's, collections, including living things that are all connected to local and global networks.

The Internet of Things is also related to the terms monitoring and controlling processes. For example, the application of monitoring and controlling system in the area of electricity utilization. This system can help us to monitor and analyse the waste of electricity. So that we can take some wise actions to rectify the condition.

Figure 2. Internet of things [2].

2.3. Components of a portable remote monitoring system

In this WPP control system, there are several components that are designed to be one so that it forms a control system and can monitor according to its function [4-6]. These components include:

2.3.1. Smart kWh Meter. The use of Smart kWh Meter functions to measure several electrical parameters such as voltage, current, power etc. The measurement result data will be displayed on the website and android application which contains several electrical parameters for a certain period of time. The specification of the smart kWh meter used in this research is as follows:

- Measured parameters: kWh, W, V, A, PF & Hz, Accuracy Class: Class 1.0
- Reference Voltage (Un): 230V AC, operating Voltage: 176 ~ 276VAC
- Rate Current (Ib) / (Imax): 5(45) A, Operating Frequency Range: 50/60 Hz + 10%
- Communication port protocol: Modbus RTU 2 wires, 4800 bps, 9600 bps, 19200 bps, 2400 bps, 1200 bps, Communication port support: Screw terminal block Max485

2.3.2. Arduino. The Arduino is an open-source single-board micro controller, derived from the Wiring platform, designed to facilitate electronic use in a variety of fields. The hardware has an Atmel AVR processor and the software has its own programming language [7].

The use of the Arduino Mega 2560 in the Monitoring and Controlling system is a data processing center for the measurement results of the smart kWh meter which will be displayed on the website and as a process for processing commands for controlling relays from the on / off command on the website.

2.3.3. Relay. The relay in the Monitoring and Controlling system functions as controlling the load by activating the contactor as a breaker or load connector. The specification of the relay used in this research is as follows [8]:

- 2-Channel relay module, Power Supply: 5 VDC
- Equipped with a high current relay, 250 VAC 10A
- Equipped with LED Relay Output Indicator for each Channel

2.3.4. MAX485 Module. The use of the MAX485 Module functions for data communication such as RS485 between the Smart kWh meter and Arduino, which converts data such as readings on the Smart kWh Meter before it is processed by Arduino. This module has a working voltage of 5 VDC.

2.3.5. Modul Wi-Fi ESP8266. The function of the ESP8266 is as an intermediary between Arduino and the website where the ESP8266 will send a signal connected to the internet from data processing or Arduino commands to the website.

2.3.6. CT-Clamp. This component serves to make accurate measurements of alternating current flowing in a conductor without the need for contact with the conductor so it is suitable for the use of portable sensors. The specification of the CT-Clamp used in this research is as follows:

- Input current: 0-20A, turn ratio: 20A:1V
- Output voltage: 0-1V, work temperature: -25°C~+70°C
- Non-linearity: ±3%, dielectric strength (between shell and output): 1000V AC/1min 5mA

3. Workflow and system design

3.1. Problem solving workflow

The work flow of the research stages are described as follows:

- The preparations stage, include literature studies on hybrid power plants, controllers, remote sensing components, and remote monitoring and controlling systems.
- Planning a web-based portable monitoring system and an android application on a wind power generator prototype.
- Selection of components for system development, including the selection of system support components, including the Arduino Mega microcontroller, power meter, smart kWh meter, contactor relay, LVDP panel
- Analysis of System Effectiveness and Accuracy, including effectiveness in monitoring, data collection and operation of safety equipment.
- Conclusion: The conclusion is the result of analysis by comparing with existing standards or installation regulations.

IOP Conf. Series: Materials Science and Engineering 1073 (2021) 012024 doi:10.1088/1757-899X/1073/1/012024

3.2. System design

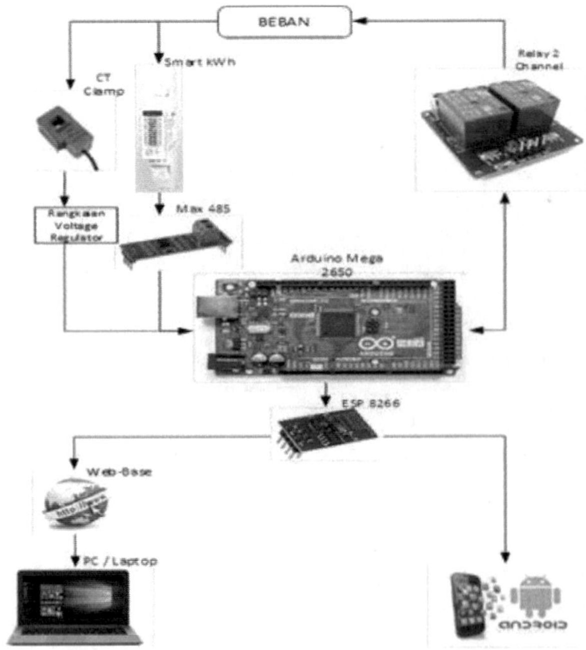

Figure 3. Block diagram of a portable remote monitoring system.

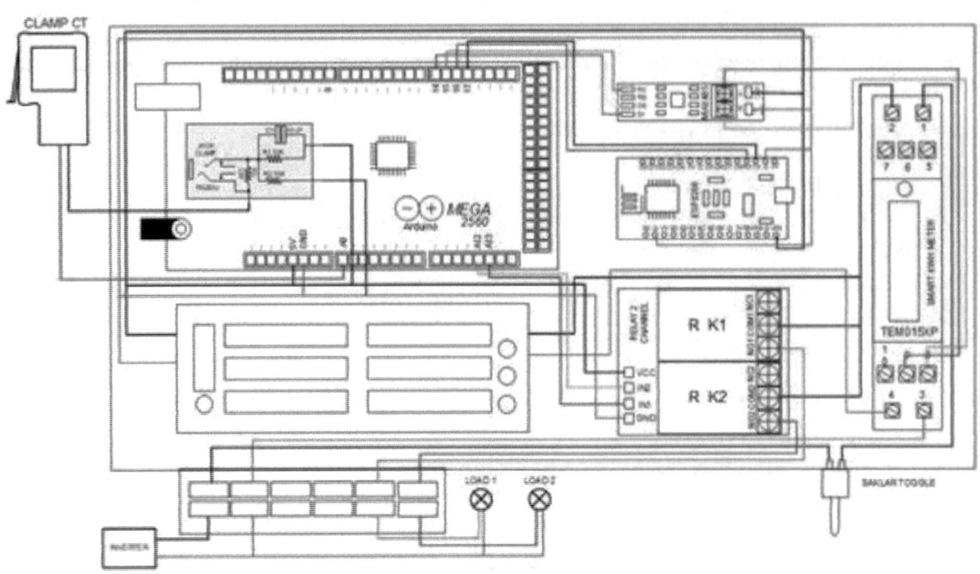

Figure 4. Schematic diagram of a portable remote monitoring system.

Figure 3 and 4 show a block diagram and a schematic diagram of a portable remote monitoring and controlling system. The measurement data is taken from the kWh meter and CT clamp, in the form of data on voltage, current, power, power factor, and frequency. Max485 functions as a data communication tool between the smart kwh meter or CT clamp and Arduino. This data will be forwarded by ESP 8266 to the internet network so that the data can be accessed via the web and the application

IOP Conf. Series: Materials Science and Engineering 1073 (2021) 012024 doi:10.1088/1757-899X/1073/1/012024

will display the measurement data from the smart kwh meter and CT-Clamp. This system can operate the system remotely, including load on / off via the web and applications with the help of relays.

4. Results and analysis

4.1. Voltage measurement

Figure 5 shows the circuit of measuring voltage on an electrical panel using a portable remote monitoring system. The results of voltage measurements and the comparison of measurement results with fixed remote monitoring equipment can be seen in the table 1.

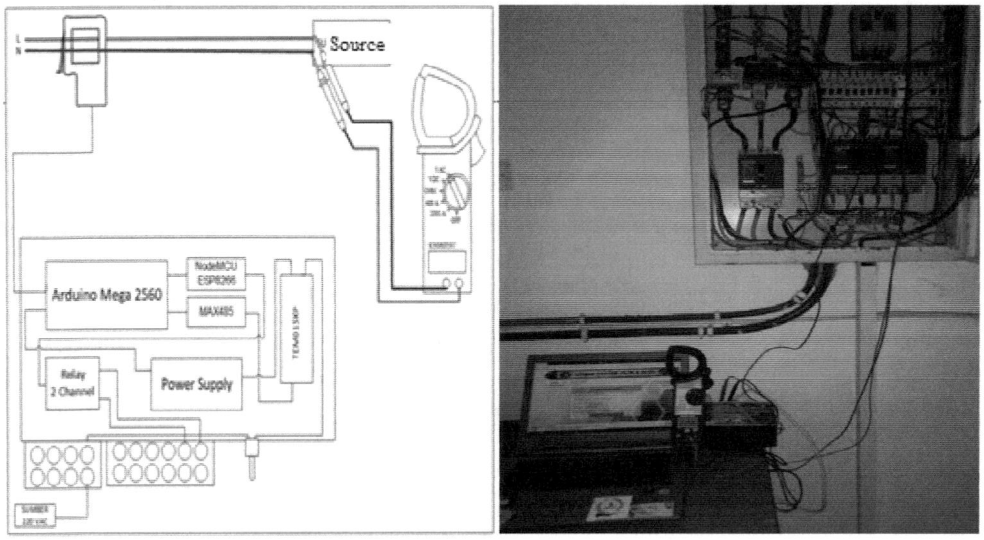

Figure 5. Circuit for voltage measurement.

Table 1. Results of voltage measurement.

No	Date	Measurement Value		Error %
		Portable Module	Fixed Module	
1	24/06/2019 09:10	224.2	226.3	0.94%
2	24/06/2019 09:15	224.1	225.3	0.54%
3	24/06/2019 09:20	224.8	226.0	0.53%
4	24/06/2019 09:25	224.5	226.1	0.71%
5	24/06/2019 09:30	224.1	226.0	0.85%
6	24/06/2019 09:35	224.3	225.4	0.49%
7	24/06/2019 09:40	224.5	225.6	0.49%
8	24/06/2019 09:45	224.3	225.4	0.49%
9	24/06/2019 09:50	224.1	225.0	0.40%
10	24/06/2019 09:55	223.4	224.6	0.54%
11	24/06/2019 10:00	223.3	224.9	0.72%
12	24/06/2019 10:05	224.0	224.7	0.54%
Average of Error (%)				0.60%

From the table 1, it can be seen that the error percentage of voltage measurements using the portable monitoring system when compared to a fixed monitoring system has a small value, around 0.6%.

IOP Conf. Series: Materials Science and Engineering 1073 (2021) 012024 doi:10.1088/1757-899X/1073/1/012024

4.2. Electrical current measurement

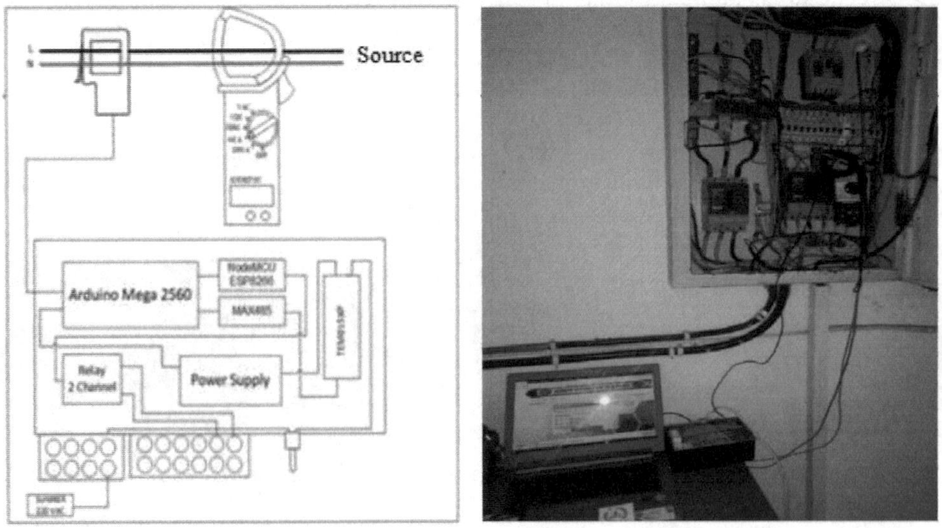

Figure 6. Circuit for current measurement.

Figure 6 shows the circuit of measuring the electrical current on an electric panel using a portable remote monitoring system. The results of electric current measurements and the comparison of measurement results with fixed remote monitoring equipment can be seen in the table 2.

Table 2. Results of electrical current measurement.

No	Date	Measurement Value		Error %
		Portable Module	Fixed Module	
1	24/06/2019 09:10	6.8	6.82	0.29%
2	24/06/2019 09:15	5.9	5.9	0.00%
3	24/06/2019 09:20	5.8	5.88	1.38%
4	24/06/2019 09:25	5.9	5.95	0.85%
5	24/06/2019 09:30	5.8	5.81	0.17%
6	24/06/2019 09:35	5.8	5.89	1.55%
7	24/06/2019 09:40	5.8	5.88	1.38%
8	24/06/2019 09:45	5.8	5.87	1.21%
9	24/06/2019 09:50	5.8	5.86	1.03%
10	24/06/2019 09:55	5.8	5.81	0.17%
11	24/06/2019 10:00	5.7	5.76	1.05%
12	24/06/2019 10:05	5.8	5.83	0.52%
Average of Error (%)				0.80%

From the table 2, it can be seen that the error percentage of electrical current measurement using a portable monitoring system when compared to the fixed monitoring system has a small percentage of errors, around 0.8%

5. Conclusion

Based on the testing and analysis that has been done, it can be concluded that:

- The main components used in the Portable Remote Monitoring System for Electrical Parameters Measurement are: A Smart kWh meter module THERA TEM015XP-D, a Max485 Module, an Arduino Mega 2560, a 2-Channel Relay Module, a Node MCU, and a CT-Clamp YHDC SCT-013-000. This equipment can facilitate monitoring of electrical parameters on the electrical panel, in addition to the relay control feature, the load can be ON / OFF when not needed. This Monitoring and Controlling System Module can be accessed via the web or android application.

- The THERA TEM015XP-D Smart kWh meter module is able to read parameters of voltage, current, active power, power factor, frequency.
- This module makes it easy to monitor electrical parameters on the electrical panel; it can be accessed via the web or android applications, and it is portable (easy to move and carry) with the error percentage less than 1%

6. Future work

After implementing a remote monitoring system for electrical parameter measurement, which is based on websites and android applications, there are some points that can be developed in the future in order to improve the quality of it. Here are some suggestions:

- Selection of the CT Clamp circuit using the OP-M amplifier so that the monitoring readings are more accurate.
- To increase the reliability of the equipment, it is better to add a safety device such as Fuse.
- Adding relays in order to control even more loads.

Acknowledgements
This work was supported by the State Polytechnic of Malang through Innovative Research Funding (2020-2021).

References

[1] Pop A A, Jurca F, Oprea C, Chirca M, Breban S and Radulescu M M 2013 Axial-flux vs. radial-flux permanent-magnet synchronous generators for micro-wind turbine application *2013 15th European Conference on Power Electronics and Applications (EPE)* (IEEE) pp 1–10

[2] Zhuga A T, Munyaradzi B and Shonhiwa C 2006 Design of alternative energy systems: A self-starting Vertical Axis Wind Turbine for stand-alone applications (charging batteries) *SPONSORS* 72

[3] Schaffarczyk A P 2020 *Introduction to wind turbine aerodynamics* (Springer Nature)

[4] Gemeasurement User Manual 16 Channel Relay

[5] Abdillah M H, Erfianto B and Wijiutomo C W Sistem Monitoring Secara Real-Time Penyimpanan Energi Listrik dari Wind Turbine Lentera Angin Nusantara (Lan) *e-Proceeding Eng.* **2** 6387–94

[6] Wihadi R B D and Respatia F F I W S 2016 Sistem Akuisisi Data Dan Monitoring Pembangkit Listrik Tenaga Bayu *Pros. SNAST* 291–300

[7] Tjahjana A *Analysis and Design of Smart Class Room Applications for Classroom Resource Management with Smart Card Technology and QT Framework*

[8] Arduino Schematic Diagram Arduino Mega

IOP Conf. Series: Materials Science and Engineering 1073 (2021) 012025 doi:10.1088/1757-899X/1073/1/012025

Application of the cascade system to increase the capacity of a pico-hydro power generation

M N Hidayat*, F Ronilaya, I H Eryk, A Setiawan and E Rohadi

Electrical Engineering Department, State Polytechnic of Malang, Indonesia

*moh.noor@polinema.ac.id

Abstract. The Pico-hydro power generation is a power plant with a scale below 5 kW. This type of power generation is suitable for areas with streams or rivers with low water levels. One area that has natural conditions like this is the Javan Langur Centre (JLC), a rehabilitation facility for one of the endangered primate species, the langur. To support the operation of JLC, a pico-hydro power plant has been built, which has a capacity of 300 watts. However, due to the water discharge is quite low, the efficiency of the generator is less than 10%, of which, the total output power from the pico-hydro generator is about 27,3 watts. While the electrical energy requirements in JLC is around 50 watts for lighting. To meet the shortage of electricity supply, it is necessary to increase the power generated by the existing pico-hydro power plant, i.e. by applying a cascade system. This system utilizes water from the existing pico-hydro power plant drainage channel as the intake flow for the new pico-hydro power plant. This system also utilizes differences in elevation or water level to make the terraces. With the application of the cascade pico-hydro system, the electrical power requirements at JLC can be met.

1. Intoduction

This research was conducted in the central Javan langur, which is approximately 10 km from the centre of Batu City, East Jaw, Indonesia. Javan Langur Centre is a rehabilitation centre for Javan Lutung, one of the primate species that is protected because it is almost extinct.

Until now, there is no electricity network included in this area. Apart from the fact that the area around this area is uninhabited, the management also keeps the environment natural so that the langur rehabilitation process is not disturbed. Because, the Javan Langur Centre management chose Genset as a source of meeting electricity needs.

IOP Conf. Series: Materials Science and Engineering 1073 (2021) 012025 doi:10.1088/1757-899X/1073/1/012025

Figure 1. Javan Langur centre area.

Around the Javan Langur Centre there are tributaries that have the potential to be a source of environmentally friendly electricity generation. The existence of this river flow can be used as a source for the Pico-hydro power plant, which is a type of power plant with a scale of under 5 kW [1][2].

In 2018, a Pico-hydro power plant was built with an electric power of 300 watts. However, this river flow rate is too low and results in a generator efficiency of less than 10%, where the output power that can be utilized is 27.3 watts.

Due to the need for electric power in the Javan Langur Centre area of around 50 watts for lighting, the fulfilment of electric power cannot be fully supplied from the Pico-Hydro Power Plant (PHPP) that has been built. To solve this problem, a pico-hydro power plant design with a staircase system was made. This system utilizes water in the existing PHPP outlet channel, which is at the top position, as an intake flow for the PHPP system which is placed in a lower position. This system makes use of differences in elevation or ground level to create terraced systems. Because the PHPP technology used in the two generation positions is identical, the output power of the PHPP terraced system is double the output power of the previous PHPP system, which is around 55 watts. With 55 watts of electric power generated by the PHPP Cascade system, the electrical energy needs at the JLC, for lighting, can be met.

2. Pico-hydro power plant

In principle, the Pico-hydro Hydroelectric Power Plant utilizes different heights and the amount of water discharge in irrigation canals, rivers or waterfalls [1]. This water flow is used to produce electrical energy using water turbines and generators.

Based on the output produced, Hydroelectric Power (PLTA) can be divided into [3]:

- Large-hydro : more than100 MW
- Medium-hydro : between 15 - 100 MW
- Small-hydro : between 1 - 15 MW
- Mini-hydro : more than 100 kW, but less than 1 MW
- Micro-hydro : between 5kW - 100 kW
- Pico-hydro : less than 5kW

This Pico-Hydro Power Plant has several advantages, such as [4]:

- The manufacturing cost is relatively cheap
- The materials are relatively cheap because they are easy to find on the market.
- Environmentally friendly because it does not use fossil fuels so that the maintenance is not too complicated.
- Its construction can be integrated with the construction of an irrigation network.
- Small in size, suitable for rural areas that have not been reached by the PLN electricity network.

2.1. *Pico-hydro power plant working principle*

The Pico-Hydro Hydro Power Plant is a hydroelectric power plant that uses an intake upstream of the river and flows water downstream through a water channel with a rather small gradient. Electric power is generated by using the height of the waterfall and the slope of the river. This power plant also uses the river flow directly by using a reservoir pool where river water is flowed into the pond through an open or closed channel by being filtered first and stored in a pond that functions to settle sand and mud as well as a reservoir. The water from the pond is flowed through a rapid pipe driving a turbine to generate electricity. The reservoir pool is equipped with several floodgates for filling or emptying when maintenance is being carried out on the reservoir pool.

The electric power generated can be calculated based on the following formula [5]:

$$P = \rho.Q.h.g$$

P is theoretical output power (watts)
ρ = density of fluid (kg / m^3)
Q = discharge of water (m^3 / s)
h = effective height (m)
g = gravity (m/s^2)

The electric power generated is the product of the height of the fall and the discharge of water, therefore the success of hydropower generation depends on the effort to get the water drop height and the large discharge effectively and economically.

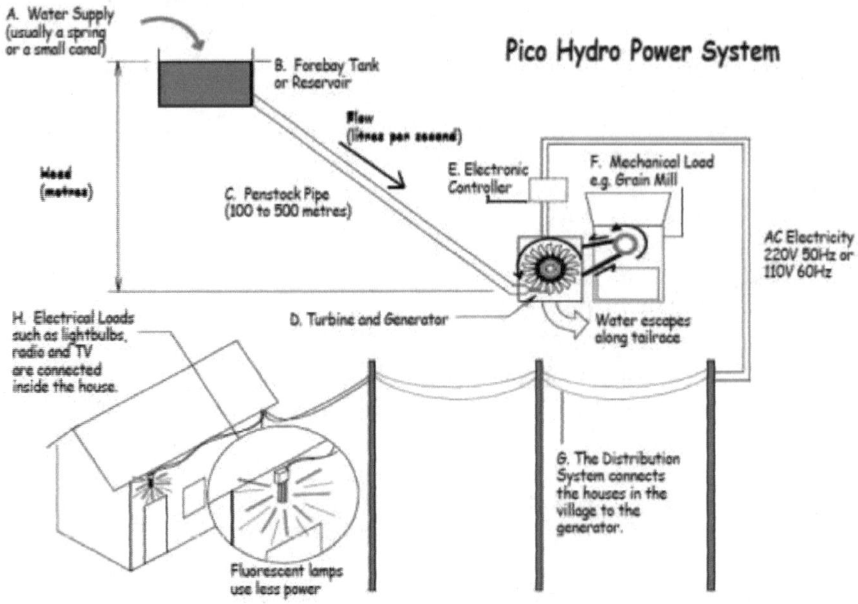

Figure 2. Pyco-hydro power plant [6].

2.2. *Components of a Pico-hydro power plant*

Some parts of a Pico-hydro Power Plant can be seen in the following explanation [6]:

2.2.1. Dam. A dam or dam is a construction built to withstand the rate of water so that water will be accommodated in a certain area.

IOP Conf. Series: Materials Science and Engineering 1073 (2021) 012025 doi:10.1088/1757-899X/1073/1/012025

2.2.2. Headrace. Headraces are water-carrying channels that follow the contour of the land to maintain the elevation of the channeled water so that water does not spread everywhere.

2.2.3. Garbage filter. Garbage filters are very useful for keeping fast pipes from becoming clogged with garbage and sand material that is carried away by the water flow

2.2.4. Head tank or forebay. The function of the soaking tub is to regulate the difference in water output between the headrace and penstock, as well as to separate impurities in the water such as sand, wood.

2.2.5. Penstock. The penstock is connected at a lower elevation to the soaking tub to accommodate the water used to turn the water turbine.

2.2.6. Generator house. In the generator house there is a sink, turbine, generator and other equipment. To maintain the safety of the equipment, the generator house is equipped with a fence and roof.

2.2.7. Hydro turbine. The choice of turbine type can be determined based on the advantages and disadvantages of the types of turbines, especially for a very specific design. At the initial stage, the selection of the turbine type can be taken into account by considering the special parameters that affect the turbine operating system, namely:

- The factor of the effective water drops height and the discharge that will be utilized for turbine operation is the main factor affecting the choice of turbine type, for example: Pelton turbines are effective for operating at high heads, while propeller turbines are very effective at operating at low heads.
- The power factor (power) desired is related to the available head and discharge.
- The speed (rotation) of the turbine will be transmitted to the generator. For example, for a direct couple transmission system between the generator and the turbine at low head, a reaction turbine (propeller) can achieve the desired rotation, while the Pelton and crossflow turbines rotate very slowly (low speed) which will cause the system to not operate.

2.2.8. Pulley system. The torque and power from the main shaft are being transmitted to the shaft of the alternator through belt pulley system. Figure 4 shows the pulley system [6].

Figure 3. Belt-pulley system.

2.2.9. Generator. A generator is an electric machine that converts mechanical power into electrical power. Electric machines can be in the form of generators and motors and based on the direction of the current, electric machines are divided into direct current electric machines and alternating current electric machines.

2.2.10. Charge controller. Charger Controller is an electronic device used to regulate direct current which is charged to the battery and taken from the battery to the load. Or it can also be said that an electronic circuit regulates the process of charging the battery (Battery Bank) and the current loading from the battery to the load.

IOP Conf. Series: Materials Science and Engineering 1073 (2021) 012025 doi:10.1088/1757-899X/1073/1/012025

2.2.11. Battery. The battery is an electric cell in which a reversible electrochemical process takes place, where the discharging process can take place in the battery, namely converting chemical energy into electric power, and the recharging process by regenerating the electrodes used, namely by passing an electric current in the opposite direction / polarity inside the cell.

2.2.12. Inverter. An inverter is a device used to convert the DC (Direct Current) input voltage into an AC (Alternating Current) output voltage. The inverter input voltage source can use batteries, fuel cells, solar power, or another DC voltage source.

2.2.13. Cable. Cable is a medium for conducting electric current. The material of this cable is various, especially as a conductor of electric current, generally made of copper and generally coated with a protector. the maximum ability of the cable to pass current continuously without causing damage to the cable is known as KHA (Current Conductivity).

2.2.14. MCB (Miniature Circuit Breaker). MCB is a component in an electrical installation that functions as a switch or a device that functions as a protector of electrical installation circuits from overcurrent. The MCB has the same function as a fuse, which will cut off the current flow of the circuit when there is an overcurrent disturbance.

3. Workflow and system design

3.1. Problem solving workflow

The method used in this study follows the order of work as mentioned below:

- Analysis of the electricity requirements required for the operational activities of the Javan Langur Centre.
- Estimation of the potential power that can be generated from the water flow around the Javan Langur Centre location to calculate the estimated capacity of turbines and generators.
- Construct civilian constructions required for turbine construction.
- If you get a generator power potential that is smaller than the load requirement, a priority scale of the load that must be supplied is arranged.
- In addition, if the potential power output of the generator is less than the load requirement, the battery requirements needed to support the energy supply during peak load are analyzed.
- The next step is to design a turbine that is suitable for the existing conditions in the Javan Langur Centre river basin. Turbine materials must be made of materials that are corrosion resistant.
- Determine the generator that matches the power and speed of the turbine designed. The generator that is planned must have high flexibility, efficiency and can rotate at low speed according to the turbine.
- Testing of generators in the lab. Electric machines Malang State Polytechnic. The test includes testing the characteristics of open circuit, short circuit generator and load characteristics.
- Turbine testing was carried out at the Javan Langur Centre.
- The design of electrical installations for the energy supply needs of the Javan Langur Centre including the protection system including the needs of batteries, charger controllers and inverters.
- Combined turbine and generator testing including zero load and full load loading. In this step, testing of the charger controller and inverter is also carried out.
- Implementation of comprehensive works including: turbine-generator installation in the constructed construction, installation of LDP panels in suitable locations, carrying out electrical installation work.

IOP Conf. Series: Materials Science and Engineering 1073 (2021) 012025 doi:10.1088/1757-899X/1073/1/012025

3.2. Block diagram of the pico-hydro power plant cascade system

The block diagram of the cascade Pico-hydro power plant system can be seen in Figure 4. From the block diagram, it can be seen that the cascade system applied to the Pico-hydro Power Plant is a combination of two Pico-hydro power generation systems in a larger system.

In the cascade system that is applied, 2 water turbines, 2 generators, and 2 charge controllers are needed, where all similar components are selected which are identical, the same in specifications and dimensions. The output of both charge controllers will be connected to a battery or accumulator. After that, it is connected to an inverter, to convert the DC voltage from the battery into AC voltage, which is then supplied to the AC load.

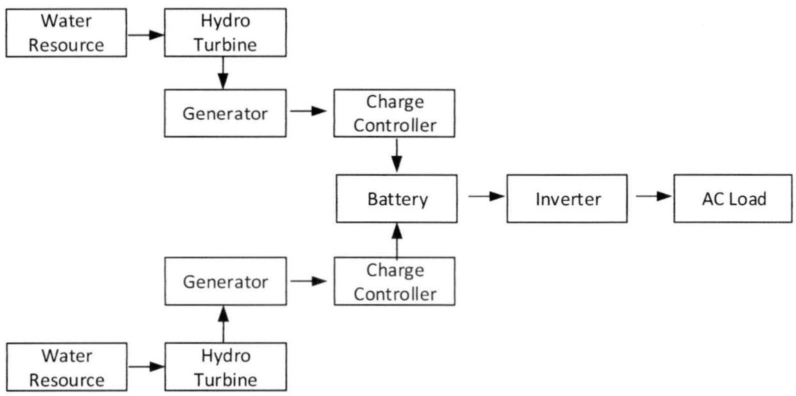

Figure 4. Block diagram of the cascade system pico-hydro power plant.

3.3. Reservoir design

The planning of the reservoir for the PHPP terraced system can be seen in the following figure 5.

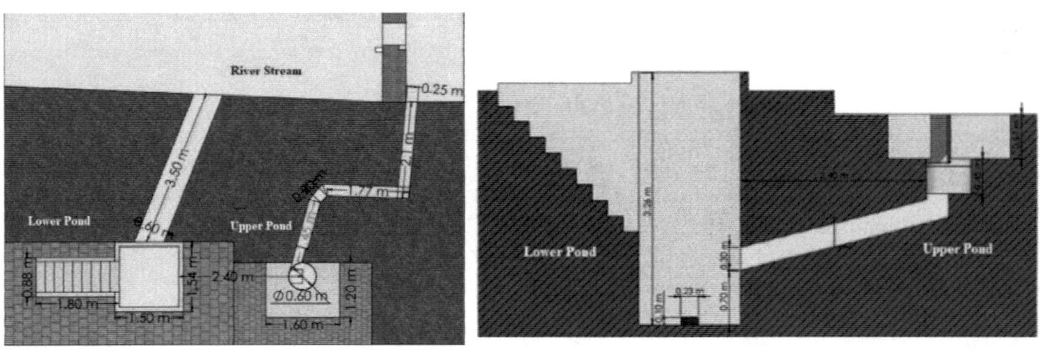

Figure 5. Pool Design (a) top view (b) side view.

From the figure 5 above, it can be seen that there are 2 PHPP systems, where 1 PHPP system is placed in the upper reservoir and 1 PHPP system is placed in the lower reservoir. Although the sizes of the two reservoirs are not the same, (1.6m x 1.2m and 1.5m x 1.5m), the types and specifications of the PHPP system used are identical, including the type and size of the turbine, pulley system, and generator specifications.

3.4. Hydro turbine design

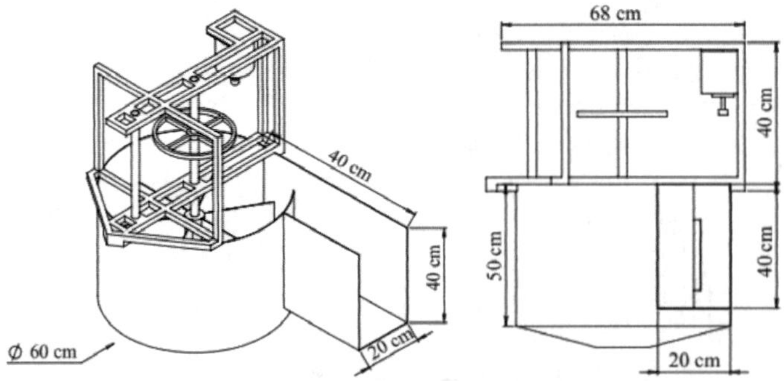

Figure 6. Hydro turbine design [8].

The water turbine design used is a portable spiral vortex hydro turbine [8] as shown above. The turbine and turbine housing have a height of 100 cm and a diameter of approximately 60 cm. The turbine intake is equipped with a water steering plate so that water entering the turbine housing will directly hit the circular wall of the turbine house. This process will produce a spiral-shaped whirlpool which has stronger power to rotate the turbine blades.

3.5. Pulley system design

The circumference of the large pulley used in this research is five times that of the small pulley; the small pulley will rotate five times for every 1-time rotation of the large pulley.

3.6. Generator specification

The generator used in the PHPP system is a 3-phase AC generator which has a nominal output power of 300 watts, a nominal voltage of 48 Vac, and a nominal rotation of 750 rpm. The generator weight is about 3.4 kg. This generator is rust resistant because the shaft is made of stainless steel and the cover is made of aluminium alloy. The dimensions of the generator can be seen in the following figure.

3.7. Charge controller and battery specification

The charge controller specifications used in this research are MPPT type charger controller with a maximum current of 20A, maximum input voltage of 48VDC, float charge 13.8V, nominal voltage 12V, charge disconnect 10.6V, charge reconnect 12.6V. While the battery used in this study has a capacity of 40 Ah and a nominal voltage of 12 V.

3.8. Inverter specification

The inverter used in this research is pure sine wave type 500 W, with a power inverter of 500 W, input: 12V (10 - 15V) DC, output: 220 - 240V AC, output frequency: 50 Hz +/- 2 Hz, output wave: pure sine wave, peak efficiency: 60 - 95%, low battery shutdown: 10.0V +/- 0.5V DC, low battery alarm: 10.5V +/- 0.5V DC, and high voltage shutdown: 15.0V + / 0.5V DC

4. Testing and analysis data

4.1. Water discharge measurement

The measurement of water discharge is carried out using a simple method, namely placing a 1.5 liter bottle in the water channel. Then note the time it took to fill the bottle to the brim. Measurements were made in 2 places, namely the river flow and the channel leading to the water turbine intake. The results of water discharge measurements can be seen in the following table, namely for river water discharge around 0.00035 m3 / s. Meanwhile, the turbine intake water discharge is around 0.00023 m3 / s.

Table 1. Water discharge measurement results.

Water Discharge of the River Stream			Water Discharge of the Turbine Intake		
Volume (liter)	Time (second)	Water Discharge (liter/second)	Volume (liter)	Time (second)	Water Discharge (liter/second)
1.5	3.67		1.5	6.39	
1.5	4.03		1.5	7.19	
1.5	5.09		1.5	6.27	
1.5	4.10		1.5	5.84	
1.5	4.27		1.5	6.29	
Average	4.23	0.00035	Average	6.40	0.00023

4.2. Generator output voltage measurement

From the measurement results, it can be seen that the turbine rotation and the generator output voltage for the two PHPP systems installed have values that are not much different. When the turbine rotation is in the range of 78 rpm, the generator rotation is around 390 rpm with a generator output voltage of about 21 volts.

Table 2. Measurement results of generator output voltage.

No	Turbine 1 Rotation (rpm)	Generator Rotation (rpm)	Generator Output (Vac)	No	Turbine 2 Rotation (rpm)	Generator Rotation (rpm)	Generator Output (Vac)
1	78.3	391.5	21.6	1	77.9	389.5	22.8
2	76.1	380.5	20.4	2	76.1	380.5	20.4
3	76.7	383.5	20.4	3	77.6	388.0	21.6
4	78.3	391.5	21.6	4	75.3	376.5	21.6
5	78.2	391.0	21.6	5	79	395.0	20.4
Average	77.52	187.6	21.12	Average		385.8	21.36

4.3. Battery charging current measurement

Table 3. Battery charging current measurement results.

Charging Current (A)		Charging Time for a 40 Ah Battery
Generator 1	0.75	53.3333
Generator 2	0.75	53.3333
Total	1.50	26.6667

From the table 3 above it can be seen that the charging currents for generator 1 and generator 2 are relatively the same, which is around 0.75 A. If you only use 1 generator (1 PHPP system), then charging the 40 Ah battery to full will take approximately 53 hours. Meanwhile, if using 2 generators (2 PHPP systems) with the cascade method, with a total charging current of 1.5 A, charging a 40 Ah battery to full will take approximately 26 hours.

4.4. Generator efficiency calculations

The calculation of generator efficiency can be seen in the following description:

Generator specifications = 3 phase, 48 volts, 300 watts. Then the nominal current value can be calculated as below:

$$I\ \text{nominal} = 300 : (1.732 \times 48) = 3.6\ \text{A}$$

To find out the value of the generator output current, it can be seen in the charging current value, which is 0.75 A. While the generator output voltage is 21 volts. Then the output power value can be calculated as below:

$$P\ \text{output} = 1.732 \times 21 \times 0{,}75 = 27{,}3\ \text{watts}$$

Then the generator efficiency value can be calculated as below:

Generator efficiency = (Actual Power: Rated Power) x 100%
= (27,3: 300) x 100% = 9,1%

4.5. Battery loading capacity calculation

Table 4. Load variations.

Load Variation			
Number of Lamps	Power (W)	Voltage (V)	Current (A)
1	10	214	0.05
2	20	214	0.09
3	30	214	0.14
4	40	214	0.19
5	50	214	0.23
1 incubator	60	214	0.28
6 lamps	110	214	0.51

Table 5. Battery loading capacity.

Battery Capacity in 1 day (Ah)	Full Load in 1 day (A)	Loading Time in 1 day (hours)	Loading Capacity in 1 day (Ah)	Battery/Loading Ratio
40	0.51	12	6.17	6.48

Based on the load variation table and the calculation of loading capacity in 1 day, it can be seen that the battery is fully charged after 18 hours and can meet the needs of electrical energy (lighting & incubator) in one day, with a ratio of battery capacity: loading capacity = 6.48.

5. Conclusion

- Measurement of water discharge in the Javan Langur Centre area shows that the water discharge in the river flow is very small, around 0.00035 m3 / s. Meanwhile, the water discharge that can be used for pico-hydro power plants is 0.00023 m3 / s. Low water discharge causes low generator efficiency. From the measurement results, it is found that the efficiency of the generator is less than 10% with the power generated around 27.3 watts.
- The 2-stage cascade system applied to the Pico-hydro power plant in the Javan Langur Centre area is able to double the power generation capacity, around 55 watts.

6. Future work

The water discharge must be enlarged to increase the rotation and torque of the water turbine. If the rotation and torque of the water turbine increases, the generator rotation will also be faster and will produce a higher output voltage. One possible option to increase the flow of water entering the turbine house is to increase the height of the existing dam, so that the volume of water collected in the dam becomes larger.

Acknowledgements

This work was supported by the State Polytechnic of Malang through Community Service Partnership Scheme (2020-2021).

IOP Conf. Series: Materials Science and Engineering 1073 (2021) 012025 doi:10.1088/1757-899X/1073/1/012025

References

[1] Zainuddin H, Yahaya M S, Lazi J M, Basar M F M and Ibrahim Z 2009 Design and development of pico-hydro generation system for energy storage using consuming water distributed to houses *World Acad. Sci. Eng. Technol.* **59** 154–9
[2] Anon Hydropower Plants
[3] Maher P and Smith N 2001 Pico hydro for village power *A Pract. Man. schemes up to* **5**
[4] Sujith K P, Tuntun K, Avinash K, Swapnil S, Kunal K and Satya P 2016 Technical and economic feasibility of Pico hydro power plants integrated to Green buildings *Int. Res. J. Eng. Technol.* **3** 77–9
[5] Roshan V R, IPE R S and George A 2016 Pico-hydro electric power generation from residential water tank *Int. J. Chem. Sci* **14** 421–6
[6] Yadav G and Chauhan A K 2014 Design and development of pico micro hydro system by using house hold water supply *Int. J. Res. Eng. Technol.* **3** 114–9
[7] Ya D A Z 2009 Pico-Hydropower system for Rural Electrification *Proceedings of the ICSE*
[8] Hidayat M N, Ronilaya F, Eryk I H and Joelianto G 2020 Design and analysis of a portable spiral vortex hydro turbine for a Pico Hydro Power Plant *IOP Conference Series: Materials Science and Engineering* vol 732 (IOP Publishing) p 12051

IOP Conf. Series: Materials Science and Engineering 1073 (2021) 012026 doi:10.1088/1757-899X/1073/1/012026

Small scale hybrid power generation for remote area electrification

M N Hidayat*, F Ronilaya, I H Eryk, S Wibowo and L Hakim

Electrical Engineering Department, State Polytechnic of Malang

*moh.noor@polinema.ac.id

Abstract. Javan Langur Centre (JLC) is a facility dedicated to the treatment of langur, an endangered primate species. To support the maintenance and rehabilitation activities, the need for electrical energy is a vital. In this facility, electrical energy is needed for lighting and baby primate's incubator with a total power of around 120 watts. There is no electricity grid included in this area and the supply of electrical energy is still very dependent on the gen-set owned by the residents which is located approximately 2 km from JLC, making it less efficient. In terms of natural energy resources, JLC has the availability of water sources that originate from small streams making it possible to develop pico-hydro technology. There has been built a pico-hydro power generation with a potential electric power that can be generated approximately 300 watts. However, the water discharge of the river stream in that area is quite low, around 0.00035 m3/s, that makes the efficiency of the power generation is 10% only. In other words, the actual electrical power that can be generated is around 30 watts. So as to meet the electrical power needs at JLC, another electrical energy source is needed. One alternative is to install solar cells or photo voltaic (PV) with a capacity of 100 watts peak. The output power of these two energy sources can be combined and regulated by applying Hybrid Power Generation technology. With the application of this technology, the need for electrical energy in the JLC area can be met.

1. Introduction

Javan Langur Centre (JLC) is the Javan Langur conservation centre, one of the primate species that is protected due to extinction, which is located 10 km from the centre of Batu City, East Jaw, Indonesia. To support daily activities, JLC has difficulties in meeting electricity needs. So far, electricity in the JLC has been supplied from the Coban Talun tourist area, which is more than one kilometer away.

Currently, a pico-hydro power plant has been developed in the area with an installed generator power of 600 watts. However, the actual power that can be generated from this generator is only around 120 watts. This is related to the very low flow of water flowing in the JLC area, which is around 0.00035 m3 / second. Low water discharge results in low turbine speed. In the end, the generator output is also low. To overcome the shortage of electricity supply in the JLC of approximately 120 watts, photo voltaic (PV) with a capacity of 100 watts peak was installed.

These two generators, namely pico-hydro and PV, are operated in a hybrid power generation system, with an output power of about 130 watts. This total power will be able to meet the electricity needs of the JLC.

2. Hybrid power plant

Hybrid Power Plant is a type of power plant that has two or more energy sources to generate electrical energy. A power plant with a hybrid system is an alternative that can be used to meet electricity needs in areas far from the existing distribution network [1].

One example of a hybrid power plant is a combination of a Pico-hydro Power Plant (PHPP) and a Solar Power Plant (SPP). In terms of system reliability, this generating combination is very beneficial, especially in areas with abundant water sources and adequate sun exposure. The Pico-hydro Power Plant can operate and produce output power 24 hours a day, as long as the water source is still available. Meanwhile, a solar power plant can operate as long as there is solar radiation, generally 10 hours a day.

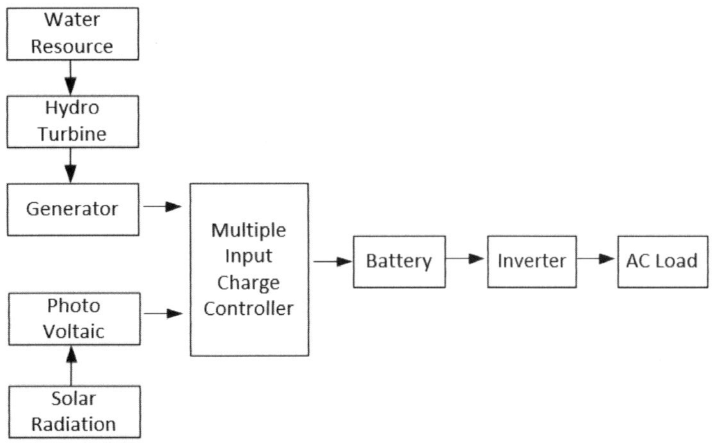

Figure 1. Block diagram of PHPP-SPP hybrid power plant.

From Figure 1, it can be seen that the hybrid generator above requires a Multiple Input charge controller, namely a charge controller that has more than one input. The input can be either AC voltage or DC voltage, depending on the type of multiple input charge controller itself. Furthermore, the output of multiple input charge controllers will be connected to the battery / accumulator. Then connected to the inverter, before finally connected to the AC load.

2.1. Solar Power Plant (SPP)

Solar Power Plant (SPP) is a power plant that utilizes solar energy sources to generate electrical energy using photovoltaic (PV) or what is often called solar cells [2-4]. Solar energy is renewable, environmentally friendly and easily converted to electricity. Even though SPP is a new, renewable energy that will never run out, this energy has one weakness, namely the unavailability of continuous energy sources because solar energy is only available during the day. So that at night, there is no electric energy that is charged into the battery [6,7].

Conventional solar cells work using the p-n junction principle, which is a junction between p-type and n-type semiconductors. When photons hit or strike a semiconductor material, such as silicon, then the material will release electrons and cause a potential difference to form between the surfaces exposed to the sun's photons [2].

IOP Conf. Series: Materials Science and Engineering 1073 (2021) 012026 doi:10.1088/1757-899X/1073/1/012026

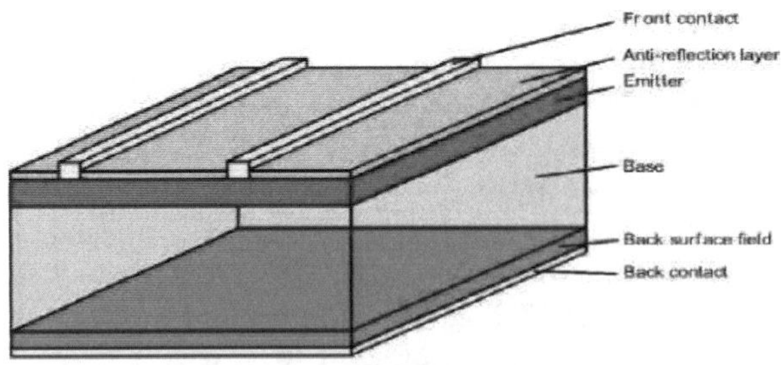

Figure 2. Inside a photovoltaic cell [2].

The maximum output power can be calculated by the formula:

$$Pmaks = Vmodul \times Imodul$$

P is the maximum power of the Solar Panel (W), V is the Maximum Voltage (V), and I is the Maximum Current (A) [2].

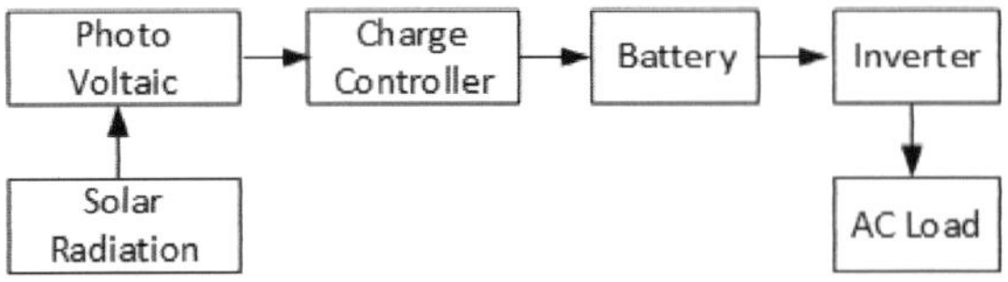

Figure 3. Block diagram of a solar power plant.

As seen in Figure 3, a Solar Power Plant utilizes PV to convert solar radiation into electrical energy. The output from the PV will be entered into the charge controller. The output from the charge controller will be used to charge the battery. Furthermore, the battery is connected to the inverter to convert the DC voltage from the battery into AC voltage which is then used to supply AC loads.

2.2. Pico-hydro Power Plant (PHPP)

The Pico-hydro Power Plant is in principle the same as other hydroelectric power plants, utilizing different heights and the amount of water discharge in the flow of irrigation channels, rivers or waterfalls [8]. This water flow is used to generate electrical energy by using a water turbine installed with a generator and the energy is stored on the battery, through a charge controller. Before it can be used to supply AC loads, the output of the battery must pass through the inverter, to convert the DC voltage into AC voltage.

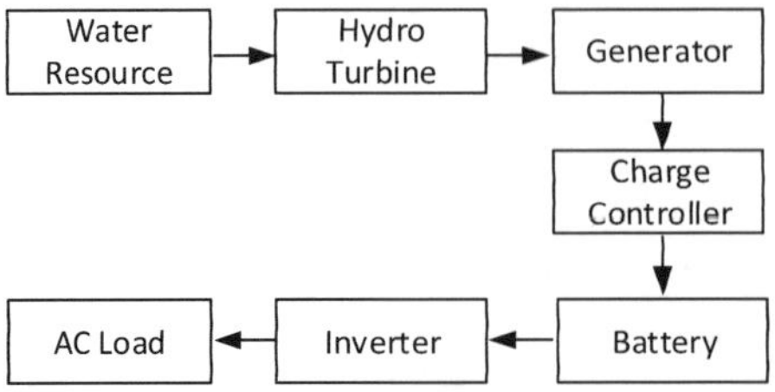

Figure 4. Block diagram of Pico-hydro power plant.

In PHPP (figure 4), the water flow model uses an intake from the river flow and flows the water through a water channel with a small slope towards a reservoir. Furthermore, water from the reservoir is flowed through a rapid pipe to the turbine housing to drive the turbine. The water turbine is coupled with a generator to generate electric power. The power (power) generated can be calculated based on the following formula [8]:

$$P = \rho.Q.h.g$$

P is the output power (watts), p is the density of the fluid (kg / m3), Q represents the water discharge (m3/s), h represents the effective height (m), and g is the gravitational force (m/s2).

3. Workflow and system design

3.1. Problem solving workflow

The method used in this research follows the order of work as mentioned below:

- Calculation of electric power requirements for lighting and incubator machines in the Javan Langur Centre area.
- Calculation of the potential electric power that can be generated from water flow and solar radiation in the Javan Langur Centre area.
- Construction of reservoirs and intake paths for pico-hydro power plants and determining the location of PV installation.
- Designing a turbine to suit the existing conditions in the Javan Langur Centre river basin.
- Build turbines and test turbines according to design.
- Determination of generator power, PV capacity, multiple input charge controller specifications, and battery capacity required in the planned hybrid generating system.
- Testing of generators in the lab. Electric machines Malang State Polytechnic. The test includes testing the characteristics of open circuit, short circuit generator and load characteristics.
- PV testing and multiple input charge controllers to determine the value of charging and discharging currents in the battery.
- The design of electrical installations for the energy supply needs of the Javan Langur Centre including the protection system including the needs of batteries, charger controllers and inverters.
- Comprehensive work implementation which includes: installation of pico-hydro and solar power plants, as well as installation of LDP panels in suitable locations,

3.2. PHPP-SPP hybrid power plant design

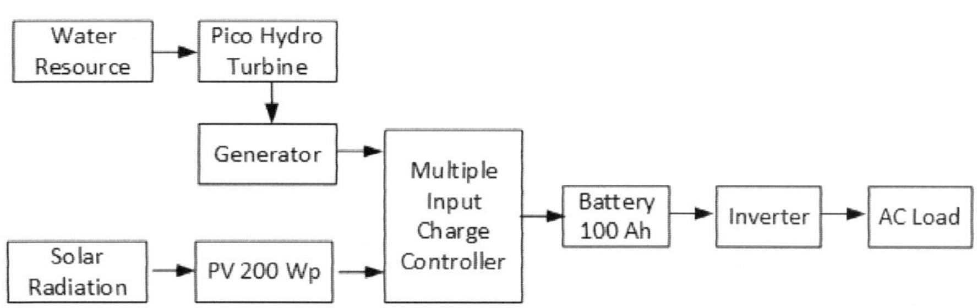

Figure 5. Block diagram of the Pico-hydro-PV hybrid power generation system at JLC.

The Hybrid Power Generation System developed at the Javan Langur Centre consists of 2 types of power plants, namely the pico-hydro power plant with a capacity of 300 watts and the solar power plant with PV with a capacity of 100 Wp.

The pico-hydro power plant has an efficiency of about 10%, so the actual power that can be utilized is 30 watts. So, to increase the supply of electrical energy, this plant is operated in a hybrid manner with PV 100 Wp to produce an output power of 130 watts, which can be used to supply lighting and incubator machines for baby langurs in the JLC area.

IOP Conf. Series: Materials Science and Engineering 1073 (2021) 012026 doi:10.1088/1757-899X/1073/1/012026

3.3. Component specifications in a hybrid power plant

3.3.1. Hydro turbine

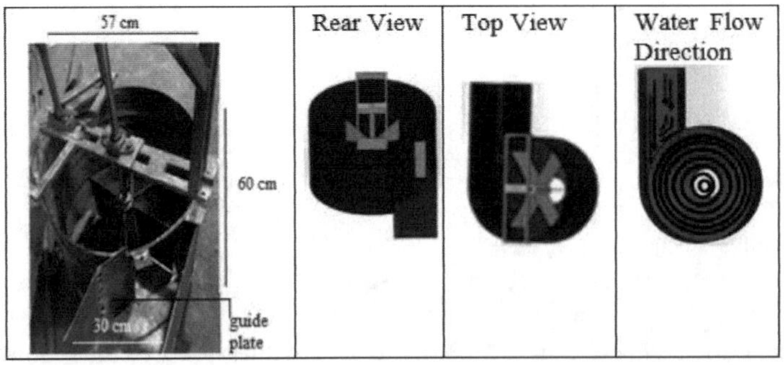

Figure 6. Water turbine design at a Pico-hydro power plant [9].

As shown in figure 6, the design of turbine house number has a water flow steering guide mounted on an iron channel. There is a drain in the middle with a diameter of 20 cm. The overall weight of the turbine and turbine house is around 22 kg.

The test results show that when water is thawed into the turbine house, the flowing water is directed to crash into the circular turbine house wall. So that a spiral whirlpool is formed immediately and the turbine rotates with a higher speed, which is around 90 rpm.

3.3.2. Generator. The generator used is a vertical axis generator, AC 3 phase, 300 W, 48 volts, and a maximum rotation of 750 rpm. This generator has a diameter of 175 mm, a height of 100 mm, a shaft diameter of 12 mm, a shaft length of 60 mm and a weight of 8 kg.

3.3.3. PV. The PV used has optimal power (peak): 100 Wp, optimal voltage: 17.8 V, optimal current: 5.62 A (Source Manual Book SOL-M24 200W).

3.3.4. Multiple input charger controller. Multiple Input Charger Controller is a charge controller which has more than one input. It may contain of two DC voltage inputs or one DC voltage and one AC voltage inputs [10]. So, in practice, it can be utilized for two different power generations, such as wind power plant and solar power plant, or solar power plant and hydro power plant.

Multiple Input Charger Controller used has specifications: rated Voltage (12/24 V), Input Power of Solar Panel (400Wp), Input Power of Pico-hydro (600W), maximum charging current of 37.5 A, Equalization Charge or the ability of multiple inputs charge controller to equalize the charging voltage of each cell battery (28.8 V +/- 1%).

The selected multiple input charge controllers also have the ability to over discharge - disconnect (21.8 V +/- 1%), Over Voltage - Shut Off (32 V +/- 1%). In addition, this device has Solar Cell Anti-Reverse Protection, which prevents backflow from the battery to the PV because it can cause the PV to break down.

3.3.5. Battery. The battery used has a nominal voltage specification of 24 volts and a capacity of 100 AH.

3.3.6. Inverter. The inverter used has specifications: rated power 500W, pure sine wave output, 220/230 V output voltage, 50 Hz output frequency, 12 V rated voltage, 52 A maximum input current, and is equipped with battery reverse protection in the form of fuse protection.

3.3.7. Load. The load supplied by the hybrid generator at the JLC consists of 6 lamps with a power of 10 watts each and an incubator machine with a power of 60 watts. So the total load supplied is 120 watts.

3.3.8. Electrical panel. The electric panel contains the electrical components mentioned above, including the multiple input charge controller, battery, and inverter, as shown in Figure 7.

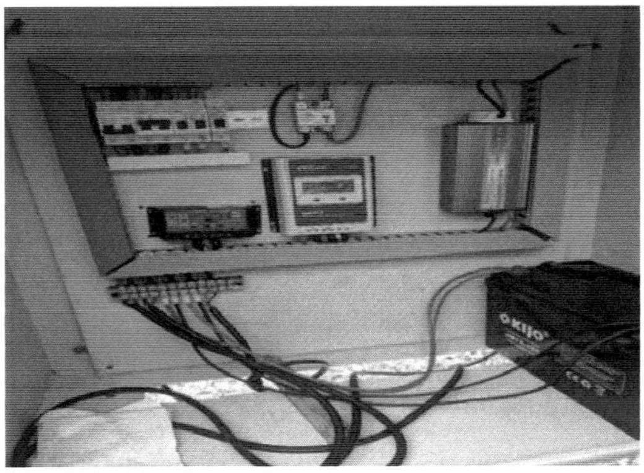

Figure 7. Electric panel hybrid generation system at JLC.

4. Testing and analysis data

4.1. Water discharge measurement

Water discharge measurements in the JLC area were carried out at 2 different locations, namely the river flow and the water turbine intake flow. Measurements are made by placing a 1.5 litre bottle in the water line and recording the time it takes to fill the bottle completely. From the measurement results, the value of river water discharge is around 0.00035 m3/s, and the turbine intake water discharge is around 0.00023 m3/s.

4.2. Output voltage measurement

Measurements were made to determine the output voltage of the 3-phase AC generator installed on the pico-hydro electric generator and the output voltage of the PV. The measurement results can be seen in the table.

Table 1. Measurement results of generator output voltage.

No	Turbine 1 Rotation (rpm)	Generator Rotation (rpm)	Generator Output (Vac)
1	78.3	391.5	21.6
2	76.1	380.5	20.4
3	76.7	383.5	20.4
4	78.3	391.5	21.6
5	78.2	391.0	21.6
Average	77.5	387.6	21.12

From table 1 it can be seen that the average turbine rotation is 77.5 rpm. After being connected through the pulley system, it produces a generator rotation of 387.6 rpm. This generator rotation will produce an output voltage of 21.12 volts. As for solar power plants, the output voltage of the PV is measured at 18 Vdc.

IOP Conf. Series: Materials Science and Engineering 1073 (2021) 012026 doi:10.1088/1757-899X/1073/1/012026

4.3. Battery charging current measurement

Table 2. Charging current measurement results.

Charging Current (A)		Charging Time for a 40 Ah Battery
Generator	0.75	53.3333
PV	1.10	36.3636
Total	1.85	21.6216

From the table above it can be seen that the charging current for the generator from the pico-hydro power plant is 0.75 A. While the charging current from the PV is measured at 1.1 A. Then the total charging current entering the battery is 1.85 A. charging current is 1.85 A, then the time required to fully charge the battery is 21.62 hours.

However, PV can only operate in the presence of solar radiation, that is, in the morning to evening, as long as sunlight is not blocked from reaching the PV surface. Therefore, in a period of 1 day (24 hours) PV is considered to be able to operate for only 10 hours out of a total of 24 hours in a day.

Table 3. Calculation of charging time.

Charging Current (A)		Charging Time in 1 day (hours)	Stored Capacity (Ah)	Charging Time for a 40 Ah Battery (days)	Charging Time for a 40 Ah Battery (hours)
Generator	0.75	24	18		
PV	1.1	10	11		
Total			29	1.3793	33.1034

From the table it can be seen that by using the PHPP and PV hybrid system, the time required to fully charge a 40 Ah battery is 1.34 days or 33.1 hours.

4.4. Generator efficiency and PV efficiency calculations

4.4.1. Generator efficiency. By referring to the generator specifications, namely 3 phase, 48 volts, 300 watts, the nominal current of the generator can be calculated:

$$I \text{ nominal} = 300 : (1.732 \times 48) = 3.6 \text{ A}$$

The generator output current can be seen by referring to the generator charging current value, which is 0.75 A. The measured output voltage of the generator is 21 volts. Then you will get the generator output power of:

$$P \text{ output} = 1.732 \times 21 \times 0{,}75 = 27{,}3 \text{ watts}$$

So that the efficiency of the generator can be calculated as below:

$$\text{Generator efficiency} = (\text{Actual Power} : \text{Rated Power}) \times 100\%$$
$$= (27{,}3 : 300) \times 100\% = 9{,}1\%$$

4.4.2. PV efficiency. Based on previous information, it is known that the installed PV capacity is 100 Wp. Meanwhile, the measured value in the battery charging process is the charging current of 1.1 A and the charging voltage is 18 volts. Then the PV output power can be calculated as below:

$$P \text{ output} = 18 \times 1{,}1 = 19{,}1 \text{ watts}$$

So that the PV efficiency can be calculated as below:

$$\text{PV efficiency} = (\text{Actual Power} : \text{Rated Power}) \times 100\%$$
$$= (19{,}1 : 100) \times 100\% = 19{,}1\%$$

5. Conclusion

- The water discharge in the JLC area is around 0.00035 m3 / s. This value is classified as very small for pico-hydro power plant applications. As a result, the water turbine rotation and generator rotation are low, which in turn results in a low generator output value. From the calculation results, the value of the generator efficiency is less than 10% with the power generated around 27.3 watts.
- Solar radiation in the JLC area can be used as a source of solar power using PV with a capacity of 100 watts peak. From the measurement and calculation results, the PV efficiency value was 19.1%.
- The Hybrid system that is applied in the JLC area combines the output of a Pico-hydro power plant and a solar power plant to meet the electricity needs in the JLC area

6. Future Work

In order to increase the output power of Pico-hydro Power Plant, it is recommended to enlarge the dimension of the waterway for turbine intake. If the water discharge higher, the rotation and torque of the water turbine increases, so that the generator rotation will also be faster and will produce a higher output voltage.

In terms of Solar Power Plant, it is recommended to install more PV's in Javan Langur Centre to produce a lot more electricity.

Acknowledgements

This work was supported by the State Polytechnic of Malang through Community Service Partnership Scheme (2020-2021).

References

[1] Syahputra R and Soesanti I 2020 Planning of Hybrid Micro-Hydro and Solar Photovoltaic Systems for Rural Areas of Central Java, Indonesia *J. Electr. Comput. Eng.* **2020**

[2] Goetzberger A and Hoffmann V U 2005 *Photovoltaic solar energy generation* vol 112 (Springer Science & Business Media)

[3] Hankins M 2010 *Stand-alone solar electric systems: the earthscan expert handbook for planning, design and installation* (Earthscan)

[4] Corporation A E 1991 *Maintenance and Operation of Stand-Alone Photovoltaic Systems* (Photovoltaic Design Assistance Center, Sandia National Laboratories)

[5] Sen Z 2008 *Solar energy fundamentals and modeling techniques: atmosphere, environment, climate change and renewable energy* (Springer Science & Business Media)

[6] Alami A H 2020 *Mechanical Energy Storage for Renewable and Sustainable Energy Resources* (Springer)

[7] Center F S E 1997 Batteries and Charge Control in Stand-Alone Photovoltaic Systems *Cocoa, Florida. EUA*

[8] Yadav G and Chauhan A K 2014 Design and development of pico micro hydro system by using house hold water supply *Int. J. Res. Eng. Technol.* **3** 114–9

[9] Hidayat M N, Ronilaya F, Eryk I H and Joelianto G 2020 Design and analysis of a portable spiral vortex hydro turbine for a Pico Hydro Power Plant *IOP Conference Series: Materials Science and Engineering* vol 732 (IOP Publishing) p 12051

[10] Nag S S and Mishra S K 2013 A multi-input single-control (MISC) battery charger for DC nanogrids *2013 IEEE ECCE Asia Downunder* (IEEE) pp 304–10

Design and analysis of the effectiveness smart home control systems based on using the Internet of Things

E Rohadi[1], H Susanto[2,*] and M Sarosa[2]

[1] Information Technology Department, State Polytechnic of Malang, Jl. Soekarno Hatta No. 9, Malang 65141, Indonesia
[2] Electrical Engineering Department, State Polytechnic of Malang, Jl. Soekarno Hatta No. 9, Malang 65141, Indonesia

*heri.susanto@polinama.ac.id

Abstract. Smart homes are a set of interconnected electrical appliances to become intelligent service systems aimed at reducing human workloads, saving electricity, and improving home security. Wireless Smart Home system enhances popularity due to flexibility, portability, and low installation cost. With the availability of technology and the internet of Things (IoT), devices can collect information from the environment according to the surrounding conditions. Raspberry Pi is a microcomputer which is a primary device that can be used to control sensors and actuators. This control can be activated from anywhere, as long as the area has an Internet connection. Using the fuzzy logic method, the controls don't need to be done manually. This method can automatically control the electronic devices in the home, using the facts in the knowledge base. The main features of the smart home are real-time, remote control, data security from intruders. Data security becomes important to access right in maintaining the confidentiality of sensitive information that may be known by the right party. Technology systems are containers created in one library, applications, and configuration files, and all that is required to be a reference to rely on.

1. Introduction

Types of work are increasingly developing based on a variety of jobs, it is undeniable that many of these workers are in service or even on duty outside the city, not only office workers but private ones, who have to negotiate related work problems, as well as workers who often mobile, which led to frequent being away from home, then leaving their homes empty and uninhabited. Usually, the homeowner will turn on the porch light of his house and other electronic equipment when he is leaving his house for a long time and the house is left empty or no one is home. Electronic equipment such as lights and tv will continue to be on during the day and night while the homeowner is away. This condition is not efficient when viewed from electricity-saving management, because all day long the electronic device will continue to run and result in an increase in the number of electricity bills at home. It is very ineffective if the homeowner has to return home just to turn on or turn off the electronic device. Also, this is done because sometimes other people think that if the electronic device does not turn on at night, it means that no one is in the house [1]. This assumption condition of the room that is left for a long time will cause the condition of the room to not be healthy, we also have to turn on the air conditioner such as the exhaust at certain times, so that the room remains fresh like a house that is maintained and inhabited. In other words, in addition to having the goal of efficiency in electricity use and maintaining the freshness

IOP Conf. Series: Materials Science and Engineering 1073 (2021) 012027 doi:10.1088/1757-899X/1073/1/012027

and health of the space and environment of the house, we also need security or prevent things that we don't want from happening to our homes when left empty. Starting from this condition, a system that can be used to control electronic equipment such as house lights, alarms, and others will be developed using the web-based internet as a medium for accessing servers at home and via the PLN power cable as communication between modules [1].

2. Literature review

2.1. Raspberry Pi
Raspberry Pi is a computer for a credit card developed in the UK by the Raspberry Pi Foundation. One of the products Raspberry Pi is the Raspberry Pi 3 Model B. Affordable compared to buy a PC for simple computing. What's more, its small size and consumption very low electrical power makes it very suitable for various purposes [2]. All Operating Systems is based on Linux. For novice users, Raspberry Pi Foundation advises to use New Out of the Box Software (NOOBS) in installing the OS on in it there is a choice of operating systems that can be installed.

2.2. FUZZY logic
Fuzzy logic has a degree of membership in the range 0 (zero) to 1 (one) and fuzzy logic shows the extent to which a value is true and the extent to which a value is false. Fuzzy logic is an appropriate way to map an input space into an output space and have a continuous value. Fuzzy is expressed in terms of membership degrees and degrees of truth. Therefore something can be said to be partly right and partly wrong at the same time [3].

2.3. Sensor Passive Infrared Receiver (PIR sensor)
Passive Infrared Receiver (PIR) Sensor is a passive sensor, where the sensor is does not radiate any inner energy detect something. PIR works with detects emitted infrared radiation by objects around it, such as the human body. Within the PIR there are sections with functions different, namely Fresnel Lens, IR Filter, Pyroelectric sensor, amplifier and comparator [4].

2.4. PHP
PHP is generally known as programming language scripts that create HTML document on the fly that is executed on the web server, an HTML document is generated of an application is not an HTML document created using a text editor or editor HTML. Also known as a programming language server side [5].

2.5. Actuator simulation software
Actuator simulation software is used show the movement of actuator and using mathematical models to replicate the behavior of an actual electronic device or circuit. Simulation software allows for modeling of circuit operation and is an invaluable analysis tool.

2.6. Relay
Relay is an electromagnetic device operated by changing a condition electrical circuit. Useful for activating other equipment by opening or close contact by giving a circuit the relay is a logic 1 or 0. One of the uses the main relay in the industrial world is to implementation of control logic in a system. As a "programming language" it is used a configuration called a ladder diagram or relay ladder logic.

2.7. DHT11
Temperature and humidity sensor is a sensor for sensing temperature and humidity objects in 1 module which has a calibrated digital output signal. This module is classified into a resistive element such as a temperature measuring device, for example, NTC. The advantage of this sensor compared to others is

that it has good quality sensing data reading, is responsive (fast in reading room conditions), and is not easily interfered with [6].

2.8. Python
Python is designed to make it easy which is very unusual to good programmers in terms of time efficiency and convenience in program development and in terms of compatibility with the system. Python can used to create tandalone programs and script programming (scripting programming) [7].

3. Design of prototype
Design and drawing of a prototype for a home system and sensor possioning.

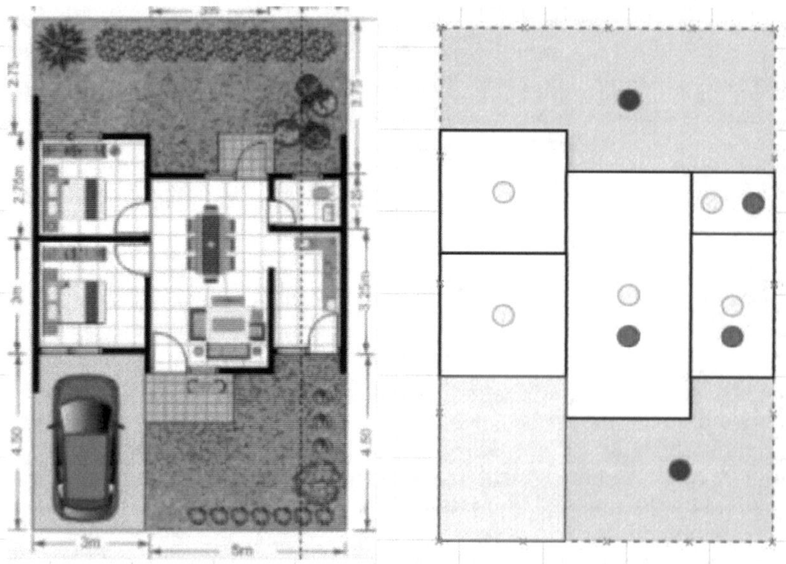

Figure 1. Prototype design for a home system.

Shows a miniature house plan consisting of 6 areas. Explanation of the control of each area in a miniature house as follows:. In the living room, living room, and kitchen, there are light and humidity sensors in the room that will control the fan lights. In the bedroom and bathroom, there are motion sensors that will control the lights. In the room, there is a motion sensor that will control the lights and a temperature sensor to control the fan. In the plant watering area, there is a water sensor and a motor to control the opening and closing of the water valve for sprinkling. This Sensor description:

3.1. Yellow color for the light control sensor
This section will control the windows for air exchange and lamps for lighting. In the lamp, the LDR sensor will detect the intensity of the light around it. If the light is bright, the light will turn off automatically The lights can also be controlled automatically and manually on the application if needed.

3.2. Red color for PIR (motion) sensor
The PIR sensor will detect the movement of objects around. If there is no object movement, the light will automatically turn off. These lights can also be controlled automatically and manually on application if needed [5].

3.3. The blue color for the plant watering sensor
This section will control the valve to adjust the water intensity to be used for watering the plants. There are 2 temperature condition parameters to control the valve, the valve will automatically shut off. If the

temperature is detected hot (with a predetermined temperature parameter), then the valve condition turns on and flushes automatically [7].

4. Methods
Operating system flow diagram.

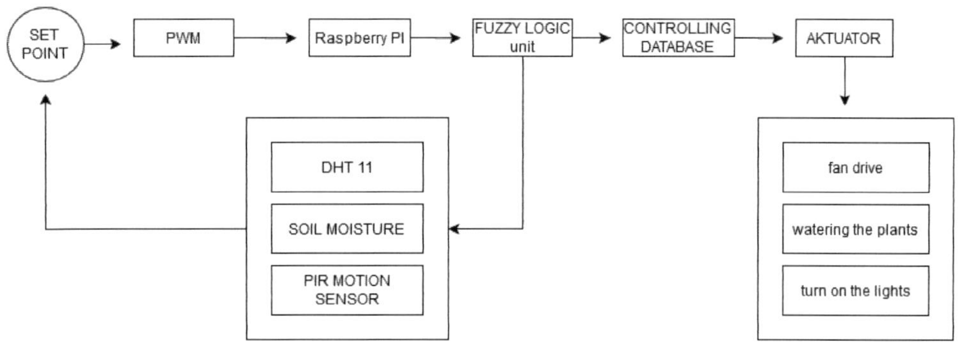

Figure 2. Process flow data looping.

Process flow sending commands to the Raspberry Pi device, starting from the button indicating activation the system will then run the function the goal, so as not to happen loading data continuously on the controlling database or wab base, which will result in a crash on the system.

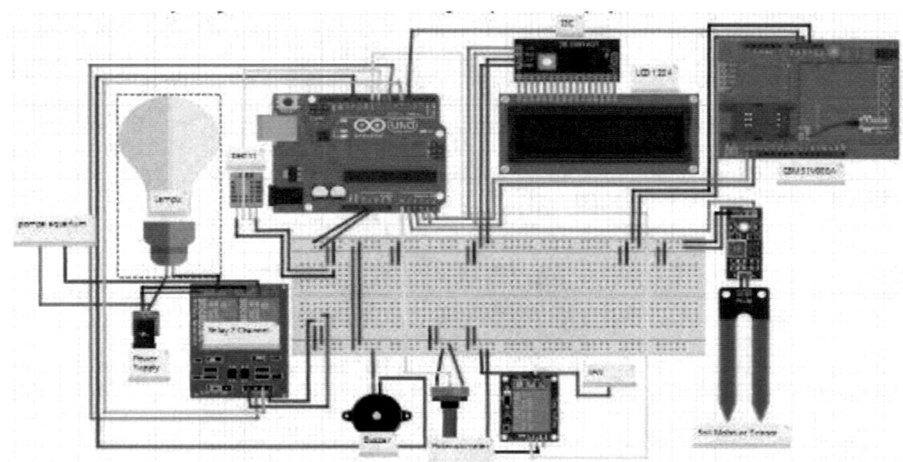

Figure 3. Blok Diagram microcontroller system.

The combined circuit scheme of several sets of hardware components into one comprehensive circuit. In this study, an automatic control system was applied to a house consisting of seven controlled areas. This control uses a light sensor, motion sensor, and temperature sensorandroid application for easy remote control. When auto mode is activated, the data is sent to the Raspberry Pi Zero microcontroller via the database [5].

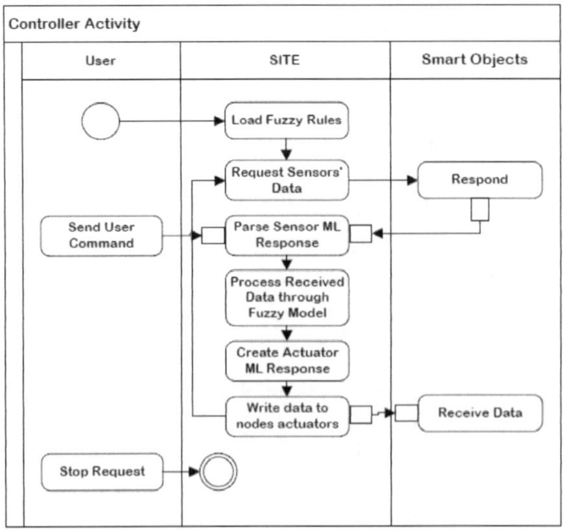

Figure 4. Blok Diagram rule sending system.

The microcontroller will process data according to the data sent by the application then the data is sent back to the application via the database. For example controlling in manual mode. The user turns on the fan control by pressing the ON switch on the application. The data is sent to the microcontroller via the database to be processed and then moves the fan blade. The data is sent back to the database .net.

Figure 5. Soil Moisture sensor.

There are 2 condition parameters to control the valve, If humidity is detected close to point 95 (with a predetermined temperature parameter), then the valve condition turns on and starts flushing automatically. The set points are in the vulnerable range between 45 da, 95 degree pie.

Figure 6. DHT11 sensor.

The sensor detects a temperature of less than 20 degrees, the sensor sends a command to the microcontroller to execute the command to the driver, which is to turn off the fan. If not, the sensor detects a temperature of 20 degrees to 30 degrees, then the sensor sends a command to the microcontroller to execute the command to the driver to turn on the fan at high speed.

IOP Conf. Series: Materials Science and Engineering 1073 (2021) 012027 doi:10.1088/1757-899X/1073/1/012027

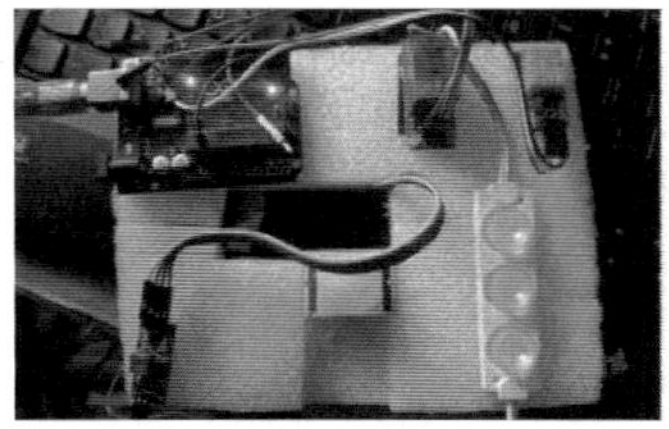

Figure 7. PIR Motion Sensor.

Indicates that the input from the light sensor and motion sensor. If in manual mode, the user can control the lights in each room with the direct ON or OFF command in the application. If not, the motion sensor detects movement (value 1), then the sensor sends a command to the microcontroller to execute the command, namely turning on the light (ON). If not, the sensor sends a command to the microcontroller to turn off the light (OFF).

5. Results

The results of the system system are divided into tests of soil temperature and humidity.

5.1. Data collection Humidity sensor on the 7 day of implementation

Soil moisture testing is done to determine whether soil moisture control is functioning properly or not. In this case, a moisture control system is used to stabilize the soil in the process of planting plants in the yard. The humidity setpoint is entered into the system with the lower set point 35 - 95.

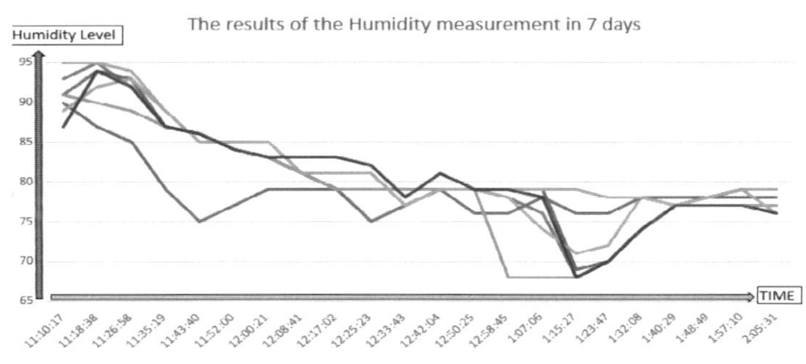

Figure 8. Implementation of data collection in the daytime.

In this case, a moisture control system is used to stabilize the soil in the process of planting plants in the yard. The humidity setpoint is entered into the system with the lower set point 45 - 95.

5.2. Data collection Temperature sensor on the 7 day of implementation

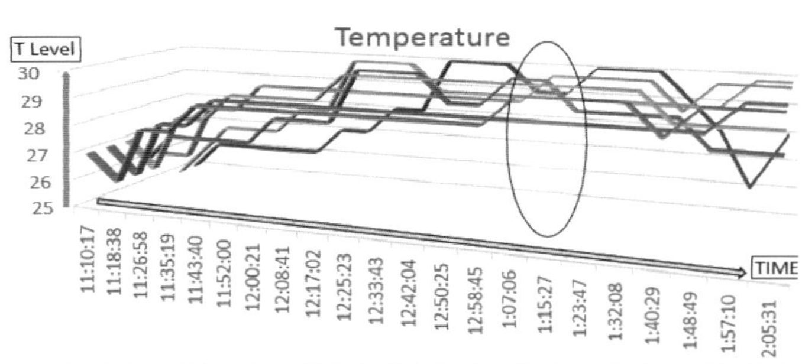

Figure 9. Implementation of data collection in the daytime.

Figure shows the temperature test by the automatic control method. The results of DHT11 control for plants can be stable at temperatures around 95^0C during the day by taking 1 state data in 10 minutes. Temperature testing is carried out to determine whether the temperature control is functioning properly or not. The temperature set point is entered in the system with a temperature between 22^0C to 27^0C.

5.3. Data collection motion sensor of implementation

The sensitivity of the Passive Infrared Receiver (PIR) sensor sends the measured data to the microcontroller with an average delay of 15.1 seconds and in general, this Passive Infrared Receiver (PIR) sensor has an effective reading range of up to 3 meters, and this sensor is very effectively used as a human detector.

6. Conclusion

In the process of controlling the smart home system, starting from soil moisture and temperature to functionality, a melting point setpoint can be achieved from 35^0C to 95^0C with a setting time of 10 minutes of data collection. The closer the motion to the PIR sensor is captured, the faster the response is generated when there is movement

Acknowledgment

We are very grateful to the Department of Information & Telecomunication Technology, State Polytechnic of Malang who has helped and provided facilities to carry out this research. With the completion of this research, the authors would like to thank Dr. Erfan Rohadi and Dr. Moechammad Sarosa as the supervisors.

References

[1] Hamdan O, Shanableh H, Zaki I, Al-Ali A R and Shanableh T 2019 IoT-based interactive dual mode smart home automation *2019 IEEE International Conference on Consumer Electronics (ICCE)* (IEEE) pp 1–2

[2] Mahmud S, Ahmed S and Shikder K 2019 A smart home automation and metering system using internet of things (IoT) *2019 International Conference on Robotics, Electrical and Signal Processing Techniques (ICREST)* (IEEE) pp 451–4

[3] Mahamud M S, Zishan M S R, Ahmad S I, Rahman A R, Hasan M and Rahman M L 2019 Domicile-an IoT based smart home automation system *2019 International Conference on Robotics, Electrical and Signal Processing Techniques (ICREST)* (IEEE) pp 493–7

[4] Estebsari A, Patti E and Barbierato L 2018 Fault Detection, Isolation and Restoration Test Platform Based on Smart Grid Architecture Model Using Intenet-of-Things Approaches *2018 IEEE International Conference on Environment and Electrical Engineering and 2018 IEEE Industrial and Commercial Power Systems Europe (EEEIC/I&CPS Europe)* (IEEE) pp 1–5

[5] Song M 2016 A Study on Business Types of IoT-based Smarthome: based on the theory of Platform Typology *J. Inst. Internet, Broadcast. Commun.* **16** 27–40

[6] Mukhopadhyay K, Sinha C, Saha H N, Rakshit S and Auddy S 2018 Smart Mirror-a Secured Application of Artificial Intelligence Recognizing Human Face and Voice *2018 IEEE 9th Annual Information Technology, Electronics and Mobile Communication Conference (IEMCON)* (IEEE) pp 1279–89

[7] Maulana H and Al-Jabari M R 2018 Development of Smart Home System to Controlling and Monitoring Electronic Devices using Microcontroller *IOP Conference Series: Materials Science and Engineering* vol 1 pp 1–3

IOP Conf. Series: Materials Science and Engineering 1073 (2021) 012028 doi:10.1088/1757-899X/1073/1/012028

Impedance matching system design using non uniform transmission line for 2.4 Ghz antenna system

N Zakaria[1,*], A Setiyawan[2], Y H Prasetyo[1], M K Wardani[1] and A W Yulianto[1]

[1] Teknik Telekomunikasi, State Politechnic of Malang, Jl. Soekarno-Hatta Malang, Malang, Indonesia
[2] Teknik Elektronika Gajayana University Malang Jl. Mertojoyo Blok L Dinoyo Malang, Malang, Indonesia

*nanak_zach@polinema.ac.id

Abstract. The Matching Impedance (MI) design method using Non-Uniform Transmission Line (N.U.T.L.) is a design method that offers more advantages over the 1/4 transformer method, the 1/4 multisection method, and the stub method. In this study, the Matching Impedance system is designed to match the WLAN antenna load with a line impedance of 50 Ohm at the 2.4 GHz. The Matching Impedance system is designed using a transmission line with a Characteristic Impedance (Zo) value that is triangular (Triangular Taper) distributed. This study also examines the effect of the number of cutting sections on a transmission line if the line with the Triangle distribution is implemented on a Microtsrip. From the test results, it is found that the number of sections used to form the Non Uniform transmission line has a significant effect on the bandwidth of the system working frequency for the required value of the Reflection Coefficient.

1. Introduction

Impedance Adjuster is an important circuit in the radio communication especially in the high frequency and microwave. Although the Impedance Adjustment field is not a new issue, its development is continuously being made to obtain an efficient and highly efficient system. One of the methods developed is by arranging a Matching Impedance using a Non Uniform Transmission Line (NUTL). Non Uniform Transmission Line is a transmission line which has a Non-Monotonous Characteristic Impedance value or changes as a function of distance ($Zo\ (l)$).

The use of matching impedance using a non-uniform transmission line is expected to improve the performance of similar devices developed from a quarter-wavelenght uniform transmission line, where to widen the working area of the device, more sections are needed [1].

2. Non uniform transmission line

The Characteristic Impedance Distribution of a Non Uniform Transmission Line is illustrated as follows [1,2].

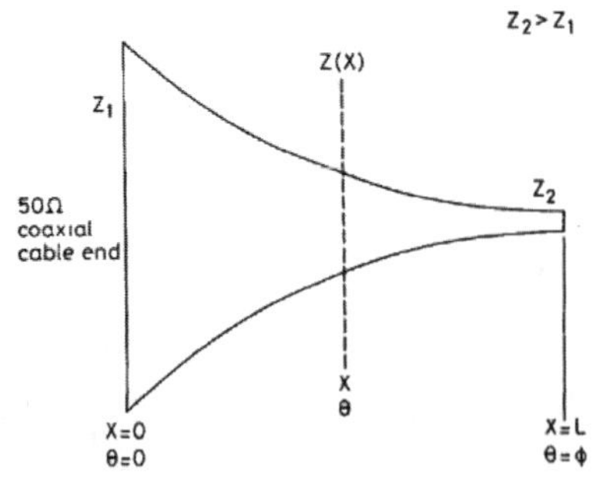

Figure 1. General model of non uniform line.

If *Z2* is the Load Impedance, *Z1* is the Input Impedance and *Z(x)* is the line's characteristics impedance as a function of distance (x), then the Reflection Coefficient on the Input side is expressed as [1]:

$$\Gamma_i = \int_0^L \exp(-j2\beta x)\left\{\frac{d\ \ln(\overline{Z(x)})}{dx}\right\} dx \quad (1)$$

where

$$\bar{Z}(x) = \frac{Z(x)}{Zi}$$

For transmission lines with a Z(x)/Zi distributed Triangular (Triangular Taper) with a mathematical model of the form [1]:

$$\frac{d}{dx} ln(\bar{Z}(x)) = \begin{cases} \frac{4x}{L^2} \ln \overline{Z_L} & 0 \leq x \leq \frac{L}{2} \\ \frac{4}{L^2}(L-z) \ln \overline{Z_L} & \frac{L}{2} \leq 0 \leq L \end{cases} \quad (2)$$

where:

$$\overline{Z_L} = \frac{Z_L}{Z_i} \quad (3)$$

then the Reflection Coefficient on the Input side can be expressed as:

$$\Gamma_i = \frac{1}{2}\exp(-j\beta L)\, ln\bar{Z}_L \left[\frac{\sin(\beta L/2)}{\beta L/2}\right]^2 \quad (4)$$

To get the value of βL as a function of Γi in a closed manner is something difficult to do so that the easiest method to design Matching Impedance using Non Uniform Channels with Triangular distribution can be used the Graph method as follows [2]:

$$\left|\frac{2\Gamma_i}{ln\bar{Z}_L}\right| = |\rho| = \left[\frac{\sin(\beta L/2)}{\beta L/2}\right]^2 \quad (5)$$

or in graphic form as stated in Figure 2. below

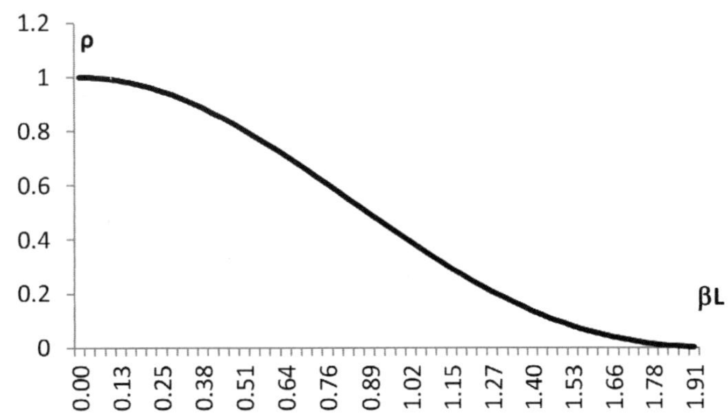

Figure 2. The coefficient of reflection vs βL.

3. Impedance adjustment system design

The design procedure for the Impedance Adjustment System using a Non-Uniform Transmission Line with a Triangular distribution can be stated as follows:

- Determine first the load impedance ratio (ZL) and the desired input impedance to obtain Equation 5.
- Determine the maximum value of the reflection coefficient ρ desired and with Graph 2 determine the value of βL.
- Find β with the equation β = (2πf√ (ε_r)) / c. By determining β and βL, the value of L (channel length) can be determined.

If the system is implemented on a microstrip or stripline board, then the value of Z(x) or the Line Characteristics Impedance needs to be expressed in the form of width (w) and strip length (l). In this study, the length line is divided into 3 (three) different specimens to obtain information on the effect of the number of division segments on the designed performance of the Matching Impedance. The number of segments divided is 4 (four) segments, 8 (eight) Segment) and 16 (sixteen) segments.

4. System testing

From the test results on 3 (three) specimens, namely: 4, 8 and 16 segments for the value $\overline{Z_L} = 2$, the working frequency is 2.4GHz, βL = 4.6 rad, the following results are obtained:

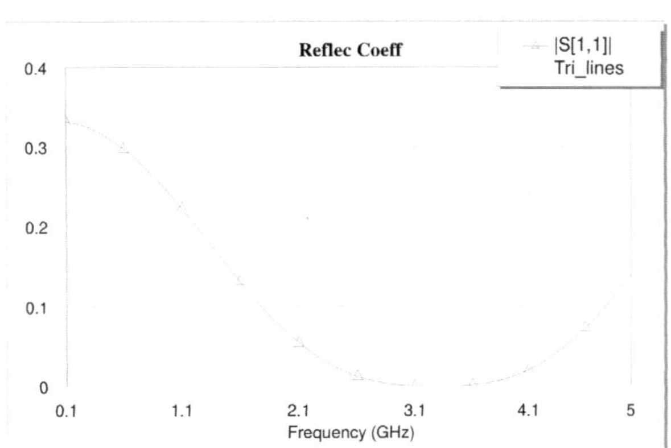

Figure 3. Segment division 4th for βL = 4.6 rad.

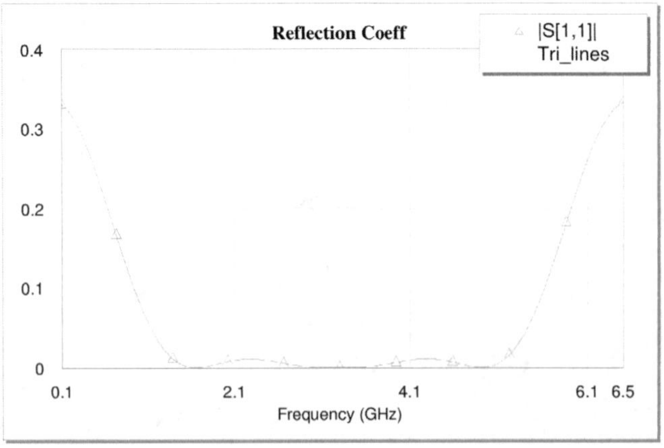

Figure 4. Segment division 8th for βL = 4.6 rad.

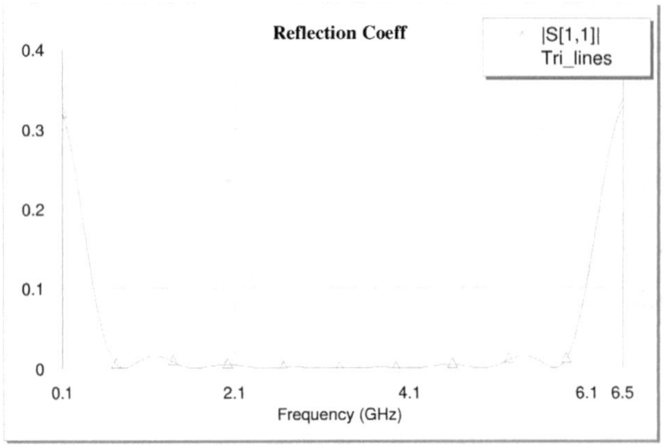

Figure 5. Segment division 16th for βL = 4.6 rad.

From Figure 3, Figure 4 and Figure 5, the width of the Matching Impedance system use can be shown as follows:

Table 1. Number of segment over working area comparison table

No	Segment	Bandwidth $\rho < 0.1$
1	4	3 GHz
2	8	4.6 GHz
3	16	5.5 GHz

In the Table 1 above, it can be seen that the more segments are divided for a fixed length L, the wider the area below the value $\rho = 0.1$ will be. This means that the smoother the Z(x) distribution, the wider the working area of the system.

5. Conclusion

- Impedance adjustment system using Non-Uniform Transmission Line with Triangular Distribution offers a wide range of system usage compared to a quarter-wave transformer system.

- The greater the segment division for a fixed βL value (the gradation level of the Characteristic Impedance distribution is getting smoother), the Matching Impedance system is designed to show a wider range of use.

References

[1] Collin 1996 *Foundations for Microwave Engineering* (New York: Mc Graw Hill)

[2] Mathur S P 1987 Design of MicroStrip Exponentially Tapered Lines To Match Helical Antenna to Standart Coaxial Lines *IEEE Proceding* **135**

[3] Urbani 2005 Synthesis of Filtering Structures for Microstrip Antennas using Orlov's Formula *Journal ETRI* **27**(2)

IOP Conf. Series: Materials Science and Engineering 1073 (2021) 012029 doi:10.1088/1757-899X/1073/1/012029

Authenticity of money using the method KNN (K-Nearest Neighbor) and CNN (Convolutional Neural Network)

C Rahmad[1,*], E Rohadi[1] and R A Lusiana[2]

[1] Department of Information Technology, State Polytechnic of Malang, Malang, Indonesia
[2] Department of Electrical Engineering, State Polytechnic of Malang, Malang, Indonesia

*cahya.rahmad@polinema.ac.id

Abstract. Nowadays, the circulation of counterfeit money is significantly increased, encouraging to conduct research related to genuine real money detection or counterfeit money based on digital imagery. Technological sophistication one of the quality of printers whose ink is very good and can print money like the original makes the layman should be more wary of money ownership. In this research conducted the authenticity of money using the method KNN (K-Nearest Neighbor) and CNN (Convolutional Neural Network). Accuracy KNN method is 87,75%. While the detection accuracy used by CNN is 96,67%. The results obtained from these 2 methods can still be improved with advanced research namely with pre production on the set and the image used. The data set used has the same exposure level, image capture angle and image size.

1. Introduction

Counterfeit money has become a serious problem in every country. The circulation of counterfeit money increased with the advancement of technology there. The more rampant circulation of counterfeit money, Bank Indonesia has designed banknotes that can be differentiated between genuine and counterfeit money. It is done by Bank Indonesia to distinguish the authenticity of money is to provide a wide variety of attributes, such as the size of the paper varied, different colors, and different patterns. With these differences, it is expected that the community can check the money manually. Manual techniques can be done by doing three things, namely seen, touched, and overlay. However, this method is not considered efficient on the amount of a lot of money because of the transaction with the money usually goes fast, while to do the checking manually takes a long time, so needed another way which is quick and efficient [1].

To overcome these problems, researchers have designed an algorithm by using the image processing as the basis of the detection system of counterfeit money. The algorithm of this system is expected can solve the problem of detection of counterfeit currency and can be used with financial transactions in the amount of a lot.

On this opportunity, the researchers have collected a variety of references related. For the example is book digital image processing and its programming techniques [2] it explains how to observe an image with the object of what was reviewed from the texture of the objects that will be observe. Perform the measurement of the texture with the contrast of colors is able to achieve an average accuracy of

74%. In distinguishing genuine and counterfeit money, Ginting [3] conducted research with edge detection using canny method. The results from the experiments performed show that the application successfully detects the sign of the water while the image of the counterfeit banknotes and a copy of the application does not successfully detect the watermark. In this study, the detection of the authenticity of the money by using 2 methods: method KNN (K-Nearest Neighbor) and CNN (Convolutional Neural Network).

Identification with the KNN method (K-Nearest Neighbor) using the color feature on the introduction of the value of banknotes. The colors provide meaningful information and values that are important in the process of describing an object [4]. The colors used are Red, Green, Blue (RGB). The process of detection of the value of banknotes using the value of the feature extraction, the Red, Green, Blue (RGB), then the result of the extraction of features used to classify the value of banknotes by the method of K-Nearest Neighbor (KNN) as the detection value of banknotes. There are several methods used for classification such as Naïve Bayes method [5], but in the Naïve Bayes method is based on the probability or conditional probability. Suyanto, in contrast with KNN method which calculates the distance to the nearest classification [6].

Research related to the detection of the value of banknotes with the KNN (K-Nearest Neighbor) have been conducted, such as research Hlaing and Gopalakrishnan [7] using texture features Gray-Level Co-Occurrence Matrix (GLCM) with a KNN method with a value of K=1. The hope of research conducted to build a model of the detection value of banknotes rupiah extraction using color feature and using the KNN method with a K value greater than the previous research to produce a good accuracy that can detect the value of banknotes with right as well as beneficial for those who need.

Identification method CNN (Convolutional Neural Network) with the cluster analysis K-Means Clustering which leads to the partitioning of N objects observations into K groups (clusters) where each observation object is owned by a group with a mean (average) closest. This method have been done by researchers Syaiful and Ahsan [8]. Application Recognition Imaging On bank notes To Know the money is counterfeit or original can show the authenticity of the money by the method of K-Means clustering using java programming on the hologram idr. 100.000. However this App can only help others in knowing the money idr. 100.000 with the description “accordance” or “not in accordance” with the results implemented on K-Means Clustering [8].

Research related to detection of banknotes with CNN (Convolutional Neural Network) is to detect the number of banknotes using the Image Processing Raspberry Pi. The method used in this research is digital image segmentation using Thresholding method with Python programming language. The results of this Study is to detect the number of banknotes using Image Processing in Raspberry Pi, where banknotes can be distinguished based on color [9].

2. A brief on related theories

2.1. KNN (K-Nearest Neigbor) method

For the research using the KNN (K-Nearest Neighbor) method can be seen in figure 1:

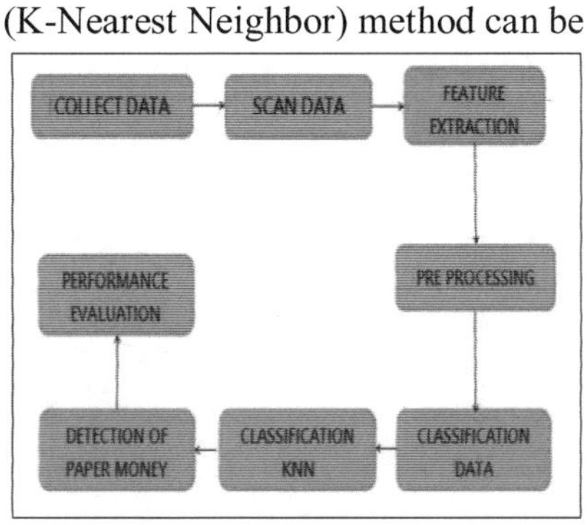

Figure 1. Process KNN method.

2.1.1. Data collection. Data used use of banknotes fractional to idr. 2000 new model, idr. 2000 old model, idr. 5000 new model, idr. 5000 old model. The number of each banknote is 10 selected from the start that the quality of the good until the quality of the crumpled because of the factor of the old or the movement of money from one hand to the other.

2.1.2. Scan data. The stage scan is done to capture the image of the banknotes using the features RGB color. Using scan the scanner tool so the authenticity and color on banknotes become more clear. On banknotes there are two different sides, namely the side of the face and the back side, so do a scan against the banknotes with two different sides. Repetition performed three times. Scan with a repetition of three times is performed to determine whether each of the stages of the scan are different have different results.

2.1.3. Feature extraction. Feature extraction is necessary to determine an image. The technique of feature extraction that is used to extract the color image RGB (Red-Green-Blue). Retrieval of color features using RGB applications in Graphic Design.

2.1.4. Preprocessing. The results the extraction of features on banknotes is the value of every feature the color Red-Green-Blue. There is the same value on the value of the feature RGB so that made the cleaning process against the same data, in addition, data processed with the draw patterns in the same class and distanced patterns of different classes so that the patterns will be cataloged and neat and separate from other classes [10].

2.1.5. Data sharing. Data Sharing used for classification on the results of detection banknotes. The model of classification takes an input learning data (training data) and testing data.

2.1.6. Classification K-Nearest neighbor. KNN (K-Nearest Neighbor) including algorithm classification is a supervised learning which classifies an object into a new class based on the training data and the attributes of the class [11]. The method of KNN (K-Nearest Neighbor) on the testing data will compare data that has been studied, namely the learning data (training data). This method is very dependent on the value of K, so one of the initial stage is to determine the value of K. Next calculate the distance from new object to data training, before the determined category of object based on the majority of value of K is made sequence data of the testing data that have the smallest distance from training data.

2.1.7. Detection banknotes. Detection is performed on the file of paper money which has been scanned as a new data to be determined banknotes is included in one of four categories class (class) who have previously carried out learning data (training data). This research uses the application of R to detect, the categories of detection of the banknotes i.e. the money 5000A to the value of banknotes idr. 5000 the old model, the 5000B to the value of banknotes idr. 5000 new model, 2000A to the value of banknotes idr. 2000 old model and 2000B to the value of the banknotes idr. 2000 new model.

2.1.8. Performance evaluation. Puidrose of performance evaluation is evaluate and analyze the model used in the study. Process of calculating the accuracy of classification results using the Confusion matrix. Confusion matrix is a table that contains the results of predicted with the true and also not true by modeling classification, this table is necessary to determine the performance of classification model.

2.2. CNN (Convolutional Neural Network) method

For the research using the CNN (Convolutional Neural Network) method can be seen in figure two:

IOP Conf. Series: Materials Science and Engineering 1073 (2021) 012029 doi:10.1088/1757-899X/1073/1/012029

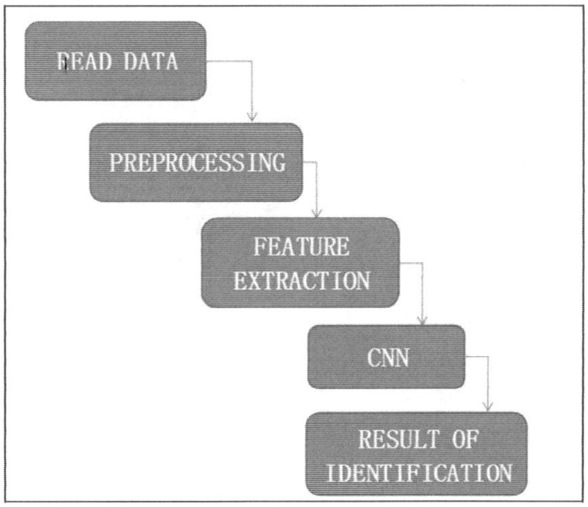

Figure 2. Process CNN method.

This research was conducted to identify the authenticity banknotes by converting color image into grayscale then the results of a new image to extraction of features in order to become a reference in detecting whether or not the original banknotes. As for the object that is used is idr. 50,000 and idr. 100,000 identify that surface texture of banknotes through the binary pattern can determine the authenticity of money accurately.
Explanation of figure 2 is:

2.2.1. Read data money. In this research, the first performed reading of the data. Process read data is a process for reading image data that has been captured, and saved in the format of *jpg.

2.2.2. Preprocessing. The stage of preprocessing is done before performing pattern recognition on image the money, at the beginning of process, the image of money will be converted to format of image grayscale (grayscale). This process to get new image with better results and is ready to be processed in the next stage.

2.2.3. Extraction of features. The next stage is performing pattern recognition. At this stage, the feature extraction of image money. Method using local binary pattern to find out the texture the surface banknotes become a reference in detection of the number value banknotes.

2.2.4. Results identification. By doing the technique of testing the accuracy by utilizing the whole data set. So the obtained result of detection genuine and counterfeit money with a level of accuracy that is able to achieve 96,67%.

3. Results and discussion

3.1. Result of KNN (K-Nearest Neigbor)
Results of research using the KNN (K-Nearest Neighbor) method is

3.1.1. Result of scan banknotes. Banknotes has two different sides, so the scan is carried out two times against the sides of retrieval feature extraction color. Examples scan paper money with three times the experiment shown in figure 3:

IOP Conf. Series: Materials Science and Engineering 1073 (2021) 012029 doi:10.1088/1757-899X/1073/1/012029

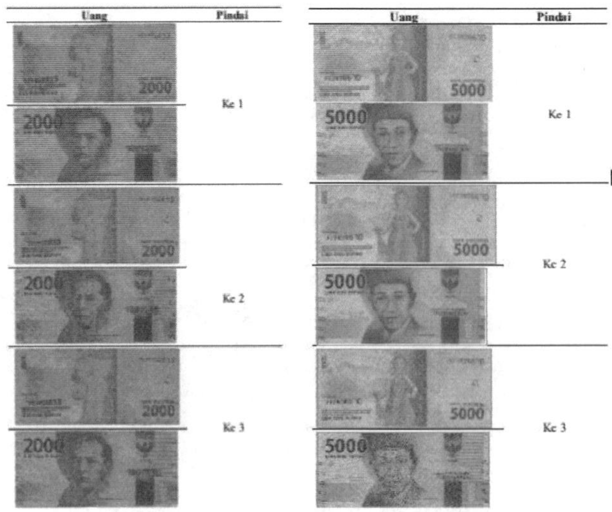

Figure 3. Result of scan banknotes.

Results of scan carried out repeated three times to produce different values for color features RGB, but there is also contained the same value between scan 1 and 2 or scan 2 and scan 3.

3.1.2. Result of feature extraction. Result of feature extraction in paper currency by issuing a value of RGB (Red, Green, Blue). The results of value color feature extraction is shown in figure 4:

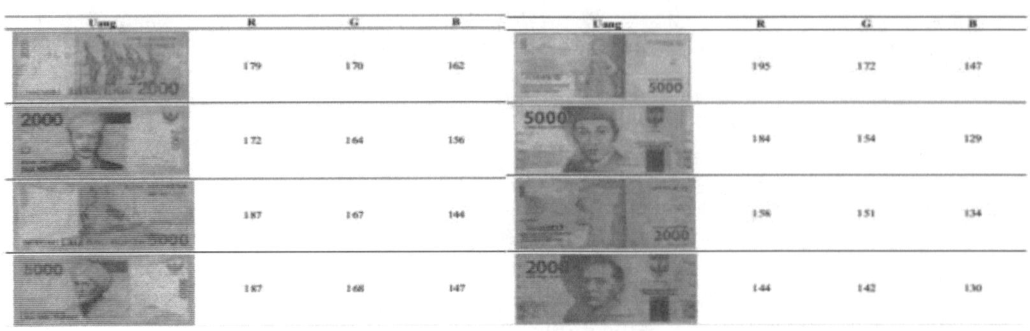

Figure 4. Result of feature extraction.

3.1.3. Result of preprocessing. Banknotes that has been scanned with the repetition of three times the same values that carried out a purge of the same value. Shown in table 1:

Table 1. Scanned with the repetition of three times the same values that carried out a purge of the same value

Banknotes 5000 old money									
	Scan Data 1			Scan Data 2			Scan Data 3		
Money	R	G	B	R	G	B	R	G	B
Money 1 back	187	167	144	188	167	145			
Money 1 front	187	168	147	188	169	147			
Money 2 back	167	144	125	168	145	126			
Money 2 front	168	145	124	168	145	125	168	146	125
Money 3 back	171	152	133	172	152	133	172	152	134
Money 3 front	174	154	135	174	155	136	175	155	136
Money 4 back	185	163	142	186	163	142			
Money 4 front	185	163	141	184	163	140	185	163	141
Money 5 back	195	172	147						
Money 5 front	195	177	153	195	177	152			
Money 6 back	163	139	119				164	140	119
Money 6 front	165	143	122						
Money 7 back	193	171	148	195	172	148	194	171	148
Money 7 front	192	173	151	193	174	151	192	173	150
Money 8 back	171	148	129	170	148	128	172	149	129
Money 8 front	170	148	128	169	147	128	170	147	128
Money 9 back	186	166	145	188	167	146	187	166	145
Money 9 front	186	165	145	188	166	146	187	165	145
Money 10 back	192	172	150				193	173	150
Money 10 front	190	173	151	190	173	152			

In table 1 there is a null value of RGB is the result of cleaning the same data. Next draw patterns in the same class and distanced patterns of a different class, shown in table 2:

Table 2. Patterns in the same class and distanced patterns of a different class.

R	G	B	Money
163	139	119	5000 A
164	140	119	5000 A
165	143	122	5000 A
167	144	125	5000 A
168	145	126	5000 A
168	145	125	5000 A
171	142	120	5000 B
172	142	119	5000 B
173	143	119	5000 B
174	141	118	5000 B
177	154	129	5000 B
177	146	122	5000 B
159	153	146	2000 A
159	146	136	2000 A
160	147	136	2000 A
160	145	132	2000 A
160	145	133	2000 A
162	147	133	2000 A
143	140	124	2000 B
144	142	130	2000 B
144	139	123	2000 B
144	142	129	2000 B
146	147	136	2000 B
148	150	137	2000 B

Data are separated into four classes (class) according to type of money that has been extracted. Here's an explanation of the some classes is 5000A: banknotes idr. 5000 old model, 5000B: banknotes idr. 5000 new model, 2000A: banknotes idr. 2000 old model and 2000B: banknotes idr. 2000 new model.

3.1.4. Results detection of banknotes. Results of testing detection value of banknotes using the 16 test data conducted on 160 training data. Process detection is done using K-Nearest Neighbor method with value of K=5. Results from testing detection using the R programming shown on figure 5:

```
               Accuracy : 0.8775
                 95% CI : (0.6977, 0.9984)
    No Information Rate : 0.25
    P-Value [Acc > NIR] : 1.141e-08

                  Kappa : 0.9167
```

Figure 5. Result detection of banknotes.

3.2. Result of CNN (Convolutional Neural Network)
Results of reasearch using the CNN (Convolutional Neural Network) method is

3.2.1. Read data money show in figure 6

Figure 6. Data money.

3.2.2. Result of conversion grayscale show in figure 7

Figure 7. An example conversion grayscale.

3.2.3. Result of feature extraction local binary pattern show in figure 8

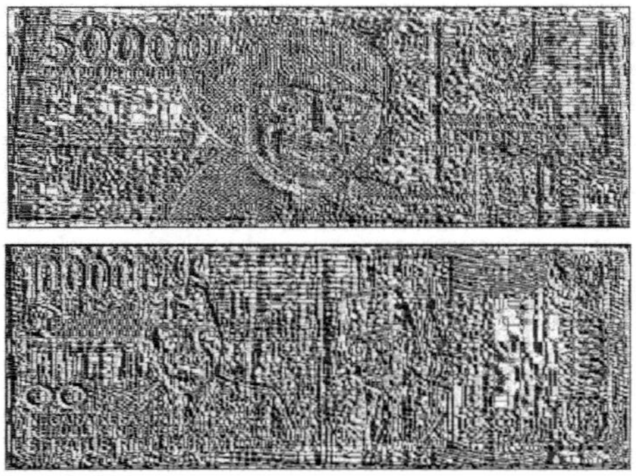

Figure 8. An example extraction local binary pattern.

3.2.4. Result of Identification

Final results obtained from testing each image of banknotes is 96,67% with image size 352 x 156 pixels. Results of identification test with the variation of accuracy different can be seen in table 3:

Table 3. Results of identification with image size 352x156.

Currency	Original Money	Counterfeit Money	Accuration
50k Ori	29	1	96,67%
50k Counterfeit	2	18	90%
100k Ori	27	3	90%
100k Counterfeit	1	23	95,8%

From the results of trials conducted in table 3, that the shaidrness in capture early through the camera smartphone greatly affect the detection of the image banknotes. Cleaning background on image the currency also affects the level of accuracy that will be obtained. Image data 50k Original, has the highest accuracy compared to image of the other because process cropping background image is done manually so that detection image of new focus to texture of surface resulting from the extraction the feature pattern on the image.

The end result without cropping the image on background image, where the accuracy of system average accuracy is 25% as seen in table 4:

Table 4. Result of identification without croping image.

Currency	Original Money	Counterfeit Money	Accuration
50k Ori	2	3	40%
50k Counterfeit	6	5	45%
100k Ori	1	3	25%
100k Counterfeit	3	1	25%

From the results of trials conducted in table 4, without process of cropping image is also very affect the value of accuracy the image money. Because the background original image of money is detected as a pattern so that authenticity of paper money difficult to detect. As on the image data 100k original and 100k false, have the same accuracy in the below average standards of accuracy, i.e. 25% as compared with the image that has been through the process of cropping. For the time testing with the 5 data is 7 seconds. So more less amount data are taken as testing sample, the faster the time of testing. There is a difference long time testing between each genuine and counterfeit money. For 50k original with 6 data testing time 8 seconds, with the low accuracy 40%. As for the 50k counterfeit with 11 data testing time 13 seconds and the accuracy is also low, namely 45%.

4. Concluding remarks

Based on the results of tests performed with two methods KNN (K-Nearest Neighbor) and CNN (Convolutional Neural Network). The accuracy obtained with the KNN method is 87,75%. While the detection accuracy with CNN is 96,67%. Of the two methods is affected when capture image at the beginning and pixel size of the image.

References

[1] Bank Indonesia 2016 *Easy To Identify The Authenticity Of Money (in Indonesia) 2016* (Jakarta: Management Department Bank Indonesia)

[2] Ponishjino P, Antony K, Kumar S and JebaKumar S 2017 Bogus Currency Authorization Using Hsv Techniques *2017 International conference of Electronics, Communication and Aerospace Technology (ICECA)* **1** 179-183

[3] Ginting E D 2009 Edge detection Using the Canny Method With Matlab To set real Money and Counterfeit Money (in Indonesia) *The Scientific Journal Information Technology Computer* **22**(3)

[4] Doush I A and Sahar A B 2017 Currency recognition using a smartphone: Comparison between color SIFT and gray scale SIFT algorithms *J. King Saud Univ. - Comput. Inf. Sci.* **29(**4) 484–492

[5] Santoso E B and Nugroho A 2019 Sentiment Analysis Of Candidates For President Indonesia, (in Indonesia) 2019 B *Jurnal Eksplora Informatika* **9**(1) 60–69

[6] Suyanto 2019 Data Mining For Classification and Data Clustering (in Indonesia). Informatic Technology Bandung

[7] Hlaing K N N and Gopalakrishnan A K 2016 Myanmar paper currency recognition using GLCM and k-NN *2016 2nd Asian Conf. Def. Technol ACDT 2016* 67–72

[8] Syaiful A M and Ahsan M 2014 Recognition Imaging Paper Money to Determine the Authenticity of Money (in Indonesia) *Journal Students Faculty of Science and Technology* **1** (5)

[9] Jalil A 2014 Control System Detection Banknotes Using Image Processing Raspberry Pi (in Indonesia) *Journal IT Stmik Handayani* **14**

[10] Sumari A D W, Sereati C O, Syamsiana I N, Wibisono M N, Abdulharis R, and Putra D H P 2020 Cognitive artificial intelligence decision support system based on knowledge growing system *International Journal of Advanced Science and Technology (IJAST)* 29 7s (Special Issue) pp 3724-3730

[11] Sumari A D W, Sereati C O, Syamsiana I N, Wibisono M N, Abdulharis R, and Putra D H P 2020 Cognitive artificial intelligence decision support system based on knowledge growing system *International Journal of Advanced Science and Technology (IJAST)* 29 7s (Special Issue) pp 3734-3743

IOP Conf. Series: Materials Science and Engineering 1073 (2021) 012030 doi:10.1088/1757-899X/1073/1/012030

Smart ecosystem for hydroponic land in the hydroponic farmers group guided by CSR PT. Otsuka Indonesia as an improved quality and quantity of harvest results

D K P Aji*, U Nurhasan, R Arianto and O D Triswidrananta

Department of Information Technology, State Polytechnic of Malang, Jalan Soekarno Hatta No.9, Jatimulyo, Kec. Lowokwaru, Kota Malang, Jawa Timur, Indonesia 65141

*deddy_kusbianto@polinema.ac.id

Abstract. Corporate Social Responsibility (CSR) is a community development program managed by PT. Otsuka Indonesia, one of which is form a hydroponic farmer community. In the activity of fostering hydroponic farmers there are several problems, namely the process of plant nursery, seeding, and harvesting are still carried out conventionally, while the assisted farmers do not have much time to supervise because they have a core job. Based on these problems, this study will focus on solving the problem of hydroponic plant supervision that can be known directly by farmers and also from the CSR of PT. Otsuka Indonesia. In completing the research, the steps carried out were based on the CRISP-DM (Cross Industry Standard Process for Data Mining) methodology. The factors used in solving the problem are the type of hydroponic plants, the acidity of the soil, the intensity of sunlight, the ambient temperature of the hydroponic plants, the water level and periodic reporting as needed in the form of a Smart Ecosystem. For this reason, it is necessary to build a device from an Arduino microcontroller based on IoT so that supervision of the hydroponic planting process can be done without having to be in hydroponic land and get reports every period of the development of hydroponic plants in the application. By using this device, it is hoped that many hydroponic farmers will use it so that during the hydroponic vegetable production process it can be in accordance with the old standards of the nursery until it is worth selling.

1. Introduction

Corporate Social Responsibility (CSR) or what is commonly referred to as Social and Environmental Responsibility (TJSL) has been regulated in Article 1 Number 3 of Law Number 40 of 2007 concerning Limited Liability Companies (Government of the Republic of Indonesia, 2007) describes the company's commitment to participate in sustainable economic development in order to improve the quality of life and a beneficial environment, both for the company itself, the local community, and society in general [1].

To fulfill the obligations of the PT Law, PT. Otsuka Indonesia makes CSR activities including: (1) Public Health (Sehat with Otsuka), (2) Community Environment (Go Green with Otsuka), (3) Community Education (Smart together with Otsuka), and (4) Disaster Nature (Rise Together with Otsuka). These CSR activities began in 2008. One of the activities of the Community Environment program (Go Green with Otsuka) is to form a hydroponic farmer community under the guidance of PT.

Otsuka Indonesia is located in Kalirejo Village, Lawang District, East Java, as many as 7 RTs from RW 04 and RW 08.

With the development of the hydroponic farming community, several problems have begun to emerge, including: market needs for hydroponic vegetables in the Lawang Market, and managing hydroponic plants that have low quality so that they only become waste for farmers.

Based on some of these problems, research in this first year will focus on solving the problem of hydroponic plant supervision that can be known directly by farmers and also from the CSR of PT. Otsuka Indonesia. The factors used in solving the problem are the type of hydroponic plants, the acidity of the soil, the intensity of sunlight, the ambient temperature of the hydroponic plants, the water level and reporting to CSR for a certain period according to needs in the form of a Smart Ecosystem. Because hydroponic farmers who are assisted by CSR PT. Otsuka Indonesia's number continues to grow, so a special hydroponic microcontroller will be built. By using a cheap microcontroller, namely using Arduino compared to using the Raspberry Pi, it is hoped that many hydroponic farmers will be interested in using it so that during the hydroponic vegetable production process it can match the old standards of the nursery until it is worth selling and the CSR of PT. Otsuka Indonesia can determine what steps to take if there is hydroponic land that does not comply with the standard.

Internet of Things, also known as IoT, is a concept that aims to expand the benefits of continuously connected internet connectivity. IoT is a concept that uses the internet as the main infrastructure network that connects certain objects [2]. As for abilities such as data sharing, remote control, and so on, including objects in the real world. For example foodstuffs, electronics, collections, any equipment, including living things which are all connected to local and global networks through embedded sensors and are always active [3]. The development of IoT for automation has been developed in various fields, in agriculture, for example. Lan-Da Van, et al. In 2019 developed a mobile-based application to monitor plant growth [4]. In this study, it was proven that an artificial intelligence could be implanted in a monitoring application to support proper analysis related to plant growth. In another study conducted by usman, et al. In 2018. That the IoT concept can also be applied to hydroponic agriculture [5]. This research focuses on controlling water circulation in hydroponic growing media of the DFT (Deep Flow Technique) type. As reinforcement of the basic research is the results of research conducted by Carlos Cambra, 2018 entitled "Smart System for Bicarbonate Control in Irrigation for Hydroponic Precision Farming [6]. In this study, Cambra monitored the availability of bicarbonate nutrients in the DFT type hydroponic growing medium.

2. Method

In completing the research, the steps were carried out based on the CRISP-DM methodology in the Figure 1 below:

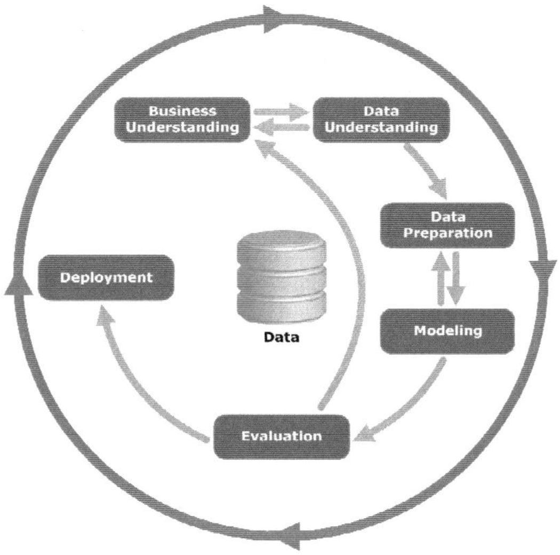

Figure 1. CRISP-DM (*Cross – Industry Standard Process for Data Mining* [7].

In 1999, sizable companies included automaker Daimler-Benz, insurance provider OHRA, hardware and software manufacturer NCR Corp and maker of statistical software SPSS, Inc. began working together to formalize and standardize approaches to data mining [8]. The result of their work is CRISP-DM, the Cross-Industry Standard Procedure for Data Mining which contains 6 stages, namely:

2.1. Business understanding
At this stage, an understanding of the ongoing business processes will be carried out. Where, to obtain this information, an interview process with the CSR PT. Otsuka Indonesia and visit the hydroponic farmer area directly so that they will get an accurate understanding of the ongoing business processes and the problems faced by the assisted farmers.

2.2. Data understanding
By knowing the problems faced by hydroponic farmers assisted by PT. Otsuka Indonesia, data will be collected which will be needed to develop solutions that can be offered, whether the data taken from the CSR of PT. Otsuka Indonesia, hydroponic farmers, and what kind of output from the system are needed by both parties so that it can become a solution that is right on target.

2.3. Data preparation
The data that has been collected will be carried out equalization or alignment both in terms of data type, data format, or deletion of data in a process so that the data used for the next stage is really the data needed and is ready to be processed.

2.4. Modelling
At this stage, modeling will be carried out from the data that has been prepared by determining the type of modeling that matches the existing data criteria. The modeling used later must have a low error rate and not have a long work process because the model will be applied to smartphones and the IoT.

2.5. Evaluation
The evaluation stage is to evaluate the modeling that has been carried out before it is implemented on the research object. At this stage, an evaluation will be carried out by building a prototype in accordance with the conditions in the field so that any mistakes can be found and can be corrected as soon as possible.

IOP Conf. Series: Materials Science and Engineering 1073 (2021) 012030 doi:10.1088/1757-899X/1073/1/012030

2.6. Deployment
At this stage, a prototype will be implemented that has been evaluated on the object of research with the actual situation in the community and training will be carried out for the use of tools, monitoring of tools, and maintenance of tools for a certain period of time.

3. Implementation
The water circulation monitoring and control system that was built aims to provide convenience to hydroponic activists in monitoring and controlling hydroponics. The initial process carried out by the system is that the sensor is placed on hydroponics to detect the value of the sensor data in accordance with the parameters needed for hydroponic mustard plants. After getting data from sensors such as temperature, humidity, pH and height of the reservoir water, it will be stored in a database and displayed on the website as the output of the monitoring process. Then the sensor data that has been obtained will be processed to determine the process of controlling the hydroponic water circulation. In this control process, the data used are temperature and humidity which will be processed using the Sugeno fuzzy method in determining the right time to flow hydroponic water. The figure 2 is the design of the tools used.

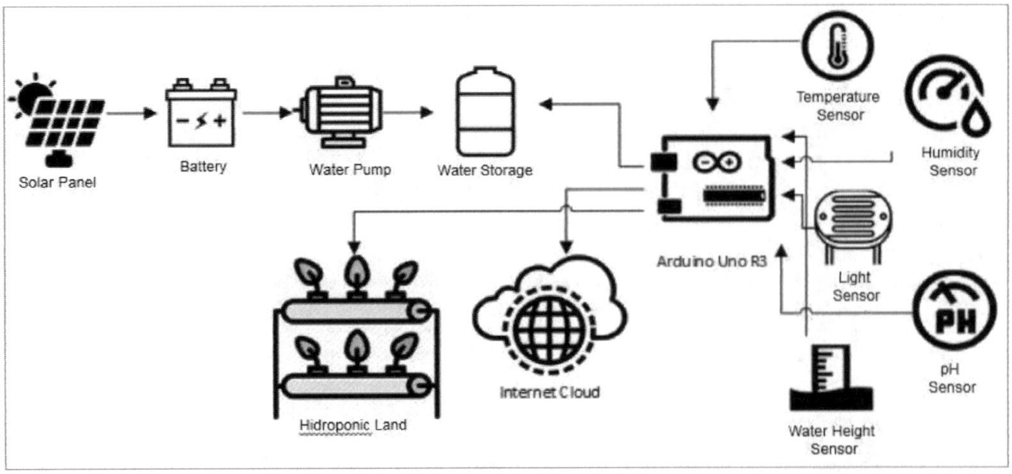

Figure 2. Device design.

Table 1. Device specifications.

Object	Information
Solar Panel	Polycrystal 100 Wp
Battery	Min. 12 Volt / 12 Ah
Water Pump	DC 12 Volt, 50 - 70 Liter / minute, Maximum height 4 - 5m, 5500 - 5800 rpm
Water Storage	Size 70-120 liters with a maximum height of 800 mm
Microcontroller	Arduino Uno R3, Temperature Sensor, Humidity Sensor, Light Sensor, pH Sensor, Water Height Sensor
Hidroponic Land	Land size between 80-100 holes, Spinach and PakChoy plant types, Height of 1 - 5 meters, Pipe size 2 "
Internet Cloud	Use the free version (ThinkSpeak)

In the figure 3, the monitoring and controlling system for hydroponic water circulation begins by taking values from several sensors. All sensor data that has been obtained is entered into the database. Temperature and humidity data will be used in the fuzzy process, then the system will turn on the water pump and turn off the water pump automatically according to the final result of the fuzzy calculation process. The results of data collection from several sensors will be displayed on the website as a parameter for monitoring plant growth elements in hydroponics. In addition, the fuzzy calculation results are also displayed along with the existing pump conditions. The following is the workflow of a fully automatic hydroponic water circulation control system using the fuzzy method:

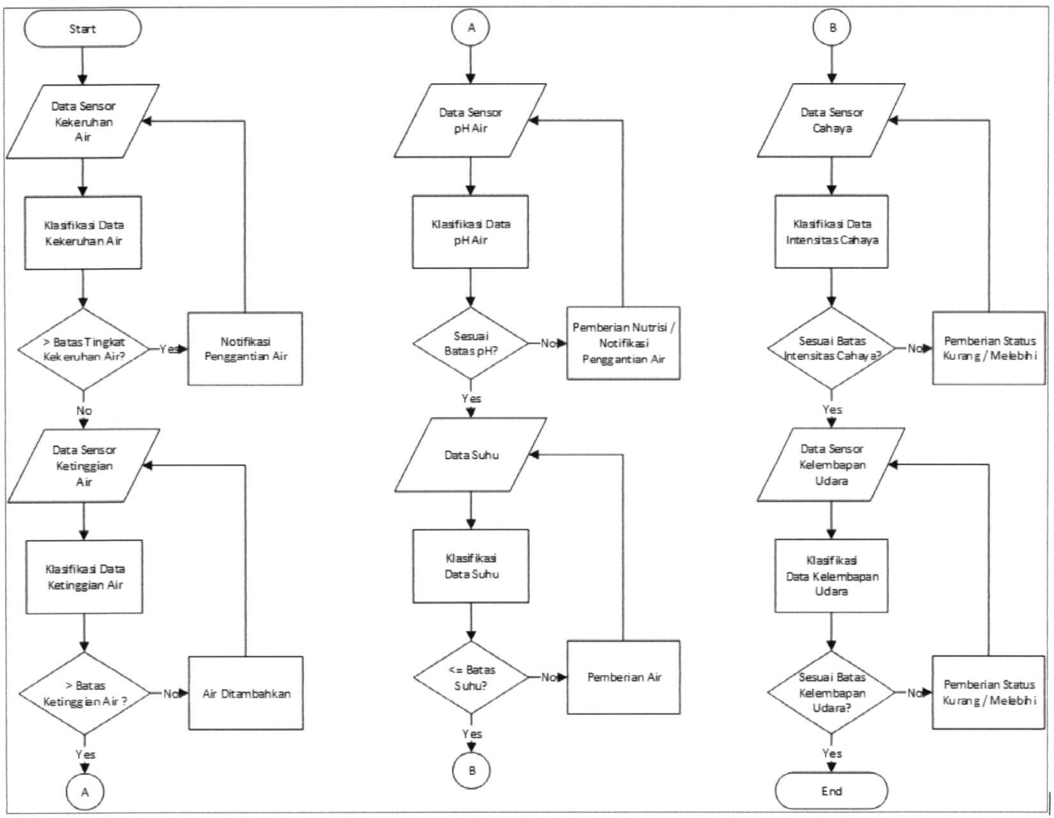

Figure 3. Fuzzy flowchart on device.

The method applied to the system of monitoring and controlling water circulation in hydroponic mustard is the Fuzzy method. The steps in calculating the fuzzy method are the first, determining the crisp value or the firm value of the values obtained by the sensor in the form of temperature and humidity values. At the fuzzification stage a membership function is obtained. Then perform the fuzzy logic operation and its implication in the form of a min function, and determine the output in defuzzification [9]. Here is a rule based on fuzzy logic.

3.1. Data obtained from the water turbidity sensor

- Starting from 4.5 - 4.58 Volts, the notification to the user is "Very Turbid" and gives a status of "Very Turbid"
- If 4.59 - 4.64 Volt, the notification to the user "Turbid" and gives the status "Turbid"
- If the result is more than equal to (≥) 4.65 Volts it will give a status of "Clean"

3.2. Data from the water level sensor in the Water Storage Container

- The water level between 0 - 350 mm will fill the water and give the status of "very less water".
- The water level between 351 - 749 mm will fill the water and give the status "Water less".

- The water level is more than 750 mm, it will give a status of "enough water" and stop water filling.

3.3. Data obtained from the water pH sensor
- Shows less than 6.0, the notification "Change Water" will appear.
- Showing more than 7.0, the notification "Add Nutrition" will appear.
- Showing between 6.0 - 7.0, it will give a notification "Normal"

3.4. Data obtained from the temperature sensor
- Shows less than 16 ° C will give the status "Cold temperature"
- Shows more than 20 ° C it will give a status of "Heat Temperature" and will turn on the water pump to drain the water.
- Showing between 16 - 20 ° C will give the status "Normal Temperature"

3.5. Data obtained from the light intensity sensor
- Showing less than 32 Lux will give you the status of "Lack of Light"
- Showing more than 108 Lux will give you the status of "Excess Light"
- Showing between 32 - 108 Lux will give you the status of "Normal Light"

3.6. Data obtained from the Air Humidity sensor
- Shows less than 40% it will give the status "Humidity Less"
- Shows more than 60% it will give the status "Excess Humidity"
- Shows between 40 - 60% it will give the status "Normal Humidity"

4. Conclusion
The results of the sensor data can control the circulation of water using a pump in hydroponics in accordance with a system that has been previously designed. Then the ultrasonic proximity sensor that is used to detect the height of the water in the pipe and reservoir can also run well can automatically start the pump when the water level in the pipe is less than 3 cm.

This research has succeeded in making a system that is used for monitoring and controlling the circulation of mustard hydroponic water using a water pump automatically with fuzzy logic through a smartphone interface.

Based on observations on the growth of mustard plants which were compared between using a monitoring system and controlling water circulation by not using the system, resulting in a significant difference in height growth results and number of leaves.

References
[1] Pemerintah Republik Indonesia 2007 Undang-Undang Republik Indonesia Nomor 40 Tahun 2007. Indonesia.

[2] Patel K K, Patel S M and Scholar P G 2016 Internet of Things-IOT: Definition, Characteristics, Architecture, Enabling Technologies, Application & Future Challenges *International Journal of Engineering Science and Computing* **6**(5) 1–10

[3] Kumar M 2017 A Review on Applications of Internet of Things *International Journal of Advanced Research in Computer Science and Software Engineering* **7**(6) 756–760

[4] Van L D, Lin Y B, Wu T H, Lin Y W, Peng S R, Kao L H and Chang C H 2019 Planttalk: A smartphone-based intelligent hydroponic plant box *Sensors* (Switzerland) **19**(8)

[5] Nurhasan U, Prasetyo A, Lazuardi G, Rohadi E and Pradibta H 2018 Implementation IoT in System Monitoring Hydroponic Plant Water Circulation and Control *International Journal of Engineering & Technology* **7**(4.44) 122-126

[6] Cambra C, Sendra S, Lloret J and Lacuesta R 2018 Smart system for bicarbonate control in irrigation for hydroponic precision farming *Sensors* **18**(5), 1333.

[7] Jensen K 2012 IBM SPSS Modeller CRISP-DM ガイド.
[8] North M 2012 *Data Mining for The Masses* (Lexington: Global Text)
[9] Purnomo R, Syauqy D and Hanafi M 2018 Implementasi Metode Fuzzy Sugeno Pada Embedded System Untuk Mendeteksi Kondisi Kebakaran Dalam Ruangan *Jurnal Pengembangan Teknologi Informasi dan Ilmu Komputer* e-ISSN 2548 964X

IOP Conf. Series: Materials Science and Engineering 1073 (2021) 012031 doi:10.1088/1757-899X/1073/1/012031

Finite element modeling and analytic algorithm for electromagnetic performance of Permanent Magnet Synchronous Generator (PMSG) 24 Slot 16 poles from modification of induction motor 0, 75kw 3 Phase

S S Wibowo[1,2,*], M Saputra[1], A H Santoso[1], N Rosidah[1] and A Arinda[1]

[1] Teknik Listrik, Jurusan Teknik Elektro, Politeknik Negeri Malang, Malang, Indonesia
[2] Departement of Electrical Engineering, State Polytechnic of Malang, Malang 65141, East Java, Indonesia

*sigi.wibowo@polinema.ac.id

Abstract. Modification of the induction motor into a permanent magnet synchronous generator is to put a permanent magnet on the squirrel cages stator to generate the magnetic pole as a replacement for the electromagnet pole produced at a separately excitation of the conventional synchronous generator called the Permanent Magnet Synchronous Generator (PMSG). The purpose of this research is to design PMSG by modification of induction motors into permanent magnet synchronous generator with 24 slots 16 pole by using radial flux and utilizing stator. This research is a software-based FEM (Finite Element Method) for design and the magnetic Material used for permanent magnet synchronous generators is Neodymium Iron Boron (NdFeB). The PMSG on this design obtains the back EMF constant for its half-cycle of 4.902 rad/s and gets a voltage of 191.455 Vdc at 375 rpm speed and when given the loading of 100 Ω at 375 rpm speed produces a voltage of 176.73 Vdc, the current of 1.77 A and the output power is 313.24 Watt with efficiency reaching 85%. The PMSG modelling is done to have a specification of 24 slots 16 poles, changes the the characteristic Induction motor with synchrounous speed of 1500 rpm becomes as low as 375 rp. So This PMSG modification can be used as wind turbine or microhydro power palnt with low speed prime mover energy.

1. Introduction

A permanent magnet synchronous generator (PMSG) is a generator where excitation of the magnetic field on the stator is provided by a permanent magnet instead of a coil that provides an electromagnetic field. Synchronous terms refer to the fact that the rotor and magnetic field rotates at the same speed, as the magnetic field is generated through a permanent magnet mechanism that is mounted shaft and induced current into a fixed or stationary armature. The Use of PMSG are the majority source of electrical energy. They are commonly used in Wind Turbine or Micro Hydro Power Palnt. Because it doesn't need power source of DC supply for excitation

Modification of the induction motor into a permanent magnet synchronous generator is to put a permanent magnet on the squirrel cages stator to generate the magnetic pole as a replacement for the electromagnet pole produced at a separately excitation of the conventional synchronous generator called

IOP Conf. Series: Materials Science and Engineering 1073 (2021) 012031 doi:10.1088/1757-899X/1073/1/012031

the Permanent Magnet Synchronous Generator (PMSG). Using a permanent magnet on the rotor will reduce weight, as the rotor construction from PMSG is smaller than the synchronous generator construction with the rotor from the excitation coil, which is one of the advantages of permanent magnets.

2. Theoritical background

2.1. Permanent Magnet Synchronous Generator (PMSG)

The Permanent Magnet Synchronous Generator (PMSG) is a generator whose excitation field is generated by permanent magnets instead of coils so that magnetic flux is produced by a permanent magnetic field [1,2]. This Generator has significant toughness, attracting researchers and is usually used in wind turbine applications [3-6].The permanent magnet synchronous generator is a rotating electric machine with a 3-phase classic stator that is like an induction Generator in general. Permanent magnets can be mounted on surfaces or embedded in Rotornya.

The permanent magnet synchronous generator (PMSG) based on the wind turbine can be easily connected to the grid via back-to-back converters. The PMSG has shown high efficiency in power enhancement and excellent performance to extract maximum power from the wind [3,4,6]. Low speed wind generators have a problem with cogging torque that could be improved via pole and slot. The small scale wind power dan microhydro power applications need a cost effective and mechanically simple generator for serving as a reliable energy source [6-8].

2.2. Magnet Permanen Noedymium (NdFeB)

Karakteristik magnet permanen yang paling tinggi saat ini adalah Neodymium Iron Boron (NdFeB), yang memiliki nilai produk energi maksimum sampai dengan 400 kJm3. Sedangkan NdFeB bonded memiliki nilai produk energi maksimum sampai dengan 200 kJm3. Neodyum Iron Boron (NdFeb) merupakan Bahan magnetic yang paling baik karena mempunyai densitas fluksi yang besar [1,2,4].

2.3. Buried magnets

In this configuration the magnets are put inside the rotor and therefore it is referred to as Interior Permanent Magnet (IPM) machine. There are many different ways in achieving interior magnet configuration. The magnets can be magnetised in radial direction as well as circumferential direction. The thickness of iron bridges between the magnets has to be designed carefully to avoid saturation. Again, the inductance in quadrature axes is different from that in direct axes direction..

2.4. Design considerations

For PMSG Axial flux coreless PMSG is regarded as the suitable generator for small turbine because of its simplicity of design and manufacture. The absence of cogging torque for coreless design eliminates any magnetic pull between stationary and rotating parts [3]. Most of the axial flux machines have two sided rotor configuration rather than the single sided rotor to balance the axial forces [4]. Coreless design reduces the mass and increases the efficiency as compared to the conventional design [5].
Generally, in axial flux machines length of the machine is much smaller compared with radial flux machines. Their main advantage is high torque density, so they are recommended for applications with size constraints especially in axial direction [9-11].

2.5. Analytic algoritm of PMSG design

The mathematical formula of the generator design is the basic formula for determining the value of some parameters that will be known on the Generator.

IOP Conf. Series: Materials Science and Engineering 1073 (2021) 012031 doi:10.1088/1757-899X/1073/1/012031

2.5.1. Calculation of rotor dimention. Calculating rotor dimensions is performed to determine the installation of magnets and water gaps to be used [12].

2.5.2. Calculation of magnetic fields. The size of the magnetic flux that goes through a field and the area of the magnetic field that will arise in the generator to obtain the maximum magnetic field.

2.5.3. Calculation of number of turn on coil. The voltage produced by the generator is very influenced from the number of turns on a coil. There must be calculations and techniques to wind a coil so that the resulting voltage is the highest result.

2.5.4. Calculation of induced voltage. In addition to magnetic flux, other factors determining the induction voltage are the winding, frequency, number of slots, and the number of phases. The calculation of induction voltage should also be adjusted with construction to be designed [13,14].

2.5.5. Calculation power and torque. The torque obtained will be a representation of the performance of the generator and will be related to the amount of input power from the generator. The power of a generator depends on the angular speed. The amount of angular speed is biased from the speed value of a generator's rotary speed. The amount of output power from the generator is worth the voltage and current [12].

$$P = D^2 L \cdot (0.5\pi^2) K_w \cdot N_s \cdot B_g \cdot ac \cdot \cos\varphi$$

Output power is the power generated by the generator in the form of electrical power. For the output power generated is

$$P_{out} = \sqrt{3} E_{ph} \cdot I_{ph} \cdot \cos\varphi$$

2.6. Finite element modellling

2.6.1. Inisialisasi dan desain geometri. Initialization and geometric design are one of the steps to provide naming and determination of materials used against the components in the generator as well as providing a thickness size for each of these components. Initialization and geometry design covers all parts of the generator components until nothing is notinialized. If something is not initialized, then the design will not be simulated. The components that need to be initialized for a quarter of the parts, namely initialization and geometric design of stators, rotors, magnets, coils, airgap stators, rotors, airboxes, and shafts [9,10,15-17].

2.6.2. Determination of Mesh. Infolityca magnet software is a software based on Fenite Element Method (FEM) [15,17,18]. FEM is a method used to solve problems about complex electromagnetic fields, so it is able to be solved by analysis models especially in parts related to the nonlinear properties of materials. This method essentially discretes the cross-section of the machine into small areas or volumes called fenite elements or meshes [19,20].

2.6.3. The boundary layer. The boundary layer is a tool on infolityca magnet software, where it is used when designing a generator 1/4 or 1/2 shape, of the full shape (360) used to reflect back the part of the generator as if in part 1/4 or 1/2 of the other part there is still the same part, so that the design of the motor can be simulated by rotating the rotor with a designated angle. Figure 6 is the giving of a boundary layer on the part of the generator [18-21].

3. Experimental dan design metodhology

3.1. Geometrical specification of induction motor

The induction motor to be modified into permanent magnet generator is using 3 phase induction motor with 24 slot 16 pole. The induction Motor has the following geometry dimensions in table 1:

Table 1. Dimensiions of Induction Motor

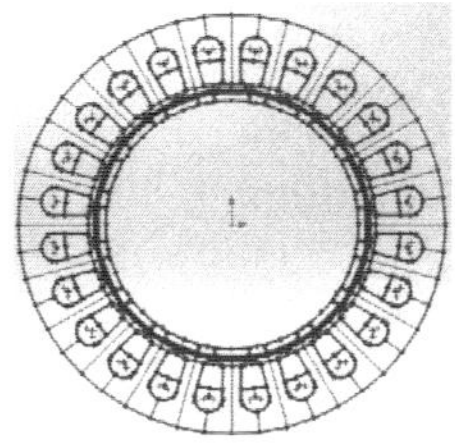

Number of slot	**-**	**24**
Diameter of outer stator	mm	121,40
Diameter of inner stator	mm	76,55
Rotor Diameter	mm	75,55
Length of rotor	mm	90
Hight of Umbrella	mm	20,5
Distance between slot	mm	6,51
Width of teeth	mm	8,10
Air gap	mm	1

3.1.1. Material data of iron core. Iron core material used in an induction motor or commonly called electrical steels is a material chosen because it has a high permeability value of materials, in which case it can increase magnetic flux density. The Material used in the induction motor is M800-50A steel. Specifications Material M800-50A steel as follows in table 2.

Table 2. Specification of iron core material.

Grade	Thickness (mm)	Density $(\mathrm{Kg/dm^3})$	Resistivity (μ Ω cm)	Core Loss (W/Kg) at 50 Hz		Magnetic Flux Density (T)	
				1,0 T	1,5 T	2500A/m	5000A/m
M800-50A	0,50	7,80	23	3,60	8,00	1,60	1,70

3.2. Analytical Algoritm on PMSG Design

In modelling of conversion from induction motor into PMSG, there are calculations based on the references from books and journals to obtain the value of the size and geometry of the generator. For the initial stage required values of the initial parameters as we want (table 3).

Table 3. Initial parameter of design.

Kecepatan Putar Rotor, $\boldsymbol{n_s}$	**375 rpm (6,25 rps)**
Medan Magnet Relatif, $\boldsymbol{B_r}$	1,2 T
Intensitas Magnet, $\boldsymbol{ac}$	12000 A/m
Faktor Belitan, $\boldsymbol{k_w}$	0,945
Faktor Daya, $\boldsymbol{\cos\varphi}$	0.85

3.2.1. Combinations of slot and pole. To determine the number of slots and poles to be used in general stator refers to the dimensions of the induction motor is 24 slots. For pole selection to be used based on the desired working frequency of the generator, the 50 Hz is then calculated using the following equation:

$$Np = \frac{120.f}{n} = \frac{120.50}{375} = 16\ Pole$$

So the number of Poles that get from the calculation above is 16 pole. After gaining the number of slots and pole is 24 slots 16 pole, the next stage calculates the slot degrees (θ_s) and degrees pole (θ_p).

$$\text{Number of slot } \theta_s = \frac{(2.\pi)}{N_s} = \frac{(2.180)}{24} = 15°$$

$$\text{Number of pole } \theta_p = \frac{(2.\pi)}{N_p} = \frac{(2.180)}{16} = 22.5°$$

3.2.2. Calculation of rotor dimensions of PMSG. The rotor used in PMSG is the rotor of the induction motor. The rotor has a diameter of 79.55 mm with Δ = Gap (mm) gap in equation 2.4 How to calculate rotor diameter:

$$\text{Diameter of Rotor (mm) } D_r = D - 2 \cdot \delta = 76{,}55\ mm - 2 \cdot 1mm = 74{,}55\ mm$$

So the required rotor diameter is 77.55 mm. The diameter will be reduced again by the magnetic dimension because the rotor will be paired with 3 stacked permanent magnets with a thickness of 2 mm. so that the rotor diameter becomes 68.55 mm after the grinding.

3.2.3. Magnetic fluks calculation of PMSG. Calculations fluks that flow on the magnet from the magnetic Pole from north to south by passing through air gap between the rotor and stator. The first step to calculate the polar area of the magnet. With a dimension of 3 stacked with each magnet measuring 30 mm x 10 mm x 2 mm, which has 16 pieces and a rotor diameter of 68.55 mm, it can be measured around the Rotor as follows:

$$\text{Circumference of rotor } K_r = \pi\, D_r = 3.14 \cdot 68{,}55 = 215{,}247\ mm$$

Then Caclculate the circumference to be magnetized :

$$K_{ri}\% = \frac{\pi D_r - D_f\, N_{Df}}{\pi\, D_r} \times 100\% = \frac{3{,}14 \cdot 68{,}55 - 6 \cdot 16}{3{,}14 \cdot 68{,}55} \times 100\% = 55\ \%$$

The percentage of magnetized area is 55% then the area of magnetic field is :

$$\text{Magnetic field area (m2) } A_{magn} = (p_m \cdot w_m)2 \cdot K_{ri}\% = (30 \cdot 10\,)2 \cdot 0{,}55 = 330\ mm^2 = 0.00033\ m^2$$

Type of magnet used is NdFeb N35 (Neodymium Iron Boron) This magnet has maximum energy product (BHmax), up to 440 kJ/m3 and relative magnetic field Br 1.2 Tesla, as well as the air gap 1mm then the calculation of magnetic field as follows:

$$\text{Maximal magnetic field (T) } B_{maks} = B_r \frac{l_m}{l_m+\delta} = 1.2\ \frac{2}{2+1} = 0{,}8\ T$$

Thus , the maximum magnetic field on one magnet is 0.8 T and at 3 stacked magnetic field maximum value of the magnetic field to 2.4 T. With maximum magnetic field = 2.4 T and the magnetic polar value of the magnet Amagn = 0.00033 m2 Then the Φ magnetic flux value can be calculated as follows:

$$\text{Magnetic Fluks (Wb) } \Phi = B_{maks}\, A_{magn} = 2{,}4 \cdot 0.00033 = 0.000792\ Wb$$

3.2.4. Calculation of number of coil. In the calculation of the number of coil, the voltage is assumed E_{ph}= 84 Vac, for magnetic flux that has already been calculated then Φ = 0.000792 Wb, because the number of slots and pole used is 24 slots and 16 pole with speed 375 rpm and frequency value f = 50 Hz, then the number of coil can be calculated as follows:

$$\text{No. of coil per winding } N_c = \frac{E_{ph}}{4{,}44 \cdot f \cdot \Phi \frac{N_s}{N_{ph}}} = \frac{84}{4{,}44 \cdot 50 \cdot 0.000792 \cdot \frac{24}{3}} = 59.7\ coil$$

For proses *rewinding* , the number of coil is rounded up to 60 coils.

3.2.5. Calculation of generator power. The active Power produced by generator can be calculated as follows:

$$P = D^2 L \cdot (0{,}5\pi^2) K_w \cdot N_s \cdot B_g \cdot ac \cdot \cos\varphi$$
$$= 0{,}07655^2 \cdot 0{,}04485(0{,}5\pi^2)0{,}945 \cdot 6{,}25 \cdot 1{,}2 \cdot 12000 \cdot 0{,}85 = 94\ Watt$$

3.3. Finite element modelling of PMSG

Modeling and determination of the material type used in each component of the generator is made using Finite Element Software based on Infolytica MagNet Software. The composed components of the magnetic permanent synchronous generator consist of a airbox, airgap, stator, rotor, magnet, coil, and shaft. Each component will be given material as in Table 4 as follows.

Table 4. Material for each component of PMSG.

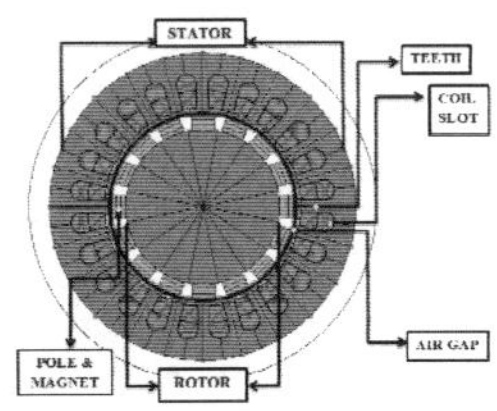

COMPONENT	MATERIAL
Air box and *Air gap*	*Air*
Coil	*Copper:5.77e7 Siemens/Meter*
Magnet	*Neodymium Iron Boron: type N35*
Rotor	M800-50A *steel*
Stator	M800-50A *steel*
Shaft	*Air*

The design results will then be simulated in order to obtain the characteristics of the PMSG. on Infolityca MagNet software can be seen in Figure 1.

3.3.1. Modelling simulation of PMSG. Before simulating a PMSG model There are several steps to be performed i.e. setting simulation parameters such as motion creation or adjusting which parts will be rotated at the specified speed and time. The results of magnetic field distribution in the form of magnetic field direction and also flux lines in the simulation of PMSG with type Neodumium Iron Boron N35 as shown in figure 1 below.

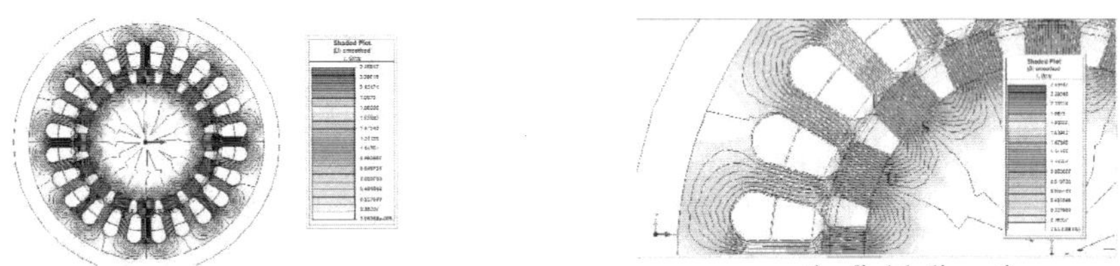

magnetic flux lines magnetic field direction

Figure 1. Magnetic field Distribution of PMSG using NdFeb 35.

Magnetic flux density is set from 0 Tesla to 2.4 Tesla. The distribution pattern shows that the highest magnetic flux density is found in the Stator umbrella section and the teeth stator marked by a red color that ranges from 2.1 Tesla to 2.4 Tesla which means magnetic flux flows in the full part or at maximum value. In green areas ,the magnetic field density value ranges at a value of 0.9 Tesla to 1.6 Tesla which means a magnetic flux that flows in a reasonable value, then followed by the outside of the teeth is characterized by a light blue area with a magnetic field density value of a range at a value of 0.16 Tesla to 0.81 Tesla in this passage of magnetic flux that flows very minimal. For the air box area, the stator air, the central part of the rotor core, and the shaft are not traversed by the magnetic flux so that this part has a small, magnetic field density value shown in white colour.

4. Results and analysis

4.1. Setting up parameter in simulation software

In the simulation will be tried various angular velocity variations. The parameters set in Simulsi are rotor angle velocity (ω), and time. The speed that will be attempted in this simulation is speed at 100 rpm, 125rpm, 150 rpm,175 rpm, 200 rpm, 225 rpm, 250 rpm, 275 rpm, 300 rpm, 325 rpm, 350 rpm, 375 rpm, 400 rpm, 600 rpm, 800 rpm, and 1000 rpm. The Rotor is part of a rotating generator. The Rotor rotates by 360 ° for a single round. The speed in rpm (rotation per minute) means that in one minute the rotor will spin once, then it will be changed in degree units per second where in one second the rotor will spin by 360 °, with the equation as follows:

$$\omega_{(250)} = \frac{250\ rpm\ x\ 360^{\circ}}{60\ sec} = 1500 deg/sec$$

The Designed generator is a permanent magnet synchronous generator with 24 Slots 16 pole, so to form one sinusoidal wave rotor only needs to rotate for per 360 °/8 = 45 °. In this simulation the rotor will be turned per 1 °, so keep in mind the time it takes for the rotor to spin for 1°.

4.2. No load simulation of PMSG

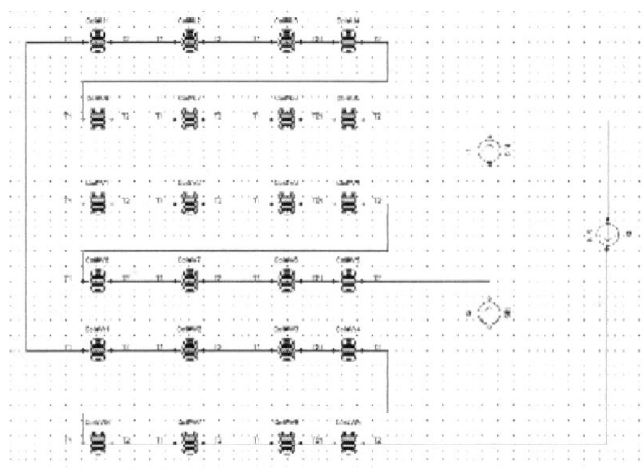

Figure 2. No load simulation circuit of PMSG.

This No Load simulation will be tested at speeds of 100 rpm until 1000 rpm. This simulation of a PMSG will result in a voltage value between phases. Where this generator is designed with 3 phases, namely phase U-V, V-W, and U-W, it will form a sinusoidal wave as in graph of speed vs Voltage.

Simulated results at a no-load 125 rpm speed will result in an inter-phase tension value. In this simulation to determine the time in which the rotor is needed when rotated per 1 ° to obtain when one wave with a speed of 125 rpm is

$$\text{Time of one step rotation} = \frac{perstep\ putaran\ (deg)}{\omega[{}^{deg}/_{sec}]}$$

$$\text{Time of one step rotation} = \frac{1(deg)}{750[{}^{deg}/_{sec}]} = 0{,}00133\ sec$$

$$\text{Time of 30 steps rotation} = 30(step).\,0{,}00133(sec) = 0{,}0399\ (sec)$$

Thus, the initial time required for the rotor speed per 1 ° of the simulation at a speed of 125 rpm is 0.00133 sec. The Rotor on the generator is rotated to 45 ° to get a single waveform, so the simulation will stop at 0.0399 sec.

In this simulation, the rotor on the generator is rotated per 1 °, with a time per 1 ° at a speed of 125 rpm of 0.00133 sec. The Rotor on the generator is rotated to 45 ° to get half the wave, so this simulation will stop at 0.0399 sec or the rotor will spin for 0.0399 sec.

Table 5. Simulation results data at a speed of 125 rpm without load.

No	Time	Flux Linkage			Tegangan Antar Fasa			DC-Voltage
	(s)	u	v	w	u-v	v-w	w-u	
1	0	0.04736	-0.02464	-0.02269	0	0	0	0
2	0.001333	0.04709	-0.01885	-0.02821	-36.38	67.84	-31.46	67.84
3	0.002666	0.045925	-0.01266	-0.03324	-44.16	67.37	-23.21	67.37
4	0.003999	0.043913	-0.00615	-0.03775	-51.16	66.12	-14.96	66.12
5	0.005332	0.041082	0.000404	-0.04148	-56.29	61.65	-5.36	61.65
6	0.006665	0.037171	0.007003	-0.04418	-63.03	55.78	7.25	63.03
7	0.007998	0.032528	0.013555	-0.0461	-67.19	50.85	16.34	67.19
8	0.009331	0.027393	0.019753	-0.04716	-68.04	43.56	24.48	68.04
9	0.010664	0.021812	0.025503	-0.04733	-68.00	35.53	32.47	68.00
10	0.011997	0.015819	0.03078	-0.04662	-67.62	27.38	40.24	67.62

The Generator rotates per 1 ° with a time of 0.00133 sec so that the recorded data is data per 1 °. The Data recorded in the simulation is an inter-phase voltage. This phase voltage is derived from the voltage of each phase, where there are phase U, V, and W. The voltage value of this phase is generated from the change of flux value of time. In table 4.7 above there is a minus voltage value, this occurs because the flux value of the generator is in and out of the rotor in the air gap. As a result, this voltage value between phases generates the following sine waves:

The output waveform shows that the rotor on the generator rotates from 0s to 0.0399 sec, indicating that the generator rotates from a rotor position of 0 ° to 45 °, resulting in half wave. The sine wave above shows that the value of t three sine waves representing the voltage between the U-V, V-W, and W-U.

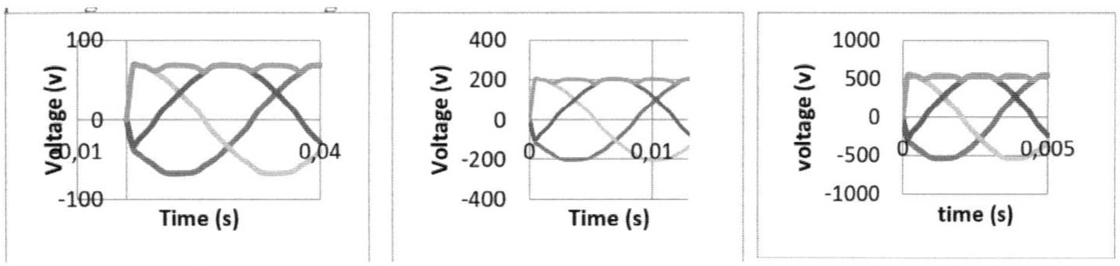

Figure 3. The output waveform of phase to phase voltage and DC output at the speed of 125 rpm.

The output voltage are rectified using three phase rectifier giving results of DC voltageThe result is shows that at no load with the average speed of 125 rpm ,PMSG will generate the peak voltage at 63.285 volts. The average voltage value will be used to calculate the EMF back constants (to) value in Equation as follows:

$$\omega = 125 \text{ rpm x } 2\,.\pi \div 60 = 13{,}02083 \text{ rad/s}$$

To obtain constant of *back EMF* (Ke) in equation 2.15 is as follows:

$$K_e = \frac{Vdc}{\omega} = \frac{63{,}285}{13{,}02083} = 4{,}902 \text{ rad/s}$$

From the results of the whole calculation dan simulation, the Ke of PMSG will produs=ces Ke at any speed as shown in table 6.

Table 6. Simulation data of noload volage of PMSG.

No.	RPM	Vdc Ave	Ke (rad/s)
1.	100	51,063	4,902
2.	125	63,825	4,902
3.	150	76,591	4,902
4.	175	89,353	4,902
5.	200	102,117	4,902
6.	225	114,892	4,902
7.	250	127,661	4,902
8.	275	140,417	4,902
9.	300	153,198	4,902
10.	325	165,956	4,902
11.	350	178,705	4,902
12.	375	191,455	4,902
13.	400	204,265	4,902
14.	600	306,396	4,902
15.	800	408,402	4,902
16.	1000	510,727	4,902

In table 6 shows that the faster the rotation of the rotor generator, the voltage of each round is getting bigger. At this no load voltage simulation produces only voltages and does not produce currents, since generators are not connected with loads. In table 6 It is known that there is a value to or Electromagnetic Force (EMF) Back constants generated due to the change in magnetic field flux in a time that remains constant shows this generator is the same generator with different speed tests.

4.2.1. The PMSG simulation results at resistive load. The equivalent circuit of PMSG with resistive loads through three phase rectifier is drawn in the software first before doing simulation process, as shown in figure 4.

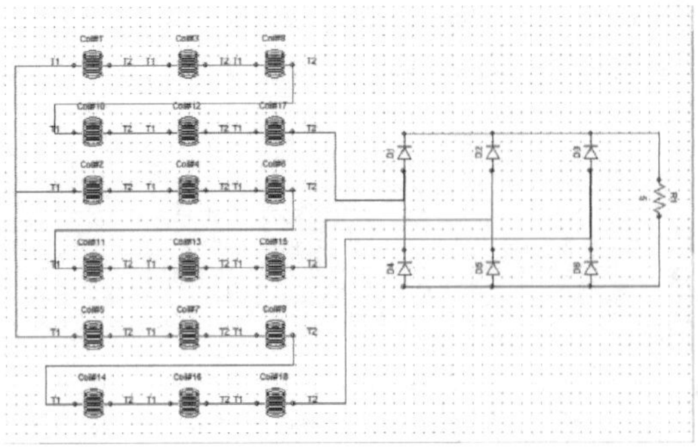

Figure 4. Simulation circuit of resistive load of PMSG.

PMSG will be given a resistor loads of between 20 Ω and 100 Ω. with a variation at a speed 100 rpm until 1000 rpm with. The Generator will get results from tests that generate multiple values at a voltage, current, torque, input power, power output ,and efficiency, as follows:

4.2.2. The Torque vs speed characterestif of PMSG. The results are then tabulated as shown ini table 7.

Table 7. Simulation data of torque vs speed with load variation.

rpm/Ω	20 ohm	40 ohm	60 ohm	80 ohm	100 ohm
100	9.84	5.92	4.32	3.438	2.88
125	11.98	7.24	5.29	4.202	3.51
150	14.01	8.54	6.25	4.97	4.14
175	15.94	9.79	7.18	5.70	4.76
200	17.77	10.99	8.08	6.43	5.37
225	19.51	12.15	8.97	7.15	5.97
250	21.24	13.27	9.83	7.85	6.56
275	22.76	14.36	10.67	8.54	7.14
300	24.47	15.41	11.49	9.21	8.39
325	25.95	16.43	12.29	9.87	8.28
350	27.51	17.41	13.07	10.52	8.84
375	28.90	18.37	13.83	11.16	9.38
400	30.35	19.46	14.66	11.87	10.00
600	38.28	26.28	19.93	16.32	13.89
800	41.01	31.17	25.24	20.56	17.50
1000	42.38	36.29	28.92	24.01	20.55

The characterestics of Torque vs spped is then drawn as as shown in Figure 5. With load variation between of between 20 Ω and 100 Ω.

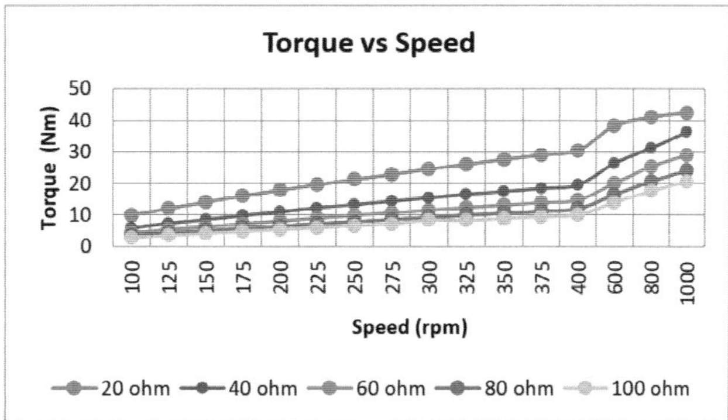

Figure 5. Characterestic of torque vs speed with load variation.

The characteristic shows in figure 5 that the more speed applied to the shaft of PMSG the more input torque produced as a primemover of PMSG. As the the load increases then the torque decreases accordingly.

4.2.3. The speed vs voltage characterestif of PMSG. PMSG will be given a resistor loads of between 20 Ω and 100 Ω. with a variation at a speed 100 rpm until 1000 rpm with. The PMSG will genereate the voltage as shown in figure 6.

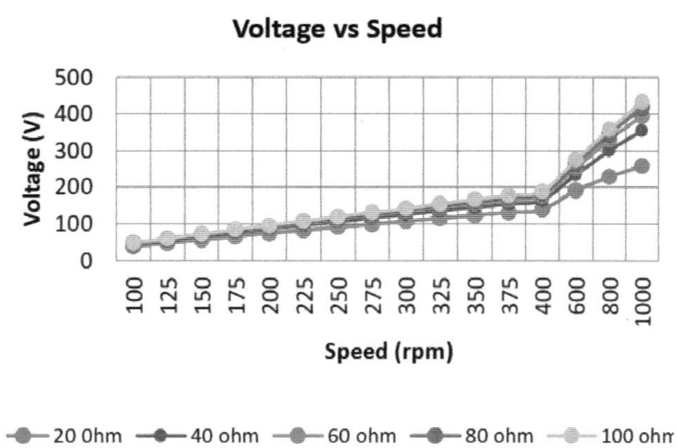

Figure 6. Characterestic of speed vs voltage with load variation.

The result shows in figure 6 that the more speed applied to the PMSG shaft , the higher the voltage will be produced. As the load is increased the voltage is ot getting down even the voltage becomes higher. This is because the more the load is increased near the nominal rating the higher the efficiency, therefore the high the output voltagr will be. From the graph shows that the lowest at 20 Ω load with speed of 100 rpm will produces the voltage of 38,2 V, whereas the highest voltage at 100 Ω with speed of 1000 rpm will produce the voltage of 433,44 V.

4.2.4. The speed vs power output Characterestif of PMSG. PMSG will be given a resistor loads of between 20 Ω and 100 Ω. with a variation at a speed 100 rpm until 1000 rpm. The PMSG will genereate the power output. The PMSG do not produce the VAR power instead it will produce WATT power because PMSG does not need excitation voltage From the graph shows that the lowest at 100 Ω load with speed of 100 rpm will produce the power output of 23,15 Watt , whereas the highest voltage at 20 Ω with speed of 1000 rpm will produce the power output of 3350,45 Watt.

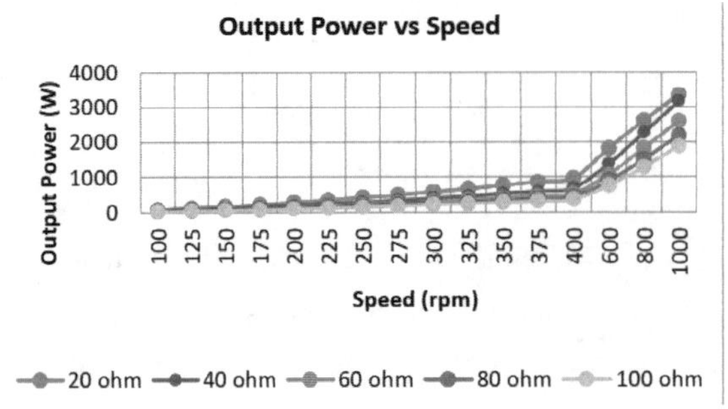

Figure 7. Characterestic of power ouput vs speed with load variation.

The graph shows that at the same load ,The output power will increases as the the speed of PMSG shat is ncreased. While the more load is applied then power output produced decreases .

4.2.5. The speed vs efficiency characterestif of PMSG. Efficiency is a measure of the level of power usage of an energy process delivered during business over a period of time. In the simulation results have been obtained the output values of voltage, and torque, so that the input power value and output power value are obtained. Then it was calculated to get the efficiency of the modified design of the PMSG created. Here's the data obtained:

Table 8. Simulation data of speed vs effivciency with with load variation.

rpm/Ω	20 ohm	40 ohm	60 ohm	80 ohm	100 ohm
100	0.72	0.78	0.79	0.80	0.80
125	0.73	0.79	0.81	0.81	0.81
150	0.73	0.80	0.82	0.82	0.82
175	0.74	0.81	0.82	0.83	0.83
200	0.75	0.81	0.83	0.83	0.83
225	0.75	0.82	0.84	0.84	0.84
250	0.76	0.82	0.84	0.84	0.84
275	0.75	0.83	0.84	0.85	0.85
300	0.76	0.83	0.85	0.85	0.85
325	0.76	0.83	0.85	0.85	0.85
350	0.76	0.84	0.85	0.86	0.86
375	0.76	0.83	0.86	0.86	0.86
400	0.77	0.84	0.86	0.86	0.86
600	0.77	0.84	0.87	0.88	0.88
800	0.77	0.85	0.88	0.89	0.89
1000	0.77	0.84	0.87	0.88	0.89

It can be found from Table 8 that the lowest efficiency of 72% is found at 20 Ω loads with a speed of 100 rpm , while the highest efficiency is 89% at 100 Ω loads with a speed of 1000 rpm

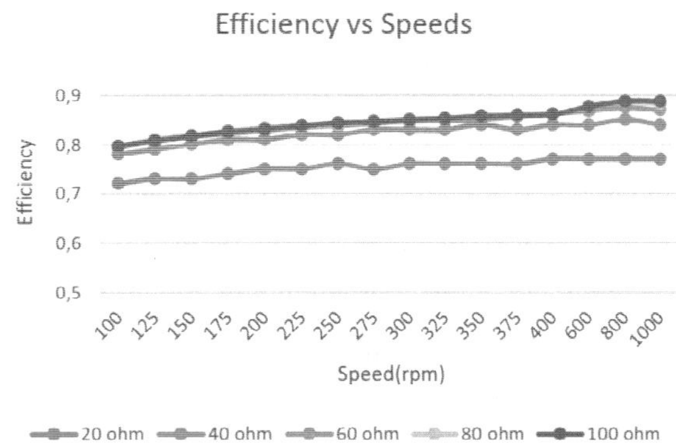

Figure 8. Characterestic of Speed vs Effivciency with with load variation.

In Figure 8 it is known that this power efficiency value indicates the level of reliability of the PMSG in converting energy. The efficiency value of this generator is fickle due to the effect of the fickle speed input value resulting in permanent magnetic synchronous generator torque also changing to a point where the value is relatively stable. And the higher the load value used, the lower the efficiency value generated. In addition in converting into electricity there are losses in the form of eddy currents and iron losses in stator and rotor materials.

4.2.6. Voltage regulation. From the simulation of the voltage test of a PMSG can be calculated the voltage regulation with the equation, as follows:

$$V_{reg} = \frac{V_{DC\ no\ load} - V_{DC\ load}}{V_{DC\ load}} \times 100\%$$

In the uni-directional voltage system, the voltage regulation value of the voltage data has been averaged as shown in the following table 9.

Table 9. Volage Regulation at the speed of 125 rpm.

	Speed of 125 rpm			Speed of 1000 rpm		
Loads	**load Voltage (V)**	**Noload Voltage (V)**	**Voltage Regulation (%)**	**load Voltage (V)**	**Noload Voltage (V)**	**Voltage Regulation (%)**
20 ohm	47,45	63,825	34	258,59	510,727	97
40 ohm	54,44	63,825	17	355,09	510,727	43
60 ohm	57,47	63,825	11	394,58	510,727	29
80 ohm	59,124	63,825	8	418,18	510,727	22
100 ohm	60,08	63,825	6	433,44	510,727	17

The data in the table 9 shows that at lower speed the voltage regulation is not as good as in high speed. At the speed of 125 rpm the increasing load is less than 50% and become worst as the load increase, Whereas the best voltage regulation is at speed 1000m rpm with lighter load of 20 Ω. As the increase the voltage regulation also decreases.

5. Conclusion

Based on the design, measurement results and analysis that have been done in this permanent magnetic synchronous generator modification design research, as follows

- The PMSG in this design from the modeling results has a specification of 24 pole 16 slots, changes the the characteristic Induction motor with synchrounous speed of 1500 rpm becomes as low as 375 rp. So This PMSG modification can be used as wind turbine or microhydro power palnt with low speed prime mover energy.

- The PMSG in this design results in an EMF back constant for half its waves of 4,902 rad/s and produce a voltage of 191,455 Vdc at a speed of 375 rpm and when given a load of 100 Ω at a speed of 375 rpm produces a voltage of 176.73 Vdc and output power of 313.24 Watts with efficiency reaching 85%.
- The variation in load and speed affects the voltage, current, torque, output power, and efficiency generated by PMSG. When the generator is charged 20 Ω at a speed of between 100 and 1000 rpm the voltage goes up.

References

[1] Bhawna G and Gagan D Y 2016 Permanent Magnet Synchronous Generator Based Energy Conversion System: A Review *International Research Journal of Engineering and Technology (IRJET)* **03**(06) 995

[2] Amir N, Hamidreza I and Yousef A B 2017 Design And Optimization Of Permanent Magnet Synchronous Generator For Use In Hydrodynamic Renewable Energy *IIUM Engineering Journal* **18**(2)

[3] Chainuson K and Montri S 2018 Low cost fabrication of permanent magnet for low speed wind turbine generators using waste motors *Songklanakarin J. Sci. Technol.* **40**(06) 1346-1353

[4] Krueasuk W, Bhasaputra P and Pattaraprakorn W 2011 Feasible Study of Restructuring Used Induction Motor for Low Speed Renewable Resource Generator *GMSARN International Journal* **05** 179 - 188

[5] de la Cruz J, Ramirez J M and Leyv L 2014 Modification of Geometric Parameters in Outer Rotor Permanent Magnet Generators to Improve THD, Efficiency, and Cogging Torque *International Journal of Emerging Electric Power Systems* **15**(5) 471-483

[6] Dai J C, Hu Y P, Liu D S and Wei J 2012 Modelling and analysis of direct-driven permanent magnet synchronous generator wind turbine based on wind-rotor neural network model. Proceedings of the Institution of Mechanical Engineers, Part A *Journal of Power and Energy* **226**(1) 62-72

[7] Sayyed T and Jagadeesh B 2015 Design and Analysis of Wind Turbine PMSG with Direct Current Vector Control *IOSR Journal of Electrical and Electronics Engineering (IOSR-JEEE)* **10**(3) 32-39

[8] Singh R and Chauhan Y K 2017 Modeling and control of PMSG based variable speed wind turbine using Proportional Resonant controller *Int. Journal of Engineering Research and Application IJERA* **7**(7)

[9] Adrian M and Mariusz K 2015 Analysis Of Integral Parameters In A Single-Phase Slotless Axial-Flux Machine Used In Small Wind Turbines Technical Transactions Electrical Engineering *Czasopismo Techniczne* (13).

[10] Mahmoudi A, Rahim N A and Hew W P 2011 Axial-flux permanent-magnet machine modeling, design, simulation, and analysis *Scientific Research and Essays* **6**(12) 2525-2549

[11] Lari H A, Kiyoumarsi A, Darijani A, Mirzaeian Dehkordi B and Madani S M 2014 Analysis and design of a permanent-magnet outer-rotor synchronous generator for a direct-drive vertical-axis wind turbine *Iranian Journal of Electrical and Electronic Engineering* **10**(4) 324-332

[12] Prof A S Shajilal and Rudra J Kumar 2017 Design of PMSG Based Wind Energy Conversion System with Unified Power Controller *International Journal of Advanced Research in Electrical, Electronics and Instrumentation Engineering* **06**(03)

[13] Acharya P, Papadakis A and Shaikh M N 2018 Modelling and Design of a 3 kW Permanent Magnet Synchronous Generator suitable for Variable Speed Small Wind Turbines *Progress In Electromagnetics Research* **85**

[14] Ebraheem M and Jyothsna T R 2018 Performance analysis of transient behaviour of PMSG model with sudden load variations *International Journal of Applied Engineering Research* **13**(10)

[15] Jian Yi Chen Chem Nayar 1998 Design and FE Analysis of an Outer-Rotor PM Generator for Directly Coupled Wind Turbine Applications *Centre for Renewable Energy Systems Technology*

[16] Mellah H and Hemsas K E 2013 Simulations Analysis with Comparative Study of a PMSG Performances for Small WT Application by FEM *International Journal of Energy Engineering (IJEE)* **3**(2)

[17] Ocak C, Uygun D and Tarimer I 2018 FEM based multi-criterion design and implementation of a PM synchronous wind generator by fully coupled co-simulation *Advances in Electrical and Computer Engineering* **18**(1) 37-42

[18] Ajay S, Harsh K S and Dr. Lata G 2017 Modelling and Simulation of Wind Turbine using PMSG *International Journal of Engineering Technology Science and Research IJETSR* **4**(3)

[19] Rajesh K, Erwan S, Laili I J and Fatihah S B 2017 design of Outer Rotor Permanent Magnet Flux Switching Machine for Downhole Application *International Journal of Applied Engineering Research* **12**(7) 1383-1388

[20] Mehrage G, Zahra N-G and Hashem O 2018 Performance Analysis of Permanent Magnet Vernier Generator under Mechanical Faults *International Journal of Industrial Electronics, Control and Optimization IECO* **1**(02) 121-131

[21] Andrzej P and Sławomir C 2019 The synchronous generator based on a hybrid excitation with the extended range of voltage adjustment in *Web of Conferences*

IOP Conf. Series: Materials Science and Engineering 1073 (2021) 012032 doi:10.1088/1757-899X/1073/1/012032

Design generator of magnetic power (Based on gap rotor-stator motor magnet)

S H Susilo*, Z Jannah, K Kasijanto and S Adiwidodo

Department Mechanical Engineering, State Polytechnic of Malang, Malang, Indonesia

*shadis172.gh@gmail.com

Abstract. The paper discusses magnetic power generators; if a conductor moves across the magnetic field, a voltage difference will occur at the ends of the conductors. The voltage rises as it approaches the area and vice versa, so electricity arises in cycles: positive - zero - negative - zero. In this study utilizing the rotation of a magnetic motor caused by the unbalance repulsive force between magnets. The design of the tool starts with designing a generator model that will be made. The design of generators is based on calculations obtained through the reference of generator design. The generator design uses magnets instead of anchor coils on the rotor (source of magnetic flux) and copper wire coils as anchor coils on the stator. This design considers rotor and stator parameters, including the distance between magnets, rotor diameter, rotor plate thickness, area of a magnetic area, number of stator coils, number of stator windings, and stator coil dimensions, and calculate generator output power. Tool testing is done to obtain data from tools that have been made. In testing the comparison of the results obtained from the study of the target to be achieved, if the results have not yet reached the target, then improvements are made to the results obtained as expected. Tests were carried out with no load and burden. The results showed that the rotor speed in the generator design could be achieved by determining the number of rotor poles, the generator voltage is influenced by the flux density produced by permanent magnets and the number of stator windings.

1. Introduction

The increasing demand for energy in line with population growth and the increasing development of science and technology encourages the development of renewable energy. Many studies have been conducted to anticipate the shortage of energy sources, starting from the utilization of energy derived from natural resources such as water, wind, geothermal, and solar. However, it is still not enough to meet the enormous energy needs, so innovations are needed to meet these needs [1–5]. As an alternative energy, it has been developed that comes from magnets that are used as electricity generators. This electric generator uses magnetic force as rotating energy, resulting in rotation. Magnets used are induction magnets or permanent magnets [6,7]. Electrical energy conversion technology is increasingly innovative to facilitate humans in doing work. The innovative electrical energy conversion technology cannot be separated from the energy source [7–9]. Energy must have a flexible value in terms of energy that can be converted to various other forms of energy; this type of energy is called electrical energy. Electrical energy is energy produced from various types of energy. A generator is needed, produced by an electric generator [10–13].

One of the most critical problems in finding solutions in Indonesia is the electricity crisis. Many natural resources are renewable and can be used to produce electrical energy [3,7,8], both natural sources such as solar heat, geothermal, wind, and water as well as physical materials such as permanent magnets, pressure differences and the effects of gravity. All of these can be utilized based on scientific analysis and experiments to obtain real results [14–17].

Countries in Asia, Europe, and America have developed electricity generation from renewable energy sources as a form of thought and concern for the increasingly troubling electricity crisis. We know that reserves of non-renewable energy sources such as oil and natural gas will be increasingly depleted to meet the world's energy demand quota while restoring them will take a long time. With the discovery of high technology by researchers, it is more comfortable and more extensive to apply, especially the optimization of energy generation systems from renewable sources[6,18–20].

The application of high technology as a step to optimize the yield of a power generation system with renewable energy is a form of compensation for the small discharge of the energy produced. We know that the release of energy generated from renewable energy generation is relatively smaller than the release of energy from non-renewable sources [21]. However, the system's optimization is expected to produce electrical energy with a discharge that is no less large, or at least the community can be independent by having a local power plant that can meet their electrical energy needs independently [22–24].

Therefore, using the concept of permanent magnets as electricity generators utilizes magnetic fields to produce rotations in electric generators. In producing the rotation of an electric generator, a permanent magnet is positioned as the stator or rotor [18].

A generator is an electrical machine that works based on motion / mechanical energy, which then converts it into electrical energy used in everyday life. The generator uses the Faraday experiment's principle, which rotates the magnet in a coil or vice versa when the magnet moves in a coil, there is a change in the magnetic flux (change in the distribution of the magnetic field) in the coil. It penetrates perpendicular to the coil so that a potential difference arises between the ends of the coil (generating electricity); this is due to changes in magnetic flux [25]. The magnetic flux can be changed by moving the magnet in the coil or vice versa by utilizing other energy sources, such as wind and water, to rotate the turbine blades to move the magnet. If a conductor moves across the magnetic field, there will be a voltage difference at the ends of the conductor [26,27]. The voltage increases as it approaches the field and vice versa so that electricity occurs in cycles: positive - zero - negative - zero (AC voltage). In a DC generator, the direction of the current when the negative voltage is reversed by a split ring mechanism so that the cycle changes to positive-zero - positive - zero (DC voltage). Many studies have been carried out on the use of permanent magnets to drive electric generators, but this research has not yet been fully achieved, so it is necessary to design an electric generator based on the magnetic motor rotor-stator gap resulting output [28–30].

2. Theory

2.1. Definition of magnets

Magnets are objects that have a magnetic field. Objects that can attract iron or steel are called magnets. Magnetic materials are widely used in equipment, including electric bells, telephones, dynamos, electric measuring instruments, compasses. Magnets are made of iron, steel, and metal alloys. Magnets are made up of tiny magnets with the same direction (arranged in an orderly manner), which are called basic magnets. Magnets have two poles, namely: north (N) and south (S). The magnetic pole is at the end of the magnet with the highest magnetic strength.

Neodymium magnets are the most potent types of permanent magnets. Neodymium magnets, also known as NdFeB, or Neo magnets, are a type of rare-earth magnet made from an alloy of neodymium metal. Tetragonal $Nd_2 Fe_{14} B$ has a very high uniaxial crystal structure of magnetocrystalline anisotropy (HA ~ 7 teslas). This compound gives the potential to have a high coercivity, that is, resistance to magnetic damage.

The magnetic characteristics of NdFeB are better when compared to other permanent magnets, such as Ferrite, Alnico, and Samarium Cobalt. The BHmax that is owned can range from 30 MGO to 52

IOP Conf. Series: Materials Science and Engineering 1073 (2021) 012032 doi:10.1088/1757-899X/1073/1/012032

MGOe. Because it has high magnetic characteristics, in its application, NdFeB magnets have small dimensions and volume. In some applications, these magnets can replace the use of Samarium-Cobalt magnets, especially those at temperatures less than 80 oC [31].

2.2. Permanent magnet electric generator

A magnetic, electric generator consists of several permanent magnets that are arranged in such a way either on the rotor or stator so that there is a repulsive force between the two that continues continuously, causing continuous rotation until the magnet is strong enough to repel and rotate the rotor.

In the 1980s, China invented magnetic field technology and succeeded in making free energy devices in which one unit measuring 150cm x 120cm x 80cm can produce about 5 KW of power. The way this free energy machine works only uses magnets as the main driving force, this tool is without batteries, without electricity, and without fuel, can move continuously at high speed, is stable and is very strong in thrust or pull so that it can play a dynamo 1000 watts to 5000 watts of electricity.

Thomas E. Bearden et al., 2013, with US patent number 6362718A have made two models of electric transformers powered by permanent magnets. This tool uses 6 watts of the electrical input to control the magnetic field path that comes out of a permanent magnet. By sending a magnetic field, first to the output coil, and then to the second output coil, and by repeatedly and quickly like ping-pong, the device can produce 96 watts of electrical output with no moving parts. Bearden refers to his device as a Moving Electromagnetic Generator (MEG), or in other words, a stationary magnet generator. Jean-Louis Naudin has duplicated the Bearden set in France. The basic principles of this tool were first disclosed by Frank Richardson from the United States in 1978. Troy Reed, also from America, has made a particular magnetic fan model that heats up when it rotates. The device absorbs the same amount of energy to rotate the fan, whether it generates heat or not.

3. Methods

The process of making this generator is done by preparing the tools and materials first. The first thing in making this generator is to make a magnetic motor to drive the generator. This process is carried out with great care and precision because the magnets' placement must be accurate so as not to lock the magnetic force. So that the rotating motion of the rotor runs typically, without any motion locking process.

Tool testing is carried out to obtain data from the tools that have been made. In testing, a comparison of the results obtained from the research is carried out against the target to be achieved, if the results have not reached the target, then the equipment is repaired until the results are as expected. The test is carried out with no load and load of 1 100 W lamp and a 350 W drilling machine connected in parallel to determine the output power of the generator that has been made. Analysis of the test result data is then carried out to obtain the optimal design.

Generators can be made by utilizing the repulsive and attractive forces between magnets. Magnets are used with neodymium material, which has very high magnetic strength. The generator is made using a motor, namely a magnetic motor. This magnetic motor uses a rotor and stator system. The unbalanced amount of repulsion and pull forces is able to move the rotor, resulting in magnetic motor rotation. The unbalanced amount of repulsion and pull forces using three magnetic groups on the rotor. Whereas in the stator, the magnetic arrangement is close together but with different distances between the magnets. Figure 1 and 2 show design of permanent magnet motor rotor and stator.

To produce rotation, the rotor and stator are close to one axis using a shaft. In the rotor part, the meeting between the non-magnetic plate and the shaft is given to place the bearing holder so that the rotor can rotate, resulting in motor rotation. The rotating magnetic motor can move the generator rotor. The arrangement of the generator rotor is a magnet-core-magnetic motor. The core is inside the copper wire coil—Single-axis magnetic power generator axis with a magnetic magnet motor. The shape of the magnetic motor is cylindrical, consisting of two plates containing a permanent magnet. The on-off of the magnetic motor depends on the distance between the plates containing the magnet. Magnetic, electric generator, the rotor is driven by a magnetic motor. The magnetic motor itself consists of a magnetic

IOP Conf. Series: Materials Science and Engineering 1073 (2021) 012032 doi:10.1088/1757-899X/1073/1/012032

arrangement configuration between two plates/cylinders. Each part of the magnetic plate has a different magnetic arrangement. The first plate has a magnetic configuration as the driving plate, while the second plate is the driven plate. The magnets' arrangement in the first plate is spaced, and the distance differs from one magnet to another, but they are regular. At the same time, the arrangement of the two magnetic plates forms a magnetic group and coincides. However, the distances between magnetic groups are the same. This permanent magnet motor is designed by making the rotor and stator parts all of the magnets with a specific configuration to produce rotation without any external energy or fuel. The rotation of this magnetic motor will rotate the core (generator rotor) so that the presence of a 16 pole copper wire will produce an electric motion. This electromotive force will produce an electric voltage.

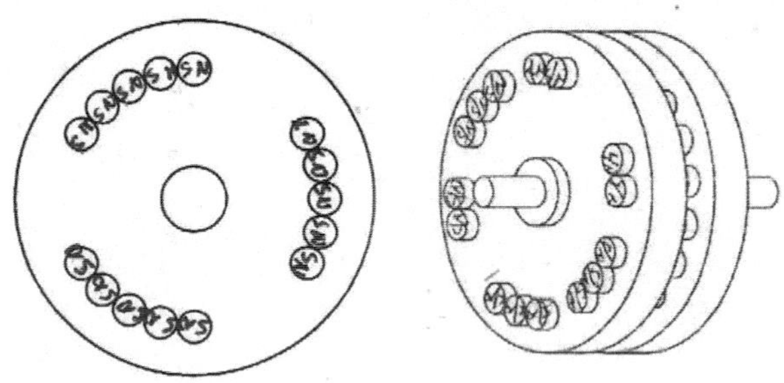

Figure 1. Design of permanent magnet motor rotor.

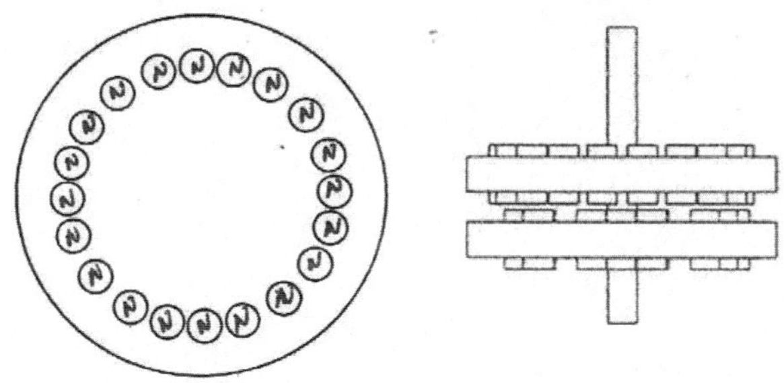

Figure 2. Design of a permanent magnet motor stator.

4. Results and discussion

The generator design designed in this research is a permanent magnet generator. The permanent magnet generator, which is designed, is a modification of the permanent magnet flywheel. This generator rotor frame is designed using a resin composite that is integrated with the flywheel. The arrangement of the permanent magnets is adapted to the modified flywheel construction. The permanent magnets used are of the neodymium type, comprising 21 magnets arranged around the stator. Replacement of permanent magnets is intended to increase the output power of the generator. The stator used is a stator that has 1 (one) number of slots with a circular winding direction of 24100 windings.

Figure 3. Permanent magnet generator and generator test circuit.

Figure 3 shows the permanent magnet generator uses an induction motor in its test. This induction motor is used as the initial drive of the generator, connected to the pulley and the v-belt. In the installation, the induction motor and generator are fitted with a pulley, which is then attached to the v-belt on the pulley. The rotating motion of the magnetic motor rotor makes the magnets in the generator rotor cut the stator magnetic field, causing an electric potential (voltage) difference in the windings.

Generator testing is carried out at a variation of the stator rotor magnetic motor gap of 10 mm - 35 mm with an increase in the distance of 5 mm to obtain accurate data. Generator testing is carried out with no load and 100 W lamp power connected in parallel to determine the maximum voltage, current, frequency, and output power, from the test results, obtained the following data.

Table 1. The test is unloaded.

Gap	Voltage (V)	Current (I)	Frequency (f)	Power (P)= V x I x cos Θ
(mm)	(Volt)	(A)	(Hz)	(W)
10	224	36	59,6	5483.52
15	224	32	58,5	4874.24
20	224	30	57,2	4569.6
25	222	27	55,2	4075.92
30	222	24	54,1	3623.04
35	218	20	53,2	2992

4.1. Relationship Gap (mm) to voltage in a unload test

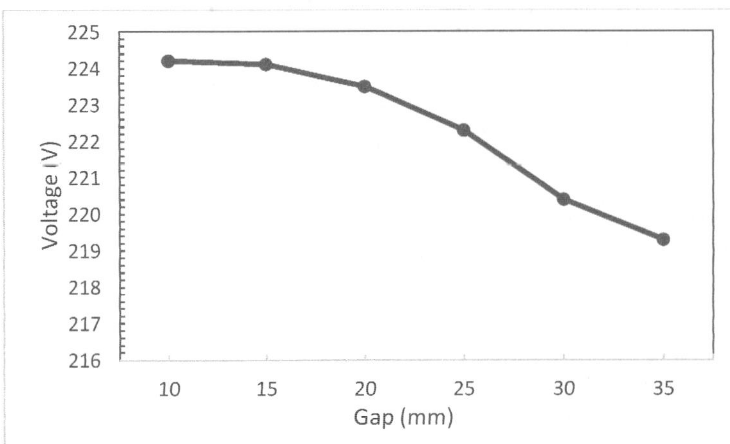

Figure 4. Relationship Gap (mm) to voltage in a unload test.

Figure 4 shows that the generator voltage has decreased in line with the increase in the stator-rotor gap (mm). The lowest voltage is obtained at a gap of 35 mm with a no-load voltage value of 218 V. When the distance gets closer between the rotor-stator motor, the voltage also increases. The voltage increases significantly according to the closer the magnetic motor stator rotor gap. The highest voltage in this unload test is obtained at a frequency of 53.2 Hz with a voltage value of 224 V. When the magnet gets closer, what happens is the higher the repulsion force, but with the tilt angle that has been set, a rotational motion will arise, and this is increasingly The closer the rotation speed accelerates. This rotational motion is also due to the magnetic levitation force. This rotational motion causes the generator rotor to rotate and produces an electromotive force. This electromotive force causes the generator output voltage to arise.

4.2. Relation of the gap to a frequency in a unload test

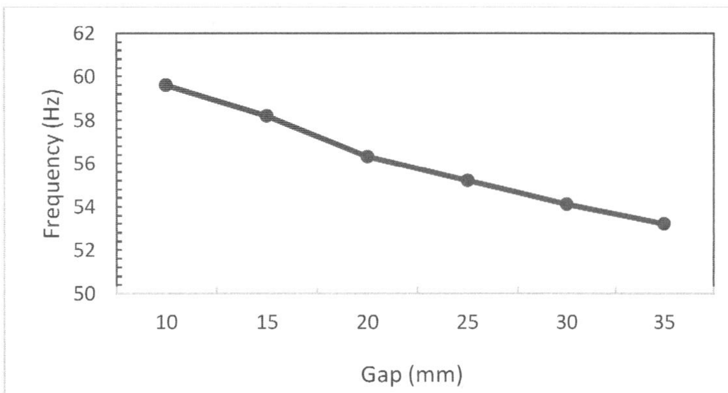

Figure 5. Relation of the gap to a frequency in a unload test.

Figure 5 shows that the generator frequency has decreased in line with the increase in the stator-rotor gap (mm). The smaller the gap between the rotor and the magnetic motor stator, the higher the generator frequency. This frequency is influenced by the period of rotation of the rotor to the generator stator. Moreover, this has to do with the number of turns and permanent magnets. The smaller the magnetic motor gab, the higher the motor rotation so that the generator rotor rotation is more significant and results in a higher magnetic flux, this causes the frequency to be greater. The highest frequency is in a 10 mm gap, with a value of 59.6 Hz.

4.3. Relationship of the gap to power in a unload test

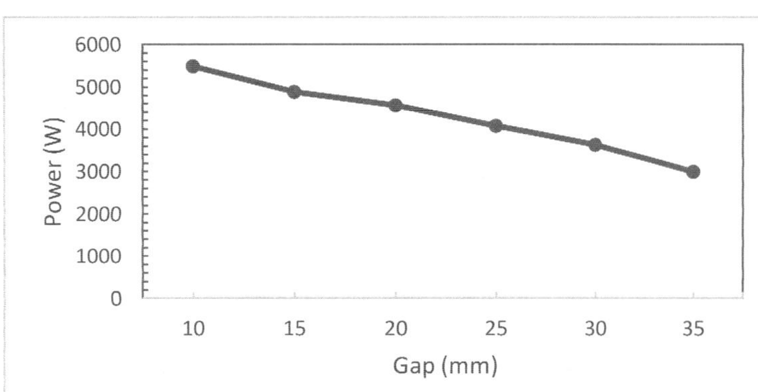

Figure 6. Relationship of the gap to power in the unloaded test.

Figure 6 shows that the generator power has decreased in line with the increase in the stator-rotor gap (mm). The lowest power is obtained at a gap of 35 mm with a no-load voltage value of 2992 W. When the gap is closed between the rotor-stator motor, the voltage also increases. The power increases significantly according to the closer the rotor-stator magnet motor gap. The highest power in this unload test is obtained at a frequency of 59.6 Hz with a voltage value of 5483.52 W. When the magnet gets closer, what happens is the greater the repulsion force, but with the tilt angle that has been set, a rotational motion will arise, and this is getting The closer the rotation speed accelerates. This rotational motion is also due to the magnetic levitation force. This rotational motion causes the generator rotor to rotate and produces an electromotive force. This electromotive force causes the generator output voltage to arise.

Table 2. Loaded testing.

Gap (mm)	**Voltage (V) (Volt)**	**Current (I) (A)**	**Frequency (f) (Hz)**	**Power (P)= Vx I x cos Θ (W)**
10	202	18	52	2981.52
15	200	16	47	2624
20	198	14	45	2273.04
25	196	12	39	1928.64
30	194	11	34	1749.88
35	190	9	30	1402.2

4.4. Relation of the gap to voltage in a unload test

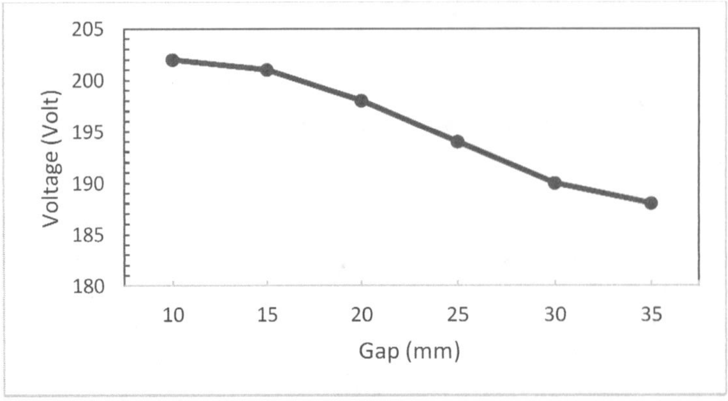

Figure 7. Relation of the gap to the voltage on a loaded test.

Figure 7 shows that the generator voltage has decreased in line with the increase in the stator-rotor gap (mm). The lowest voltage is obtained at a gap of 35 mm with a voltage value with a load of 190 V. When the distance is getting closer between the rotor-stator of the motor, the voltage also increases. The voltage increases significantly according to the closer the magnetic motor stator rotor gap. The highest voltage in this loaded test is obtained at a frequency of 52 Hz with a voltage value of 202 V. When the magnet gets closer, what happens is the higher the repulsion, but with the tilt angle that has been set, a rotational motion will arise, and the closer it is to accelerate. Rotation round. This rotational motion is also due to the magnetic levitation force. This rotational motion causes the generator rotor to rotate and produces an electromotive force. This electromotive force causes the generator output voltage to arise.

4.5. Relation of the gap to current in a loaded test

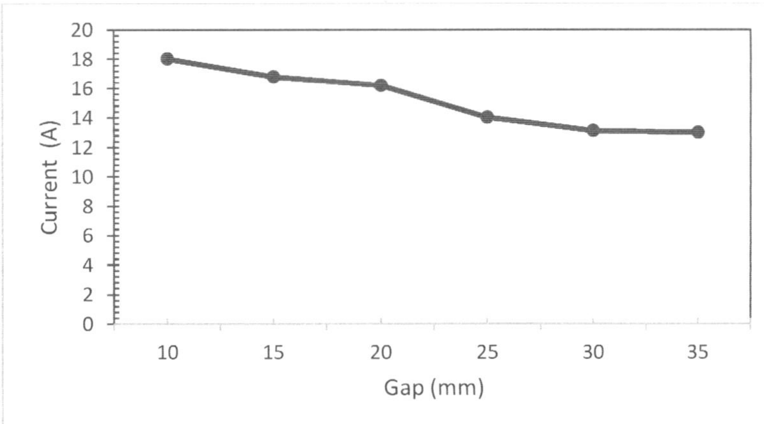

Figure 8. Relation of the gap to current in a loaded test.

Figure 8 shows that the generator frequency has decreased in line with the increase in the stator-rotor gap (mm). The smaller the gap between the rotor and the magnetic motor stator, the higher the generator frequency. This frequency is influenced by the period of rotation of the rotor to the generator stator. Moreover, this has to do with the number of turns and permanent magnets. The smaller the magnetic motor gab, the higher the motor rotation so that the generator rotor rotation is bigger and results in higher magnetic flux. This causes the frequency to be greater. The highest frequency is in a 10 mm gap, with a value of 52 Hz.

4.6. Relation of the gap to a frequency on a load test

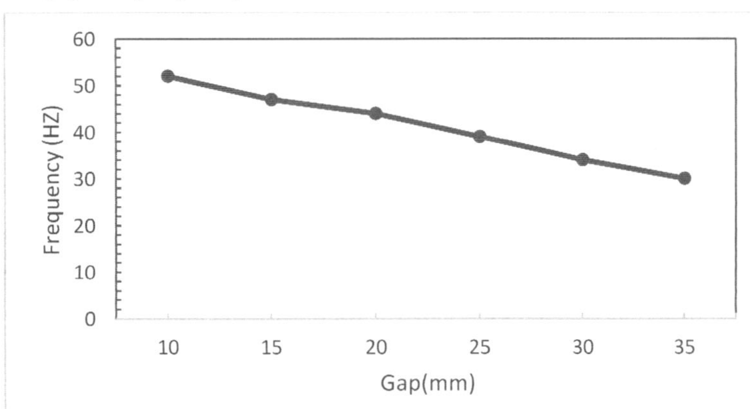

Figure 9. Relation of the gap to a frequency on a load test.

Figure 9 shows that the generator frequency has decreased in line with the increase in the stator-rotor gap (mm). The smaller the gap between the rotor and the magnetic motor stator, the higher the generator frequency. This frequency is influenced by the period of rotation of the rotor to the generator stator. Moreover, this has to do with the number of turns and permanent magnets. The smaller the magnetic motor gab, the greater the motor rotation so that the generator rotor rotation is bigger and results in greater magnetic flux. This causes the frequency to be greater. The highest frequency is in a 10 mm gap, with a value of 52 Hz.

4.7. Relationship gap (mm) to power (watts) on a load test

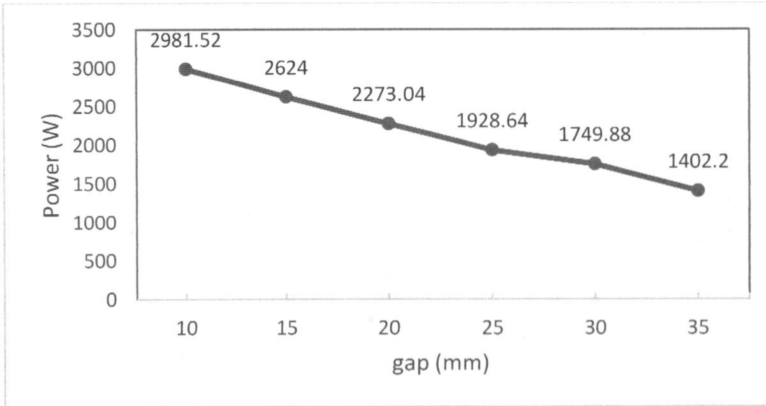

Figure 10. Relationship gap (mm) to power (watts) on a load test.

Figure 10 shows that the generator power has decreased in line with the increase in the stator-rotor gap (mm). The lowest power is obtained at a gap of 35 mm with a no-load voltage value of 1402.2 W. When the distance between the rotor-stator motor voltage also increases. The voltage increases significantly according to the closer the magnetic motor stator rotor gap. The highest power in this load test is obtained at a frequency of 52 Hz with a voltage value of 2981.52 W. When the magnet gets closer, what happens is the higher the repulsion force, but with the tilt angle that has been set, a rotational motion will arise, and the closer the rotation accelerates. Rotation. This rotational motion is also due to the magnetic levitation force. This rotational motion causes the generator rotor to rotate and produces an electromotive force. This electromotive force causes the generator output voltage to arise. The power value increases as the stator rotor gap get closer. The load test found that the power value is lower than the unloaded test at the same gap distance. This happens because some of the generated voltage loses voltage on the load. The addition of load when the generator is inactive condition will generally make the generator rotational speed decrease. This will also have an impact on decreasing the frequency and decreasing the generator output voltage. The reason is that the load served will generate a current, which then passes through the stator winding resulting in a magnetic field in the opposite direction to the rotating field of the rotor so that the rotation of the rotor becomes restrained. However, in the research that has been carried out using this permanent magnet generator, when the load is applied when the generator is an inactive condition, the rotating speed of the rotor remains stable. The load on the generator is too small so that the current flowing in the stator winding only creates a small magnetic field and is unable to withstand the rotating speed of the rotor.

5. Conclusion

In this designed permanent magnet generator, the resulting output is influenced by several variables. The variables in question are the number of magnets and the strength of the resulting magnetic field, the number of a stator winding poles, the number of stator turns, the air gap between the stator and the rotor, the closer the air gap is, the higher the generator output will be—the more the number of turns and the number of poles, the higher the generator output voltage. The stronger the magnetic field, the higher the generator output voltage will be, this can be done by adjusting the air gap between the stator and the rotor and using a type of permanent magnet with high magnetic field strength. However, the stronger the generator's magnetic field will affect the generator's rotational speed. The permanent magnet generator that is made and tested in this study is capable of producing a minimum output voltage of 190 V at a 35 mm gap under load conditions. The highest output voltage is 202 V at a frequency of 59.6 Hz with an electric current of 18 A with a gap of 10 mm under load conditions. The output power of the generator increases in proportion to the closer the gap is. The frequency of the generator increases in proportion to the increase in the gap, the closer the stator rotor of the magnet motor is, the resulting frequency will also increase, and vice versa.

References

[1] US Department of Energy 2015 *Quadrennial Technology Review: An Assessment of Energy Technologies and Research Opportunities* (Increasing Efficiency of Building Systems and Technologies) 38
[2] Hudson C R and Badiru A B 2008 Energy Systems In *Operations Research Applications* 148-176 CRC Press.
[3] Gielen D, Saygin D and Rigter J 2017 *Renewable Energy Prospects: Indonesia, a REmap analysis*
[4] Owusu P A and Asumadu-Sarkodie S 2016 A review of renewable energy sources, sustainability issues and climate change mitigation *Cogent Engineering* **3**(1) 1167990
[5] World Energy Resources 2016 *World Energy Counc* 6–46
[6] Susilo S H, Jannah Z, Mustofa L D and Effendhi Y 2020 Design of permanent magnet motor," *IOP Conf. Ser. Mater. Sci. Eng.* **732**(1)
[7] Shettigar Y P, Akshay B N and Abhishek R 2018 Electrical energy amplifying generator *IOP Conf. Ser. Mater. Sci. Eng.* **376**(1)
[8] Grover M, Lohith Kumar B and Ramalla I 2014 The Free Energy Generator *Int. J. Sci. Res. Publ.* **4**(1) 2250–3153
[9] Vesperman G 2014 02 Electrical Energy Innovations 1590
[10] Chan Y, Kantamaneni R and Allington M 2015 Study on energy efficiency and energy saving potential in industry and on possible policy mechanisms *ICF Consulting Limited, London. Tratto il giorno* **6**(08)
[11] Brauer H, Ziolkowski M, Porzig K and Toepfer H 2011 Modelling and simulation of a simple homopolar motor of Faraday's type *Facta universitatis-series: Electronics and Energetics* **24**(2) 221-242
[12] Matizamhuka W 2018 The impact of magnetic materials in renewable energy-related technologies in the 21st century industrial revolution: The case of South Africa *Adv. Mater. Sci. Eng.*
[13] Emetere M E, Okoro U, Etete B and Okunbor G 2016 Free energy option and its relevance to improve domestic energy demands in southern Nigeria *Energy Reports* **2** 229–236
[14] Dong W, Li Z, Sun H and Zhang J 2020 Modeling and analysis of a deflection type permanent magnet synchronous wind generator system *Energies* **13**(6)
[15] Hidayat M N, Santoso R, Syamsiana I N, Setiawan A and Djulihenanto S 2019 Design and analysis of a radiant charger using 5 coils and 5 poles of neodymium magnet as a rotor drive *J. Phys. Conf. Ser.* **1402**(3)
[16] Alagh Y K 2006 India 2020 Policy Energy Review *J. Quant. Econ.* **4**(1) 1–14
[17] IRENA 2019 *Innovation landscape for a renewable-powered future: Solutions to integrate variable renewables*
[18] Basford L 1968 a.C. Generators and Motors *electricity* (877) 184–194
[19] DOE 2015 Quadrennial Technology Review an Assessment of Energy Technologies and Research Chapter 3: Enabling Modernization of The Electric Power System *Dep. Energy united states*
[20] Isenberg G, Edinger R and Ebner J 2002 Renewable energies for climate benign fuel production *Environmental Science and Pollution Research* **9**(2) 99-104
[21] Jonathan E and Rucker J E 2005 Generator for Naval Applications
[22] Patrick S L 2002 The Motionless Electromagnetic Generator , the Active Vacuum
[23] Rodrigues L 2008 Wave power conversion systems for electrical energy production *Renew. Energy Power Qual. J.* **1**(6) 601–607
[24] Jurasz J, Canales F A, Kies M Guezgouz and A Beluco 2020 A review on the complementarity of renewable energy sources: Concept, metrics, application and future research directions *Sol.*

Energy **195** 703–724
[25] Supardi A and Dzikri N M 2019 A prototype of low power permanent magnet generator for renewable energy applications *AIP Conf. Proc.* **2114**
[26] Fremerey J K 2019 Permanent Magnet Bearings
[27] David A P 2019 Electro-Magnetic Induction: Free Electricity Generator *SSRN Electron. J.*
[28] Farrok O, Ahmed K, Tahlil A D, Farah M M, Kiran M R and Islam M 2020 Electrical power generation from the oceanic wave for sustainable advancement in renewable energy technologies *Sustainability* **12**(6) 2178
[29] Irhas and Suryaningsih R 2014 Study on wave energy into electricity in the south coast of yogyakarta, Indonesia *Energy Procedia* **47** 149–155
[30] M O N Motors 2009 MOTORS & GENERATORS 1–27
[31] Barragán-Escandón A, Ruiz J M O, Tigre J D C and Zalamea-León E F 2020 Assessment of power generation using biogas from landfills in an equatorial tropical context *Sustain.* **12**(7), 1–18

IOP Conf. Series: Materials Science and Engineering 1073 (2021) 012033 doi:10.1088/1757-899X/1073/1/012033

Mini PC based parking location determenation in POLINEMA

M Kusumawardani, N Suharto and S Wirayoga*

Electrical Engineering Department, State Polytechnic of Malang, Malang, Indonesia

*yoga.septriandi@polinema.ac.id

Abstract. Motorbikes and cars are the most widely used means of transportation by students, lecturers, and employees at Polinema. 14 motorcycle parking locations are available with a total capacity of approximately 6605 motorcycles and 18 car park locations with a capacity of around 330 vehicles. With a total of 12,226 students and 530 permanent lecturers and assuming 50% of them bring vehicles to campus at the same time, it can be estimated that the real need for parking and the difficulty in finding a parking space that is currently empty. The purpose of this study is to provide information to parking users about the location that can be occupied by the vehicle. The method used is to install 2 infrared sensors in 8 parking locations. Every 2 minutes the sensor sends data to the server, and then the server processes and provides information about the parking space vacancy. Information is displayed through displays displayed at each parking location in POLINEMA. The results of the work of the system made in this study can be seen on a monitor connected to the server which will make it easier for motorists who want to park their vehicles in the POLINEMA area. Measurements taken to see system performance are delay and the success rate of the system. Of all the measurements, the largest delay average was 1.9 seconds and the smallest delay average was 0.636 seconds. for the success rate it has the highest 100% and for the lowest is 80%.

1. Introduction

Based on data found in the Higher Education Database (PDPT), Malang State Polytechnic (Polinema) has 12,411 students registered up to 2019. The number of permanent lecturers is 530 people. The population will increase if it is added by the number of non-permanent lecturers, educational staff and administration. If it is assumed that 50% of them use transportation in the form of motorbikes and cars, it can be estimated that the need for parking spaces that need to be provided. On a campus area of approximately 13 hectares, Polinema provides 14 parking locations for motorbikes with a total capacity of about 6605 motorbikes. As for cars, parking lots are prepared in 18 locations with a total capacity of 330 standard-sized four-wheeled vehicles. With the large number of vehicles, it is often difficult for drivers to find out the position of an empty parking lot [1]. Motorists need to circle several parking locations before finally finding a spot. This condition is common, especially when big events are held on campus or events that involve the attendance of almost all lecturers and employees simultaneously. Based on the description above, this research proposes the title Mini PC Based Parking Location Determenation in POLINEMA. Several parking lots have become nodes and are equipped with sensors, mini PCs, and microcontrollers. All nodes are connected to each other to form a Manet. Sensing data will be obtained and then provide information about empty parking spaces. This information is displayed through displays in each parking lot.

2. The methods and descriptions of system

2.1. Methods

In this study, a web with a MySQL database was designed [2] that can manage the existence of a parking space, determine parking rates and direct the position of the vehicles that will park according to the block easily without being guided by parking attendants, both 4-wheeled and 2-wheeled vehicles. The model used is an infrared sensor (proximity sensor) and internet of things (IoT) [3], in this case, a chip that is programmed and placed in each parking lot so that it can transmit digital information signals both to the server and to the parking user's gadget. This research is intended for indoor parking locations, so it has not yet discussed outdoor locations. For network used, Mobile Ad Hoc Network (MANET) is a temporary network formed by several mobile nodes without an administrative center and wired infrastructure. In MANET [4,5], mobile nodes that are connected to wireless on the network can move freely and also act as a router. The nodes in this network are responsible for finding and handling routes to each node on the network. The nodes that wish to communicate with the fixed host must pass through the gateway first.

2.2. System design

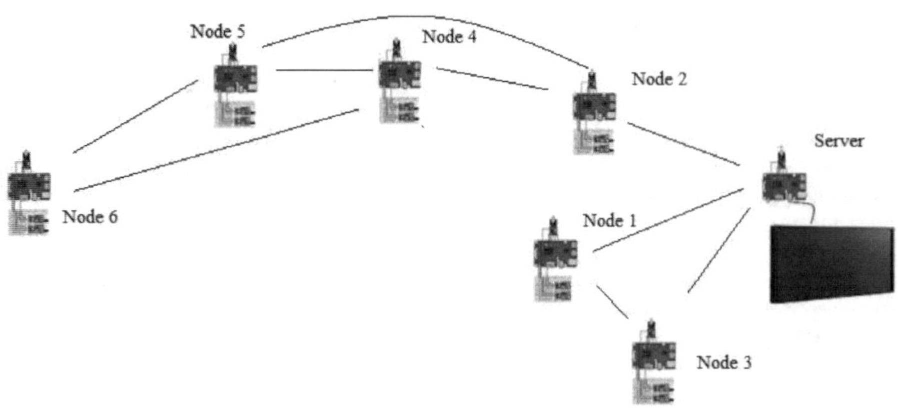

Figure 1. System design.

Figure 1 shows the design of the Implementation of Manet (Mobile Ad Hoc Network) as Network Infrastructure in Determining the Position of Empty Parking Areas in Polinema. This system consists of several parts which are:

- The first stage is to install the putty software for testing the AODV protocol [6].
- The second stage is the installation of the AODV routing protocol on the Ubuntu operating system (OS). If during the installation process there are no errors, the AODV protocol installation process is successful. However, if there is an error in the installation process, then re-install the AODV protocol.
- The third stage is to configure the AODV protocol in the MANET network, namely by building a MANET network consisting of 6 nodes, then configuring the AODV routing protocol and activating the AODV protocol on each node. After all the nodes are connected, the AODV protocol is tested in implementation.
- The fourth stage is to test the AODV protocol so that it produces data to be able to analyze the protocol performance by implementation.
- The fifth stage is the analysis of the work system between devices including programs that can run, between devices can communicate, between devices can send data to each other, the

distance that can be reached for communication between devices, delay, packet loss and throughput that occur in implementation.

- The last stage is making conclusions from the results of the research that has been done.

2.3. System flowchart
Determination of procedures and parameters in the system used in this study can be explained through the research flow diagram below in Figure 2.

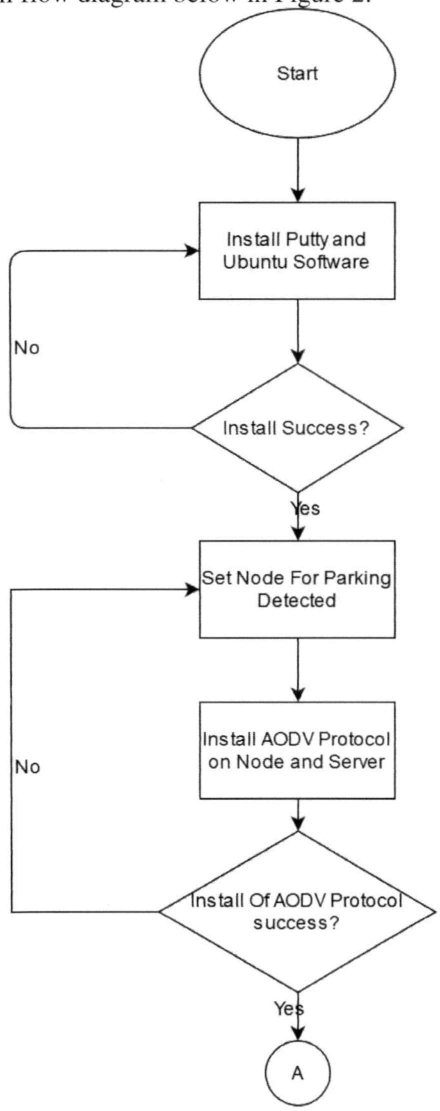

Figure 2. System flowchart.

3. Results and discussion

3.1. The implementation of web design
This section will show the results of a simple web design that will be displayed in the built local network. In this web interface, there are 6 parking areas fitted with 6 nodes, where each node covers one parking area. To view the parking web that is built can be seen in Figure 3.

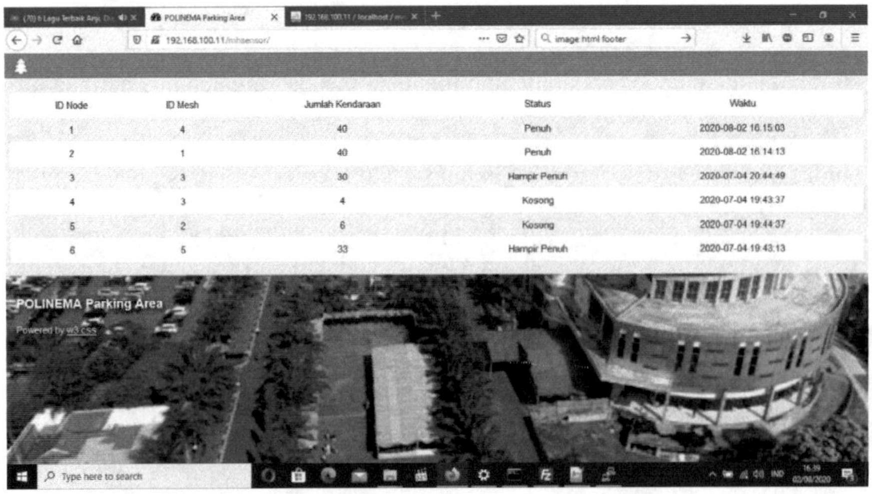

Figure 3. Web design.

In the web view above, the status has 3 types, namely full, almost full and empty. For a full condition, it is a condition where "Jumlah Kendarann" has reached the maximum value set in this system, which is 40. While almost full is a condition where the parking lot contains about 30 <= "Jumlah Kendaraan" <40. Meanwhile, for an empty condition is if "Jumlah Kendaraan "<30. Or can be seen on Table 1.

Table 1. Status criteria.

"Status" Criteria	**Numbers of "Jumlah Kendaraan"**
Penuh	40
Hampir Penuh	30<= Jumlah Kendaraan<40
Kosong	Jumlah Kendaraan <30

3.2. The system testing

The results of system testing are carried out to observe the performance of the AODV protocol on the MANET network that has been built with delay parameters and success rates. The first test that will be carried out here is testing the success rate of the node sending data to the server. This needs to be done because this system is real time so that it is always updated even though the condition of the node is very far away which is overcome by building a MANET network system. The results of the success rate can be seen in Table 2.

Table 2. The results of the success rate.

No.	Testing	Node					
		1	2	3	4	5	6
1	Testing Number 1	V	V	V	V	V	V
2	Testing Number 2	V	V	V	V	V	V
3	Testing Number 3	V	V	V	V	V	V
4	Testing Number 4	V	V	V	V	V	V
5	Testing Number 5	V	V	V	V	V	X
6	Testing Number 6	V	V	V	V	V	V
7	Testing Number 7	V	V	V	V	V	V
8	Testing Number 8	V	V	V	X	V	X
9	Testing Number 9	V	V	V	V	X	V
10	Testing Number 10	V	V	V	V	V	V
	Succes Rate	100%	100%	100%	90%	90%	80%

Table 2 above shows the test results of each node made in terms of sending data to the server. For node 1, node 2, and node 3 have a perfect success rate of 100%. Whereas for node 4 and node 5 it has a success rate of 90%, this is because out of 10 tests it has experienced one failure, this is because the distance of nodes 4 and 5 to the server is quite far, which is around 500-600 meters which makes nodes 4 and 5 ride with nodes. another as much as one node to reach the server. Meanwhile, node 6 has a success rate of 80% because it has failed twice when it was tested 10 times. This is the lowest success rate because node 6 is indeed the farthest node from the server about 850 m from the server. To go to server node 6, you need to ride another node with two nodes. Next is the delay table shown in Table 3.

Table 3. Delay table.

No.	Testing	Node Delay (Seconds)					
		1	2	3	4	5	6
1	Testing Number 1	0.51	0.62	0.85	1.31	1.41	1.83
2	Testing Number 2	0.63	0.64	0.98	1.27	1.43	1.84
3	Testing Number 3	0.56	0.73	0.97	1.25	1.51	1.86
4	Testing Number 4	0.56	0.65	0.88	1.33	1.46	1.87
5	Testing Number 5	0.81	0.92	1.02	1.35	1.42	-
6	Testing Number 6	0.51	0.66	1.05	1.41	1.45	1.98
7	Testing Number 7	0.72	0.69	0.94	1.38	1.43	1.92
8	Testing Number 8	0.67	0.85	1.02	-	1.46	-
9	Testing Number 9	0.65	0.65	1.01	1.42	-	1.97
10	Testing Number 10	0.74	0.82	0.98	1.44	1.52	1.93
	Delay Average	0.636	0.723	0.97	1.351111	1.454444	1.9

From the table 2 above, it can be seen that node 6 has the largest average delay with a delay of 1.9 seconds, this is because the distance between node 6 and the server is very far. Meanwhile, node 1 has the lowest average delay with a delay value of 0.636 seconds, this is because node 1 is the closest distance to the server.

4. Conclusion

From the results of the above discussion, it can be written that the manet network that was built to cover a large area was successfully implemented to support the parking monitoring system in the POLINEMA area. Based on the test results, the success rate of sending data can be seen that node 1, node 2, and node 3 have a perfect success rate of 100%. Meanwhile, node 4 and node 5 have a success rate of 90% and node 6 is 80%. From this data it can be seen that the further the location of the sensor from the server also affects the success rate of sending data to the server. For delay, the smallest average delay value is owned by node 1 with a delay value of 0.636 seconds while for the largest average delay is owned by node 6 with a delay value of 1.9 seconds. From this value, it can be seen that the farther the node is from the server, the more time it takes to send it.

Acknowledgment

The Authors would like to thank The State Polytechnic of Malang for the support to attend this conference.

References

[1] Costa Á, Rocha C and Melo S 2014 Parking management policies and the effectiveness of public policy solutions *Procedia-Social Behav. Sci.* **111** 965–73
[2] Gonzalez C V and Jung G 2019 Database SQL Injection Security Problem Handling with Examples *2019 International Conference on Computational Science and Computational Intelligence (CSCI)* (IEEE) pp 145–9
[3] Xu D, Wang Y, Xiong B and Li T 2017 MEMS-based thermoelectric infrared sensors: A review *Front. Mech. Eng.* **12** 557–66
[4] Rishiwal V, Agarwal S K and Yadav M 2016 Performance of AODV protocol for H-MANETs *2016 International Conference on Advances in Computing, Communication, & Automation (ICACCA)(Spring)* (IEEE) pp 1–4
[5] Rathod N and Dongre N 2017 MANET routing protocol performance for video streaming *2017*

International Conference on Nascent Technologies in Engineering (ICNTE) (IEEE) pp 1–5
[6] Paranavithana P and Jayakody A 2017 Compromising AODV for better performance: Improve energy efficiency in AODV *2017 6th National Conference on Technology and Management (NCTM)* (IEEE) pp 201–4

IOP Conf. Series: Materials Science and Engineering 1073 (2021) 012034 doi:10.1088/1757-899X/1073/1/012034

Design of HHO generators as producers of water fuel (HHO generator product analysis based on electric current and catalyst)

Z Jannah* and S H Susilo

Department of Mechanical Engineering, State Polytechnic of Malang, Malang, Indonesia

*zahra172.gh@gmail.com

Abstract. One of the fossil fuels is petroleum fuel. Petroleum is a non-renewable natural resource. Various aspects of life have felt its effects, where fuel oil has begun to experience scarcity. The higher price of fuel oil should make us realize that the amount of oil reserves has begun to run low. The purpose of this study was to determine the effect of electrodes, current strength, and concentration of NaHCO3 on the HHO gas generator output. HHO electrode generator material made of aluminum and brass. The electrode's shape is rectangular with a size of 250 mm x 400 mm, with a gasket seal. The test begins with the HHO gas pressure test. The independent variable used is aluminum and brass electrode variations. The electric current given is 20A, 30A, 40A, and the amount of 5-25% NaHCO3 catalyst. Then proceed with analyzing the pressure obtained on the HHO gas flow rate. To test the HHO gas flow rate carried out with a duration of 1 to 3 minutes in each treatment, the measurement of the discharge using a gas regulator to determine the flow rate of H2 and O2 produced. The results showed that the lowest gas discharge value was on aluminum electrodes with a current of 20A, i.e., 0,00033554 m3/s, while the highest discharge value on brass electrodes with a current of 50A was 0,001657 m3/s. It shows that the better the metal element contained in the electrodes and the higher the current applied, the greater the flow of H2 and O2 gas flow.

1. Introduction

The biggest energy needs that occur in Indonesia today is the use of fossil fuels. One of the fossil fuels is petroleum fuel. Petroleum is a non-renewable natural resource. Various aspects of life have felt its effects, where petroleum fuels have begun to experience scarcity. The higher price of fuel oil should e able to make us realize that the amount of oil reserves has begun to run low. The purpose is that efforts need to be made, such as finding and developing alternative energy sources [1]–[5]. One alternative energy is the use of hydrogen gas (H_2), which is environmentally friendly and can save the use of fuel oil in vehicles [6].

Hydrogen is an element that is widely available in nature, but not in the form of gas but in the form of compounds, for example, in water and coal. The purpose is to get hydrogen gas that is not available in a free state, and it can be done by industrial production so that the final price of the hydrogen depends on the production results that have been used. To obtain or produce hydrogen can be done by the water electrolysis method [2,6–8]. The hydrogen is different from energy from fossils. Fossil energy is very abundant in nature. Besides abundant existence, fossil energy can also be mined. It is very different from hydrogen, which must be produced first using electrolysis of water [3,9,10].

Hydrogen gas is an alternative energy source that is environmentally friendly so that it can be used as a fuel mixture for motor vehicles. The use of hydrogen for motor fuel oil is still rarely done [8,11, 12]. The use of hydrogen as an alternative to motor fuel oil can be used to conserve fossil fuels. Hydrogen can be used in fuel cells and has advantages, among others, can provide high energy efficiency, does not have soot particles that can harm humans, and has a low noise level. Gupta and Pant [13], Susilo and Anis [14], and Sudrajat et al. [15], states that the use of hydrogen for motor vehicle fuel oil can reduce harmful exhaust emissions such as CO_2, which has decreased by 59.93%. This figure is high enough for hydrogen to be said to be an excellent material to replace or serve as a mixture of motor vehicle fuel.

2. Theory

Colli et al [16] and Susilo et al [17] researched the electrodes dimensions, which stated in his research that the production of HHO gas by wet type HHO generators using the water electrolysis method produces the best generators at 1 mm electrode thickness. The electrodes used are stainless steel with $NaHCO_3$ (Sodium Bicarbonate) catalyst in the electrolyte solution. The HHO power used was 59.11 Watt resulting in an HHO gas production of 0,00054 kg/s, and the efficiency value of the HHO generator was 9.42%. Tranggono et al [18] and Morse [19] also stated in his research in making electrolyzer that only uses two variations of the electrode area, namely the front and rear surfaces only so that the level of accuracy and current flow is hampered. This shows that further research is needed on the effect of the extent of the electrodes that are directly in contact with the electrolysis system so that the hydrogen gas product produced matches the energy expended [20].

One way to save energy is to produce hydrogen, which can be obtained using the electrolysis method or through HHO technology. Production of HHO gas as much as possible is expected to reduce fuel prices and scarcity. This is supported by the discovery of Stanley Meyer, which is about fuel made from water used in his VW frog so it can go at speeds of up to 160 km/h using only 3 liters of water [21,22]. Previous research conducted only focused on the catalyst, shape, and extent of the electrodes on the production of hydrogen gas produced without analyzing the HHO generator made. Nylund et al [23] and Praveen and Sethumadhavan [24] state the hydrogen is not available to free in the earth, it is produced from an industrial production process so that the final price is determined by the production process used. To get or produce hydrogen gas can be done using the water electrolysis method. Stainless steel metal has a fairly good corrosion resistance because it has a chromium oxide (CrO) content. In the study using stainless steel log, a modification was made to the electrodes used in the form of the addition of catalytic metals such as Fe, Co, and Ni because they have similar properties and sizes with stainless steel metals. Hydrogen production can also be done by steam reforming of hydrocarbons [26–28]. In addition, hydrogen can be produced from the reaction of metals and water with the help of a catalyst. One of them is hydrogen produced from aluminum metal, which is reacted with water using a NaOH catalyst [29,30].

This generator has a working principle of water electrolysis, which is to flow DC through an electrolyte using an electrode. It can cause changes in electrical energy into chemical energy (redox reactions), causing water molecules to break down into hydrogen and oxygen. HHO Generators have two types: wet type and dry type [31].

3. Experimental methods

The object of research, is the HHO generator, several variables given in the experimental study include variations of aluminum and brass electrode materials, electric current 20A, 30A, 40A, and the amount of 5-25% NaHCO3 catalyst mixed into the solution, and the dependent variable, i.e., the discharge is converted from the HHO gas flow pressure. From the HHO generator testing process, HHO gas pressure data can be obtained and then converted to HHO gas discharge. Acetylene regulator is used to measuring the value of the resulting gas pressure. On the aluminum electrode, 15 times of research were carried out using a catalyst variation of 5% (200 g), 10%, (400 g) 15% (600 g), 20% (800 g) and 25% (1000 g) catalyst used is NaHCO3. Pressure data obtained will be processed into discharge (m ^ 3 / s).

3.1. HHO generator design

The generator design in this study was a prototype HHO generator with dimensions of 250 mm x 400 mm with electrodes made of aluminum and brass, it show figure 1. While the experimental setup show figure 2.

Figure 1. HHO generator design.

3.2. Experimental setup

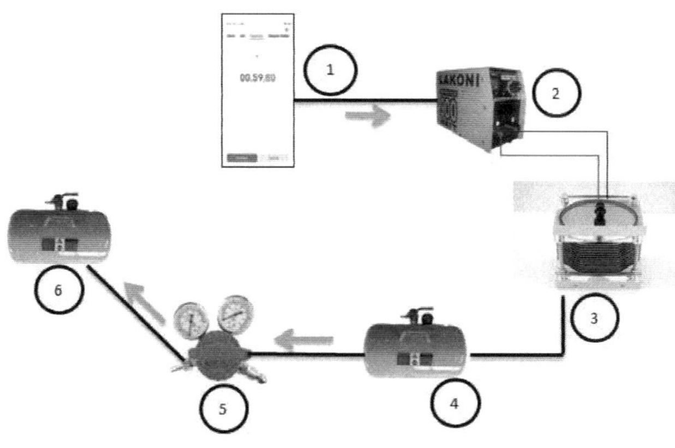

Figure 2. Research scheme.

Where: 1. Stopwatch, 2. DC inverter, 3. HHO Generators 4. HHO Gas Cylinders 1, 5. Asetylin Regulators, 6. HHO Gas Cylinders 2.

IOP Conf. Series: Materials Science and Engineering 1073 (2021) 012034 doi:10.1088/1757-899X/1073/1/012034

4. Results and discussion

4.1. HHO gas discharge with 5% NaHCO3

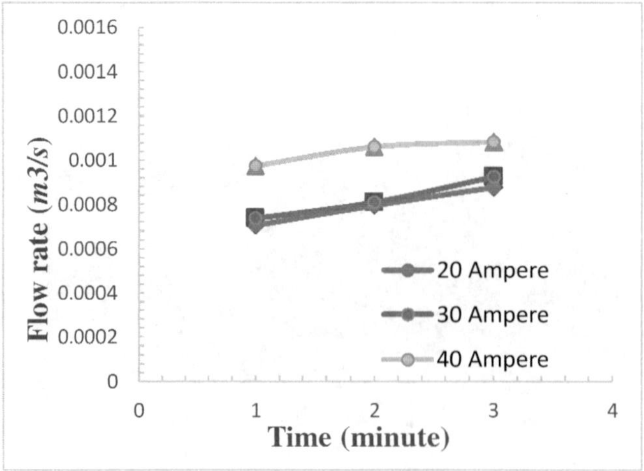

Figure 3. HHO gas discharge with 5% NaHCO3.

Figure 3 shows an increase in HHO gas discharge when the HHO generator is given a voltage source with a current of 40 Amperes with a time of 1 minute, that is, produces a Debit of 0.0009725 (m^3/s) and within 3 minutes the discharge reaches 0.001082 (m^3/s). It shows that the longer the HHO generator gets electricity to supply, the higher the HHO gas produced. The lowest HHO gas discharge is obtained at an electric current of 20 Amperes with a time of 1 minute that is equal to 0.0007038 m^3/s. With 5% NaHCO3 content, the maximum HHO gas discharge on electric current is 40 Amperes. Thus, the time and electric current are very influential on the amount of HHO gas production, because the longer and the higher the current is given to the HHO generator, the reduction and oxidation reaction (redox) will run more optimally. In this research, sodium bicarbonate ($NaHCO_3$) acts to accelerate the reaction. When sodium bicarbonate is added to water, it separates sodium cations (positively charged) and hydroxide anions (negatively charged oxygen and hydrogen). Sodium bicarbonate is easily soluble in water and will produce heat (exothermic). From the results of the above study, it can be concluded that the best electric current to produce the amount of HHO gas discharge is an electric current of 40 amperes. It is because the greater the source of electric current, the faster the reaction process is carried out. The results of testing the measurement of discharge from HHO generators using variations in the electrode material, the amount of $NaHCO_3$, and the electric current are given to the HHO generator showed a significant effect on the rate of HHO gas production. The better the electrode material used, the higher the HHO gas discharge generated. The more amount of $NaHCO_3$ given, the higher the HHO gas discharge generated. Furthermore, the higher the electric current used, the rate of HHO gas production will be even higher. Thus providing variations in the electrode material, the amount of $NaHCO_3$, and electric current influences the rate of HHO gas production (discharge), but the amount of $NaHCO_3$ has a significant effect on the amount of HHO gas production rate when HHO gas is given a power source for 3 minutes, while for electrode material and the electric current influences the speed of the H_2O electrolysis process, the better the metal content of the electrodes and the higher the current is given, the faster the electrolysis process carried out by the HHO generator.

4.2. HHO gas discharge with NaHCO3 10%

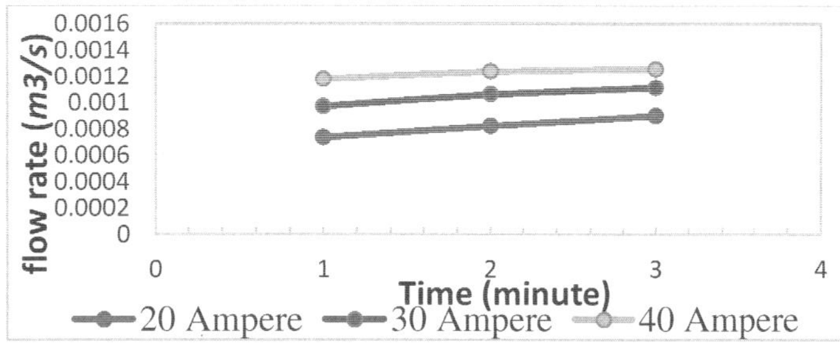

Figure 4. HHO gas discharge with NaHCO3 10%.

Figure 4 shows the increase in each current given for 1 minute to 3 minutes. The largest increase in HHO gas discharge within 3 minutes occurred at a current of 20 Amperes in the amount of 0.0001649 m^3/s. The largest HHO gas discharge is obtained at a current of 40 Amperes with a time of 3 minutes of 0.001255 m^3/s, the lowest HHO gas discharge produced is at a current of 20 Amperes with a flow of 0.0007351 m^3/s. It shows that the higher the electric current is given to provide electricity to the HHO generator, the greater the HHO gas produced. With 10% NaHCO3 content, the maximum HHO gas discharge when using a current of 40 Amperes. Thus it can be said that the time and electric current affect the amount of HHO gas production, because the longer and the greater the current given to the HHO generator, the reduction and oxidation reaction will run more optimally, and the more $NaHCO_3$ levels are given, the reaction reduction and oxidation will proceed quickly too. From the results of the above study, it can be concluded that the best electric current to produce the amount of HHO gas discharge is an electric current of 40 amperes. It is because the greater the source of electric current, the faster the reaction process is carried out. The results of testing the measurement of discharge from HHO generators using variations in the electrode material, the amount of NaHCO3, and the electric current are given to the HHO generator showed a significant effect on the rate of HHO gas production. The better the electrode material used, the higher the HHO gas discharge generated. The more amount of $NaHCO_3$ given, the higher the HHO gas discharge generated. Furthermore, the higher the electric current used, the rate of HHO gas production will be even higher. Thus providing variations in the electrode material, the amount of $NaHCO_3$, and electric current influences the rate of HHO gas production (discharge), but the amount of $NaHCO_3$ has a significant effect on the amount of HHO gas production rate when HHO gas is given a power source for 3 minutes, while for electrode material and the electric current influences the speed of the H_2O electrolysis process, the better the metal content of the electrodes and the higher the current is given, the faster the electrolysis process carried out by the HHO generator.

4.3. HHO gas discharge 20 amperes brass electrodes

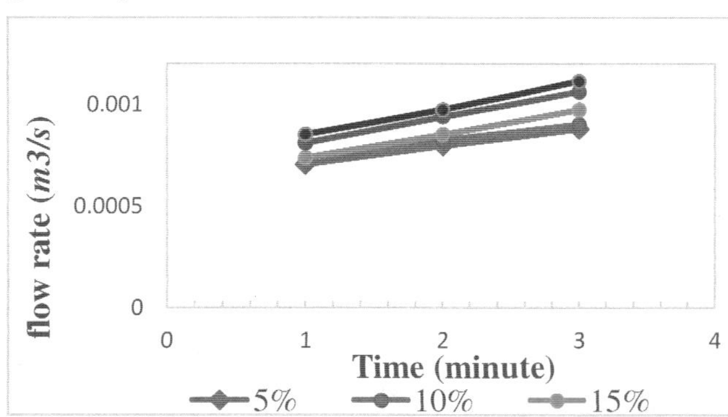

Figure 5. HHO gas discharge with $NaHCO_3$ 10%.

Figure 5 shows the increase in each current given for 1 minute to 3 minutes. The largest increase in HHO gas discharge within 3 minutes occurred at a current of 20 Amperes in the amount of 0.0001649 m^3/s. The largest HHO gas discharge is obtained at a current of 40 Amperes with a time of 3 minutes of 0.001255 m^3/s, the lowest HHO gas discharge produced is at a current of 20 Amperes with a flow of 0.0007351 m^3/s. It shows that the higher the electric current is given to provide electricity to the HHO generator, the greater the HHO gas produced. With 10% $NaHCO_3$ content, the maximum HHO gas discharge when using a current of 40 Amperes. Thus it can be said that the time and electric current affect the amount of HHO gas production, because the longer and the greater the current given to the HHO generator, the reduction and oxidation reaction (redox) will run more optimally, and the more NaHCO3 levels are given, the reaction reduction and oxidation (redox) will proceed quickly too. From the results of the above study, it can be concluded that the best electric current to produce the amount of HHO gas discharge is an electric current of 40 amperes. It is because the greater the source of electric current, the faster the reaction process is carried out. The results of testing the measurement of discharge from HHO generators using variations in the electrode material, the amount of $NaHCO_3$, and the electric current are given to the HHO generator showed a significant effect on the rate of HHO gas production. The better the electrode material used, the higher the HHO gas discharge generated. The more amount of $NaHCO_3$ given, the higher the HHO gas discharge generated. Furthermore, the higher the electric current used, the rate of HHO gas production will be even higher. Thus providing variations in the electrode material, the amount of $NaHCO_3$, and electric current influences the rate of HHO gas production (discharge), but the amount of $NaHCO_3$ has a significant effect on the amount of HHO gas production rate when HHO gas is given a power source for 3 minutes, while for electrode material and the electric current influences the speed of the H_2O electrolysis process, the better the metal content of the electrodes and the higher the current is given, the faster the electrolysis process carried out by the HHO generator.

4.4. HHO gas discharge 20 amperes brass electrodes

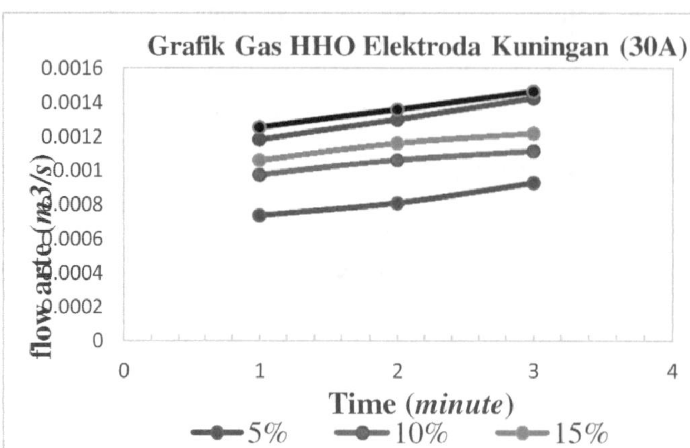

Figure 6. HHO gas discharge 30 ampere brass electrodes.

Figure 6 shows an increase in HHO gas discharge within 3 minutes for each given $NaHCO_3$ level ranging from 5-25% on the brass electrode. The addition of $NaHCO_3$ by 5% within 3 minutes produced a gas discharge of 0.000925 m^3/s. It shows an increase in HHO gas discharge from time to time and from each addition of NaHCO3, unlike the aluminum electrodes. The lowest HHO gas discharge was obtained with a 5% $NaHCO_3$ level in 1 minute of 0.000735 m^3/s. On the addition of 25%, $NaHCO_3$ debit gas produced large enough to reach 0.001462 m^3/s within 3 minutes. From the graph above, it can be concluded that the addition of 25% $NaHCO_3$ with 30A electrical current to the brass electrode is still better in terms of the amount of discharge generated, in addition to the addition of $NaHCO_3$ will accelerate the electrolysis process. Based on these data, the gas discharge produced in the H2O electrolysis process is best shown by the addition of $NaHCO_3$ at 25% (1000gr). The results of testing the measurement of discharge from HHO generators using variations in the electrode material, the amount of $NaHCO_3$, and the electric current are given to the HHO generator showed a significant effect on the rate of HHO gas production. The better the electrode material used, the higher the HHO gas discharge generated. The more amount of $NaHCO_3$ given, the higher the HHO gas discharge generated. Furthermore, the higher the electric current used, the rate of HHO gas production will be even higher. Thus providing variations in the electrode material, the amount of $NaHCO_3$, and electric current influences the rate of HHO gas production (discharge), but the amount of $NaHCO_3$ has a significant effect on the amount of HHO gas production rate when HHO gas is given a power source for 3 minutes, while for electrode material and the electric current influences the speed of the H_2O electrolysis process, the better the metal content of the electrodes and the higher the current is given, the faster the electrolysis process carried out by the HHO generator.

4.5. Discharge Gas HHO 40 amperes Brass Electrodes

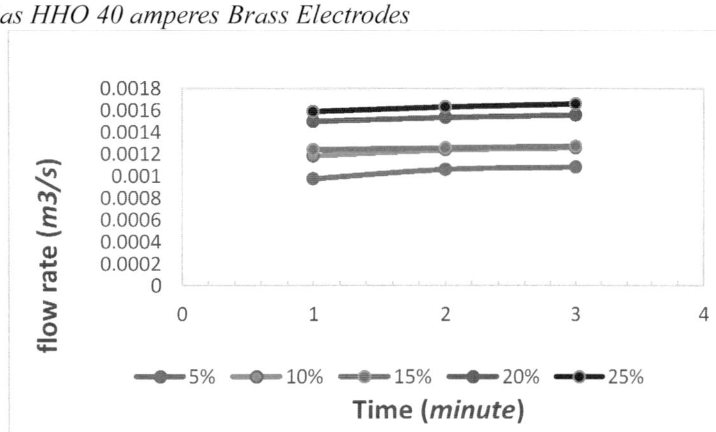

Figure 7. HHO gas discharge 40 ampere brass electrodes.

Figure 7 shows an increase in HHO gas discharge within 3 minutes for each given $NaHCO_3$ level starting from 5-25% on the brass electrode. The addition of $NaHCO_3$ by 5% within 3 minutes resulted in a gas discharge of 0.001082 m^3/s. It shows an increase in HHO gas discharge from time to time, as well as from each addition of $NaHCO_3$, unlike the aluminum electrodes. The lowest HHO gas discharge was obtained with 5% $NaHCO_3$ levels in 1 minute of 0.000972 m^3/s. The addition of 25% $NaHCO_3$ debit gas produced is large enough to reach 0.001657 m^3/s within 3 minutes. However, 10% and 15% $NaHCO_3$ only slightly increased HHO gas discharge by 0.000018 m^3/s. From the graph above, it can be concluded that the addition of 25% $NaHCO_3$ with 40A electric current on the brass electrode is still better in terms of the amount of discharge generated, and the addition of NaHCO3 will accelerate the electrolysis process. Based on these data, the gas discharge produced in the H_2O electrolysis process is best shown by the addition of NaHCO3 at 25% (1000gr). The results of testing the measurement of discharge from HHO generators using variations in the electrode material, the amount of $NaHCO_3$, and the electric current are given to the HHO generator showed a significant effect on the rate of HHO gas production. The better the electrode material used, the higher the HHO gas discharge generated. The more amount of $NaHCO_3$ given, the higher the HHO gas discharge generated. Moreover, the greater the electric current used, the rate of HHO gas production will be even higher. Thus providing variations in the electrode material, the amount of $NaHCO_3$, and electric current influences the rate of HHO gas production (discharge), but the amount of $NaHCO_3$ has a significant effect on the amount of HHO gas production rate when HHO gas is given a power source for 3 minutes, while for electrode material and the electric current influences the speed of the H_2O electrolysis process, the better the metal content of the electrodes and the higher the current is given, the faster the electrolysis process carried out by the HHO generator.

5. Conclusion

Based on the results, the following conclusions can be drawn.

- Giving variations of electric current 20 amperes and 30 amperes has a significant influence on the discharge of HHO gas produced, especially on the most significant current. The provision of massive electrical currents makes the HHO gas discharge faster, but too large currents also make the generator heat up quickly when using poor electrode materials such as aluminum.
- The variation in the amount of $NaHCO_3$ has not had a significant effect on the discharge of HHO gas produced, especially on aluminum electrodes. However, the largest HHO gas discharge was obtained from the research conducted at the amount of 25% $NaHCO_3$ (1000gr).
- The most gas flow rate or HHO gas discharge can be produced from brass electrodes with 25% $NaHCO_3$ (1000gr) with a 40-ampere electric current source equal to 0.001657 (m^3/s).

References

[1] Goodger E M 1994 Fuels and combustion *Mech. Eng. Ref. B.* 1994
[2] European Commission 2015 State of the Art on Alternative Fuels Transport Systems in the European Union 128 2015
[3] Staffell I, Scamman D, Abad A V, Balcombe P, Dodds P E, Ekins P and Ward K R 2019 The role of hydrogen and fuel cells in the global energy system. *Energy & Environmental Science* **12**(2) 463-491
[4] Ajanovic A and Haas R Economic and Environmental Prospects for Battery Electric- and Fuel Cell Vehicles: A Review *Fuel Cells* **19**(5)515–529
[5] BOC 2011 Hydrogen offers the perfect balance between low carbon mobility and climate protection
[6] Bossel U and Eliasson B 2002 Energy and Hydrogen Economy *Eur. Fuel Cell Forum, Lucerne* 36
[7] Yuvaraj A L and Santhanaraj D 2014 A systematic study on electrolytic production of hydrogen gas by using graphite as electrode *Mater. Res.*, **17**(1) 83–87
[8] Susilo S H and Jannah Z 2020 Effect of Electrodes , Electric Currents , And Nahco 3 Concentration Against Hho Pressure Generator Effect of Electrodes , Electric Currents , And Nahco 3 Concentration Against Hho Pressure Generator 3–9 2020.
[9] Braithwaite D 2019 Beyond Fossil Fuels: Indonesia's fiscal transition GSI REPORT
[10] Rusdianasari, Bow Y and Dewi T 2019 HHO Gas Generation in Hydrogen Generator using Electrolysis," *IOP Conf. Ser. Earth Environ. Sci.*, **258** (1)
[11] Walter L 2001 Hydrogen Use in Internal Combustion Engines *Hydrog. Fuel Cell Engines* 3–1
[12] Barrera J, Paya A, Mang N and Siss T 2013 Hydro-Gen
[13] Gupta R and Pant K 2008 Fundamentals and Use of Hydrogen as a Fuel *Hydrog. Fuel* 2–32
[14] Susilo S H and Anis U 2020 The Effect of Magnet Strength and Engine Speed on Fuel Consumption and Exhaust Gas Emission for Gasoline Vehicle **17**(3) 18–25
[15] Sudrajat A, Mayfa Handayani E, Tamaldin N and Kamal Mat Yamin A 2018 Principle of generator HHO hybrid multistack type production technologies to increase HHO gas volume *SHS Web Conf.* **49**
[16] Colli A N, Girault H H and Battistel A 2019 Non-precious electrodes for practical alkaline water electrolysis *Materials (Basel)* **12**(8) 1–17
[17] Susilo S H, Suparman S, Mardiana D and Hamidi N 2016 The effect of velocity ratio study on microchannel hydrodynamics focused of mixing glycerol nitration reaction *Period. Polytech. Mech. Eng.* **60**(4) 228–232
[18] Tranggono A, Salim A, Yudha R G P, Lawu K S, and Yusuf A 2019 Performance Characteristics between Titanium and Stainless Steel Materials as Electrodes on Dry Cell Type HHO Gas Generator **4**(3) 563–566
[19] Morse J Hho aviable means or reducing emissions."
[20] Frumkin H, Hess J and Vindigni S 2009 Energy and public health: The challenge of peak

petroleum *Public Health Rep.* **124**(1) 5–19

[21] Nagashima M 2018 *Japan's Hydrogen strategy and its economic and geopolitical implications études de l'Ifri monica nagasHima Centre for Energy*

[22] Water T, Vinegar, and Soda B The best HHO Electrolyte (catalyst) for HHO Gas Generator ?," vol. 2.

[23] Nylund N O, Aakko-Saksa P, and Sipilä K 2008 *Status and outlook for biofuels, other alternative fuels and new vehicles* **2426**

[24] Praveen K and Sethumadhavan M 2017 On the extension of XOR step construction for optimal contrast grey level visual cryptography *2017 Int. Conf. Adv. Comput. Commun. Informatics, ICACCI 2017* 219–222.

[25] Van Kranenburg K, Schols E, Gelevert H, De Kler R, Van Delf Y and Weeda M 2016 Empowering the chemical industry. Opportunities for electrification *Tno-Ecn*, 32

[26] Sapountzi F M, Gracia J M, Fredriksson H O and Niemantsverdriet J H 2017 Electrocatalysts for the generation of hydrogen, oxygen and synthesis gas *Progress in Energy and Combustion Science* **58** 1-35

[27] T. H. E. Role *et al.*, "University of Ghana http://ugspace.ug.edu.gh University of Ghana http://ugspace.ug.edu.gh," no. 10388417, 2018.

[28] Hamaad A S A A, Tawfik M, Khattab S and Newir A 2017 Device for Using Hydrogen Gas as Environmental Friendly Fuel for Automotive Engine (GREEN & ECO H 2) *Procedia Environ. Sci.* **37** 564–571

[29] Givirovskiy G, Ruuskanen V, Ojala L S, Lienemann M, Kokkonen P and Ahola J 2019 Electrode material studies and cell voltage characteristics of the in situ water electrolysis performed in a pH-neutral electrolyte in bioelectrochemical systems *Heliyon*, **5**(5) e01690

[30] Singh S, Jain S, Venkateswaran P S, Tiwari A K, Nouni M R, Pandey J K and Goel S 2015 Hydrogen: A sustainable fuel for future of the transport sector. *Renewable and sustainable energy reviews* **51** 623-633

[31] Esposito D V 2017 Membraneless Electrolyzers for Low-Cost Hydrogen Production in a Renewable Energy Future *Joule* **1**(4) 651–658

IOP Conf. Series: Materials Science and Engineering 1073 (2021) 012035 doi:10.1088/1757-899X/1073/1/012035

Analysis and comparison of nonlinear control for DC/DC buck converter in PV system

L N Palupi[1,*], T Winarno[1], E Mandayatma[1] and L Ardhenta[2]

[1]Department of Electrical Engineering, State Polytechnic of Malang Soekarno Hatta St. 9 Malang, Indonesia
[2]Department of Electrical Engineering, Brawijaya University, MT Haryono St. 167 Malang, Indonesia

*lucky.nindya.palupi@polinema.ac.id

Abstract. The use of DC power systems is increasingly widely used, so we need a system to convert DC power systems, one of which is a buck converter. The buck converter system requires a controller so that the output voltage generated is as desired without any disturbance when the buck is operating. This study examines improvements in buck converter performance using backstepping and SMC controllers. Backstepping controller and SMC are able to maintain stability where it can be shown by looking at the output voltage value of the buck converter. The results of a comparison of the simulation implementation of the backstepping controller and the SMC show that the recovery time of the SMC is faster than backstepping. From this comparison it is also seen that the amount of SMC voltage deviation is smaller than backstepping.

1. Introduction

In this modern era, almost all electronic device systems require a DC source. DC voltage is widely used in low-voltage electronic systems to high-voltage electrical systems. Including for DC electric motors, maximum power tracker (MPPT), battery charging, and so on. Therefore, the use of a DC power supply system requires a system that is able to convert DC voltage from one DC voltage level into another DC voltage level. In the industrial world, DC conversion systems are widely used to improve system efficiency [1]. There are different types of DC voltage conversion designs that have different capabilities. The types of energy conversion devices include a boost converter, buck converter, buck-boost converter, zeta converter, cuk converter, DC-DC bidirectional and others [2].

DC-DC converters have a DC voltage input and an output DC voltage. The output voltage can be greater or less than the input voltage. The buck converter is used to convert a large input voltage into a smaller output voltage. This type of converter is very suitable to be used to provide voltage to a load where the load requires a smaller voltage than the DC source it has. In the voltage converter that converts a DC quantity to another DC quantity, there is a step called regulating which has the function of regulating the output voltage, increasing or decreasing the value of the voltage. The output DC voltage of PV system depends on irradiation and ambient temperature such that it has unstable output voltage. Therefore, DC voltage does not only have to be regulated by the equipment we install. So, we need a DC voltage control, so that the equipment we install works in accordance with its capabilities [3-7].

From the description above, there are problems that occur with the buck converter. First, it is difficult to control the current because it has a large enough ripple. The second is the transient output voltage at

IOP Conf. Series: Materials Science and Engineering 1073 (2021) 012035 doi:10.1088/1757-899X/1073/1/012035

the start of the system which causes the length of time Ts (Time Settling). This transient symptom cannot be eliminated, but its impact can be minimized by controlling the PWM on the buck converter. In the buck converter, control can be done with various controlling techniques such as PID, linear quadratic regulator (LQR), sliding mode controller (SMC), backstepping, and many more controlling techniques that can be used [8-11].

In this paper, a study of buck converter control was carried out using backstepping controllers and sliding mode controllers. Mathematical model of buck converter is used to help learning and understanding the operating principles of the buck converter [12]. The backstepping control was chosen because the output of the buck converter with backstepping control was displayed using the parameters listed in the journal, which has a fast output voltage setting time and low overshoot [13]. As a comparison, this research uses Sliding Mode Controller (SMC) control. The SMC was chosen because it was able to force the system output to move according to its surface path [14], besides that in the research written by Uçak and Günel [15] the output of the buck converter using the SMC was shown with interference only with the load. In this research, a simulation of disturbance testing was carried out in the form of a load change, input voltage change, and the change of voltage reference. The purpose of comparing the two nonlinear controls is so that we can see which of the two controllers is best used to regulate the output voltage of the buck converter. In this study also examine the durability these two controls kind of interferences. The interferences are given to compare which method is the most efficient for buck converter.

2. Mathematical model of buck converter

Buck converter is a power electronic converter with DC input voltage and DC voltage output where the output voltage value can be smaller than the source voltage. This converter is included in a switching type converter which works by opening and closing electronic switches. This converter uses inductors, diodes, capacitors and a MOSFET or IGBT which functions as an electronic switch. The buck converter circuit is shown in Figure 1.

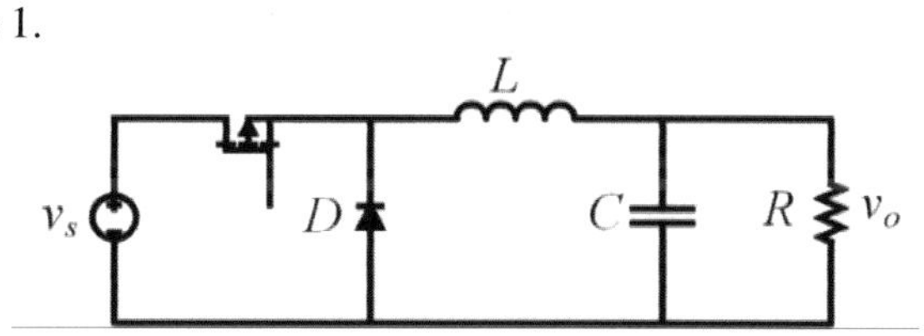

Figure 1. Block diagram of proposed system.

When the switch is closed, the diode is reverse bias as shown in Figure 2. In accordance with Kirchhoff's voltage law, the loop passes through the closed source, inductor, and switch. By analyzing the voltage and current that passes through the inductor, it can be determined by the kirchoff voltage law and kirchoff current law, which the calculation will also be used to make the equation of inductor current () and capacitor voltage (V_C) as,

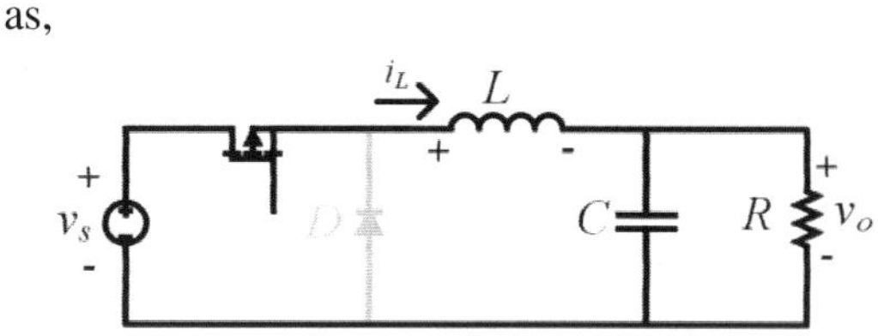

Figure 2. The equivalent circuit when the switch is ON.

$$\frac{di_L}{dt} = \frac{V_{in} - V_C}{L}$$

IOP Conf. Series: Materials Science and Engineering 1073 (2021) 012035 doi:10.1088/1757-899X/1073/1/012035

$$\frac{dv_C}{dt} = \frac{1}{C}(i_L - \frac{V_C}{R})$$

When the switch is open as shown in Figure 3, the inductor current does not change instantaneously, the diode is biased forward to provide a path for the inductor current. It is assumed that the output voltage Vo is constant.

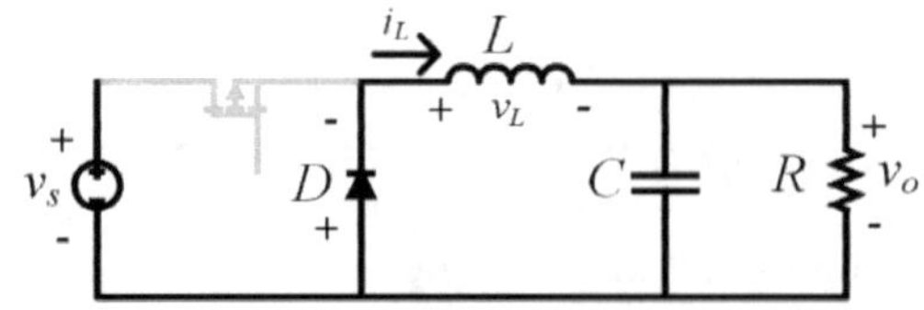

Figure 3. Block diagram of proposed system.

The value of the voltage and current that passes through the inductor when the switch is open can be determined by the following equation.

$$\frac{di_L}{dt} = -\frac{V_C}{L}$$

$$\frac{dv_C}{dt} = \frac{1}{C}(i_L - \frac{V_C}{R})$$

By using state space averaging modeling, the converter equation is simplified as equation,

$$\dot{x} = \begin{bmatrix} 0 & -\frac{1}{L} \\ \frac{1}{C} & -\frac{1}{RC} \end{bmatrix} \times \begin{bmatrix} i_L \\ V_C \end{bmatrix} + \begin{bmatrix} \frac{d}{L} \\ 0 \end{bmatrix} \times [\text{Vin}]$$

So that we can get the state space average value of the buck converter current ($\dot{X}_1$) and voltage ($\dot{X}_2$).

$$\dot{X}_1 = -\frac{X_2}{L} + \frac{dV_{in}}{L} \quad (1)$$

$$\dot{X}_2 = \frac{X_1}{C} - \frac{X_2}{RC} \quad (2)$$

3. Sliding mode control design

In order to obtain the desired output voltage, it is necessary to make an error function in SMC, the error function can also be interpreted as sliding surface. The error function is based on the current output of buck converter. In designing sliding surface, current, voltage and power can be used, but in reality, power is difficult to control so the reference that is often used is current. The proposed sliding surface is given by,

$$\sigma = K_1 e + \dot{e}$$

where e is the error of inductor current and K_1 is positive gain for sliding surface. The derivation of sliding surface is derived to reach its equilibrium system.

$$\dot{\sigma} = -K_1\dot{X}_1 + \frac{1}{L}\dot{X}_2 \quad (3)$$

The equivalent control U_{eq} is obtained as (4) by substituting (2) into (3).

$$U_{eq} = \frac{1}{V_{in}K_1C}x_1 - \frac{x_2}{V_{in}}\left(-\frac{1}{K_1CR} + 1\right) \quad (4)$$

The first step to deriving U_n is to employ the Lyapunov function.

$$V = \frac{1}{2}\sigma^2$$
$$\dot{V} = \sigma\dot{\sigma}$$

By considering a positive definite Lyapunov function, it is defined

$$\sigma\left(\frac{K_2 V_{in} U_n}{L}\right) < 0$$

The signal of natural control U_n is determined as

$$U_n = K_3\left(\frac{L}{K_2 V_{in}}\right) sign(\sigma)$$

The input control U is defined by $U_{eq} + U_n$ and, therefore, it can be obtained as,

$$U = \frac{1}{V_{in} K_1 C} X_1 + \frac{V_d}{V_{in}}\left(-\frac{1}{K_1 CR} + 1\right) + K_3\left(\frac{L}{K_2 V_{in}}\right) sign(\sigma)$$

4. Results and analysis

The research was conducted to see the output voltage generated by the buck converter, then given a controller so that the regulated voltage is better as shown in Figure 4. In buck converter, there are several parameters that must be known and set. The buck converter specifications studied are given in Table 1.

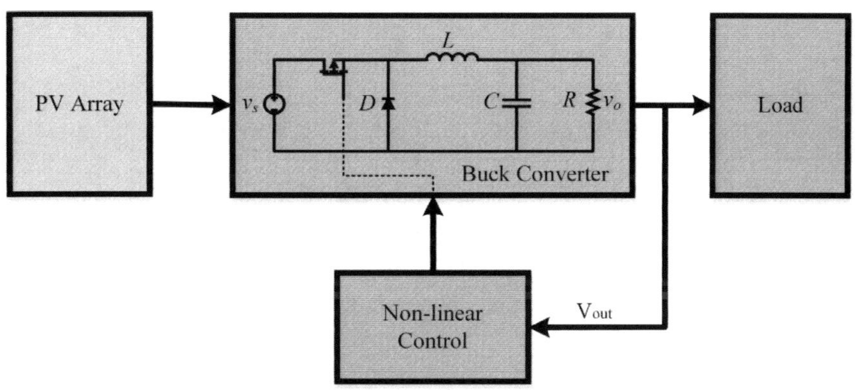

Figure 4. Block diagram of proposed system.

Table 1. Buck converter specifications.

Parameter	Symbol	Value
Input Voltage	V_{in}	15-24 V
Inductor	L	2 mH
Load Resistor	R	30 Ω
Capacitor	C	6 μF
Frequency	f	15 kHz

Table 1 shows the specifications and values used. The input voltage uses a value of 15 V, which is adjusted to the value of most photovoltaic outputs, which is around 12 V to 14.7 V. The output voltage generated by the buck converter is 5 V, this voltage is commonly used in battery charging systems starting from the battery. to the device.

In order to find out the success rate of the plant and controller in producing an output voltage that is in accordance with what is required by the load and to know the resistance of the controller when there

IOP Conf. Series: Materials Science and Engineering 1073 (2021) 012035 doi:10.1088/1757-899X/1073/1/012035

is a change in input voltage, a change in the loading side, and a change in reference voltage. The proposed controller is compared by backstepping controller [13] as a nonlinear control and the parameters are defined in Table 2.

Table 2. Parameters of controller.

Types of controllers	Parameters		
Backstepping	C_1=3000	C_2=10000	
Proposed Method	K_1=100	K_2=10	K_3=100

4.1. Input voltage change

In this test the input voltage will be varied to be greater than the initial input voltage, the magnitude of the change is 0%, 25%, 50%, 75% and 100% while the reference voltage and load are made constant. Then we look at the output of the two control systems as presented in Table 3.

Table 3. The buck converter responses when the input voltage changes.

No	V_{in} (V)	R (Ω)	V_{ref} (V)	Backstepping		SMC	
				t_{rec} (s)	ΔV (V)	t_{rec} (s)	ΔV (V)
1	15			-	-	-	-
2	18,75			-	-	-	-
3	22,5	30	5	-	-	-	-
4	26,25			-	-	-	-
5	30			-	-	-	-

Based on Table 3, the results of the backstepping and SMC controllers are able to maintain a steady state even though the input voltage increases. It can also be seen from the t_{rec} and ΔV values where there is no change in the output voltage. For more details, the test results are shown in Figure 5.

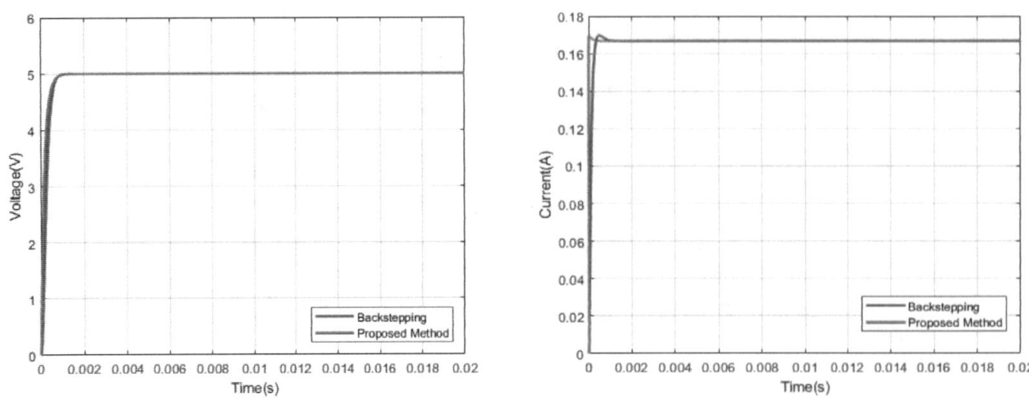

Figure 5. Output voltage and output current when input voltage is changed into 30V.

4.2. Load resistor change

In this second test, the load resistor will vary, while the input voltage and reference voltage are kept constant. The performances are presented in Table 4, the results of the backstepping controller is not able to maintain quickly a steady state when the load conditions increase. It can also be seen from the t_{rec} and ΔV values where the changes that occur in the output voltage are able to return to steady conditions with relatively small t_{rec} and ΔV values.

Table 4. The buck converter responses when the load resistor changes

No	V_{in} (V)	R (Ω)	V_{ref} (V)	Backstepping		SMC	
				t_{rec} (s)	ΔV (V)	t_{rec} (s)	ΔV (V)
1		30		0	0	0	0
2		37,5		0,0012	0,305	0	0
3	15	45	5	0,0013	0,501	0	0
4		52,5		0,0013	0,638	0	0
5		60		0,00128	0,739	0	0

Figure 6 depicts the output of the buck converter with backstepping control and SMC which is given interference in the form of a change in the value of the load by 100%. The system with backstepping control has an overshoot at 0.01 seconds of 5.97 V and has a time recovery 0.00114 s and using SMC has no overshoot.

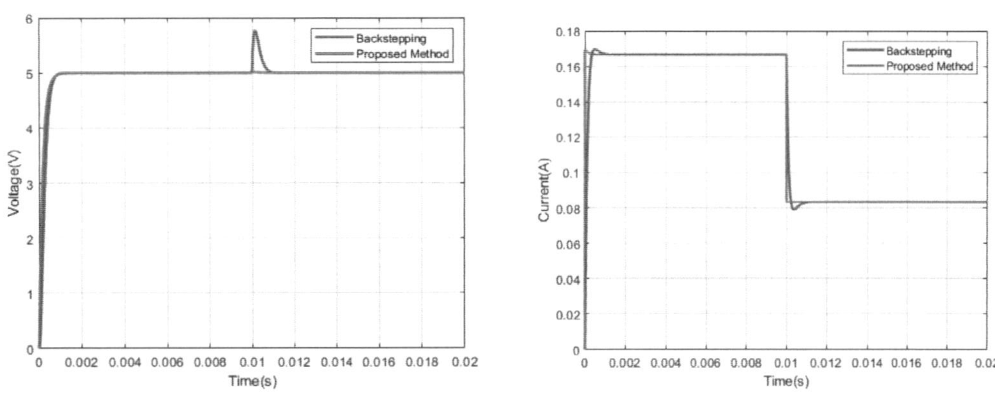

Figure 6. Output voltage and output current when load resistor is changed into 60Ω.

4.3. Voltage reference change

In order to determine the performance of the two types of nonlinear controllers used, a third test is carried out. The last examination is conducted on this system by changing the reference voltage of the buck converter under constant load and input voltage. From the results of Figure 7, it can be seen that the time required for a system with an SMC controller to reach the desired value is faster than a system with a backstepping controller. The change in reference voltage in 0.01 seconds makes the backstepping controller have a very large voltage overshoot or deviation. This shows that using the backstepping controller causes the system to become unstable for about 0.006 seconds.

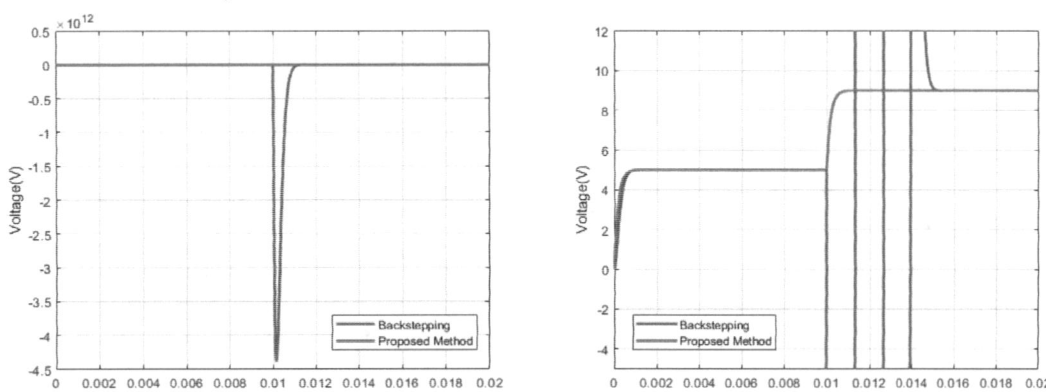

Figure 7. Output voltage and output current when reference voltage is changed into 9V.

5. Conclusion

In uncontrolled condition, the buck converter is not able to generate the desired output voltage and it is not robust against the interferences. By using backstepping and SMC, the buck converter when there is a change in input voltage is able to generate the output voltage as a reference and is able to eliminate overshoot. Voltage deviation occurred in backstepping control when load is changed. It is less able to follow the reference voltage compared with SMC. Backstepping control is not able to reach the reference voltage when it changes. It is proved that Buck Converter with SMC generates controlled output current and constant output voltage.

References

[1] Kordkheili R A, Yazdani-Asrami M and Sayidi A M 2010 Making DC-DC converters easy to understand for undergraduate students In *2010 IEEE Conference on Open Systems (ICOS 2010)* 28-33 IEEE

[2] Mohan N, Undeland T M and Robbins W P 1995 *Power Electronics: Converters, Applications, and Design* (New York: Wiley)

[3] Zhang X, Yin C and Bai H 2017 Fixed-boundary-layer Sliding-mode and Variable Switching Frequency Control for a Bidirectional DC–DC Converter in Hybrid Energy Storage System *Electric Power Components and Systems* **45**(13) 1474-1485

[4] Fathabadi H 2018 Plug-in hybrid electric vehicles: replacing internal combustion engine with clean and renewable energy based auxiliary power sources *IEEE Transactions on Power Electronics* **33**(11) 9611-9618

[5] Lepanov M and Rozanov Y 2013 Multifunctional regulator based on SMES and power electronic converter for increase of power quality and power supply reliability *4th International Conference on Power Engineering Energy and Electrical Drives* 1387-1391

[6] Mukundan C N and Jayaprakash P 2016 A new binary hybrid cascaded H-bridge multi level converter for power quality applications *Proc. IEEE 7th Power India International Conference* (PIICON) 1-6

[7] Sahoo M and Kumar K S 2014 High Gain Step Up DC-DC Converter for DC Micro-Grid Application *Information and Automation for Sustainability (ICIAfS) 2014 7th International Conference* 1-6

[8] Ardhenta L, Kuswiradyo P and Subroto R K 2019 Model direct adaptive control of buck converter by using MRAC *International Journal of Innovative Technology and Exploring Engineering* 1108-1112

[9] Ardhenta L and Subroto R K 2020 Application of direct MRAC in PI controller for DC-DC boost converter *International Journal of Power Electronics and Drive Systems* 851-858

[10] Subroto R K, Ardhenta L and Maulana E 2017 A novel of adaptive sliding mode controller with observer for DC/DC boost converters in photovoltaic system *5th International Conference on Electrical, Electronics and Information Engineering* (ICEEIE) 9–14

[11] Subroto R K, Ardhenta L and Lian K L 2019 Observer Based Adaptive PI Sliding Mode Controller for Cuk Converter *6th International Conference on Instrumentation, Control, and Automation (ICA)* 82-87

[12] Tan R H and Hoo L Y 2015 DC-DC Converter Modeling and Simulation using State Space Approach. *Energy Conversion (CENCON), 2015 IEEE Conference*. Malaysia: Electrical & Electronics Engineering Department UCSI University 42-47

[13] Sureshkumar R and Ganeshkumar S 2011 Comparative study of proportional integral and backstepping controller for buck converter *in Proc. Int. Conf.Emerging Trends Electr. Comput. Technol. Tamil Nadu, India* 375–379

[14] Vijay K O and Sriramalakshmi P 2016 Comparison between Zeta Converter and Boost Converter using Sliding Mode Controller *International Journal of Research & Technology (IJERT)*. India: School of Electrical Engineering, Vit University pp 368-373

[15] Uçak K and Günel G Ö 2020 An adaptive sliding mode controller based on online support vector regression for nonlinear systems *Soft Computing* **24**(6) 4623-4643

IOP Conf. Series: Materials Science and Engineering 1073 (2021) 012036 doi:10.1088/1757-899X/1073/1/012036

Energy Database Management System (EDBMS)-based data acquisition audit for electricity savings analysis

I N Syamsiana*, S S Wibowo, M F Hakim, I Ridzki, and R Firjatoellah

Department of Electrical Engineering, State Polytechnic of Malang, Malang 65141, East Java, Indonesia

*ikanoersyamsiana@polinema.ac.id

Abstract. State Polytechnic of Malang (POLINEMA) has taken a part of a joint action to energy savings. The load profile of POLINEMA has specificity in the intensity of electricity energy (IKE) consumption due to it charachterictic as a vocational higher education. This specificity has caused POLINEMA to have many laboratories and workshops in each building. Energy savings can be done if it is supported by an energy consumption data audit system. The data will be calculated and compared to IKE standart. For that purpose, this research is aimed to design and implement an Energy consumption monitoring device based on Enegry Database Management System (EDBMS). In this research, the energy audits were carried out on the lighting and cooling systems of seven buildings in POLINEMA, namely AB, AC, AD, AE, AF, AG, and AH buildings. After auditing of each building is as follows, the IKE value in the AB building is 30.5 kWh /m2/Year, the AC building is 7.5 kWh/m2/Year, the AD building is 15.2 kWh/m2/Year, the AE building is 24.3 kWh/m2/Year, AF building at 71 kWh/m2/Year, AG building at 79.5 kWh/m2/Year, and AH building at 22.6 kWh/m2/Year. In conclusion is Energy consumption in POLINEMA is classified as efficient where the standard of a building for Commercial/Education is 240 kWh/m2/Year.

1. Introduction

Increasing the number of students each year will trigger the construction of new buildings as new classrooms. With the addition of new buildings, of course there will be additional new facilities to support learning process. Furthermore, it's increased the consumption of electrical energy. The electricity system at the State Polytechnic of Malang (Polinema) is centered at the UPT.PP office which functions as a controller and monitors electrical energy. Currently, UPT.PP has two LVMDPs that supply each group of buildings. And as a back up from the PLN source, there are two generators that will be operated by the operator when occurs a blackout by PLN.

Recently, energy saving becomes an issue of the world because of the increasing electricity needs of all aspects of life. Indonesian government through mandatory circular number 302.E / 07 / DJE / 2010 issued by the Ministry of Energy and Mineral Resources (KESDM) obliges users of electric and non-electric energy equivalent to 6000 tons of oil or 69780 MWh per year to conserve energy [1].

One of the steps to conserve energy is to conduct regular energy audits. An audit can be done if the data electricity consumption is already collected. The data collection is carried out by measuring the required energy conventionally not in short time. Even, it requiring a professionals person in the retrieval of data in each building will be energy audits. The collection of energy consumption data that was done by means of conventional measurement way needs energy, budget, and time. Even, it needs professional worker in data taking at each building which will be audited. The problem is the conventional measurement cannot be used to carry out the energy audit regularly. To cope with such problem, an EDBMS has been designed as a facility the data collection in terms of accuracy, easiness, and retrievability.

The designed EDBMS integrates 7 Polinema's buildings. For this purpose, an energy monitoring tool is installed in each building's Single Distribution Panel (SDP). It is equipped with a current transformer to get current data from each line (R.S.T) which is then connected to Power Monitoring (PM) and transmitted to the web. By this way, the auditing worker can get the required data easily. The energy audit consists of lighting system and cooling system audits. Electrical standard for Energy Consumption Intensity (IKE) for Office sad commercial Educations in Indonesia is 240 kWh/m^2/year based on energy audit set for ASEAN and APEC countries. If from the analysist the energy used is not according to the regulation, then energy saving will be carried to fit the IKE standard.

2. Material and methods

2.1. Energy Database Management System (EDBMS)

EDBMS is an application software that enables user to define, create, collect the electrical energy data, and control the access to database [2]. It is desigend to monitor the electrical energy unit at POLINEMA and uses cloud as the data storage central from the monitored buildings [3]. The kWh usage data in each building is taken through PM and delivered to the cloud storage. The collected data will be analyzed to calculate the value of the Energi Consumption Intensity in each building. There are five main components of the designed EDBMS, namely:

a) Hardware which can be in form of one personal computer/laptop, one mainframe, and computer network.
b) EDBMS software, namely a web-based application program.
c) Database which contains data and meta-data, as well as database structure called scheme.
d) Procedure that shows the impact of instructions and rules to the design, and the use of database.
e) Humans who are related to the system such as database designers, application developers, and end-users.

2.2. Energy audit

Audit energy is a mandatory and a first step to identify the energy saving potentials. The results of the audit will be used as a reference and information regarding the appropriate steps to run energy efficiency program. It is also the basis for determining the efficiency target as a reference in constructing action plan which consists of energy saving recommendations [4]. The procedure of energy audit includes planning, base data collection, tools testing data, data analysist, no-cost or low-cost recommendation, capital investation, implementation plan, and reporting.

2.3. Energy Consumption Intensity (IKE)

IKE is a terminology used to state the amount of energy usage per gross squared meter of a building at a certain range time. The energy usage can calculated if the followings are known [4]:

a) Details of the building area and the total building area (m^2).
b) Building energy consumption per year (kWh/year).
c) Building IKE (kWh/m^2/year).
d) Building energy cost (Rp./kWh).

According to ASEAN-USAID research in 1987, the electrical IKE in Indonesia is as follows [5].

- IKE for offices (commercial) : 240 kWh/m^2/year.
- IKE for shopping centers : 330 kWh/m^2/year.
- IKE for hotels/apartments : 300 kWh/m^2/year.
- IKE for hospita.ls/industries : 380 kWh/m^2/year

IOP Conf. Series: Materials Science and Engineering 1073 (2021) 012036 doi:10.1088/1757-899X/1073/1/012036

Some calculations are needed to obtain IKE, kWh average or occupancy rate, room factor value (RK), light flux of room total (Q total), lighting strenght, and required AC capacity as shown in (1) to (6).

$$Average\left(\frac{kWh}{month}\right) = \frac{Total\ kWh\ in\ 1\ Year}{The\ Number\ of\ Months\ in\ 1\ Year} \tag{1}$$

$$Occupancy\ Rate = \frac{The\ Number\ of\ Usage\ Time}{The\ Number\ of\ Available\ Time} x100\% \tag{2}$$

$$RK = \frac{2x(room\ width) + room\ lenght}{6\ x\ (room\ height)} \tag{3}$$

$$Total\ Light\ Flux = The\ Number\ of\ Lamps\ (n) x\ Lamp\ Lumen \tag{4}$$

$$Lighting\ Strenght = \frac{Total\ Light\ Flux\ x\ Lighting\ Efficiency\ (\%)}{Room\ Area} \tag{5}$$

$$Required\ AC\ Horse\ Power = \frac{Lenght\ x\ Widht\ x\ Height}{3} x500 \tag{6}$$

3. Results and disscussion

Electrical energy needs at Polinema is supplied by PLN sources that are included in the electricity tariff class S3/TM. It is supplied with 2 transformers which each one is 500kVA. The electricity system at Polinema is centered in the UPT.PP which function as a controller and monitors electricity energy. It is also supplied with 2 genset which each a capacity is 250kVA. The generator is activated by the operator if in case of a blackout.

The distribution system at Polinema is divided into 2 LVMDP (Low Voltage Main Distribution Panel). LVMDP 1 is loaded as much as 795,827 VA and LVMDP 2 is loaded in the amount of 801,221 VA. So that the total power is 1,597,048 VA. The total installed power looks beyond the power provided. This can happen because the installed power is not used at the same time, for example PJU lights that only light up at the night.

3.1. Analysis of energy audit

The energy consumption data below is obtained based on data retrieval using the Energy Database Management System. Figure 1 shows the energy consumption data of each building.

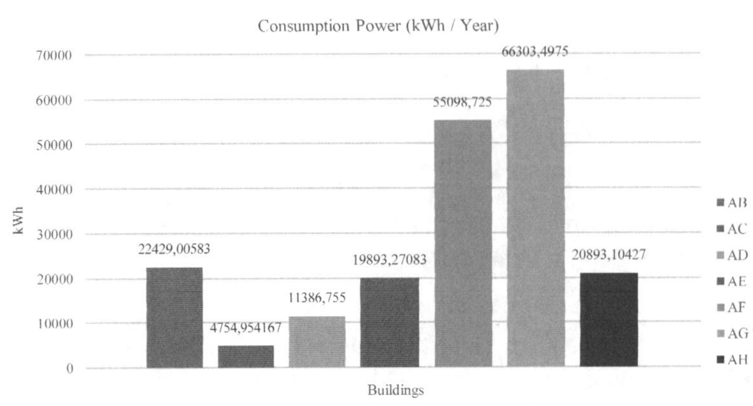

Figure 1. The energy consumption data in each monitored building.

From the figure 1 it is used to obtain data to calculate IKE in the lighting system. The light intensity for each area in Polinema building is calculated according to the light installed. Then it will compare with the national lighting standards contained in SNI 03-6197-2000. Table 1 shows the results of energy audits in 7 buildings that have been monitored based EDBMS.

Table 1. Calculation of average consumption power (kWh) and energy intensity.

Buildings	Consumption Power (kWh/Year)	Total Month per year	Average of Consumption Power (kWh/month)	Wide (m2)	EI (kWh/m2)	EI Standard (kWh/m2)
AB	22680,58	12	1890,048333	736	30,81600543	240
AC	5145,86	12	428,8216667	640,8	8,03036829	240
AD	11475,12	12	956,26	752	15,25946809	240
AE	20685,69	12	1723,8075	817	25,31908201	240
AF	54886,46	12	4573,871667	776	70,72997423	240
AG	64084,26	12	5340,355	834,2	76,82121793	240
AH	16987,86	12	1415,655	924,96	18,36604826	240

The table show EI of some buildings in Polinema is lower than EI standart. That means the building is very efficient because it used natural lighting.

The lighting in each classroom as shown in the table 2 which shows the light intensity is not according to the standart. Based on field observations, each class has sufficient room lighting during the day, so that the installation of lights in each room is adjusted according to the needs. Besides that, the classrooms at Polinema are only used during the day. To conduct an audit of the cooling system in Polinema, will be calculated to the install AC (Air Conditioner) then compared to the applicable standards.

Table 2. The calculation of lumen building AB in second floor.

No	Room	Dimension (m)			Area (m2)	The Number of Lamps		Ql Lamp Lumen (lumen)		Qt Total Light Flux Installed (lumen)	Rk	Efficiency	E Calculation (lux)	E Standard (lux)	Noted
		p	l	t		TL	CF	TL	CF						
1	Computer Lab 1	8	7,75	3	62	6	-	4100	-	24600	1,31	0,52	206,32	500	< SNI
2	Computer Lab 2	8	7,88	3	63	6	-	4100	-	24600	1,32	0,52	203,05	500	< SNI
3	Auditorium	15,6	8	3	125	8	-	4100	-	32800	1,76	0,57	149,57	300	< SNI
4	Classroom V	6	8	3	48	4	-	4100	-	16400	1,22	0,51	174,25	250	< SNI
5	Classroom VI	5,88	8	3	47	4	-	4100	-	16400	1,22	0,51	177,96	250	< SNI

Table 3 shows the calculations for air conditioning requirements in the AB building, it can be seen that many of the air conditioners installed are not in accordance with the cooling required. Polinema can still make savings from the cooling system, if the classroom has sufficient air ventilation.

Table 3. The calculations for air conditioning requirements in the AB building.

No	Room	Dimension(m)			The Number of AC installed	P (W)	Capasity of AC installed (PK)	BTU/hour installed	BTU/Hour required
		p	l	t					
1	Classroom 1	8	7,75	3	1	1472	2	18000	31000
2	Classroom 2	8	7,88	3	1	1472	2	18000	31500
3	Lecturer room 1	8	7,88	3	1	1472	2	18000	31500
5	Lecturer room 2	8	8	3	2	2944	2	36000	32000

4. Conclusion

This paper has proposed an Energy Database Management System to be used as a data collector. It is used to facilitate periodic energy audits. The conclusions that can be sumarized based on the results of the energy audit are:

- The value of the Energy Intensity (EI) of some Polinema Buildings are still within standard.
- The lighting system has a lower standard based on SNI 03-6197-2000.

References

[1] Unknown n.d. Begini Proses dan Manfaat Audit Energi Listrik [Online] Available at: https://sewatama.com/proses-dan-manfaat-audit-energi-listrik/
[2] Connolly T and Begg C 2004 Database Solutions (England: Addison Wesley)
[3] Syamsiana I N, Setiawan A and Saputra M 2020 Design of Energy Database Management System (EDBMS) the Unit 2 Outgoing Transformer at State Polytechnic of Malang *IOP Conf.Series: Materials Science and Engineering* **732**(2020) 012052
[4] Anonim 1 2000 *SNI-03-6196-2000. Prosedur Audit Energi Pada Bangunan Gedung* (Jakarta: Badan Standardisasi Nasional)
[5] Deringer J and Busch J 1992 *Building Energy Conservation Project* (Jakarta: ASEAN – USAID)

IOP Conf. Series: Materials Science and Engineering 1073 (2021) 012037 doi:10.1088/1757-899X/1073/1/012037

Improving teleoperation robots performance by eliminating view limit using 360 camera and enhancing the immersive experience utilizing VR headset

S Wibowo*, I Siradjuddin, F Ronilaya, and M N Hidayat

Department of Electrical Engineering, State Polytechnic of Malang, Jl. Soekarno Hatta No. 9, Malang, Indonesia

*sapto.wibowo@polinema.ac.id

Abstract. A teleoperated robotic system offers a rapid deployment for performing tasks in dangerous environments. An effective teleoperation requires a proper feedback system such as a vision system using cameras. Despite its effectiveness of using camera, the limited field of view (FoV) become one of the main burdens of the operation. To deal with this problem, some researchers used wider-view camera or multiple cameras. In contrast, this research proposed to achieve a better teleoperation by eliminating FoV limit by using 360 camera. This research also aimed to improve the immersive experience by utilizing virtual reality (VR) headset. The proposed framework then developed to accommodate all the requirements and then evaluated by some numerous trials by some participants. An AprilTag marker was placed on the robot and tracked by a ceiling mounted camera to record the trajectory data. The obtained trajectories then measured its performance by using a standard performance metrics. The experiments demonstrated that the proposed method achieved lower collision area, closer to desired ideal path and smoother than using a standard camera as the feedback system. However, as the 360-video streaming induced higher latency, the proposed approach was slower. The measurements justified that the proposed approach achieved a better teleoperation especially for mission critical tasks.

1. Introduction

Often, some works need to be done in dangerous environments, such as inspecting exploded nuclear power plant, getting some samples in viruses' highly-contaminated areas or other dangerous situations. Sending human for such tasks is not a good option and the better alternative is to use robots to do the job. Full autonomous robots are an ideal solution for performing these tasks since autonomous robots capable to operates without the need of human [1]. To operate fully autonomous, the robots have to be able to recognise obstacle, know passable routes, planning the motion paths, create surrounding maps, localize itself within the environment, explore the environment efficiently and manipulate the objects as instructed. Those capabilities require complex algorithms such as SLAM, path planning, motion planning, object recognition, pose estimation, etc, which is non-trivial [2]. Hence, as those algorithms best work on a particular situation, there is no general-purpose autonomous algorithm that can be used for any environments without some significant adjustments. Different environments require non-trivial and time-consuming algorithm adjustments which make it difficult to implement promptly in different complex missions. According to Mahmoud Zadeh et all, the full autonomy is still hard to realize for complex mission [3].

A considered approach for rapid deployment of robots on a dangerous, yet complex missions is to use human teleoperated robotic system. In this case, the robots are operated by human operators in safe area, which might miles apart from the robots. By inducing a human experience, the system takes the benefit of their intelligence that capable to promptly adapt arbitrary situations and perform complex tasks [4]. An effective teleoperation requires a proper feedback system such as LIDAR and vision system. Since images consist rich information that can be easily interpreted by human to build situational awareness, a vision system using camera(s) become a popular feedback system for teleoperated robotic systems. In this case, the operator observes the video streaming to perform the tele operation. However, despite its effectiveness of using camera, the limited field of view (FoV) of the camera become one of the main constrains of the tele operation. The limited FoV decrease the situational awareness and affect the performance of the operation [5].

To deal with this problem, some researchers used wider view or multiple cameras [5–9]. In contrast with those approaches, this research proposed to achieve a better teleoperation by eliminating FoV by using 360 camera. The 360 cameras practically can be seen to have no limited field of view as it can capture every single point around the cameras. This feature is potential to eliminate the narrow FoV problem. While the 360 camera can deliver full environment information, displaying the 360 video in the base station can be a problematic. Displaying as a flat image require mental adjustment for the operator to understand the scene as the display will not be natural [10]. To deal with this, the proposed method utilized virtual reality (VR) headset to display the 360 video streaming. With this approach, the operator can see the scene in more natural way and it will also improve the immersive experience as the operator feel they are on the location. This approach also can potentially improve the performance of the teleoperation.

2. System design

To enable teleoperation of robots, a framework has to be developed to accommodate some features, such as:

- Provides a connection between robot to base station via local area network. The communication media must be wireless to ensure unconstrainted motion;
- The base station should be able to accept robot control command from the operator, using a controller. In this case, the controller was a game pad;
- The control command then should be able to be transmitted to the robot via local area network;
- The robot should be able to receive this control command and performs motion as instructed;
- In the same time, the robot must be able to capture image around the robot and send it as video streaming to the base station via local area network;
- The base station should be able to receive the video stream and display it on the screen or in the VR headset;
- The protocol for control command as well as for video streaming has to be selected carefully to minimize latency.

The developed framework can be seen in Figure 1.

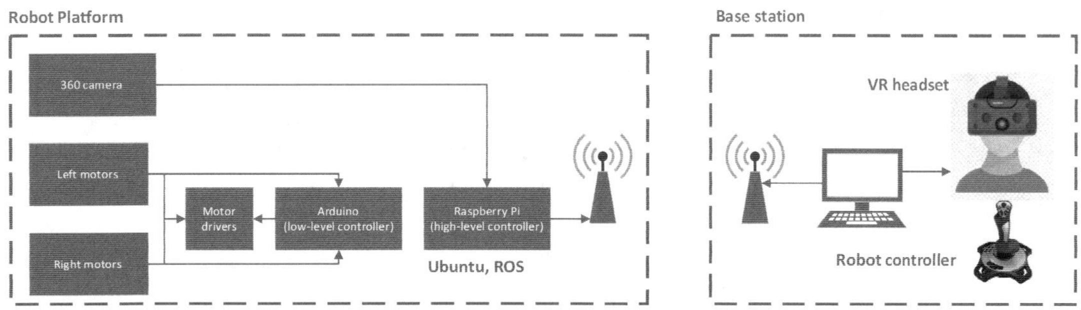

Figure 1. The framework design of teleoperated robotic systems.

The robot platform used in the experiment was 4 wheeled, and configured as differential wheel mechanism. Each of the 4 wheels has its own motor and those 4 motors were grouped into 2 groups: left motors and right motors. Using this configuration, only 2 motor drivers were needed. The 2 motor drivers were connected to an Arduino Nano which provided PWM signal for each group. The Arduino Nano (served as low-level controller) was connected to the Raspberry Pi as high-level controller through serial UART communication. The communication between Arduino Nano and Raspberry Pi used ROS serial protocol.

The Raspberry Pi as high-level controller used Ubuntu 16.04 as its operating system. The Ubuntu operating system managed the Wi-Fi connection to the WLAN so the robot can connect to the base station. The Raspberry Pi also implemented Kinetic Robotic Operating System (ROS). The ROS provides robot command communication using standard velocity command messages (*cmd_vel*). The velocity command was received from base station which also implemented ROS. In the same time, the Raspberry Pi captured image from web camera and send it to base station using standard ROS image transport class. The base station subscribed to the compressed image transport to reduce latency. While webcam video streaming was done using ROS image transport, the 360 streaming was done by a custom plugin for the Ricoh Theta V 360 camera. The plugin provided WebRTC web server that then accessed from Oculus Quest VR browser.

3. Experiment methods

We evaluated these designs with 28 trials by 7 participants. Each trial done by teleoperating robots on a predefined mission. A ceiling mounted camera tracked an AprilTag marker placed on the robot to precisely identify the 2D pose of the robot within the arena. The ceiling mounted camera capable to capture the whole arena and to ensure the accuracy, the ceiling mounted camera was previously calibrated using a checkerboard. The trajectories of each trial were recorded in rosbag and then analysed its performance in Matlab using a standard performance metrics as introduced by (Munoz Ceballos, Valencia Velasquez, & Ospina, 2010). While the full performance metrics are:

1) Mission success: number of successful missions
2) Collisions: number of collisions per mission
3) Path length: distance travelled to accomplish the task
4) Obstacle clearance: minimum and mean distance to the obstacles
5) Robustness in narrow spaces: number of narrow passages successfully traversed
6) Smoothness of the trajectory
7) Time taken to accomplish the task

Some metrics are un-relevant, such as mission success (all trials were always successful), "robustness in narrow spaces" (no "narrow spaces") and "obstacle clearance" (no obstacles in the arena setup). In final, we measured the performance only using 4 metrics: 1. Collision area; 2. Path length (distance travelled to accomplish the task); 3. Smoothness of the trajectory; 4. Time taken. To minimize the effect of learning, each participant was trained to get familiar controlling the robot. The robot and arena can be seen in Figure 2.

IOP Conf. Series: Materials Science and Engineering 1073 (2021) 012037 doi:10.1088/1757-899X/1073/1/012037

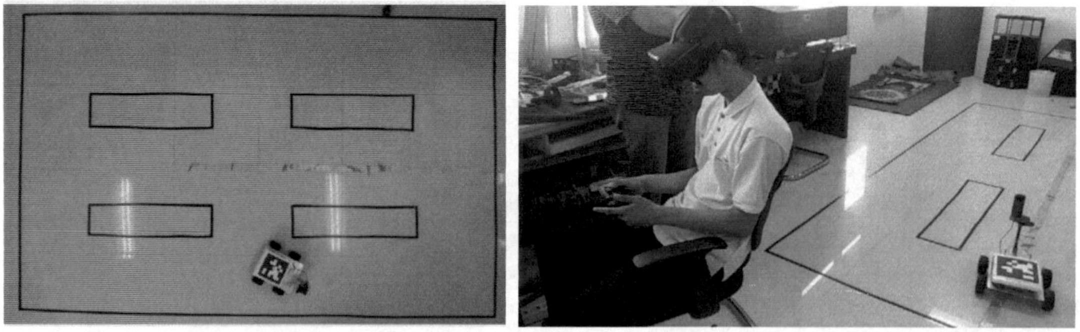

Figure 2. The robot arena (left) and a participant performed a trial in teleoperating the robot(right).

The example results of trajectories, plotted in Matlab can be seen in Figure 3.

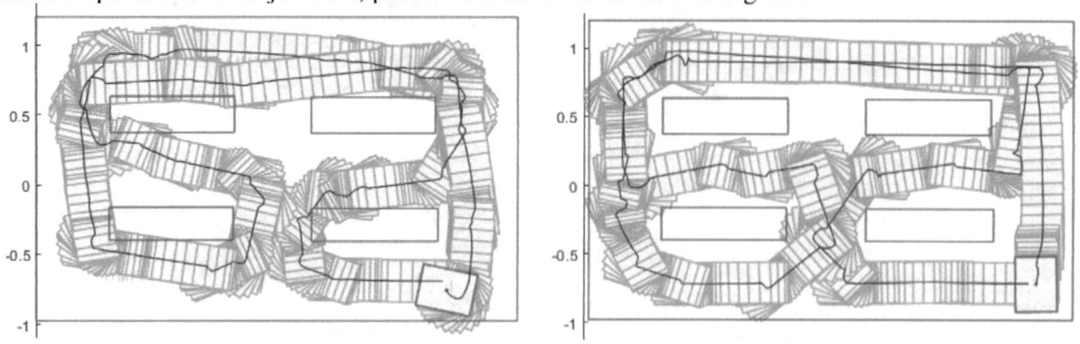

Figure 3. The obtained trajectories plotted and analysed in Matlab.

4. Results and discussions

The recorded trajectories were analysed in Matlab and the results then plotted as box plot. The box plot provides more insight to the data as it can display the median, 25th – 75th percentiles, maximum-minimum data and outliers (if any). Since it can display median, percentiles and mix-max data, the boxplot can be used to measure accuracy (from the median) and precision (from the dispersion of dataset).

4.1. Collision area

The boxplot of the collision area of teleoperating robot using webcam and 360 camera is presented in Figure 4 (left). While the median of the data both webcam and 360 camera was very close, the boxplot shows the teleoperation using 360 camera was significantly more precise as shown by a lower deviation. It shows the webcam operation was harder to avoid collision and the teleoperation depends significantly by the skill of operator. In contrast, using 360 camera as feedback is more consistent and the median was also lower (0.255 m^2) compared to webcam (0.268m^2).

4.2. Deviation from ideal path

The trials should follow the ideal path, which is the middle of the way path. However, most of the teleoperation will not perfectly follow the desired path and it induces longer taken path. Again, the boxplot shows the teleoperation achieved more precise and more accurate. The accuracy can be seen from the median of deviation that scored 0.996m on 360 camera teleoperation, compared to 1.06m on webcam teleoperation as shown in Figure 4 (right). The precision is significantly better as the deviation of 360 camera operation was lower.

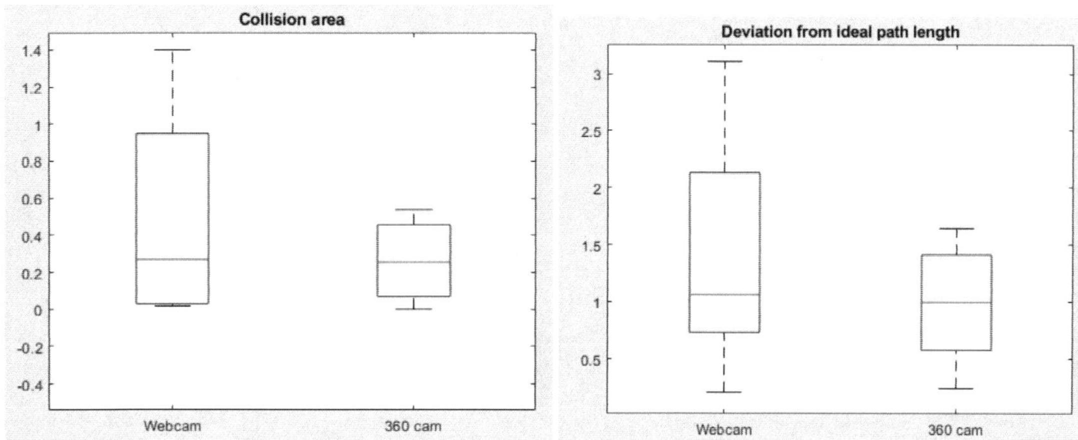

Figure 4. The area of collision in m^2 (left) and the deviation from ideal path length in m (right).

4.3. Smoothness

The comparison of smoothness both teleoperation modes is presented in Figure 5. While 360 camera feedback achieved significantly smoother trajectories (lower medians), however the consistency can be seen similar with webcam feedback.

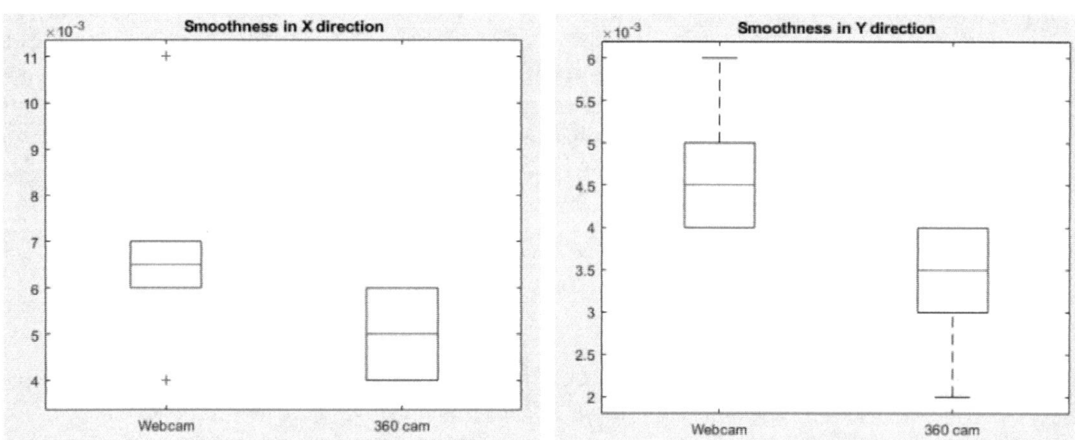

Figure 5. The boxplot of smoothness position both in X and Y directions.

4.4. Time

While aforementioned performance metrices shows that 360 camera feedback system is superior than webcam camera feedback, the last performance metric shows an opposite result. The operation time of 360 camera feedback system took significantly longer than webcam at 74.8 seconds (median) and 61.7 seconds (median) respectively. Teleoperation using 360 camera also more inconsistent in time as shown by larger percentiles. This phenomenon took place due to 360 camera streaming had higher latency. In this case, the operator cannot force the robot to move faster as the real robot position and the video streaming can be unsynchronized. The operator then unwillingly to drive the robot slower. Some disturbance such as intermittent connection also forced the operator to wait until the video streaming getting normal which induced larger time deviation.

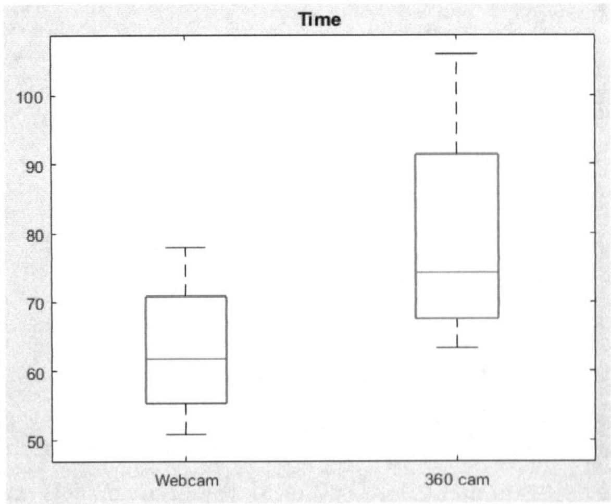

Figure 6. The boxplot of dataset related to time taken to accomplish the missions.

5. Conclusions

The experiment demonstrated that the developed framework successfully accommodated the teleoperation. The experiments also proved that teleoperating robot using limited FoV camera (webcam) achieved bigger error from ideal path, while teleoperating robot using 360 camera was closer to the ideal path. A robot teleoperated using webcam tend to collide the obstacles, while teleoperated robot using 360 can better avoid the obstacle. Using 360 camera combined with VR headset also achieved a smoother trajectory. However, due to higher latency of 360 video streaming, the operation time of the proposed method was longer than traditional approach. Therefore, the proposed framework demonstrated good for teleoperating robot in mission critical task while the typical approach (using limited FoV camera) is better in time critical task.

Acknowledgments

Authors wishing to acknowledge financial support from the State Polytechnic of Malang

References

[1] Katyal K D, Brown C Y, Hechtman S A, Para M P, McGee T G, Wolfe K C, Murphy R J, Kutzer M D, Tunstel E W, McLoughlin M P and Johannes M S 2014 Approaches to robotic teleoperation in a disaster scenario: From supervised autonomy to direct control *IEEE International Conference on Intelligent Robots and Systems* pp 1874–1881

[2] Mostafa S A, Ahmad M S and Mustapha A 2019 Adjustable autonomy: a systematic literature review *Artif. Intell. Rev.* **51**(2) pp 149–186

[3] Mahmoud Zadeh S, Powers D and Zadeh R 2019 Augmented Reactive Mission Planning Architecture: Augmented Reactive Mission and Motion Planning Architecture *Cognitive Science and Technology* pp 95–107

[4] Asami R, Sawai Y, Sato N, Morita Y, Endo T, and Matsuno F 2016 Teleoperation system with virtual 3D diorama for moving operation of a tracked rescue robot *International Conference on Advanced Mechatronic Systems, ICAMechS* **0** pp 90–95

[5] De León J, Garzón M, Garzón D, Narváez E, Del Cerro J, and Barrientos A 2016 From Video Games Multiple Cameras to Multi-robot Teleoperation in Disaster Scenarios *Proceedings - 2016 International Conference on Autonomous Robot Systems and Competitions, ICARSC* pp 323–328

[6] Keyes B, Casey R, Yanco H, Maxwell B, and Georgiev Y 2006 Camera Placement and Multi-Camera Fusion for Remote Robot Operation *IEEE International Workshop Safety, Security,*

Rescue Robot

[7] Seo S H, Rea D J, Wiebe J, and Young J E 2017 Monocle: Interactive detail-in-context using two pan-and-tilt cameras to improve teleoperation effectiveness *RO-MAN 2017 - 26th IEEE International Symposium on Robot and Human Interactive Communication* **2017** pp 962–967

[8] Nielsen C W, Goodrich M A, and Ricks R W 2007 Ecological interfaces for improving mobile robot teleoperation *IEEE Transactions on Robotics* **23**(5) pp 927–941

[9] Hedayati H, Walker M, and Szafir D 2018 Improving Collocated Robot Teleoperation with Augmented Reality *Proceedings of the 2018 ACM/IEEE International Conference on Human-Robot Interaction* pp 78–86

[10] Walker M E, Hedayati H, and Szafir D 2019 Robot Teleoperation with Augmented Reality Virtual Surrogates *ACM/IEEE International Conference on Human-Robot Interaction* **2019** pp 202–210

IOP Conf. Series: Materials Science and Engineering 1073 (2021) 012038 doi:10.1088/1757-899X/1073/1/012038

Design locomotion for automatic guided vehicles using double ackerman inverted method (2WD-4WS-1A)

A M Karim*, B Setiawan, and I Siradjuddin

Electrical Engineering Departement, State Polytechnic of Malang. Jl. Soekarno-Hatta No. 9, Malang 65141, Indonesia.

*karim.arifmafatia@polinema.ac.id

Abstract. This paper introduces a robot vehicle designed for the same forward or reverse maneuver specifications. The vehicle has two two-wheel drives with a Double Ackermann Inverted (2WD-4WS-1A) steering method for Automatic Guide Vehicle (AGV). In the batch processing system industrial estate, AGV is used for transportation in the component warehouse area. In the steering system, Ackermann's mechanics are described as optimizing the maximum maneuvering radius of the AGV, based on the turning center point. AGV drive is designed with four wheels, two drives and one steering mechanics. The vehicle's mechanical steering system is designed with a system geometry analysis to control the turning radius of the vehicle. The Ackermann AGV steering method that has been implemented will be tested for its maneuvers. The results of the study were measured empirically to determine the efficiency and optimization of the system in the AGV design. The efficiency of steering control using only 1 steering wheel as 4 wheel control and small optimal radius on the AGV will increase maneuverability compared to using 2 WD.

1. Introduction

There is a growth and development progress in human life on technology towards automation. One of the identifications such as dirty, dangerous, and boring is something that is often encountered [1]. The introduction of robots, flexible programming can simplify various repetitive tasks in various conditions. Although the history of industrial robots can be traced back to the 1930s, the first industrial programmable robots are believed to have been commissioned by Unimation in the early 60s [2]. Better performance and stability appear to be at the forefront of the auto industry looking for unconventional approaches to power distribution and maneuverability. Four wheel steering (4WS) is an advanced control technique that can improve characteristics. Compared to traditional two-wheel steering (2WS), the four-wheel steering system directs the front and rear wheels individually during cornering. Four-wheel steering can improve handling stability, increase active vehicle safety, and allow the vehicle to turn at a much smaller turning radius [3]. The ideal condition of the steering system can be approached by analyzing the Ackerman condition, which is a condition where the turning angle given by the driver to the steering wheel is the same as the turning angle experienced by the vehicle.

2. Literature review

2.1. Ackerman steering

Today most vehicles use a two-wheel steering mechanism as their primary handling system. However, the efficiency of the two-wheeler was shown to be lower [4] and the 2WS system requires more turning radius compared to the 4WS system which requires more space to turn the vehicle [5].

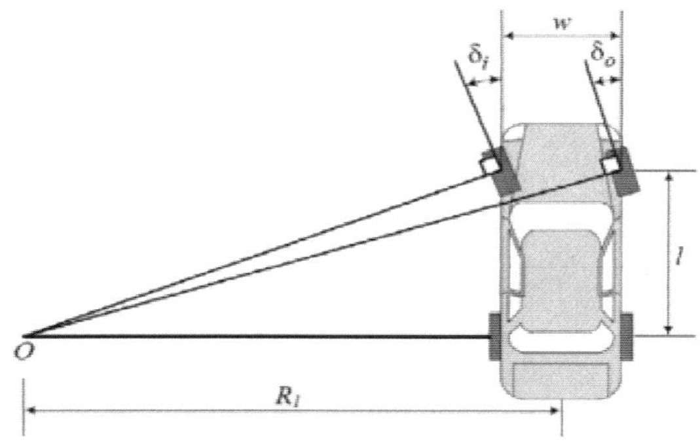

Figure 1. Model 2WS 2WS [1,5,6].

2.2. Four wheel steering

The 4WS is also called rear-wheel steering or all-wheel steering where the four-wheel symmetrical steering system as shown in figure 2. shows that the front and rear wheels outside and inside are turning at the same angle. So that this condition results in the shortest possible turning radius for the vehicle because the perpendicular lines of the wheels meet at the center line of the wheel base [5].

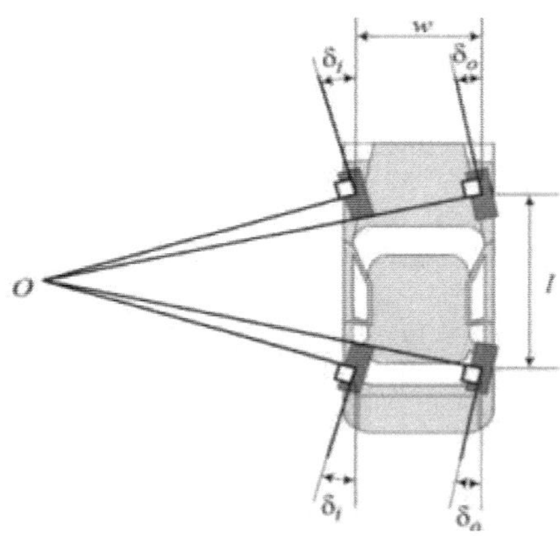

Figure 2. Four Wheel Steering [5].

3. Methods

The research method is basically a scientific way to obtain information with specific purposes and benefits. The steps in this research method can be seen in the research design in the following explanation:

3.1. Design of double ackermann inverted AGV locomotions

In this research, the manufacture of an AGV prototype consisting of 4 wheels, 2 motors forward and backward on the front side and a steering system to change the angle of the four wheels as a change in the direction of motion of the AGV. Angle the front and rear wheels in one command and reverse the polarity of the angle. Steering motor position in the middle of the longitudinal AGV, can operate optimally.

3.2. The AGV steering and chassis design

Steering is a mechanical control that is able to adjust the position of the wheels where the four wheels are set. Steering is designed in such a way as shown in figure 3. To maintain effectiveness and flexibility in controlling the position of the wheels of the wheel, several component parts composed of threads are used as rotating wheels. In addition, there are cylinder pistons and cross joints to help drive elasticity in the steering.

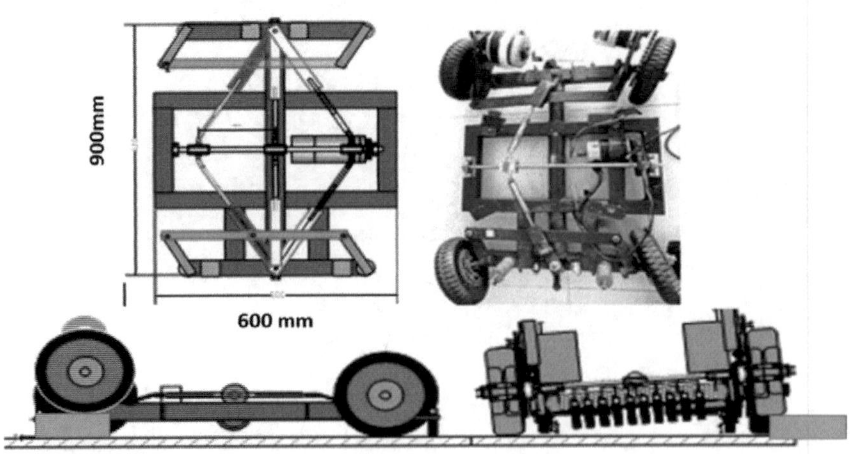

Figure 3. AGV steering system models (2WD-4WS-1A).

The frame design is made flexible in such a way as to minimize sleep on one of the wheels. So that it appears that one side of the wheel segment will be lifted if there is an uneven surface side.

3.3. Data collection method

The identification of the AGV steering uses a maximum limit with an angle of 30°, which uses the double ackerman inverted method. The results of the calibration are shown in the table below.

Table 1. Identification of AGV steering.

Trial to -	Front Wheel Angle (deg)	Rear Wheel Angle (deg)
1	0	0
2	5	-5
3	10	-10
4	15	-15
5	20	-20
6	25	-25
7	30	-30

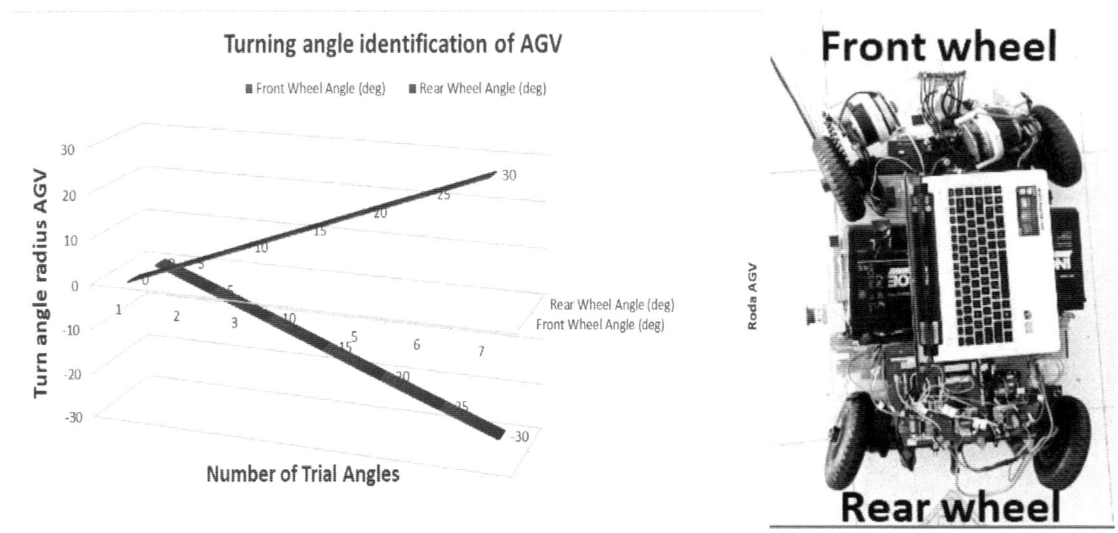

Figure 4. Steering graphs against angles.

In Table 1, it can be seen that the angles generated from the front wheel and rear wheel produce an inverse result from the position of the front wheel. The motor on the steering control adjusts the angle results obtained by the AGV.

4. Results and discussion

This test is carried out to identify, in graphical form, sensor signals, motor and steering drivers, current sensors and encoders. The test is carried out in a closed room with a tiled platform with the results below.

4.1. Round with 30° angle (PWM KA 50 and KI 25)

Table 2. Identification of rotation with a 30° angle (PWM KA 50 and KI 25).

	V. Right IBT Conversion	V. Left IBT Conversion	V. IBT Steering Conversion	I. Right IBT	I. Left IBT	I IBT Steering
1	2.32	2.48	2.48	1	0.1	0.12
2	1.97	2.38	2.46	2.89	0.63	0.2
3	1.99	2.41	2.47	2.74	0.48	0.18
4	1.97	2.4	2.47	2.86	0.52	0.17
5	1.99	2.41	2.47	2.77	0.49	0.17

The current used in the right motor driver (PWM Ka) produces a greater value than the left motor (PWM ki). In the table above the experiment uses PWM, with a right motor speed of 50 and a left motor 25 pwm. With a radius of 30° angle. Markers are used to mark the results of the AGV movement.

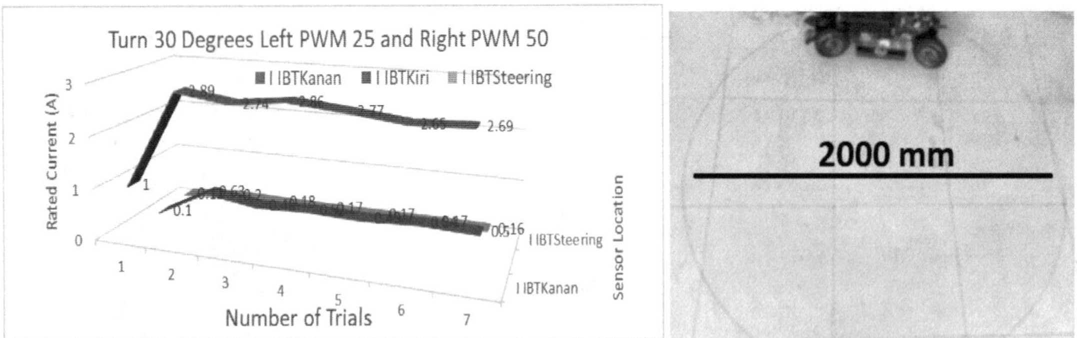

Figure 5. Response of a 30° angle radius on pwm Right 50 and Left 25.

With this radius, the results obtained with an AGV turning diameter of 2000 mm. Markers are used to mark the results of the AGV movement. Markers are used to mark the results of the AGV movement.

4.2. Round with 30° angle (PWM KA 100 and KI 25)

Table 3. Round identification with 30° Angle (PWM KA 100 and KI 25).

	V. Right IBT Conversion	V. Left IBT Conversion	V. IBT Steering Conversion	I. Right IBT	I. Left IBT	I IBT Steering
1	2.42	2.47	2.61	2.61	0.44	0.16
2	2.41	2.47	2.49	2.49	0.5	0.14
3	2.44	2.47	2.62	2.62	0.32	0.15
4	2.44	2.47	2.65	2.65	0.32	0.14
5	2.41	2.47	2.69	2.69	0.49	0.15

The use of current consumption is directly proportional to the PWM used in both motors. In the table above the experiment uses a PWM with a right motor speed of 100 and a left motor pwm of 25. With an angle radius of 30°.

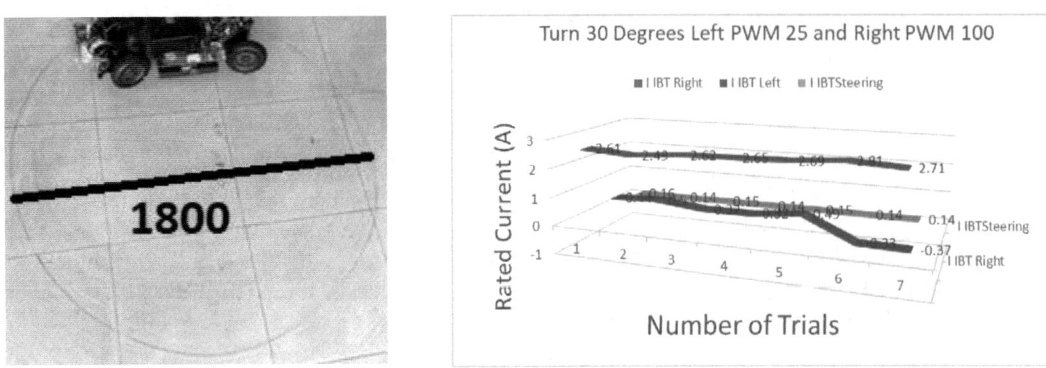

Figure 6. Response of an angular radius of 30 at AGV.

With this radius, the results obtained with the AGV turning diameter of 1800 mm. Markers are used to mark the results of the AGV movement. Markers are used to mark the results of the AGV movement.

4.3. Round with 20° angle (PWM KA 50 and KI 25)

Table 4. Identification of rotations with angles of 20 ° (PWM KA 50 and KI 25).

	V. Right IBT Conversion	V. Left IBT Conversion	V. IBT Steering Conversion	I. Right IBT	I. Left IBT	I IBT Steering
1	2.02	2.47	2.44	2.62	0.17	0.3
2	2.01	2.46	2.44	2.65	0.22	0.3
3	2	2.46	2.44	2.71	0.24	0.3
4	2	2.47	2.44	2.71	0.19	0.3
5	2.03	2.45	2.44	2.57	0.25	0.3

The use of current consumption is directly proportional to the PWM used in both motors. In the table above, the experiment uses PWM motor ka 50 and pwm motor ki 25. With a radius of 20°.

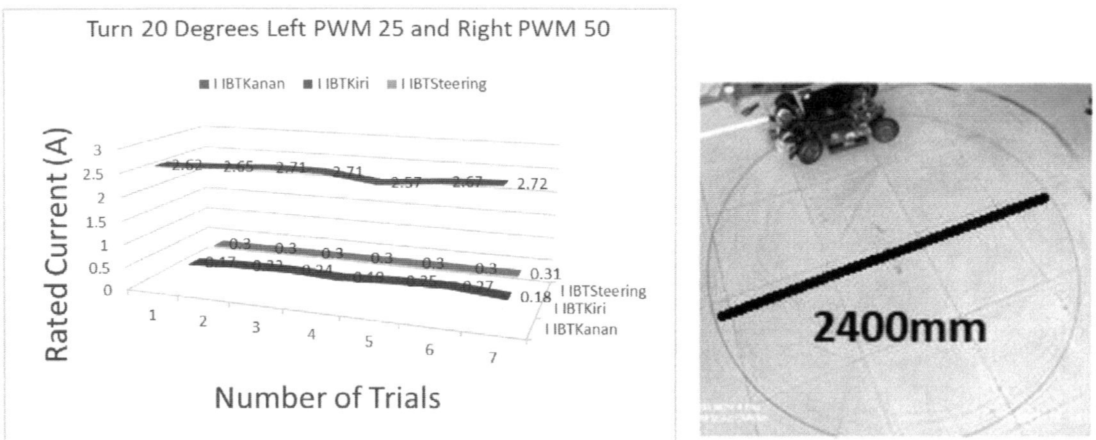

Figure 7. Respons radius angle 20° pada AGV.

With this radius, the results obtained with an AGV turning diameter of 2400 mm. Markers are used to mark the results of the AGV movement. Markers are used to mark the results of the AGV movement.

5. Conclusion

From the discussion and analysis that has been carried out in this study, it can be concluded, with an angle of 30 ° with the double Ackerman inverted steering method when the maneuver results in the turning radius of the AGV, it can be smaller when testing on the track. AGV with the double inverted (2WD-4WS-1A) method can operate optimally. The efficiency of steering control using only 1 steering as 4 wheel control and small optimal radius on the AGV will increase maneuverability compared to using 2 WD.

Acknowledgments

The author would like to thank the Ministry of Research, Technology, and Higher Education who funded this research with a master's thesis research scheme. This research was conducted at the Master of Applied Science in the Department of Electrical Engineering in 2020, Malang State Polytechnic. We are also grateful for providing facilities in the process of doing so.

IOP Conf. Series: Materials Science and Engineering 1073 (2021) 012038 doi:10.1088/1757-899X/1073/1/012038

References

[1] Klindworth K E 2017 Dynamic Modelling For The Path Tracking Control Of A Four-Wheel Independent-Drive, Four-Wheel Independent-Steer Autonomous Ground Vehicle (North Dakota State University)

[2] Unimate - The First Industrial Robot n.d. Robotics Online. [Online]. Available: https://www.robotics.org/joseph-engelberger/unimate.cfm [Accessed: 02-Jun-2020]

[3] Schein, Victoria L 2016 Optimal Vehicle Control ofFour-Wheel Steering (Northampton, Massachusetts)

[4] Sundar S 2018 Review of Design and Fabrication of four wheel Steering system *International Journal of Recent Trends in Engineering & Research (IJRTER)* **04**

[5] Gohil K, Bhavesh 2018 Optimization of steering system for four wheel Vehicle *Proceedings of ARSSS International Conference*

[6] Singh B P 2015 Advanced Four Wheel Steering System *International Journal of Research in Engineering & Advanced Technology [IJIERT]* **3**(2)

IOP Conf. Series: Materials Science and Engineering 1073 (2021) 012039 doi:10.1088/1757-899X/1073/1/012039

Development of smart energy meter to measure energy saving of dimmable LED panel light

A Parastiwi, R A Prasojo*, M N H Adzani and H K Safitri

Electrical Engineering Department, Politeknik Negeri Malang, Malang, Indonesia

*rahmanazisp@polinema.ac.id

Abstract. Conserving energy is important now more than ever. Dimming lamps to the required illuminance is one way to reduce the energy usage. Technological advancement in the form of new materials, sensors, and apparatus allows us to reduce the energy usage of an equipment. To obtain a clear understanding of consumer energy usage, a smart meter which is able to monitor and store energy usage characteristics can be a useful tool to have. This paper discusses the development of multi-points smart energy meter using ESP32, ZMPT101B voltage sensor, and SCT-013 current sensor. The measurements result of the meter developed has been validated using commercial energy meter. A 35 W dimmable LED panel light benefitting dimming function in QH7938 chip has also been constructed. The dimming characteristics were obtained, with the output luminous flux of 5644 to 950 lumen, and power input of 35 to 5.32 watt. Subsequently, the energy saving case scenario in 2.5 x 2.8 x 3 meters home office with various natural lighting condition were also presented.

1. Introduction

Climate change is becoming a more and more critical issue, and many belief that efficient use of energy is an essential approach in greenhouse gas emission reduction [1]. In attempt to reduce fossil energy, households are responsible for 15-20% of energy usage. The use of electricity in household could be reduced by changing the behaviour of the user. This might be done by promoting the energy-efficient products and services [2]. Lighting is one of the main loads in household, and the study in making power reduction is interesting topic.

Several previous studies have discussed energy saving aspects of lighting loads. Study in Khan [3] compares techno-economic performance of different light sources, which concluded to propose a widespread use of LED lamps. LED lights are electrical devices that produce light using one or more light-emitting diodes (LEDs). LED lights have a life span several times longer than equivalent incandescent lamps, and are significantly more efficient than most fluorescent lamps. In addition, LEDs are considered as flicker-free when compared to their predecessor, fluorescent lamps [4].

In its development, many LEDs are implemented as smart lighting systems because of the flexibility of their control. Smart lighting is a lighting technology designed for energy efficiency. This concept includes equipment with high efficiency and automatic controls that make adjustments based on room conditions or lighting requirements. In that matter, this study focusses on the development of dimmable LED panel light in attempt to save lighting energy consumption. Effects of LED control to the reduction of power needed, lumen output, and its power factor will be investigated. Finally, an experiment in home

IOP Conf. Series: Materials Science and Engineering 1073 (2021) 012039 doi:10.1088/1757-899X/1073/1/012039

office setup were conducted to exhibit the possibility of dimmable LED panel to further be developed into a smart lighting system.

2. Methods

The study starts with the development of multi-point smart meter. The meter was validated using commercial energy meter, as well as utility energy meter. A dimmable LED panel lamp were also developed. The LED panel dimming characteristics will be measured. The device then will be installed at home office with various natural lighting condition to simulate power reduction scenario. Fig 1 shows the flowchart of methodology of this study.

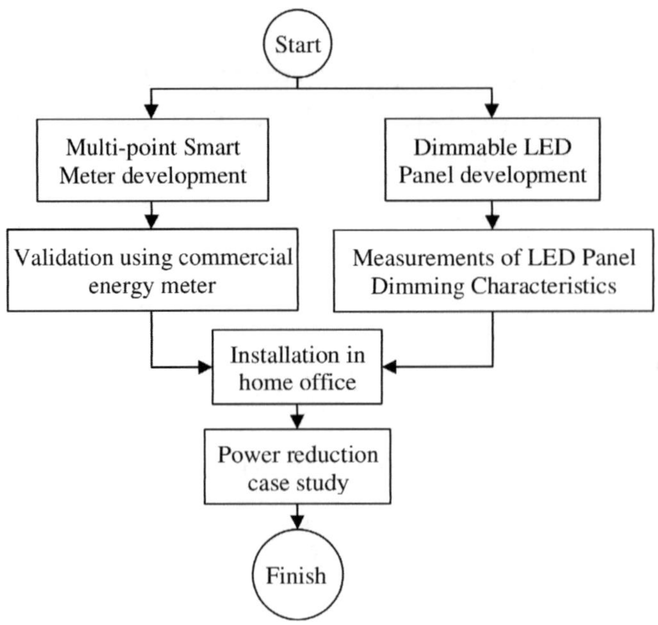

Figure 1. Flowchart of methodology of the study.

2.1. Smart energy meter design

Several other studies has developed devices to monitor energy meter [5–9]. This project is using two sensors such as ZMPT101B for AC voltage measurement and SCT-013 10A/1V for current measurement. The data from this sensor is sent to smartphone over Wi-Fi features on board from ESP32 that connected to internet. The platform used to make this monitoring application is Blynk which can be used for free. Fig 2 shows the block diagram of this project.

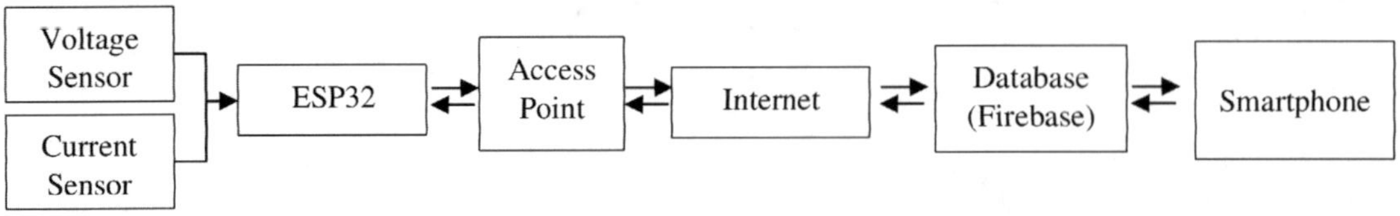

Figure 2. Block diagram of smart energy meter proposed.

Fig 3 is a schematic of signal conditioning for SCT-013 10A/1V current sensor. This schematic explains the installation of the sensor to ESP32.This sensor require a burden resistor which can be read below, and it is needed to use resistor value for R1 and R2 between 10k-470k. The ceramic capacitor 104 is used to filter noise from input. The sensor is CT clamp with 2000 turns and should be calibrated with formula to calculate the calibration value for emonlib library used.

$$\textit{Calibration number} = \textit{number of turns of CT Clamp / Burden Resistor} \quad (1)$$

An output from CT clamp need to be used with a burden resistor. The burden resistor completes or closes the CT secondary circuit. The burden resistor value is chosen to provide a voltage proportional to the secondary current. The burden resistor value needs to be low enough to prevent CT core saturation. The system voltage is getting from the supply using 3.3.VDC.

$$\textit{Burden resistor} = (\textit{system voltage} / 2.0) / ((\textit{I\_RMS} * 1.414) / \textit{CT Turns}) \quad (2)$$

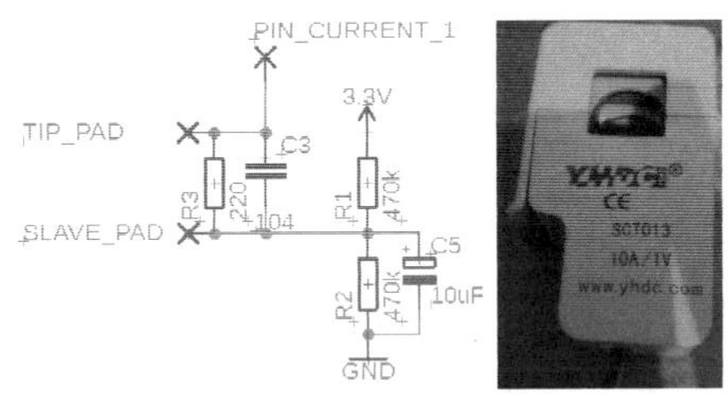

Figure 3. SCT-013 signal amplifier conditioning.

2.2. Dimmable LED panel light design

A 35-watt LED panel light has been constructed from 35 high power LED connected in series. The dimmer was constructed from QH7938 chip. Fig 4 shows the schematic of QH7938 obtained from the datasheet. The characteristics of the developed LED panel will be measured, such as lumen, power usage, and also power factor. The constructed LED lamp were then be installed in a home office with the natural light available. An experiment will be conducted to maintain illuminance in workspace to be within the limit, regarding various natural light condition.

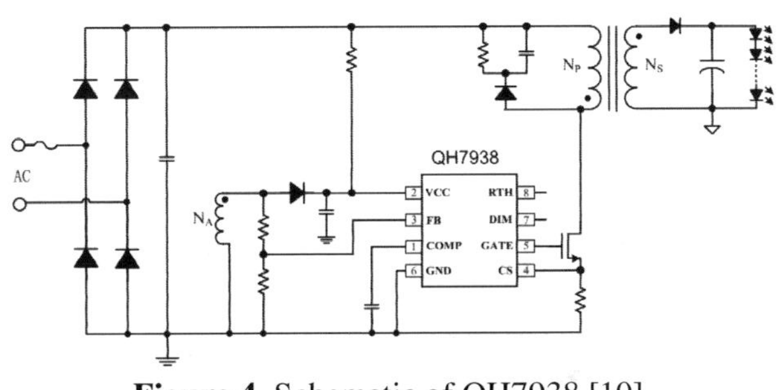

Figure 4. Schematic of QH7938 [10].

Fig 5 is a schematic of home office. The LED panel was installed in 2.5 x 2.8 x 3 m home office setup. The workspace is situated 1.75 m above the floor, and 2.55 m to the LED panel. There is a window where the natural light comes into the room. The illuminance is to be kept within the limit. Table 1 shows the residential building minimum illuminance and colour rendering recommendation. A value of 250 lux is used as limit in home office experiment.

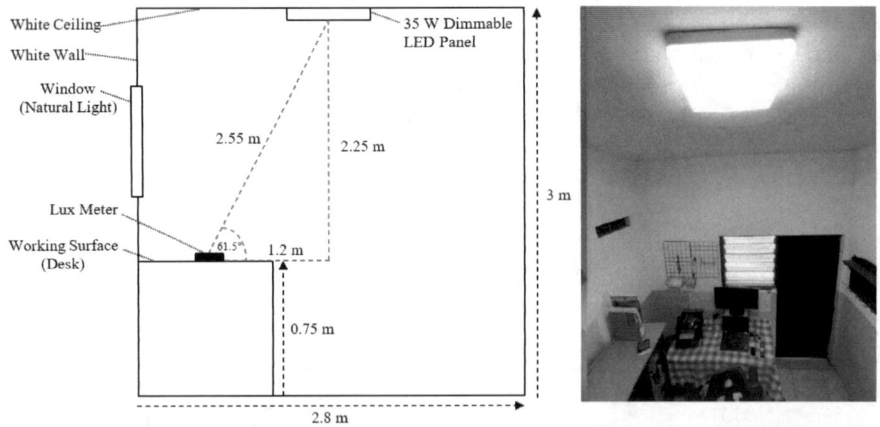

Figure 5. Schematic of home office dimmable LED panel installation and measurement.

Table 1. Residential building minimum illuminance and colour rendering recommendation [11].

Room Function	Illuminance (lux)	Colour Rendering
Terrace	60	1 or 2
Living Room, dining room, bed room	120-250	1 or 2
Office	**120-250**	**1**
Bath room	250	1 or 2
Kitchen	250	1 or 2
Garage	60	3 or 4

3. Results and discussion

Smart energy meter and LED dimmable panel has been developed. This section presents the results and discussion of the proposed system.

3.1. Smart energy meter

Fig 6 shows the proposed energy meter device, which constructed from ESP32, ZMPT101B AC voltage sensor, SCT-013 10A/1V current sensor. The measurement of AC voltage is done by ZMPT101B and the current flow from SCT-013 10A/1V. The values of power factor, apparent power, real power, and energy in Watt-hour is obtained from calculations through the code from library of the program.

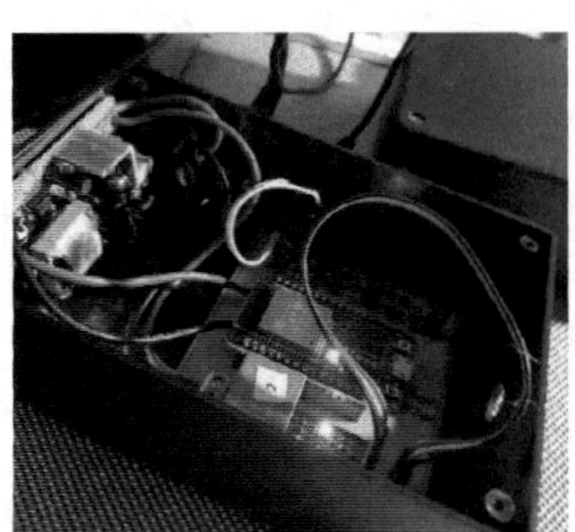

Figure 6. The developed smart energy meter.

Fig 7 shows the measurement value that was sent to the Blynk App, and can be monitored via a smartphone with some add-ons like to set the set point of the maximum energy value, and the notification can be sent to the smartphone that the energy has reached the maximum value. The energy value should be reset if the maximum value is reached.

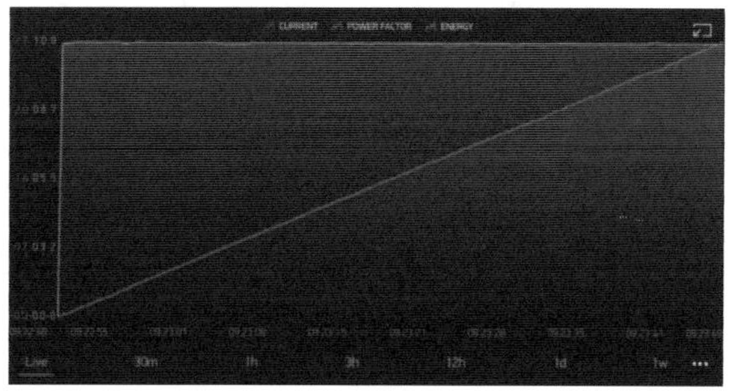

Figure 7. Current, Power Factor, Energy with load graph.

Fig 8 shows the measurement of the energy meter (including AC voltage, current, energy in Watt-hours, power factor, power and frequency) and the Blynk App with no load, with load, and when the energy threshold passed. The commercial energy meter can measure very low current with no issues. The Blynk App shows the small tolerable difference in error values between the power analyzer and the hardware on the AC voltage value. Table 2 shows the comparison of commercial energy meter and the proposed smart energy meter. It shows some differences in measurements.

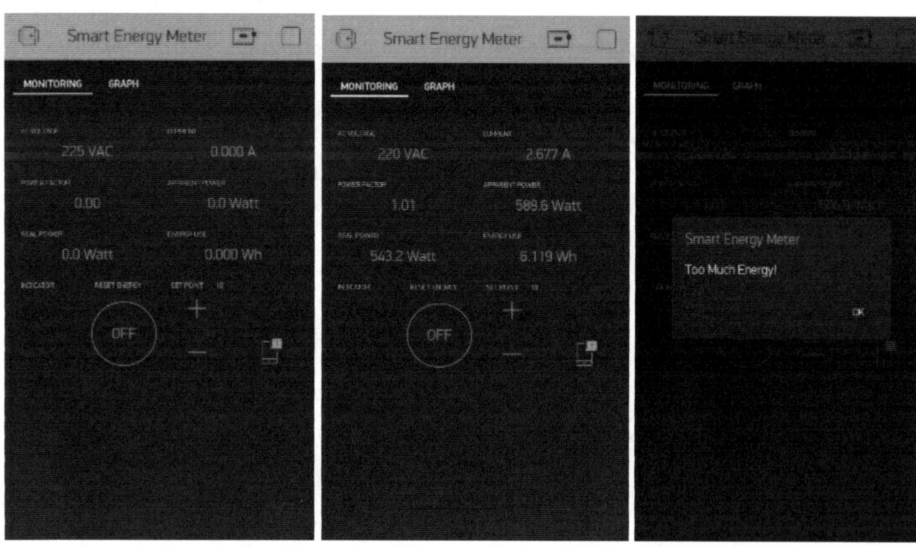

Figure 8. Current, Power Factor, Energy with load graph.

Table 2. Measurement comparisons of different loads.

No	Load	Commercial Energy Meter					Proposed Smart Energy Meter				
		V	I	PFa	Fq (Hz)	Pwr	V	I	PFa	Apparent Power	Real Power
1	Kettle	220	2,68	1	50	586	220	2,68	1,0	589	543
2	Drill	221	1,2	0,9	50	260	219	1.23	1,0	292	223
3	Blower	222	2,7	0,9	50	300	220	2,7	1,0	516	504
4	Solder	219	0,118	0.93	50	23,9	220	0,15	0,8	33	21

3.2. Dimmable LED panel light

After development of smart meter, a dimmable LED panel light were constructed. Fig 9a shows the dimming characteristics of LED panel constructed. The maximum power consumption is 35.6 watt, with the measured output of 5644 lumen. The minimum power of the dimmable LED panel is 5.35 watt, with measured output of 950 lumen. The output shows linear relationship between power consumption with lighting output.

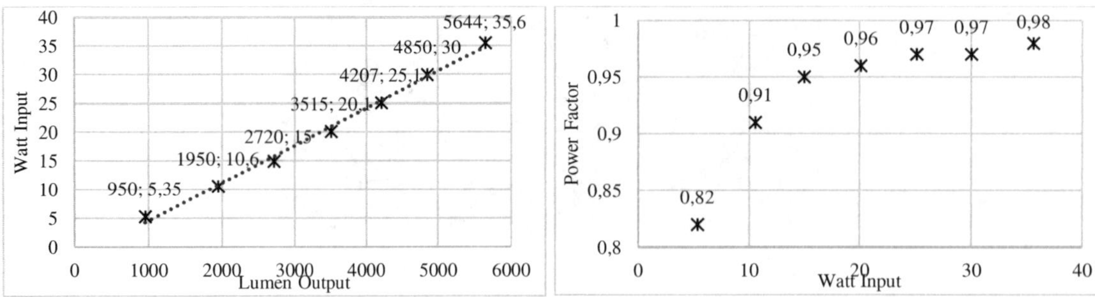

a **b**

Figure 9. a) LED panel power input versus centre spot lumen output dimming characteristics; **b)** LED panel power factor versus power input dimming characteristics.

Fig 9b shows the power factor characteristics versus power consumption. The power factor of LED panel in brightest setting is 0.98. The power factor decreases when the lamp is dimmed. The lowest setting of the LED panel causes the power factor to be drop to as much as 0.82.

After obtaining the characteristics, the LED panel lamp was installed to the home office setup. The ceiling and wall paint white with a window act as natural light source. The lux meter is placed on worksurface, and the natural light is varied. The LED was kept to light the work surface at the predefined illuminance, 250 lux. Fig 5 shows the home office experimental setup.

Fig 10 shows the power needed for LED panel to maintain illuminance at the work surface to be at 250 lux. The natural light is varied from 115 to 200. The experiment resulted in the variance of LED panel light needed to maintain the 250 lux as shown by Fig 10. A linear relationship can be obtained, with the correlation coefficient R^2 0.9903. This result is interesting to be used in future study, to develop self-dimming smart LED panel.

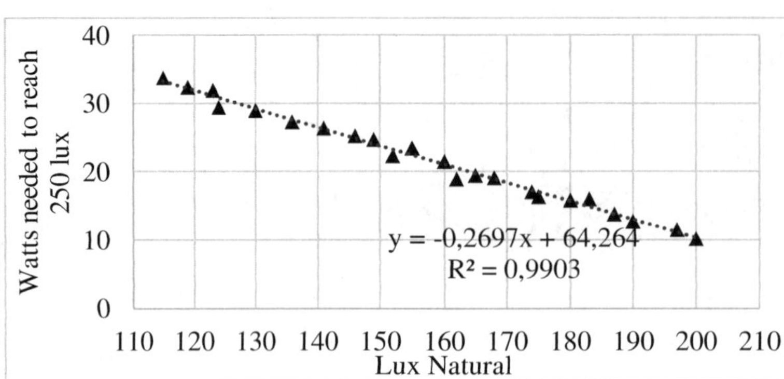

Figure 10. LED power needed to reach 250 lux in various natural lighting condition

4. Conclusion

From the research experiment that has been done, it can be concluded that the dimmable LED panel in attempt to save electrical power has been successfully implemented. A 35 W dimmable LED panel light

IOP Conf. Series: Materials Science and Engineering 1073 (2021) 012039 doi:10.1088/1757-899X/1073/1/012039

benefitting dimming function in QH7938 chip has been constructed. The dimming characteristics were obtained, with the output luminous flux of 5644 to 950 lumen, and power input of 35 to 5.32 watt. Subsequently, the power reduction case scenario in 2.5 x 2.8 x 3 meters home office with various natural lighting condition were presented. A dimming curve with high correlation coefficient of 0.99 were obtained. This result shows the ability of the system to be used in the future study, to develop self-dimming smart LED panel.

References

[1] Allcott H 2011 Social norms and energy conservation *J. Public Econ.* **95** 1082–95

[2] Steg L 2008 Promoting household energy conservation *Energy Policy* **36** 4449–53

[3] Khan N and Abas N 2011 Comparative study of energy saving light sources *Renew. Sustain. Energy Rev.* **15** 296–309

[4] Grigoryeva S, Baklanov A, Titov D, Sayun V and Grigoryev E 2017 Analysis energy efficiency of automated control system of LED lighting *2017 Int. Sib. Conf. Control Commun. SIBCON 2017 - Proc.* 3–7

[5] Syamsuri T U, Buwono H and Amalia R N 2019 Aplikasi Mikrokontroler dalam Sistem Kontrol dan Monitoring Energi Listrik *J. ELTEK* **17** 107–19

[6] Xia T, Liu C, Zheng X, Lei M, Wang S and Wang Y 2018 Research on Field Testing Method of Digital Energy Meters Based on Digital Reference Meter with High Accuracy *2nd IEEE Conf. Energy Internet Energy Syst. Integr. EI2 2018 - Proc.* 1–9

[7] Barman B K, Yadav S N, Kumar S and Gope S 2019 IOT Based Smart Energy Meter for Efficient Energy Utilization in Smart Grid *2nd Int. Conf. Energy, Power Environ. Towar. Smart Technol. ICEPE 2018* 1–5

[8] Chaudhari S, Rathod P, Shaikh A, Vora D and Ahir J 2018 Smart energy meter using Arduino and GSM *Proc. - Int. Conf. Trends Electron. Informatics, ICEI 2017* 2018-January 598–601

[9] Syamsiana I N, Setiawan A and Saputra M 2020 Design of Energy Database Management System (EDBMS) on the unit 2 outgoing transformer at the State Polytechnic of Malang *IOP Conf. Ser. Mater. Sci. Eng.* **732**

[10] Anon Datasheet of QH7938 Dimmable primary side feedback single-stage active PFC LED constant current chip 1–9

[11] Badan Standar Nasional Indonesia 2001 SNI 03-6575-2001 tentang Tata Cara Perancangan Sistem Pencahayaan Buatan pada Bangunan Gedung 1–32

IOP Conf. Series: Materials Science and Engineering 1073 (2021) 012040 doi:10.1088/1757-899X/1073/1/012040

Low-cost Transverse Electromagnetic (TEM) cell design for radiated emission measurement

M Y Hariyawan*, R S Darwis and B A Habibi

Department of Electrical Engineering, Politeknik Caltex Riau, Pekanbaru, Indonesia

*yanuar@pcr.ac.id

Abstract. The increasing use of electronic and electrical equipment that continuously emits radiated emissions in the industrial, military, communications, household, and medical fields has increased dramatically over the past few decades. Radiated emission can interfere performance of other electronic devices around it. There is a demand for electric devices to comply with electromagnetic compatibility (EMC) standards before being marketed, requiring manufacturers to carry out EMC testing before devices enter the market. Transverse Electromagnetic (TEM) cells are usually used by designers to ensure that electronic devices comply with EMC standards. TEM cells for radiated emission measurement are widely used because they are cheap and time-saving before full compliance testing is carried out, which expensive for one-time screening. The main problems of commercial open TEM cells are relatively high price, weight, handling problems and portability. In this study, low-cost TEM cells were designed that could be used to test radiated emission generated by electronic devices. The fabricated TEM cells can operate up to 960 MHz and it can be used to test radiated emission generated from small electronic devices. It has better performance and more sensitive than branded TEM cells in the radiated emission test of a mobile phone.

1. Introduction

Electromagnetic transverse cells (TEM) are equipment used to test various radio frequencies (RF) produced by electronic devices. This equipment was first introduced in 1974 by Myron L. Crawford to create a uniform electromagnetic field in a shielded environment [1]. In general, TEM cells are closed rectangular coaxial lines with tapered regions at each end. TEM cells are designed to spread the single TEM method on all measurement frequencies within the bandwidth. TEM cells consist of three main parts: connectors, transition parts, and rectangular coaxial lines. There are various types of TEM cells, which type to use depends on the application, frequency of interest, and size of the device being tested. This equipment can be used for E-field probe calibration, E-field immunity radiation test, and emission radiation measurement. Electromagnetic interference (EMI) testing for small RF devices in a laboratory environment, usually using TEM cells [2,3].

The main problems of commercial open TEM cells are relatively high price, weight, handling problems and portability. In addition, conventional TEM cell construction requires expensive copper/ ferrous sheet metal. As an important note, TEM cells made of sheet metal generally experience a loss of skin effect [4].

Various methods have been proposed in the design of TEM cells for multiple purposes. TEM cells built using 3D printers and wrapped in aluminum tape can operate up to 1.6 GHz [1]. The distribution of EM fields in cells, especially in the transverse section, does not seem to have much effect in high

IOP Conf. Series: Materials Science and Engineering 1073 (2021) 012040 doi:10.1088/1757-899X/1073/1/012040

order mode [5]. Open TEM cells with conductors in cells are tapered using multi-step and stepwise linear tapering methods designed to produce a standard EM field from 1 MHz to 1 GHz [2]. A small strip-line TEM cell can be used to measure EMC on IC components, which cannot be done by TEM cells in general [6]. Development of an open broadband TEM cell was likewise acquired and explored [7].

An evaluation of the distribution of the EM field in closed and open TEM cells is presented to determine whether the field distribution influenced by the presence of ferromagnetic and dielectric materials [8]. Characterization of DUT radiation emission with different orientations to electromagnetic field polarization in open TEM cells is presented through simulation and measurement [9].

An aluminum-based low-cost TEM cell was developed and fabricated in this study. It can operate up to 960 MHz. TEM cell design results can be used to test radiated emission generated from small electronic devices such as mobile phones. Fabricated TEM cells can measure emissions emitted by cell phones. It can respond EMI in the frequency range 170-270 MHz, 400-510 MHz, 630-740 MHz, and 870-960 MHz. TEM cells designed can measure.

2. TEM cell design and fabrication

The flow of TEM Cell design can be seen in Figure 1. The research stages began by studying the specifications of TEM cells to be designed by considering return loss, VSWR, and matching impedance. The design results are simulated to ensure that the design results are by the research objectives. The next process is the TEM Cell fabrication. Fabrication of TEM Cell test results is carried out to ensure that parameters such as return loss, VSWR, and matching impedance match the simulation results. If the test results match the results of the simulation fabrication process is carried out.

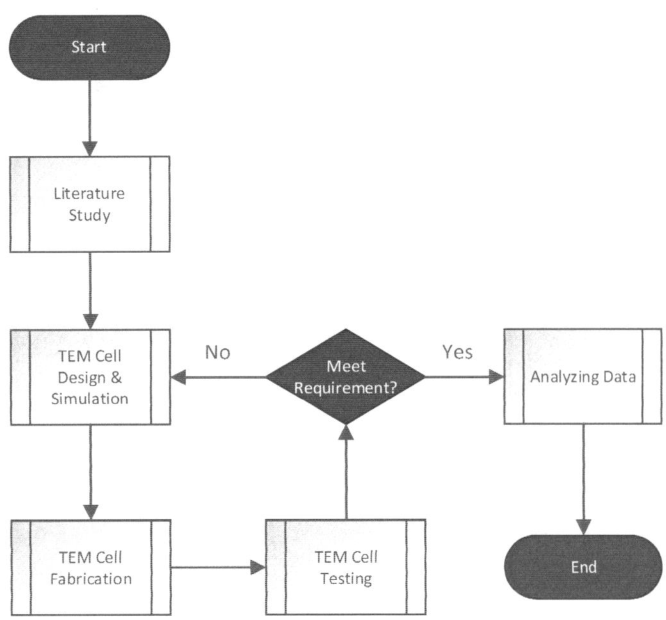

Figure 1. Research flow.

The material used for the design of this TEM cell is aluminum to make TEM Cells that are easier to fabricate and more cost-effective. TEM cells are built following the specifications of branded TEM Cells, as in Table 1. To see the performance of TEM cells, the parameters observed in this simulation are return loss, VSWR, propagation of power in the cell, and matching impedance in the specified frequency range.

The simulation results are then analyzed, whether the simulation results meet the specified standards or not. If the simulation results do not show the des. If the simulation results show that the results comply with the established criteria, then the design of the TEM cell that is designed can be implemented to the TEM cell construction at a higher stage. The matching impedance of the septum is calculated using the following equation [4]:

$$Z_O = \frac{94,15}{\sqrt{\varepsilon_r}\left[\frac{w}{2d(1-\frac{t}{2d})} + \frac{C_f}{0,0885\ \varepsilon_r}\right]}$$

where:
εr = the medium inside the cell relative dielectric constant (1 for air)
w = the septum width (m)
t = the septum thickness (m)
d = distance between the septum and upper wall
2d = the cell height (m)
Cf = fringing capacitance per unit length (0.053 pF/cm).

Table 1. Dimensions of the TEM cell.

View	Name	Dimensions
Top	Length (un-tapered)	350 mm
Top	Length (tapered)	165 x 2 mm
Cross-section	Width (outer Shield)	200 mm
Cross-section	Width (Septum)	126 mm
Cross-section	Septum-outer shield	80 mm x 2
Side	Tapering angle with respect horizontal	≈30°
Side	Height (Septum)	50 mm
Side	Height (TEM cell)	100 mm

Table 2. Illustration of dimensions of TEM cell parts.

View	Dimension	View	Dimension
Top and bottom	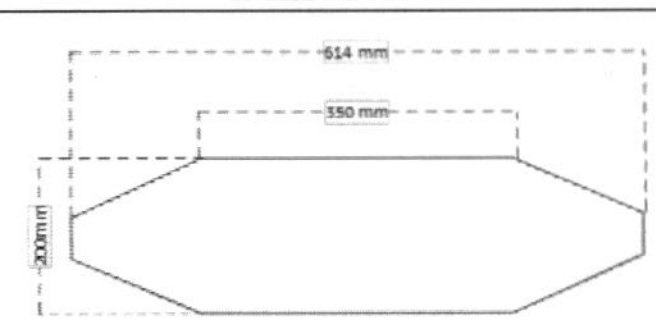	Tapered septum	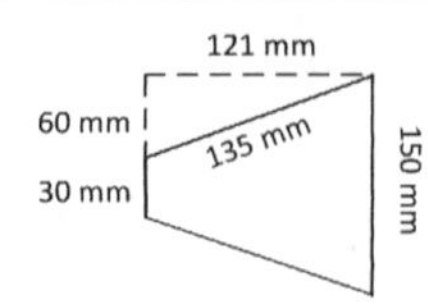
Tapering side	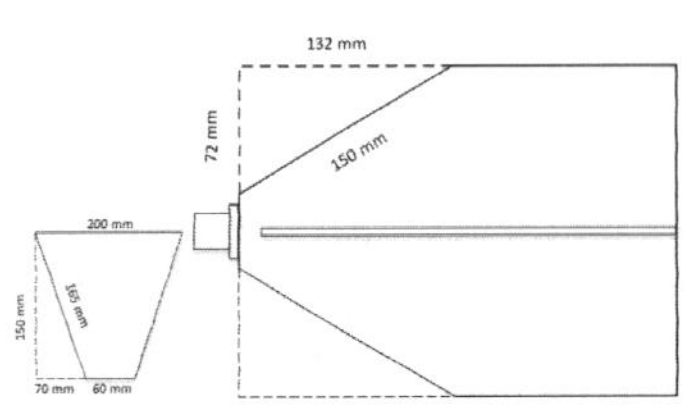	Probe section	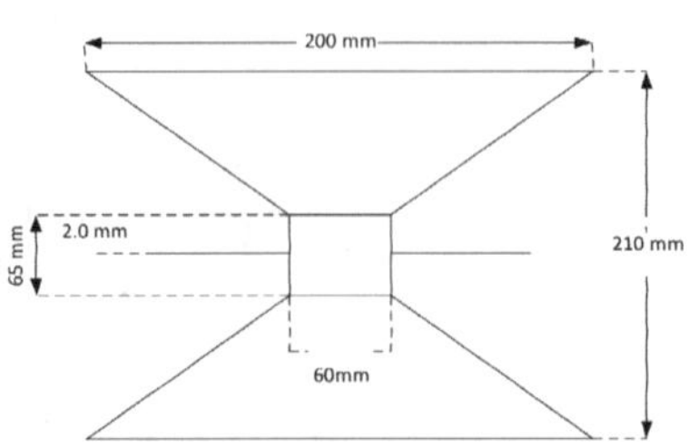
Septum	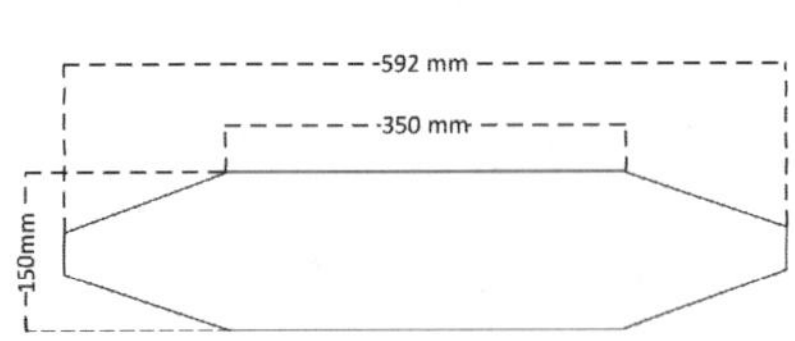	Front and back	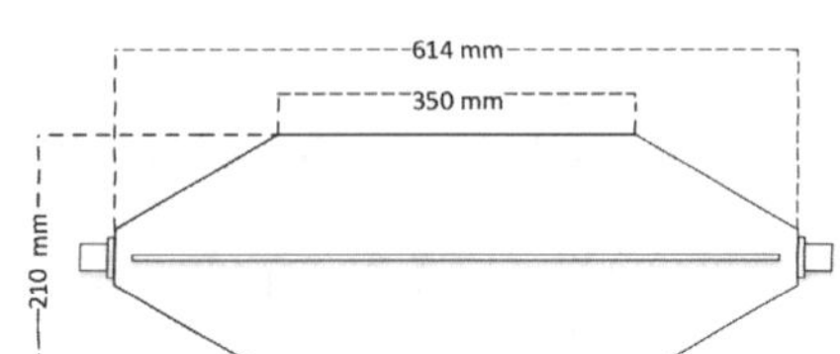

Before a TEM Cell simulation is done, the whole TEM cell's dimensions must be determined first. Three parts of TEM Cell must be determined by the size of the field (length, height, and thickness of the aluminum plate): the top and bottom of the TEM cell, the middle part of the TEM Cell called the septum and the left and right side (the probe) of the TEM Cell. Based on the predefined specifications, by Table 1, the dimensions of some parts can be seen in **Error! Reference source not found.**.

Visually the 3D image of the TEM Cell design can be seen in Figure 2. Based on the design results' dimensions, the VSWR and Return Loss obtained can be seen in **Error! Reference source not found.** and **Error! Reference source not found.**. From **Error! Reference source not found.** several frequency ranges are not in line with expectations. In the frequency range of 600 MHz -700 MHz, VSWR ranges from 2 to 2.6, and in the frequency range of 800 MHz - 950 MHz, VSWR ranges from 2.3 to 4.6. In **Error! Reference source not found.**, the return loss value is quite good, although several frequency ranges have a poor return loss quality. In the frequency range of 600-710 MHz, the return loss ranges from -7 dB to -9 dB, and in the frequency range 790-950 MHz, the return loss ranges from -5 to -9 dB. With a satisfactory simulation result, the next step is the fabrication process and continued with TEM cell fabrication testing. Fabricated TEM cells are shown in **Error! Reference source not found.** and the TEM cell test set up is shown in **Error! Reference source not found.**. The fabricated TEM cells were modified at the termination section (port 1 and 2), and there were four nylon-based supports between the septum and the bottom of the TEM cells.

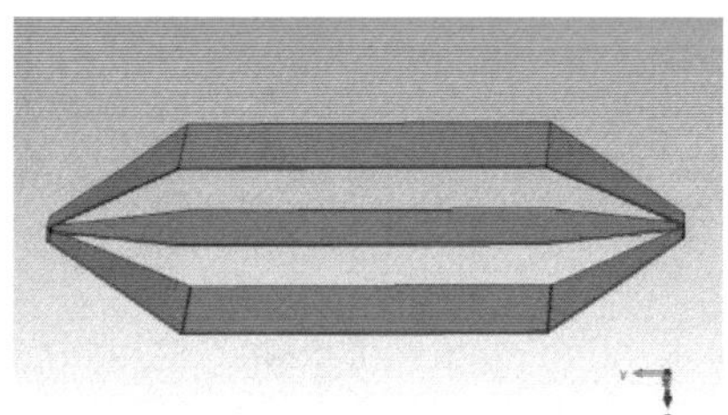

Figure 2. TEM cell 3D visual simulation.

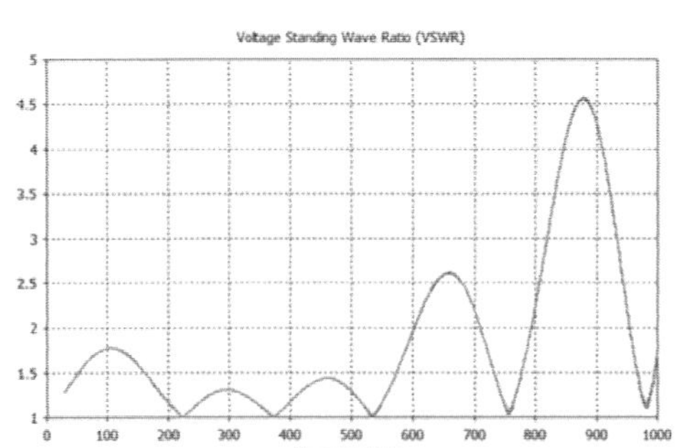

Figure 3. VSWR simulation results.

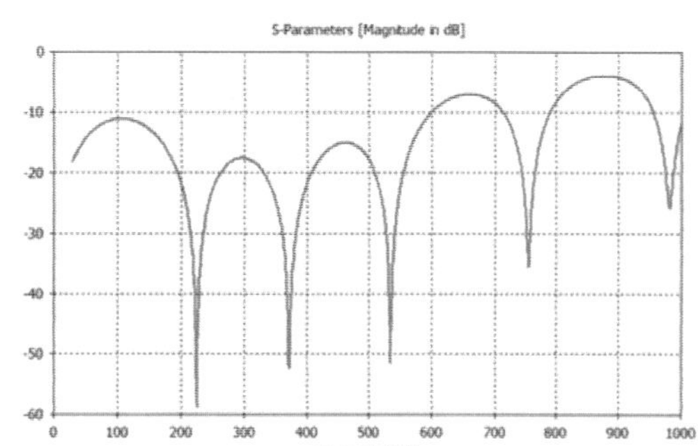

Figure 4. Return loss simulation result.

Figure 5. Fabricated TEM cells.

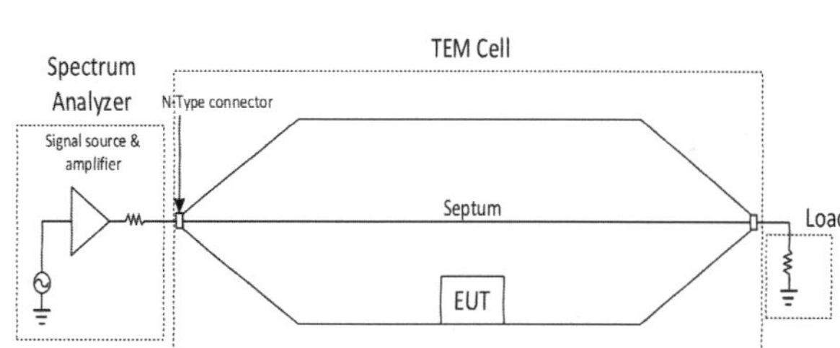

Figure 6. TEM cells test set up.

3. Data analysis

3.1. Characteristic testing

The parameters of the voltage standing wave ratio (VSWR), scattering, and Return Loss of fabricated TEM cells are tested. The return loss test results can be seen in Figure 7. From the measurement results, several frequency ranges are above -10 dB. At a frequency of 50 - 180 MHz, a return loss value of -7.5 dB, - 6 dB at 280 - 405 MHz, -6.5 dB at 505 - 630 MHz, -6 dB at 750 - 870 MHz, and -8 dB is obtained at a frequency of 960 - 990 Mhz. This means, at some point in the frequency range, the septum does not receive full power, or it can be said that some of the energy is reflected. However, if seen in general, the fabricated return loss has a broader frequency range than the simulation. The frequency range of the simulation results has a narrower frequency range when compared to fabrication.

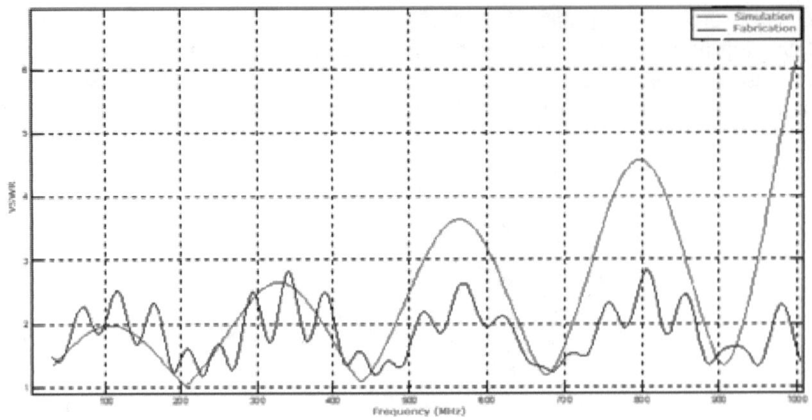

Figure 7. VSWR simulation vs fabrication.

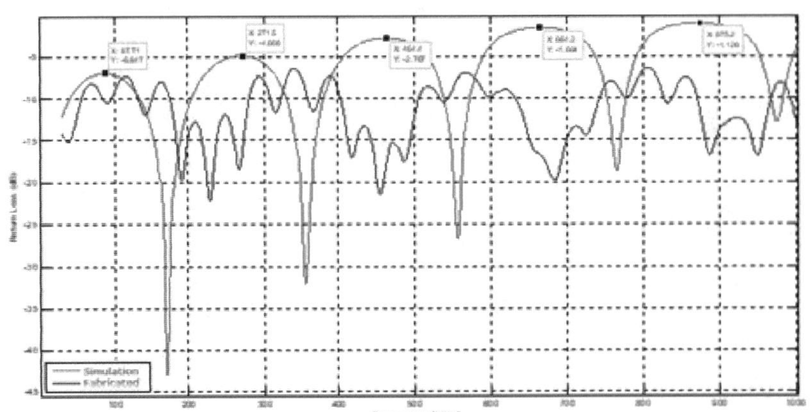

Figure 8. Return loss simulation vs fabrication.

Voltage Standing Wave Ratio (VSWR) is defined as the ratio of the maximum voltage and minimum voltage standing waves on a transmission line. If the load impedance does not match the transmission line impedance, part of the incoming wave energy on the load will be reflected. It creates a reflected wave on the line, which interferes with the transmission power. This interference produces standing waves whose magnitude depends on the magnitude of the reflection power.

The results of measuring the VSWR value can be seen in Figure 7, which shows a significant difference. In general, the fabrication results have a better VSWR than the simulation results. The reflected waves are caused by the power losses reflected from the load (dummy load). While the frequency shift caused by a change in the size of the tapered portion of the TEM Cell. The difference in the size of the tapered part of the TEM Cell is carried out so that the integration between the upper and lower parts of the TEM Cell (outer) can firmly integrate with the septum part of the TEM Cell (inner).

Compared with the simulation results, the TEM Cell fabrication's measurement results show a better return loss value, as shown in Figure 8. From the simulation results, there are several points whose return loss values are above -10 dB. In the simulation, at a frequency of 271 MHz, the return loss value is -4.8 dB, while the measurement results obtained a lower return loss value, which is around -7.2 dB. The frequency shift is probably caused by the TEM Cell manufacturing process, where the terminal ports 1 and 2 widen the size from 2.5 x 4 cm to 5 x 5 cm. Changes were made to simplify the process of unifying the septum and outer (top and bottom) parts of the TEM Cell.

3.2. Radiated emission testing

Radiated emission test was carried out by 2 DUTs (device under test) (EMI source and mobile phone) to test the performance of the fabricated TEM Cells. The test results compared with branded TEM Cells in the Transmission Media Laboratory of the Polytechnic of Caltex Riau.

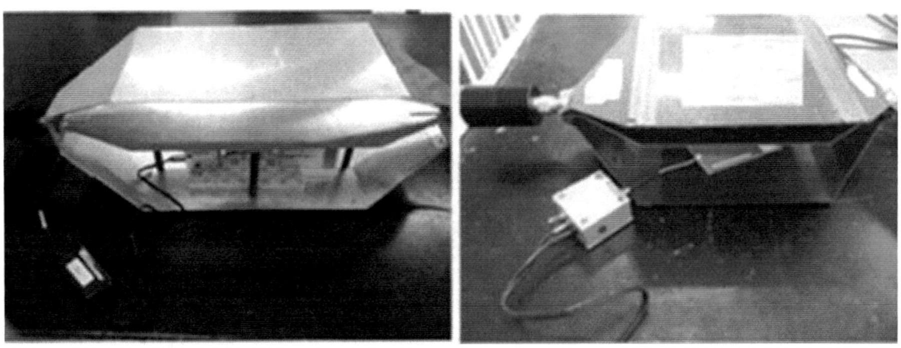

Figure 9. EMI source radiated emission test set-up: Fabricated vs. branded.

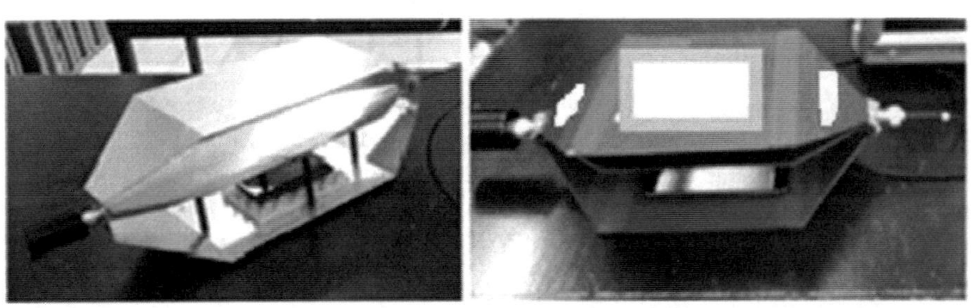

Figure 10. Mobile phone radiated emission test set-up: Fabricated vs. branded.

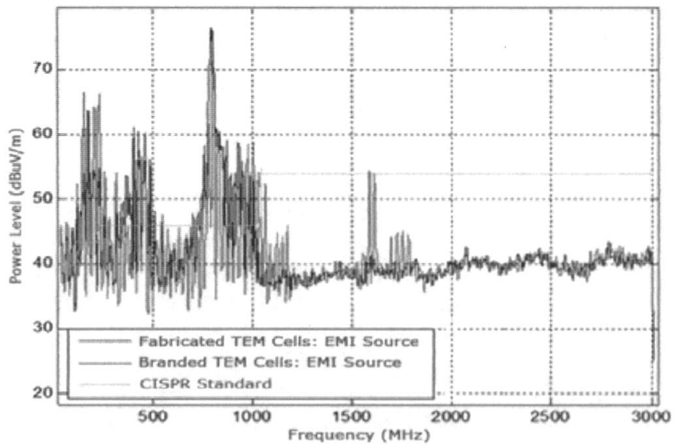

Figure 11. Radiated emission generated by EMI source: Fabrication vs branded.

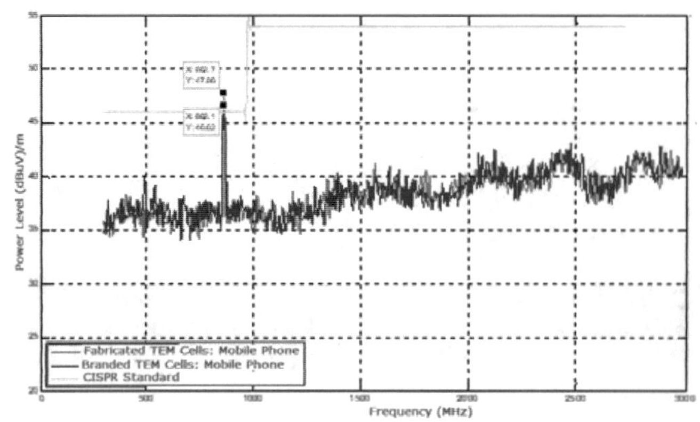

Figure 12. Radiated emission generated by mobile phones: Fabrication vs branded.

Performance testing of fabricated TEM cells is carried out by measuring the emission signal generated by an electronic device and comparing it with the performance of branded TEM cells made from mixed materials. EMI source and mobile phones are used as devices under test (DUT). The Radiated Emission test generated by EMI source and mobile phone is shown in Figure 9 and Figure 10. When the EMI source is set to produce an emission signal at a frequency of 10 MHz, the radiated emission test results are obtained, as shown in Figure 11. The fabricated TEM Cell can capture radiation signals that are not captured by branded TEM cells. It proves that the fabricated TEM cell is more sensitive in capturing emission signals. According to the CISPR 22 class B standard regulations, radiation emitted by the EMI Source passes through the radiation standard at several frequencies, 40 dBμV/m (30 - 88 MHz), 43.5 dBμV/m (88 - 216 MHz), 46 dBμV/m (216 - 960 MHz).

The radiation emission test produced by a mobile phone when making a call is shown in Figure 12. It can be seen that the radiation emission captured on the fabricated TEM cells has a higher power level when compared to branded TEM cells. It also proves that the fabricated TEM cells are more sensitive when compared to branded TEM Cells. The radiation emission test results on the mobile phone show that the shielding on the mobile phone cannot reduce the emission it produces accurately.

4. Conclusion

The designed, low-cost TEM cell can be used to test the radiation emission generated by electronic devices. From the VSWR and Return Loss measurements, the designed TEM cells have better characteristics compared to the simulation results. The fabricated TEM cells have better performance and more sensitive than branded TEM cells in the radiated emission test of a mobile phone. It can operate up to 960 MHz and it can be used to test radiated emission generated from small electronic devices such as mobile phones.

References

[1] Al Takach A, Ndagijimana F and Al-husseini M 2018 3D-Printed Low-Cost and Lightweight TEM Cell *in 2018 International Conference on High Performance Computing & Simulation (HPCS)* 2018–2021

[2] Arezoomand M, Meybodi M K and Noori N 2016 Design of a TEM Cell Using Both Multi-Step and Piecewise Linear Tapering *in 2016 8th International Symposium on Telecommunications (IST)* 571–574

[3] Ho T H, Lin H N, Ko P N, Huang Y C, Hsiao H C and Li J K 2019 Design and characteristic calibration of TEM cell for IC radiation EMC test *IEEE Electr. Des. Adv. Packag. Syst. Symp.* 10–12

[4] Satav S M 2008 Do-it-Yourself Fabrication of an Open TEM Cell for EMC Pre-compliance *IEEE EMC Society Newsletter* 66–71

[5] Paffi A, Liberti M, Fratta F, Merla C, Pinto R and Lovisolo G 2012 A TEM Cell System for In Vivo Exposure at 2 . 45 GHz *in 2012 6th European Conference on Antennas and Propagation (EUCAP)* 1099–1101

[6] Wen L, Yalin G and Jin L 2016 Three New Strip-Line TEM Cells in EMC Test *in 2016 IEEE International Conference on Electronic Information and Communication Technology (ICEICT)* 497–500

[7] Chechetkin V, Korotkov A and Kozlov V 2019 Investigation of the Characteristics of the TEM Cell Prototype *in SIBIRCON 2019 - International Multi-Conference on Engineering, Computer and Information Sciences, Proceedings* 241–244

[8] Koohestani M, Ramdani M, Besnier P, Levant J L and Perdriau R 2019 Perturbations of Electric and Magnetic Fields Due to the Presence of Materials in TEM Cells *IEEE Trans. Electromagn. Compat.*, **62**(4) 1–0

[9] Koohestani M, Ramdani M, Lafon F, Moreira A A and Perdriau R 2020 Impact of Field Polarization on Radiated Emission Characterization in an Open TEM Cell *IEEE Trans. Instrum. Meas.* **69**(9) 1–1

IOP Conf. Series: Materials Science and Engineering 1073 (2021) 012041 doi:10.1088/1757-899X/1073/1/012041

Solar panels output optimization using Phase Change Material (PCM) and heatsink applied in open-pit mining

S Widodo[1,2], A Taqwa[3] and R Rusdianasari[3,*]

[1] Applied Master of Renewable Energy Engineering, Politeknik Negeri Sriwijaya, Jalan Srijaya Negara, Palembang, 30139 Indonesia
[2] Maintenance of Mining Support and Production Equipment, PT. Bukit Asam Tbk, Jalan Parigi No. 1 Tanjung Enim, 31716 Indonesia.
[3] Renewable Energy Engineering Department, Politeknik Negeri Sriwijaya, Jalan Srijaya Negara, Palembang, 30139 Indonesia.

*rusdianasari@polsri.ac.id

Abstract. Solar energy can be utilize to power a system located far from the electricity sources such a mobile tower applied in open-pit mining in PT Bukit Asam Tbk. However, the tropical weather in an open-pit mining can increase the surface temperature of a Photovoltaic (PV) panel, lead to overheating and the reduction of the efficiency and the reduced electricity power. This condition requires a cooling system to reduce the temperature. This paper presents the application of Phase Change Material (PCM) and heatsink as the cooling system for a PV panels. The PCM installed are in two sizes aluminum hollows, a 0.0025 m^3 and 0.00625 m^3. The experiment was conducted in 2 batches, 14-29 February 2020 for the installed 0.0025 m^3 aluminum hollow, and 1-14 March 2020 for 0.00625 m^3. The application of a 0.00625 m^3 can reduce the PV panel's surface temperature up to 3-5°C. The generated current using PCM 0.00625 m^3 is 0.034 A for I_{sc} and 0.014 A I_{load} higher than without cooling system. The application of a aluminum hallow with the size of 0.00625 m^3 gives 9-20% efficiency, and the efficiency gives 8-15%. The experimental results show that the use of a hollow aluminum with a size of 0.00625 m^3 can reduce the surface temperature of the PV panel and increase the power and efficiency of the PV panels.

1. Introduction

The industrial revolution has increased the need for energy sources for both business and non-commercial activities. As the scientists have predicted, the availability of widely used fossil fuels are becoming depleted by time. In Indonesia, 95% of the energy consumption still uses fossil fuels, and about 50% of the fossil energy is oil, whereas the rest is gas and coal. Consumption of fossil fuel leads to environmental pollution and greenhouse CO. Among other floods, drought and global warming are impacts of these pollutants. This condition leads people to find energy alternatives that are more natural and easy to replenish. The new renewable energy trend is expected to improve the quality of life for people [1-6].

However, electricity generation by solar energy is effected by many external factor including sun position relative to the photovoltaic (PV) panel position, overheating and shading [6-9] and the overheated PV surface [10-18]. Recent studies using phase change materials (PCMs) have attracted many researchers to the surface of solar cells as a cooling system [19-21].

This paper discusses the comparative analysis of PV panels installed in a mobile tower used in open-pit mining. The first panel is equipped with a PCM and heat sink as a cooling device using an aluminum hallow with a size of 0.0025 cm^3, and the second panel uses an aluminum hallow with a size of 0.00625 m^3. The third PV panel is installed without a cooling system as a benchmark for the cooled PV panel. This research took place in the South Sumatra Open-Pit mine. The compared PV panels supply electricity to the mobile tower used in the pit. It is critical that this renewable technology continues to grow due to its high potential for use in everyday life [22,23].

2. Proposed method

This paper proposes the use of PCM for a passive cooling system combined with a heat sink. Three PV panels are installed, two cooling panels with a different hollow aluminum size and the third one is normally installed. The aluminum hollow sizes considered in this study are 0.0025 m^3 and 0.00625 m^3.

2.1. Phase Change Material (PCM)

PCM is a phase change material that allows thermal energy to be absorbed, processed, and released. PCM is also referred to as a latent heat storage unit. The use of certain PCMs depends very much on the temperature of the phase change, where latent heat can be used as a whole. When properly applied to a system, PCM can make considerable energy savings. When PCM reaches temperature changing phase (melting temperature), PCM absorbs large amounts of heat at almost constant temperatures. PCM continues to absorb heat until all materials are converted into liquid phases, without significant temperature changes. If the heat temperature decreases around the environment, the liquid PCM begins to freeze and release the stored latent heat. Large amounts of PCM are available in any temperature range up to 190°C. Therefore, the use of paraffin wax in solar cells is one alternative for reducing the surface temperature of too high solar panels. The heat sink is used as a heat transfer metal to help spread heat to the PCM. Thus, the expected surface area of the object's heat transfer is faster than the cooling process.

2.2. Proposed cooling system

This research is implemented for PV panels to supply mobile towers in PT. Bukit Asam Tbk which is located in Tanjung Enim, South Sumatra. The PT. Bukit Asam Tbk open-pit mining site and its surrounding area is 7700 ha. The region is located at 103°45′ –103°50′ East Longitude and 3°42′30′ –4°47′–30′ South Latitude. Mobile tower shown in Figure 1 is required to allow data communication from a wireless network on a vast mining site, but traditional energy supply for the mobile tower is not feasible. Solar energy is suitable for this type of situation because it receives plenty of solar rays.

This mobile tower is critical to the implementation of the FMS (Fleet Management Systems), which enables the transmission of data on speed units, working tonnage units, heavy-duty equipment conditions, fuel and working conditions, and the disposal or front of the excavation process. Updated data is sent to the dispatcher or operator. The PV panels to power the mobile tower is shown in Figure 2.

Figure 1. The mobile tower deployed in the open-pit mine.

Figure 2. The PV panels to power mobile tower deployed in the open-pit mine.

IOP Conf. Series: Materials Science and Engineering 1073 (2021) 012041 doi:10.1088/1757-899X/1073/1/012041

2.3. PV panels efficiency
Efficiency is the parameter used to show the effectiveness of PV panel application using a proposed method. Efficiency is the ratio between the output produced and the energy input received by the PV panel. Therefore, the efficiency of a PV panel is given by

$$\eta = \frac{\rho_{max}}{\rho_{in}} x100\%, = \frac{I_{mp} \cdot V_{mp}}{P_{in}} x\ 100\%, = \frac{I_{sc} V_{oc} FF}{P_{in}} x\ 100\%, \quad (1)$$

where *Imp* and *Vmp* are the generated current and voltage when the PV panel supplies a load, and *Isc* is the current during short circuit and without load, and *Voc* is the voltage measured during open circuit without any load attached.

3. Results and discussion

3.1. Experimental setup
The experiment was carried out using three solar panels installed on the mobile network. The PV panels used are monocrystalline types with 50 WP (W-peak) each. Two panels installed using PCM paraffin and heatsink systems with 0.0025 m^3 of hollow aluminum size measurements were taken from data (14-29 February 2019) and from 1-14 March 2020 to 0.00625 m^3 of hollow aluminum. Another PV panel is installed without the use of a PCM and heatsink system. Data is collected every day from 9.00 a.m. to 3.00 p.m. The circuit is equipped with a thermal camera to detect the temperature of each device in detail. The solar charger is used to limit the incoming and outgoing currents of the battery as a source of energy from installed solar cells.

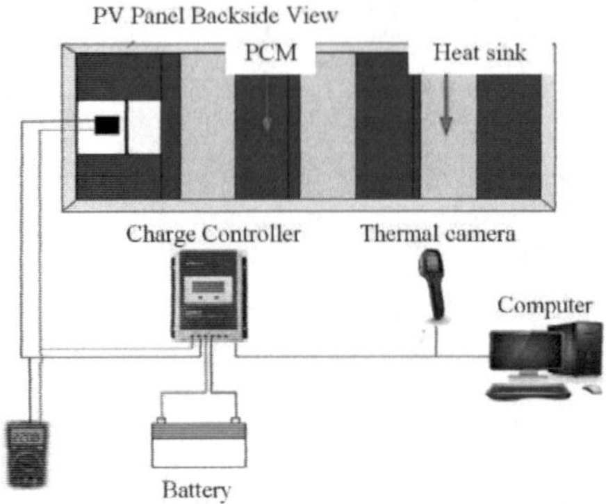

Figure 3. Experimental setup.

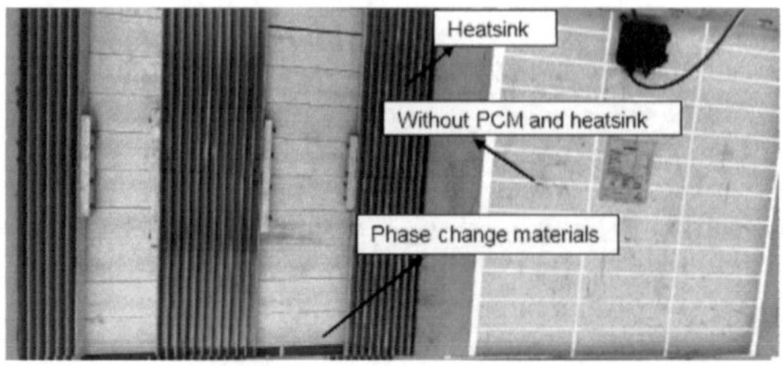

Figure 4. The installation of PCM and heat sink compared to the panel without cooling system.

Figure 3 shows the experimental setup of the PV panels attached to the mobile tower shown in Figure 1. Figure 4 shows the back of the PV panel when the cooling system is installed (left), and the cooling system is not installed (right). Both PV panels are installed side by side; therefore, both panels received the same amount of irradiance.

3.2. Experimental results
The experiment was initially scheduled for 30 days, but the weather was challenging in February and March when South Sumatra had medium cloudy and sunny conditions.

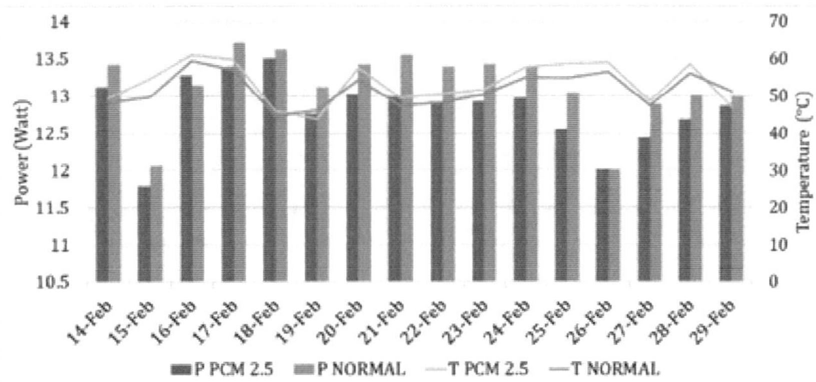

Figure 5. Temperature effect on output power with the application aluminum hollow 2.5 m^3.

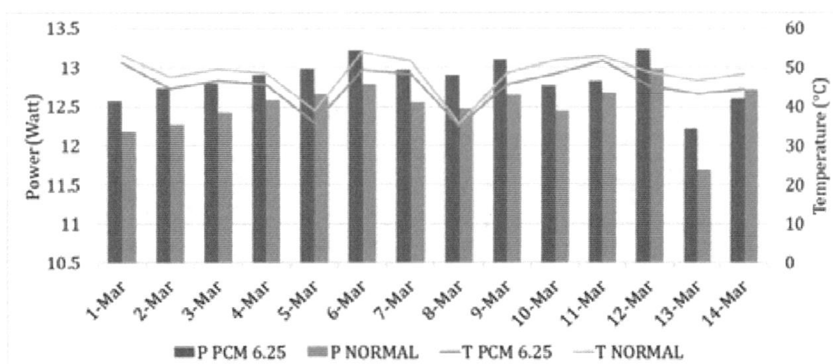

Figure 6. Temperature effect on output power with the application aluminum hollow 6.25 m^3.

Figure 5 shows the temperature conditions that could affect the output power generated by the PV panel. Weather conditions are observed every day in order to compare the effect of the temperature measured with the application of a 0.0025 m^3 aluminum Hallow PCM system and heat sink compared to the condition without the use of a 0.0025 m^3 Hallow aluminum PCM system and heat sink. Power generated by a 0.0025 m^3 PCM system (P PCM 2.5) is less than normal power without a 0.0025 m^3 PCM system (P NORMAL). The temperature measured using the 0.0025 m^3 PCM system (T PCM 0.0025) is higher than the temperature without using the 0.0025 m^3 PCM system (T NORMAL).

Figure 6 shows the comparison of the produced power between the application of the PCM cooling system and the 0.00625 m^3 aluminum hallow heat sink and the standard application of the PV panel. T Normal is defined as temperature without a cooling system, and T PCM 0.00625 m^3 is the temperature at which the cooling system is used. P PCM 0.0025 is the power produced with the application of 0.00625 m^3 PCM cooling system. The cooling system reduces the PV panel temperature to 3-5°C for the application of a 0.00625 m^3 aluminum hollow. Therefore, the overheated temperature can be properly reduced by using a PCM with a 0.00625 m^3 hallowed aluminum heat sink. Figures 7 and 8 show the cooling system's effect when using a 0.0025 m^3 and 0.00625 m^3 aluminum hallow. The application of an aluminum hallow with a size of 0.0025 m^3 is not sufficient to reduce the overheated PV panel.

The power generated by the use of a cooling system with an aluminum hollow of 0.00625 m^3 is higher than without a cooling system, while the power generated by the use of a cooling system with an aluminum hollow of 0.0025 m^3 is lower than the power generated without a cooling system. The average temperature with a hollow aluminum cooling system of 0.0025 m^3 PCM is 53°C and without 51°C and with a hollow aluminum cooling system of 0.00625 m^3 PCM is 45°C and without 48°C.

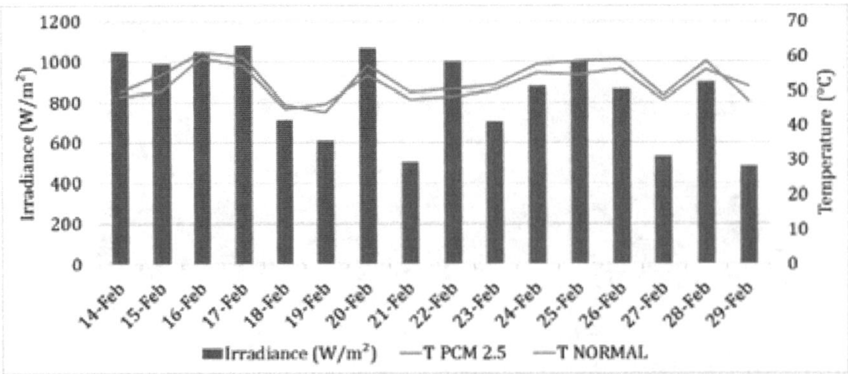

Figure 7. Irradiance effect on temperature using 0.0025 m^3 hollow aluminum.

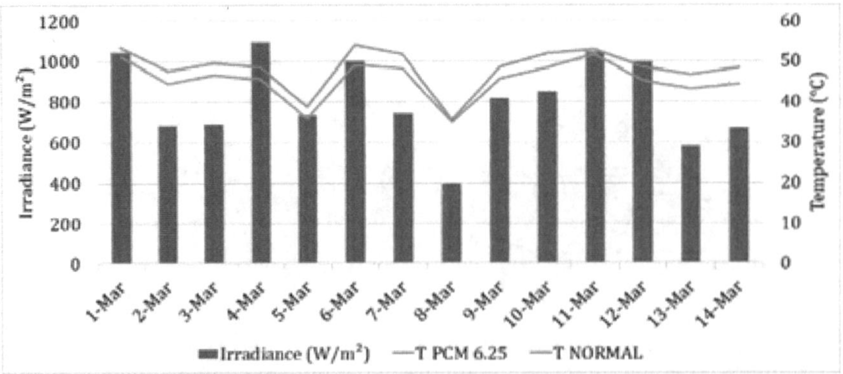

Figure 8. Irradiance effect on temperature using 0.00625 m^3 hollow aluminum.

Figures 7 and 8 that weather determines the irradiance value generated on the PV solar cell. This condition means that the lower the solar cells' irradiance, the lower the current produced by the solar cells.

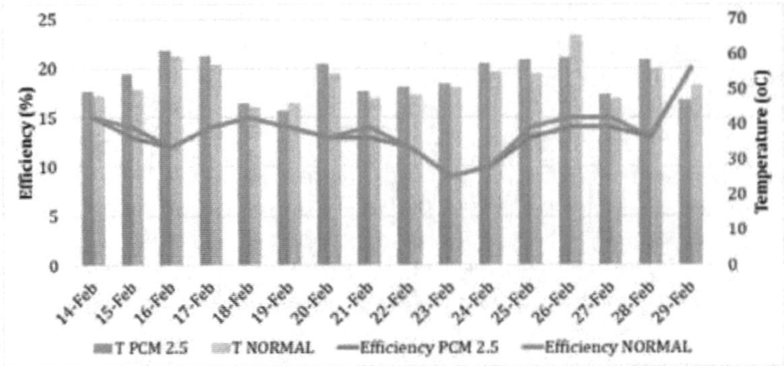

Figure 9. Comparison of efficiency using a 0.0025 m^3 PCM system and without cooling system.

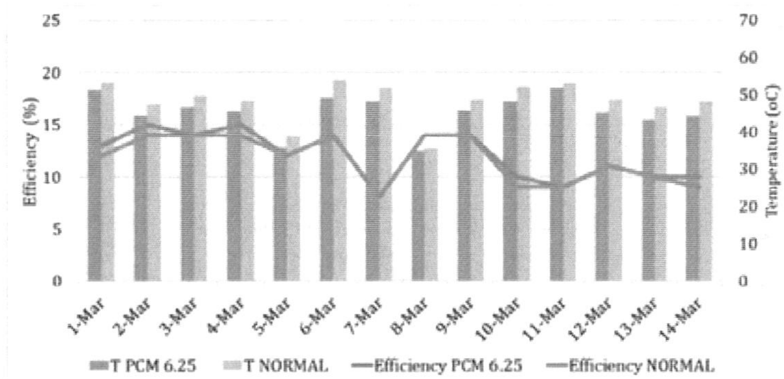

Figure 10. Comparison of efficiency using a 0.00625 m3 PCM system and without cooling system.

Figures 9 and 10 provide the efficiency comparison of the cooling system's performance and without the cooling system, where Efficiency PCM 0.0025 is the efficiency of application PCM 0.0025 m^3 and Efficiency PCM 0.00625 is the efficiency of application PCM 0.00625 m^3. Figure 10 shows that using an aluminum hallow with a size of 0.0025 m^3, the highest efficiency is 20%, and the lowest is 9%. The efficiency obtained by using a cooling system with an aluminum hallow solution of 0.00625 m^3 is 15% the highest and 8% the lowest.

The experimental results show that the application of a hollow aluminum with a size of 0.0025 m^3 is not sufficient to optimize the performance of the PV panel to power the PT Bukit Asam Tbk mobile tower. The application of a hollow aluminum with a size of 0.00625 m^3 can reduce the surface temperature of the PV panel and increase the power and efficiency of the PV panels.

4. Conclusion

This paper presents the application of PCM and heatsink as the cooling system for a PV panels. The PCM installed are in two sizes aluminum hollows, a 0.0025 m^3 and 0.00625 m^3. The experiment was conducted in 2 batches, 14-29 February 2020 for the installed 2.5 m^3 aluminum hollow, and 1-14 March 2020 for 0.00625 m^3 aluminum hollow. The application of a aluminum hallow with the size of 0.0025 m^3 gives 9-20% efficiency, and the efficiency gives 8-15%. The experimental results show that the application of a hollow aluminum with a size of 0.0025 m^3 is not sufficient to optimize the performance of the PV panel to power the PT Bukit Asam Tbk mobile tower. The use of a hollow aluminum with a size of 0.00625 m^3 can reduce the surface temperature of the PV panel and increase the power and efficiency of the PV panels.

References

[1] Ploetz R, Rusdianasari R and Eviliana E 2016 Renewable Energy: Advantages and Disadvantages in *Proceeding Forum in Research, Science, and Technology (FIRST)*

[2] Yuliandi R B, Dewi T, and Rusdianasari 2018 Comparison of Blade Dimension Design of a Vertical Wind Turbine Applied in Low Wind Speed E3S Web of Conferences EDP Sciences **68** 01001

[3] Bow Y, Dewi T, Taqwa A and Rusdianasari 2018 Power Transistor 2N3055 as a Solar Cell Device 2018 *International Conference on Electrical Engineering and Computer Science (ICECOS) IEEE* 327-332

[4] Yudha M H, Dewi T, Risma P, and Oktarina Y 2018 Life cycle analysis for the feasibility of photovoltaic system application in Indonesia *IOP Conference Series: Earth and Environmental Science* **124** 012005

[5] Arissetyadhi I, Dewi T, and Kusumanto R D 2020 Experimental Study on The Effect of Arches Setting on Semi-Flexible Monocrystalline Solar Panels *Kinetik: Game Technology, Information System, Computer Network, Computing, Electronics, and Control. KINETIK* **5**(2)

[6] Dewi T, Risma P, and Oktarina Y 2019 A Review of Factors Affecting the Efficiency and Output of a PV system Applied in Tropical Climate presented at the *IOP Conference Series: Earth and Environmental Science* **258** 012039.

[7] Edward A, Dewi T, Rusdianasari 2019 The effectiveness of Solar Tracker Use on Solar Panels to The Output of The Generated Electricity Power *IOP Conference Series: Earth and Environmental Science* **347**(1) 012130

[8] Hamdi B R D M, Dewi T, and Rusdianasari 2019 Performance Comparison of 3 Kwp Solar Panels Between Fixed and Sun Tracking in Palembang-Indonesia *IOP Conference Series: Earth and Environmental Science* **347**(1) 012131

[9] Zhafarina I N, Dewi T, and Rusdianasari 2018 Analysis of Maximum Power Reduction Efficiency of Photovoltaic System at PT. Pertamina (Persero) RU III Plaju VOLT: Jurnal Ilmiah Pendidikan Teknik Elektro 3(1) 19-25.

[10] Bow Y, Syakdani A, Taufik M and Rusdianasari 2020 Effect of Drying Airflow Rate on H2O Mass Evaporated on Banana Chips Drying using Photovoltaic Solar Panel *Journal of Physic: Conf.* **1500** (012015)

[11] Tabaei H and Ameri M 2018 Improving the effectiveness of a photovoltaic water pumping system by using booster reflector and cooling array surface by a film of water in Proceeding Iranian *Journal of Science and Technology Transactions of Mechanical Engineering* **39** 5160

[12] Chumpolrat K, Sangsuwan V, Udomdachanut N, Kittisontirak S, Songtrai S, Chinnavornrungsee P, Limmanee A, Sritharathikhun J, and Sriprapha K 2014 Effect of Ambient Temperature on Performance of Grid-Connected Inverter Installed in Thailand, Hindawi Publishing Corporation

[13] Duffie A and Beckman W A 2013 *Solar Engineering of Thermal Processes 1st Ed* (John Wiley & Sons, Inc.) pp 776-778

[14] Kamkird P, Ketjoy N, Rakwichian W and Sukchai S 2012 Investigation on Temperature Coefficients of Three Types Photovoltaic Module Technologies under Thailand Operation Condition *Procedia Engineering, Elsevier* **32** 376-383

[15] Sasmanto A A, Dewi T, and Rusdianasari 2020 Eligibility Study on Floating Solar Panel Installation over Brackish Water in Sungsang, South Sumatra *EMITTER International Journal of Engineering Technology* **8**(1)

[16] Junianto B, Dewi T, and Sitompul C R 2020 Development and Feasibility Analysis of Floating Solar Panel Application in Palembang, South Sumatra *Journal of Physics: Conf. Series 3nd Forum in Research, Science, and Technology Palembang, Indonesia*

[17] Kurniawan A, Taqwa A, and Bow Y 2019 PLC Application as an Automatic Transfer Switch for on-grid PV System; Case Study Jakabaring Solar Power Plant Palembang *Journal of Physics: Conf.* **1167** 012026

[18] Harahap H A, Dewi T, and Rusdianasari 2019 Automatic Cooling System for Efficiency and Output Enhancement of a PV System Application in Palembang *Indonesia Journal of Physics: Conf.* **1167** 012027

[19] Kasali S O, Ordonez-miranda J, and Joulain K 2020 Conductive thermal diode based on two phase-change materials International *Journal of Thermal Sciences* **153** 106393

[20] Selvam C, Manikandam S, Khrisna N V, Lamba R, Kaushik S C, and Mahian O 2020 Enhanced thermal performance of a thermoelectric generator with phase change materials *International Communications in Heat and Mass Transfer* **114** 104561

[21] Singh P, Khanna S, Becerra V, Newar S, Sharma V, Mallick T K, Hutchinson D, Radulovic J, and Khusainov R 2020 Power improvement of finned solar photovoltaic phase change material system *Energy* **193** 116735

[22] Taqwa A, Rusdianasari, Budiman, Kusumanto R D, and Dewi T 2018 Synchronization and Application of IoT for on Grid Hybrid PV-Wind System 2018 *International Conference on Applied Science and Technology (iCAST) IEEE* 617-621

[23] Dewi T, Risma P, Oktarina Y, Taqwa A, Rusdianasari, and Renaldi H 2020 Experimental analysis

on solar powered mobile robot as the prototype for environmentally friendly automated transportation *Journal of Physics: Conference Series, International Conference on Applied Science and Technology (iCAST on Engineering Science) Bali, Indonesia*, 1450 012034.

IOP Conf. Series: Materials Science and Engineering 1073 (2021) 012042 doi:10.1088/1757-899X/1073/1/012042

Automatic irrigation monitoring system based on photovoltaic solar energy with fuzzy logic

Y Ariyanto, R Wakhidah, H F Utari* and B Harijanto

Information Technology Department, State Polytechnic of Malang, Malang, Indonesia

*hanifahfanidyautari@gmail.com

Abstract. Indonesia's food self-sufficiency program requires the use of paddy fields with rice as the main food. That irrigation infrastructure problem also needs efficiency mitigation of cost impact using fuel oil in operation manually of engine irrigation. An automatic photovoltaic irrigation system can solve the problem of deficiency irrigation infrastructure, especially in the dry season as a more efficient power supply for irrigation infrastructure. In this research, implemented the automatic photovoltaic irrigation system is combined with the sunlight tracking system to get power supply, according to the direction of the sun. The irrigation pattern requires the suitability of soil moisture, water level, and power supply voltage as variables for the application of an automatic photovoltaic irrigation system. An automatic photovoltaic irrigation system is equipped with a website to monitor the irrigation variables. The Irrigation variables data is obtained from sensor readings then sent by NodeMCU to the website and calculates the suitability of irrigation variables as parameters using fuzzy logic to control automatic irrigation. The conclusion from this study was obtained after testing the system with the system running well, fuzzy logic parameters run with variables value of voltage > 9.9 Volt, water level < 2cm, and soil moisture < 69%.

1. Introduction

Paddy agriculture produces rice as the dominant food commodity in Indonesia's food self-sufficiency program. The area of rice fields in Indonesia in 2018 was 7,105,145 ha. However, 50% of the agricultural land does not have adequate irrigation infrastructure [1]. Generally, the rice fields that experience this problem are rainfed. So that independent efforts developed by farmers to meet irrigation needs in the fields, especially during the dry season.

Rainfed rice fields rely on rainwater for irrigation. During the dry season, farmers use rivers or wells as the closest source of water for irrigation. This water source requires lifting with a water pump because it is located at an elevation below the surface of the rice fields. The water pump requires a source of electrical energy. The alternative used is a portable water pump with an energy source of fuel oil (BBM) or LPG gas cylinders. However, the burden of fuel costs in using a portable pump is an obstacle in its operation. So that the growing use of irrigation pump technology with a more effective and efficient solar panel electric energy source.

Researches continue to be developed on the agricultural side, especially accompanied by the use of solar energy. The researcher develops automated irrigation that monitors and maintains the desired soil moisture content via automatic watering using ATMEGA328P microcontroller and soil moisture sensor on the Arduino Uno platform is used to implement the control unit. IoT with Thing speaks transmission is used to keep the farmers updated about the status of sprinklers. but this system needs improvement at power management sides like photovoltaic technology [2]. Photovoltaic Solar energy technology offers an alternative to irrigation system development. Research on the Solar Water Pumping System Using the IoT Monitoring System designed the concept of an AC and DC water pump system using solar panel power sources with battery backup [3-6]. Seeing this condition, it is necessary to develop a solar panel irrigation automation system using a microcontroller to control the irrigation water pump. The development of solar panel technology to capture sunlight according to the direction of the rays is also needed so that the solar panels can maximize power for irrigation pumps.

The amount of water needed for the growth process of rice plants is still not known with certainty. This causes farmers to give excessive water to paddy fields. Excessive water supply can lead to wasteful use of water. The way to maintain water availability without causing waste is to adjust the height of water inundation. The system of rice intensification (SRI) method of applying a minimum inundation height of 1-2 cm can save water use without reducing rice production [7]. The scale of soil moisture in agriculture also influences irrigation. Soil moisture is usually indicated by a sensor with three types of moisture, dry soil, humid soil, and wet [5,8].

In this study, built a system to monitor solar panel irrigation automation using fuzzy logic as a control of the irrigation pump status to suit irrigation needs with soil moisture parameters and water level and power source voltage.

2. Methods

Irrigation automation system using fuzzy as logic methods. Fuzzy Sugeno type is used to control the status of the irrigation pump based on parameters of voltage, water level, and humidity. The flow diagram of the Sugeno fuzzy method is shown in Figure 1.

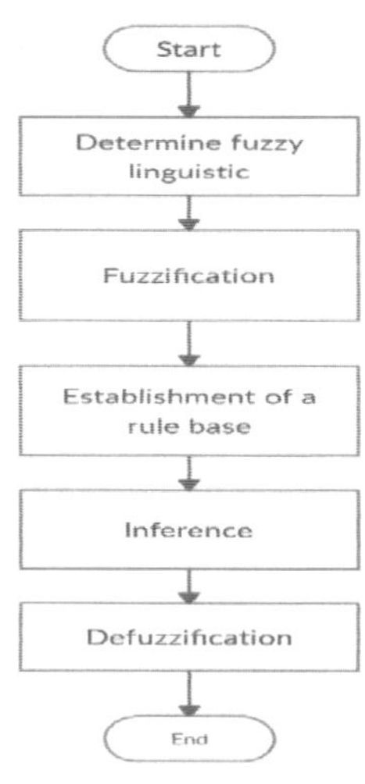

Figure 1. Sugeno fuzzy flowchart.

The fuzzy method consists of fuzzification, inference to defuzzification [9]. Sugeno fuzzy stages are shown as follows:

- Determine the linguistic value.
- Fuzzification

The fuzzification stage is changing the information from the input data input from the sensor to the fuzzy linguistic set data [10]. Fuzzification of membership degrees for humidity is shown in Figure 2, Fuzzification of membership degrees for water is shown in Figure 3, Fuzzification of membership degrees for voltage is shown in Figure 4.

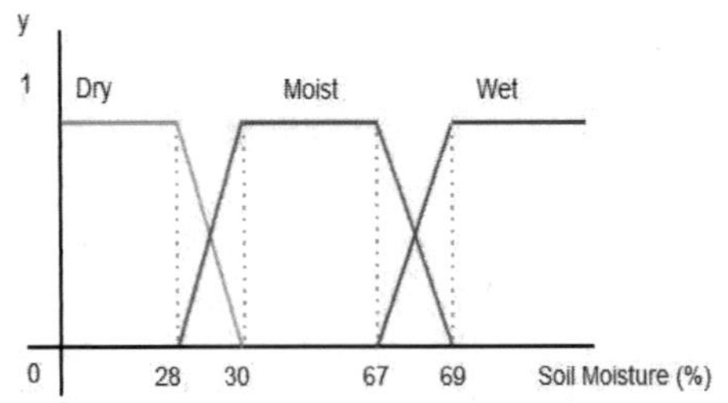

Figure 2. Membership degrees of humidity.

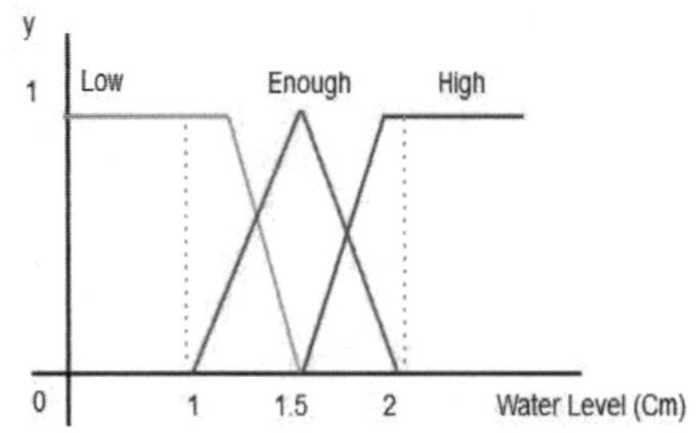

Figure 3. Membership degrees of water.

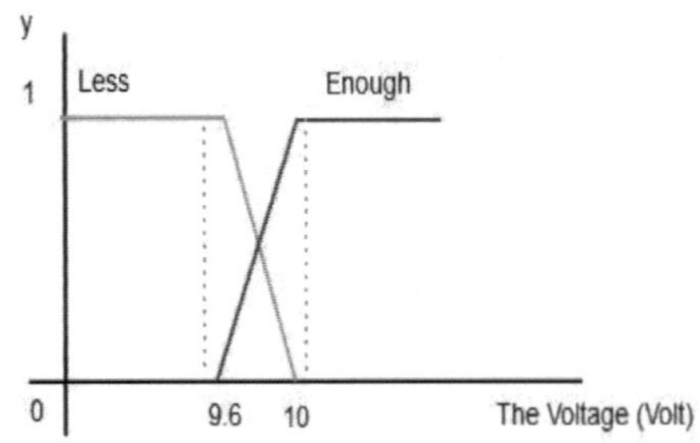

Figure 4. Membership degrees of voltage.

- The formation of a fuzzy knowledge base (Rule in the form of IF ... THEN).
- Inference, at this stage using the MIN implication function to obtain the α-predicate value for each rule (α1, α2, α3,…αn).
- Defuzzification, at this stage defuzzification, takes input in the form of αpredicate and z values of each rule. Defuzzification is done by calculating the center of singleton value, namely the sum of the multiplication between the membership value and the singleton value then divided by the total membership value. When the defuzzification result is> = 0.5, the pump output condition will turn on [11].

3. Design system

In the implementation of the system, the software and hardware requirements of the solar panel irrigation automation system using the fuzzy method are shown in table 1.

Table 1. Software and hardware requirement.

No	Software	Hardware
1	Windows Operating System 7/8/10	NodeMCU Microcontroller, DT-Avr Unoduino
2	Arduino IDE	Voltage sensor ina219, LCD, soil moisture sensor, water level sensor, servo motor
3	Sublime text editor, Xampp, Codeigniter	The solar panel voltage specification is 12V 10W 833mA, 12v 3.4aH battery, Solar charger controller, Water pump specification 12V/350mA, Water flow 240 L/H, maximum thrust maximal 350 cm, ADS1115 (Analog extender), relay.

Implementation in the form of miniature rice fields with a scale of 1: 24000 measuring $0{,}098m^2$ that is 35 x 28 cm. The water reservoir is placed with a height difference of 50 cm. A miniature rice field design is shown in Figure 5.

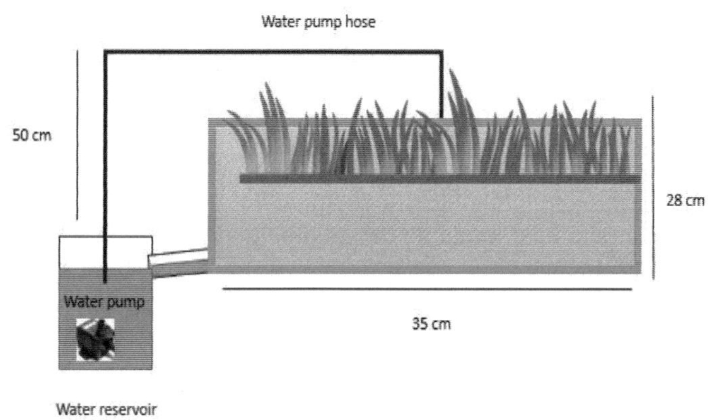

Figure 5. Miniature rice field design.

This system overview explains how the operation of the system functions as a whole shown in Figure 6.

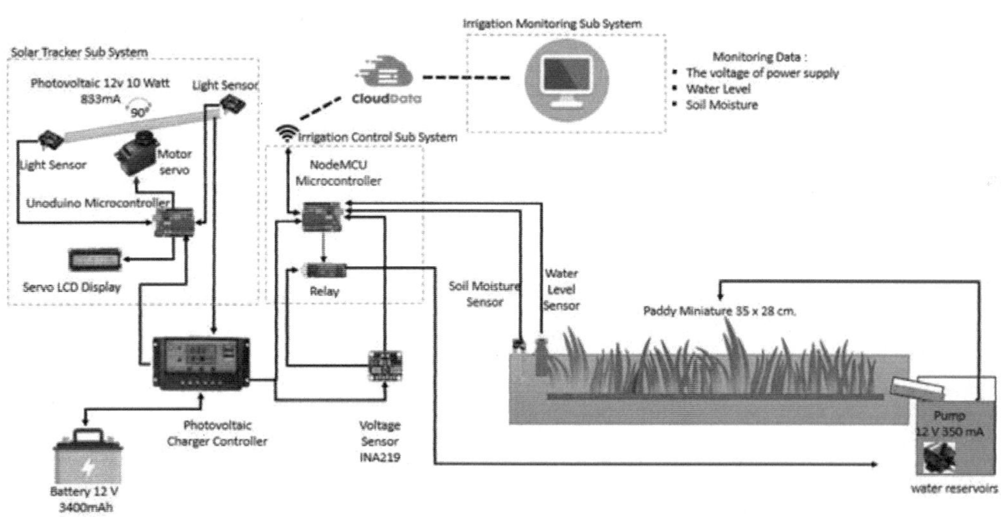

Figure 6. Irrigation automation and monitoring system implementation design.

4. Results and discussion

Testing of the irrigation automation system using the fuzzy method on 1-month old rice plants. Testing the Sugeno fuzzy algorithm calculation on the website system is shown in Figure 7.

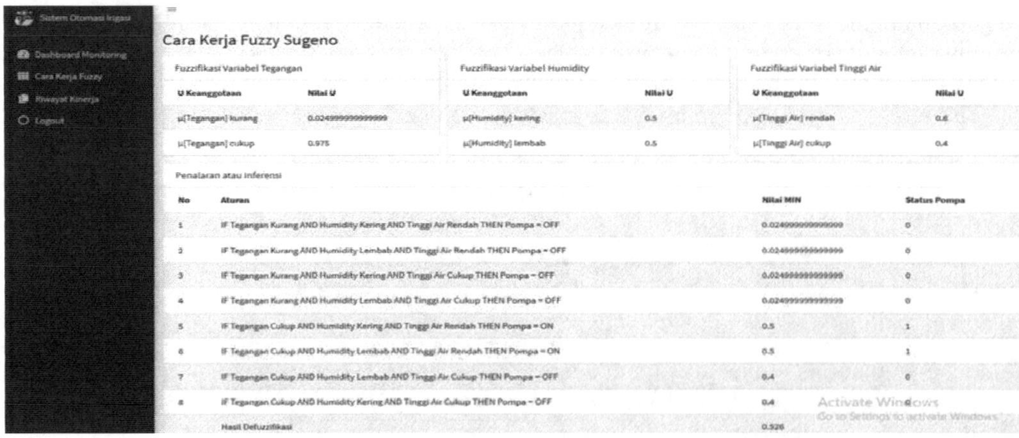

Figure 7. Sugeno Fuzzy calculations in the website.

The irrigation automation system test shows the received sensor data with the pump status output and the irrigation system status, and the results of the pump condition test are sent to the database. The test of the irrigation automation system from the data in the database is shown in table 2.

Table 2. Sensor data in the database.

No	Date Time	Voltage Sensor	Humidity Sensor	Water level sensor	Pump Status	Irrigation system status	Pump condition result
1	2020-06-15 11:34:39	10.8	50	1.02	ON	OFF	OFF
2	2020-06-15 11:35:11	10.78	50	1.04	ON	ON	ON
3	2020-06-15 11:35:49	10.84	51	1.07	ON	ON	ON
4	2020 06-15 11:36:22	11.24	51	1.13	ON	ON	ON
5	2020 06-15 11:36:55	11.09	51	1.12	ON	ON	ON
6	2020 06-15 11:37:25	11.08	51	1.08	ON	ON	ON
7	2020 06-15 11:37:57	10.75	51	1.1	ON	ON	ON
8	2020 06-15 11:38:34	10.89	51	1.14	ON	ON	ON
9	2020 06-15 11:39:12	10.88	51	1.17	ON	ON	ON
10	2020 06-15 11:39:47	10.82	51	1.2	ON	ON	ON
11	2020 06-15 11:40:24	10.82	51	1.29	ON	ON	ON
12	2020 06-15 11:40:54	11.05	51	1.41	ON	ON	ON
13	2020 06-15 11:41:24	11.34	51	1.7	OFF	ON	OFF
14	2020 06-15 11:41:54	11.38	50	1.58	OFF	ON	OFF
15	2020 06-15 11:42:30	11.13	49	1.55	OFF	OFF	OFF

Solar tracker system testing to show the movement of solar panels. The solar tracker system test is shown in Figure 8. The implementation of the rice field prototype measuring 35 x 28 cm. As well as a miniature water source with a size of 40 x 35 cm as a prototype of a well/river as a water source in the rice fields. Panel box measuring 30 x 20 x 12 cm as a place for placing Unoduino and NodeMCU microcontrollers, INA219 voltage sensors, relays, and batteries on the inside. Then on the outside wall-mounted LCD solar tracker and solar charger controller. As support for solar panels with a solar tracker system, a pipe with a height of 65 cm and a width of 35 cm is used. rice field prototype is shown in Figure 9.

Figure 8. Solar tracker trial.

Figure 9. Rice field prototype.

Testing of the irrigation automation website monitoring system is shown in Figure 10.

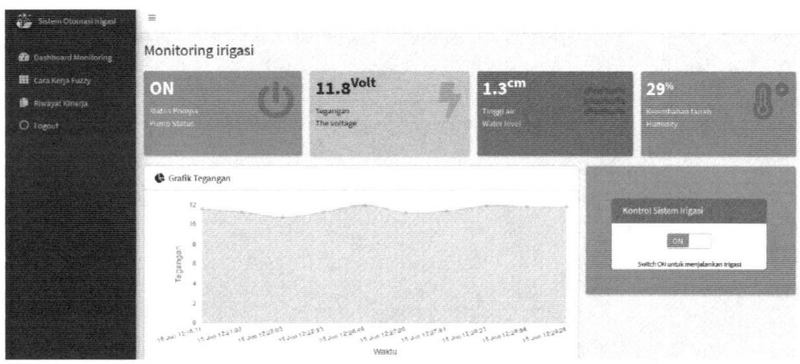

Figure 10. Irrigation automation website monitoring system.

5. Conclusion

From the test results, the solar tracker system can move the solar panel servo to follow the sun's motion automatically based on the light intensity received by the light sensor which is then processed by the Unoduino microcontroller. The degree of movement of the servo rotation is 0-90 degrees. Solar panels can charge batteries with a voltage of 11-12 Volts. The NodeMCU microcontroller reads and sends the INA219 voltage sensor data, soil moisture, and water level as parameters for the water pump status on the irrigation automation system website. Irrigation automation system based on Sugeno fuzzy calculations in making decisions to control the condition of the pump can work with parameters of voltage> 9.6 Volt, water level <2 cm, and soil moisture <69%. The irrigation automation system can work remotely and monitor it via the website. Sending sensor data through REST communication with HTTP sending protocol with a period of adding data every 30 seconds.

Acknowledgment

The authors would like to thank the Indonesian Directorate General of the Higher Education (DIKTI) and State Polytechnic of Malang who have supported this research project.

References

[1] Ruslan K 2019 Improving Indonesia's Food Statistics through the Area Sampling Frame Method
[2] Rawal S 2017 IOT based smart irrigation system *Int. J. Comput. Appl.* **159**(8) 7–11
[3] Veena R, Shyamala G, Bhavyashree N, and Murthy R 2019 Solar Water Pumping System Using IOT Monitoring System *Perspect. Commun. Embed. Signal-processing - PiCES* **2**(10)
[4] Orozco-Lugo A G 2020 Monitoring of water quality in a shrimp farm using a FANET *Internet of Things* 100170
[5] Kalia P and Ansari M A 2020 IOT based air quality and particulate matter concentration monitoring system *Mater. Today Proc.*

[6] AlRikabi H T S, Nasser K W, and Alaidi A H M 2020 The application of wireless communication in IOT for saving electrical energy *Int. J. Interact. Mob. Technol.* **14**(1) 152–160
[7] Uphoff N, Kassam A, and Harwood R 2011 SRI as a methodology for raising crop and water productivity: productive adaptations in rice agronomy and irrigation water management *Paddy Water Environ.* **9**(1) 3–11
[8] Rani M U and Kamalesh S 2014 Web based service to monitor automatic irrigation system for the agriculture field using sensors *in 2014 International Conference on Advances in Electrical Engineering (ICAEE)* 1–5
[9] Daud M, Handika V, and Bintoro A 2018 Design and realization of fuzzy logic control for Ebb and flow hydroponic system *Int. J. Sci. Technol. Res.* **7**(9) 138–144
[10] Manjunatha P, Verma A K, and Srividya A 2008 Multi-sensor data fusion in cluster based wireless sensor networks using fuzzy logic method *IEEE Reg. 10 Colloq. 3rd Int. Conf. Ind. Inf. Syst. ICIIS 2008* 1–6
[11] Nurhasan U, Prasetyo A, Lazuardi G, Rohadi E, and Pradibta H 2018 Implementation IoT in System Monitoring Hydroponic Plant Water Circulation and Control *Int. J. Eng. Technol.* **7**(4.44) 122

IOP Conf. Series: Materials Science and Engineering 1073 (2021) 012043 doi:10.1088/1757-899X/1073/1/012043

A development of an Arduino pure sine wave inverter for a small scale off-grid solar PV system

F Ronilaya*, S Ilmawati, M Huda, W Anistia, I N Syamsiana and M N Hidayat

State Polytechnic of Malang, Soekarno-Hatta No. 9 Malang, 65141, Indonesia

*ferdian@polinema.ac.id

Abstract. This paper presents the implementation of Arduino Nano microcontroller for a single-phase pure sine wave inverter, which can convert DC voltage to AC voltage at high efficiency and low cost. Solar-powered electricity generation is being favored nowadays as the world increasingly focuses on environmental concerns. The designed inverter converted DC voltage into AC voltage for a small-scale off-grid solar PV system suitable for electrification in remote areas, pollution-free, and inexpensive. Its inverter uses a sinusoidal pulse width modulation technique and a simple circuit, consisting of only 2 MOSFET switches and 1 MOSFET driver. The H-bridge inverter's output is applied to a step-up transformer with a dual coil input and a single-coil output, and hence, we can create positive and negative sides of the wave. Mitigate a voltage noise; a capacitor is parallelly installed at the secondary side of the transformer. Several simulations are performed to verify the effectiveness of the designed inverter using Proteus software, and continued hardware implementation. Based on some experiments we have done, the designed inverter produces a 230 V r.m.s 50 Hz sine wave with very low harmonics distortion. The highest efficiency was obtained using 2200nF / 400V of the filter capacitor, and the smallest voltage regulation gained using 2200nF / 400V of the filter capacitor when compared with other capacitors. The proposed system is economical, efficient, and reliable and can be used for small scale power applications.

1. Introduction

Electricity is one of the greatest inventions man has ever made, due to its very important role in socio-economic and technological development [1]. The need for electrical energy is increasing every year and also an increase in the government's construction of power plants. The power plant relies on a thermal generator that can produce a large amount of power. On the other hand, thermal generators have several drawbacks, including air pollution, noise, waste, exhausted fuel, and expensive initial investment. Therefore, we need the alternative way to overcome these problems by utilizing renewable energy. One sustainable renewable energy, solar energy, is converted into electrical energy using the Solar PV System.

Solar PV Systems generate DC from sunlight, which can be directly used for DC electrical equipment or to recharge batteries. Solar PV system based off-grid has several advantages, including suitable for electrification in remote areas and pollution-free. Solar PV Systems can be applied to public street lighting systems, electricity needs for boats, recreational vehicles, electric cars, camera detectors, and remote monitors.

IOP Conf. Series: Materials Science and Engineering 1073 (2021) 012043 doi:10.1088/1757-899X/1073/1/012043

Besides, Solar PV Systems are dependent on solar radiation. The current produced is direct current (DC). Thus the Solar PV System requires AC, a converter from DC to AC is needed, an inverter. The inverter converts the DC from the energy storage unit then supplied into the connected AC load.

There are square wave inverters, modified square wave inverters, and pure sine wave inverters. Among the three types of inverters, the most efficient and suitable for all loads is the pure sine wave inverter [2]. It is an inverter that has an output voltage with a pure sine waveform. This type of inverter can provide a voltage supply to inductor loads or electric motors with excellent power efficiency. This paper documents the design of an Arduino-based pure sine wave inverter, focusing on a small scale solar PV system. The inverter's various applications are solar electrical systems, Remote homes, Telecommunications, Computers, Tools, Security applications, Mobile power, Monitoring equipment, Emergency power, and lighting

2. A review of sine wave inverter

Single-phase inverters are of three types i.e. square wave inverter, modified sine wave inverter, and pure sine inverter. The pure sine inverter is studied in this paper. The square wave or modified sine wave; these two types of inverters are cheaper and are not suitable for delicate electronic devices [3]. The output of pure sine wave inverter is a near-perfect sine wave. Pure sine wave inverters have fewer power losses and less heat generation [4]. The sine wave has minor harmonic distortion resulting in a very clean supply. It makes it most suitable for running electronic systems such as microwaves ovens, computers and motors and other sensitive equipment without causing problems or noise. Pulse Width Modulation (PWM) technique is best for sine wave generation.

A pure or true sine wave inverter changes or converts the DC supply into a near-perfect sine wave. The sine wave has minimal harmonic distortion, which results in a very clean supply [5]. It makes it suitable for working electronic systems such as computers, motors, and microwave ovens, and other sensitive equipment without causing problems like noise. Things like mains battery chargers also run better on pure sine wave converters. Ideally, the output waveforms of an inverter should be sinusoidal. However, the waveforms of efficient inverters are non-sinusoidal and contain specific harmonics [6]. Due to the availability of high-speed power semiconductor devices, the harmonic contents present in the output voltage can be minimized significantly by using switching techniques [7].

Table 1. Comparison of inverter techniques [8].

Comparison of Inverter Techniques			
Parameter	Square wave inverter	Modified sine wave inverter	Pure sine wave inverter
Output waveform	Square wave	Square except that the output goes to zero volts for a time before switching positive or negative wave.	Sine wave
Harmonic distortion	More	Moderate	Less
Heat generation	More	Moderate	Less
Design complexity	Simpler	Less complex design than pure sine wave inverter	More
Cost	Less expensive	Moderate expensive	More expensive
Suitability	Not appropriate to delicate electronic devices	Not appropriate to delicate electronic devices	Appropriate to delicate electronic devices

3. System model of an Arduino-based inverter design

Figure 1, the block diagram below explains that the sources used to make pure sine wave inverters are batteries, batteries, or a 12V DC power supply. This voltage will be converted into AC voltage. In the process of converting DC voltage to AC voltage using a switching circuit, using two pairs of MOSFETs with each side using four parallel MOSFET in order to obtain SPWM waves back and forth from switching the two MOSFET pairs. In the MOSFET switching process, the MOSFET driver IC will be driven by the type IR2110. The MOSFET driver circuit will drive the MOSFET, which has been ordered by the Arduino nano microcontroller. It gives a signal to the MOSFET Driver circuit then forwarded to the MOSFET, where the two pairs of MOSFET have different signal values, where one MOSFET will be worth one and the other one MOSFET will be worth 0 so that an alternating SPWM signal will be obtained. Reverse, which will produce an AC voltage. The inverter output is in the form of an AC voltage, which will then enter the step-up transformer. This step-up transformer serves to increase the voltage from 12V AC to 220V AC. After the transformer output voltage, a mylar capacitor is installed to reduce the ripple, and the resulting wave is sinusoidal. Then the inverter will be tested by turning on the lamp.

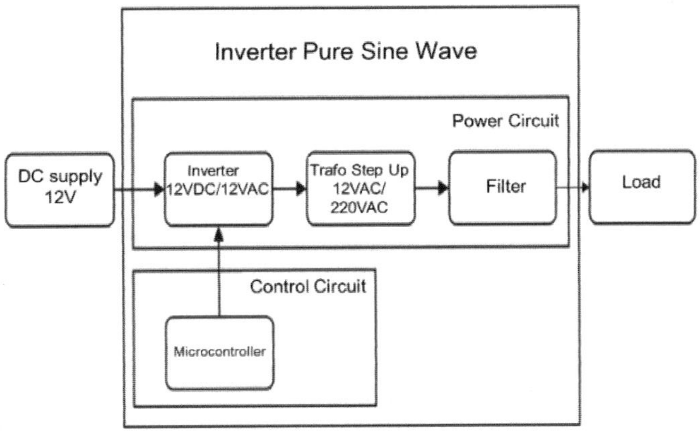

Figure 1. Block flow diagram of inverter design.

3.1. Software design

In this study, Arduino IDE software for Windows is used, with serials, namely series 1.8.5.0. This software is built-in Arduino, which is useful in compiling the source code for programming the Arduino board. The Arduino IDE software can generate the setting SPWM signal in the pulse configuration. There are two types of SPWM are arranged to obtain alternating waves to produce AC voltage [9]. In the coding PIN D9 and D10 are set as output because these ports include ports that support PWM signals as shown in the coding:

```
DDRB = 0b00000110; // Set PB1 and PB2 as outputs.
```

For a frequency setting of 50 Hz, shown in the coding:

```
ICR1    = 1600;         // Period for 16MHz crystal, for a switching frequency
of 100KHz for 200 subdevisions per 50Hz sin wave cycle.
```

After the programming process is uploaded, then it will be integrated with hardware to see the SPWM output signal.

3.2. Hardware design

In designing the Proteus software, it aims to simulate an inverter circuit to make it easier to manufacture the hardware. In the Proteus simulation that has been made, the inverter circuit concept is converted from DC to AC then the voltage generated is stepped up using a transformer. The transformer used is a CT step-up transformer 12V to 220V. From the simulation, the inverter circuit can be seen in the figure 2 below.

IOP Conf. Series: Materials Science and Engineering 1073 (2021) 012043 doi:10.1088/1757-899X/1073/1/012043

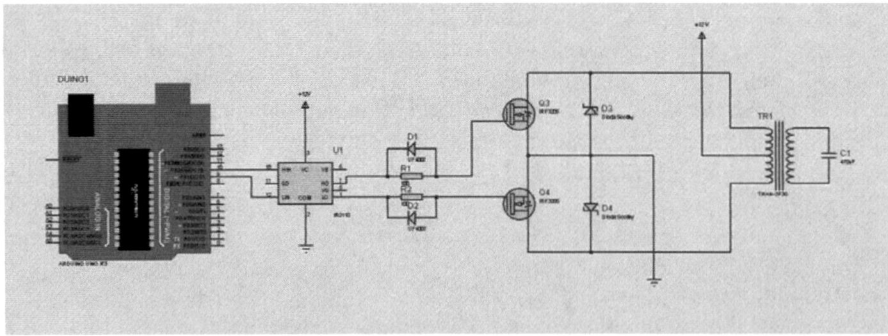

Figure 2. Inverter circuit in proteus software.

In the picture above, it is explained that the manufacture of this inverter uses two pairs of MOSFET for the switching process, which produces AC voltage. Where to set the PWM control, the picture above to get a PWM signal, then the output voltage from the switching process will be stepped up by using a transformer, the switching output voltage is higher than the input voltage.

The other software design is the design using Eagle software. In this design, the schematic and PCB paths will be used for the hardware. This design is obtained from the correct topology, which is then continued with the schematic design and PCB so that hardware work can be done immediately. The Eagle software consists of 2 types, namely the circuit schematic and the PCB board. The schematic circuit used in the Eagle software can be seen in the Figure below and the PCB board image can be seen in the figure 3 below.

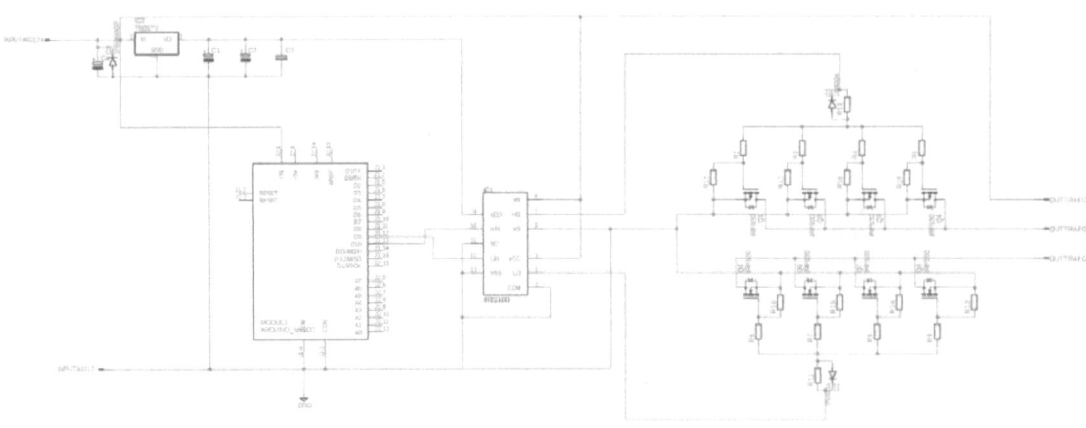

Figure 3. Inverter schematic circuit in eagle software.

The picture above is a schematic of an inverter using the Eagle software. In this circuit scheme, it is more specific than using Proteus software, where there is a control circuit in the form of an Arduino Nano Microcontroller and a MOSFET Driver circuit, and there is a diode that will reverse the back current that will pass through the MOSFET, no short circuit, and safe. Besides, the power circuit uses 2 pairs of MOSFET with each MOSFET pair, there are four MOSFET arranged in parallel, so that the number of MOSFET used is eight MOSFET to slow down the temperature rise of the MOSFET. In this circuit, the hardware for the inverter is made.

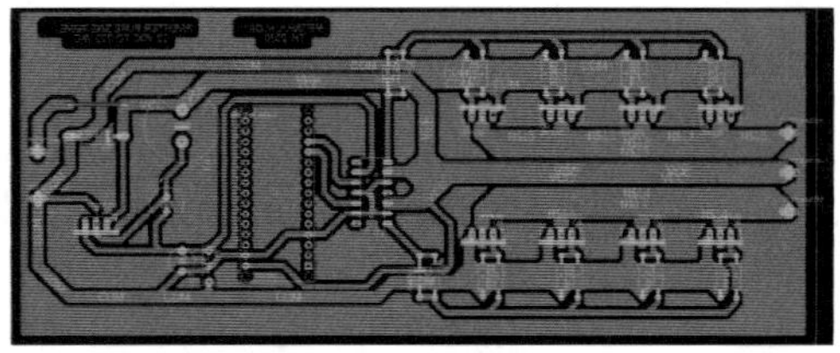

Figure 4. Inverter board schematic circuit in eagle software.

Figure 4 above is the board that will be printed onto the PCB, so that connection paths can be made between components so that hardware can be made inverter. In designing the hardware system, it integrates the inverter topology that has been made, which is then made a schematic circuit in the eagle software, which will be made a schematic board and printed in PCB form. After that, the components are integrated into the PCB and soldered. In this hardware, there are several parts in making an inverter, namely a switching driver circuit, an inverter or switching circuit, a step-up transformer circuit, and a filter circuit. Tools and materials in making pure sine wave inverters can be seen on table 2.

Table 2. Tools and materials in making pure sine wave inverters.

Elements of the Design	Quantity
Arduino Nano	1
Modul Step Down LM2596	1
MOSFET IRF3205	8
MOSFET Driver IR2110	1
Dioda 6A05MIC	1
Dioda UF4007	2
Elco 4700 uF/35 Volt	1
Capacitor non polar 100nF	1
Capacitor Milar 470nF	1
Capacitor Milar 1000nF	1
Capacitor Milar 2200nF	1
Resistor 10 ohm	8
Resistor 100 ohm	2
Resistor 1000 ohm	8
IC Regulator LM7805	1
Transformator 20 A	1
Post Binding Connectors	1
Fan DC 12 Volt	1
Heatsink	1

In designing the inverter hardware, the first stage is the inverter. The inverter used is the push-pull topology. The working principle of push-pull is that when there is a voltage source in the form of a DC voltage, which will be converted into AC voltage, then there are two switches that will be used, namely S1 and S2. At first, switch S1 will close, a voltage will appear on the load, where the electric current will flow from the negative source to S1 then to load 1 then to the positive source. Then S1 will be opened while S2 will close so that a gap appears between the loads, thus current will flow from the negative source to load and then to the positive source. If done continuously will result in AC voltage. The switch used is a N type MOSFET. The MOSFET used is a MOSFET

IRF3205 with input from the inverter in the form of a 12V DC voltage. In the N-type MOSFET, the gate output voltage value is the same as the gate input voltage. How the MOSFET works when the condition is ON where the MOSFET condition is active or there is a current flowing from the MOSFET to the drain leg to the source leg so that the Vds value is more than 0. While the OFF condition or no current flows from the MOSFET to the drain leg then to the source leg, so that the value of Vds equals 0.

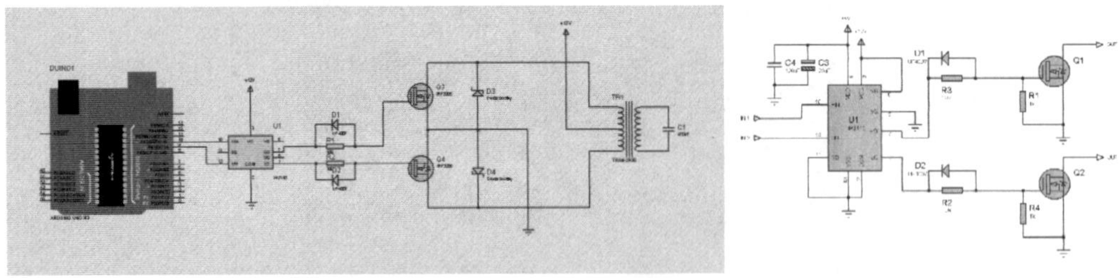

Figure 5. (a) The push-pull inverter circuit topology, (b) IR2110 MOSFET driver circuit.

Figure 5 (a), it explains that the circuit converts DC voltage into AC voltage with a frequency of 50 Hz. The MOSFET used is the IRF3205 type because the ability of the Vds MOSFET has a maximum value limit of 55 Vds and Id 110 A. Figure 5 (b) shows a series of IR2110 MOSFET Drivers which will switch 2 pairs of MOSFET that are made which produce high and low signals from the program created. Based on the datasheet related to the MOSFET IR2110 Driver circuit, it will generate an SPWM signal for the MOSFET of the microcontroller used. The microcontroller on the system used is an Arduino Nano. Based on the datasheet related to Arduino Nano pins, this system uses pins D9 and D10 to produce high and low signals because the SPWM setting uses OC1A and OC1B to set high and low signals, which can be seen the Arduino Nano pin configuration. Only a capacitor is used, because the transformer also functions as an inductor for the filter.

The following is the calculation for the ideal transformer [10]:
Inverters are made with a power of 1000 Watts. It can be calculated based on the following equation:

$$P = V \times I \rightarrow I = \frac{P}{V} \quad (1)$$

If the input voltage is 12 V on the primary side and the desired power is 1000 Watts, then:

$$I = \frac{P}{V} = \frac{1000}{12} = 83.3\ A \quad (2)$$

As for the output on the secondary side with an output of 220 V then:

$$I = \frac{P}{V} = \frac{1000}{220} = 4.5\ A \quad (3)$$

A center tap transformer is needed which is capable of delivering a current of 83.3 A. However, the transformer used in this test is 20 A:

$$P = V \times I \rightarrow 12 \times 20 = 240\ W \quad (4)$$

Thus, in this prototype, the maximum power that can be generated is 240 Watt.

4. Results and analysis

Testing the Pure Sine Wave Inverter without load to find out that the Pure Sine Wave Inverter can work adequately, producing sinusoidal waves at no-load conditions. The inverter will be tested in the form of a waveform and an RMS voltage value. The test was carried out by varying the capacitance value of the capacitor filter with a capacitor variation of 470 nF / 400V, 1000 nF / 400V, and 2200 nF / 400V.

4.1. No-load test

In testing the inverter output waveform at no load, the tools used are Laptop, DC 12V 10A Power Supply, Hantek-6022BE Oscilloscope, 1A Stepdown Transformer. Since the oscilloscope only has a maximum limit of 35V AC, the inverter output side is connected to a stepdown transformer so that the oscilloscope is not damaged.

4.1.1. Inverter output waveform of no-load test. In the figure 6, it can be explained that, to test the inverter using a voltage source in the form of a 12V / 10A DC Power Supply. The power supply is connected to the inverter input, and then the inverter output is connected to a stepdown transformer. The output from this stepdown transformer is connected to the oscilloscope to see the inverter wave output, which can be seen from the laptop.

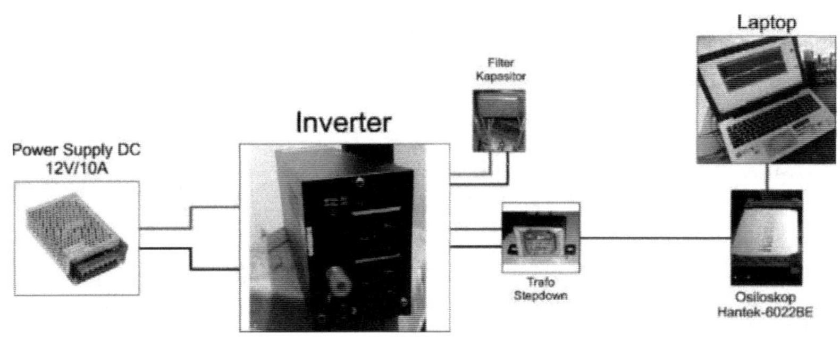

Figure 6. Output wave test without load and filter capacitors.

In the figure 7, it can be seen that the filter capacitor is installed at the inverter output. The variations of the capacitors used are 470 nF / 400V, 1000 nF / 400V, and 2200 nF / 400V.

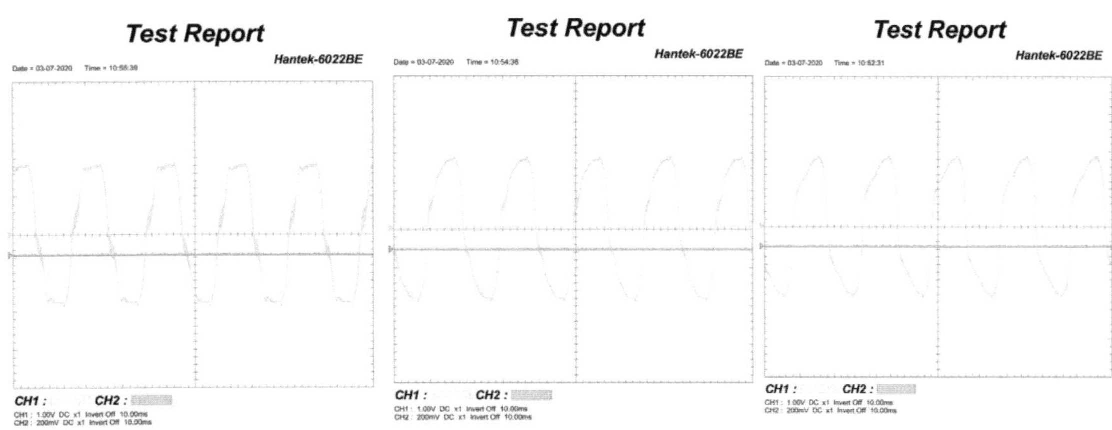

Figure 7. (a) No-load 470nF / 400V (b) No-load 1000nF / 400V (c) No-load 2200nF / 400V.

The inverter output waveform without load and filter capacitor, the waveform is still a lot of ripple noise. The output waveform of the inverter without load, and with a 470nF / 400V filter capacitor, the waveform has become a sinusoidal wave, but at the peak of the wave, there is still a little ripple noise so that the peak waveform is slightly cut off. The no-load inverter output waveform and with a 1000nF / 400V filter capacitor, the waveform has become a better sinusoidal waveform than using a 470nF / 400V capacitor. The waveform has less ripple noise when the wave moves upward and downwards. The no-load inverter output waveform using a 2200nF / 400V filter capacitor becomes a sinusoidal wave better than using a 470nF / 400V capacitor. Its waveform almost has no ripple noise, and only the waveform is slightly cut off when the wave goes up and the time the wave goes down.

4.1.2. Inverter output RMS voltage of no-load test. In testing the measurement of the value of the inverter output RMS voltage at no load, the tools used are the DC 12V / 10A Power Supply and the True RMS Multimeter.

Table 3. The RMS voltage measurement results.

Without filter capacitors (Volt)	Using a filter capacitor (Volt)		
	470 nF	1000 nF	2200 nF
297,1	232,7	228,5	226

Table 3, the measurement results of the inverter output RMS voltage value without load and filter capacitor, and with the filter capacitor, the RMS voltage value is different. At the time without the capacitor, the voltage is very high at 297.1 volts. Then after installing the filter capacitor, the inverter output voltage drops. The higher the capacitor value, the smaller the inverter output RMS voltage [11]. It is because of the more significant the capacitor value, the smaller the ripple noise at the inverter output.

4.2. Load-test

Testing of the Pure Sine Wave Inverter with a load aims to determine the performance of the Pure Sine Wave Inverter with the effect of the load. The loads used are bulb lamps with a power of 5 Watt, 10 Watt, 15 Watt, 25 Watt, and 100 Watt, as well as LED lamps with 3 Watt, 5 Watt, 7 Watt, 9 Watt, and 11 Watt. The inverter will be tested in the form of a waveform, RMS voltage value, efficiency, and regulatory voltage. The test was carried out by varying the capacitance value of the capacitor filter with a capacitor variation of 470 nF / 400V, 1000 nF / 400V, and 2200 nF / 400V. The equipment used for testing loaded inverters includes laptops, DC 12V 10A power supply, Hantek-6022BE oscilloscope, 1A stepdown transformer, True RMS multimeter, and bulb and LED lighting loads.

4.2.1. Inverter output waveform with load. In testing the output waveform of the loaded inverter, the tools used are Laptop, DC 12V 10A Power Supply, Hantek-6022BE Oscilloscope, 1A Stepdown Transformer, and bulb and LED lamp loads. Since the oscilloscope only has a maximum limit of 35V AC, the inverter output side is connected to a stepdown transformer so that the oscilloscope is not damaged.

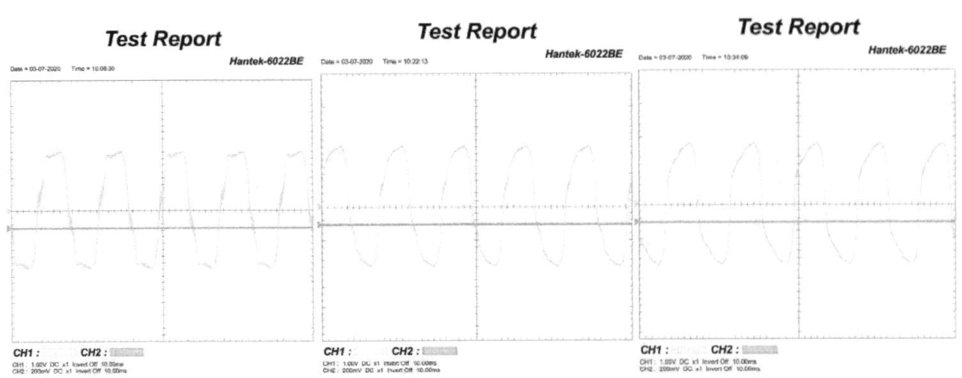

Figure 8. (a) with-load 470nF / 400V (b) with-load 1000nF / 400V (c) with- load 2200nF / 400V.

When the capacitor filter is installed, when the inverter output is connected to the load of the bulb lamp, the output voltage increases as the capacitor value increases, until the highest load is 100 Watt, the output voltage decreases as the capacitor value increases. When the inverter output is connected to the load of the LED lamp, the output voltage decreases with the rated capacitor 470nF / 400V and 1000nF / 400V, and the output voltage rises to the rated capacitor 2200nF / 400V.

4.2.2. Inverter output RMS voltage with load. It can be seen on table 4.

Table 4. Inverter output RMS voltage with load.

Load		Trafo output voltage (Volt)			
Load variation	Watt	No capacitor	Using capacitor		
			470 nF	1000 nF	2200 nF
	5	258	215	215	219
	10	246	210	211	215
	15	234	204	206	209
Bulb	25	229	202	204	207
	100	107	142	141	140
	3	294	226	224	225
	5	228	223	222	224
	7	286	221	220	223
LED	9	230	220	219	222
	11	275	219	218	221

The measurement results of the RMS voltage value of the inverter output loaded without the filter capacitor, and the filter capacitor, the value of the RMS voltage is different. In the absence of a capacitor, when the inverter output is connected to the bulb lamp's load, the voltage decreases with the magnitude of the lamp power. Moreover, when the inverter output is connected to the LED lamp's load, the voltage is erratic, fluctuates, and when the power of the LED lamp is 3 Watt, 7 Watt, and 11 Watt, the voltage drops gradually. Furthermore, when the power of the 5 Watt LED lamp and 9 Watt LED, the voltage is far from the previous lamp load. It can be concluded that the voltage decreases with the amount of load power. Moreover, the capacitor value that makes the most significant output voltage is the capacitor with a value of 2200nF / 400V.

4.3. Efficiency inverter

In simple terms, efficiency can be defined as the percentage of the amount of power released and the amount of power taken. The following equation can calculate efficiency:

$$\eta = \frac{P_{out}}{P_{in}} \times 100\% = \frac{V_{out} \times I_{out}}{V_{in} \times I_{in}} \times 100\% \quad (5)$$

It is necessary to find data on input voltage, input current, output voltage, and output current in order to find efficiency. In measuring the value of efficiency, the tools used are DC 12V / 10A Power Supply and True RMS Multimeter.

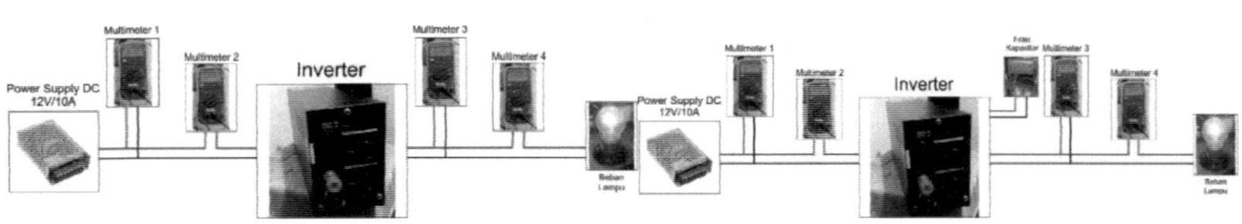

Figure 9. (a) Without filters capacitor (b) with filters capacitor.

In the figure 9 it can be explained that, to test the efficiency of the inverter using a voltage source in the form of a 12V / 10A DC Power Supply. The power supply is connected to the inverter input, then there are four multimeters, each of which has its function. Multimeter 1 is connected in parallel on the input side to measure the input voltage, and multimeter 2 is connected in series on the input side to measure the input current, Multimeter 3 is connected in parallel on the output side to measure the output voltage, multimeter 4 is connected in series on the output side to measure current the output. Moreover, it can be seen that a capacitor filter is attached to the output of the inverter. The variations of the capacitors used are 470 nF / 400V, 1000 nF / 400V, and 2200 nF / 400V.

Table 5. Efficiency of inverter.

Watt Lamp	Efficiency			
	No capacitor	470nF	1000nF	2200nF
5	45%	26%	28%	30%
10	51%	35%	36%	40%
15	56%	41%	43%	47%
25	57%	43%	45%	49%
100	43%	54%	54%	53%
3	16%	8%	10%	12%
5	86%	19%	18%	20%
7	38%	19%	21%	23%
9	91%	26%	29%	31%
11	49%	34%	33%	36%
average	53%	31%	32%	34%

Table 5 show the results of measuring the efficiency of the inverter without filter capacitors and filter capacitors, the efficiency values differ. At the time without capacitors and at bulb lamp loads, the efficiency value increases with increasing load power. Until the highest load is a 100 Watt bulb lamp, the efficiency value decreases due to the very high rising input current. At the time without capacitors and at a load of LED lamps, the efficiency values fluctuate and fall. The efficiency is low at the load of 3 Watt, 7 Watt, and 11 Watt LED lamps. Moreover, the efficiency is high at the load of 5 Watt, and 9 Watt LED lamps.

When the capacitor is installed, bulb as the load, the efficiency value increases with increasing load power. When the load on the LED lamp, the efficiency value also continues to increase with increasing load power. When compared, the efficiency is higher when the load is bulb lamps than LED lamps. The highest efficiency is when using a 2200nF / 400V filter capacitor rather than other filter capacitors.

It can be concluded that the inverter's efficiency is higher when without a capacitor filter than using a filter capacitor. It is because when the filter capacitor is installed, the filter capacitor will become a burden to the inverter results in the current on the input side getting bigger [12]. When a capacitor filter is installed, the highest efficiency is when using a 2200nF / 400V filter capacitor than other filters. It is because of the more significant the filter capacitor, the smaller the current on the input side [13].

4.4. Voltage regulation. It can be shown on table 6.

Table 6. Voltage regulation.

Variety of load	Watt	2200nF/400V		
		VNL	VFL	VR
Bulb	5	226	219	3%
	10	226	215	5%
	15	226	209	8%
	25	226	207	9%
	100	226	140	61%
	3	226	225	0,4%
	5	226	224	1%
	7	226	223	1%
LED	9	226	222	2%
	11	226	221	2%

The calculation results of the Voltage Regulation output of the inverter without filter capacitors and with filter capacitors, the value of the calculation of Voltage Regulation is different. When the circuit does not use a capacitor, and the bulb is used as a load, the voltage regulation value continues to increase as the power from the load increases. Moreover, the highest is the voltage regulation value for a 100 Watt bulb lamp load. Then at the load of the LED lamp, the voltage regulation value does not fluctuate.

When a capacitor filter is installed, the voltage regulation value increases continuously as the load power increases [14]. When loaded with a 100 Watt bulb lamp, the highest voltage regulation value is up to 60% more. When a 100 Watt bulb lamp installed, the voltage drops mainly so that the voltage regulation is high. The smallest voltage regulation value is when using a filter capacitor of 2200nF / 400V when compared to other capacitors.

It can be concluded that the value of voltage regulation is small when using a filter capacitor compared to without using a filter capacitor. Furthermore, the smallest voltage regulation value is when using a 2200nF / 400V filter capacitor.

5. Conclusion

A microcontroller-based inverter circuit that generates a pure sine waveform was developed and tested. The Arduino Nano microcontroller pure sine wave inverter was designed and simulated with Proteus Software, programmed with Arduino IDE Software. It was then constructed on a printed circuit board using Eagle Software. When tested, the circuit produced a 230 V r.m.s 50 Hz sine wave with very low harmonics distortion, which makes it suitable for small-scale electronic devices. From the test carried out and obtained, the system performed according to the design specification. The Inverter efficiency without a filter capacitor is higher than using a filter capacitor. The installed filter capacitor turns as a resistance of the inverter, which increases the input current [15]. The Voltage regulation using filter capacitors is smaller than without using filter capacitors. The highest efficiency was obtained using 2200nF / 400V of the filter capacitor and the smallest voltage regulation gained by using 2200nF / 400V of the filter capacitor when compared with other capacitors. The proposed system is economical, efficient, and reliable and can be used for small scale power applications.

References

[1] Wang Y, Ronilaya F, Chen X, and Roskilly A P 2013 Modelling and simulation of a distributed power generation system with energy storage to meet dynamic household electricity demand *Applied Thermal Engineering*, **53**(2) 312-324

[2] Jana J, Saha H, and Bhattacharya K D 2017 A review of inverter topologies for single-phase grid-connected photovoltaic systems. *Renewable and Sustainable Energy Reviews*, **72** 1256-1270

[3] Ronilaya F and Miyauchi H 2014 A Load Frequency Control in an Off-Grid Sustainable Power System Based on a Parameter Adaptive PID-Type Fuzzy Controller *International Journal of Emerging Electric Power Systems* **15**(5) 429–441

[4] Doucet J, Eggleston and Shaw J 2007 DC/AC pure sine wave inverter *PFC Worcester Polytecnic Institute*

[5] Ghalib M A, Abdalla Y S and Mostafa R M 2014 Design and implementation of a pure sine wave single phase inverter for photovoltaic applications *AMERRICAN SOCIETY FOR ENGINEERING EDUCATION, ASEE* 1-8

[6] Alhamrouni I, Salem M, Rahman N H and Awalin L 2019 Design of single phase inverter for photovoltaic application controlled with sinusoidal pulse width modulation *Indonesian Journal of Electrical Engineering and Computer Science* **15**(2) 620-630

[7] Novak M, Belany P, Bolf A, Roch M and Bracinik P 2018 Off-grid model of photovoltaic system with real time monitoring function In *2018 IEEE International Energy Conference (ENERGYCON)* 1-5

[8] Islam M, Nuzhat F, Nadim M and Hoque N 2012 *AC sine wave generation by using SPWM inverter* (Doctoral dissertation, BRAC University)

[9] Kersting W H, Phillips W H, and Carr W 1999 A new approach to modeling three-phase

transformer connections *IEEE Transactions on Industry Applications* **35**(1) 169-175

[10] Koutroulis E, Chatzakis J, Kalaitzakis K and Voulgaris N C 2001 A bidirectional, sinusoidal, high-frequency inverter design *IEE Proceedings-Electric Power Applications* **148**(4) 315-321

[11] Alexa D, Sîrbu A and Lazar A 2006) Three-phase rectifier with near sinusoidal input currents and capacitors connected on the AC side *IEEE Transactions on Industrial Electronics* **53**(5) 1612-1620

[12] Lu D D C, Iu H H C, and Pjevalica V 2008 A single-stage AC/DC converter with high power factor, regulated bus voltage, and output voltage *IEEE Transactions on Power Electronics* **23**(1) 218-228

[13] Zhao J J 2009 Filtering capacitor feedthrough assembly *U.S. Patent No. 7,502,217*

[14] Kjaer S B and Blaabjerg F 2003 Design optimization of a single phase inverter for photovoltaic applications *In IEEE 34th Annual Conference on Power Electronics Specialist* **3** 1183-1190

[15] Ronilaya F 2018 A double stage micro-inverter for optimal power flow control in grid connected PV system *2018 International Conference on Information and Communications Technology (ICOIACT), Yogyakarta* 808-813

Optimisation solar and wind hybrid energy for model catamaran ship

B Setiawan, E S Putra*, I Siradjuddin and M Junus

Electrical Engineering Department, State Polytechnic of Malang. Jl. Soekarno-Hatta No. 9, Malang 65141, Indonesia

*ekananda_putra@polinema.ac.id

Abstract. Based on WindSat data from 2004 to 2014, Indonesia has an average wind speed of above 8m/second at sea. In addition, the energy potential of the sun is around 4.8 kWh/m^2. A mechanical model ship with a catamaran type will be installed photovoltaic module and wind turbine generator which are equipped with microprocessor to control the angle of the wind turbine blades, the energy generated by the PV and wind turbine generator will be regulated by the dual input buck boost converter using the PID method to regulated voltage to 14 Volt used to charge the battery. The power stored in the battery will be used to drive the ship model propulsion system. This paper discusses about pv performance, generator performance, dual input buck boost calculation, and simulation. From testing the energy obtained is 774WH. From the simulation of the dual input buck boost converter circuit, the setling time is 11ms when using the PID control system and 35ms when using PI. Energy from the sun and wind can be combined into one with a dual input buck boost converter so that the energy can be absorbed is more optimal.

1. Introduction

Indonesia is the largest archipelago country in the world which consists of around 17,058 islands with a coastline length of 81,000 km. The sea area in Indonesia reaches 5.8 million km^2 or about 75% of the total area of Indonesia. Therefore, ship transportation is one of the most important components.

The wind speed map based on the WindSat data, in Indonesia has wind speeds from 5m/s to 10m/s or greater than the minimum wind speed to drive wind turbine generator which is 4m/s [1]. Apart from wind energy, Indonesia has solar energy potential. Based on solar radiation data collected from 18 locations in Indonesia, solar radiation in the Western Region of Indonesia is around 4.5kWh/m^2/day with a monthly variation of around 10%, and in Eastern Indonesia around 5.1 kWh/m^2/day with a monthly variation of around 9%. Thus, Indonesia's average solar radiation potential is around 4.8kWh/m^2/day with a monthly variation of around 9% [2].

From the data on the potential for wind and sunlight in Indonesia. This study intends to determine the ability of the two energies to drive the catamaran ship propulsion system. This study makes an empirical model of catamaran ships by adding PV and wind turbine generators as energy.

IOP Conf. Series: Materials Science and Engineering 1073 (2021) 012044 doi:10.1088/1757-899X/1073/1/012044

2. Literature review

2.1. Catamaran ship
Catamaran ship is a twin hull ship, where the two hulls are connected with a strong deck construction and stretches on it to withstand large bending moments and shear forces and work towards the midline (Center line) ship. The shape of the catamaran hull on various ships is not the same. There are many models of catamaran body shapes, but in general there are three basic forms of catamarans [3,4].

- Symetric
- Asymmetric with a straight inside
- Asymmetric with a straight outside

2.2. Wind turbine generator
Wind turbines are alternative energy technologies that are able to convert wind energy into electrical energy [5]. There are 2 types of wind turbines, namely horizontal vertical axis wind turbine (HAWT) and vertical axis wind turbine (VAWT). The working principle of the HAWT wind turbine is based on the lift force of the wind energy while the VAWT wind turbine is based on the drag force that occurs due to wind movement [6,7]. HAWT is widely used for higher production volumes that require a large investment and occupy more space for installation compared to VAWT. VAWT requires a low cost investment and less space for installation compared to HAWT [6,7].

2.3. Photovoltaic module
Photovoltaic Module is a semiconductor consisting of a diode p-n junction, which when exposed to sunlight will create electrical energy that can be utilized, this energy conversion is called the photoelectric effect [8].

2.4. Dual input buck boost converter
The dual-input DC-DC converter is a development of power electronics technology where two DC-DC converters are combined either in parallel or in series to supply the same load from several different sources interconnected. Before using a multi-input DC-DC converter, a separate converter for each source is usually used, the output of each converter is then managed by DC Bus so that the output voltage is the same and stable [9-11].

2.5. PID controller
PID is a controller system to determine the precision of a system with the feedback on the system [12]. Set point is a certain condition to be achieved. The sensor functions to detect the plan output and convert it into the same quantity as the set point unit. The sensor output is a feedback signal which will be subtracted from the set point resulting in an error signal. The goal of the control is to make the output of the plan equal to the set point, which means that the error signal is (or is close to) zero [13,14].

2.6. Ziegler Nichols method
Ziegler Nichlos' PID tuning method is experimental (assuming the model plan is not yet known). There are two ways of tuning the PID with this method, namely [12,14].

2.6.1. Ziegler Nichols 1 method. The tuning method using this method is by looking at the response plan first by giving input unit steps and then a curve will be obtained in which there are variables for further calculations.

2.6.2. Ziegler Nichols 2 method. This method is used for Plans which have an oscillatory response when given a certain Kp value. The method of tuning the PID with this method is to give a certain Kp value until the system oscillations per period are found

IOP Conf. Series: Materials Science and Engineering 1073 (2021) 012044 doi:10.1088/1757-899X/1073/1/012044

3. Methods
In this study utilizing renewable solar and wind energy as a driving force for catamarans as explained in Figure 1 below

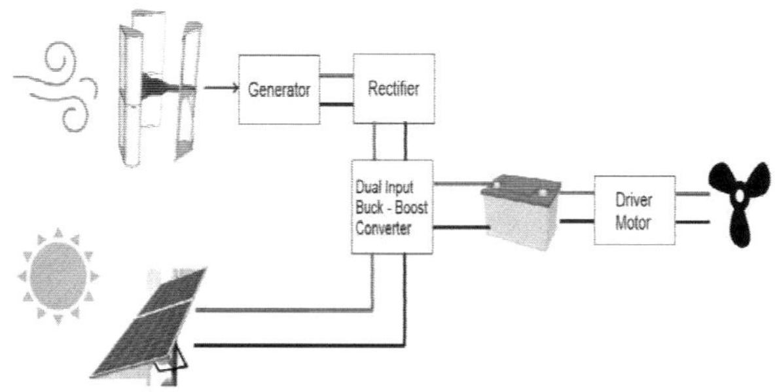

Figure 1. Diagram block system.

From the block diagram above, the mechanical design ship is mounted by PV and Wind Turbine as an energy source for motor. The voltage generated by the PV and the wind turbine generator goes to the dual input buck boost converter for the regulation process so that the voltage generated is stable at 14V and the energy absorbed is optimal.

Figure 2 is a ship's mechanical design with the specifications Length of Arc 170 cm, Beam 100 cm and Depth 32 cm Depth. At the top of the ship an array of PV will be installed along the deck of the ship and the Wind Turbine is placed at the middle of the ship described in the figure 3.

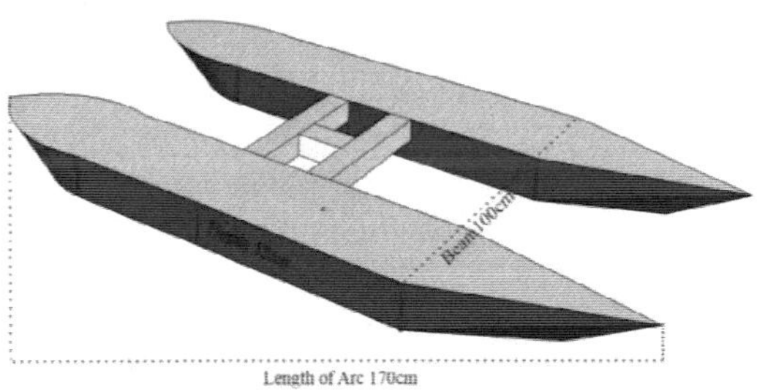

Figure 2. Ship mechanical design

Figure 3. Ship mechanical design with PV and wind turbine generator.

The dual input buck boost converter circuit is described in the figure 4 below.

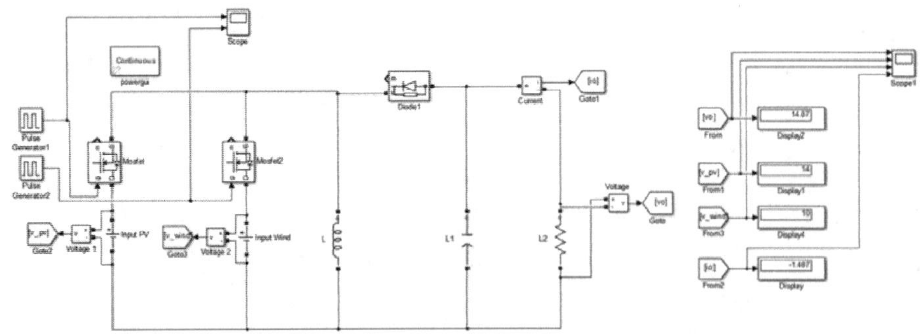

Figure 4. Dual input buck boost converter circuit.

4. Results

4.1. Generator

Measurement of generator performance is done by directly coupling the generator using a drill and the generator is given a battery load. The RPM is obtained from the tachometer. The results of generator performance measurements are shown at the charts below.

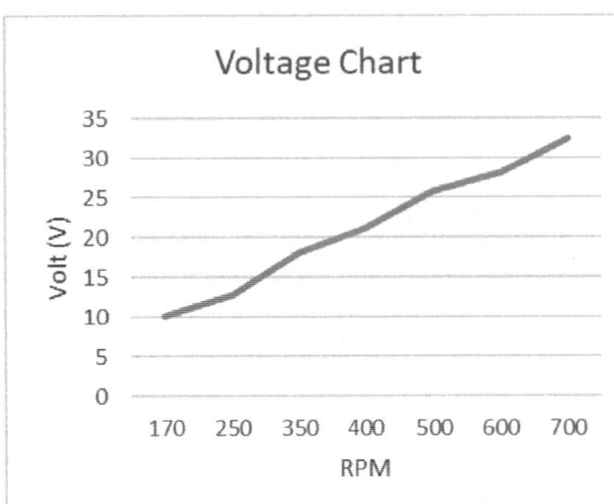

Figure 5. Voltage chart of generator.

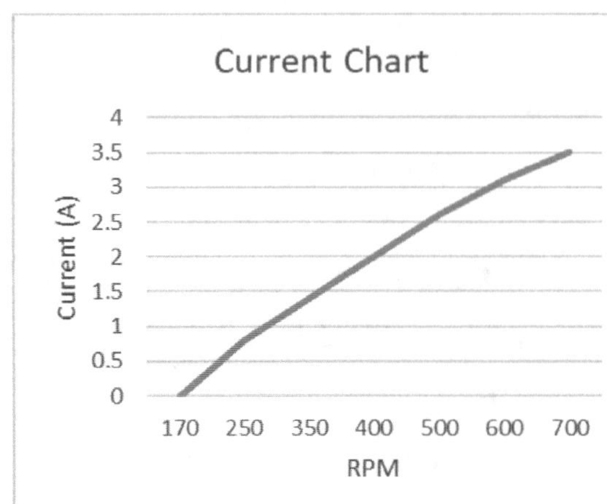

Figure 6. Current chart of generator.

From the measurement results of the minimum RPM generator performance that can be used for charging a battery of around 170 RPM which produces a voltage of about 10.2 Volts and a charging current of about 0.2 Ampere.

IOP Conf. Series: Materials Science and Engineering 1073 (2021) 012044 doi:10.1088/1757-899X/1073/1/012044

4.2. *Photovoltaic module*

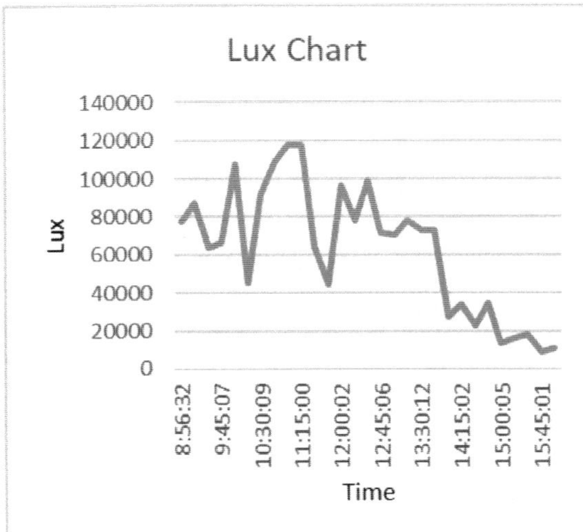

Figure 7. Lux chart of PV.

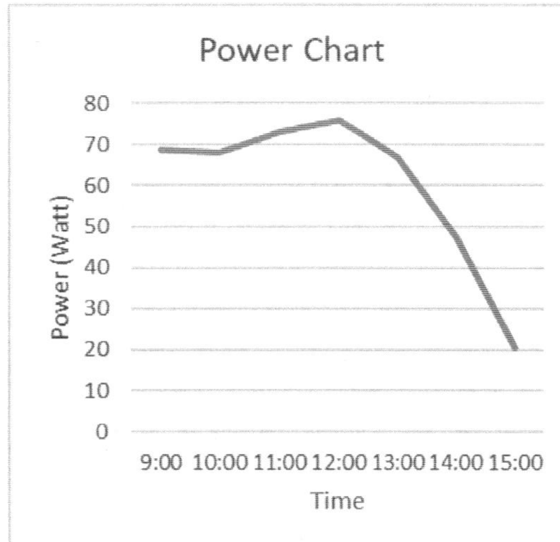

Figure 8. Power chart of PV.

The chart above shows the lux value in the time range 08.56 to 15.50 hours. The highest lux value was obtained at 11.15, which is around 110,000 lux and the average "lux at the time of testing was around 66,365 lux. The highest power is obtained around 75 Watt. Meanwhile, the average power during the testing period is around 60.07 Watt.

4.3. *Wind turbine generator*

Wind turbine generator testing is carried out for 24 hours and the results are as shown in the chart below.

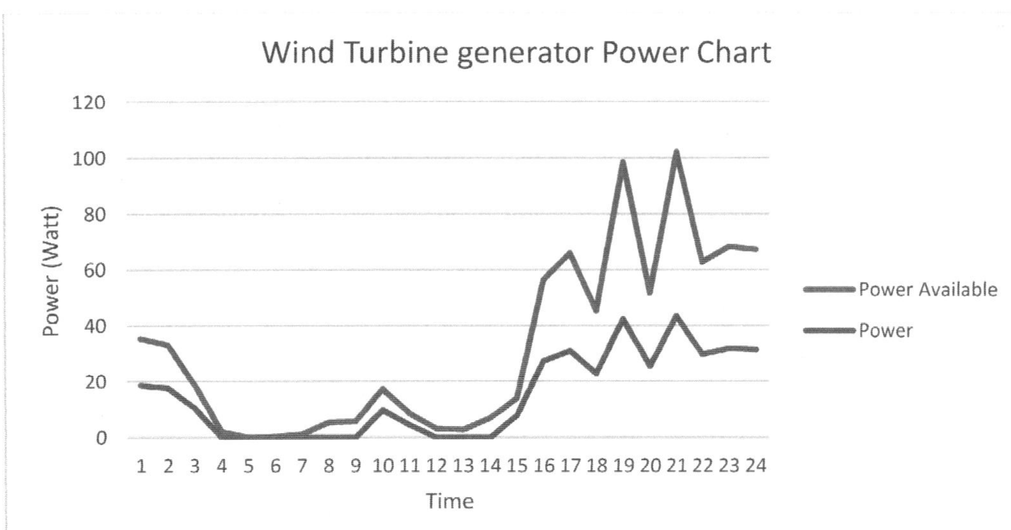

Figure 9. Wind turbine generator power chart.

From chart above, the maximum power that can be released by the wind turbine is around 40 Watts. The total power that can be generated by the wind turbine is 354Wh.

IOP Conf. Series: Materials Science and Engineering 1073 (2021) 012044 doi:10.1088/1757-899X/1073/1/012044

4.4. Simulation dual input buck boost converter

After identifying the performance of PV and Wind Turbine, the PV and Wind Turbine parameter values are obtained which will be used for the calculation of buck boost. From the calculation results have obtained the inductor value of 656uH and capacitor 200uF. The PID calculation uses the Ziegler Nichols type 2 method and the results are as shown in Figures 10 and 11 below.

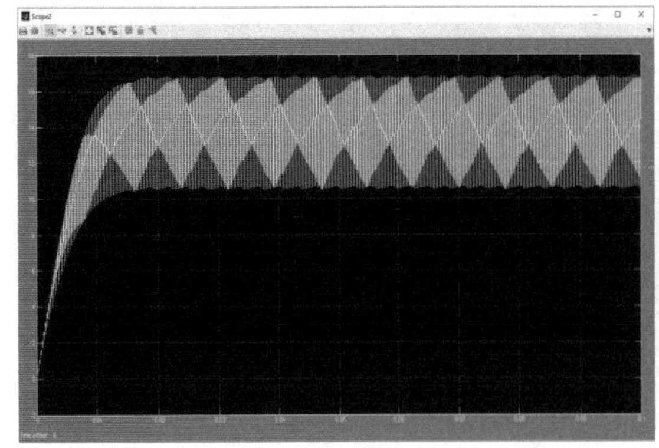

Figure 10. Signal output using KP 0.495 KI 528 and KD 2112

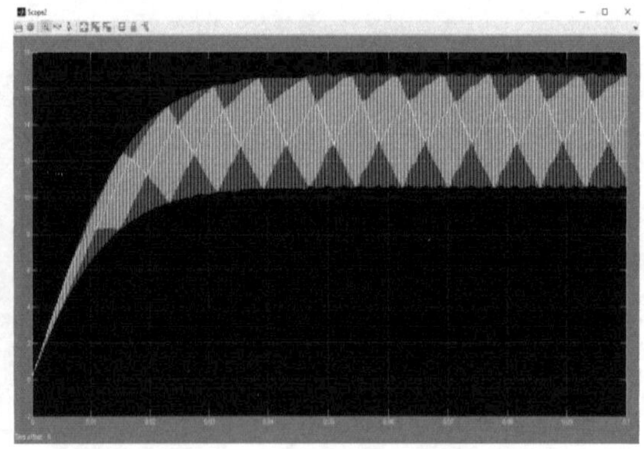

Figure 11. Signal output using KP 0.495 and KI 237.6

From the signal output image above using the KP value 0.495, KI 528, and KD 2112, the result is a rise time of 8ms while the setting time is 11ms. For a maximum oscillation voltage of 16.8 Volts and a minimum of 10.5 Volts. The RMS value is in accordance with the 14 Volt set point, which is 14,119 Volts.

From the signal output image above using the KP value 0.495, and KI 237.6, the result is a rise time of 20ms while the setting time is 35ms. For a maximum oscillation voltage of 16.8 Volts and a minimum of 10.5 Volts. The RMS value is in accordance with the 14 Volt setpoint, which is 14,119 Volts.

5. Conclusion

The conclusion from this research is that the energy produced by PV and wind turbine generators is 774Wh. To optimally absorb the two energies, can use a dual input buck boost converter circuit with a PID control system that can combine the two energies to produce a fixed voltage. From the simulation when using the PI control system, the setting time is 35ms, while when using the PID, the setting time is 11ms.

References

[1] Dida H P, Suparman S and Widhiyanuriyawan D 2016 Pemetaan Potensi Energi Angin di Perairan Indonesia Berdasarkan Data Satelit QuikScat dan WindSat *Jurnal Rekayasa Mesin,* **7**(2) 95-101

[2] Widayana G 2012 Pemanfaatan Energi Surya *Jurnal Pendidikan Teknologi dan Kejuruan* **9**(1) 37-46

[3] Chrismianto D, Arswendo B and Sobirin Y 2014 Pengaruh Variasi Bentuk Hull Kapal Catamaran Terhadap besar Hambatan Total Menggunakan CFD *KAPAL* **11**(2) 99-106

[4] Muk-Pavic E, Chin S and Spencer D 2006 Validation Of The CFD Code Flow-3D For The Free Surface Flow Around The Ship's Hulls *Annual Conference Of The CFD Society of Canada*

[5] Promdee C and Photong C 2016 Effects of Wind Angles and Wind Speeds on Voltage Generation of Savonius Wind Turbine with Double Wind Tunnels in *International Electrical Engineering Congress, IEECON2016*, Chiang Mai, Thailand

[6] Pane I E 2017 Optimasi Perancangan Turbin Angin Vertikal Tipe Darrieus Untuk Penerangan Jalan Tol in *Seminar Nasional Sains dan Teknologi*

[7] Vivek C, Gopikrishnan P, Murugesh R and Mohamed R R 2017 A Review on Vertical and Horizontal Axis Wind Turbine *International Research Journal of Engineering and Technology (IRJET)* **4**(4) 247-250

[8] Rohmana R A 2016 Analisa Performansi dan Monitoring Solar Photovoltaic System(SPS) Pada Pembangkit Listrik Tenaga Surya Di Tuban Jawa Timur *Junral Teknik POMITS* 1-8

[9] Komarudin A 2014 Desain dan Analisis Proporsional Kontrol Buck-Boost Converter Pada Sistem Photovoltaik *Jurnal ELTEK* **12**(2) 78-89

[10] Sundareswaran K, Kuruvinashetti K S, Sankar H B P and Nayak P S 2013 Output Voltage Control of Dual Input Buck-Boost Converter in *IEEE Innovative Smart Grid Technologies-Asia (ISGT Asia)*, Bangalore, India

[11] Arfianto D F, Asfani D A and Fahmi D Pemantauan, Proteksi, dan Ekualisasi Baterai Lithium-ionTersusun Seri Menggunakan Konverter Buck-Boost dan LC Seri dengan Kontrol Synchronous Phase Shift *Jurnal Teknik ITS* **5**(2) 122-127

[12] Aisuwarya R and Hidayati Y 2019 Implementation of Ziegler-Nichols PID Tuning Method on Stabilizing Temperature of Hot-water Dispenser in *16th International Conference on Quality in Research (QIR): International Symposium on Electrical and Computer Engineering*, Padang, Indonesia

[13] Jamal Z 2015 Implementasi Kendali PID Penelaan Ziegler-Nichols Menggunakan Mikrokontroller *Jurnal Informatika* **15**(1) 81-88

[14] Patra A 2014 PID Controller Tuning using Ziegler-Nichols Method for Speed Control of DC Motor *International Journal of Scientific Engineering and Technology Research* **3**(13) 2924 - 2929

IOP Conf. Series: Materials Science and Engineering 1073 (2021) 012045 doi:10.1088/1757-899X/1073/1/012045

Solar panel performance monitoring using Wireless Sensor Network in State Polytechnic of Malang

M Junus*, N Hidayati and R A Wijayanti

Department of Electrical Engineering, State Polytechnic of Malang, Malang, Indonesia, 65141

*mochammad.junus@polinema.ac.id

Abstract. The electricity supply used in State Polytechnic of Malang (Polinema) is still fully supported by PLN by utilizing the electricity source of the Electric Steam Power Plant (PLTU) with coal as a fuel. As the use of electricity in Polinema's environment is very large, this has an impact on the expensive cost that must be incurred every month. The use of solar power as a new source of electricity that utilizes the nature of renewable energy can be used as an alternative electricity in Polinema. Supported by the use of solar panels with relatively affordable price and can be used as a long-term investment, it is expected that the use of electricity from PLN can be replaced by this solar power plant (PLTS). In this research, a prototype of electrical energy monitoring device produced by solar panel using several sensors connected in one network or commonly called Wireless Sensor Network (WSN) was designed and made. WSN was built using ESP32 microcontroller which is assembled with a voltage and current sensor which then the readings from the sensors will be sent wirelessly to the router to proceed to the server with internet facilities. Information from the sensors are being continously stored and monitored, so user can see the results of WSN reading from any location connected to the internet (real time).

1. Introduction

Currently, the need for electricity use in the Malang State Polytechnic (Polinema) is very high especially used for activities in the classroom, practicum activities in the laboratory and also office activities from morning to evening. Until now, the electricity needs in Polinema are fully supported by PLN by utilizing a power source from the Steam Power Plant (PLTU) in Paiton [1].

The use of PLTU itself has its drawbacks, that it still requires fuel of coal, whose availability can be exhausted, expensive investment and also requires a lot of cooling water, so it must be placed in a location close to many water sources [2].

The use of solar power as a new source of electricity that utilizes the nature of renewable energy can be used as an alternative to electrical energy in Polinema. Being in the equatorial region, the potential for solar energy in Indonesia reaches an average of 4.8 kilowatt hours per square meter per day (kWh / m2 / day), equivalent to 112,000 GWp [3]. With an average sun shining 6 to 8 hours per day, it can meet the needs of electricity production in solar panels, which the ideal length of irradiation is 4 to 5 hours per day [4]. Supported by the use of solar panels at a relatively affordable price and can be used as an investment in the long term [5-6], it is hoped that the use of electricity from PLN can be replaced by Solar Power Plant (PLTS).

However, the design of PLTS cannot be done carelessly, but must meet several factors that determine the performance of solar panels. The amount of output power generated from the conversion of sunlight energy into electrical energy is determined by several environmental conditions such as sunlight intensity, temperature, direction of sunlight and the spectrum of sunlight. Environmental conditions that always change every time can cause the output power of the solar panels to also fluctuate [7].

In this research, a prototype of a monitoring tool for electrical energy produced by solar panels will be designed and built using several sensors connected in a network or commonly called a Wireless Sensor Network (WSN) which is then connected to a microcontroller to obtain data such as voltage and current as well as take advantage of the Wi-Fi facility available as a medium for sending and receiving the data in real time. The performance of solar panel can be determined by monitoring its output parameters such as voltage, current, power and efficiency output. If the results of these monitoring are known, the appropriate energy consumption and electrical load can be well planned.

2. Methods

This research applies the Wireless Sensor Network (WSN) concept to monitoring the solar panel parameters such as voltage and current values generated from the solar panel and also the temperature around. WSN itself at least consist of node sensor to get data from the environment, gateway or sink to collect the data from node sensor and base station to store the data. The topology used in the WSN system is a star topology, means each node sensor is directly connected to the sink.

Every sensor node is sensing or monitoring the environment, in this research sensor nodes are sensing the voltage and current value from solar panel and also temperature around solar panel. These monitoring results are sent to sensor gateway/sink. After several times, sink will deliver the monitoring result data to the base station. At the base station, the data will be processed into another data desired by the user.

2.1. Solar panel monitoring system

Based on WSN architecture above, here is in Figure 1 shown the architecture of solar panel monitoring system implemented in this research:

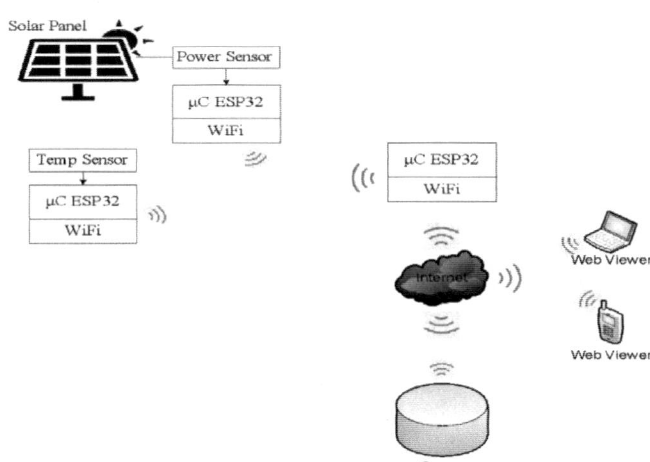

Figure 1. Architecture of solar panel monitoring system.

Note:

a. Node Sensor = Voltage Sensor and Current Sensor ACS712 connect with ESP32µC
b. Node Sensor = Temperature Sensor BME280 connect with ESP32µC
c. Gateway/Sink = ESP32µC
d. Base Station = Server (http://niaja.id)

Figure 1 explains the overall research diagram that consists of node sensor around solar panel for sensing voltage, current and temperature sensor. Node sensor send the data to sink through Wi-Fi. Sink also send the data to base station or server through Wi-Fi. At base station, these data will be performed in number and also in chart, so user can access it from mobile phone or laptop anywhere in real time.

ESP32μC is a single 2.4 GHz Wi-Fi-and-Bluetooth combo chip designed with the TSMC ultra-low-power 40 nm technology. It is designed to achieve the best power and RF performance, showing robustness, versatility and reliability in a wide variety of applications and power scenarios. ESP32μC is designed for mobile, wearable electronics, and Internet-of-Things (IoT) applications. It features all the state-of-the-art characteristics of low-power chips, including fine-grained clock gating, multiple power modes, and dynamic power scaling [8].

From research diagram shown in figure 1 above, the system is implemented as shown in Figure 2, Figure 3 and Figure 4 as follow:

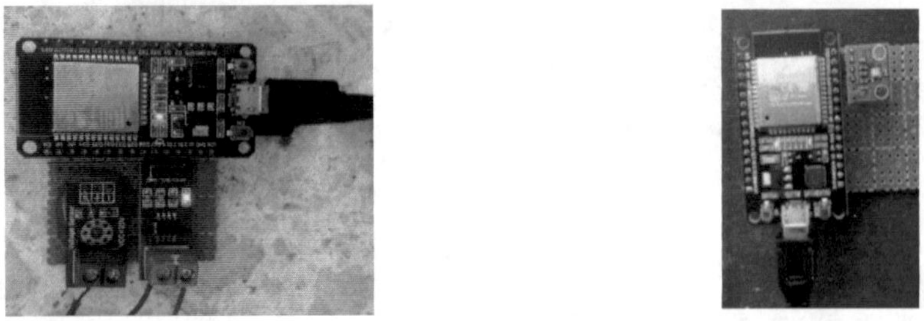

(a) **(b)**

Figure 2. Configuration of node sensors: power sensor (a) and temperature sensor (b)

From Figure 2 above, node sensor (a) consists of voltage sensor and current sensor that connected to ESP32μC. This node sensor is used to retrieve voltage and current data generated by the solar panel. Node sensor (b) consists of temperature sensor that also connected to ESP32μC. This sensor is used to retrieve temperature data around solar panel. Temperature around solar panel is important to monitor because higher temperature around can decrease the solar panel efficiency about 0.055% /°C [9].

Figure 3. Configuration of sink

Figure 4. Configuration of solar panel monitoring system.

From Figure 3 above, we also can see that sink is also using ESP32μC so the communication between node sensor and sink can be done more easily, by utilizing the MAC address on each ESP32 board. After node sensors send the power and temperature data to sink, sink will collect all data and then send them to base station or server through Wi-Fi.

In Figure 4, the whole system is shown. The solar panel is connected to a battery with a capacity of 12 mAh, so the electrical energy obtained from the sunlight can be stored in the battery.

3. Results

The monitoring results from sensors can be accessed at address http://niaja.id/esp-data-daya.php for power and efficiency of solar panel, http://niaja.id/esp-data-temp.php for temperature around solar panel and also http://niaja.id/esp-chart.php for every data in chart. Here in Figure 5 (a), (b) and (c) show the all result of monitored data:

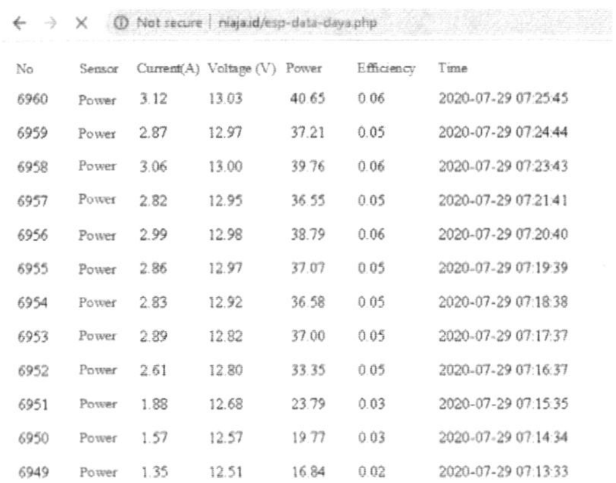

No	Sensor	Current(A)	Voltage (V)	Power	Efficiency	Time
6960	Power	3.12	13.03	40.65	0.06	2020-07-29 07:25:45
6959	Power	2.87	12.97	37.21	0.05	2020-07-29 07:24:44
6958	Power	3.06	13.00	39.76	0.06	2020-07-29 07:23:43
6957	Power	2.82	12.95	36.55	0.05	2020-07-29 07:21:41
6956	Power	2.99	12.98	38.79	0.06	2020-07-29 07:20:40
6955	Power	2.86	12.97	37.07	0.05	2020-07-29 07:19:39
6954	Power	2.83	12.92	36.58	0.05	2020-07-29 07:18:38
6953	Power	2.89	12.82	37.00	0.05	2020-07-29 07:17:37
6952	Power	2.61	12.80	33.35	0.05	2020-07-29 07:16:37
6951	Power	1.88	12.68	23.79	0.03	2020-07-29 07:15:35
6950	Power	1.57	12.57	19.77	0.03	2020-07-29 07:14:34
6949	Power	1.35	12.51	16.84	0.02	2020-07-29 07:13:33

(a) Monitoring result: Power and Efficiency data

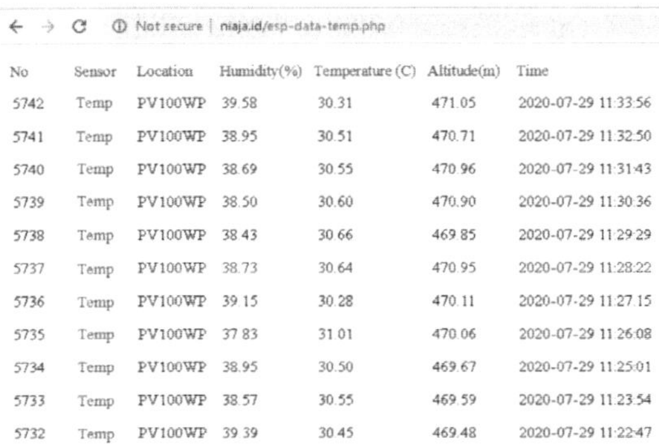

No	Sensor	Location	Humidity(%)	Temperature (C)	Altitude(m)	Time
5742	Temp	PV100WP	39.58	30.31	471.05	2020-07-29 11:33:56
5741	Temp	PV100WP	38.95	30.51	470.71	2020-07-29 11:32:50
5740	Temp	PV100WP	38.69	30.55	470.96	2020-07-29 11:31:43
5739	Temp	PV100WP	38.50	30.60	470.90	2020-07-29 11:30:36
5738	Temp	PV100WP	38.43	30.66	469.85	2020-07-29 11:29:29
5737	Temp	PV100WP	38.73	30.64	470.95	2020-07-29 11:28:22
5736	Temp	PV100WP	39.15	30.28	470.11	2020-07-29 11:27:15
5735	Temp	PV100WP	37.83	31.01	470.06	2020-07-29 11:26:08
5734	Temp	PV100WP	38.95	30.50	469.67	2020-07-29 11:25:01
5733	Temp	PV100WP	38.57	30.55	469.59	2020-07-29 11:23:54
5732	Temp	PV100WP	39.39	30.45	469.48	2020-07-29 11:22:47

(b) Monitoring result: Temperature data

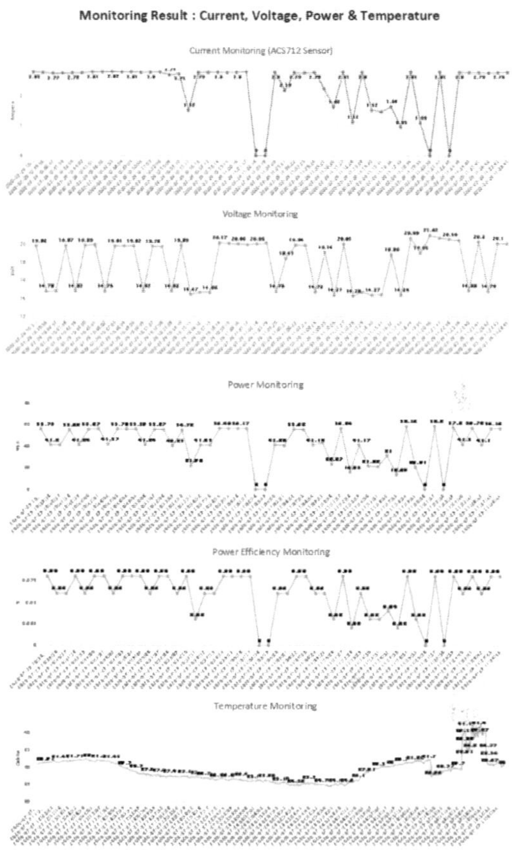

(c) All monitoring result in chart

Figure 5. All results of monitored data.

From the Figure 5 above, it is shown that the maximum power voltage generated by the solar panel is 14.63 V, while the specifications of the solar panel said that the maximum power voltage at normal temperature (25°C) is 16.8V [9]. This difference is due to measured temperature around the solar panel is around 31°C. This is in accordance with the characteristics of the solar panel that higher the temperature around the solar panel will decrease the voltage value. The efficiency of solar panel is about 12%, by using incident radiation flux about $1000 W/m^2$.

4. Testing and analysis

Testing process was carried out to ensure that the results obtained by WSN were correct, this includes voltage testing, current testing, temperature testing and communication between nodes and sink. Voltage and current testing were done by measure the voltage and current value using digital multimeter, so the result between multimeter and node sensors can be compared. Here in Figure 6 shows the voltage testing:

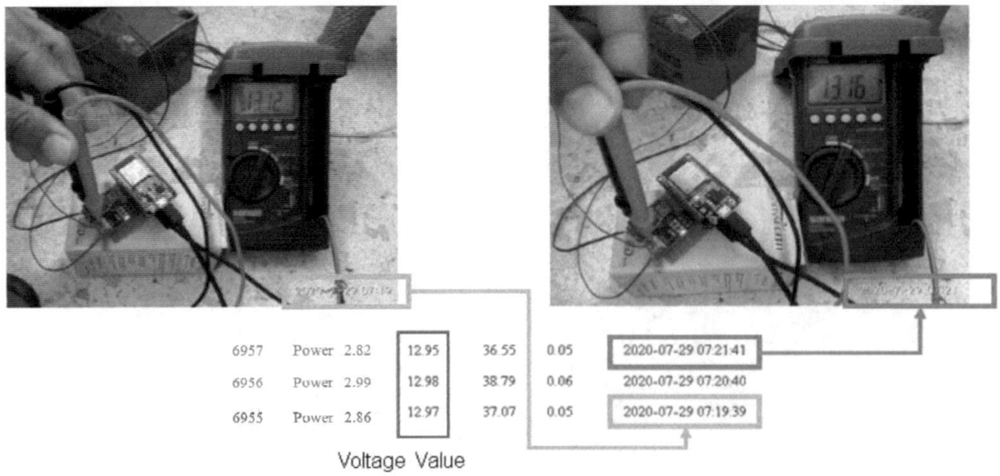

Figure 6. Voltage testing.

From the Figure 6 above, voltage data from digital multimeter has higher value about 0.15 – 0.21 V than value from voltage sensor, with percentage error is about 0.91%.

Temperature testing was also done by measure the temperature value using digital thermometer, so the result between thermometer and node sensor can be compared. Here in Figure 7 shows the temperature testing:

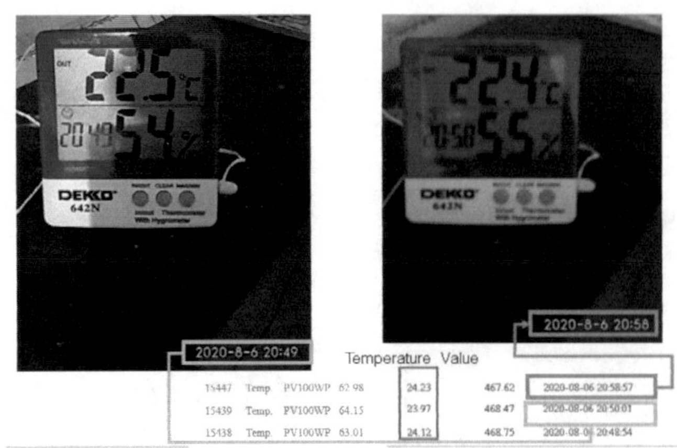

Figure 7. Temperature testing using digital thermometer

From the Figure 7 above, temperature data from digital thermometer has different value about 1.62° – 1.83° lower than value from temperature sensor, with percentage error is about 1.73%.

Communication between node and sink testing was done through serial monitor using IDE Arduino, shown in Figure 8 below:

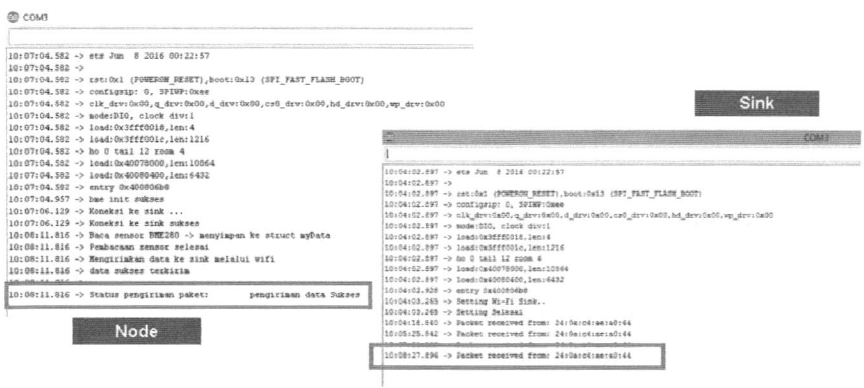

Figure 8. Communication testing between node and sink.

From Figure 8 above, connection between node and sink is successfully done and after that nodes send the data to sink through Wi-Fi. About 16 seconds after, sink successfully received the data.

5. Conclusion

From the results of testing and analysis in this research, several conclusions can be drawn as follows: 1) Wireless Sensor Network are successfully configured to monitoring the efficiency of solar panel. 2) Sensor nodes that consist of voltage, current and temperature sensors are able to sensing data but the data taken has different values measured manually (still in normal value). 3) Sensor nodes are able to communicate with sink by utilized their MAC address. 4) Data taken form Wireless Sensor Network are real time and can be accessed from anywhere.

References

[1] Paiton Energy 2020 https://www.paitonenergy.com/id_ID/informasi-teknik/ accessed on January 13th, 2020

[2] Kartawijaya S 2020 Pembangkit Listrik Tenaga Uap. https://adoc.tips/apa-itu-pltu- pembangkit-listrik-tenaga-uap-pltu-adalah-pemba.html accessed on January 13th, 2020

[3] Sekretariat Jenderal Dewan Energi Nasional. 2016 *Indonesia Outlook Energy 2016. Online Version*

[4] Janaloka 2015 *Potensi Energi Matahari di Indonesia* https://janaloka.com/potensi-energi -matahari-di-indonesia/ accessed on January 14th, 2020

[5] Sharma H, Haque A and Jaffery Z A 2018 An Efficient Solar Energy Harvesting System for Wireless Sensor Nodes *IEEE ICPEICES* 461 - 464

[6] Alippi C and Galperti C 2008 An Adaptive System for Optimal Solar Energy Harvesting in Wireless Sensor Network Nodes *IEEE Trans. On Circuits and Systems* **6** 1742 - 1750

[7] Fachri M R, Sara I D and Away Y 2015 Pemantauan Parameter Panel Surya Berbasis Arduino secara Real Time *Jurnal Rekayasa Elektronika* **11** 123 – 128

[8] ESP32 Series Datasheet Version 3.4. 2020 https://www.espressif.com/sites/default/files/documentation/esp32_datasheet_en.pdf accessed on January 13th, 2020

[9] Kenika Panel Surya NPS100W Specifiation https://www.kenika.com/index.php?route=product/product&product_id=1050 accessed on March 23th, 2020

IOP Conf. Series: Materials Science and Engineering 1073 (2021) 012046 doi:10.1088/1757-899X/1073/1/012046

Android based Wireless Sensor Network (WSN) mobile application on humidity and temperature environmental monitor using CC2650 sensor tag

R A Asmara[1,*], I Siradjuddin[2], and R D Romadhona[3]

[1] Dept. Computer Science, State Polytechnic of Malang, Malang, Indonesia
[2] Dept. Control and Robotic, State Polytechnic of Malang, Malang, Indonesia
[3] Dept. Computer Science, State Polytechnic of Malang, Malang, Indonesia

*rosa.andrie@polinema.ac.id

Abstract. This research proposed environmental sensing humidity and ambient temperature or object use CC2560 Texas instrument sensor tag device. This device will be implemented on a mobile app to detect surrounding humidity and temperature environment, hoping that it will be used to monitor the user's daily activity. The sensor will transmit raw data to a central data collector mobile app using Bluetooth Low Energy (BLE) communication and configured with the other same device CC2650 sensor tag in a field with Wireless Sensor Network (WSN). Temperature and Humidity sensing was doing with distance to testing the accurate and sensitivity about the sensor device and the sensor is going work well when the time sampling in second time, the humidity and temperature can follow and change with real time sampling until 30s start from 30.12 and humidity start from 85.81 percent because of specification of the sensor device data transfer very fast wit Bluetooth Low Energy / multi-protocol 2.4 GHz and wireless MCUs with 352-kB flash and 80-kB SRAM, so the device get very little energy, low cost, and have good durability can access data and transmit for a long time.

1. Introduction

The future of technology, internet will be most usable around 26 until 50 billion in 2020 and 100 billion internets connect around the world [1]. We can connect to the other people more easily use any device, mobile or computer sensor in our life [2,3]. In the future the climate will change randomly, research or journal about concept of environmental sensing use IoT to solve the problem in our environment. We need to care to each other about healthy. Some earlier research came study about climate problems, unique ecology, and how the weather is changing [4].

While we do our activity, we can connect sensors to the other places, example the information from hospital can access with other device or user on somewhere else like public transportation, and they will know about humidity and temperature at hospital in that time. Sensor device will contribute the information from another place if we still connect with them. In this part IoT wireless sensor network is very important to sharing about information sensor to the other users and help prepare if they want to go outside or somewhere else.

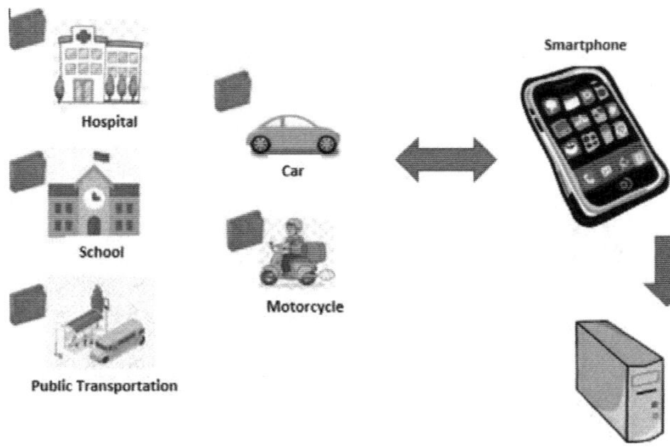

Figure 1. Concept of mobile app environmental sensing.

Device will connect by smartphone Bluetooth and use internet connection from smartphone to send humidity and temperature data, however the location and position place from device can detect and tracking in our web hosting use longitude and latitude from GPS smartphone access permission. Then data will be collected in firebase to processing, grouping or classification the data based on standard health percentage with visual human interface.

This Paper describes a challenge create a simple and friendly user sensor system especially in Malang, Indonesia that can be operated in school, public transportation, public places and another places with sensor tag CC2560 device and use mobile app android to collected data humidity and temperature more save energy to preventing the global warming.

2. Sensor tag CC2560

In order to provide an executable for small system sensor devices we use sensor tag cc2560 was designed by Texas instrument. The sensor device is use for detect humidity and temperature data in somewhere else. Connected to a smartphone Bluetooth the sensor transmits raw data and collect data to forward to the server or firebase. We focus on the following requirement sensor device first is compatibility, sensor must small and easy to bring while we do our activity, second is functionality, sensor must connect with smartphone is wireless connection like Bluetooth, last is durability, long lasting battery and low-cost energy must be worn on sensor device.

Figure 2. Hardware sensor tag CC2560.

2.1. Specification sensor tag module 1

The device generally must simple, long lasting battery, low energy cost, and fast enough to develop and transfer data cause the program from android studio always increase and update every time. that why we choose sensor tag CC2650. there is specification for hardware sensor module from Texas instrument about sensor tag CC2650.

IOP Conf. Series: Materials Science and Engineering 1073 (2021) 012046 doi:10.1088/1757-899X/1073/1/012046

Sensor tag CC2560 Texas instrument has 5x4 cm dimension so tiny and the weight just 28gr, it so wearable and mobile easy to carry out. Sensor Tag is based on the CC2650 wireless microcontroller (MCU), offering 75 percent lower power consumption than previous Bluetooth products [5].

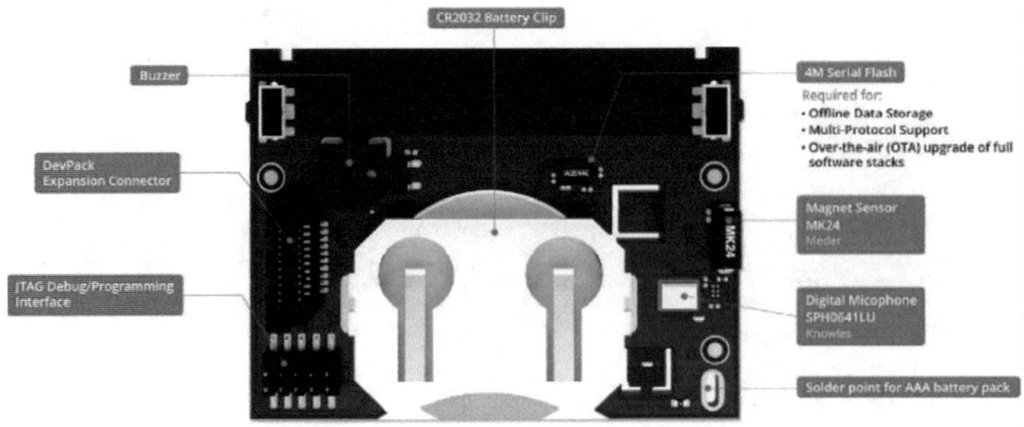

Figure 3. Sensor tag CC2560 module specification 1.

2.2. Specification sensor tag module 2

This allows the Sensor Tag kit to be battery powered, and offer years of battery life from a single coin cell battery [6]. Cloud connectivity lets you access and control your Sensor Tag kit from anywhere. Pin-to-pin compatible with the Simple Link Bluetooth Low Energy / multi-protocol 2.4 GHz CC2642R and CC2652R wireless MCUs with 352-kB Flash and 80- kB SRAM [7].

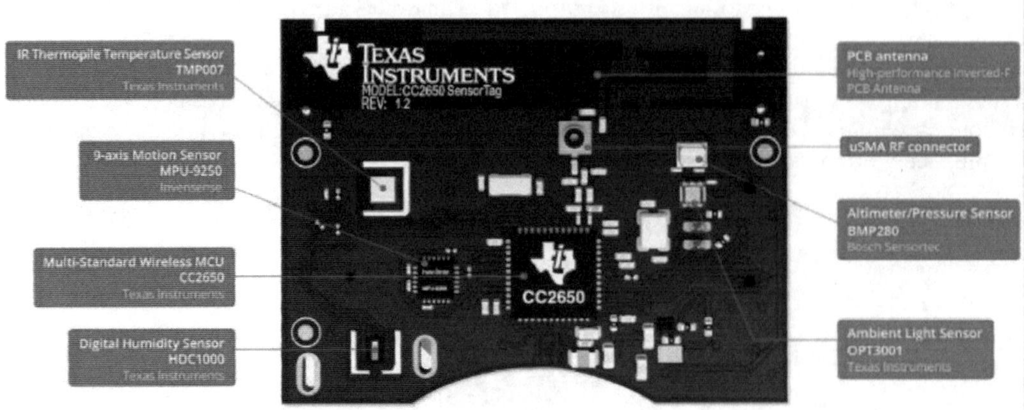

Figure 4. Sensor tag CC2560 module specification 2.

3. Application

We have developed application system using android studio for collect data temperature and humidity from sensor tag CC2560. We try with 1 device sensor tag to detect the sensor and can open them in mobile application. Sensor connect with smartphone use Bluetooth low energy (BLE) work properly they can send data or transmit raw data temperature and humidity sensing. In android studio we change and developed UUID data sensor to number of temperature and humidity sensor. There are the programs to detected sensor use android studio.

IOP Conf. Series: Materials Science and Engineering 1073 (2021) 012046 doi:10.1088/1757-899X/1073/1/012046

Figure 5. Import bluetooth connection.

Figure 5 is declared import Bluetooth connection and permission in application to use Bluetooth connectivity with sensor tag cc2560. Bluetooth from smartphone must be always on while data transmit occur.

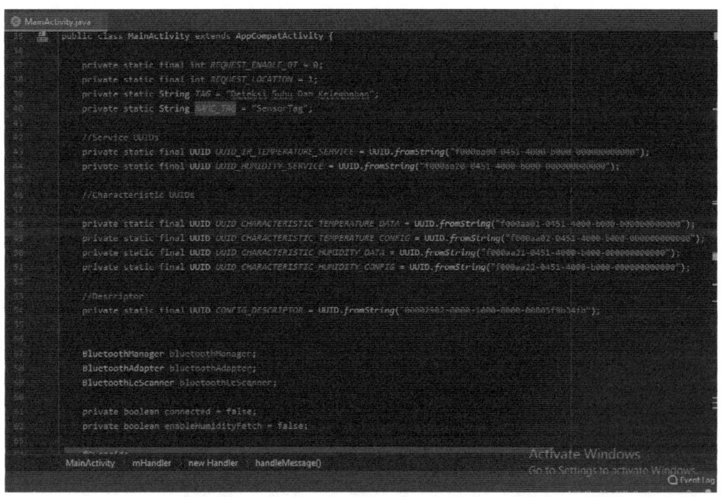

Figure 6. UUID binary data sensor.

Figure 6 is configuring data transfer in mobile app from sensor tag cc2650 and change UUID binary data to temperature and humidity number. It detected the sensor and collect the data temperature and humidity sensor.

Figure 7. Bluetooth low energy is on.

Figure 8. Android mobile application.

Figure 9. Permission app to bluetooth connection.

Figure 10. Detection temperature and humidity.

After temperature and humidity have been detected, we improve the location where the device will be found it, so we create a script to show longitude and latitude from the smartphone who connected with sensor tag Bluetooth Low Energy. With android application monitoring Sensor and location can push to firebase with internet connection of the smart- phone so the smartphone should connect the internet always to send data to firebase and we can collect the data from that. Here the figure about location access in application below.

Figure 11. Location access.

IOP Conf. Series: Materials Science and Engineering 1073 (2021) 012046 doi:10.1088/1757-899X/1073/1/012046

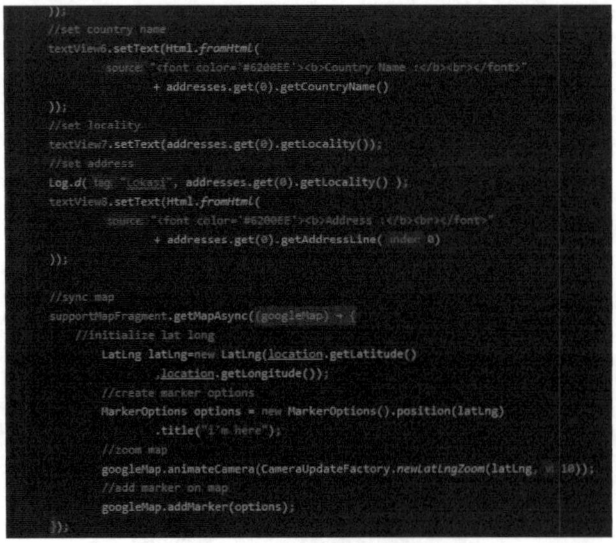

```
));
//set country name
textView6.setText(Html.fromHtml(
        source: "<font color='#6200EE'><b>Country Name :</b><br></font>"
                + addresses.get(0).getCountryName()
));
//set locality
textView7.setText(addresses.get(0).getLocality());
//set address
Log.d( tag: "Lokasi", addresses.get(0).getLocality() );
textView8.setText(Html.fromHtml(
        source: "<font color='#6200EE'><b>Address :</b><br></font>"
                + addresses.get(0).getAddressLine( index: 0)
));

//sync map
supportMapFragment.getMapAsync((googleMap) -> {
    //initialize lat long
    LatLng latLng=new LatLng(location.getLatitude()
            ,location.getLongitude());
    //create marker options
    MarkerOptions options = new MarkerOptions().position(latLng)
            .title("i'm here");
    //zoom map
    googleMap.animateCamera(CameraUpdateFactory.newLatLngZoom(latLng, v: 10));
    //add marker on map
    googleMap.addMarker(options);
});
```

Figure 12. Location access program.

Program 5 is algorithm to allow access location and detection of latitude and longitude from GPS smartphone. They send and push the data location to the firebase use internet connection and position of user on smartphone.

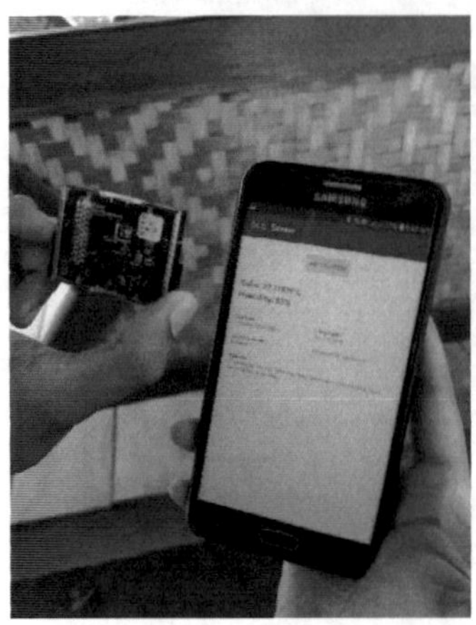

Figure 13. Device connected with smartphone.

Figure 13. show about device sensor tag cc2560 Bluetooth Low Energy (BLE) connected with the application BLE sensor in android smartphone, there are current location, current data sensor in real time and actual temperature and humidity. The data will always push to firebase with internet connection and Bluetooth from smartphone.

4. Experiment and measurement

Sensor testing from sensor tag cc2560 device is accurate and the time is following with data sensor read. We try and search about the sensitivity and delay time data transfer to firebase data cloud in google console. The sensor read humidity and temperature from 5m distance to smartphone will be connected with sensor tag cc2560.

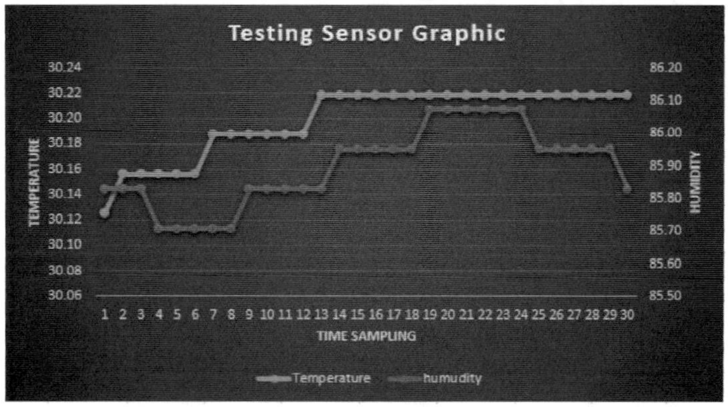

Figure 14. Graphic sensor.

The graphic show humidity and temperature in axis y right and left to read the value of data they are going together with time sampling in 30 second in below graphic and there are not delay time also the accurate from the sensor is good. Temperature start from 30.12 and humidity start from 85.81 percent, temperature is steady at 13 second time sampling in 30.22 °C while humidity reaches 85.80 percent. we try this in outdoor or field so the humidity and temperature can change in any condition.

Figure 15. Application layout with real time.

Figure 15 is application layout and there are data real time with date, month, hours, minutes until second so user can know about the sensor when take the data and transmit data from sensor tag trough the smartphone use internet connection to cloud or firebase google console. Then we collect the data in firebase to memories and save them. The figure below explains how the data will be save and memorize at firebase. The function of firebase is to store data that has been transferred from a smartphone connected to CC2560 tag sensor and clustering data into easy to read data. In next research we using bayes or KNN algorithm to clustering raw data.

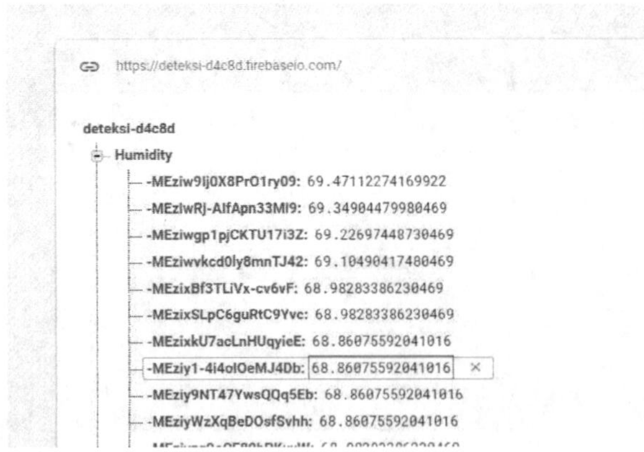

Figure 16. Humidity data in firebase.

Figure 17. Temperature data in firebase.

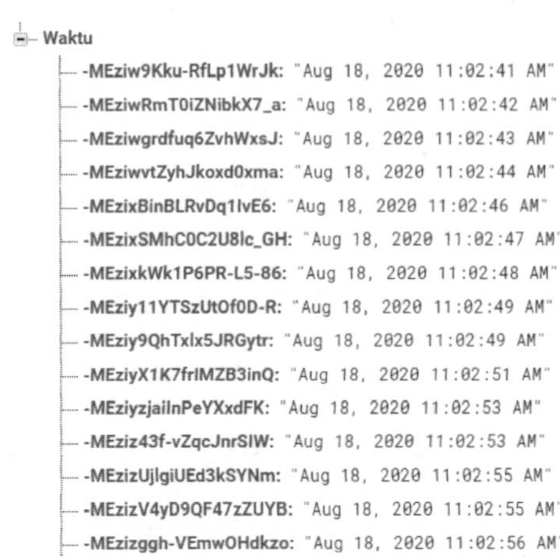

Figure 18. Real time in firebase.

5. Conclusion

This paper presented abstracting environmental sensing to detect humidity and temperature in mobile application use android studio to developed the program with sensor tag CC2560 as sensor device. The specification of the sensor is work properly and can access data transfer quickly with BLE multi-protocol 2.4 GHz and wireless MCUs with 352-kB flash and 80-kB SRAM. So, they get consume the energy is very small and low-cost energy. Durability from sensor also good and long lasting to get data for a long time and we can connect with GPS from smartphone to know about the condition or surrounding in another place.

References

[1] Wu F, Wu T and Yuce M R 2019 Design and im- plementation of a wearable sensor network system for iot-connected safety and health applications *IEEE 5th World Forum on Internet of Things (WF-IoT)* pp 87–90

[2] Mois G, Folea S and Sanislav T 2017 Analysis of three iot-based wireless sensors for environmental monitoring *IEEE Transactions on Instrumentation and Measurement* **66** 8 pp 2056–2064

[3] Mois G, Sanislav T and Folea S C 2016 A cyber-physical system for environmental monitoring *IEEE Transactions on Instrumentation and Measurement* **65** 6 pp 1463–1471

[4] Corbellini S, Ferraris F and Parvis M 2008 A system for monitoring workers' safety in an un- healthy environment by means of wearable sensors *IEEE Instrumentation and Mea- surement Technology Conference* pp 951–955

[5] Vishwakarma S K, Upadhyaya P, Kumari B and Mishra A K 2019 Smart energy efficient home automation system using iot *4th International Conference on Internet of Things: Smart Innovation and Usages (IoT-SIU)* pp 1–4

[6] Andre Chase O, das Neves Carvalho A, do Socorro Silva Andrade E, Souza de Almeida J F and da Costa Junior C T 2018 Environmental mea- surement technology: An approach to the amazo- nian environment *IEEE Latin America Transactions* **16** 4 pp 1036–1041

[7] Joris L, Dupont F, Laurent P, Bellier P, Stoukatch S and Redout´e J 2019 An autonomous sigfox wireless sensor node for environmental monitoring *IEEE Sensors Letters* **3** 7 pp 01–04

IOP Conf. Series: Materials Science and Engineering 1073 (2021) 012047 doi:10.1088/1757-899X/1073/1/012047

Experimental analysis of temperature, light intensity, and humidity on rooftop standalone solar power plant

W Widjanarko*, N Alia, A Dani and F A Perdana

Department of Mechanical Engineering, State Polytechnic of Malang, Malang, Indonesia

*widjanarko@polinema.ac.id

Abstract. Rooftop standalone solar technology is a new renewable power generation solution worldwide and can reduce carbon dioxide levels in the environment. However, several factors affect the performance of the power plant, it is highly influenced by weather parameters. This paper shows the design of the rooftop standalone solar power plant and the influence of temperature, light intensity, and environmental humidity on it. The designed system consists of a 200WP photovoltaic, 1260Wh LiFePO4 battery, and an MPPT solar charge controller 40A. This standalone power plant was applied to supply the energy of the home lighting, with 20Watt the total of power. The result shows the design of the rooftop standalone solar power plant. Furthermore, there is an influence of temperature, light intensity, and environmental humidity on the LiFePO4 battery input power.

1. Introduction

Indonesia is a country located in the equator that gets abundant sunshine throughout the year, which has a very large potential for solar energy with average daily insolation of 4.5-4.8 KWh/m^2/day [1]. Solar energy is a renewable source of energy that is gaining ground because of the benefits it offers [2]. A Photovoltaic (PV) system is a power generation system that produces electrical power by harnessing solar radiation and converting it into electricity [3]. Sunshine is a renewable energy source of solar electricity that can provide electricity continuously. Furthermore, it is also a clean and environmentally friendly energy, to reduce gas emissions and greenhouse [4–7].

The off-grid system is a stand-alone system that not relating to the gird facility [8,9]. Especially, the system that are not connected to the main electrical grid are referred as an off-grid photovoltaic system. Off-grid system can generate the power and run the appliances by itself, it was called a mini-grid or a standalone system. If the distinct living and spread population in the vast area or in the in the districts where they do have no access to the electricity, the off-grid electrification system is viable for the remote this areas [10]. Electrical energy in the stand-alone photovoltaic system produced through the Solar photovoltaic panels needs to be stored or saved on the battery bank because the requirement from the load can be different from the solar panel output [11].

A rooftop solar power plant is a photovoltaic system that has its electricity-generating solar panels mounted on the rooftop of a residential or commercial building [12]. A photovoltaic (PV) system consists of a photovoltaic, MPPT Solar Charge Controller, battery bank, and some electronic elements for power conditioning. The urban environment provides a large number of empty rooftop spaces and can inherently avoid potential land use and environmental concerns. The space on the rooftop can harness solar radiation and converting it into electricity by applying the A rooftop photovoltaic power station [13]. On the other hand, the owners of rooftop photovoltaic power station have the authority to incorporate the solar electricity into the home grid electricity [14].

The output power generated by photovoltaics on the rooftop standalone power plant depends on many aspects [15]. There are from environmental weather condition, such as air temperature [16,17], relative humidity [18,19] and light intensity [20]. Moreover, PV material, parasitic resistance, inverter efficiency, dust, module orientation, geographical location, cable thickness also give affect the performance of photovoltaic [21].

2. Methodology

Weather parameter: light intensity, relative humidity, air temperature, were measured intermittently and simultaneously with solar illuminance and output voltage from solar charge controller to storage energy system. The development of a rooftop standalone solar power plant and data logger monitoring system consists of some parts and software. The rooftop standalone power plant consists of 1) polycrystalline Photovoltaic 200WP, 2) MPPT Solar Charge Controller 40A, 3) LiFePO4 Battery 12V 100Ah. Then, the data logger monitoring system consists of microcontroller ATMEGA 328 (Arduino UNO), current sensor ACS758 50A, voltage sensor, DHT11 sensor, BH1750 sensors, RTC DS3231, and SD Card module. Figure 1 shows the experimental layout of the rooftop standalone solar power plant and data logger monitoring system.

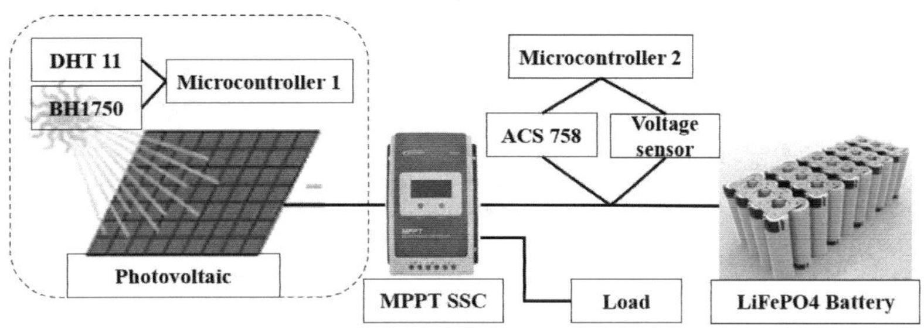

Figure 1. Experimental design of rooftop standalone solar power plant and data logger monitoring system.

2.1. Microcontroller

Microcontroller ATMEGA 328 (Arduino UNO) has 14 digital input/output pins (of which 6 can be used as PWM outputs), 6 analog inputs, a 16 MHz ceramic resonator, a USB connection, a power jack, an ICSP header, and a reset button. It was used to develop the data logger and monitoring system in this rooftop standalone power plant [22].

Figure 2. Location to applied rooftop standalone power plant.

The development of this rooftop standalone power plant is focused as a source of energy for home lighting. The selected location for applied this rooftop standalone power plant is on KH. Hasyim Asyhari Street, Gg. Bandar Kidul, Mojoroto, Kediri (Figure 2), which has coordinates -7.831849, 112.002152. Measurement of all parameter tests was taken at an interval of 24 hours for a 10 days 22-31 July 2020. During the measurements shown in figure 1, the relative humidity and air temperature were measured using the DHT 11 sensor. The light intensity was measured using the BH1750 sensor. The output voltage from the MPPT solar controller to battery was measured using a voltage sensor that consists of a voltage divider. Then, the current from the MPPT solar charge controller to the battery was measured using current sensor ACS758 50A. All data measurements were real-time recorded, therefore RTC DS3231 and SD card module were added into the data logger system.

```
File  Edit  Format  View  Help
30.07.2020,08:40:21,78.00,25.70,9439.00
30.07.2020,08:40:21,78.00,25.70,9454.00
30.07.2020,08:40:21,78.00,25.70,9447.00
30.07.2020,08:40:21,78.00,25.70,9453.00
30.07.2020,08:40:22,78.00,25.70,9451.00
30.07.2020,08:40:22,78.00,25.70,9460.00
30.07.2020,08:40:22,78.00,25.70,9460.00
30.07.2020,08:40:22,78.00,25.70,9455.00
30.07.2020,08:40:22,77.00,25.70,9458.00
30.07.2020,08:40:23,77.00,25.70,9473.00
30.07.2020,08:40:23,77.00,25.70,9469.00
30.07.2020,08:40:23,77.00,25.70,9460.00
```

Figure 3. Real-time data recorded of relative humidity, air temperature and light intensity from microcontroller 1.

```
File  Edit  Format  View  Help
28/7/2020 10:38:3,19.78,4.87
28/7/2020 10:38:3,19.85,4.90
28/7/2020 10:38:3,19.85,4.90
28/7/2020 10:38:3,19.82,4.92
28/7/2020 10:38:4,19.82,4.90
28/7/2020 10:38:4,19.80,4.90
28/7/2020 10:38:4,19.85,4.90
28/7/2020 10:38:4,19.87,4.91
28/7/2020 10:38:4,19.78,4.89
28/7/2020 10:38:5,19.85,4.90
```

Figure 4. Real-time data recorded of voltage and current from microcontroller 2.

The example of data recorded that recorded on the SD card is shown in Figure 3 and Figure 4. From figure 3, the data real-time of relative humidity, air temperature, and light intensity was recorded on microcontroller 1. Furthermore, figure 2 is a real-time data from microcontroller 2 that were recorded the voltage and current from the MPPT Solar Charge controller to the battery. All of the data recorded were saved in a micro SD card with a txt extension. Subsequently, data were converted to excel and analyze using origin software.

3. Result and discussion

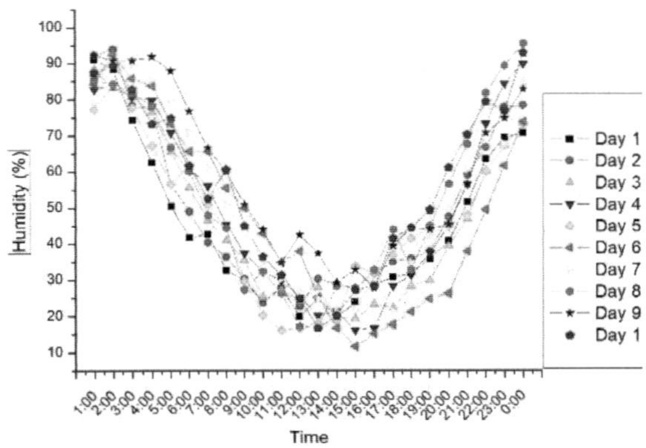

Figure 5. Graph of the relative humidity.

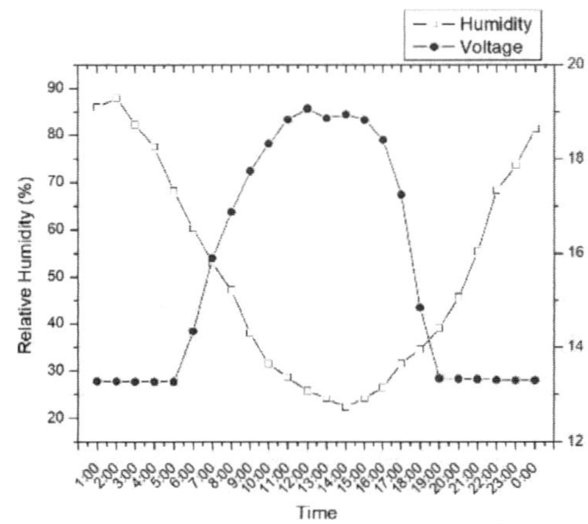

Figure 6. Comparison of the rate of the relative humidity and voltage.

Figure 5 shows the relative humidity as long as 10 days. The data were recorded by real-time using the data logger. From the data can be seen that the humidity during the test is fairly good and in normal condition. At night day, the highest humidity can reach to be 80%. While during the day it can reach 18% of humidity. Then, Figure 6 shows the comparison of the rate of humidity and the rate of voltage from the MPPT solar charge controller into the battery as long as a 10-day test. During the test, the lowest humidity and the maximum voltage occur between 10.20 AM to 14.10 PM.

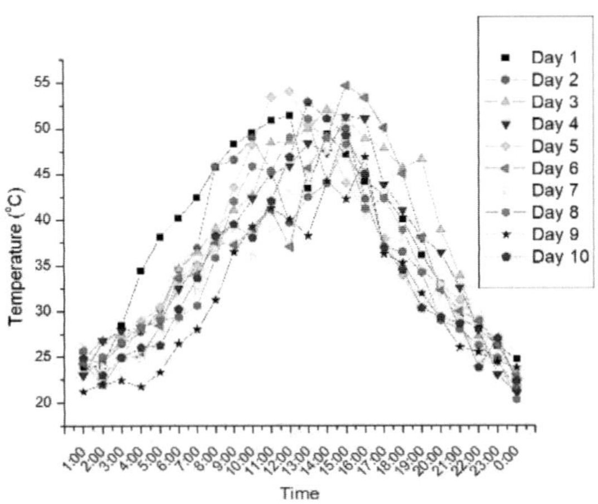

Figure 7. Graph of the air temperature.

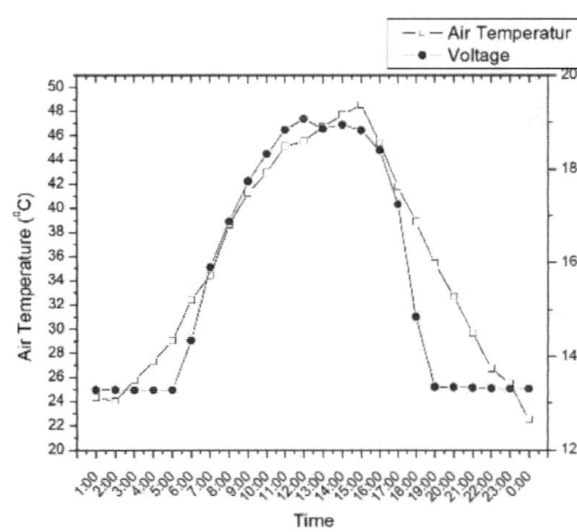

Figure 8. Comparison of the rate of air temperature and voltage.

Figure 7 shown the air temperature as long as a 10-day test. The air temperature during the test is fairly good and in normal condition. The maximum air temperature during the test is 57,5°C, then the minimum air temperature during the test is 21.8°C. Figure 8 shows the comparison of the rate of the air temperature and the rate of voltage. The graph shows that temperature has a significant effect on the power production of photovoltaic.

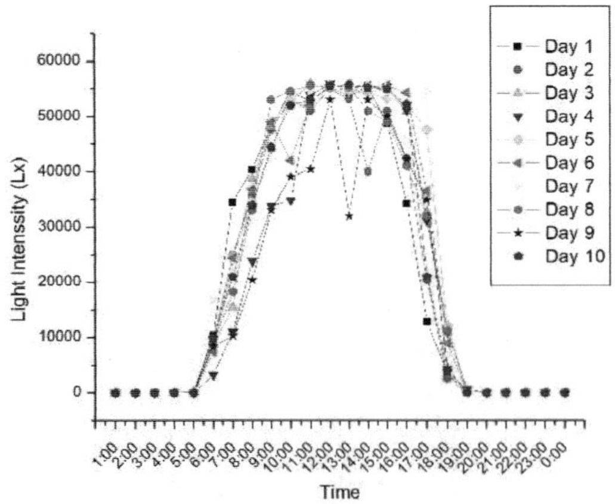

Figure 9. Graph of light intensity.

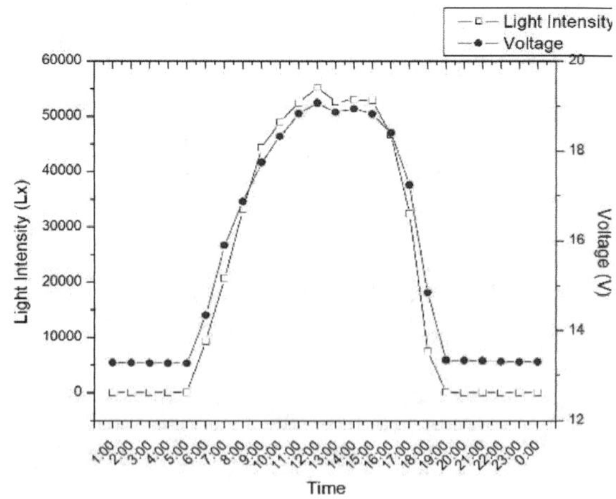

Figure 10. Comparison of the rate of light intensity and voltage.

Figure 9 shown the graph of the light intensity as long as the 10-day test. The maximum light intensity during the test is 55765 Lx that occur between 10.46 AM to 14.38 PM. Figure 10 shows the comparison of the light intensity and the rate of voltage. The graph show that light intensity has a significant effect on the power production of photovoltaic.

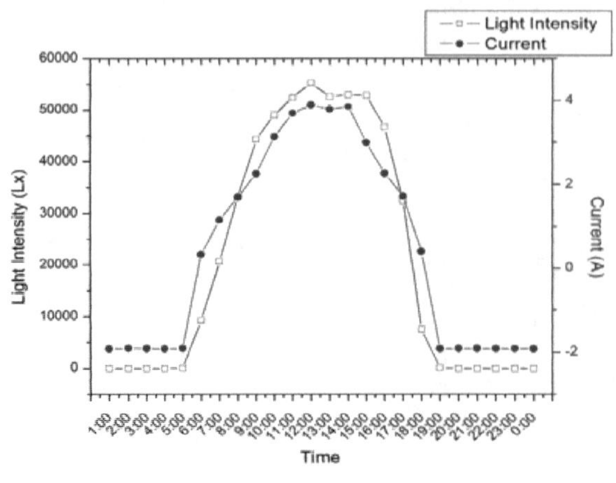

Figure 11. Comparison of the light intensity and current.

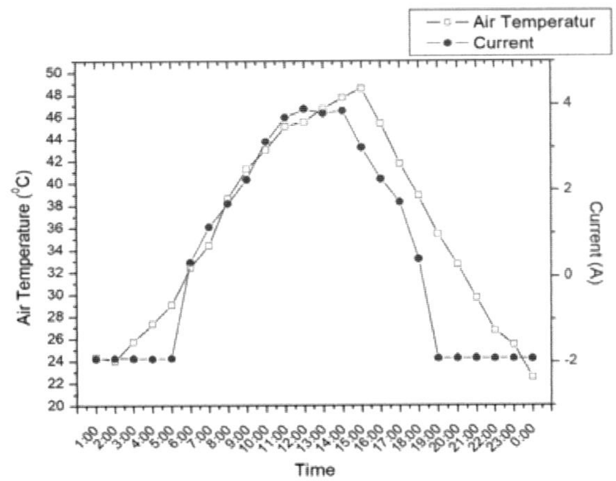

Figure 12. Comparison of the air temperature and current.

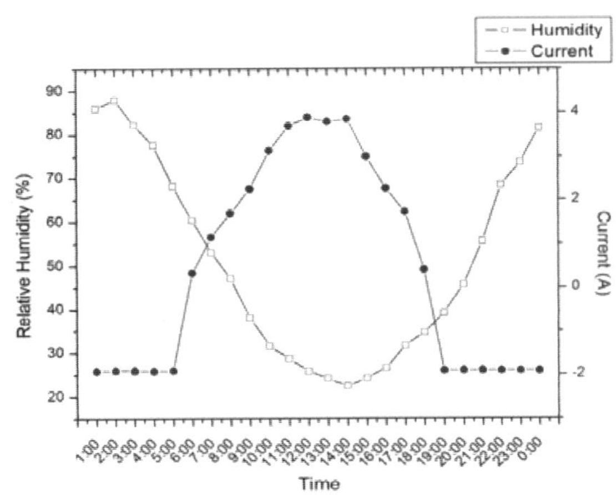

Figure 13. Comparison of the relative humidity and current.

Figure 11 shows the comparation of the rate of light intensity and current. The maximum current that produce by photovoltaic occur between 11.12 AM to 14.18 PM. Figure 12 shows the comparison of air temperature and the current. From the graph, a direct correlation is observed between air temperature rise and current increase. Figure 13 shows the comparation of the of relative humidity to changes the current from the photovoltaic. The graph shows if the relative humidity increase, so the current that produce will be decrease.

4. Future work

Future work will focus on:

- Wireless data logger software
- The specific lithium ion battery (LiFePO4) with cooling system
- Smart BMS that include on the battery
- The rooftop standalone power plant connects with smart home system

5. Conclusion

The real-time performance and monitoring of rooftop standalone solar power plant located on the rooftop of a boarding house in Kediri district are carried out by developing a hardware model of 200WP polycrystalline photovoltaic, MMPT solar charge controller 40A, and LiFePO4 battery 12V 100Ah. Moreover, the data logging monitoring system was developed using microcontroller Arduino UNO (Arduino UNO), current sensor ACS758 50A, voltage sensor, DHT11 sensor, BH1750 sensors, RTC DS3231, and SD Card module.

The light intensity affects the energy output of the rooftop standalone solar power plant. If the light intensity increases by 100 Lux, the energy output will increase by 4Wh assuming a constant temperature. During the day, the temperature affects the energy output of the rooftop standalone solar power plant. If the temperature increases by 1 degree, the energy output of the rooftop standalone solar power plant will increase 0.321 Wh, assuming constant light intensity. In the morning, afternoon, and evening, humidity does not affect the energy output of the rooftop standalone solar power plant.

Acknowledgment
This research was founded by Politeknik Negeri Malang University grant through Penelitian Riset Terapan with the funding scheme grant no. DIPA-023.18.2.677606/2020.

References

[1] Hardianto H 2019 Utilization of Solar Power Plant in Indonesia: A Review *Int. J. Environ. Eng. Educ.* **1** 3 pp 1–8
[2] Gielen D, Boshell F, Saygin D, Bazilian M D, Wagner N and Gorini R 2019 The role of renewable energy in the global energy transformation *Energy Strateg. Rev.* **24** pp 38–50
[3] Nwaigwe K N, Mutabilwa P and Dintwa E 2019 An overview of solar power (PV systems) integration into electricity grids *Mater. Sci. Energy Technol.* **2** 3 pp 629–633
[4] Duzenli M, Kocar G and Eryasar A 2018 a Review of Floating Solar Power Plants *From Sci. Ed.* pp 281–288
[5] Adejuyigbe S B, Bolaji B O, Olanipekun M U and Adu M R 2013 Development of a solar photovoltaic power system to generate electricity for office appliances *Eng. J.* **17** 1 pp 29–39
[6] Mariano J D A, Campos H M, Tonin F S, Junior J U and Junior E F C 2016 Performance of photovoltaic systems: Green office's case study approach *International Journal of Energy and Environment* **7** 2 123
[7] Gul M, Kotak Y and Muneer T 2016 Review on recent trend of solar photovoltaic technology *Energy Exploration & Exploitation* **34** 4 485-526
[8] Ishaq M, Ibrahim U H and Abubakar H 2013 Design of an off grid photovoltaic system: A case study of Government Technical College, Wudil, Kano State *International Journal of Technology Enhancements and Emerging Engineering Research* **2** 12 175-181
[9] Widjanarko W, Nugroho P W, Dani A and Alia N 2019 Studi Implementasi Small Plts Off Grid Berbasis Baterai Lifepo4 Pada Rumah Tinggal Daya Tenaga Surya 200 W *Jurnal Teknologi* **13** 2 10-14
[10] Feron S 2016 Sustainability of off-grid photovoltaic systems for rural electrification in developing countries: A review *Sustainability* **8** 12 1326
[11] AlShemmary E N A, Kadhom L M and Al-Fahham W J 2013 Information technology and stand-alone solar systems in tertiary institutions *Energy Procedia* **36** 369-379
[12] Journal 2016 *Feasibility for Small Scale Rooftop Solar Photovoltaic System in Heritage Buildings of* 12591–12603
[13] Freitas S, Catita C, Redweik P and Brito M C 2015 Modelling solar potential in the urban environment: State-of-the-art review *Renewable and Sustainable Energy Reviews* **41** 915-931
[14] Moraitis P, Kausika B B, Nortier N and Van Sark W 2018 Urban environment and solar PV performance: the case of the Netherlands *Energies* **11** 6 1333
[15] Kazem H A and Chaichan M T 2016 Effect of environmental variables on photovoltaic performance-based on experimental studies *International Journal of Civil, Mechanical and Energy Science (IJCMES)* **2** 4 1-8
[16] Amajama J 2016 Effect of air pressure on the output of photovoltaic panel and solar illuminance

(or intensity) *International Journal of Scientific Engineering and Applied Science (IJSEAS)* **2** 8 139-144

[17] Njok A O and Ogbulezie J C 2018 The Effect of Relative Humidity and Temperature on Polycrystalline Solar Panels Installed Close to a River *Physical Science International Journal* 1-11

[18] Amajama J and Oku D E 2016 Effect of Relative humidity on Photovoltaic panels' Output and Solar Illuminance/Intensity *Journal of Scientific and Engineering Research* **3** 4 126-130

[19] Tijjani J, Hamza B and Igwenagu S 2020 The Effect of Humidity and Temperature on the Efficiency of Solar Power Panel Output in Dutsin-Ma Local Government Area (LGA), Nigeria *Journal of Asian Scientific Research* **10** 1 1-16

[20] Kane S N, Mishra A, Dutta A K and Sen P 2012 PREFACE: International Conference on Recent Trends in Physics (ICRTP 2012) *JPhCS* **365** 1 011001

[21] Fouad M M, Shihata L A and Morgan E I 2017 An integrated review of factors influencing the perfomance of photovoltaic panels *Renewable and Sustainable Energy Reviews* **80** 1499-1511

[22] Sudhan R H, Kumar M G, Prakash A U, Devi S A N U R and Sathiya P 2015 Arduino ATMEGA-328 microcontroller *International journal of innovative research in electrical, electronics, instrumentation and control engineering* **3** 4 27-29

IOP Conf. Series: Materials Science and Engineering 1073 (2021) 012048 doi:10.1088/1757-899X/1073/1/012048

Voltage control for variable speed wind turbine using buck converter based on PID controller

R I Putri*, I N Syamsiana, M Rifa'i and F Aditya

Department of Electrical Engineering, State Polytechnic of Malang, Jl. Soekarno Hatta No. 9, Malang, Indonesia

*ratna.ika@polinema.ac.id

Abstract. Currently, energy diversification is needed to overcome the limitations of fossil energy as a fuel for electricity generation. In Indonesia, one source of renewable energy that has great potential is wind energy. Wind energy conversion system is an environmentally friendly electricity generator with unlimited availability. However, the electrical energy produced depends on the fluctuating wind speed so that it produces a fluctuating voltage. Variable speed wind turbine is one of the most widely used wind turbine types because it has better efficiency but requires control to overcome the fluctuations in the voltage generated. This paper presents voltage control on a prototype variable speed wind turbine with a permanent magnet synchronous generator (PMSG) and buck converter. The output voltage generated by the PMSG will be rectified by a three-phase uncontrolled rectifier. Voltage control is done through the duty cycle settings on the buck converter that is connected to the battery for energy storage. PID controller is used to set the duty cycle of PWM pulses with a switching frequency of 31.5 kHz. The Propotional Integrator Differensiator (PID) controller is embedded in the microcontroller. Based on the test results, the resulting output voltage can be held constant at 14V by providing variations in wind speed changes which are simulated through PMSG speed changes. Transient responses with the fastest settling time of 1.2 s were obtained with the PID controller parameter values Kp = 0.6, Ki = 0.15 and Kd = 0.6.

1. Introduction

The use of wind energy as electrical energy generator is increasingly being used because it is environmentally friendly energy, free of CO2 emissions and unlimited availability [1-2]. Wind turbine technology continues to be developed to improve the efficiency of wind energy conversion. Wind turbines are needed to convert wind energy into electrical energy. One type of wind turbine system that is widely used is the variable speed wind turbine (VSWT). Compared to a fixed speed wind turbine, VSWT has higher efficiency. However, the electrical energy produced is highly dependent on fluctuated wind speed so that the electrical energy produced also fluctuates [2-5]. To overcome the fluctuation of electrical energy produced, VSWT is connected to a battery as energy storage. In addition, constant voltage control is needed so that battery charging becomes more efficient.

To improve the performance of the controller, a power electronic configuration for VSWT is required. The use of the dc-dc converter is widely used in wind turbine systems to increase efficiency [6-7]. Buck converter, boost converter and buck-boost converter are types of dc-dc converters which are usually used in wind turbine systems. Two-phase buck converter has been applied to the small scale wind turbine system and produces good performance. Two single buck converters are connected in parallel so as to produce an output power twice as large as a single buck converter at currents below 4 A [8]. A buck-boost converter is also used in PMSG wind turbines and is connected to a DC link to maintain a constant supply voltage. Based on the simulation results, this converter provides low cost and good quality power conversion but has not been implemented in the prototype [9-10].

In this paper, voltage control is designed using a PID controller for a wind turbine system using PMSG. The voltage control will maintain the buck converter output voltage at 14 V through the duty cycle setting sent to the buck converter switching component. The PID controller is embedded in the microcontroller which sends a pulse width modulation (PWM) signal to the buck converter circuit through the MOSFET driver circuit.

2. Methods

Variable speed wind turbine in this paper consists of permanent magnet synchronous generator (PMSG), 3 phase uncontrolled rectifier, buck converter, driver converter, voltage and current detector, microcontroller and battery as energy storage. Figure 1 shows block diagram of variable speed wind turbine. PMSG will convert wind energy captured by the blade to be 3 phase voltage. PMSG has rating power of 500 Watt. 3 phase voltage will be changed to DC voltage by 3 phase uncontrolled rectifier. The greater the wind energy captured, the greater the PMSG rotation will be as well as the resulting output voltage. The rotation of the turbine causes the generator to induce and emits a three-phase AC voltage with a value depending on the rotation. The output voltage of the PMSG will be rectified by an uncontrolled rectifier circuit and the filter becomes a DC voltage. The voltage limit to be processed is 20-100 volts. If the voltage is less than or more than this limit, the relay will break the connection between the PMSG and the rectifier. The voltage that has been rectified will enter the Buck Converter. The microcontroller will adjust the output voltage of the buck converter to a constant 14V through the duty cycle setting of the PWM signal sent to the driver circuit. The 0utput voltage of buck converter is depended by duty cycle value. The duty cycle setting uses a PID controller based on the battery voltage detected by the voltage detector as feedback. PID controller is embedded on microcontroller. The driver circuit functions as an interface between the microcontroller and the buck converter which will match the output voltage of the microcontroller and the MOSFET on the buck converter of 12V.

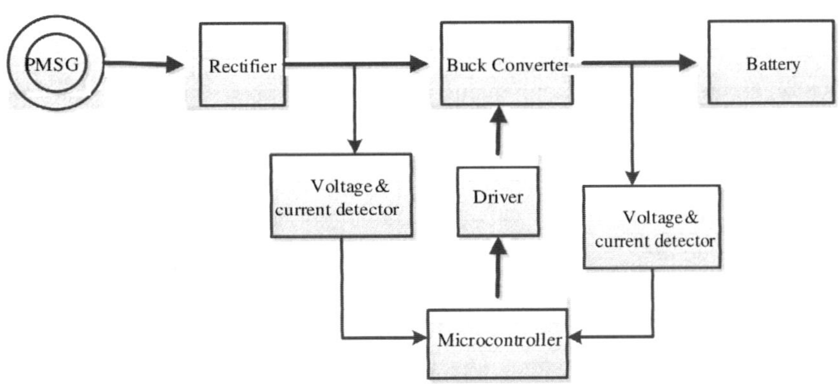

Figure 1. Block diagram of proposed variable speed wind turbine.

2.1. Design of voltage detector

Voltage detector will detect battery and rectifier voltage and generate output voltage that will send to microcontroller. Figure 2 shows voltage detector circuit that used voltage divide circuit.

IOP Conf. Series: Materials Science and Engineering 1073 (2021) 012048 doi:10.1088/1757-899X/1073/1/012048

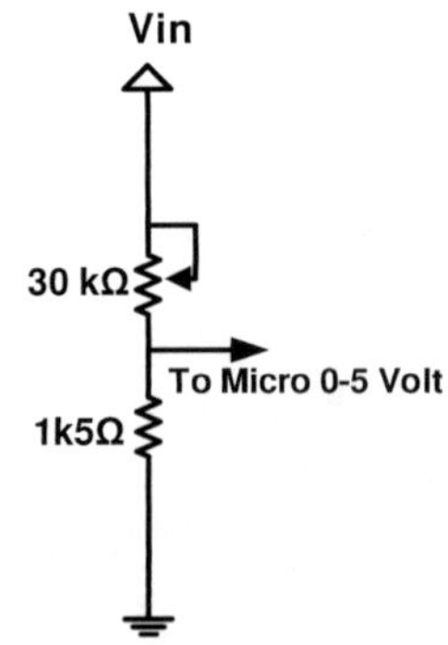

Figure 2. Voltage detector circuit.

The voltage detector circuit is used to detect the PMSG output voltage and the buck converter output voltage to be displayed on the LCD via the microcontroller or as feedback to the controller. The voltage sensor circuit uses a voltage divider system with two resistors connected in series to convert a maximum voltage of 100 Volts to 5 volts so that it can be read by the microcontroller via the ADC pin. The resistors used have maximum power of 0.5 Watt so that maximum current and voltage are 5 mA and 100 V respectively. The resistance value used can be determined by:

$$\frac{V_{out}}{V_{in}} = \frac{R1}{R1 + R2} \quad (1)$$

Where V_{out} is the output voltage of voltage detector with maximum value of 5 V. V_{in} is the output voltage of PMSG with maximum value of 100 V. R_1 and R_2 is resistance value in voltage detector circuit. The total resistance is 30 kΩ, so that R_1 and R_2 is 28.5 kΩ and 1.5 kΩ respectively.

2.2. Design of buck converter

Buck converter is a type of switching converter that produces an output voltage that is lower than the input voltage. The buck converter can be operated in 2 modes, namely continuous conduction mode and discontinuous conduction mode. In continuous conduction mode, the current will continue to flow through the inductor, or in other words, the current in the inductor never reaches zero. Whereas in discontinuous conduction mode, the current flowing through the inductor will be zero for a certain period of time. The inductor value selected will determine which mode to use. The buck converter circuit consists of a MOSFET as a switching component, inductor, capacitor, and diode, as shown in Figure 3.

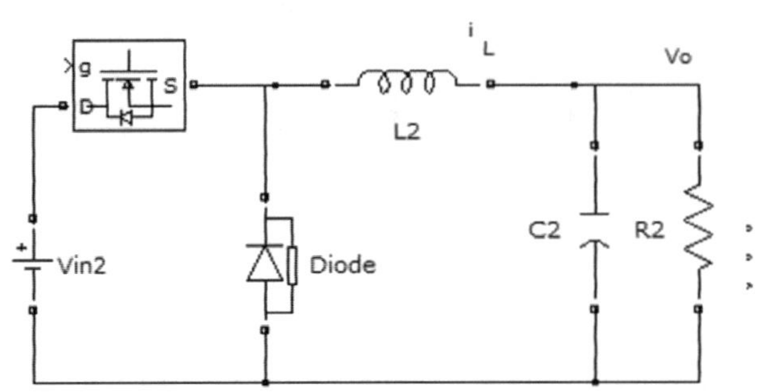

Figure 3. Buck converter topology.

In this article, the buck converter input voltage is the 3 phase rectifier output voltage. The buck converter circuit used is a modification of the basic buck converter circuit. The buck converter output voltage depends on the duty cycle of the PWM signal sent by the microcontroller to the Mosfet (M1). The lower the duty cycle, the lower the output voltage will be. The diode in Figure 3 is replaced by Mosfet 2 (M2) by providing a PWM signal with a duty cycle opposite to M1, as shown in Figure 4. When M1 gets a high pulse, M2 will get a low pulse. When M1 gets a high pulse and M2 gets a low pulse, the current will flow to the inductor and fill the capacitor. Otherwise, the inductor current will decrease and the input will be disconnected with the circuit.

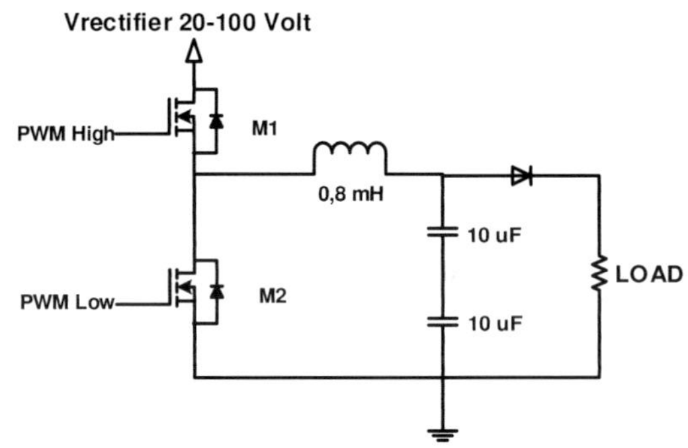

Figure 4. The Proposed buck converter.

The switching frequency of the buck converter is designed at 31.5 KHz, the output voltage ripple is 4% and the maximum output current is 10A. Value of inductor (L) and capasitor (C) can be determined by:

$$L = \frac{1}{f} \times (Vin - Vout) \times D \times \frac{1}{\Delta Vo} \quad (2)$$

$$C = \frac{\text{Vout} \times (1-\text{D})}{8 \times \text{L} \times \Delta\text{Vo} \times \text{f}^2} \quad (3)$$

Where f is switching frequency, D is duty cycle, and ΔV is voltage ripple. The inductor value used is 0.8 mH with a copper coil diameter of 0.9mm and two 10 μF capacitors are connected in series.

2.3. Design of PID controller

The PID controller is used to determine the duty cycle of the PWM signal sent to the buck converter driver circuit so that the buck converter output voltage can be maintained constant at 14V. The determination of the PID controller parameter value uses the Ziegler Nichols I method which can be determined by:

$$\text{Kp} = 1.2 \text{ x} \frac{T}{L} \quad (4)$$

$$\text{Ti} = 2 \text{ x L} \quad (5)$$

$$\text{Td} = 0.5 \text{ x L} \quad (6)$$

Where L is delay time and T is settling time from open loop response. Based on value Kp, Ti and Td, Ki and Kd can be determined. In this paper, the value of Kp is 0.6, Ki is 0.15 and Kd is 0.6.

3. Experiment results

The performance of the prototype variable speed wind turbine with voltage control was tested experimentally. Figure 5 (a) shows the measurement results of the PMSG output voltage at a rotating

IOP Conf. Series: Materials Science and Engineering 1073 (2021) 012048 doi:10.1088/1757-899X/1073/1/012048

speed of 800 Rpm. The PMSG output voltage is an AC voltage of 200Vpp. The greater the rotational speed, the greater the PMSG output voltage. The PMSG output voltage is rectified into a DC voltage signal by a 3 phase uncontrolled rectifier. The comparison of the measurements results of the rectifier and PMSG output voltage is shown in Figure 5 (b). Measurement of PMSG and rectifier output voltage used an oscilloscope.

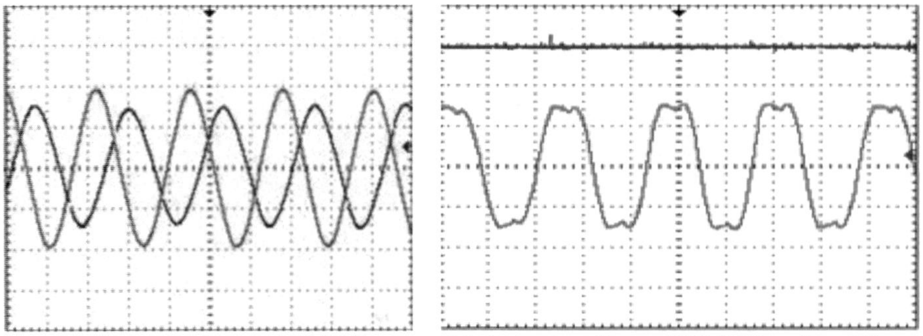

(a) Output voltage of PMSG (b) Output voltage of PMSG and rectifier

Figure 5. Measurement result of PMSG and rectifier output voltage.

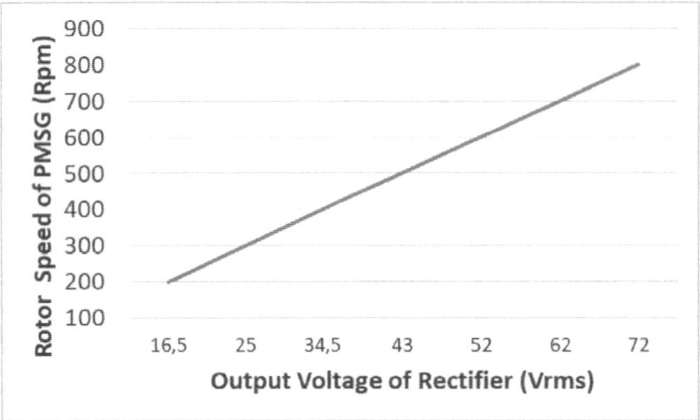

Figure 6. Relationship between rectifier output voltage and rotor speed of PMSG.

Figure 6 shows relationship between rectifier output voltage and PMSG rotor speed. The rectifier output voltage is proportional and linear to the rotor speed of the PMSG. The greater the rotation of the rotor, the greater the output voltage generated by the rectifier. Figure 7 shows measurement result from voltage detector with input voltage range between 5 V – 50 V. The average error in testing of the voltage sensor is 1.66%.

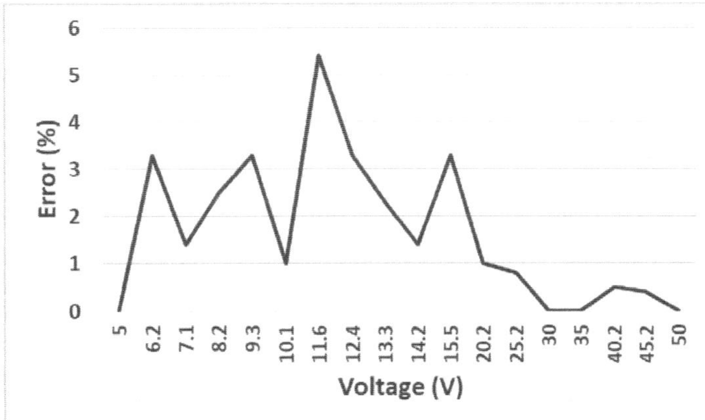

Figure 7. Measurement result of voltage detector.

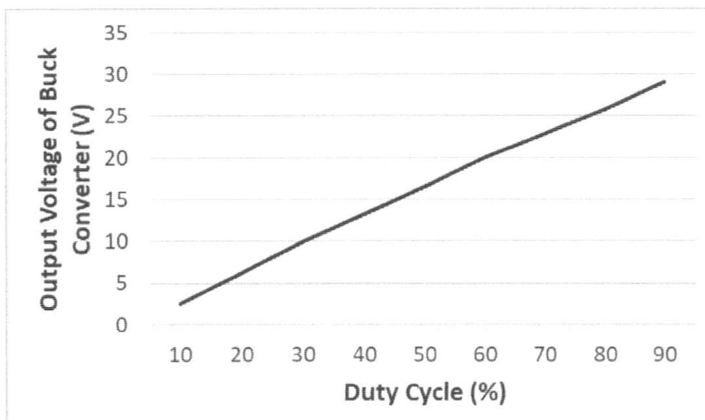

Figure 8. Measurement result of buck converter output voltage.

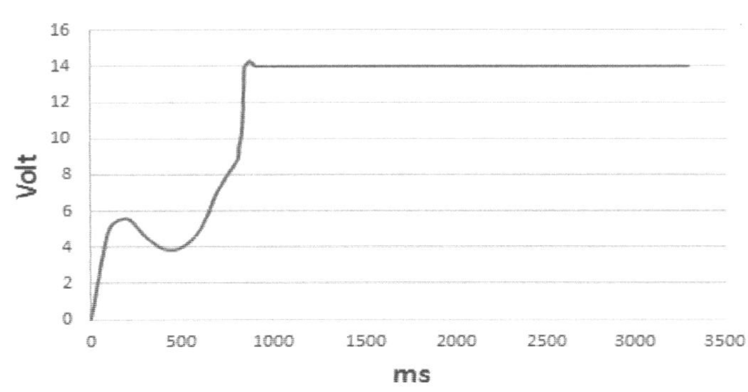

Figure 9. Output voltage of buck converter with PID controller.

Figure 8 shows the measurement results of the buck converter output voltage to changes in duty cycle. The greater the duty cycle, the greater the output voltage of the buck converter. The buck converter output voltage test is done by giving a change in duty cycle and the input voltage is kept constant at 37.5V. The PID controller can work well maintaining a constant voltage of 14V, as shown in Figure 9.

The system response has no maximum overshoot with a settling time of 1.2 s and a steady state error of 0.3%.

4. Conclusion

Voltage control for variable speed wind turbine using buck converter based on PID controller has been described in this paper. The prototype uses PMSG with a power rating of 500 Watt. The voltage generated by the wind turbine system is highly dependent on wind speed. The wind turbine system is connected to a battery, so to maintain a battery voltage of 14V, voltage control is needed. Voltage control using a PID controller embedded in the microcontroller. Based on the test results, the PID controller can work properly. With the controller parameter values Kp = 0.6, Ki = 0.15 and Kd = 0.6, the resulting transient response does not have a maximum overshoot with a settling time of 1.2 s and a steady-state error of 0.3%.

Acknowledgments

Authors wishing to acknowledge financial support from the Ministry of education and culture, Indonesia through funding for superior basic research of universities.

References

[1] Putri R I, Rifa'i M, Pujiantoro M, Priyadi A and Purnomo M H 2017 Fuzzy MPPT Controller For Small Scale Stand Alone PMSG Wind Turbine *ARPN J. Eng. Appl. Sci.* **12** 188–93

[2] Cheng M and Zhu Y 2014 The state of the art of wind energy conversion systems and technologies: A review *Energy Convers. Manag.* **88** 332–47

[3] Hui J and Jain P K 2012 Power management and control of a wind energy conversion system (WECS) with a fuzzy logic based maximum power point tracking (MPPT) *IECON 2012-38th Annual Conference on IEEE Industrial Electronics Society* (IEEE) pp 5966–71

[4] Xia Y, Ahmed K H and Williams B W 2012 Wind turbine power coefficient analysis of a new maximum power point tracking technique *IEEE Trans. Ind. Electron.* **60** 1122–32

[5] Rosyadi M, Muyeen S M, Takahashi R and Tamura J 2011 Transient stability enhancement of variable speed permanent magnet wind generator using adaptive pi-fuzzy controller *2011 IEEE Trondheim PowerTech* (IEEE) pp 1–6

[6] Tawfiq K B, Mansour A S, Ramadan H S, Becherif M and El-Kholy E E 2019 Wind energy conversion system topologies and converters: Comparative review *Energy Procedia* **162** 38–47

[7] Korai A, Denecke J, Torres J L R and Rakhshani E 2019 New control approach for blackstart capability of full converter wind turbines with direct voltage control *2019 IEEE Milan PowerTech* (IEEE) pp 1–6

[8] Sejati P, Suzuki H and Yasuno T 2016 Two-phase buck converter for battery charger of small scale wind turbine power system *2016 International Electronics Symposium (IES)* (IEEE) pp 244–7

[9] Patel R, Arora K and Katiyar S 2016 Control Analysis of PMSG Based Wind Energy Conversion System Using Buck-Boost Converter *2016 Second International Conference on Computational Intelligence & Communication Technology (CICT)* (IEEE) pp 395–402

[10] Suwarti, Mulyono and Wahyono 2017 Application of voltage controller with buck-boost converter model PLTB (the Gorlov turbine) to maximize power output *Int. Semin. Appl. Technol. Inf. Commun* Putri, R.I, Rifa'i M, Pujiantara M, Priyadi A, and Purnomo M.H, " Fuzzy MPPT Controller For Small Scale Stand Alone PMSG Wind Turbine", ARPN Journal of Engineering and Applied Sciences **12**(1)

IOP Conf. Series: Materials Science and Engineering 1073 (2021) 012049 doi:10.1088/1757-899X/1073/1/012049

Control plastic pellet feeding double extruders for 3D printing symmetric bilateral

B Setiawan[1], A M Damayanti[1,*], I Siradjuddin[1], W Widjanarko[2], D C Permatasari[1] and V U Audiana[1]

[1] Electrical Engineering Department, State Polytechnic of Malang, Indonesia
[2] Mechanical Engineering Department, State Polytechnic of Malang, Indonesia

*annisa_maulidia@polinema.ac.id

Abstract. Three Dimensional Printing (3DP) is one of the technologies that support industry 4.0. One of the common 3DP technology is FFF (fused filament Fabrication), the technology is also known as Fused Deposition Modelling (FDM). In this case, extrusion thermoplastic through a hot nozzle with certain melting temperature. The extrusion machine (extruder) will melt the plastic pellet and push the pellet through a heating zone that has a different temperature and pushed out by a screw to the nozzle to be printed into objects. This research aims, to make a control innovation for plastic pellet feeding double extruders. Those extruders for a bilateral 3DP symmetrically to print an object with a size of 2000x2000x2000mm using FDM method. The research simulation results showed that the extruder melts the plastic pellets with a temperature of melting is 200°C, temperature controlled by PID. In this research using 4mm plastic pellets diameter polyethylene high density (HDPE). The double extruders is controlled by two stepper motors Y- and Y+ axis with speed 70mm/s, on 2mm nozzle diameter. The extruder can be produce 110mm melted plastic filaments in 250 steps.

1. Introduction

Three Dimensional Printing (3DP) technology is one of the technologies that revolutionize the design and manufacturing of a new product [1]. The advantage of using 3D printers to create prototyping is that it can create prototypes in a short time and cost-effective comparison of conventional prototypes. This rapid prototyping machine becomes a vital tool in the industrial world. But for industries in Indonesia have not been used because the price of the machine is relatively expensive for developing industries in Indonesia [2].

The most famous and inexpensive 3D Printing technology is FFF (fused filament Fabrication) Technology is also known as Fused Deposition Modelling (FDM), its working principle is thermoplastic extrusion through a hot nozzle at melting the plastic layer by layer [3]. Extruder (extrusion machine) has a function to melt the plastic pellets and then be processed through the heating zone. In the extrusion process occurs transfer event (conveying) resin from one point to another using screw, then melting and pressing. Heat temperature comes from the coil mounted around the screw. Once the movement of the material towards the end, there is a pressure increase through the small holes while the screw is continuously pressing. To produce good Quality 3D Printing products must be set and controlled appropriately [4].

2. Literature review

2.1. 3D printing
Additive manufacturing or 3D Printing is a process in which the material is merged under the control of the computer to create a three-dimensional object, with materials added (such as liquid molecules or powder grains combined), 3DP used in prototype making. Objects can be either form or geometry and are usually generated using the digital data model of the 3D model derived File Manufacture additive (AMF). There are several techniques used, such as stereolithography (STL) or Fused Deposit Modelling (FDM). The printing builds a three-dimensional object of a CAD model or computer-aided AMF file, usually by adding layers by layers in succession [5].

2.2. CAD software
Computer-Aided Design is a computer program to draw a part of a product or a total product. The products that want to be depicted can be represented by lines as well as symbols that have certain meanings has become more important for virtual technology for advance [6].

Rapid Prototyping (RP) is a term used to describe a technology that can create three-dimensional objects in a single process directly through CAD. RP simplifies the complexity of the 3D manufacturing process by reproducing it into a series of the limited thickness of 2D shapes or layers and adding both together. RP technology is growing with varied methods [7].

2.3. Simplify 3D
Three dimensional (3D) software was used to digitally slice the samples and create the G-Code. G-Code is the basic language used to control CNC machines and it is the ultimate guide for CNC Programming. For 3D printers, G-Code contains the command to set the stepper motor movement. Create G-Code by Slicing 3D design files using the Simplify 3D app and storing them. The saved files will be converted to G-Code files and used by 3D printers as a reference for creating 3d objects [8].

2.4. Marlin firmware
Marlin firmware is used by the microcontroller and the test phase for controlling the actuators [9]. Marlin supports many designs of 3D printer platforms [10]. Marlin was made for 3D printer enthusiasts who used to be direct, reliable, and adaptable 3D printer firmware. Marlin is a firmware that can be run directly using an inexpensive 8-bit Atmel AVR microcontroller. Marlin version 2.x can be used on 32-bit microcontrollers. This chip sits at the center of the popular Arduino/Genuino open-source platform. The reference platform for Marlin is the Arduino Mega 2560 with RAMPS 1.4.

2.5. Extruders
The extruder is a process machine to carry out the extrusion process including mixing materials and printing. Extrusion is a combination treatment of pressure, friction, and temperature processes in a screw that moves at the same time. The extruders are used for the blown film, pipe processes, and blow molding where the discharge temperatures need [11]. Most of the 3D FFF Printers belong to the direct extruder type. In this type of extruder, the moving parts include the nozzle, heater, and motor, which attract and direct the filament from the cold end to the hot-end directly. For the Bowden type extruder, the driving part is hot-end and that is the only difference between the direct extruder and the Bowden [12]. Bowden extruder The FFF printer has a flexible cable connected between the separate sections. This cable is known as Bowden cable and is sometimes also referred to as a Teflon tube (depending on the material name). Thermoplastic filaments in the tube to obtain feedback from the moving material to the hot-end [13].

IOP Conf. Series: Materials Science and Engineering 1073 (2021) 012049 doi:10.1088/1757-899X/1073/1/012049

3. Methods

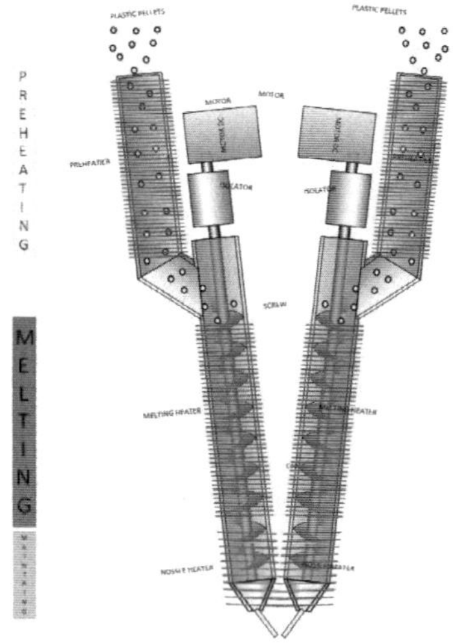

Figure 1. Mechanic design of double extruders.

Double extruder mechanical design using a screw with plastics pellets feeding method shown in Figure 1. The extruders process plastic pellet by melting the material applying some heaters. The melted plastic will be press to the nozzle using screw. And then, the melted plastic will be printed to form an object by 3D printing CNC. Figure 1 consists of two extruders. An extruder has an inlet/hopper as input for plastic pallet that will flow the pellet in to heated 200°C chamber of the extruder. Hopper is formed to accommodate plastic pallet for processing; screw to push the plastic pallets to flow into the nozzle; heater consists of a pre-heater, melting heater, and a nozzle heater for melting plastic pallets; and nozzle as output melted plastic whose diameter of melted plastic depends on the nozzle.

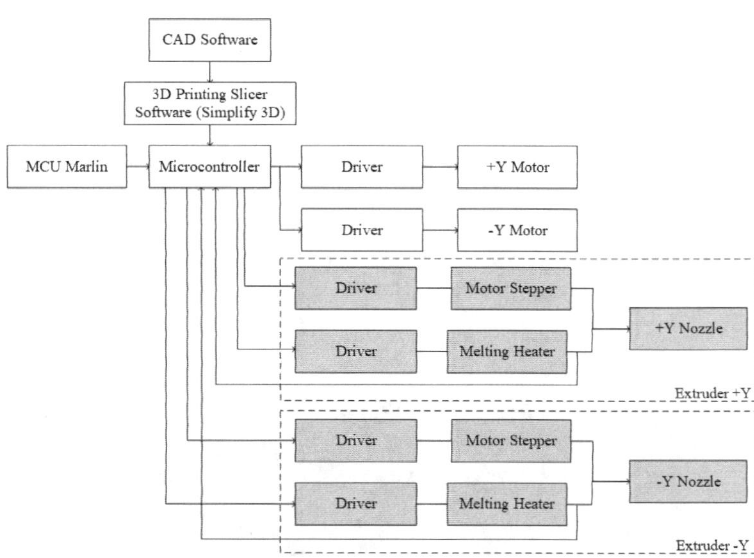

Figure 2. Double extruders block diagram.

Figure 2 shows block diagram how to print the 3DP object with double extruders. Create a design on the CAD software and then save and draw in the process of slicing in simplified 3D to generate G-code. Next, it's read by a microcontroller with marlin firmware to drive the +Y & -Y motor and +Y & -Y nozzle trough the command in the driver.

4. Simulation and results

4.1. Temperature Control for each extruder's heaters

Temperature control is used to stabilize the temperature in the melting heater with the set point.

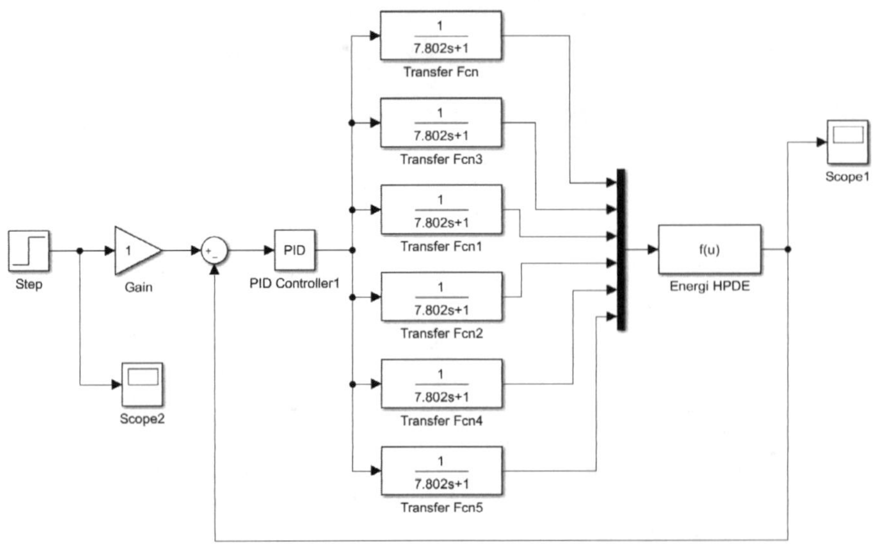

Figure 3. Temperature control simulation.

Figure 3 shows the temperature control simulation using PID control. Temperature control melting plastic pellets process using PID controller used to adjust the melting point of the 6 heaters, with a temperature sensor feedback. In the simulation, PID auto-tuning is used to obtain optimal control. From this simulation, temperature control was used to stabilize temperature on the melting plastic pellets process in each extruder. Plastic pellet using High-Density Polyethylene (HDPE), where the required melting temperature is 200°C. For the temperature control to be stable to produce melted plastic.

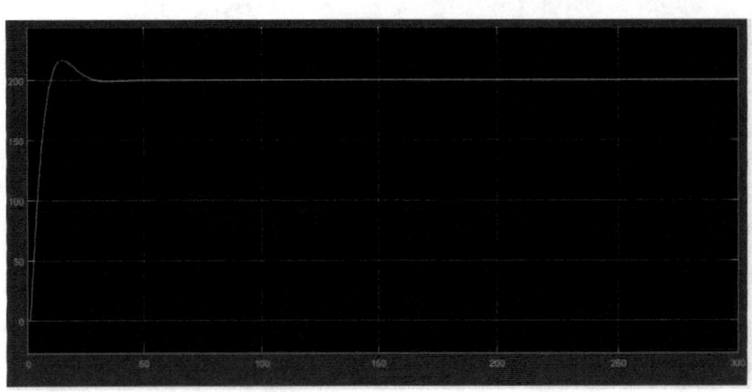

Figure 4. Melting temperature response.

IOP Conf. Series: Materials Science and Engineering 1073 (2021) 012049 doi:10.1088/1757-899X/1073/1/012049

From the temperature control simulation with HDPE plastic pellets, the resulting graph is shown in Figure 4, stable temperature control is 200°C and overshoot is 1%. The temperature control results have shown the desired stability.

4.2. Stepper Motor Control

Figure 5 shows a stepper motor simulation, a stepper motor speed setting of 70mm /s. In this simulation, the hybrid stepper motor must be adjusted according to the specifications of the motor to be used, motor Nema 23.

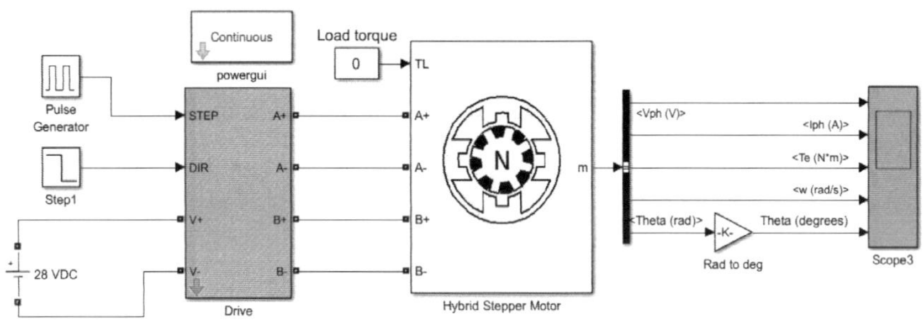

Figure 5. Stepper motor simulation.

From this simulation, the values of voltage phase, current phase, torque, speed, and angle of movement of the motor position shown in figure 6.

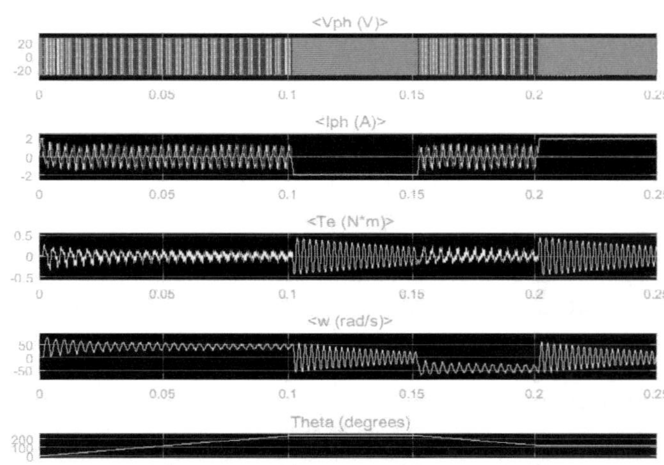

Figure 6. Stepper motor response.

From figure 6, it can be seen from the graph above that the system is optimal. Furthermore, it can be seen the simulation results by testing the system directly.

Stepper motor testing through the Pronterface application, by changing the value of "length", step/mm, and stepper motor input pulses. At pronterface, the speed stepper motor is fixed at 70 mm/s and the melted plastic length setting is 100mm, then changing the step/mm to see the actual discharge of the HDPE plastic pellets produced. This is intended to see the relations between motor pulses and the melting of HDPE plastic pellets. Shown in Table 1.

Table 1. Pulses of stepper motors with melted plastic pellets.

Step/mm	Time (s)	Length of Filament (mm)	Pulse	Frequency (Hz)
250	85	110	25000	289
220	94	98	22000	255.1
200	100	80	20000	232

Table 1 shows the relations between melting length HDPE plastic with motor pulses is directly proportional, the higher the motor pulse, the longer the HDPE plastic melt. And the higher the frequency, the longer the plastic melt.

Table 2. Stepper motor speed with filament length.

Stepper Speed (mm/s)	Time (s)	Length of Filament (mm)	Pulse	Frequency (Hz)
100	71	81	20000	331.1
90	82	83	20000	297.6
80	92	80	20000	265.2
70	99	84	20000	232

In table 2, the pulse is fixed at 200 pulses and change the speed of the stepper motor (mm/s) to see the actual length of the filament that comes out. This is to determine the relations between stepper motor speed and HDPE plastic melt length. From the table 2, it can be seen the lower speed of the stepper motor, the shorter the results of the melting filament.

5. Conclusion

From the simulation results temperature melting point plastic pellets is stable at 200°C, with an overshoot is 1%. And the higher the motor stepper steps, can produce more length plastic pellet. The extruder can produce 110mm melted plastic filaments in 250 steps.

References

[1] Noorani R 2017 *3D printing: technology, applications, and selection* CRC Press
[2] Dahlan M, Gunawan B and Hilyana F S 2017 Rancang Bangun Printer 3D Menggunakan Kontroller Arduino Mega 2560 *Prosiding SNATIF* 105-110
[3] Xiaoyong S, Liangcheng C, Honglin M, Peng G, Zhanwei B and Cheng L 2017 Experimental analysis of high temperature PEEK materials on 3D printing test. In *2017 9th International conf. on measuring technology and mechatronics automation (ICMTMA)* 13-16 IEEE
[4] Samikannu R and Poonamallie B A 2011 Design of synthetic optimizing neuro fuzzy temperature controller for dual screw profile plastic extruder using labview *Journal of Computer Science* **7**(5) 671-677
[5] Share E C Robert Bosch Tool Corp 2019 *Temperature control system for 3D printer extrusion head* U.S Patent 10 500 831
[6] Maiti C K 2017 *Introducing Technology Computer-Aided Design (TCAD): Fundamentals, Simulations, and Applications* CRC Press
[7] Cooper K 2001 *Rapid prototyping technology: selection and application* CRC press
[8] Sahputra B, Hakim R and Saputra I Studi Parameter Temperatur Nozzle Dan Base Plate pada Material ABS sebagai Bahan Baku 3D Printer Terhadap Hasil Kehalusan

Permukaan

[9] Quatrano A, De S, Rivera Z B and Guida D, 2017 Development and implementation of a control system for a retrofitted CNC machine by using Arduino. *FME Transactions* **45**(4) 565-571

[10] Moore S B, Glisson W B and Yampolskiy M 2017 Implications of malicious 3D printer firmware

[11] Spalding M A and Womer T W, 2017 Mitigating gels in polyethylene products produced using grooved-bore, single-screw extruders *SPE-ANTEC Tech Papers 63*

[12] Duong T H, Jaksic N I, DePalma J L, Ansaf B, Daniel D M, Armijo J and Galaviz M 2018 G-code Visualization and Editing Program for Inexpensive Metal 3D Printing *Procedia Manufacturing 17* 22-28

[13] Abilgaziyev A, Kulzhan T, Raissov N, Ali, M H, Match WK and Mir-Nasiri N 2015 Design and development of multi-nozzle extrusion system for 3D printer *International Conference on Informatics, Electronics & Vision (ICIEV)* 1-5 IEEE

IOP Conf. Series: Materials Science and Engineering 1073 (2021) 012050 doi:10.1088/1757-899X/1073/1/012050

Feedback remover on the sound system using inverse phase method

A Hariyadi*, R H Y Perdana, N Hidayati and Y Ratnawati

Electrical Engineering Department, State Polytechnic of Malang, Indonesia

*aadhariyadi@gmail.com

Abstract. Currently, almost all activities carried out by all parties are inseparable from the sound system. With the sound system, it is not necessary to speak loudly. There are several types of sound systems or models, ranging from the cheapest to the most expensive, with the usual quality up to good. However, there is a problem that can be obtained by a sound system that is of good quality or the usual quality, namely feedback from the speaker to the mic. Feedback will be very annoying because of the entry of sound from the speaker to the mic, which should be avoided. By using the inverse phase method and narrow bandpass filter, the simulation results can eliminate the feedback signal that is 100%. However, there are side effects on the band that occurs feedback, which is the reduction in the sound signal by the feedback power level that occurs between 30% - 100% that are in the frequency band range.

1. Introduction

Sound processing systems, or better known as sound systems, have become a device inherent in humans today, starting from television, mobile phones, home theatre etc. Large-scale sound systems are used when there are events such as weddings, meetings, meetings and so on. However, sometimes there is a small but very annoying problem, namely the emergence of feedback. Feedback occurs due to incorrect placement of sound equipment or improper sound system settings. So that the waves emitted by the loudspeaker go back into the microphone and enter back into the sound system, it is repeated continuously so that the signal will get bigger. With the feedback signal getting bigger, it will disturb the listener even more. Currently, the process of eliminating the feedback signal is still using manual methods, namely by finding the frequency of the feedback signal and reducing it.

The latest technology in sound systems already uses digital signal processing [1,2]. With the implementation of digital signal processing, it will further improve the quality of the sound signal. The use of simulations in digital signal processing is carried out to simplify calculations [3]. By using the Fourier transformation, which is the basis of digital signal processing, the signal value from the frequency domain will be obtained [4]. The frequency domain is the basis for the filter circuit design. Filters that are often used in a sound system are low pass filters, bandpass filters and high pass filters. The old type sound system still uses analogue filters, while the new sound system uses digital filters [5]. The advantage of digital filters is that the processing of values is easier and more precise. Because the filter is digital, the data that is processed is also in the form of digital data with the processing hardware, namely FPGA [1,6]. In digital signal processing, there is a method for removing a signal, which is called the inverse phase. This method uses an input signal with the phase reversed and then added to the input signal again so that the output signal is an addition to the input signal with a reversed signal.

IOP Conf. Series: Materials Science and Engineering 1073 (2021) 012050 doi:10.1088/1757-899X/1073/1/012050

In this study, a simulation is proposed to eliminate feedback that occurs in a sound system, by using the inverse phase method and combined with the use of bank filters. Hopefully, it can eliminate the feedback signal accurately.

2. Methods

The design method used in this study uses six filters, consisting of 1 low pass filter, four bandpass filters, and one high pass filter. Filter manufacturing begins using the order 64, 126, 190 and 254. The system design in this study is shown in Figure 1.

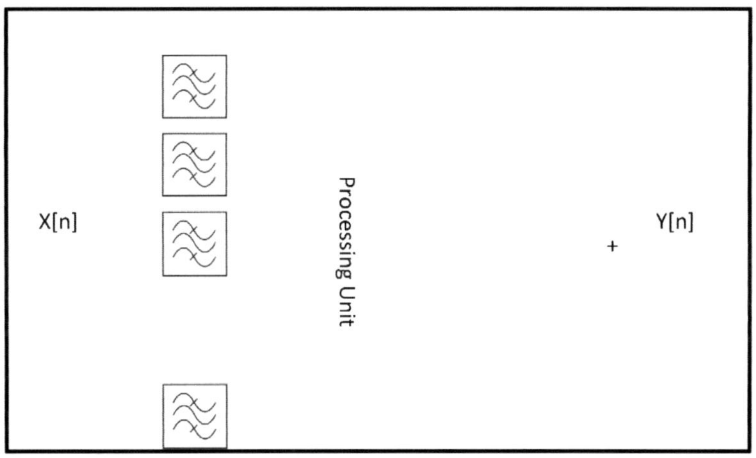

Figure 1. Design system.

Figure 1 shows that the x [n] input signal will enter the filter bank with a different cut off frequency. With a speech signal bandwidth of 20-20KHz, the bandwidth used in the filter design is 2KHz wide. Then the signal after passing through the filter series will enter the processing unit. In this section, the signal that contains feedback will be selected, and the phase will be reversed using the inverse phase method. At the end of the signal that has been reversed will be added again with the input signal. The mathematical process in this study is shown in formula 1.

$$Y[n] = X[n] + X_1[n]' \dots X_n[n]' \quad (1)$$

In the formula 1, Y [n] is the output of the system, with the input being X [n] and X1 [n] 'is the input signal from the first filter to Xn [n]' which is the input signal from the nth filter that has been an inverse phase. The filter design process used is shown in Figure 2. Figure 2 is a picture showing six bandpass filters arranged to form a filter bank.

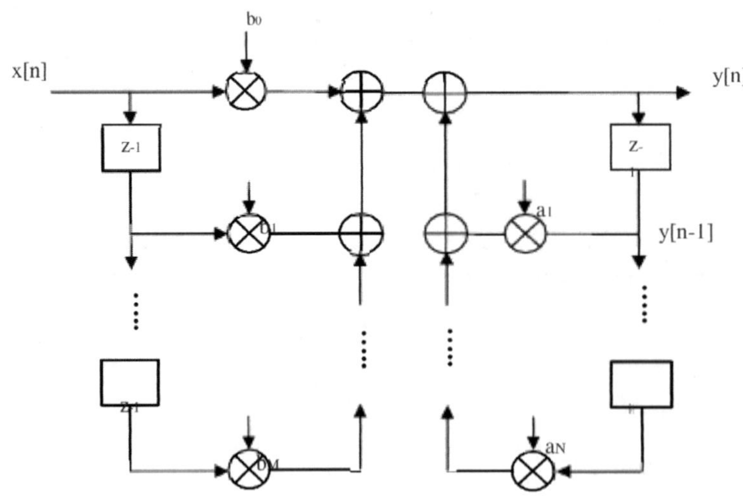

Figure 2. Bandpass filter bank.

3. Results and discussion

Filters are made with orders ranging from 64, 126, 190 and 254. The results of filter design are shown in Figure 3 (a) for filters with order 64, Figure 3 (b) for filters with order 126, Figure 4 (a) for filters with an order: order 190 and filter with order 254 in figure 4 (b).

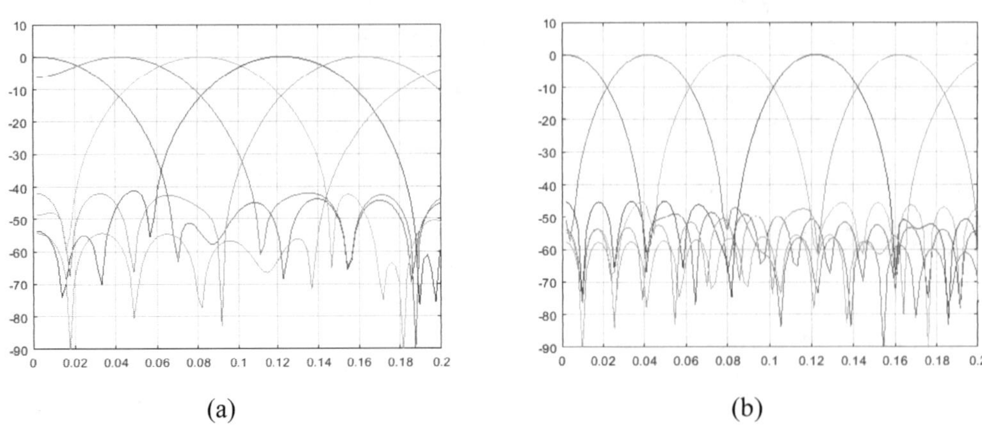

Figure 3. (a) Filter with order 64, (b) Filter with order 126.

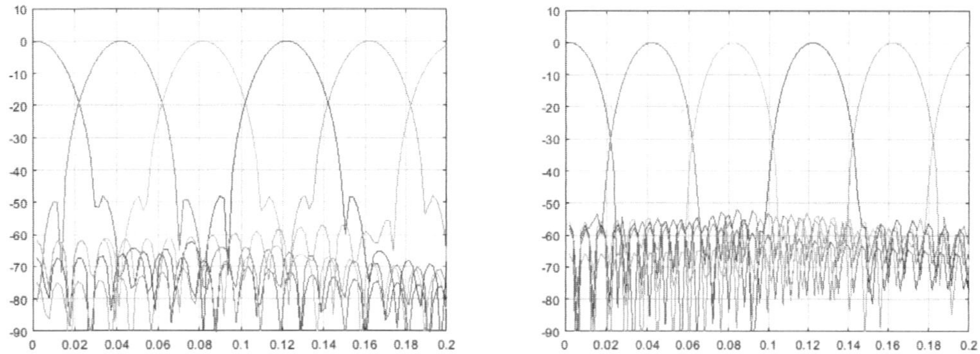

(a) (b)

Figure 4. (a) Filter with order 190, (b) Filter with order 254.

In a filter with a small order, it can be seen that the resulting filter response is not too sharp, but on orders above 190, the filter response is getting sharper. The process of designing and testing filters with higher orders will burden the processor because the mathematical process is very long and long. Filters are designed from 0Hz to 20KHz with six filter banks.

The system testing process is carried out by entering a speech signal that has been mixed with a feedback signal. Speech signals in the time domain and frequency domain are shown in Figures 5 (a) and (b). Furthermore, the feedback signal is shown in figures 6 (a) and (b).

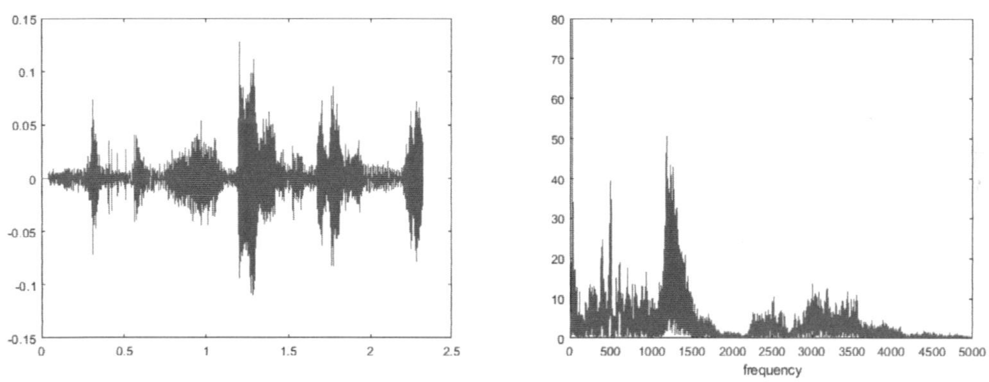

(a) (b)

Figure 5. Voice signal (a) time domain, (b) frequency domain.

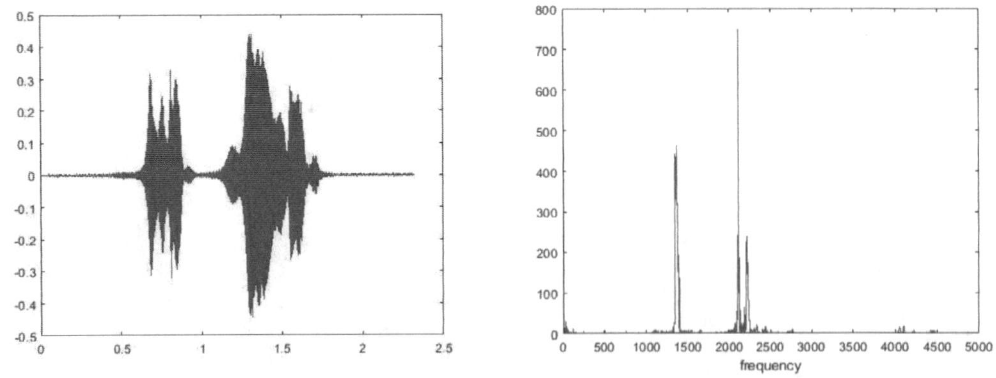

(a) (b)

Figure 6. Feedback signal (a) time domain, (b) frequency domain.

In Figure 5, the time domain signal has a duration of 2.4 seconds and has a frequency ranging from 0Hz to 5KHz. Whereas in Figure 6, the feedback signal has a duration of about 1.3 seconds with a frequency of 1.3KHz and 2.1KHz to 2.3KHz. The result of adding the two signals produces a new signal, as shown in Figure 7.

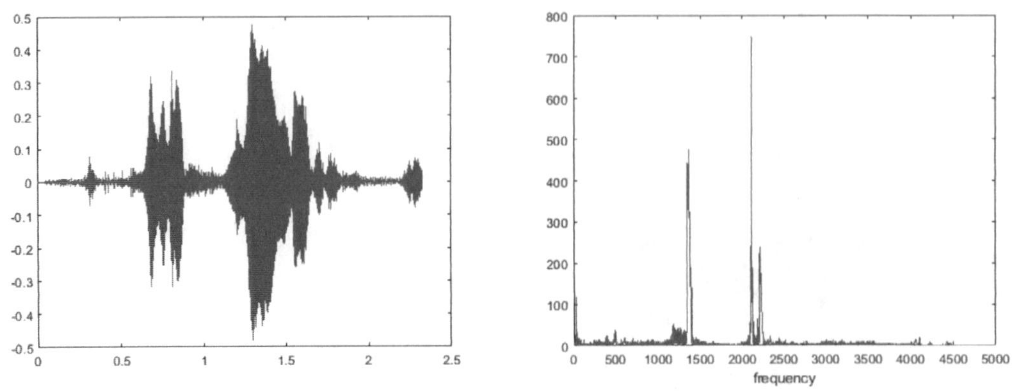

(a) (b)

Figure 7. Voice signal + Feedback signal (a) time domain, (b) frequency domain

It is shown in Figure 7, a speech signal that has added a feedback signal. The feedback signal has a very large amplitude compared to the speech signal so that in the frequency domain, it can be seen that the speech signal has a very small amplitude.

The results of the phase reversal process on the filtered signal can be seen in Figure 8.

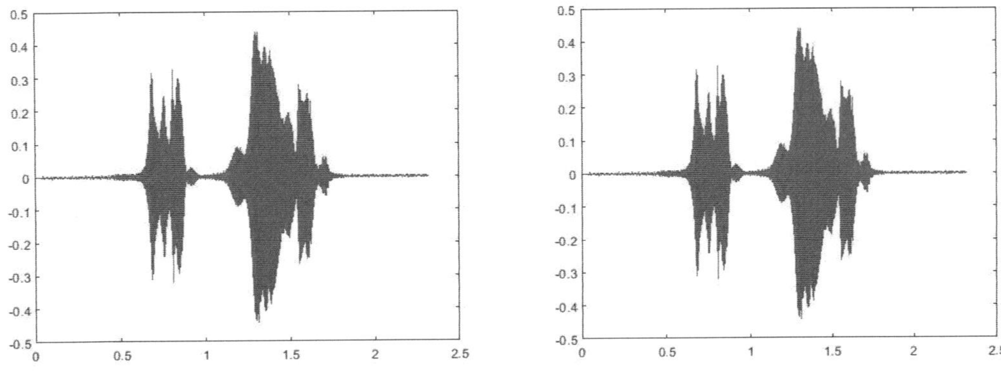

(a) (b)

Figure 8. (a) Feedback signal after filter bank, (b) Inverse signal.

In Figure 8 (a), a filtered feedback signal is shown then the phase of the signal is inverted to produce the signal shown in Figure 8 (b). not visible difference from the inverse signal or not, but when enlarged you will see that the signal phase has been reversed.

After obtaining the inverse signal, then adding to the original signal in the form of a speech signal is still mixed with the feedback signal, resulting in an output signal y [n] that has no feedback in it. The y[n] signal is shown in figure 9.

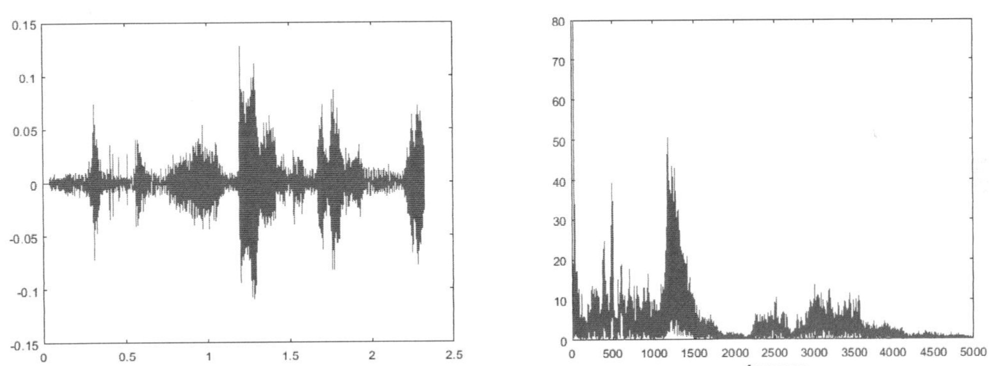

(a) (b)

Figure 9. y[n] signal (a) time domain, (b) frequency domain.

Figure 9 shows an output signal that has no feedback signal in it. It appears that the feedback signal is no longer visible in the signal in the frequency domain. However, the process of removing this feedback signal has a weakness. Namely, the original signal is also weakened at the frequency where the feedback signal is 30% -100% of the original signal.

4. Conclusion

From the experiments that have been done, it can be concluded that using the inverse method to eliminate feedback on the sound system is running very well. With a percentage of the accuracy, eliminating the feedback signal reaches 100%. However, based on the test, it was found that the speech signal attenuation is at the same frequency as the feedback signal, so that the speech signal has

IOP Conf. Series: Materials Science and Engineering 1073 (2021) 012050 doi:10.1088/1757-899X/1073/1/012050

attenuation by 30% -100%. Overall, the use of the inverse method is successful in eliminating the feedback contained in a sound system.

References

[1] Lv M, Gao T, Yan D and Qiao L 2020 Design of Audio Signal Analyzer Based on MCU and FPGA *J. Phys. Conf. Ser.* **1544**

[2] Ye Y, Wang K, Hu W, Li H, Yang K, Sun L and Chen Z 2019 A Wearable Vision-To-Audio Sensory Substitution Device for Blind Assistance and the Correlated Neural Substrates *J. Phys. Conf. Ser.* **1229**

[3] Zhou N, Quan X and Wang Y 2019 Research on Digital Signal Processing Experiment System Based on Virtual Instrument Technology *J. Phys. Conf. Ser.* **1288**

[4] Hidayat D, Syafei N S, Wibawa B M and Tumbelaka B Y 2018 Fourier transform of high frequency ultrasonic waves propagated with a transmission mode *J. Phys. Conf. Ser.* **1080**

[5] Thabit A A 2020 Design and Software Implementation of Analog and Digital Filters for Coherent Receivers under Noisy Channels *IOP Conference Series: Materials Science and Engineering* vol 765

[6] Xin W, Zhi-Qiang H, Da-Qing J, Zhi-Xiao S, Yu Z and Yong L 2019 Several Implementation Methods of Signal Processing Algorithm Based on FPGA *IOP Conference Series: Materials Science and Engineering* vol 565

IOP Conf. Series: Materials Science and Engineering 1073 (2021) 012051 doi:10.1088/1757-899X/1073/1/012051

Design and optimization of digitalization device of temperature control system using PID Ziegler-Nichols control in chemical engineering laboratory

C Wiharya*, H Sungkowo, A S Suryandari and B I Kurniawan

State Polytecnic of Malang, 9 Sukarno Hatta street, Malang

*chandra.wiharya@polinema.ac.id

Abstract. The Temperature Control Module is a tool that is useful for adjusting the temperature of a liquid or solution so that the temperature of a liquid is as determined. A device that uses a heat exchanger as a place to combine temperature, so that the temperature of a liquid with another liquid will meet there so that the temperature of the liquid in the heat exchanger can be changed. From the temperature control practicum module prototype in the chemical laboratory, it is known that in the open loop test with hot water input temperature above 90 °C and cold water input temperature 28 °C an output value of 69 °C is obtained. This value is stable but slowly decreases. By providing feedback and the controller in the form of sensors and data will be processed by the PLC, then a value that corresponds to a set of values is obtained, namely 60 °C, but the output temperature is unstable because the output graph is still fluctuating and unstable. Based on experimental calculations and PID calculations using the Ziegler Nichols method, the Kp = 30, Ti = 77.95, and Td = 19.489 were obtained. By providing the PID algorithm calculated by the Ziegler Nichols method, it can increase the output temperature to 58 °C. This is due to the addition of the PID gain to be more substantial so that the system becomes unstable.

1. Introduction

The electric power control system is used as a remote control. Arrangements are usually made in the SCADA control room. Many things can be controlled in the field of learning such as motors, valves and other electrical equipment. One example of the use of control systems in field learning is found in the Chemistry Laboratory of State Polytechnic of Malang AQ in the student practicum module. Examples of practicum modules are pH control, valve control, temperature control and others. The practicum module is very important because it supports the learning process and trains students' skills.

The problem in the practicum module is that the practicum module used by practicum students is damaged so that teachers and students cannot use the module and interfere with the learning process. Proportional Integral Derivative (PID) is a feedback mechanism controller that is usually used in industrial control systems. Output the PID controller is the sum of the outputs proportional controller, integral controller output.The characteristics of the PID controller are strongly influenced by the large contribution of the three parameters P, I and D [1]. In the Temperature Control prototype module you can use PID to calculate the error value as the difference between the desired set point and the measured process variable so that by determining the PID value the error value can be reduced.

2. Literature review

2.1. Heat transfer and flow rate

2.1.1. Energy balance. The flow in the gap is completely closed, so that the energy equilibrium can be used to determine the varying fluid temperature and the total value of convection heat transfer Qconv depending on the mass flow rate [2]. If the changes in kinetic energy and potential energy are neglected, then the changes in heat energy and working fluid will have a significant effect. So that the energy balance depends on 3 variables, which can be formulated as follows [3-6].

$$Q = m_c \,.\, C_{p,c} . \Delta T_c = m_h \,.\, C_{p,h} . \Delta T_h \qquad (1)$$

Q = heat transfer rate (W)
m_c = mass flow through the tube (kg / s)
m_h = mass flow through the shell (kg / s)
C_p = coefficient of heat transfer (Kj / kg.K)
ΔT_c = temperature difference of cold fluid through the tube (oC)
ΔT_h = temperature difference of hot fluid through the shell (oC)

2.2. Control system

A control system is the process of regulating or controlling one or several plants to be within a certain price range. A factory in a control system is a set of equipment that may consist of only a few parts of a machine that work together, which are used to carry out a certain operation. In the application of the control system there is also interference in the form of signals that tend to have a detrimental effect on the system output price.

2.2.1. PID Ziegler Nichols. To get the appropriate PID value used several methods, including the *Auto Tuning* method on MATLAB software, *Ziegler Nichols* method and *try and error* method [7,8]. In previous studies, it was found that the fastest response to show temperature stability was the Ziegler and Nichols method [9]. Ziegler and Nichols suggest setting the Kp, Ti and Td parameter values based on the formula shown in Table 1.

Table 1. Ziegler Nichols adjustment rules based on strengthening the Kcr and the critical Pcr period (Second method).

Control Type	Kp	Ti	Td
P.	0.5 Kcr	~	0
PI	0.45 Kcr	$\frac{1}{1,2}\,Pcr$	0
PID	0.6 Kcr	0.5 Pcr	0.125 r

3. Methodology

3.1. Time and place of research

The research site is in the Control System Laboratory and Laboratory of Chemical Engineering in the State Polytechnic of Malang. The time of this research takes place in December 2019 to May 2020.

3.2. Object of research

In the process of working on this thesis, the object of research is the change in temperature at the valve output under uncontrolled or open loop conditions, valve control with PID and without PID.

3.3. Temperature control prototype testing flow chart

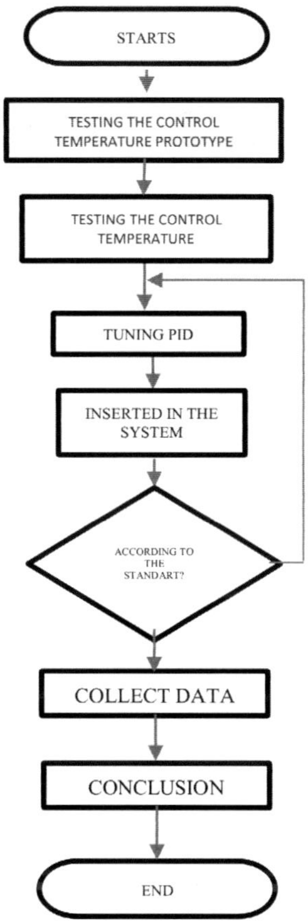

Figure 1. Temperature control prototype testing flow chart.

4. Discussion

4.1. Input, process, output parameters on temperature control

4.1.1. Input parameters. The component for input to Temperature Control is Thermocouple. So the parameter on the input component is temperature, the temperature will be detected by a thermocouple and then it will be converted to voltage so that it can be processed on the PLC. Values read on the PLC range from 0 to 5000. Input temperature used for hot water input is 90 °C and cold water temperature is around 28 °C if according to equation 2.1 a value of 59 °C is obtained for the output temperature if it is assumed that the mass and heat values considered 1.

4.1.2. Process parameters. The process component of Temperature Control is the Programmable Logic Controller (PLC), which is used as control of this prototype. PLC accepts the value of the input which is in the form of voltage and then used as decimal or hexadecimal numbers. This value will be processed using a ladder program. The value that is in the PLC can be sent in the form of current or voltage to activate the output according to user wiring [10].

4.1.3. Output parameters. The output component in the Temperature Control is a control valve, the valve is controlled by the voltage generated by the PLC, the signal voltage rating is 0-10V, the greater the voltage sent by the PLC, the greater the hole that is opened by this valve. For output components in the form of relays only logic 1 and 0 which means on or off, the PLC will send a signal in the form of a voltage to make the relay active on the digital output. This relay is useful for activating pumps that require a 220V supply so that relay contacts are needed so that the pump can be activated and controlled by a PLC.

4.2. Analysis of open loop, close loop and close loop testing using PID in the temperature control prototype

4.2.1. Open loop temperature control experiment analysis

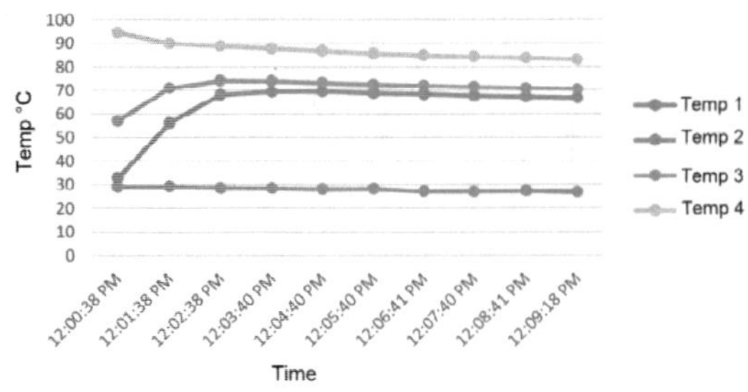

Figure 2. Open loop output input output graph.

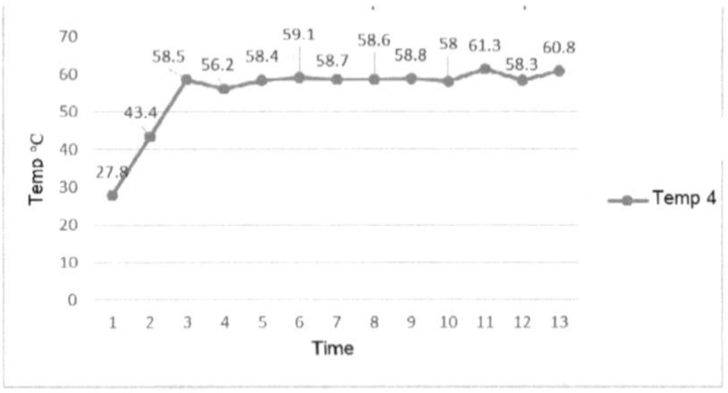

Figure 3. Graph of output close loop output without PID.

Based on Figure 2, the heat temperature used as input in the heat exchanger will continue to decrease from the original 94 °C to 83 °C. The temperature of the cold water input is initially 26 °C but after entering the heat exchanger the temperature will change to 60 °C. When the hot temperature for the input is valued at 90 °C, it can change the temperature of the cold water that was originally 28-26 °C to a temperature of 69 °C can even be 70 °C. The peak time needed to reach the peak value from the first overshoot is 3 minutes. The maximum peak value (maximum overshoot) is 69.2 °C, if formulated in a percentage of 6.3%. ESS of open loop testingi.e. 13.3%.

4.2.2. Close loop trial analysis without PID. Based on Figure 3 the temperature experiment data graph above the time needed to rise to reach the final value of the response which is about 1 minute (rise time). The peak time needed to reach the peak value from the first overshoot is 3 minutes. The maximum peak value (maximum overshoot) is 61.3 °C, if formulated in a percentage of 2.83%.ESS from close loop testing without PID which is 5%.

4.2.3. Close loop testing analysis using PID

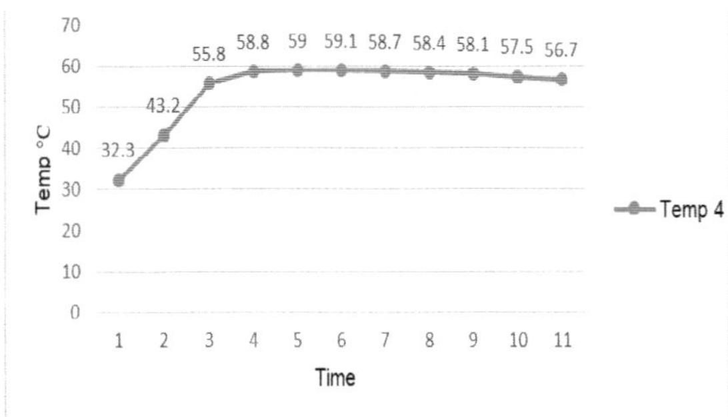

Figure 4. Graph of output close loop output using PID.

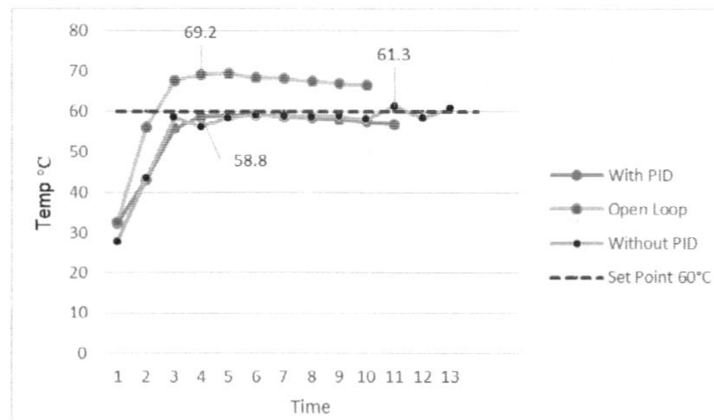

Figure 5. Test graph open loop, close loop without PID, close loop using PID.

Based on the experimental data chart table, the temperature of 30 °C will rise to near set value. The output temperature will be stable at 58 °C, using PID the output temperature cannot reach the set value that is 60 °C. Based on the table graph of temperature experiment data above the time needed to rise to reach the final value of the response which is about 2 minutes (rise time).ESS from close loop testing using PID which is 2.3%.

4.2.4. Graph analysis of open loop trial, close loop without PID, close loop using PID. Based on Figure 5, during the open loop experiment the output temperature will continue to rise to a temperature of 69.2 °C. When the close loop experiment without PID, the temperature fluctuates so that it exceeds the setpoint value of 61.3 °C. By using a PID system (with PID), the system becomes better because the resulting output temperature is more stable in the range of 58 °C with the specified setpoint is 60 °C, not only that if the system is given a PID, the steady state error is smaller than the other experiment is 2.3%. If the system is not given a PID (without PID) configuration, the output value can exceed the setpoint of 61.3 °C or experience an overshoot of 2.83%. When the system is given a PID, the output value is stable close to the setpoint value and does not experience overshoot. The time to output reaches a 60 °C setpoint value is almost the same, that is, it takes about 3 to 4 minutes to reach the 60 °C setpoint.

4.3. PID tuning with the Ziegler Nichols method

To determine the value of PID in a system whose equation is unknown, testing is needed to get the value of the equation in the system then a multiple method approach is used to obtain the appropriate PID value. Tests carried out on a prototype system Temperature Control by providing a variable temperature input then see the response from the heat exchanger output. Here are the results of calculating the PID value based on testing.

$Kp = 0.02418$
$Ki = 0.0031$
$Kd = 0.0124$

Adding the Kp value to 30 makes the system more stable.

Figure 6. Response graph using the Ziegler Nichols method before and after the addition of Kp 30.

5. Conclusion

Based on the formulation of the problem in research, it can be concluded that:

The Temperature Control prototype consists of input, control, and output components, each component has a parameter, for the input component that is changing the temperature to DC voltage, then the voltage is processed into hexadecimal or decimal numbers so that it can be processed by the PLC, after the number is processed then the number The DC voltage will be made again by 0-10 V so that it can be used to activate the PLC output. The temperature value of the heatexchanger temperature is calculated based on 59 °C using hot water input temperature of 90 °C and cold water of 28 °C.

Based on experimental calculations and PID calculations using the Ziegler Nichols method, the values of Kp = 30 Ti = 77.95 and Td = 19.448 were obtained. By entering the PID value, the system will run more stable than not using the PID algorithm.

Based on the three experiments, namely Open Loop, Close Loop without PID, Close Loop with PID, the output results obtained are as follows:

- When the Open Loop test uses hot water input temperatures of 90 °C and cold water input temperatures of 28 °C, an output value of 69 °C is obtained, this value exceeds the setpoint of 60 °C, with a time of about 4 minutes. mthe output temperature is overshooted by 6.3%. In addition, the Steady State Error value is also large at 13.3%.
- When testing the Close Loop without PID using a hot water input temperature of 90 °C and cold water input temperature of 28 °C, an output value of up to 60 °C is obtained, this value has reached a setpoint of 60 °C but the temperature is still less stable due to an increase temperature so that it experiences an overshoot of 2.83%. In addition the value of Steady State Error is also large at 5%.
- When testing the Close Loop using PID using a hot water input temperature of 90 °C and cold water input temperature of 28 °C, an output value of 58 °C is obtained. Steady State error when using PID is 2.3%, and does not experience overshoot like an experiment without PID.
- If giving a PID to the system will make the system more stable, reduce Steady State Error by 2.7% and also eliminate overshoot.

The SCADA design on the Temperature Control prototype was created using Cx-Supervisor software. In this SCADA there are several menus such as monitoring that function to monitor which components are active and also see the temperature in the system. Apart from SCADA monitoring, it also has a controller feature that functions to run programs, enable or disable pumps and PID configuration. There is also a trend graph that functions to see the increase or decrease in temperature in the system and there is a menu for storing temperature data so that the data will be stored in Microsoft Excel in the form of a table.

References

[1] Arindya R 2017 PID Control Tuning for Process Controllers *Journal of Electrical Engineering Mercu Buana University* **8**(2) 109-110

[2] Stoecker W F and Jones J W 1982 *Refrigeration and Air Conditioning* (New York: McGraw Hill)

[3] Cengel Y A 2003 *Heat Transfer: A Practical Approach Second Edition* (New York: McGraw Hill)
[4] Holman J P 1997 *Heat Transfer, language transfer: Jasjfi* E, Ir. M.Sc. Sixth Edition (Jakarta: Erlangga)
[5] Incropera F P 1996 *Foundational Heat and Transfer* (Canada: John Wiley and Sons)
[6] Kuppan T 2000 *Heat Exchanger Design Handbook* (New York: Maercel Deker Inc.)
[7] Ogata K 1995 *Automatic Control Techniques* E. Leksono, Penerj. (Bandung: Erlangga) vol 1
[8] Ogata K 1995 *Modern Control Engineering* E. Leksono, Penerj. (Bandung: Erlangga) vol 2
[9] Firmansyah R D, Wahyunggoro O and Cahyadi A I 2015 Sistem Kendali Suhu menggunakan PID Ziegler-Nichols dan Fuzzy Logic *CITEE 2015*
[10] Aji W S and Junaidi A 2017 Room Temperature Control System With PID Algorithm Using PLC OMRON CP1E-NA20DR-A *Journal of Electrical Engineering Study Program Ahmad Dahlan University* **3**(2) 47-54

Microcontroller-based control system design using step counter method in drilling fixture

A Faizin*, H Wicaksono and N P Sari

Department of Mechanical Engineering, State Polytechnic of Malang, Jl. Soekarno Hatta, No. 9, Malang 65141, Indonesia

*akhmad.faizin@polinema.ac.id

Abstract. Automation in the manufacturing process in the industry today is absolutely needed. Advances in technology has changed the characteristics of the manufacturing industry. The development of robotics, visualization technology and collaborative automation has increased the development of the manufacturing industry, enabling automation to be applied not only in the mass production process but also in low volume production. This is done to pursue the benefits of investing in industrial automation machines: productivity, accuracy, safety, cost savings, quality improvement, and repeatability. Industrial automation is the main key in following the development of technology into industry 4.0. This study aims to propose a new pneumatic system using the microcontroller by implementing the step counter methods. In this research, it is used to control a drilling fixture that has 3 (three) DACs. The Arduino microcontroller system used in this study to perform the step counter methods. Arduino type Mega 2560 control system has 10 inputs and 10 outputs for the 5/2 directional control valve (5/2 directional control valve – DCV. Through the application of this method, the preparation of microcontroller-based control circuits can be done in a structured and easier way, both the steps of preparation, control, and troubleshooting.

1. Introduction

Manufacturing automation processes in many industries use pneumatic systems. The pneumatic system is a system that produces work using compressed air media. The examples of manufacturing systems are flanging fixtures, drilling fixtures, etc. The use of jigs and fixtures in the manufacturing process can be controlled by various types of control systems, such as pneumatics, electro pneumatics, PLC, and microcontroller-based control systems. With compressed air, it can be used to drive and control the manufacturing system automatically [1]. The composition of the pneumatic system consists of compressor, air service unit, the control circuit, and actuators. As a pressure generating unit, the compressor sucks air from the atmosphere through the suction filter and compresses it. The increased temperature of the air due to compression is cooled by the cooling unit. Then the compressed air flows into the storage tank, which is equipped with a condensate drain, a manometer, and a pressure limiting valve. Through a series of piping, compressed air is channeled to the distributor for use. Then the compressed air is passed through the air service unit, to be filtered, regulated by pressure, and lubricated with atomized oil. Finally, the compressed air is passed valve 2/2 directional control valve (2/2-DCV), which serves as a channel open/close (on/off) to the pneumatic circuit. Pneumatic circuit is divided into signal elements, control elements, and driving elements. The signal element is usually used 3/2-DCV. This valve functions to control the direction of flow which can be activated by spring, manually, mechanically, pneumatically, or electrically. This valve can be adjusted as normally close/open [2]. The control element functions to control the actuator's movement forward/backward or rotating left/right, usually using 4/2-DCV, 4/3-DCV, 5/2-DCV, or 5/3-DCV. In the use of pneumatic systems in industry, the arrangement of these valves into a pneumatic control circuit design, is a very important step. Furthermore, compressed air is used to drive actuators in the form of a single acting cylinder (SAC), double acting cylinder (DAC), or a pneumatic motor [3]. To control the movement, a control element is used, which functions to regulate the flow rate of the air flow into/out of the actuator, so that it can affect the speed of forward/backward or left/right rotation. This control element is a flow control valve combined with a one-way valve. This valve combination is called a one-way flow control adjustable valve or throttle valve. The installation of this valve uses a metering-out system. This method can be used to adjust the actuator's speed [4].

This study aims to implement microcontroller-based control system in a pneumatic drilling fixture. This study offers a new control approach using the step counter method on the microcontroller.

2. Literature review

2.1. Pneumatic based control

In its application, the pneumatic system can be used to carry out the manufacturing process, as a tool for the production process, as well as material handling equipment. Some examples of pneumatic system applications include flanging fixture, drilling fixture, milling fixture, and material handling equipment. The automation process in drilling machines involves various pneumatic devices, pneumatic systems, and some electrical and electronic devices. By implementing automation strategy, its resulting in reduced cycle time, costs, improved product quality and increase in productivity [5]. By considering a particular task of drilling holes, it is found from the present investigation that by the process of conversion the output rate increases four times as compared to that of conventional type of machines [6].

2.2. Step counter method

In principle, the step in designing a pneumatic control circuit using the step counter method is to arrange the power line (pressure line) according to the number of steps. The power line is formed based on the output signal from the 3/2 DCV valve combined with the AND Gate and arranged in a row [7]. Figure 1 shows an example of the formation of a power line step counter method. In a circuit, usually the number of power line could be two (2), 3 (three), 4 (four) power line, and so on. The power line arrangement is numbered sequentially according to the step or step number. In the step counter method, a power line consists of 3 (three) input signals and 3 (three) output signals. The input signal is: Prepare (P), meaning that this signal is prepared from the previous step. Reset (R), meaning that this signal is reset from the next step. Confirm (C), meaning that this signal is confirmed by a confirming signal. The output signal is: Prepare (P), this means that this signal prepares from the next step. Reset (R), meaning that this signal resets from the previous step. Set (S), this means that this signal sets a specific work [8].

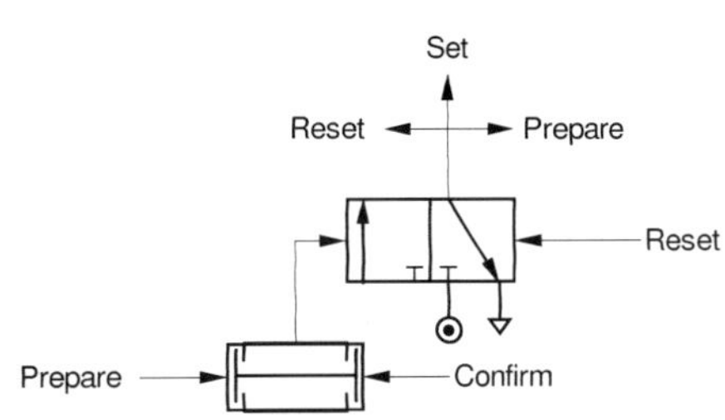

Figure 1. Step counter power line method.

3. Research methods

The steps taken in this study are:

- Literature Study of Drilling Machine: At this stage, preparation for activities such as literature search, collection of library sources, and journals is carried out. The collected literature and journals are studied and understood, in order to obtain the theoretical basis or necessary literature. Literature review conducted includes the process of strangulation and drilling.
- Design Step Counter Methods: Identification of actuator movements. Design of sensors at the location used. Making a step diagram. Arranging the power line as needed using the step counter method. System controlled pneumatic circuit depiction. The depiction of a series of controlling systems. The address arrangement in microcontroller adjusted in this step.
- Simulation Study: the simulation focused on the programming code. The step counter method translated into programming language. This step also includes the dry run test of the coding.
- Making a circuit: The arrangement of pneumatic components according to the pneumatic circuit drawn using the cascade and step counter methods. The completed circuit is connected to a power source. Movement test. Actuator speed settings as needed.
- State Diagram Check: Making a step diagram (state diagram) based on the pneumatic circuit that has been made. The resulting step diagram check is compared with the designed step diagram as needed. Troubleshooting, if there are problems.
- Conclusion: Conclusions are drawn based on the results obtained from the comparation between circuit design using the cascade method and the step counter method.

In the circuit pneumatic system can be divided into two main groups, the system is controlled and the control system. Figure 2(a) shows the application of the pneumatic system. This system has 3 DACs, which work sequentially A forward (A+), A backward (A-), B forward (B+), B backward (B-), C forward (C+), and C backward (C-). The work sequence is further illustrated by a step diagram in Figure 2 (b).

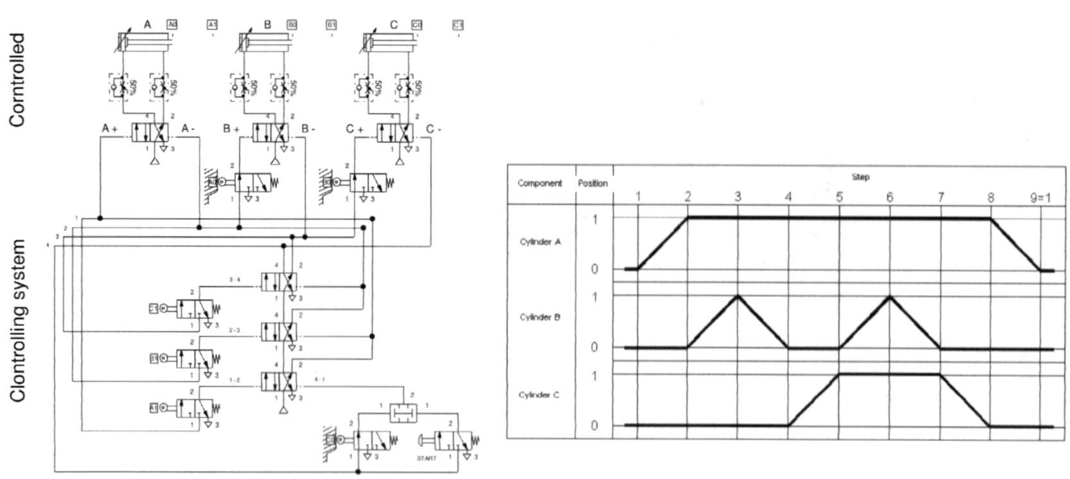

(a) (b)

Figure 2. (a) Pneumatic system (b) Step diagram.

In this study, the preparing of the circuit uses the step counter method that is commonly used in the preparation of pneumatic control circuits. Through the application of this method, the arrangement of relay circuits can be done in a structured and easier way, in the steps of preparation, control, and troubleshooting [9]. The number of relays used is very dependent on the amount of output available. In the drilling fixture used in this study, 6 relays are needed for the control circuit and 3 relays for driving the directional control valve (DCV) 5/2 directional control valve.

There were 8 steps in this installation, thus resulting in 8 block of step counter programs. In this case there are 8 steps: A+ B+ B- C+ B+ B- C- A-. Then the feedback signal diagram can be arranged. (cylinder A is moving forward, cylinder B is moving forward, cylinder B is moving backward, cylinder C is moving forward, cylinder B is moving forward, cylinder B is moving backward, cylinder C is moving backward, and cylinder A is moving backward).

4. Results and discussion

4.1. Design of control system

The design of the control system can be seen from address list in Figure 3, the relay module blocks is connected to digital pins of Arduino. Figure 4 shows the diagrams of sensor, and LCD.

INPUT			OUTPUT	
Description	Address		Address	Description
START	PIN52	**A**	PIN23	**Cylinder A forward**
RESET	PINRESET	**R**	PIN24	**Cylinder A backward**
a_0	PIN30	**D**	PIN25	**Cylinder B forward**
a_1	PIN31	**U**	PIN26	**Cylinder B backward**
b_0	PIN32	**I**	PIN27	**Cylinder C forward**
b_1	PIN33	**N**	PIN28	**Cylinder C backward**
c_0	PIN34	**O**		
c_1	PIN35			

PIN LCD	ADDRESS
PINRS	**PIN8**
PINE	**PIN9**
PIND4	**PIN4**
PIND5	**PIN5**
PIND6	**PIN6**
PIND7	**PIN7**

Figure 3. Input and output addressing.

Figure 3 depict the pin distribution as the installation guidance. Input a_0 to c_1 connected to the limit switches. The limit switches use as the sensor device that indicates the position of each piston. The **0** subscript indicates the piston in the starting position, meanwhile the **1**subscript indicates the piston in the main position (forward). The PIN23 – PIN28 used to control the relay that connected to the solenoid valve. The reset button directly connected to the reset pin in Arduino. This method simplifies the reset action in the coding, which the reset pin will trigger the reset command in Arduino [10]. The basic principle of designing a relay control system using step counter method is similar to the design of pneumatic circuit, namely in one work cycle divided into several steps according to the actuator step. Each step consists of 3 input signals (prepare, confirm, reset) and 3 output signals (prepare, confirm, set). In the control circuit there is always a START button to start the process and RESET to return the position of the actuator to its original position. In addition, another function of the RESET button is to turn off all STEP unless the last STEP is activated. To facilitate troubleshooting, in the relay assembly process, the input signal and the output signal for each STEP connected to each relay terminal must use the same sequence.

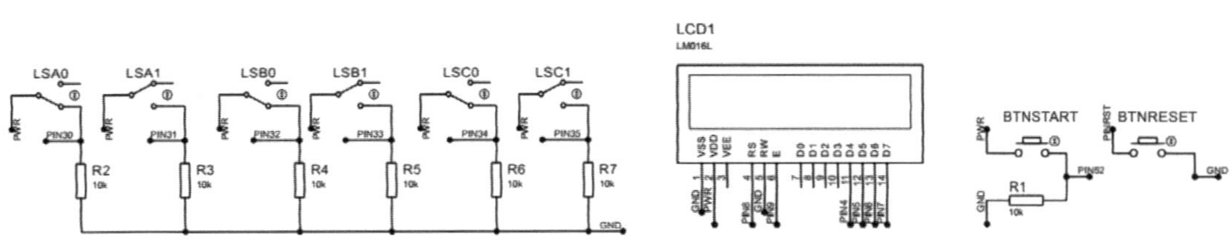

Figure 4. Input sensor and LCD schematic diagram.

The limit switches equipped with 10k Ohm resistor. The resistor is used to stabilize the input signal. In this installation the read command is using the pull up resistor method to enhance the accuracy and response of the signal. The start button also uses the pull up resistor with the same method.

IOP Conf. Series: Materials Science and Engineering 1073 (2021) 012052 doi:10.1088/1757-899X/1073/1/012052

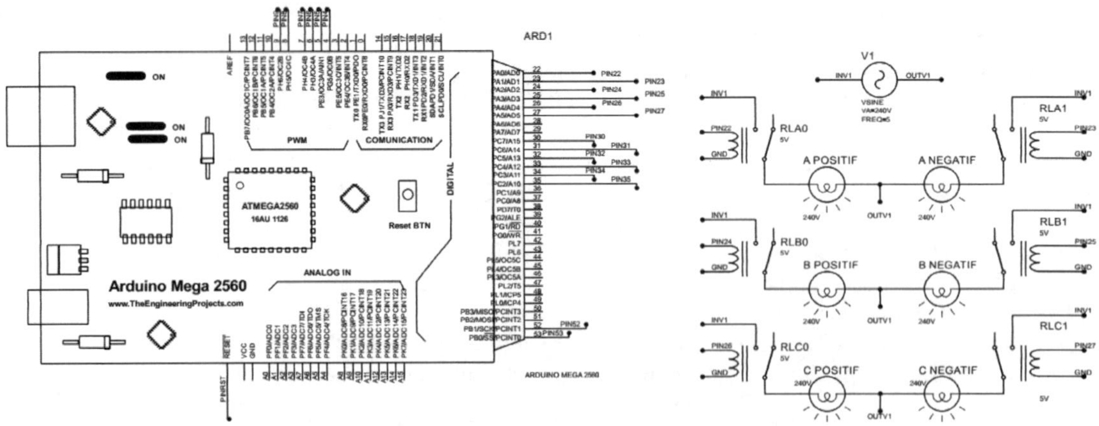

Figure 5. Output schematic diagram.

The relay schematics diagram as shown in the figure 5, is using the normally open type. When the output pin gives the signal, the relay will conduct the electricity. The current in this study is 220 V to electrify the solenoid valve. When the solenoid is charged, the magnetic field causes the solenoid rod to lift, moving the spool and allowing air to pass through to the one side of the piston chamber. This solenoid mechanism has a relatively quick response. This air flows moves the piston to the other side. Figure 6 depicts the Arduino Mega 2560 pin addressing. Each desired pin used to give the variable names in the program code.

4.2. Step counter method implementation

The algorithm in this study is as follows:

- Pressure area 1 is confirmed by a0 and Start and then move the Cylinder A forward.
- Pressure area 2 is confirmed by a1 and then move the Cylinder B forward.
- Pressure area 3 is confirmed by b1 and move the Cylinder B backward.
- Pressure area 4 is confirmed by b0 and move the Cylinder C forward.
- Pressure area 5 is confirmed by c1 and move the Cylinder B forward.
- Pressure area 6 is confirmed by b1 and move the Cylinder B backward.
- Pressure area 7 is confirmed by b0 and move the Cylinder C backward.
- Pressure area 8 is confirmed by c0 and move the Cylinder A backward.

These algorithms then encoded in the program using the counter variable as the pointer. In this program the “stepcounter” variable is beginning with 0 value. As seen in Figure 6(a) the initial program code shows when the start button is triggered the “stepcounter” variable changed in to 1 defines that the loop has started. In this installation the LCD will show the time different between each step. Thus, the millis() function is used to determine the initial time of the loop.

```
lcd.setCursor(0, 0);
  lcd.print("READY");

  if (digitalRead(btStart) == HIGH) {
    time = millis();
    stepcounter = 1;
    lcd.clear();
    lcd.setCursor(0, 0);
    lcd.print("STEP1 MULAI");
    Serial.println("Start");
    delay(500);
  }
```

(a)

```
if (stepcounter == 1) {
    Start = millis();
    while ((digitalRead(lsA1) == LOW)) {
      digitalWrite(rA0, HIGH);
    }
    digitalWrite(rA0, LOW);
    Stop = millis() - Start ;
    Serial.println("Proses Step 1 = ");
    Serial.print(Stop / 1000);
    Serial.println(" detik");
    lcd.clear();
    lcd.setCursor(0, 0);
    lcd.print("Step2");
    lcd.setCursor(0, 1);
    lcd.print(Stop / 1000);
    stepcounter = 2;
  }
```

(b)

Figure 6. (a) Initial stepcounter program code (b) Stepcounter code block.

In every step of the algorithm, the value of "stepcounter" variable will be changing depends on the step number as seen in Figure 6 (b). These block of code shows the lines of code that runs in every steps. The script begins with the millis() function to initialize the time counter. The "while" statement used to run the cylinder A move forward (relay A0) to be active until the limit switch in a_1 is giving the signal. Right after the piston activate the a_1 limit switch, the relay A0 will be inactive (LOW). The time calculation is beginning in the end of the step. This program also using the serial communication through the computer's port to log the data. The LCD shows the current time interval and the stepcounter state. After every data collected, the "stepcounter" variable is change into next step.

This step counter code block is implemented in each step, only the state number that makes it different. After the last step, the "stepcounter" variable will be continuing to be 1 again. Thus, makes these procedure runs continuously until the reset button is pressed. With the implementation of these code, the variation of the step can be changed easily by the simple adjustment of the coding. The time counter which displayed in serial monitor and LCD can help to detect the ineffective installation thus can be the parameter for the maintenance activity.

5. Conclusion

The conclusion in this study is as follows:

- The stepcounter method is an effective way to enhance the effectivity of a control system. This method can be widely use in every application to guarantee the desired algorithm. By implementing the stepcounter method in the drilling machine, the processes will be run automatically.
- Troubleshooting by using the microcontroller require less time than the other method. This is due to the replacement of controlling hardware that used in the other method, by only using the line code.
- Applying microcontroller in the drilling fixture is a unique method to simplify the hardware requirements. Various sensor can be implemented to promotes the maintenance activity. This resulting in the less cost of production. The microcontroller implementation can enhance adaptivity to the stepcounter algorithm changes.

References

[1] Ahsan Q, Grosvenor R and Prickett P 2006 A distributedon-line system for process plant monitoring *Proc. Of Mech.E, Part E.* **220** 61-77

[2] Prsic D, Fragassa C, Nedic N and Pavlovic A 2019 Describing function of the pneumatic flapper-nozzle valve *Mechanical Systems and Signal Processing* **124** 696-714

[3] EUGENIO BRUSA 2015 *Mechatronics Principles, Technologies, and Applications* (New York: Nova Publisher)

[4] Mahalik N P 2003 *Mechatronics Principles, Concepts, and Applications* (New Delhi: McGraw Hill)

[5] Kumar A, Baig M A A and Lathe R 2015 Automation of a Drilling Machine using Pneumatic Devices

[6] Vijaya V G 2013 Design and Fabrication of Sequencing Circuit with Single Double Acting Cylinder *International Journal of Innovative Research in Science, Engineering and Technology* **2**(8)

[7] Sovaiala G, Radoi R, Matache G and Pavel I 2015 Determining the Response of a Pneumatic System with Medium and High-Pressure Actuators to Step Signal *Hidraulica* (1) 46

[8] Lukman T, Godena G, Gray J, Heričko M and Strmčnik S 2013 Model-driven engineering of process control software–beyond device-centric abstractions *Control Engineering Practice* **21**(8) 1078-1096

[9] Papoutsidakis M, Chatzopoulos A and Tseles D 2018 Hydraulics and Pneumatics: A Brief Summary of their Operational Characteristics *Hydraulics and Pneumatics* **5**(10)

[10] Blum J 2019 *Exploring Arduino: tools and techniques for engineering wizardry* (John Wiley & Sons)

IOP Conf. Series: Materials Science and Engineering 1073 (2021) 012053 doi:10.1088/1757-899X/1073/1/012053

Design of flow control system based on PLC armfield pressure control module in chemical engineering laboratory

P H Suharti[1,*], H M Kristiana[2] and R N Amalia[2]

[1] Chemical Department, State Polytechnic of Malang, East Java, Indonesia
[2] Electrical Department, State Polytechnic of Malang, East Java, Indonesia

*profiyanti@polinema.ac.id

Abstract. A control system in industry has an important role to increase the effectiveness and efficiency in manufacture process. Laboratory of Chemical Engineering Department in State Polytechnic of Malang has several tools to develop competence in automation process control system, one of them is the pressure control module. This module aims to control the rate of compressed air flow coming out from the compressor. Nonetheless, the control system used is still analog. In order to improve the performance of module, rejuvenation has created by PLC that is optimized by PID controller method. This rejuvenation process is carried out in several stages, such as designing PLC hardware and software then testing and analyzing the system. There are four stages in testing and analyzing the system, both testing and analyzing the open loop system without tank and with tank, PID close loop with tank and without tank. Based on the results, this study indicated that the close loop system with PID has a better response than open loop system, which is able to improve the errors and transient responses so that the system is more stable and fit with set point value. The PID parameter that produces a good response when used in the pressure control module is Kp = 100 Ki = 1 and Kd = 0. This parameter is able to produce a system measurement value of 80% with error of 0%.

1. Introduction

Chemical Engineering Laboratory of State Polytechnic of Malang has several teaching modules, one of those modules is the pressure control module. The module is used as learning for Chemical Engineering students to figure out the flow rate of compressed air that comes out from the tank or without tank by applying the open loop and close loop systems (with feedback). Furthermore, analyse some parts of air pressure when conducting experiments using tanks. The problem of the system is the controller itself as the main control in the module. The controller is broke which cannot accept input from the sensor results or providing a set point signal, so the module cannot be used properly and data retrieval still uses a recorder. PLC (Program Logic Controller) is one of the technological developments in controller field. The use of PLC in the pressure control module plays an important role as the main control in the module, therefore when using PLC, users can use the module manually and automatically. To improve the working principal of PLC, the writer put PID as an extra controller [1]. PID control consists of P (Proportional), I (Integral) and D (Derevative) which aims to accelerate and produce the best system response. The pressure control module uses PID to calculate the error value as the desired set point difference [2]. From these problems and descriptions, a PLC-based controller was made and evaluated the comparison of Open Loop experiments using tank and Close Loop trials using tank or without tank on the compressed air control module that was upgraded by the Proportional, Integral, and Derevative algorithm (PID).

2. Methods

2.1. Schematic of process control system

Both models that contained in PCT 10 are manual and automatic mode. In relatively slow process, the controller is able to analyse the process which connected with controller, calculate the appropriate PID amount and set it automatically [3]. Such operations are referred to auto-tune or self-optimize and can be activated as follows. Set Point pressure is determined via HMI. From pressure value that has been read, the pressure data will be processed by controller then be used as a comparison with set point, which given the control results to the system [4]. The error value obtained by the controller is processed by the PID control output from the system and adjusted to the pressure set point value determined from the reading.

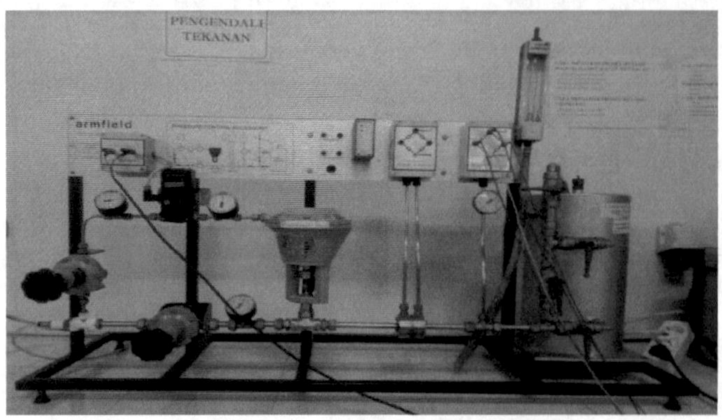

Figure 1. Pressure control module armfield PCT 14.

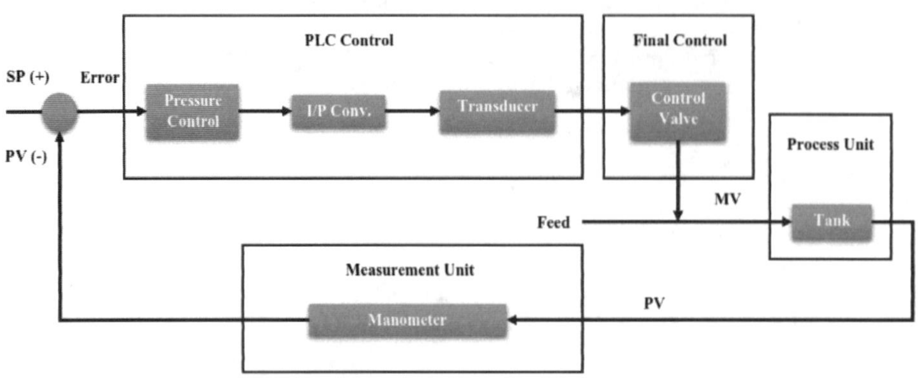

Figure 2. Process block diagram.

2.2. OMRON CP1E-NA20DR-A PLC wiring design

PLC is used as the main controller to control the holistic system. The Omron CP1E-NA20DR-A PLC type used is a power voltage configuration of 100 to 240 V AC, 12 inputs and 8 outputs. The I / O configuration of the Omron CP1E-NA20DR-A used is shown as in Figure 3 below. The flow of electricity from 220 voltage source flows to L1 PLC which functions as power from the PLC, then the neutral source is connected to the L2 PLC terminal as a neutral source.

The analog input from PLC is connected to transducer sensor, I IN0 is coupled with V IN0 and connected to sensor phase. Furthermore, COM 0 is connected to sensor neutral. The analog output and I Out 0 are connected to IP Converter phase and COM 0 is connected to the IP Converter. The START button enters the PLC input 01 address and the STOP button enters the PLC address 02 which connected to the PLC source.

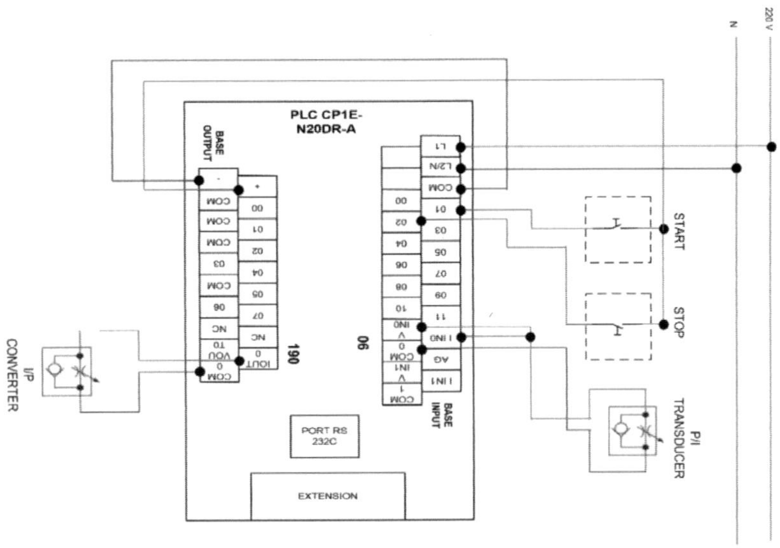

Figure 3. Wiring I/O PLC Omron CP1E-NS20DR-A.

2.3. *Omron CP1E-NA20DR-A PLC program design*
Program design on PLC Omron CP1E-NA20DR-A uses special software that user friendly program, CX Programmer. This software has been used to program the PLC Omron CP1E-NA20DR-A using ladder diagram programming language [5].

3. Analysis and testing
The analysis is carried out in order to determine whether the pressure module control system has been able to function as expected by testing the circuit tool. Testing tools in form of testing pressure module control system components, Open Loop testing and Close Loop testing with PID to prove the results of the design made.

3.1. *Testing of open control with tanks (Open Loop)*
The Open Loop test is carried out without providing feedback to the system. When the set value is entered, the response from the output will be analyzed.

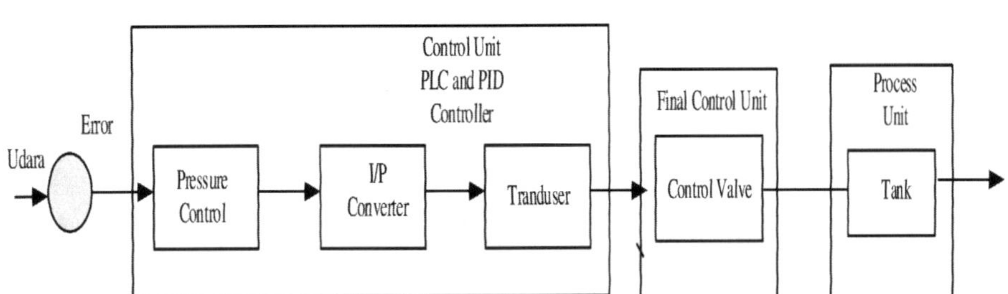

Figure 4. Block diagram of Open Loop system with tank.

3.1.1. Test results and Open Loop testing graphs using tank

Table 1. Test results with 0-100% PO tank.

%PO	%PV	P4(Psi)	P2(Psi)	% PVideal	Δ%Pvideal
0	100	8.5	3	100	0
10	97	8	4	90	7
20	86	7.7	5	80	6
30	80	7.1	6.2	70	10
40	67	6.6	7.5	60	7
50	52	6	8.7	50	2
60	37	5.1	10	40	3
70	22	4.2	11	30	8
80	6	3	12,3	20	8
90	2	1.5	13.5	10	8
100	0	0	15	0	0

3.1.2. Open Loop experiment analysis using tank. When conducting an open loop experiment on pressure control through the tank, a significant change according to response of graph will show briefly. The graph shows decrease gradually when the initial set point settings change to steady state until subsequent set point. This phenomena can be happen because the air pressure that flowing through the tank is compressed and slows down the output of air pressure that is released, so the pressure will flow out slowly [6].

3.2. Automatic Close Loop testing mode using PID with tank

This system is applied in automatic mode because this experiment observes the amount of response time, error on a given setpoint. The test is carried out as follows:

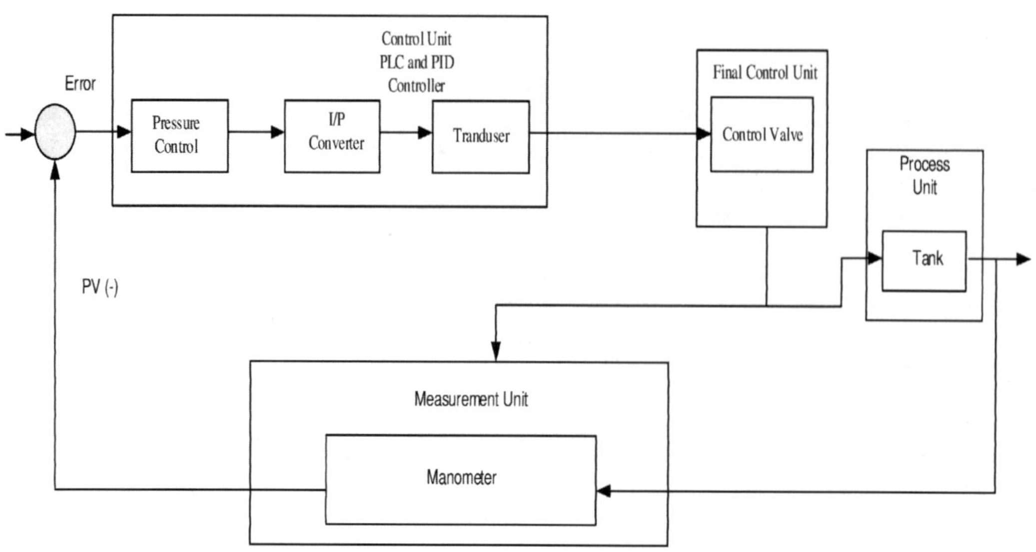

Figure 5. Close Loop circuit diagram with PID using a tank.

3.2.1. Test results and Close Loop testing graphs using a tank

Table 2. Close Loop testing using a tank with variable Kp.

Parameter	Kp = 80 Ki = 0	Kp = 90 Ki = 0	Kp = 100 Ki = 0	Kp = 110 Ki = 0
Risetime (ms)	50	48	40	40
Settling Time	0.02	0.02	0.02	0.02
Error Average (%)	2,5	2,5	2,5	2,5

Table 3. Close Loop testing using a tank with variable Ki.

Parameter	Kp = 10 Ki = 1	Kp = 100 Ki = 2	Kp = 100 Ki = 3	Kp = 100 Ki = 5
Risetime (ms)	50	52	50	50
Settling Time	0.02	0.02	0.02	0.02
Error Average (%)	0	0	0	1,25

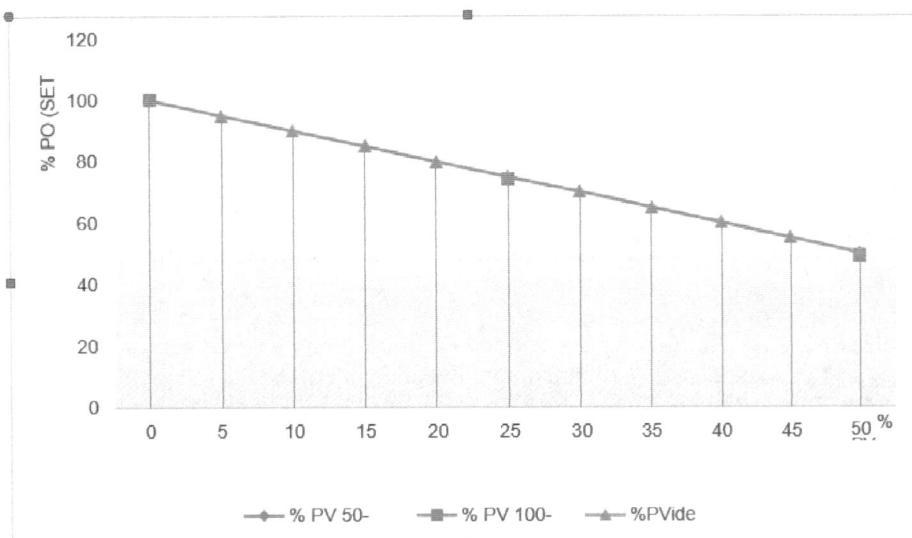

Figure 6. Comparison change of% PO to% PV in a Close Loop system using a tank.

3.2.2. Closed Loop experiment analysis using a tank. This system can be obtained that the work tool system is in accordance with the target value of% PV Testing is close to% PVideal or% PV Testing =% PO =% PVideal, because the comparison of changes in the% PV value between 50-100 and 100-50 obtained data with a difference of 1 to 0 points . This happens because there are errors in the equipment which are quite small and these errors have been processed by feedback in the form of measurements and the PID system [7].

4. Conclusion

In operating this tool, there are 2 experimental practicums, open loop for manual mode and close loop for automatic mode, in each practicum there are two methods, without and using tank respectively.

IOP Conf. Series: Materials Science and Engineering 1073 (2021) 012053 doi:10.1088/1757-899X/1073/1/012053

- Open loop: The main target of testing is the value of% PV =% PV testing -% PVideal because in open loop system there is no feedback and corrections toward output error value in order to fix the system. So this system is suitable for manual mode and the value from% PO to% PVideal becomes the reference. When the system using a tank has the advantage of being able to easily provide a% PO (set point) value for valve openings and has the disadvantage of not maximizing the% PV test results against% PVideal as to produce a difference of% PVmax = 16 points.
- Close loop: The main target for testing is a value of% PO =% PV because in close loop system there is feedback and corrections to output error value on the system using PID controller. This system is suitable for automatic mode so the reference of the target is the value of% PO =% PV and% PVmax = 0.

Furthermore, when system using a tank has the advantage of being able to easily provide a% PO (set point) value for the valve opening, by changing the values of Kp, Ki and Kd will greatly affect the results of the process output. The drawback of this method is % PV test results against% PVideal are not maximal, resulting in a difference in value of% PVmax = 0 points. For further research, an automatic system with SCADA will be built, then HMI can be displayed according to the real conditions.

References

[1] Gozali R M 2005 Desain Kontrol Pid Dengan Metoda Tuning Direct Synthesis Untuk Pengaturan Kecepatan Motor Dc *Teknoin* **10**(4) 283-293

[2] Wicaksono H and Pramudijanto J 2004 Kontrol PID Untuk Pengaturan Kecepatan Motor DC Dengan Metode Tuning Direct Synthesis *Jurnal Teknik Elektro* **4**(1) 10-17

[3] Hardiansyah R 2018 Kendali Posisi Linear Actuator Berbasis PID Menggunakan PLC *Journal of Applied Electrical Engineering* **2**(1) 12-17

[4] Suyanto S and Yulistyawan D 2007 Otomatisasi Sistem Pengendali Berbasis PLC Pada Mesin Vacuum Metalizer Untuk Proses Coating (Studi Kasus Di Pt. Astra Otoparts, Tbk-Divisi Adiwira Plastik, Bogor) *GEMATEK (Jurnal Teknik Komputer)* **9**(2)

[5] Lukito E S 2018 *Pengembangan Media Pembelajaran PLC (Programmable Logic Controller) Studi Kasus Aplikasi Sorting Machine Pada Mata Kuliah PLC Di Prodi Pendidikan Teknik Elektronika Fakultas Teknik Universitas Negeri Yogyakarta* (Skripsi, Yogyakarta: Universitas Negeri Yogyakarta)

[6] Amaroh H M 2017 *Kontrol Kecepatan Putar Motor Pada Proses Pembuatan Arum Manis Dengan Metode PID* (Skripsi, Malang: Politeknik Negeri Malang)

[7] Hermawan W n.d. Simulasi Kontrol PID Berbasis PLC (Programmable Logic Controller) dan HMI (Human Machine Interface) Pada Sistem Automatic Well Test *Jurnal Universitas Jayabaya*

IOP Conf. Series: Materials Science and Engineering 1073 (2021) 012054 doi:10.1088/1757-899X/1073/1/012054

The information security strategy of Bogor's smart city to deal with threat in cyber space

M A Madjid*, T Legionosuko and E G Samudro

Universitas Pertahanan, Bogor, Indonesia

*adnan.madjid@idu.ac.id

Abstract. The increasing quality and quantity of cyber-attacks and the demand for the information security revolution are also followed by the threat in developing Smart City of Bogor. In assuring these problems, it is crucial for the City Government of Bogor to develop and information security strategy which is capable to protect public privacy data and critical information of Smart City Bogor. The research was conducted with a qualitative method by using a grounded theory approach at the critical infrastructure of Smart City Bogor, which is the Department of Transportation, Department of Health, Department of Fire Rescue and Department of Communication and Information. The result showed that the information security as a core strategy which only created by the readiness of the legal ware as a causal factor in forming the knowledge and information security awareness in the preparation of priority program and development of the command, control, communications, computers, intelligence, surveillance and reconnaissance (C4ISR) systems of Smart City Bogor which became the basis of the formulation of Smart City Bogor's information security theory.

1. Introduction

In Indonesia, the development of districts and cities as smart cities are based on the concept of the Ministry of National Development Planning (*Bappenas*), which is the City of the Future of Indonesia: Sustainable and Competitive Cities, that is expected to be realized in 2045. The concept of Indonesia's smart cities in the future is supported by the three pillars of sustainable and competitive cities, namely safe and comfortable decent cities, green cities that have conservation and slavery resilience, and smart cities that are competitive and technology-based using applications that use 6 (six) main dimensions of smart cities namely smart governance , smart society, smart economy, smart environment, smart life and smart mobility in the daily life of city residents supported by the latest information and communication technology that is expected to be one of the efforts to solve various urban problems that are getting increased with growth of Indonesian residents who live in urban areas. Furthermore, based on data from the Central Statistics Agency (BPS), the Ministry of National Development Planning (PPN) / National Development Planning Agency (*Bappenas*) estimated that in 2045 as many as 82.37% of Indonesia's population will live in urban areas [1].

This can have a direct impact on the level of security and well-being of the urban population which ultimately affects Indonesia's national security as a whole, especially in aspects relating to public security and human security.

Flowing from this explanation, it can be understood that the use of information and communication technology in the development of smart cities is carried out in a system that has interoperability capabilities in which all functions of government systems and public services are integrated together through a system mechanism capable of monitoring (surveillance) in a comprehensive and real time manner. The various activities of the city and its citizens are able to be recognized (reconnaissance) and provide awareness of the situation of the various activities through the city control center (control) which is then intelligently communicated (communications) in a network system (computers) that are mutually integrated with various authorized institutions to do the actions deemed necessary quickly and precisely (command). The system mechanism that is able to assist smart city governments in making decisions based on comprehensive and real time situation awareness as stated, is actually a decision support system based on the framework of the Command, Control, Communications, Computers, Intelligence, Surveillance and Reconnaissance (C4ISR) which has been used by the United States Department of Defense and has been adapted at the federal, departmental and bureau level within the United States Government [2].

The Development of smart cities by utilizeing a decision guide system based on the C4ISR System is currently being developed by Bogor City which is also on of the cities that develops the concept of smart cities in Indonesia for the sake of improving government performance maximally, quickly and accurately in serving public services according to the needs of the residence as well as for the safe, prosperous and sustainable city development. Through the development and management of the city which is supported by the use of information and communcation technology that is packgaed in a decision guide system based on the C4ISR system framework, the Bogor City has succeeded in increasing effetiveness and efficiency in the implementation of government functions and public services which are expected to help improving the quality of life of Bogor City residents. This is proved by the leadership of the Mayor of Bogor City, Bima Arya, who successfully received the Gatra 2019 award for the category of Infrastructure “Smart City – Development Innovation”. The award was given by the President Director of Gatra at IPB International Convention Center, 1st December 2019 [3].

In addition, the concept of a smart city that is dynamic and focuses on innovation, solutions and optimal use of human resources and technological resources is an inevitable choice for the development of the city of Bogor. The e-Government Movement in Bogor City, which began to emerge from 2007 and experienced a boom in 2015, became an embryo in the planning of the construction of the Smart City of Bogor City. Therefore, the work program of the Movement Towards 100 Smart City Indonesia is very relevant to the direction of the Bogor City development goals [4].

Based on the description above, the potential threats and challenges may come to the Bogor City Government in developing an information security strategy in cyberspace that is based on a complete information security architecture, starting from the stage of determining the priority programs of the city to the stage of preparing and developing the C4ISR system. It as outlined in the security policy information in cyberspace and actualized through operations and information security assessment in cyberspace. It is aimed to to realize the the goal of developing Bogor Smart City as a world-class city that is comfortable for all and is one of the non-military defense forces of the the Republic of Indonesia in facing the threat of asymmetric warfare of technological dimension. Based on these problems, the researcher is interested in being able to reveal more deeply in the hope of providing benefits in the form of contributing information security strategies in cyberspace which can be used as a reference in the development of the Bogor Smart City.

2. Research methods

This research was conducted using a qualitative research methodology through a systematic and fundamental grounded theory approach. Glasser and Strauss in Yusuf stated that grounded theory is a general methodology for developing theories through qualitative research conducted systematically and fundamentally. The theory is built based on data collected about a phenomenon that is the focus of research [5]. While Gunawan explained that the characteristics of qualitative research with a grounded theory approach were the comparison between data from various categories and the use of different samples to maximize their similarities and differences [6].

Based on these explanations, this research begins with determining research problems, research questions, theoretical sampling, data collection, analysis and codification of data (open coding, constant comparison, axial coding, selective coding) and theory formulation, as presented as Figure 1 [7].

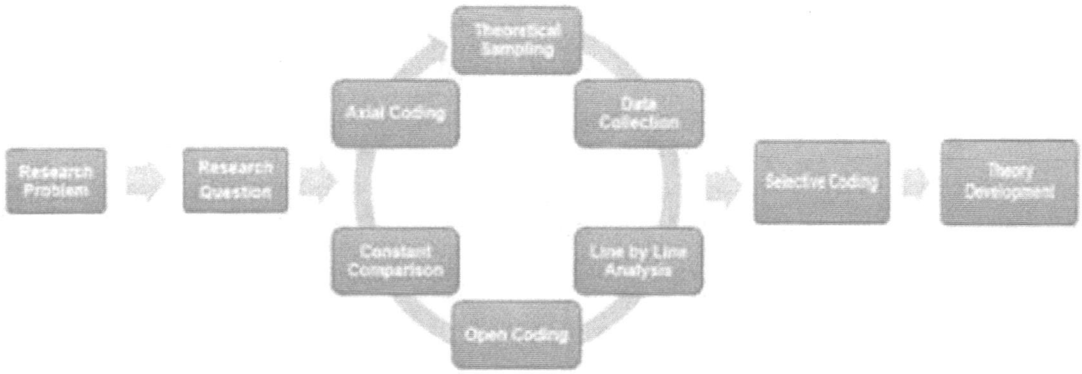

Figure 1. Grounded theory research step.

3. Data analysis

3.1. Data interpretation

Based on the results of the analysis conducted, there are several categories that are the findings of the research, namely information security as the main category, the rules (legal ware), knowledge, information security awareness, vulnerability of priority programs, weaknesses in preparation and also development of the C4ISR system, and translation of information security. The findings of the research are described as follows:

3.1.1. Information security (the main category). In the development of the Bogor as Smart City, information security in the form of public data and critical information of the Smart City Bogor has not been placed as an important aspect that must be placed as a short-term priority that must be immediately prepared by the Bogor City Government in developing its territory as a smart city. Utilization of information and communication technology in the preparation, development and integration of various information systems as one of the means and resources used to achieve these objectives has the potential to become a cyber-weapon that threatens public security and human security for citizens of Bogor City, especially those relating to public data security, critical information to the life safety of the citizens of the Bogor as Smart City. In addition, the preparation of steps that are not accurate and systematic that actually leads to wasteful use of the budget that has the potential to cause regional losses. Finally, the low quality of information security in cyberspace has the potential to become a barrier to the realization of the vision of creating Bogor as a city that is comfortable for all.

3.1.2. The rules (legal ware). In the development of the Bogor Smart City, the absence of a set of rules (legal ware) regarding information system governance and information security governance both at the city government policy level and at the operational level of programs and activities in the regional apparatus organizations studied, has become a causal factor for vulnerability priority programs of the city and regional apparatus organizations. The weak levels of resource preparation and successful development of information systems in each of the regional apparatus organizations studied and the integration of the information systems within the framework of the C4ISR Smart City Bogor system and the inability of the Bogor City Government to translate information security in the development of the Bogor Smart City, both in the form of risk assessment and information security operational in the Bogor

IOP Conf. Series: Materials Science and Engineering 1073 (2021) 012054 doi:10.1088/1757-899X/1073/1/012054

Smart City. This is indicated by the lack of knowledge and information security awareness of the regional apparatus organizations under study due to the absence of rules as a guideline.

3.1.3. Knowledge. In terms of knowledge about information system governance and information security governance, the state civil apparatus in the regional apparatus organizations studied, do not yet have full and accurate knowledge. The understanding is based on the explanations of the resource persons, among them stated by the resource person at the Bogor City Fire Department who stated that the shortage of personnel was lack of technological understanding so that when the continuity of knowledge did not go well. While the Department of Communication and Information Agency in general explained that currently the number of human resources who understand the Information-Technology in the Bogor City Government is very limited and failure in the development of information systems in several regional apparatus organizations, particularly those investigated, among others due to the limited human resources in the Bogor City Government who understand the technology of information.

3.1.4. Information security awareness. Similar with the level of knowledge about information system governance and information security governance, awareness of the importance of information security in the development of the Bogor Smart City is very low. This understanding is also based on the explanation of the informants in the regional apparatus organizations studied. At the Transportation Office, it was stated that in terms of information security including the internet network, the Transportation Office had not yet focused on this matter and it was entirely left to the Communication and Information Agency. While at the Health Office, it was stated as a form of information security in the use of information technology in Bogor Home Care telemedicine services, the Health Service only think about how quickly patients are handled, not thinking about the security of that information. In addition, the Bogor City Fire Service stated that an information security analysis had not been carried out, bearing in mind that it was still new and its existence had been very beneficial for them.

3.1.5. Vulnerability of prior programs. In the development of the Bogor Smart City, a priority program vulnerability analysis is carried out on priority programs at the city level as well as at the regional level organization, especially those that utilize information and communication technology. At the priority program at the city level as referred to is the Implementation Program of IT Governance and a Convenient City Program for All. Whereas at the level of the organization of the regional apparatus, especially the organization under study, the priority programs referred to are:

- Bogor Traffic Program (One way system, Area Traffic Control System) by the City Transportation Agency
- Bogor Home Care Program by the Bogor City Health Office.
- Fire mapping and Regional Management Program by Bogor City Fire Service.
- One Data Program and Big Data Program by Bogor City Communication and Information Agency.

Based on the analysis carried out on these priority programs, an understanding is obtained that all priority programs are not based on an information security analysis that takes into account aspects of confidentiality, integrity and availability which ultimately actually places the development of the Bogor Smart City put on a fragile foundation.

3.1.6. Preparation and development of information system. In preparing resources for the development of information systems in the Bogor Smart City, it was analyzed through 7 (seven) main resources in each of the regional apparatus organizations studied, namely:

- Human resources
- Software
- Hardware

IOP Conf. Series: Materials Science and Engineering 1073 (2021) 012054 doi:10.1088/1757-899X/1073/1/012054

- Network resources
- Physical infrastructure
- The set of rules
- Source of budget

The results of the analysis provide an understanding that except for the Bogor City Communication and Information Agency, the readiness of all regional apparatus organizations under study is limited as a system user and does not yet have a readiness as a manager especially as an information system developer. While the Bogor City Office of Communication and Information, although it already has the resources and ability to develop information systems independently, the lack of knowledge and information security awareness as a basis for developing information systems has caused the regional apparatus organizations not to have the readiness as intended. Furthermore, along with the weak level of resource readiness and the success of the development of information systems in the regional apparatus organizations, it has direct implications for the integration of the information system within the framework of the C4ISR system of Bogor Smart City. The results of observations at the Bogor Smart City War Room provide understanding that the percentage of integrated information systems in the smart city control center does not reach 10% of the overall regional apparatus organization in the Bogor City Government.

3.1.7. Translation of information security. Analysis of the translation of information security in the development of the Bogor Smart City provides an understanding that the Bogor City Government has not been able to translate it both in information security operations to protect the confidentiality, integrity and availability of privacy data belonging to city residents as well as critical information of the Bogor Smart City, particularly in the C4ISR system and information security risk assessment in each stage of the development and management of the Bogor Smart City. This understanding is based on the explanation of the speakers at the organization of the area in the research. At the Department of Transportation, it was stated that the operational standards and procedures for managing and securing the Bogor City Adaptive Traffic Control System and the Bogor City Traffic Monitoring and Control Room have not yet been prepared, including scenarios for handling cyber incidents. As for the Bogor City Office of Communication and Information along with the lack of a set of rules regarding information security, of course there is no important matter related to information security in Bogor City. Besides that, for the time being, information security is still translated in terms of the type of information that belongs to the Bogor City Government that can be published or that cannot be published to the public.

3.2. Discussion

Further analysis of information security in Bogor Smart City gives an understanding of the relationship between categories which is then used as a basis for developing an information security strategy theory that can be used as a reference to be applied in the development of Bogor Smart City. The relationship between these categories is presented in Figure 2, as follows:

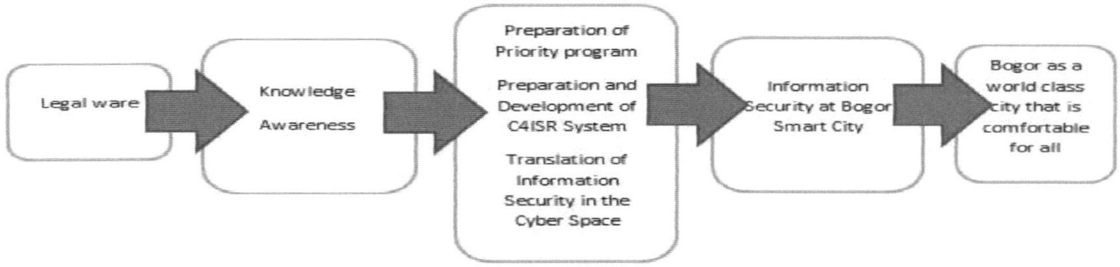

Figure 2. The relationships of the resources.

Flowing from this analysis it can be argued that Bogor Smart City's information security strategy is: "A series of protections on public data and critical information of Bogor Smar City in dealing with threats and challenges in the cyber space based on the rule of law (legal ware) as a causal factor forming knowledge (knowledge) and information security awareness in the preparation of priority programs as well as the preparation and development of the Bogor Smart City C4ISR system in order to realize Bogor as a world-class city that is comfortable for all.

4. Conclusions

Based on the results of research and discussion, several conclusions can be made as follows:

- The absence of a set of rules is a causal factor of the lack of knowledge and information security awareness which subsequently causes vulnerability to the Bogor Smart City priority program, the weak preparation and development of the Bogor Smart City C4ISR system and the inability of the Bogor City Government to translate information security for the protection of public data and critical information of Bogor Smart City.
- To realize information security in the Bogor Smart City, it is important for the Bogor City Government to have an information security strategy that is based on a set of rules that can guide the development of the Bogor Smart City.

Acknowledgement

This research was supported by the informants from the the institutions in Bogor City, which are Department of Transportation, Department of Health, Department of Fire Rescue and Department of Communication and Information and all of our colleagues' lecturer in assisting the research

References

[1] Nasional K P P and NASIONAL B P P 2015 Pengembangan Kota Cerdas di Indonesia *Makalah disampaikan dalam Konferensi e-Indonesia Initiative dan Smart Indonesia Initiatives Forum ke-1* vol 15

[2] Thomas R, Beamer R A and Sowell P K 2000 Civilian application of the DoD C4ISR architecture framework: A treasury department case study *5 th International Command and Control Research and Technology Symposium*

[3] Berita Bogor Pemkot Bogor Raih Penghargaan Inovasi Smart City

[4] Diskominfotandi 2017 *Exceutive Summary Masterplan Smart City*

[5] Yusuf A M 2016 *Metode penelitian kuantitatif, kualitatif & penelitian gabungan* (Prenada Media)

[6] Imam G 2013 Metode Penelitian Kualitatif Teori dan Praktik *Jakarta Bumi Aksara*

[7] Choi Y-J, Choi H-B and O'Donnell M 2018 Disaster reintegration model: a qualitative analysis on developing Korean disaster mental health support model *Int. J. Environ. Res. Public Health* **15** 362

Prediction of student graduation accuracy using decision tree with application of genetic algorithms

A Maulana*

Department of Telecommunications, Semarang State Polytechnic, Indonesia

*lana_ooh@ymail.com

Abstract. Many of students who can't complete the study promptly is a problem that needs to be faced with a fast and effective way. In the student's education system is an important asset to the note the graduation rate of students promptly. Predicting graduation is very important for the institution to determine the strategic policies for the institution. The scope of the study is limited in performing accurate comparison between studies that only use Genetic Algorithm method and the research-based Genetic Algorithm using forward selection. Here we can conclude that the value of the highest accuracy is there in the Genetic Algorithm-based method Forward selection. Thus the Genetic Algorithm method with the selection of attribute-based Forward Selection is the best method for solving a problem in the prediction accuracy of graduation.

1. Introduction

The Graduation is one of the assessment items in the college accreditation process. The assessment elements for graduates in a university will refer to the national standard SNPT for higher education PERMEN DIKBUT number 3 of 2020. So that if students graduate academically on time it will help assess the accreditation of a college. Graduating on time is very important and this is a priority policy issue, even the passing rate is considered one of the institutional effectiveness [1].

The value of graduate accreditation plays an important role in accreditation, so every university tries to make various efforts so that its students can graduate on time.

The Diploma (D3) education level is included in the scope of higher education, which is one of the basic requirements in finding a job, where universities will prepare qualified and skilled undergraduate candidates who have skills in their fields. Of course, in achieving the degree in the Associate Expert (AMD) field, it takes a normal time for 3 years [2].

2. Literature review

Verry Riyanto, Abdul Hamid, Ridwansyah 2019 "*Prediction of Student Graduation Time Using the Best Algorithm*" The results were obtained from the study of academic achievement during the study period, architecture, and appropriate weights of the NN, DT, and SVM methods [3]. Of the three models tested and experimented with, the highest value was SVM 85.18% to occupy the second position with a value of 84.96% occupied by DT, and the smallest accuracy value was NN with a value of 84.68%.

Joseph Stephen Bassi, Emmanuel Gbenga Dada, Alkali Abdulkadir Hamidu and Mshelia Dauda Elijah 2019 "Students Graduation on Time Prediction Model Using Artificial Neural Network" The results showed that artificial neural networks are a promising tool [4].

Agus Romadhona, Suprapedi, H. Himawan, 2017 "Timely student graduation prediction by age, gender, and achievement index using decision tree algorithm" From this research, it was found that the prediction of students who entered the study period of students who entered was influenced by age, gender, GPA semester 1 to GPA 4 semesters, and the most influential was the 4th-semester student GPA on time with a gain value of 0.340 from all attributes [5].

Susi Mashlahah 2013 "Prediction of student graduation uses the decision tree method with the application of the C4.5 algorithm" The more sample data used, the greater the correctness of the prediction [6].

Eko Prasetiyo Rohmawan, 2018 "Predicting student graduation on time uses the decision tree and artificial neural network methods" The use of data mining with the Decision Tree method has an accuracy of 74.51% and an Artificial Neural Network has a higher accuracy of 79.74% which can be applied in predicting student graduation on time [7].

3. Research methods

3.1. Research purposes

In this study, the data used were data from students of the D-3 Computer Engineering Study Program, Harapan Bersama Polytechnic of Tegal, the purpose of this study was to find and find the best algorithm patterns with student data objects based on several determining variables to get the level of accuracy in predicting graduate accuracy students using the Decision Tree method and research that will be carried out with the Application of Genetic Algorithms.

The data in this study were taken from graduation data in 2019 as many as 306 students with initial Nim attributes, gender, marital status, occupational status, social studies scores 1 to 4 occupation, and parents' income. The research stages are as follows:

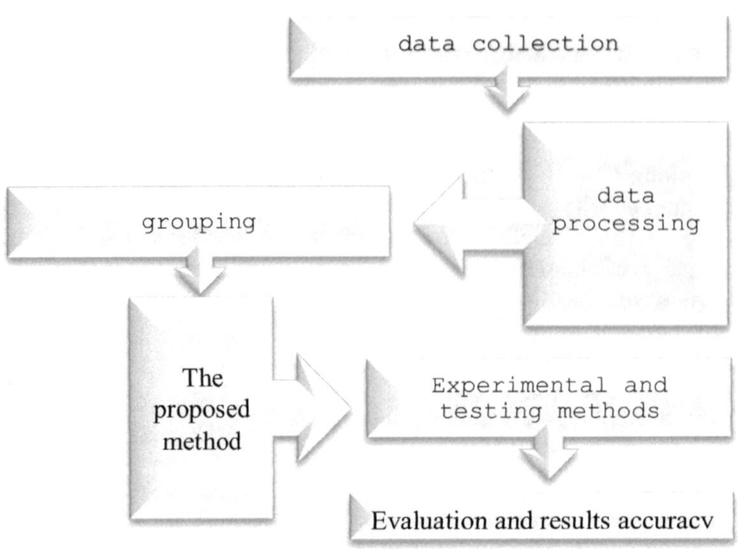

Figure 1. Research stages.

3.2. Testing phase

The research was conducted using the decision tree method with the application of genetic algorithms, the object of research taken from students of the D-3 Computer Engineering Study Program at the

IOP Conf. Series: Materials Science and Engineering 1073 (2021) 012055 doi:10.1088/1757-899X/1073/1/012055

Harapan Bersama Polytechnic Year 2016 in the form of data from Microsoft Excel then transformed into the Rapid Miner tool by doing 10 fold cross-validation to get one prediction about the student's on-time graduation to get the best accuracy rate results. As shown in the table below:

Table 1. Table of Attributes of data for graduates of the Harapan Bersama Tegal Polytechnic.

Atribut	Information
NIM	Student ID Number
Gender	Gender
Marital status	Marital status
Job Status	Students or parents with Working or not working status
Income	Total rupiah
IPS1	Semester Achievement Index 1
IPS2	Semester Achievement Index 2
IPS3	Semester Achievement Index 3
IPS4	Semester Achievement Index 4
Label	Be on time Not on time

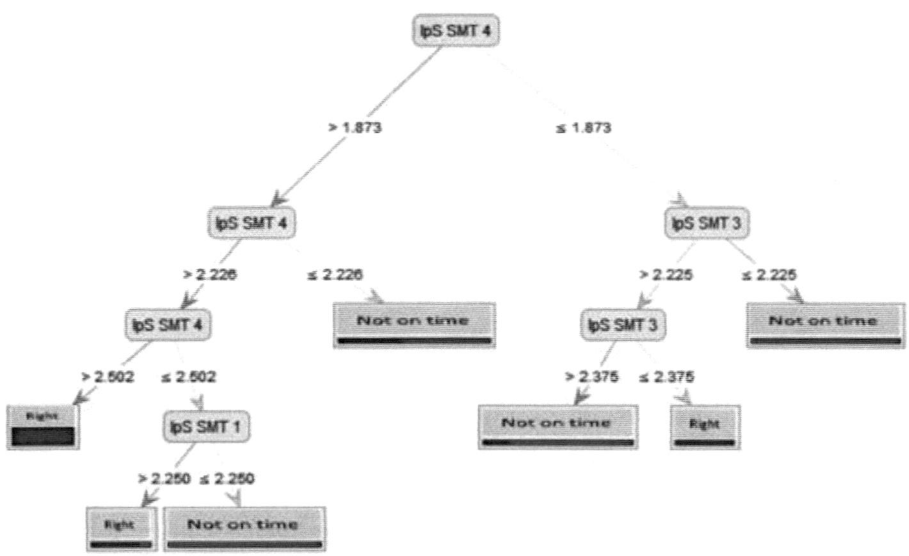

Figure 2. Decision Tree Model of student graduation data.

From the Decision Tree in Figure 2. obtained rules for predicting student graduation data on time or not on time. By using the Decision Tree Algorithm, so that you get a rule that will be used for decision making on new data.

In conducting classification, this method can be evaluated based on criteria such as the level of accuracy, speed, reliability, scalability, and interpretability. Then after the data is processed, the level of accuracy can be tested to see the performance of the Decision Tree algorithm

From the results of testing the data for graduate students in 2019 above, it is evident that the test results of the Decision Tree Algorithm method have a high accuracy value.

IOP Conf. Series: Materials Science and Engineering 1073 (2021) 012055 doi:10.1088/1757-899X/1073/1/012055

4. Conclusion

Based on the results of the data accuracy analysis, the graduation data of the students of the Harapan Bersama Tegal Polytechnic using the Decision Tree with the Genetic Algorithm Application to predict student graduation. The resulting model can determine the best method in predicting student graduation.

References

[1] Depkumham 2010 *Peraturan pemerintah RI no 66 Tentang Perubahan atas peraturan Pemerintah no 17 tahun 2010 tentang pengolahan dan penyelenggaraan pendidikan* www.djpp.depkumham.go.id, Jakarta

[2] Mashlahah S 2013 Prediksi kelulusan mahasiswa menggunakan metode decision tree dengan penerapan algoritma C4. 5

[3] Riyanto V, Hamid A and Ridwansyah R 2019 Prediction of Student Graduation Time Using the Best Algorithm *Indones. J. Artif. Intell. Data Min.* **2** 1–9

[4] Bassi J S, Dada E G, Hamidu A A and Elijah M D 2019 Students Graduation on Time Prediction Model Using Artificial Neural Network *IOSR J. Comput. Eng.* **21** 28–35

[5] Romadhona A, Suprapedi S and Himawan H 2017 Prediksi Kelulusan Mahasiswa Tepat Waktu Berdasarkan Usia, Jenis Kelamin, Dan Indeks Prestasi Menggunakan Algoritma Decision Tree *Cyberku J.* **13** 8

[6] Mashlahah S, Yaqin M A and Faisal M 2013 Prediction of Students Graduation Using Decision Tree Method with the Implementation of Algorithm C4. 5 *IEESE Int. J. Sci. Technol.* **2** 1

[7] Rohmawan E P 2018 Prediksi Kelulusan Mahasiswa Tepat Waktu Menggunakan Metode Desicion Tree Dan Artificial Neural Network *J. Ilm. Matrik* **20** 21–30

The development and analysis of an information system for incentives publication of reputable indexed international journal and national journal (Case study State Polytechnic of Malang)

E Rohadi[1], A Amalia[2,*], D K Kusumah[1], G A A Fahrezi[1], W I Sabilla[1] and I Siradjuddin[2]

[1]Department of Information Technology, State Polytechnic of Malang, Jl. Soekarno Hatta No. 9 Malang 65141, Indonesia
[2]Department of Electrical Engineering, State Polytechnic of Malang, Jl. Soekarno Hatta No. 9 Malang 65141, Indonesia

*amalia@polinema.ac.id

Abstract. The research activities in The State Polytechnic of Malang addressed to achieve the output as articles, intellectual properties, and product innovations. The incentives are given to the researchers who publish articles in reputable publishers due to increasing the articles in both numbers and quality. In this work, an information management system of publications is developed and analysed. The system is integrated into the staff database in order to relate with yearly staff performances. In the system, articles are categorized according to classification as International and National publications. Furthermore, both the International and National Categories are classified as a reputable and accredited publisher, respectively. The proposed system has tested by administrators in each unit and randomly staffs as respondents to fill the questioner. The result shows that the management system can be operated by an administrator in each unit and the staffs do not have difficulty to use the proposed system. It means the proposed system can be applied in The State Polytechnic of Malang.

1. Introduction

Three Pillars of Higher Education is one of the goals that must be achieved and carried out by every college in Indonesia. The essence of the Three Pillars of Higher Education consists of 3 points, namely: Education and Teaching, Research and Development, and Community Service. For this reason, the Three Pillars of Higher Education is the responsibility of all elements contained in college. Not only students, but lecturers, and the various academics involved [1].

In general, the developed countries in the world have very high research and development culture. Rapid technological developments, innovations, and even cutting-edge problem-solving products, are born out of research and development. Research and development are something this nation must always do if it wants to progress and develop [2]. Carrying out research and development will have an impact on the advancement of the economy, education, social and other sectors in society [3]. Lecturers as teaching staff are required to fulfil the Three Pillars of Higher Education. One of them is research and development whose output is in the form of journals or scientific papers.

Currently, State Polytechnic of Malang applies a policy of giving incentives to lecturers for journal publications they produced. This policy is given in order to meet the number of publications produced in accordance with the work contract between the director and the ministry. Furthermore, this policy is also related to the Three Pillars of Higher Education which must be fulfilled by lecturers. Incentives are considered as one of the essential factors to encourage workers to make significant efforts and work more efficiently. This is because the incentive and reward system direct workers' capabilities into more efficient in their work to achieve the institution's goals [4-5].

In practice, the submission of incentives for lecturers for the articles or journals they made is still not well accommodated. In this case, the accommodation means that no information system can store the data related to the incentive application. Lecturers who want to apply for incentives for articles or journals they made must submit data via google form without knowing whether the data has been processed or not by authorized staff, which then the data stored in the form of a Microsoft Excel file. This is time-consuming and inefficient in terms of data management. Based on that problem, in this research, the development and analysis of an information system for incentives publication of reputable indexed international journal and national journal (case study State Polytechnic of Malang) proposed.

This information system is about submission of incentives for article or journal publications that have been done by lecturers at the State Polytechnic of Malang. All data will be stored and managed properly with this system. Article or journal categories can be seen by the lecturer so that they can become a reference later. The number of incentives given depends on the article or journal category chosen by the lecturers. This system is expected to be able to help related units as well as lecturers who wish to propose incentives for the articles or journals that have been made.

2. The descriptions of system

This web-based information system is designed to make it easier for lecturers to submit the articles or journals they made. The submitted articles or journals are intended to get incentives, where the amount of incentives received depends on the article or journal category. The journal category consists of international journals and national journals. International journals are divided into reputable publishers such as Scopus and ISI Thomson Reuter and other publishers such as Copernicus, DOAJ, and CABI. Providing incentives is very important, the absence of appropriate incentives can have a negative impact on employee performance. It may also weaken their workplace productivity, which decreases in the chances of achieving promising institutional goals.

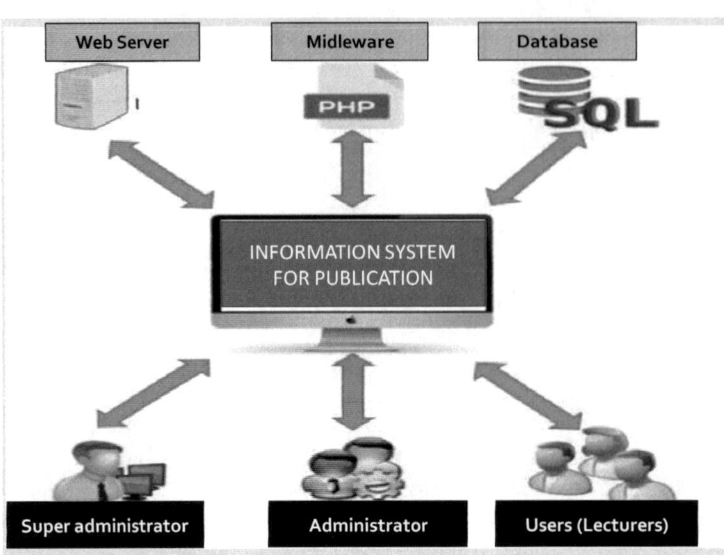

Figure 1. System architecture.

Figure 1 shows the system architecture of this system. This system has three access rights: admin, super admin, and user and managed by the related unit. In this system, the admin can see the articles entered by the lecturers and can also change the articles status starting from being processed until the process is complete. Thus, the related unit does not need to do this manually but has been systemized [6]. Super admin can see data from admin, user data (lecturers), and data from lecturers' articles uploaded. Lecturers, as users, can input article data to be submitted to get incentives into the system by entering article data and uploading article documents and can monitor the article submission process through the status given by the admin [7-8].

The incentive submission process starts with the user submitting their article or journal in the system, then selecting the journal category and then uploading all the data requested by the system. The data that has been uploaded by the user will be stored in the system database. Then the admin will verify the data that has been uploaded by the user. After completing the verification process, the admin will change the status of "submission" to "processed" or "accepted" if the data uploaded by the user meets the predetermined criteria. Users with accepted status can print the related data, which will later prove that the submission of article incentives has been verified and accepted by the related unit's admin.

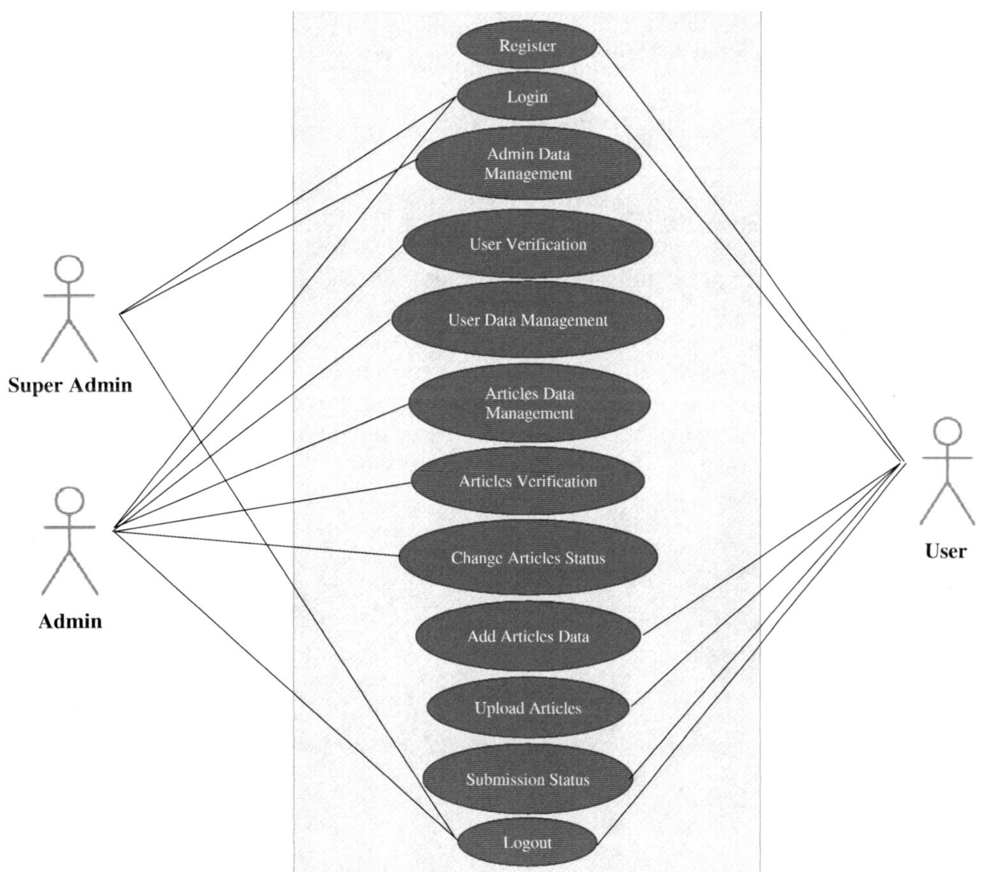

Figure 2. Use case diagram.

Figure 2 shows the use case diagram of this information system. Use case diagrams are UML diagram models used to describe the expected functional requirements of a system. Use case diagrams are used to briefly explain who uses the system and what they can do [9-10]. In this information systems for incentives publication of reputable indexed international journal and national journal, there are three actors including super admin, admin, and user.

3. Results and discussion

Testing is carried out on all system functionalities from the super admin, admin, and user pages. This is done to ensure all features of this system can run smoothly and are satisfactory for users [6]. The super admins have three features, such as login, admin data management, and logout. Meanwhile, admins have seven features, such as login, user verification, user data management, article data management, article verification, changing article status, and logout. Lastly, users (lecturers) have six features: register, login, add article data, upload articles, view submission status, and logout. All system features have been tested and can works properly as shown in Table 1 below.

Table 1. Functional test results.

Actors	Number of Features	Status
Super Admin	3	Works Properly
Admin	7	Works Properly
Users (Lecturers)	6	Works Properly
Main Page Login/Logout		**Status**
Login Super Admins, Admins and users		Works Properly
Logout Super Admins, Admins and users		Works Properly
Register Page		**Status**
Users (Lecturers) Register		Works Properly

For the purpose of this study, user satisfaction testing on this system is also carried out by using a questionnaire distributed to the admin and the lecturers. A total of 55 respondents filled out the distributed questionnaires consisting of 9 respondents as super admins, 13 respondents as admins from 7 departments and 1 unit, and also 33 users (lecturers). From a total of 9 respondents as super admins, 8 respondents stated that this system is excellent, and one is good. Meanwhile, a total of 13 respondents as admins, 10 respondents stated that this system is excellent, and 3 are good. Lastly, 28 out of 33 users (lecturers) stated that this system excellent, 4 good, and 1 fair, respectively. The questionnaire results can be seen as shown in Figure 3 below.

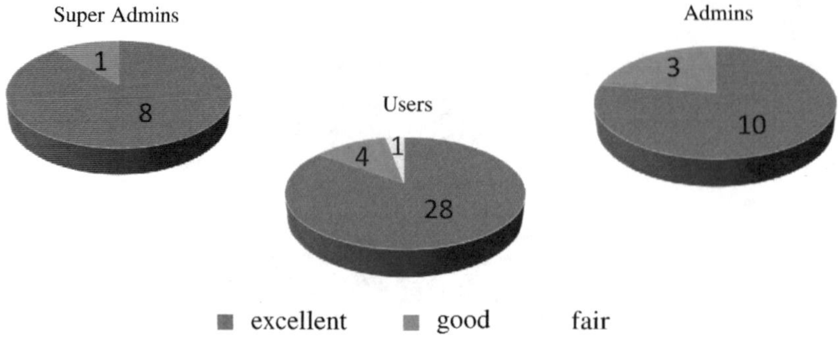

Figure 3. Questionnaire results.

Figure 4 shows the user view of this system. On this page, the user can download two files, a personal data form, and a scientific publication statement letter, which must be filled in. After that, it is uploaded back into the system when filling in new article data. Article uploading can be done on the article upload page, where on this page, the user is required to fill in all available fields. The fields in the topic include the name of the article, the name of the journal, criterion 1 (international or national journal), criterion

2 (the type of publisher), URL of SINTA, URL of Scopus, URL of articles, contributions, indexing agencies, upload article file, upload personal data, and upload a scientific publication statement letter. After all fields are filled in, user can submit the data to store the data in the system database. The process of articles that have been submitted can be monitored by looking at the status of the article on the system.

Figure 5 shows the user responds of the system performances. The system tested by 50 lecturers as respondents. Most of performance categories have good responds except the difficulties. It is occurred by the old lecturers have difficulties to operate the systems by themselves. The proposed of information system development is realize the effectiveness data transactions. On the other hand, the information technology needs the capability in order to operate as their requirements [8].

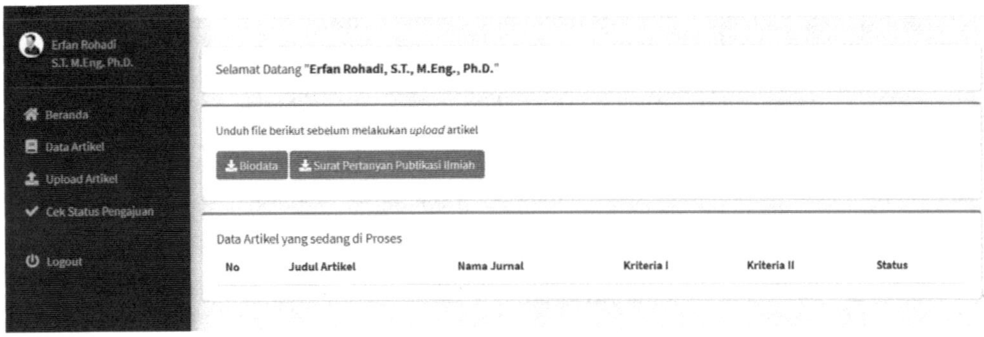

Figure 4. User view.

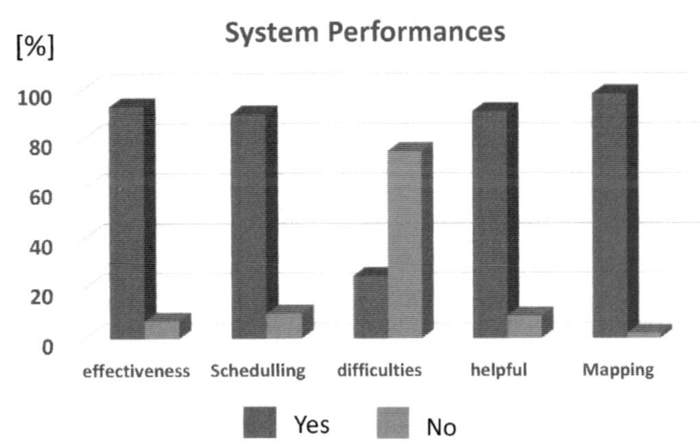

Figure 5. Respond of the users.

4. Conclusion

The development and analysis of an information systems for incentives publication of reputable indexed international journal and national journal (case study State Polytechnic of Malang) has been presented. The proposed system has been tested by admins in each unit and randomly staffs as respondents to fill the questionnaire. The results show that an administrator can operate the system in each unit, and the staffs do not have difficulty using the proposed system. As a result, the proposed system can be applied in The State Polytechnic of Malang.

Acknowledgment

The Authors would like to thank The State Polytechnic of Malang that provide funding and support for this research.

References

[1] Ayllón S, Alsina Á and Colomer J 2019 Teachers' involvement and students' self-efficacy: Keys to achievement in higher education *PloS one* **14** (5) e0216865

[2] Sharma E 2019 A review of corporate social responsibility in developed and developing nations *Corporate Social Responsibility and Environmental Management* **26** (4) pp 712-720

[3] Zafar M W, Shahbaz M, Hou F and Sinha A 2019 From nonrenewable to renewable energy and its impact on economic growth: the role of research & development expenditures in Asia-Pacific Economic Cooperation Countries *Journal of cleaner production* 212 pp 1166-1178

[4] Ponta L, Delfino F and Cainarca G C 2020 The role of monetary incentives: bonus and/or stimulus *Administrative sciences* **10** (1) 8

[5] Ahmad I, Danish R Q, Ali S A, Ali H F and Humayon A A 2019 A Comparative Study of Banking Industry Based on Appraisal System, Rewards and Employee Performance *SEISENSE Journal of Management* **2** (1) pp 1-11

[6] Soegoto D S and Oktady D A 2018 Information System Design of an Inventory Online Website *IOP Conference Series: Materials Science and Engineering* **407** 1 p 012025

[7] Börner K, Klavans R, Patek M, Zoss A M, Biberstine J R, Light R P, Larivière V and Boyack K W 2012 Design and update of a classification system: The UCSD map of science *PloS one* **7** p.e39464

[8] Payam H, Samira G and Masoud A 2017 Impact of Information Technology on Lifestyle: A Literature Review and Classification *Int. Journal of Virtual Communities and Social Networking (IJVCSN)* **9** pp 1-23

[9] Khurana N, Chhillar, R S and Chhillar U 2016 A Novel Technique for Generation and Optimization of Test Cases Using Use Case, Sequence, Activity Diagram and Genetic Algorithm *JSW* **11** (3) pp 242-250

[10] Syed M U G, Gogadi S, Nadendla B and Han Y 2018 Qualitative Software Engineering and Event-B Modeling for Communication and Use Case Diagrams *Int. Conf. on Computational Science and Computational Intelligence (CSCI)* pp 841-845

IOP Conf. Series: Materials Science and Engineering 1073 (2021) 012057 doi:10.1088/1757-899X/1073/1/012057

Securing the electronic medical record by implementing Advanced Encryption Standard (AES) on the information system of a health service place

C Rahmad[1], A R Syulistyo[1,*] and A D W Sumari[1,2,3]

[1]Department of Information Technology, State Polytechnic of Malang, Jl. Soekarno-Hatta No. 9, Malang 65141, East Java, Indonesia
[2]Cognitive Artificial Intelligence Research Group (CAIRG), Department of Electrical Engineering, State Polytechnic of Malang, Jl. Soekarno-Hatta No. 9, Malang 65141, East Java, Indonesia
[3]Faculty of Defense Technology, Indonesia Defense University, IPSC, Sentul, Bogor, West Jawa, Indonesia

*arie.rachmad.s@polinema.ac.id

Abstract. Medical records are very crucial in health services all over the world. This is due to not only to record individual health conditions but also for getting the most appropriate medical services at a local health facility when in mobile conditions. By having a look at the medical records, the local health workers can know the medical history and give their best service in treating the patient. Conventional medical records commonly are written manually and are not flexible for the mobile environment. Therefore, the electronic medical record (EMR) has become mandatory to be implemented to cope with such problems. Another important thing that requires attention is the patient has the right to know the contents of the medical record, especially when having health services at a different hospital or other health services when at mobile conditions. Accessing the EMR from the mobile environment brings a crucial matter, namely, how to secure the contents of EMR and ensure that the access can only be given to the legitimate owner fit as well as other legitimated actors such as the related doctors and the EMR officials. Based on the arisen problem, the contribution of this research is to make the implementation of encryption algorithms that will be used on Secure Electronic Medical Record (SEMR) systems using secure and fast encryption techniques by using Advanced Encryption Standard (AES) on the information system of a health service place.

1. Introduction

The advancement of information and communication technology (ICT) along with the introduction of Industrial Revolution 4.0 technologies a couple of years ago does not always have a big impact on health services such as hospitals primarily community health services or Public Health Center or Puskesmas [1]. Providing the best services to patients, medical records are one of the keys [2]. The medical record is a compilation of facts about the health condition and illness of a patient which includes documented data about the present and past illness condition, and written medication that has been and will be carried out by a health worker [3]. This is very important when patients change public health care places so that it makes it easier for medical personnel at that place to find out about the patient's medical history. The facts show that medical records of the patient on several hospitals and Public Health Center are still handled conventionally [4,5]. Meaning, the records are still stored in hard-form folders and not in digital format.

On the other hand, the dynamic mobility of the community allows people to seek treatment at the hospital or Puskesmas where they are at that time. These situations make medical records a very important document and access to medical records is very crucial in health care services [3]. This mobile condition happens for everyone who travels a lot whether in-country or abroad. Conventional medical records that are stored in hard-form folders are no longer able to support this condition. Therefore, medical records that are stored in electronic form have become mandatory since they have to comply with the recent technologies and in order to enhance the health services. Since then Electronic Medical Record (EMR) became a standard for storing health services given to patients as well as recording their health conditions.

There is one important thing that has to get serious attention, namely patients have the privilege to know the contents of the medical records, especially when receiving health services from hospitals or Puskesmas. One solution to dealing with this problem is to create an information system that stores medical records electronically and has a standard so that it can be easily integrated between hospitals and Puskesmas. Accessing the EMR from the mobile environment brings a crucial matter, namely, how to secure the contents of EMR and ensure that the access can only be given to the legitimate owner o fit as well as other legitimated actors such as the related doctors and the EMR officials. However, the application of EMR made medical personnel doubt, but after being passed the Law on Information and Electronic Transactions (UU ITE) Number 11 of 2008 has provided answers to such doubts [6]. The ITE Law has provided opportunities for EMR implementation. Based on the arisen problem, the contribution of this research implements an encryption algorithms that will be used on Secure Electronic Medical Record (SEMR) systems using secure and fast encryption techniques by using Advanced Encryption Standard (AES).

2. Background study

2.1. Electronic Medical Record (EMR)

EMR is a digital version of a patient's medical record which consists of a patient's treatment from certain care delivery organizations (CDO) such as hospital or community health service by a particular medical worker such as general doctor, specialist doctor, and dentist. It also contains the doctor's diagnosis and drugs given to the patient as well as other related medical information. EMR is said only documented and stored at CDO such hospitals or community health services especially in remote areas that are very far from the closest hospitals in their areas. Meanwhile, Electronic Health Record (EHR) is said that the MR file can be brought by the patient and can be accessed by other CDOs that deliver health care to the patient. It is like the MR is attached to the patient wherever he/she goes. EHR is more flexible that EMR but anything that is relaxed will invite threats.

As has been known, MR is an old kind of health data recording which has been carried out for a long time in the health care world. MR is a very important private health document not only for the patient him/herself but also for the caring doctor as well as CDO that delivers health care. The introduction of information technology brought positive impacts to EH/MR where it was recorded by hand-written and put in the folders, now it can be done electronically and stored in an electronic manner. In the era of cyberspace, medical or health information becomes an important target for certain actors and it can be used to do harmful things. Another thing that has become big attention is the possibility of unauthorized actors to access EM/HRs database and make them as ransom. Therefore protecting EM/HRs for this case will be considered as very important.

2.2. System Usability Scale (SUS)

The System Usability Scale (SUS) was created by John Brooke in 1986 as a "quick and dirty" questionnaire. This questionnaire allows practitioners to quickly and easily assess the usability of a particular product and services [7]. Originally this tool consisted of 10 questionnaire items with five response scales with a minimum score of 1 which means strongly disagree and a maximum value of 5 which means strongly agree. The three steps for calculating the SUS score are as follows [8]:

- For each of the odd numbered questions, subtract 1 from the score.
- For each of the even numbered questions, subtract their value from 5.
- Take new values which you have found, and add up the total score. Then multiply this by 2.5.

The calculation results that you have a maximum score of 100.

2.3. Advanced Encryption Standard (AES)

Medical image record should be encrypted to ensure privacy and integrity, and it was recommended to use AES [9]. AES has been introduced by NIST in 2000 which has 4 main operational blocks, such as:

- Substitute byte transformation: An S-box is used to substitute each data block byte with another block.
- Shift transformation of rows: Each row of the state matrix is given a cyclic shift to the right side according to its location.
- Mix Transformation of Columns: It is a matrix multiplication operation where each column of the state matrix is multiplied by that of the fixed matrix.
- Add Round Key Transformation: XOR operation is performed between the new state matrix and the round key one.

AES algorithm based on electronic codebook is simplest to implement however, the same block of plaintext will always generate the same block of ciphertext, this can give clues for attackers. So, in this research, we choose AES encryption based on cipher block chaining (CBC) mode which illustration can be seen in Figure 1 [10].

CBC Encryption: $C_1 = CIPH_K(P_1 \oplus IV);$
$C_j = CIPH_K(P_j \oplus C_{j-1})$ for $j = 2 \ldots n$.

CBC Decryption: $P_1 = CIPH^{-1}_K(C_1) \oplus IV;$
$P_j = CIPH^{-1}_K(C_j) \oplus C_{j-1}$ for $j = 2 \ldots n$.

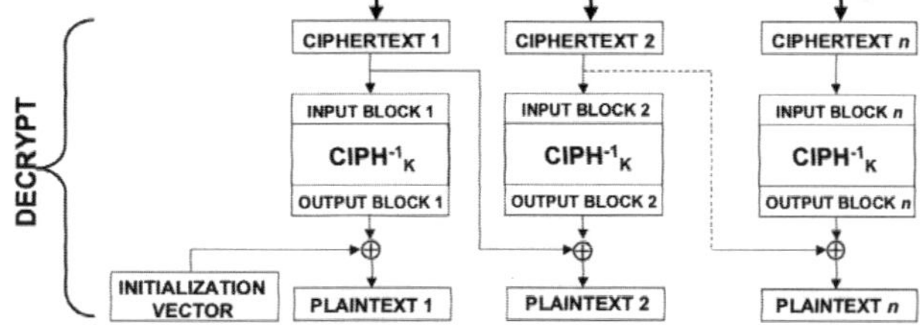

Figure 1. AES cipher block chaining mode structure.

IOP Conf. Series: Materials Science and Engineering 1073 (2021) 012057 doi:10.1088/1757-899X/1073/1/012057

3. Proposed method

This research consists of 3 stages, namely information system development, encryption algorithm implementation, and the last integrating encryption algorithm and information system. The encryption algorithms are used to encrypt digital images only. Three stages of this research are given in the following illustration.

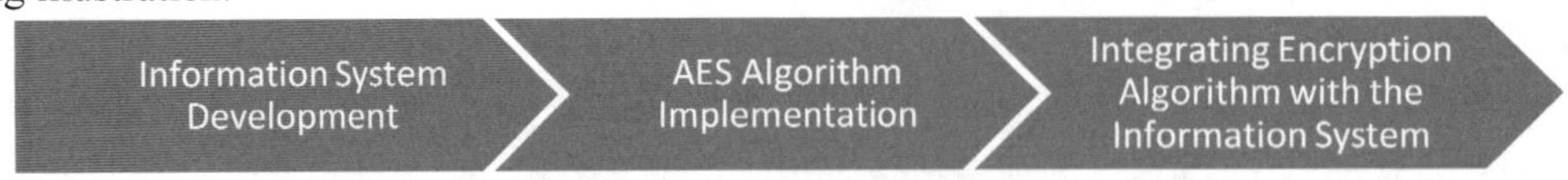

Figure 2. Research stages.

3.1. Information system development

The information system was mainly developed by using PHP language. The information system has several functions, including online queuing and recording medical records. In recording medical records in the digital image section, an encryption algorithm will be integrated to encrypt digital image data.

3.2. Advanced Encryption Standard (AES) algorithm implementation

This research focuses on encrypting digital image data using the AES algorithm. The illustration on the x-rays image is as follows. The image is taken from several open sources on the internet and the size of the image used for the experiment varies.

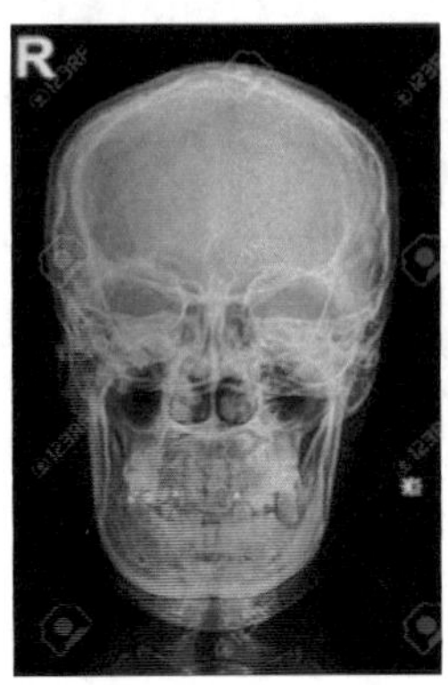

Figure 3. Head X-rays.

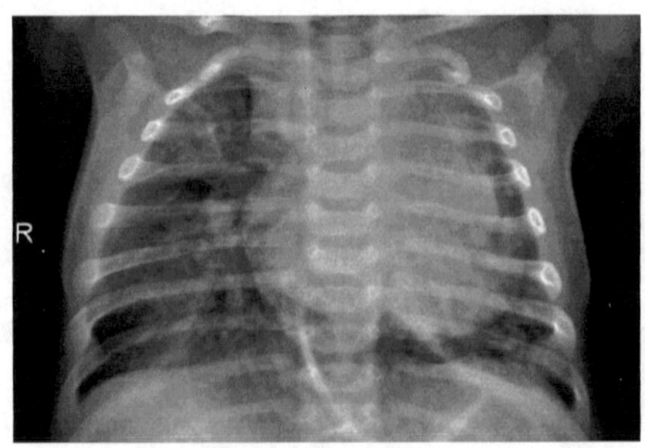

Figure 4. Chest X-rays.

4. Results and discussion

4.1. Usability measurement of information system

The information system which has develop can be seen on figure 5 and 6. After the development of information system then we measure the usability of this system by using system usability scale. The usability scale which used on this paper consist of 7 questions, such as:

- The information system has no benefit for the user
- The information system provides the promised online queue feature
- The information system is difficult for users to access
- The information systems can be operated comfortably without problems
- The information systems can produce data information accurately as needed
- The information systems can solve long time problems in patient queues
- The information system user interface is good

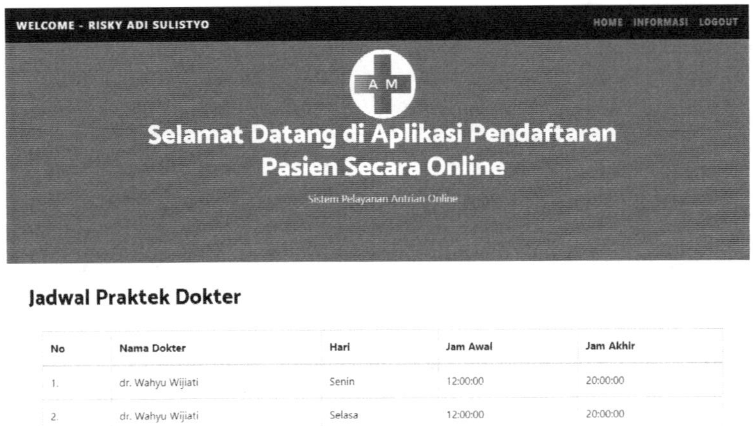

Figure 5. Illustration patient queue system.

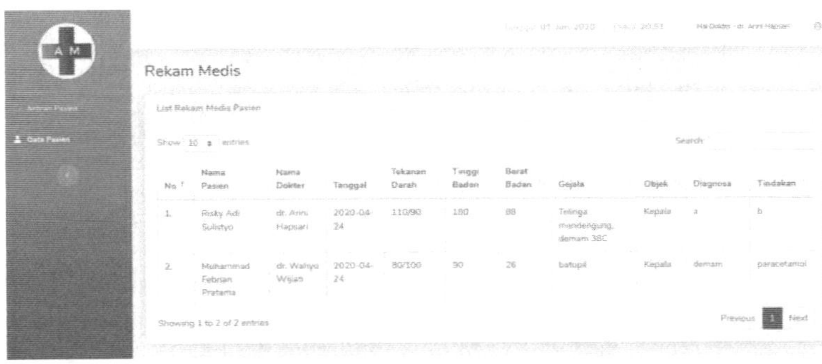

Figure 6. Illustration patient medical records management.

The SUS which already designed and distributed to 3 users then we recapitulate the questionnaire. We get score 69 of the overall usability score of the system. Based on the usability score, these scores above the average however, this system should be improved to get a better score. This can be caused due to the lack of the information system flow or the information provided to the user that is not precise so that the user feels moderate about the application being developed.

4.2. Implementation Result of Advanced Encryption Standard (AES)

Encryption of digital images has been implemented using the PHP language. The steps to encrypt medical images by converting medical images into strings with the base64 function in PHP. Then the string is encrypted using the AES algorithm in Figure 1. Based on the evaluation results, the AES algorithm can perform encryption and decryption with an average speed of 1MB / sec. Encryption illustration can be seen in Figure 7. The results of the decryption can be converted back into an image with the decode_base64 function in PHP.

Upload berhasil!
Link: 16296185-skull-x-rays-image.jpg
Encrypt Text =
/9j/4AAQSkZJRgABAQEASABIAAD//gA7Q1JFQVRPUjogZ2QtanBlZyB2MS4wICh1c2luZyBJSkcgSlBFRyB2NjIpLCBxdWFsaXR5ID0gOTUK/9sAQwAKBwcIBwYKCAgICwoKCw4YEA4N
encrypt :1.1084098815918 seconds
Encrypt Result =
OgODPJndRF/3I80vdALIWOpMLdLzD3WM2KEFDycax5B6JaBE8nAOCrh77HbLG4X9y0eAzxEwCkb3G8OYUUc5M75Dvg262AjFCrfiGFA8OUWavJ9kTZ3rrc/IaKPvT85Ixa3s3cfMa8r1gyHn
Decrypt Result =
/9j/4AAQSkZJRgABAQEASABIAAD//gA7Q1JFQVRPUjogZ2QtanBlZyB2MS4wICh1c2luZyBJSkcgSlBFRyB2NjIpLCBxdWFsaXR5ID0gOTUK/9sAQwAKBwcIBwYKCAgICwoKCw4YEA4N
time decrypt :1.1922659873962 seconds

Figure 7. Illustration encryption result.

IOP Conf. Series: Materials Science and Engineering 1073 (2021) 012057 doi:10.1088/1757-899X/1073/1/012057

5. Conclusion

Medical record documentation is essential to provide good quality services for patients, accurate and timely to reduce the risk of malpractice. And to help health care providers evaluate, plan patient handling, and maintain continuity of treatment. This research succeeded in developing an information system for recording medical records that has been integrated with an encryption algorithm so that only users who have access can view medical records.

References

[1] Min J, Kim Y, Lee S, Jang T W, Kim I and Song J 2019 The fourth industrial revolution and its impact on occupational health and safety, worker's compensation and labor conditions *Safety and health at work* **10**(4) 400-408

[2] Ajami S, Ketabi S and Torabiyan F 2015 Performance improvement indicators of the Medical Records Department and Information Technology (IT) in hospitals *Pakistan Journal of Medical Sciences* **31**(3) 717

[3] Bali A, Bali D, Iyer N and Iyer M 2011 Management of medical records: facts and figures for surgeons *Journal of maxillofacial and oral surgery* **10**(3) 199

[4] Tang P C, Fafchamps D and Shortliffe E H 1994 Traditional medical records as a source of clinical data in the outpatient setting In *Proceedings of the Annual Symposium on Computer Application in Medical Care* (p. 575) American Medical Informatics Association

[5] Vaidotas M, Yokota P K O, Negrini N M M, Leiderman D B D, Souza V P D, Santos O F P D and Wolosker N 2019 Medication errors in emergency departments: is electronic medical record an effective barrier? *Einstein (São Paulo)* **17**(4)

[6] P. R. INDONESI 2019 "https://kominfo.go.id," [Online] Available: https://web.kominfo.go.id/sites/default/files/users/4761/UU%2019%20Tahun%202016.pdf. [Accessed 24 August 2020

[7] Anonim 2020 "https://www.usability.gov," [Online]. Available: https://www.usability.gov/how-to-and-tools/methods/system-usability-scale.html. [Accessed 24 August 2020].

[8] Thomas N 2020 Usability Geek [Online]. Available: https://usabilitygeek.com/how-to-use-the-system-usability-scale-sus-to-evaluate-the-usability-of-your-website/. [Accessed 25 August 2020].

[9] P Patil, P. Narayankar and e. al., "A Comprehensive Evaluation of Cryptographic Algorithms: DES, 3DES, AES, RSA and Blowfish," in International Conference on Information Security & Privacy, Nagpur, India, 2015.

[10] Veness C 2020 Movable Type Scripts [Online] Available: https://www.movable-type.co.uk/scripts/aes.html. [Accessed 24 August 2020]

IOP Conf. Series: Materials Science and Engineering 1073 (2021) 012058 doi:10.1088/1757-899X/1073/1/012058

Object detection on dental x-ray images using deep learning method

D Suryani, M N Shoumi*, and R Wakhidah

Department of Information Technology, State Polytechnic of Malang, Jl. Soekarno-Hatta No. 9, Malang 65141, East Java, Indonesia

*milyun.nima.shoumi@polinema.ac.id

Abstract. Radiological examination has an important role in determining the diagnosis of dental problems and making decisions about the right type of treatment according to the case indications. Dental x-ray is a medical procedure for taking pictures of the inside of the mouth using radiation fluid, where the results are used diagnostically to help the dentist see the entire structure of the jaw bone and teeth, and dental problems that cannot be seen directly. Dental radiographic interpretation, which is generally performed by dentists, is a time-consuming and error-prone process due to high variations in tooth structure, low experience levels, and fatigue factors experienced by dentists. The workload of a dentist and the occurrence of misdiagnosis can be reduced by the existence of a system that can automatically interpret the x-ray results. To overcome these problems, a model will be developed to be able to detect objects in the dental panoramic x-ray images using Mask R-CNN, one of the methods in Deep Learning. Deep Learning is an artificial intelligence function that modelled the workings of human brain in processing data and creating patterns for use in decision making. With the detection of objects in panoramic x-ray image automatically, it is expected to save time, improve the quality of dental care, and also the quality of diagnosis made by dentists.

1. Introduction

In the world of dentistry, radiological examination has an especially important role in determining the diagnosis and making the right type of treatment decision according to the indication of the case. Dental x-rays or dental radiography are medical procedures to take pictures of the inside of the mouth using radiation fluids, where the results are used diagnostically to help dentists see the entire structure of the jawbone and teeth, as well as dental problems that cannot be seen directly by the eye. Without dental X-rays, many dental problems go undiagnosed. One of the most performed radiographic techniques is the panoramic radiograph technique. A panoramic x-ray can provide a complete picture of the upper and lower jaw at once, as well as various adjacent anatomical structures. Doctors perform this procedure to find out disorders in the mouth. This procedure can also be used to plan treatment for dentures, braces, tooth extractions, or dental implants.

Dental radiographic interpretation, which is generally performed by dentists, is a time-consuming and error-prone process due to high variation in tooth structure, low level of experience, and the large number of dental X-rays that need to be checked every day by the dentist, which can cause dentists to become exhausted. leading to a misdiagnosis or an inaccurate diagnosis, which in turn can hinder treatment. A dentist's workload and the occurrence of misdiagnoses can be reduced by having a system that can automatically interpret x-ray results. To overcome this problem, a model that can detect objects in a panoramic x-ray image of teeth will be developed using the Mask R-CNN method, which is one of the methods in Deep Learning. The Mask R-CNN method is proven to produce high accuracy in recognizing objects in an image with input data in the form of an annotated image. The object to be detected is a restoration object. Dental restoration is a treatment to repair cavities or damaged teeth, as well as dental repair treatment after endodontic treatment, to restore the normal function and appearance of teeth. Tooth restoration after endodontic treatment is the key to successful treatment. With the automatic detection of objects on panoramic x-rays, it is expected save time, improve the quality of dental care, and the quality of diagnosis made by dentists.

2. Background

2.1. Deep learning

The term deep learning was first introduced to the machine learning community by Rina Dechter in 1986 [1] and was introduced to the realm of artificial neural networks by Igor Aizenberg in 2000 [2], in the context of Boolean threshold neurons. Deep learning is a subset of machine learning and it is a function of artificial intelligence that imitate the way of human brain processes data and creates patterns for use in decision making. Deep Learning is a network that can learn unsupervised from unstructured or unlabelled data. In deep learning, a learning model performs classification tasks from images, text, or sounds. The model is trained using large labelled datasets and a neural network architecture that contains multiple layers. The deep learning scheme can be seen in Figure 1. Each layer is treated separately and trained sequentially. When the previous layer has been trained, the new layer will be trained using the encoding from the previous layer's input.

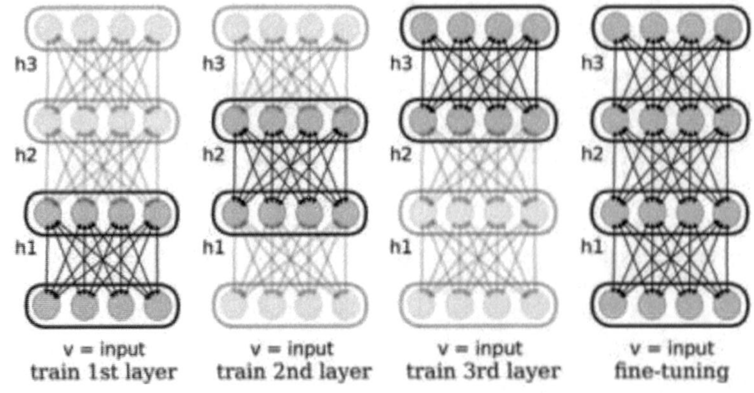

Figure 1. The deep learning scheme [3].

Most deep learning models are based on Artificial Neural Networks, and the most popular type is Convolutional Neural Network (CNN). CNN combines features learned from input data and uses 2D convolutional layers to make this architecture particularly suitable for processing 2D data, such as images. CNN eliminates the need for manual feature extraction, eliminating the need to identify the features used to classify images. CNN works by extracting features directly from images. This automated feature extraction makes deep learning models highly accurate for tasks such as object classification and detection. One type of CNN that has high accuracy in detecting an object is Mask R-CNN.

2.2. Mask R-CNN

Region-Based Convolutional Neural Networks (R-CNN), is a family of convolutional neural network models designed for object detection, developed by Girshick [4]. There are four main variations of the R-CNN approach. The salient aspects of each variation can be summarized as follows:

- R-CNN: The bounding box is proposed by a "selective tracing" algorithm, which is each stretched, and features extracted via a deep convolutional neural network, before the final set of object classifications is created with linear SVM.
- Fast R-CNN: Simplified design with single model, bounding box is still defined as input, but the merge area of interest layer is used after CNN in to combine regions and the model predicts class and region labels of interest directly.
- Faster R-CNN: Added Regional Proposal Network that interprets extracted features from deep CNN and learns to propose areas of interest directly.
- Mask R-CNN: Extension of Faster R-CNN which adds an output model to predict the mask for each detected object.

The Mask R-CNN Model introduced in 2017 [5] is the latest variation of the family model and supports object detection and object segmentation. The R-CNN Mask algorithm builds on the Faster R-CNN architecture with two main contributions:

- Replacing the ROI Pooling module with a more accurate ROI Align module
- Incorporate additional branches from the ROI Align module

This additional branch receives the ROI Align output and then feeds it into the two CONV layers. The output of the CONV layer is the mask itself. The R-CNN Mask architecture can be visualized in Figure 2.

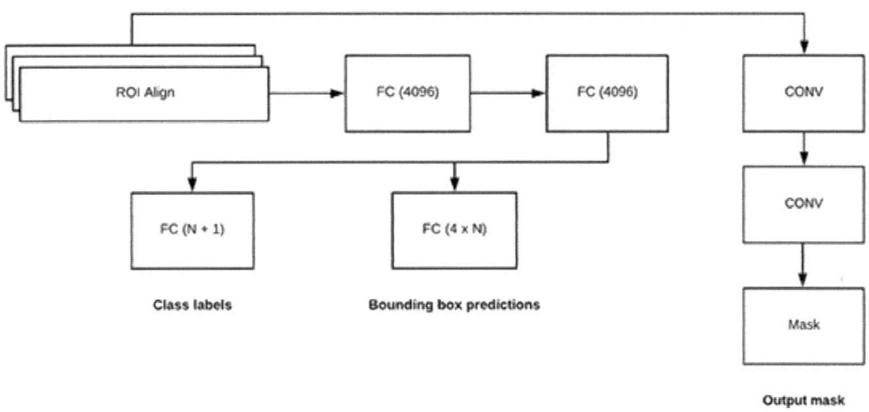

Figure 2. Mask R-CNN architecture [6].

3. Research methodology

3.1. System architecture

The architecture of object detection system in dental images using mask R-CNN can be seen in Figure 3.

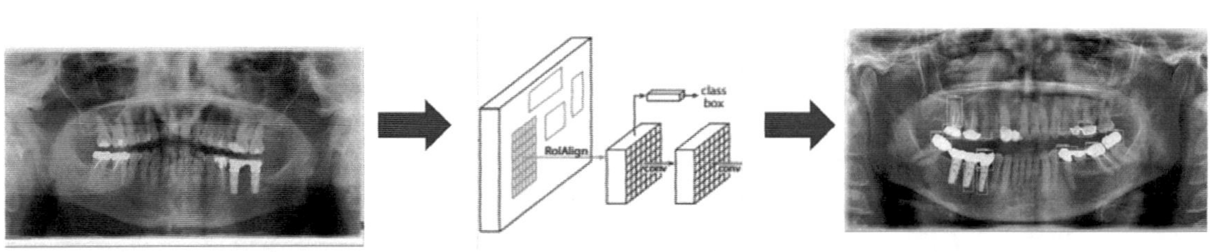

Figure 3. System architecture of object detection in dental images.

The first step is to take the images and extract the features using the ResNet 101 architecture [7]. These features act as an input for the next layer. Then take the feature map obtained in the previous step and apply the Region Proposal Network (RPN). This stage basically predicts whether an object is present in that region (or not) and get a map of the region or feature map that the model predicts contains multiple objects. The next step is to apply a pooling layer and convert all the regions to the same shape. Then this region is passed through a fully connected network so that the class labels and bounding boxes are predictable. For all predicted regions, Intersection over Union (IoU) will be calculated with the ground truth box. The IoU calculation is as follows:

$$IoU = Area\ of\ intersection\ /\ Area\ of\ unity \tag{1}$$

If the IoU is greater than or equal to 0.5, it is considered as a region of interest. If not, the region will be ignored. It is implemented for all regions and then selects only one set of regions which the IoU is greater than 0.5. After having a RoI based on the IoU value, the next process is to add a mask branch to the existing architecture. This returns the segmentation mask for each region containing the object. This returns a mask with a size of *28x28* for each region which is then increased for inference. The model has segmented all objects in the image. This is the final step in Masks R-CNN where it performs mask prediction for all objects in the image.

4. Results and discussion

4.1. Dataset
The data used in this study are panoramic x-ray images of teeth that have been annotated using the open-source Visual Object Tagging Tool (VoTT) software. This data set consisted of anonymous and de-identified panoramic dental X-rays from 116 patients, taken at the Noor Medical Imaging Centre, Qom, Iran [8]. Subjects covered a wide variety of dental conditions from healthy, to partial and complete toothless cases. The dataset is annotated with one class, which is restoration, and it is divided into two parts, 110 images for the training process and 6 images for the testing process.

4.2. Detection results
Experiments were carried out using a GPU with a number of training steps per epoch of 100, the learning rate was 0.006, and the minimum value for detection confidence was 0.9. The pre-fit R-CNN Mask model on the MS COCO object detection dataset can be used as a starting point and then adjusted to a specific dataset, in this case, a dental x-ray image. The model is trained on the output layer (heads) with a layer on top to speed up the learning process, using 5 epochs. Detection results for 6 testing data can be seen in Figure 4.

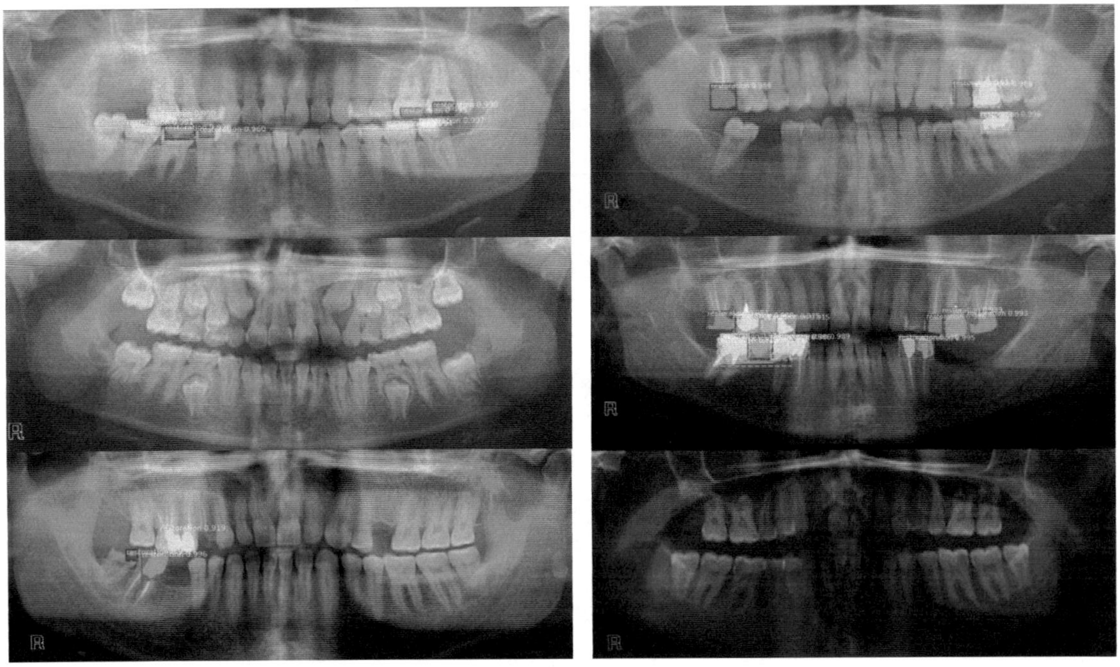

Figure 4. Experiments results on testing image.

4.3. Analysis
Model performance for object detection is evaluated using mean absolute precision, or mAP. The bounding box is predictable so it can be determined whether the bounding box prediction is good or not based on how well the predicted and actual bounding boxes overlap. It can be calculated by dividing the area of the overlap by the total area of the two bounding squares, or the intersection divided by the union, it is known as the "intersection over union," or IoU. The perfect bounding box will have IoU of 1.

IoU is a standard for assuming positive predictions from the bounding box. Precision refers to the percentage of bounding boxes predicted correctly (IoU> 0.5) of all predicted bounding boxes. Meanwhile, recall is the percentage of bounding boxes that are predicted correctly (IoU> 0.5) of all objects in the photo.

5. Conclusion
The object detection model on panoramic x-ray images of teeth has been developed and has successfully detected the location of the restoration object in the data. Only one type of object in the panoramic tooth image was detected, namely tooth restoration. This is due to the difficulty in finding a panoramic dental image dataset that contains several objects, such as dental implants and endodontic treatments. The results of object detection in the panoramic tooth image show confidence level values in the range of 0.91 to 0.96. This research can be developed for several different objects in one image (multiple object). So, it is necessary to search for deeper datasets for further research development.

References
[1] Dechter R 1986 Learning while searching in constraint-satisfaction problems. University of California, Computer Science Department, Cognitive Systems Laboratory

[2] Aizenberg I, Aizenberg N N, and Vandewalle J P 2000 *Multi-Valued and Universal Binary Neurons: Theory, Learning and Applications.* Springer Science & Business Media

[3] Ludovic A, Sébastien R, Sylvain C, and Hélène P 2011 An Introduction to Deep Learning.

European Symposium on Artificial Neural Networks (ESANN), Bruges, Belgium
[4] Girshick R, Donahue J, Darrell T, and Malik J 2014 Rich feature hierarchies for accurate object detection and semantic segmentation. *Proceedings of the IEEE conference on computer vision and pattern recognition*
[5] He K, Gkioxari G, Dollár P, and Girshick R 2018 Mask R-CNN. arXiv:1703.06870v3 [cs.CV]
[6] Rosebrock A 2018 Deep Learning For Computer Vision With Pyhton. Pyimagesearch.
[7] He K, Zhang X, Ren S, and Sun J 2016 Deep residual learning for image recognition, in: *Proceedings of the IEEE Conference on Computer Vision and Pattern Recognition*, pp 770–778.
[8] Abdi A and Kasaei S 2020 Panoramic Dental X-rays With Segmented Mandibles. Mendeley Data, v2. DOI: 10.17632/hxt48yk462.2.

Public service innovation through the application of mobile-based smart city concepts in Sukun sub-district, Malang City

D K P Aji*, R Ariyanto, F Rahutomo, A Ismail and K S Batubulan

Information Technology Department, State Polytechnic of Malang. Jl. Soekarno Hatta No.9, Malang 65141, West Java, Indonesia

*deddy_kusbianto@polinema.ac.id

Abstract. Public service provides services to the public carried out by the government whose services can be in the form of goods, services, or administration. The sub-district is a public service provider from the application stage to the publication phase of the document. Where people need not go to several agencies to take care of licensing documents, but enough to come to the district in their respective regions. Districts as the spearhead of government administrators must carry out their functions as the embodiment of excellent public service. However, in reality, public services at the sub-district level are of poor quality. Less responsive, less informative, and convoluted bureaucracy is often a major problem in service. Even though in the current technological era, the sub-districts should adopt technology for their activities. Innovation in using smart city concept technology in sub-districts is the key to solving existing problems because integrating entities in the smart city concept can facilitate the exchange of information, speed of data exchange, and rapid decision-making by service officers the public. So that the public services organized by the sub-district become more quality.

1. Introduction

Sukun sub-district has an area of 2,655.19 ha which divided into 11 villages [1]. With such a large population, the local government's public services in this district included in the category of excellent service. But the reality is that limited facilities, infrastructure, and human resources are the main problems in maintaining this public service. For example, the number of residents is not comparable to the number of service employees. To makes the people who need the assistance have to be willing to queue for a few hours to get the intended service and limitations of facilities and infrastructure that have not fully supported.

Based on the above description, there are several problems faced by Sukun Sub-District of Malang City. Among others is Less responsive, where responses to various complaints, aspirations, and expectations of the community are often slow or even ignored. Less informative where information should be conveyed to the public, delayed, or even not to the public. And bureaucratic where the service is carried out through various levels, thus causing the completion of the service too long [2].

The above problems solved through several systems aimed explicitly at specific purposes such as information systems and integrated service systems. But the increasing system in an instance will increase the difficulty, especially for the community in using the system. Integration is currently a trending word in IT [3]; one of its applications is smart city concepts. Smart City as a concept of urban planning integrated with a wide range of development and combined with the development of information and communication technology with the objectives among others, creating habitable, advanced and modern urban planning and development, improving regional productivity and economic competitiveness and building the foundation of Indonesia smart nation [2]. The concept of intelligent cities believed to be the solution to urban development in the area. Smart City designed to increase the productivity of the people who live in it so that the result of the structuring and management of the city done by utilizing information and digital technology optimally in all aspects—starting from building management system, environmental quality management, and public service.

In this article, the author tries to develop a public service information system with the concept of a smart city that integrates several services that can be accessed and used directly by the public in a mobile-based application. Mobile-based apps are the most appropriate application used as a connecting device between the Sukun sub-district government and the community. It hoped that this information system can help people access services easily and quickly.

2. Methods

The system development method used is the waterfall method. The waterfall method is a classic model that is systematic, sequential in building software. Here are the stages contained in the waterfall method that can be seen on figure 1, namely:

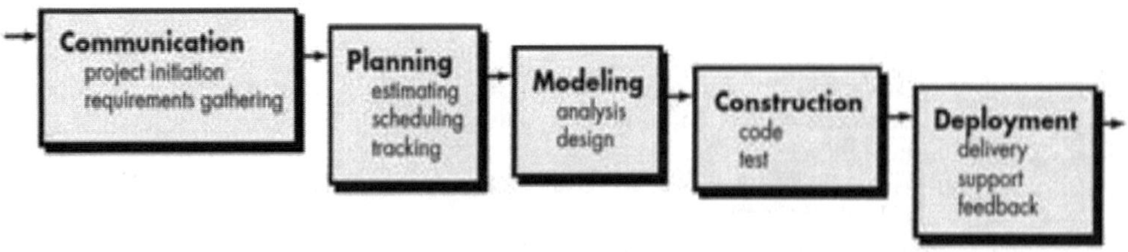

Figure 1. Waterfall software development model [3].

2.1. Communication

This step is an analysis of the needs of the software and the stage to conduct data collection by meeting with customers and collecting additional data in journals, articles, and the internet.

2.2. Planning

The planning process is a continuation of the communication (analysis requirement) process. This stage will generate a user requirement document or can be said as data related to the user's desire in the creation of software, including the plan to be carried out

2.3. Modelling

This modeling process will translate the requirements of need into a predictable software design before coding is created. The process focuses on designing data structures, software architectures, interface representations, and procedural details (algorithms). This stage will generate a document called software requirement.

2.4. Construction

Construction is the process of creating code. Coding or coding is a translation of a design in a language that can be recognized by a computer. The programmer will translate the transaction requested by the user. This stage is the real stage in working on software, meaning the use of the equipment will be maximized in this stage. After coding is complete, testing of the system has been done. The purpose of testing is to find errors to be corrected.

IOP Conf. Series: Materials Science and Engineering 1073 (2021) 012059 doi:10.1088/1757-899X/1073/1/012059

2.5. *Deployment*
This stage can be said to be final in the creation of a software or system. After doing the analysis, design, and encoding, the user uses the complete order. Then the software that has been created should be done maintenance periodically.

3. Program development

3.1. *Changes in business processes in institutions*
From the application to be created, the implementation will change existing business processes, business process changes can be seen in the following image:

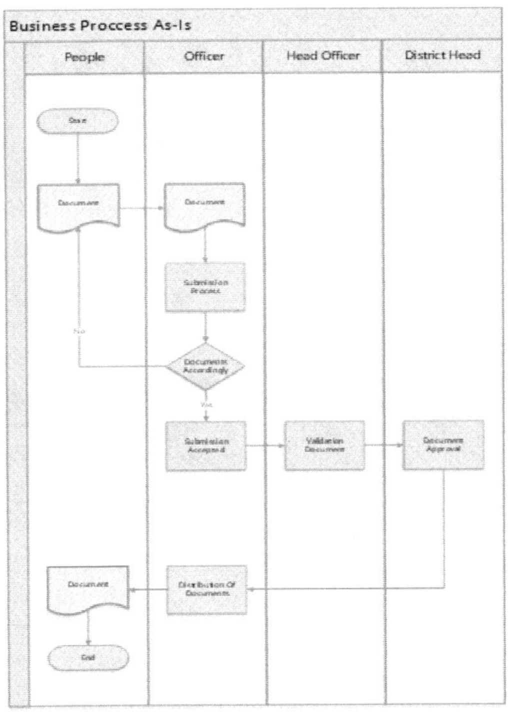

Figure 2. Business Process As-Is.

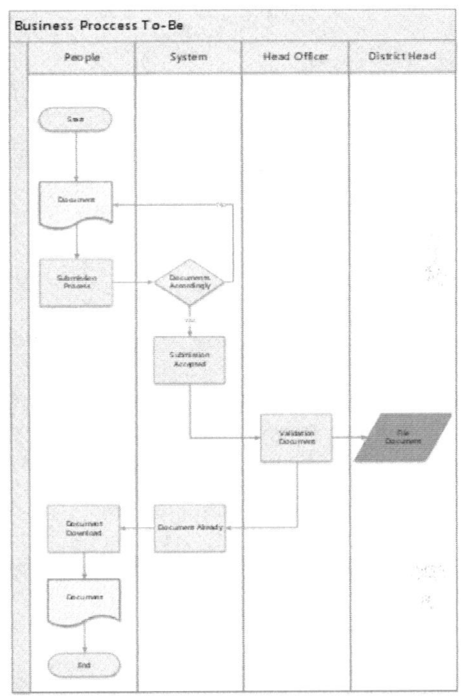

Figure 3. Business Process To-Be.

The above picture explained that Business Process As-In (figure 2) as a previous business process wherein the processing of licensing documents will involve many parties that will rely heavily on the working officer. In contrast, in Business Process To-Be (figure 3) as a business process submitted, the method of processing documents will no longer depend on the officer ultimately, but instead return to the public, the officer only validates and approves the text and subsequently the system that will process it. In addition to speeding up preparing documents, this change can make it easier for the community to complete the process independently.

3.2. *Application architecture design*
The design of public service application architecture with Mobile Based Smart City Concept consists of the first user or user using the android platform that accesses the system with the internet network using the webserver. Then, the web server will display data or information retrieved from the database. The second user or user is the officer who plays in the government, who will later provide feedback or input. Presented by the first user. Here's the design of the app architecture, shown on figure 4:

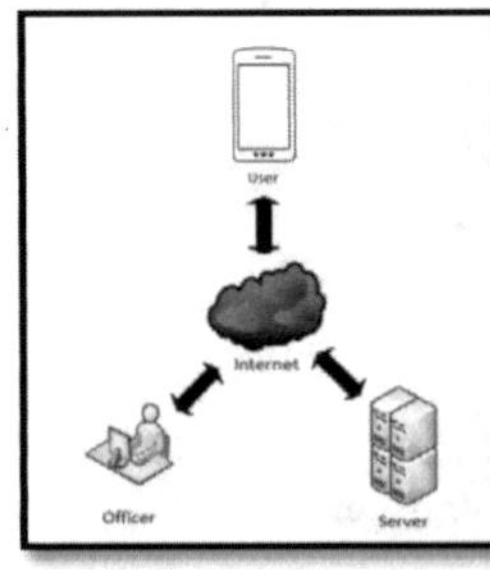

Figure 4. Application architecture.

3.3. *Concept design*

The concept of a smart city that existed only in developed countries is currently widely adopted by cities in Indonesia. For example, the Provincial Government of DKI Jakarta, Surabaya City Government, and Bandung City Government implemented a smart city. As a major city, especially the Sukun sub-district, Malang needs a breakthrough in supporting its government's performance to realize a sound governance system. Several ways have done by improving infrastructure, transportation, and other public services. But its nature is still limited and passive because there is no direct communication between the government and the community [4]. Therefore, it needs to designed in a system that can bridge the community and government in interacting. In addition to providing information directly to the community, the city can also participate in development through this system.

3.4. *Program implementation*

After doing the analysis and design stage, then the next step is the implementation of the program. The testing of this program aims to implement the design that has been created. The results of the application are as follows on figure 5 and figure 6:

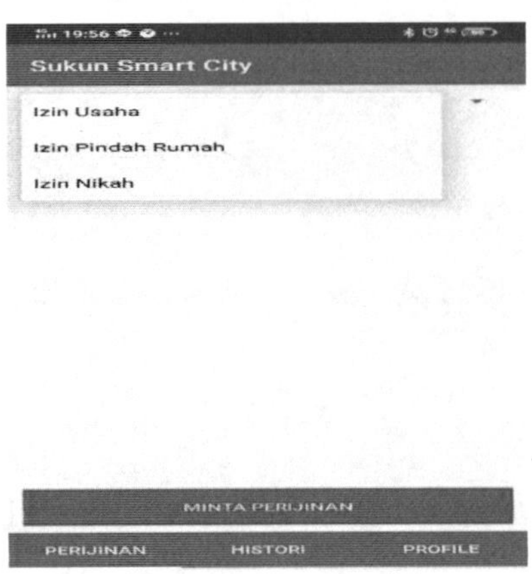

Figure 5. User layout menu.

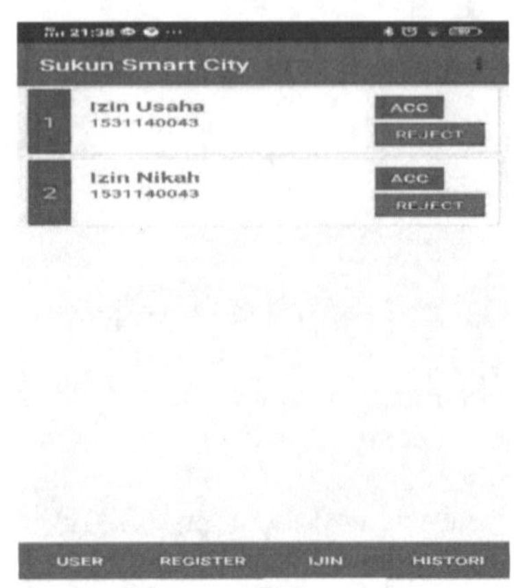

Figure 6. Admin option menu.

The image above is the implementation of a system where there are admin logins and logins allocated by the population to function for security in the data in the order. Furthermore, the registration page intended for residents who enter personal data and the desired documents. After registering, the resident will immediately recorded into the Smart City form. On the last way, the admin will validate the resident's permit is complete and appropriate or not. Furthermore, the user can track the complainant's status so that there is no need to come to the sub-district to wait, and the results of the submission of documents can directly viewed in the user's hand phone.

4. Results and discussion

After the system has created, it has implemented in the sub-district office. In this study, we only focus on making cellular-based applications as an existing information system. From the implementation, results obtained the following results:

Get positive results based on the User Acceptance Test conducted with the sub-district. Then from the results of application testing carried out using the white box method, 80% of the system runs as desired, and 20% gets revisions for file size and file naming. For naming, we include encryption, so there will be no duplicate names and files. Based on the system's implementation above, the system will be developed again for other features added according to community needs.

5. Conclusion

With the use of this application, this public service can meet the needs of the civil service in the district that no longer need to come and queue at the area to get information and application procedures especially licensing applications; also, users can even know to what the application process is so that there is not much direct interaction with the sub-district.

References

[1] BPS Kota Malang 2020 *Kecamatan Sukun Dalam Angka 2019* (Malang: BPS)
[2] Sinambela L P 2011 *Reformasi Pelayanan Publik* (Jakarta: Bumi)
[3] Pratama I P A E 2014 *Sistem Informasi dan Implementasinya* (Bandung: Informatika)
[4] Talis Z 2019 Pengaruh Pelaksanaan Good Governance Terhadap Pelayanan Publik (Studi Pada Kantor Kelurahan Temas, Kecamatan Batu, Kota Batu) *Jurnal Ilmu Sosial dan Ilmu Politik* **VIII**(2) 178-185

IOP Conf. Series: Materials Science and Engineering 1073 (2021) 012060 doi:10.1088/1757-899X/1073/1/012060

Web application implementation of Android programming learning assistance system and its evaluations

Y W Syaifudin[1,*], N Funabiki[1], M Kuribayashi[1], M Mentari[2], P Y Saputra[2], Y Yunhasnawa[2] and F Ulfa[2]

[1] Dept. Electrical and Communication Engineering, Okayama University, Okayama, Japan

[2] Dept. Information Technology, State Polytechnic of Malang, Malang, Indonesia

*p52m5jeb@s.okayama-u.ac.jp

Abstract. Currently, Android has been installed on more than 70 percent of the smartphone devices in the world, which has significantly increased the need for Android-based application engineers. Many IT departments in universities and professional schools have included the Android-based programming into the main subjects. To improve Android programming educations, we have developed Android Programming Learning Assistance System (APLAS), a platform to accommodate the students learning Android programming independently. APLAS adopts the test-driven development method to realize automatic validations of students' answers to the programming assignments. Previously, APLAS has been implemented as a desktop offline application. In this paper, we present a Web application implementation to manage the learning activities online. This implementation consists of three main parts, namely, Validator, Web Interface, and Database System. To evaluate the effectiveness, we offered three learning topics, Basic UI, Basic Activity, and Advanced Widgets, into this Web application, and asked 60 students in Indonesia to solve them. The results showed that the Web application has well been accessed and smoothly utilized by them. The reliability of the validator program was confirmed by handling more than 180 submissions.

1. Introduction

Mobile applications have become one of the main fragments in the global economy after enormous achievements of smartphones. Then, mobile application programmers become one of the most popular and highly demanded jobs in the IT sector. Currently, Android has been installed on more than 70 percent of the smartphone devices in the world. The statistics shows that Google Play is the leading application store and provides more than 2.5 million applications in the 1st quarter of 2020 [1]. With this demand, many professional schools and universities have carried out the mobile programming courses, especially for Android application developments.

To improve the effectiveness of Android programming by students, we have proposed the Android Programming Learning Assistance System or APLAS [2]. APLAS provides a self-learning platform with automatic validations of the students' answers for Java-based Android programming, by adopting a test-driven development method. Unit testing is applied to validate the requirement specifications in the source code from a student. For unit testing in Android application, JUnit and Robolectric are adopted.

In previous implementations, students must solve the learning assignments using Android Studio and validate source codes by themselves using given test codes [3]. Then, the students report the validation results and source codes to the teachers. In this case, the teachers must check the validity of the answers one by one. APLAS has offered 13 learning topics with 8-10 tasks for each. Each task contains one or two test codes. If a teacher has 50 students, he/she must execute at least 400 test codes for one learning topic. This situation makes the teacher difficult to handle the students with many learning topics.

In this paper, we present the web application implementation with the automatic validation process for APLAS. This web application can be used by the students to obtain the learning materials, manage the assignments, and submit the results. In the server, the validator program runs in the background to detect new submissions and validate them automatically. The result of the validation process can be accessed on the web interface by the students and the teachers.

To evaluate the effectiveness, we implemented three learning topics, Basic UI [3], Basic Activity [4], and Advanced Widgets [5]. Then, we asked 60 students in Indonesia to solve the assignments and submit the results to the web application. The validation process has taken a long time relatively, depending on the number of test files, test methods, and types of unit testing. The results confirmed the reliability of the validator program that has validated than 180 submissions.

The rest of this paper is organized as follows: Section 2 studies related works in literature. Section 3 explains the overview of APLAS. Section 4 presents the model of web application. Section 5 describes the online validation method. Section 6 presents the evaluation results of web application and validation process to 60 students with its discussion. Finally, Section 7 concludes this paper with future works.

2. Related works

In 2012, Bosnić et. al. introduced ORVViS, a module in the Moodle learning management system, to validate source codes from students [6]. However, it did not support the grading system, including checking the source codes according to the given specification or assignment from teacher. It focuses mainly on syntax.

In 2013, Funabiki et al. proposed a Web-based Java Programming Learning Assistant System (JPLAS) for Java programming language self-learning [7]. JPLAS adopts JUnit for unit tests to validate source codes as assignment answers from students. In this system, a student writes a source code by reading the statement and the test code, then it will be tested at the server automatically.

In 2014, Yulianto and Liem introduced a source code analyzer to grade programming assignments automatically [8]. Their study produced an application called SCAGrader consisting of a Java engine application and web-based user interface.

In 2016, Staubitz et.al introduced CodeOcean as a web-based platform to practice programming course exercises [9]. It is designed to be used in programming MOOCs. Also, it allows teachers to plan programming exercises, which can be graded automatically using unit tests as a quality measurement.

Khan et al. presented AUTOGRADER in 2019, a tool to decide the correctness of programming assignments automatically [10]. It also examines the semantically different execution paths the codes and the reference implementation. This tool successfully decreases the teacher's efforts in writing test cases and delivers the grading more flawless.

3. Review of Android Programming Learning Assistance System (APLAS)

In this section, we review APLAS in our previous studies [2,5].

3.1. System overview

APLAS is a students' self-learning platform for Android programming learning with automatic validation features [3]. To validate source codes from students, APLAS adopts a Test-driven Development method based on Java unit testing. The developed project is validated by executing the given test codes. To validate the correctness of the program codes and the UI components in an Android project, APLAS adopts JUnit and Robolectric for integration testing. On these tools, APLAS provides a set of learning assignments to be solved by students individually.

3.2. System specification
APLAS implements a learning platform for the students to solve the learning tasks on PCs. Using Android Studio, a student can build a specific Android application to solve learning tasks and validate the built projects by him/herself using the given test codes. Android Studio offers tools to facilitate Android developments to work on tasks using the combinations of Java and XML.

3.3. Learning model
In APLAS, the learning materials consist of several topics. Each topic represents one learning assignment. To cover the wide contents in Android application programming subjects, we grouped the learning topics into four stages: User Interface, Interactive Application, Content Provider, and Service Interaction. For a topic, students will get a package consisting of guide documents, test codes, and supplement files. The test code is applied to validate the code written in Android project.

4. Proposal of web application for APLAS
In this section, we explain the design and implementations of the proposed web application for APLAS.

4.1. System architecture
The web application for APLAS adopts a client-server model, and has two main sections, a client section and a server section, as illustrated in Figure 1. The client section interacts with students, contains the learning platform, and the web client. The server section contains the validator and the web server. The web server accesses the database to manage the data.

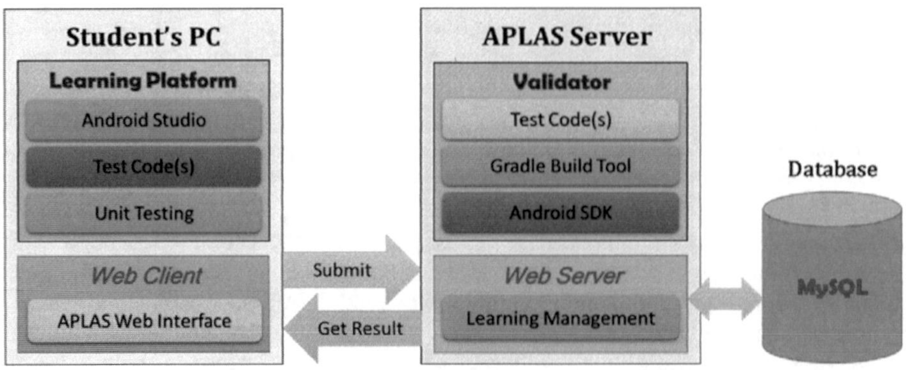

Figure 1. Improved system architecture of APLAS.

4.2. Learning platform
The learning platform is the part used by the students to solve the assignments on PCs. This part adopts *Android Studio* as the most popular *Integrated Development Environment (IDE)*, to create *Android* applications based on Java programming language. The project created by the student will be validated using given test codes that will run on *JUnit* and *Roboletric* manually. The student can get results directly in *Android Studio* when the execution of a test code finished.

4.3. Web application
To facilitate the students in learning managements and online assignment validations, the web application for *APLAS* provides the interface to students in Figure 2. The interface assists students in using the learning packages. After they finish a topic, they can submit the results of an assignment to this web application. Then, the system will validate them automatically, and record the learning progress in the database.

The validation process will be carried out by the validator, a thread program that validates any new submission from a student. The detailed validation results will be recorded in the database and will be displayed on the web interface. If the validation results are incorrect, the messages will explain the incorrectness. Then, students should improve the answer codes and resubmit them to the web application. Students can monitor the achievements of their learning and track the history of their submissions.

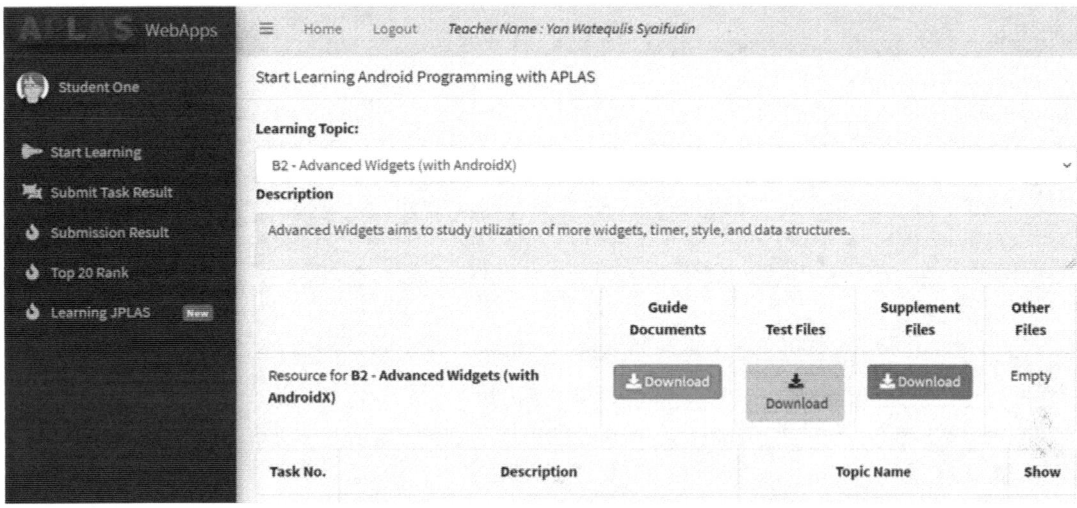

Figure 2. The web interface for students to access learning materials.

4.4. *Validator*

Validator is the background process written in Java and has the functions to detect new submissions from students and to validate each submission. The validation process is achieved by executing the given test codes using Gradle [11] in the form of an Android project. Gradle will integrate the Android project with the testing tools (JUnit and Robolectric) and Android SDK library. The validation results will be stored in the database so that it can be accessed from the web application. Each validation process will record the execution time, error messages (if any), and execution error messages (if occur).

5. Online validation model

In this section, we describe the student's learning process and the online validation model of their answers submitted using the web application.

5.1. *Learning process*

There are three steps for students to learn Android programming in APLAS using the web application, as shown in Figure 3. First, they have to open the web interface on a web browser and access the "Start Learning" page. As shown in Figure 2, they must choose a topic and download the learning materials that contain guide documents, test files, supplement files, and other files (if available). Then they use Android Studio to work with the assignment.

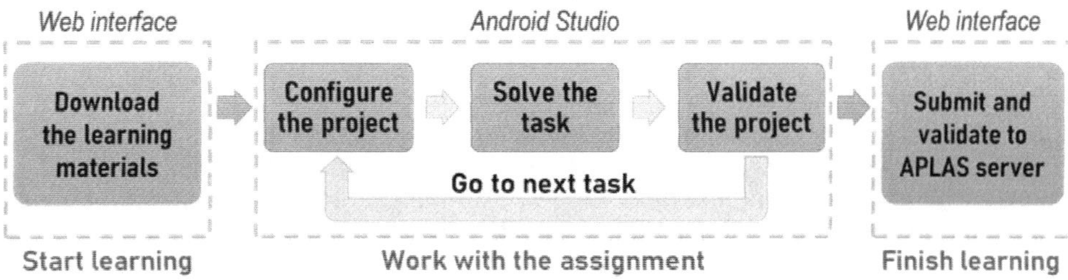

Figure 3. Learning process using web application.

IOP Conf. Series: Materials Science and Engineering 1073 (2021) 012060 doi:10.1088/1757-899X/1073/1/012060

A topic in APLAS contains several tasks that must be solved one by one sequentially, to finish the assignment for one topic. A student must follow the instructions in the guide document provided for each task. The student has to check the correctness of the answer using the provided test code(s) until he/she can solve all the tasks in the topic. Finally, he/she must submit the results and upload the necessary files on "Submit Task Results" page in the web application.

5.2. Validation method
After submitting the assignment results, the student has to wait for the results of the validation process in the APLAS server. As shown in Figure 4, there are five steps to validate a new submission. First, the validator receives a new submission from a student. Then, it prepares a folder that contains the Android project files associated with the submission. In this step, all the uploaded files will put in the folder. After the folder is completed, the validator will execute the test codes automatically using the Gradle command prompt, and get the results of validation. The results will be submitted into the database. The student can access the results using the web application in "Submission Results" page.

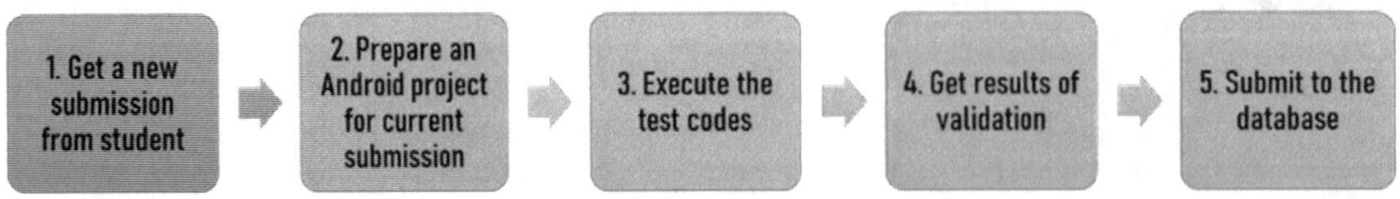

Figure 4. The validation process for a new submission by validator.

6. Evaluation
In this section, we clarify the effectiveness of the web application by evaluating it to students online.

6.1. Evaluation setup
To perform online implementation, we chose randomly 60 undergraduate students of a university in Indonesia that were taking a mobile programming course in an IT department. We asked them to download the learning materials for the three topics, Basic UI, Basic Activity, and Advanced Widgets. Then, they solved them on their PCs using Android Studio. Here, we provided two versions for each topic, Android support library version and AndroidX library version. The students can use one of them independently. For each student, we gave three days to solve three assignments. If some students failed to solve all or some of the assignments, we gave the second chance to solve and submit the results again one week later.

In this evaluation, we used the server that runs on Ubuntu Linux with VMWare virtual machine. This virtual server is with Processor AMD Opteron(TM) Processor 6238 4 Core 2.6 GHz, RAM 11 GB, Storage 100 GB, and internet address "aplas.polinema.ac.id".

6.2. Evaluation results
In general, the results for the three topics show that finally, all the students can pass all the tasks, as shown in Table 1. Only in solving Advanced Widgets topic, two students failed in the first chance. However, they can pass it in the second chance. The observation of students in accessing the web application shows that there is no serious problem on the registration, login, downloading, learning materials, submissions of assignment results, and getting validation results.

6.3. Discussion on evaluation results
All the students took various time to solve one learning topic from 63 minutes until 96 minutes on average. If compared with previous studies [2,5], the students' performances on online learning are faster than the results on in-class learning. Moreover, all the submissions can be validated by the validator successfully. It took time from 176 seconds until 294 seconds on average. It confirmed the effectiveness

of the online validation method. The detail of validation results can be accessed by students on the web application, including the messages to explain the incorrectness of the failed tasks.

In performance evaluations, we summarize the conditions that affect execution time:

- the number of test files and test methods has the big impact on the execution time,
- the number of assertion commands in a test file has the small impact on the total time, and
- the initial method of Robolectric takes around 85% of the total execution time.

Table 1. Validation results of students' submissions in three topics.

Topic	First chance		Second chance		Total		Avg. solving time	Avg. time to validate
	#Passed	#Failed	#Passed	#Failed	#Passed	#Failed		
Basic UI	60	0	0	0	60	0	96 min.	233 sec.
Basic Activity	60	0	0	0	60	0	63 min.	176 sec.
Advanced Widgets	58	2	2	0	60	0	78 min.	294 sec.

7. Conclusion

This paper presented the design and implementation of the web application with the online validation for Android Programming Learning Assistance System (APLAS). The students can download the learning materials from the web interface, manage the learning process, and submit the assignment's answers at any time, where the validator will validate it automatically. The evaluation results confirmed the effectiveness and reliability of our proposal. In future works, we will implement the web interface for the teachers, improve the accessibility of the web application, and the reliability of the validator.

References

[1] Clement J 2020 Google Play: number of available apps 2009-2020 *Statista GmbH*

[2] Syaifudin Y W, Funabiki N, Kuribayashi M, and Kao W C 2019 A proposal of Android Programming Learning Assistant System with implementation of basic application learning *International Journal of Web Information Systems* **16**(1) 115-135

[3] Syaifudin Y W, Funabiki N, and Kuribayashi M 2019 Learning model for Android programming learning assistant system *in Proc. IEICE General Conference 2019, Tokyo*

[4] Syaifudin Y W, Funabiki N, and Kuribayashi M 2019 An implementation and evaluation of basic activity topic for interactive application stage in Android programming learning assistance system *in Proc. FIT 2019, Okayama, Japan*

[5] Syaifudin Y W, Funabiki N, and Kuribayashi M 2020 An implementation and evaluation of Advanced Widgets topic for interactive application stage in Android programming learning assistance System *in Proc. 8th Int. Conf. Inf. and Edu. Tech., Okayama*

[6] Bosnić I, Mihaljević B, Orlić M, and Žagar M 2012 Source code validation and plagiarism detection - technology-rich course experiences in *Proc. 4th Int. Conf. Comp. Supp. Edu., Porto*

[7] Funabiki N, Matsushima Y, Nakanishi T, Watanabe K, and Amano N 2013 A Java programming learning assistant system using test-driven development method *IAENG Int. J. Comp. Sci.* **40**(1) 38-46

[8] Yulianto S V and Liem I 2014 Automatic grader for programming assignment using source code analyzer in *Proc. Int. Conf. Data. Soft. Eng., Bandung, Indonesia*

[9] Staubitz T, Klement H, Teusner R, Renz J, and Meinel C 2016 CodeOcean - a versatile platform for practical programming exercises in online environments in *Proc. Glo. Eng. Edu. Conf., Abu Dhabi*

[10] Khan I A, Iftikhar M, Hussain S S, Rehman A, Gul N, Jadoon W, and Nazir B 2020 Redesign and validation of a computer programming course using inductive teaching method *PLOS ONE* **15**(6) 1-21

[11] DiMarzio J 2017 *Beginning Android programming with Android Studio* (John Wiley & Sons) 2017

IOP Conf. Series: Materials Science and Engineering 1073 (2021) 012061 doi:10.1088/1757-899X/1073/1/012061

Developing vocabulary card base on Augmented Reality (AR) for learning English

I F Rozi*, E Larasati and V A Lestari

Informatics Engineering Study Program, Information Technology, Politeknik Negeri Malang

*imam.rozi@polinema.com

Abstract. The application of technology in the world of education can affect the quality of students in understanding learning material because technology is one of the important elements to assist in the teaching and learning process. One educational application that can be developed is Augmented Reality (AR) technology. AR is one of the learning media that can be used to combine virtual objects or objects into a real environment with the virtual world to be able to interact in realtime. English is a foreign language that must be mastered. In learning English, students need to be given adequate vocabulary learning that can be provided through reading or listening activities. The importance of adequate vocabulary in English to support the smoothness of oral communication, where there are 2 things that inhibit the ineffectiveness of oral communication namely students do not have adequate vocabulary mastery and do not know how to use vocabulary in their spoken language. Effective and interactive learning media will facilitate students in learning vocabulary. This research an instructional media based on augmented reality vocabulary cards will be made for learning English vocabulary. Learning English vocabulary developed is a basic vocabulary consisting of object recognition and greeting. The learning media will display real objects and sounds, so that the object looks real and the ease of pronunciation of English consonants correctly.

1. Introduction

Augmented Reality technology can be used in language teaching and learning, where it brings real-life objects that can create convenience in learning, so it can affect the quality of students in understanding learning material because technology is one of the important elements to assist in teaching and learning [1]. One of the challenges in learning is in the process of mentoring students when learning, so this makes teachers have to take advantage of a tool with new technology in teaching, one of which is by using Augmented Reality (AR) [2]. The use of instructional media that is less interactive in the learning process causes students to not understand the material to be delivered. The more interesting the media used and communicative, the more students will be interested in learning the subject matter. Some advantages in using learning media in the teaching and learning process are that it can increase desires and interests, motivation and stimulation of learning activities and even bring psychological influence on students [3]. One of the benefits of technological development is that it can be used in a more creative, effective and efficient learning process [4]. One educational application that can be developed is Augmented Reality (AR) technology. Augmented Reality, AR has a wide range of applications in teaching and learning contexts, one of which is the use of AR applications in the form of 3D objects that combining real and virtual that seem to coexist with the same real world [5]. AR is the incorporation of objects that exist in cyberspace into the real world in the form of two dimensions or three dimensions that can be touched, seen, and heard [6].

Foreign language learning is the process of learning a language that is not used as a communication language in one's environment [7]. The foreign language that must be mastered is English, so that in early childhood education until college is introduced and taught English. Effective oral communication skills are very important for students of English as a foreign language. To support the effectiveness of oral communication requires 3 elements, namely language, knowledge, and support strategies [8] . Students need to be given adequate vocabulary mastery that can be provided through reading or listening activities. The importance of mastering adequate vocabulary to support the smoothness of oral communication, where there are 2 things that inhibit the ineffectiveness of oral communication: first, students do not have adequate vocabulary mastery, and second, students do not know how to use vocabulary in their spoken language [9]. In previous studies, researchers have made mobile-based English learning media applications [8]. In connection with this in this study, the author will create an English learning application, is vocabulary by utilizing AR technology so that it is more effective and interactive. Effective and interactive learning media will facilitate students in learning [10].

This research an instructional media based on augmented reality vocabulary cards will be made for learning English vocabulary. Learning English vocabulary developed is the basic vocabulary of object recognition. In this English vocabulary learning application, a voice is also added that makes it easier to pronounce vocabulary correctly. The learning media will display real objects and sounds making it easier for students to learn and understand English vocabulary both in correct pronunciation and writing.

2. Methods

The main objective of this research is to design mobile-based AR applications, which can be used as vocabulary learning independently. This vocabulary card application is aimed at elementary level students and is expected to make it easier to remember and learn English vocabulary. This research can be categorized as a developmental study, in which media learning developed in this study uses Augmented Reality technology. The development method used in this study was developed by Sutopo [11]. Multimedia development is chosen based on the steps of the method, where each stage is explained in detail and specifically. There are 6 stages used in the Multimedia Development Method, are; concept, design, material collection, assembly, testing, distribution and implementation. Stages of multimedia development methods can be seen in Figure 1.

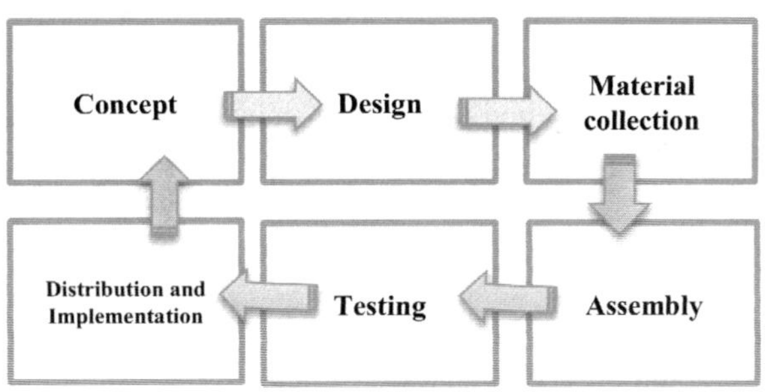

Figure 1. Multimedia development method.

3. Results and discussion

3.1. Concept

Augmented Reality technology in learning media has a positive impact on the understanding of learning materials for users [12].This is the basis for developing Augmented Reality applications as alternative learning media. Table 1 is a description of the concept of augmented reality application that was developed.

Table 1. Description of the application concept.

The Title	Vocabulary Card
Types of Applications	AR-based learning application
Types of users	Users are identified application or intended for elementary school students, and therefore its contents range to the age of 6-12 years.
Content Application	The content is the basis vocabulary consisting of nouns, the introduction of the limbs, animals and fruits.
Pictures	2D image file format is .JPG, while 3D is. .FBX taken from the internet and making 3D & 2D models.
Audio	Using .MP4 format

3.2. Design

The design process undertaken is to make markers in the form of 2D drawings and 3D object models. 2D images are used as AR markers, while 3D object models are used as 3D AR modeling output. The tools used to create 2D and 3D designs are blender. Besides making with the blender application, 3D AR object models are also obtained from the internet. At this stage also recorded sound with the .MP4 format used as audio sound when the target marker is run. Figure 2 is an example of the design process.

Figure 2. The example design process created 3D object models.

3.3. Material collection

The materials needed in making this application consist of 2D and 3D images. 2D images are designed by themselves in the form of a flip card, while there are 3D images obtained from several sources on the internet with some modifications and make your own designs using a blender.

Content. The vocabulary taken was obtained from literature studies and interviews with English teachers in elementary schools, so that 70 vocabularies consisted of 70 basic vocabularies are 30 nouns, 20 fruits, and 20 animals. The following is a 2D image used as a marker for the English vocabulary card application which can be seen in Table 2.

Table 2. The example picture used as a marker.

No	Name	Format	Picture
1	Car	.JPG	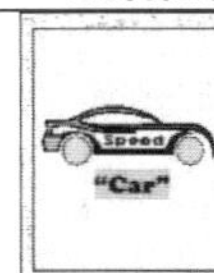
2	Bicycle	.JPG	
3	Bird	.JPG	
4	Bowl	.JPG	
5	Broom	.JPG	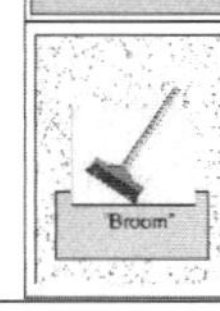

Audio. The developed audio is obtained from dubbing by English lecturers for each vocabulary in either an English word or its meaning in Indonesian. The results are then saved in .MP3 format.

3.4. Assembly

After completing the AR vocabulary card making process that is used for markers, 3D objects, and audio, the next is the aurasma process. In this process upload 2D marker images in 3D format to the database. The database used is vuforia. Next, import the database into Unity as a marker so that it can be recognized by AR. Image upload process marker vuforia database can be seen in Figure 3.

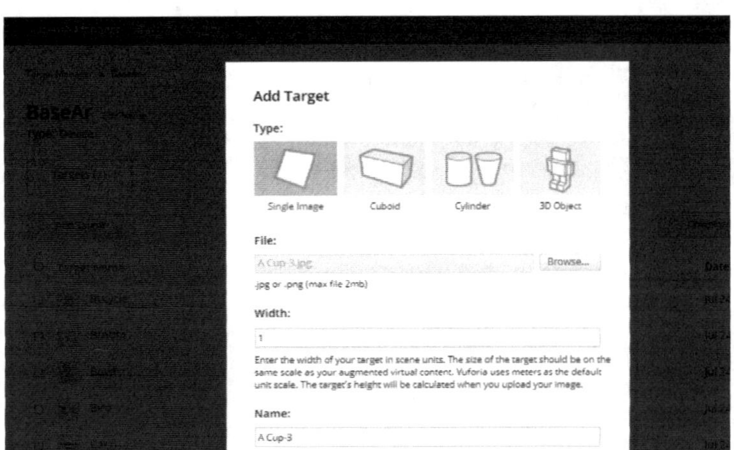

Figure 3. Image Upload 2D marker to vuforia.

After the marker is ready, the next step is to import the 3D object image into unity so that when the marker is used, the 3D object will appear. The next process is to upload the audio that has been made, so that in addition to appearing 3D objects also appear the sound that contains the English vocabulary and its meaning. Figure 4 is the process of making AR with Unity. Aurasma process is carried out for all vocabulary used in this study, amounting to 77 vocabularies. After the Aurasma process is complete for all English vocabulary words, it will then be built in the form of .APK, so that it can be installed on Android for use by users.

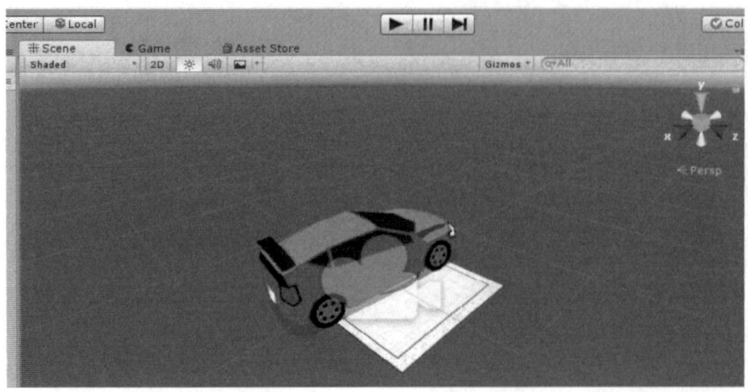

Figure 4. Making AR with unity.

3.5. Testing

The first test is carried out with the aim of verifying whether the application is running properly according to the previously made design. Tests are carried out on markers that have been made using a mobile device. The test results show that each marker can display Augmented Reality content properly and clearly which can be seen in Figures 5. Distance testing shows good results at a distance of 10-60 cm with normal indoor lighting conditions. The results of distance testing can be seen in Table 3. At a distance of 70 cm Augmented reality images are sometimes visible and invisible.

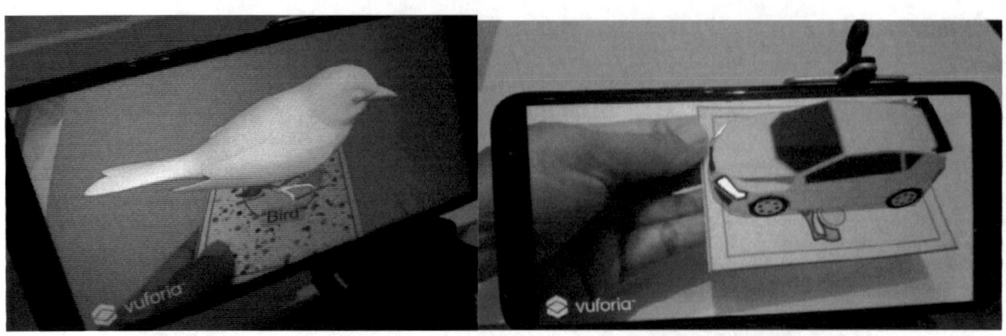

Figure 5. The example display of Augmented Reality marker test results.

Furthermore, the second phase of testing will be carried out on students in elementary and junior high schools, but due to the conditions of the Covid 19 pandemic it cannot be done.

Table 3. The results of the Augmented Reality distance test.

No	Distance	Status
1	10 cm	AR Content appear
2	20 cm	AR Content appear
3	30 cm	AR Content appear
4	40 cm	AR Content appear
5	50 cm	AR Content appear
6	60 cm	AR Content appear
7	70 cm	Sometimes appear, sometimes not
8	80 cm	AR cannot appear

3.6. Distribution
Augmented Reality markers are printed on a flipcard. This is because to make it easier in the learning process because it is simpler than in the form of a book, besides that, it can also be easily carried anywhere with a smaller form.

4. Conclusion
In this research, a mobile-based AR application was created to assist students in learning English vocabulary either independently or in class. The AR marker is designed in the form of a flipcard so that it is simpler than a book. Vocabulary cards in the form of Flip Cards can be used as one of the main learning media in learning English, especially for elementary school students or who are just learning to speak English. Augmented reality can be used to make learning media more attractive in learning English vocabulary. Further research will be made an English vocabulary game which is expected to help students in remembering and learning vocabulary through games.

References
[1] Solak E and Cakır R 2015 Exploring the effect of materials designed with augmented reality on language learners' vocabulary learning *J. Educ. Online* **13**(2) 50–72
[2] Martín-Gutiérrez J, Fabiani P, Benesova W, Meneses M D, and Mora C E 2015 Augmented reality to promote collaborative and autonomous learning in higher education *Comput. Human Behav.* **51** 752–761
[3] Arsyad 2007 *Media Pembelajaran* (Jakarta: PT Raja Grafindo Persada)
[4] Küçük S, Yilmaz R M, and Göktaş Y 2014 Augmented reality for learning english: Achievement, attitude and cognitive load levels of students *Egit. ve Bilim*, **39**(176) 393–404
[5] Taskiran A 2019 The effect of augmented reality games on English as foreign language motivation *E-Learning Digit. Media* **16**(2) 122–135
[6] Wu H K, Lee S W Y, Chang H Y, and Liang J C 2013 Current status, opportunities and challenges of augmented reality in education *Comput. Educ.* **62** 41–49
[7] Chen C M and Chung C J 2008 Personalized Mobile English Vocabulary Learning System Based on Item Response Theory and Learning Memory Cycle *Comput. Educ.* **51**(2) 624–645
[8] Rohani S, Suyono A, and Rozi I 2019 Designing a Mobile Application for Autonomous Learning of English *Int. Conf. Adv. Educ. Humanit. Lang.*
[9] Leaver B L, Ehrman M, and Shekhtman B 2005 *Achieving Success in Second Language Acquisition* (Cambridge, UK,)
[10] Yingprayoon J 2015 Teaching Mathematics using Augmented Reality *Proc. 20th Asian Technol. Conf. Math.* **20** 384–391
[11] Sutopo A 2003 *Multimedia Interaktif dengan Flash* (Jakarta: Graha Ilmu)
[12] Cascales A, Pérez-López D, and Contero M 2013 Study on parents' acceptance of the augmented reality use for preschool education *Procedia Comput. Sci.*, **25** 420–427

IOP Conf. Series: Materials Science and Engineering 1073 (2021) 012062 doi:10.1088/1757-899X/1073/1/012062

Detecting body parts from natural disaster victims using You Only Look Once (YOLO)

S Moechammad[1,*], R Cahya[2] and A N A Berkah[1]

[1] Telecommunication and Information Technologies, Malang State Polytechnic, Soekarno-Hatta No. 9 Malang, Indonesia
[2] Computer Science and Engineering, Malang State Polytechnic, Soekarno-Hatta No. 9 Malang, Indonesia

*msarosa@polinema.ac.id

Abstract. Natural disasters are unpredictable events in both the location and the time of occurrence. Natural disasters can cause property loss and can even be claimed by life. To reduce the amount of losses, the handling of rapid evacuation should be conducted by the SAR team to help victims of natural disasters. But in fact, there are a lot of obstacles in the evacuation process. Starting from the difficulty of searching the victim's body, the difficulty of the terrain reached until limited equipment needed. In this study designed the body detection system of natural disaster victims using image processing where the shooting of victims was carried out using drones aiming to help find victims in a difficult or prone location when reached directly by humans. Background of the problem, in this research proposed a development method for the detection of victims of natural disaster that aims to help the SAR team as well as natural disaster volunteers in the search for victims who are in a difficult to reach place. The method used by You Only Look Once (YOLO) uses the Python programming language associated with image processing. From the research has been obtained accuracy detection object disaster victims with good accuracy. Based on the experiments that have been done obtained a good accuracy value of 95.49% with epoch of 12000.

1. Introduction

Indonesia is one of the countries that has a natural structure consisting of a meeting of tectonic plates namely Eurasia, Indo-Australia and Pacific. It makes Indonesia has a very complex geological condition. This condition resulted in many areas in Indonesia which are prone to natural disasters such as tsunamis, volcanic eruptions, and earthquakes. In addition to natural disasters caused by geological factors, Indonesia also often experienced disasters triggered by natural damage caused by human deeds such as flood and landslide [1]. Based on data from the National Disaster Management Agency (BNPB) in 2019 there was a disaster of 1911 events. Many areas in Indonesia are prone to disasters. Almost all of the disaster-prone areas have high levels of population. During the disaster, one of the important things is the evacuation process of disaster victims. But at the field, there are often obstacles in the evacuation process. Currently, the evacuation process of disaster victims is still no precise technology, making it difficult for the SAR team in the process.

Image processing is a growing field in the scope of research where the observation system in real time increases the opportunity in researchers to develop new research on a wide range of problems [2]. The processing is done that detects objects that are implemented to detect the existence of certain objects

IOP Conf. Series: Materials Science and Engineering 1073 (2021) 012062 doi:10.1088/1757-899X/1073/1/012062

[3]. Object detection relates to identifying objects in the real world such as people, animals, and objects. Object detection algorithms use various image processing applications to extract the desired parts of the object [4]. Object detection can be implemented to detect natural disaster victims such as patterns of human body [3].

YOLO is one of the single stage detection methods. By dividing the input image into an area and predicting the coordinates of the boundary box and the probability of the class for each region, that is, converting object detection problems into regression issues, which can actually achieve end-to-end detection [5]. YOLO is one of the fastest object detection methods with good performance and high accuracy [6].

In this research, using the YOLO method, will be designed the body detection system of natural disaster victims using image processing where the victim's image is performed using a drone aiming to help find victims in difficult locations when reached directly by humans.

2. Study literature

2.1. Related work

There are several references used as state of the art on research to be conducted. The first research is on detection of natural disaster victims. This research designed the system to detect natural disaster victims using the Convolutional Neural Network (CNN) method and implemented on Raspberry Pi that can detect natural disaster victims through streaming cameras placed in UAV. With CNN method obtained 100% accuracy with distance object 1-4 m using Mobile-Net SSD model [3].

The second study discusses a new approach to object detection and tracking using the adaptive background subtraction method. Research discusses the subtraction background used to recognize human movements such as bending, running, walking and jumping. The average level of introduction of movement generated was 97% [2].

Third research on object detection algorithms for video surveillance. This research designed object detection that includes skin detection, color detection, and face detection in video surveillance. The accuracy generated in this application is 95% [4].

The fourth study discusses object detection and tracking using the Deep Learning and OpenCV methods. This research discusses the implementation of deep learning to detect objects in real time. The resulting accuracy is 99% [7].

The fifth study is the detection of natural disaster victims using You Only Look Once (YOLO). This research has a focus on the detection of natural disaster victims with conditions where the victim's body should be visible. With a 100-training data, it generates an accuracy of 89% in literacy to 3000 [8].

Novelty in this research I developed a detection system with the focus of every part of the body disaster victims with conditions where the victim's body should be partially visible using the You Only Look Once (YOLO) method.

2.2. Algorithm You Only Look Once (YOLO)

YOLO is an object recognition method based on the Convolutional Neural Network (CNN) [9]. There are 3 phases of YOLO to detect an object. These stages include:

- Enlarged input image.
- YOLO is a method that divides the image input into a grid of S × S size. To get a bounding box, it will be the convolution of the image input. Bounding boxes have x coordinates, y coordinates, width (width), height, and confidence score. In the bounding box, the normalization is 0 to 1. The x and Y coordinates are normalized to adjust the corresponding top-left point. The height and width will be normalized according to the image size. Illustrations of YOLO are shown in Figure 2.
- After bounding box and confidence has been obtained next, the process of Operation Convolutional Neural Network on each bounding box.

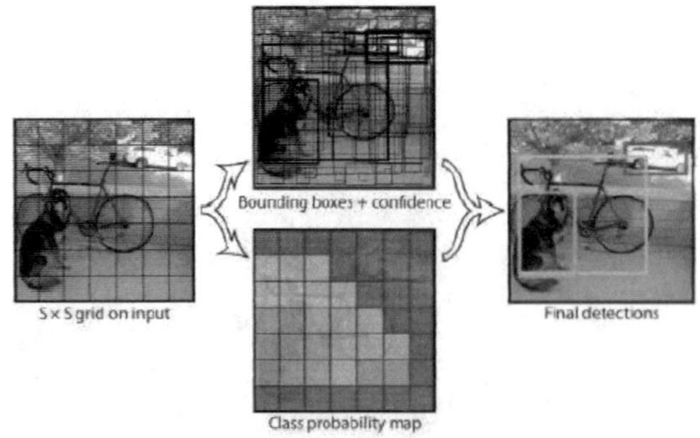

Figure 1. Illustration *You Only Look Once* (YOLO) [1].

2.3. Library OpenCV

OpenCV is a library of programming functions used for computer vision [10]. Its purpose is to make the computer has human-like capabilities. This Library is created for programming using the C/C++ language. The features of OpenCV libraries include basic image processing (filtering, edge detection, color conversion, histogram), can detect moving objects, camera calibration, and object recognition [11].

3. Methods

The proposed system and methods will be outlined in the steps as reference as the analysis framework for the introduction of disaster-affected objects. The following applied object detection diagram flow will be described in Figure 2.

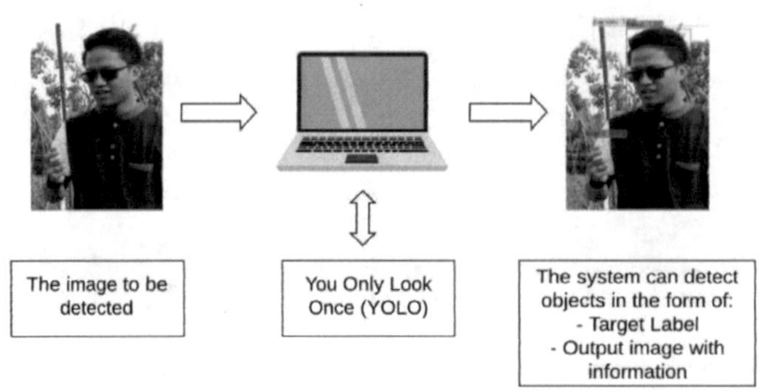

Figure 2. Object detection diagram flow.

Based on the diagram flow in figure 2, when the image is used as input from the network model, the output of the network model will be counted according to the parameters and model structure. The output of the model contains the image category information, coordinating information corresponding to the bounding box, and according to the output information. Value, allows the computer to display a box selectively to detect targets.

There are several steps done in the process of detecting objects using the YOLO method. The following stages of disaster victim's detection using the YOLO method shown in Figure 3.

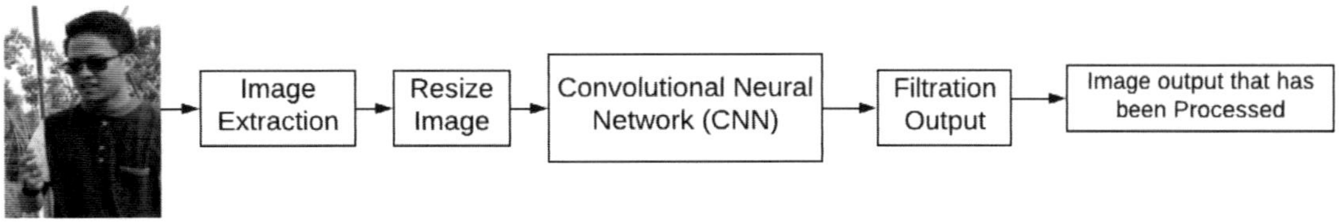

Figure 3. Stages of disaster victim detection using the YOLO method.

Based on Figure 4, the first stage of disaster victim's detection is initiated by extracting a single image from the image that has been taken. Further resize image where image output resize image is input that will be processed using CNN. The Convolutional Neural Network (CNN) processes the input image. The process that will be carried out include the prediction of coordinates and the position of the bounding box that contains the object, the probability of bounding box containing the object, and the probability of each object in the bounding box that belongs to the specified class. Then the output image of the CNN process carried out the filtration process to determine a more specific object. So, in the output video there is information on the object name on each of the detected objects.

4. Results

In this research, the platform used in this paper is Google Colaboratory platform (Google Colab), CUDA 10.1, cuDNN 8.0, OpenCV 4.1.2 and GPU Nvidia Tesla K80 and the Darknet Network framework. Data used by 279 images consisting of 244 training data and 35 data testing. Training data is conducted with Epoch 1000 until epoch 12000.

Tests were conducted with True Positives (TP), False Positive (FP), and False Negatives (FN) data retrieval. TP is the human object of the image successfully detected by the YOLO algorithm. FP is a incorrectly detected object as human. FN is when the algorithm cannot recognize the human object in the image. From TP, FP and FN values can be obtained accuracy, Precision, and Recall values with the following formula:

- Accuracy $= \frac{TP}{TP+FP+FN}$
- Precision $= \frac{TP}{TP+FP}$
- Recall $= \frac{TP}{TP+FN}$

Test results can be seen in table 1.

Table 1. Evaluation metrics of YOLOv4.

EPOCH	1000	2000	3000	4000	5000	6000	7000	8000	9000	10000	11000	12000
TP	156	251	259	266	267	271	269	275	275	272	272	275
FP	71	51	26	21	20	27	23	16	17	14	14	11
FN	121	26	18	11	10	6	8	2	2	5	5	2

Based on table 1, the highest TP value on the number of epochs 12000 is 275. The lowest FP and FN values are also on a total of 12000 epoch which are respectively 11 and 2. The more the number of epoch then TP value is higher, the value of FP and FN is lower. Thus, the body object detection of the best disaster victims when the epoch reaches 12000. Accuracy, Precision, and Recall value recognition objects displayed on table 2.

Table 2. Accuracy, precision, and recall value.

Epoch	Accuracy	Precision	Recall
1000	44.84%	68.72%	56.32%
2000	76.51%	83.11%	90.61%
3000	85.48%	90.88%	93.50%
4000	89.26%	92.68%	96.03%
5000	89.90%	93.03%	96.39%
6000	89.14%	90.94%	97.83%
7000	89.67%	92.12%	97.11%
8000	93.86%	94.50%	99.28%
9000	93.54%	94.18%	99.28%
10000	93.47%	95.10%	98.19%
11000	93.47%	95.10%	98.19%
12000	95.49%	96.15%	99.28%

Based on table 2, the best accuracy, Precision, and Recall values are the 12000 epochs with a value of 95.49%, 96.15%, and 99.28% respectively.

5. Conclusion

In this paper, we present the YOLO algorithm for detection of natural disaster victims ' bodies. Datasets are used as many as 279 images in which 244 images for training data and 35 images for data testing. Training data is carried out until the epoch of 12000. Based on the experiments that have been done obtained a good accuracy value of 95.49%. But there are several factors affecting the detection result, i.e. the background object in the image, the position of the object, and the height/distance of the object.

For further research is expected to develop to solve the problem in order to achieve more efficient results in assisting the SAR team in evacuating natural disaster victims.

Acknowledgment

This research was conducted at the Dept. Computer Science and Dept. Telecommunication and Information State Polytechnic Malang and is intended for conference publication journals. The author would like to thank Faried, Rizky, Muna, Putra and others for assisting in the preparation and testing of my research. and also, to Mr. M. Sarosa and Mr. Cahya R. as mentoring lecturers who always provide solutions to problems.

References

[1] Shianto K A, Gunadi K and Set E 2019 Deteksi Jenis Mobil Menggunakan Metode YOLO dan Faster R-CNN *Jurnal Infra* **7**

[2] Angelo K M 2018 A novel approach on object detection and tracking using adaptive background subtraction method *International Conference on Computing Methodologies and Communication* (Chidambaram, India)

[3] Raghunandan A, Raghav P and Aradhya H R 2018 Object detection algorithms for video surveillance applications *International Conference on Communication and Signal Processing (ICCSP)* pp 0563-0568

[4] Ramadhan M I and Prihandoko P 2017 Penerapan Data Mining untuk Analisis Data Bencana Milik BNBP Menggunakan Algoritma K-Means dan Linear Regression *Jurnal Informatika dan Komputer* **22** pp 57-65

[5] Darma M D 2018 Sistem Monitoring Kursi Pada Ruangan Kelas Dengan Metode Convolutional

Neural Network (CNN) (Doctoral dissertation, Universitas Islam Negeri Sultan Syarif Kasim Riau)
[6] Sikumbang S and Suryadi K 2015 Human Detection Menggunakan Metode Histogram Of Oriented Gradients (Hog) Berbasis Open_Cv *Jurnal Pendidikan Teknik Elektro* **4** 2
[7] Jeong H -J, Park K -S and Ha Y –G 2018 Image Preprocessing for Efficient Training of YOLO Deep Learning Networks *IEEE International Conference on Big Data and Smart Computing* pp 635-637
[8] Muna N 2019 *Deteksi Korban Bencana Alam menggunakan You Only Look Once (YOLO)*
[9] Zufar M and Setiyono B 2016 Convolutional Neural Networks untuk Pengenalan Wajah secara Real-Time *Jurnal Sains Dan Seni ITS* **5** pp A72-A77
[10] Nurfita R D 2018 *Implementasi Deep Learning Berbasis Tensorflow untuk Pengenalan Sidik Jari* (Surakarta)
[11] Hartawan D R, Purboyo T W and Setianingsih C 2019 Disaster Victims Detection System Using Convolutional Neural Network (CNN) Method *IEEE International Conference on Industry 4.0, Artificial Intelligence, and Communications Technology*

Designing real-time research data portal of JTI POLINEMA

G Budiprasetyo*, Y Yunhasnawa, M Mentari and D C Pratama

Dept. of Informatics Engineering, State Polytechnic of Malang, Malang, Indonesia

*gunawan.budi@polinema.ac.id

Abstract. Research culture in Department of Information Technology Polinema is always developed continuously every time. In addition, many research titles that are critical of environmental problems are beginning to be seen. The Information Technology Department has an outstanding focus on issues, namely smart systems which include smart education, smart living, smart healthy, smart city and smart tourism. Research resources in the Information Technology Department Polinema are competing to realize that research focus. For this reason, efforts are needed to increase the availability of facilities that are not only material. The fast response can be solved by providing Real-Time Research Data Portal. The contribution of this paper is to design of middle ware on web portals that have high scalability capabilities for providing large data facilities and fast access. Real-time big data processing is the path that will be described in the architecture that will be created and adapted to the needs of the Department of Information Technology Polinema. The results of the latency evaluation test indicated that RabbitMQ allowed us to enforce concurrency, thereby reducing overall RTC by up to 50% compared to traditional HTTP.

1. Introduction

Research in Indonesia has gradually increased over time. The Indonesian government also provides a lot of support in the form of annual grants which provide research opportunities for Indonesian researchers. Researchers, namely lecturers in the industrial era 4.0 and disruptive innovation who open the need to collaborate and be open in sharing research data each other [1]. This results in the need for direction and governance in an effort to increase research. Entering the industrial era 4.0, namely the era of digitalization, there are main and very important areas in its implementation, namely the field of Information Technology (IT). One of the industries that uses IT support, especially the Internet of Things (IoT) tool in carrying out all existing activities [2]. And it turns out that real time processing supported by IoT and big data also improves services better and has an impact on increasing certain collaborations.

Back to the discussion of research in Indonesia, especially in higher education institutions like State Polytechnic of Malang (Polinema), especially the Information Technology Department (JTI), apparently needs the same amount of money in fulfilling research facilities. Real time data access for each research group, such as in Julian and Surya [3] for the Computer Network, Architecture and Data Security group research which explains about industry in droves using integrated into a single integrated network and also IoT which requires assurance [4]. data security whose reality must be accessed in real time; research group computer vision requires real time data access as well as Pan et al [5] on real time surgery has performed accurate detection and tracking of corneal contours; a research group of artificial intelligence according to Ionescu et al [6] which carried out route tracking with low power consumption for robots in real time; Information Systems research group looks at research Cheung et al [7] on simulation of real time operating performance in information technology (IT) equipment of data centers; and research group multimedia real time access games used in various applications from then until now [8]. The needs and work of various kinds of group research will be better if they are collected into a centralized source at JTI-Polinema in the form of a real time web portal according to the needs in this digitalization era. JTI-Polinema does not yet have a web portal regarding real time research data. Therefore, this study aims to design a Real-time Research Data Portal JTI POLINEMA. Flexible communication systems and models are needed, because the application is naturally separated in layers and geographically. An application that has flexible communication with technology is needed to address real time data access. The data portal designed based on Hadoop using the RabbitMQ message broker to handle large amounts of data access because access is done in real-time.

2. Related work

Data portals are becoming a trend to meet current data collaboration needs. Many countries and institutions have created data portals like Koppe and Schafer [9] on the German Marine Research Data Portal which provides "one-stop-shop" condition. In addition, technology concentration is also carried out to improve the performance, efficiency and productivity of an organization using a Web-based portal [10]. Application of data storage and data processing in real time [11] on an integrated platform that builds Hadoop and Spark cluster environments that have the principle of reading data interactively can cause bottlenecks so that data cannot be accessed and uploaded quickly on the data portal. Therefore, real-time data portals require additional technology that can solve these problems. Such as processing data streams in parallel and distributed environments [12] handled by the framework as parallel processing center and message broker [13] to obtain smooth processing, fault tolerance and efficient. Message broker systems commonly used in distributed systems that perform parallel and real time processing are Kafka, RabbitMQ, ActiveMQ, etc. A study on the comparison of two types of message brokers [14] namely RabbitMQ and Kafka proved that in the basic setup consisting of one node file, single channel, single partition and unreplicated RabbitMQ performance outperformed ActiveMQ. In addition, a literature study regarding RabbitMQ and distributed system. The developer implemented RabbitMQ's clustering capabilities to improve reliability [15]. Exchange messages in bulk, getting the best results from Advanced Message Queuing Protocol (AMQP) [16]. The performance of a single queue drops significantly when the queue is mirrored across the cluster for fault tolerance (at least 10% -20% better than full mirrored) [17]. Challenge to achieve the real-time capability using Map Reduce which open source and provides off-the-shelf high scalability that can significantly shorten the processing time for big data [18]. RabbitMQ Faster generates a message from the broker (when the client receives the message) [19]. In terms of latency, both systems (Kafka and RabbitMQ) were able to provide low latency results (i.e. an average / median of about 10 ms) [20]. Technologies for processing parallel data streams and highlighting problems and experiences, scale well [21]. When a large number of users send requests to the web application at the same time, the results will be more stable if using RabbitMQ as a message-oriented middleware rather than the REST API communication method [22]. RabbitMQ as a load balancer can share the workload evenly, thereby reducing the latency time of the Naive Bayes Classification by 30.3% [23]. A distributed system on a real-time research data portal requires basic support in the form of a collection of software tools like Apache Hadoop [11].

Research Data portal carries a complex data handling architecture in the form of web observatories (WO) so that researchers can be involved with each other in the research data that has been carried out or that has been produced [24]. During its development [2], data access is needed in real-time in the era of digitalization. This has an impact on the need for quick and precise decision making.

Real-time data or message access so that the infrastructure is made first so that it can interact with each other on a distributed system [25]. This can be handled using a message broker like RabbitMQ. RabbitMQ is a message broker that can control the sending and receiving of messages [15]. Rabbit MQ is a message queuing system used in middleware for distributed systems that implements Advanced Message Queuing Protocol (AMQP) [17].

3. JTI-Polinema real time research data portal design

For real-time research data designing portal JTI-Polinema, a prototype is needed to prove the latency of data access using a message broker.

3.1. Hadoop with traditional HTTP

The real-time data portal access traditionally shown in Figure 1 can be accessed by users from the front-end side via browse and will be forwarded as soon as the HTTP request is received. Then the request continued on the back-end REST access with Spring (support Java-based applications). Rest response will be given after Spring instruction is performed. Because the data is estimated to be accessed in real-time which allows for complex data transactions and operations, middleware is needed to handle it. After going through the middleware layer, the client requests large data will be continued on HDFS to be divided into fixed nodes.

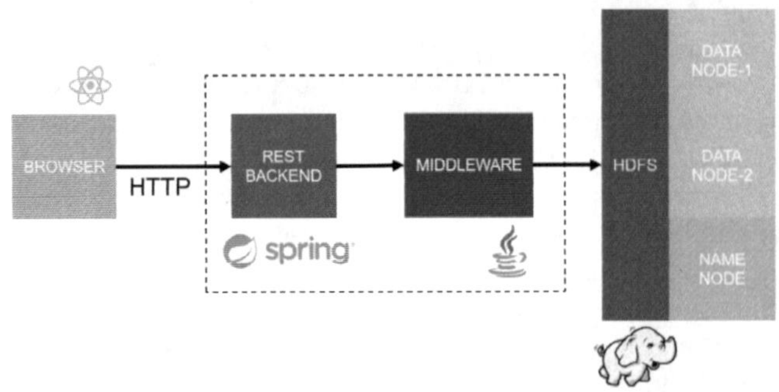

Figure 1. Technology without message broker.

3.2. Design real-time research data portal of JTI POLINEMA

JTI Polinema's real-time research data portal prototype design is shown in Figure 2. However, it is realized that there are not only many transaction data problems that occur in real-time data access, but also the possibility of bottlenecks. Therefore, it is necessary to add AMQP Message Broker, especially RabbitMQ. RabbitMQ will manage the sending of requests from the client, then it will be handled with middleware on each RabbitMQ that has been installed so that it is expected to create speed of sending requests and also get uncaught in the middle of sending data process in HDFS.

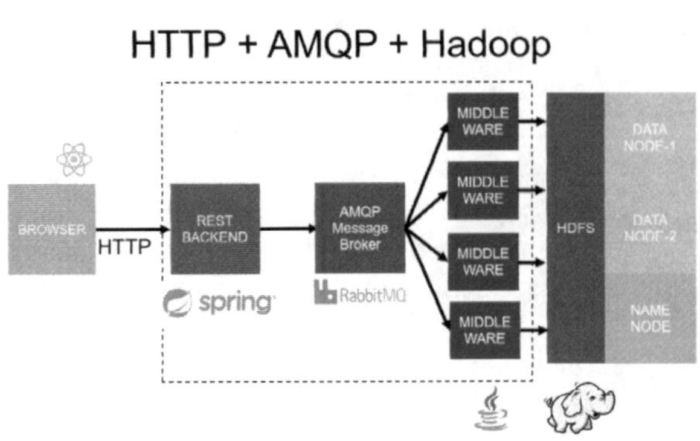

Figure 2. Technology with message broker.

4. Research and discussion

To determine the performance evaluation on the design prototype made, a performance evaluation (PE) test was carried out on five (5) Different Sizes, namely: 1 KB, 10, KB, 100 KB, 1000 KB, 10000 KB and also Four (4) Different Tasks: Store Only, MD5, Base 64, Word Count.

4.1. Performance evaluation (HTTP) over performance evaluation (HTTP and RabbitMQ)

The performance evaluation of the four types of tasks and five types of sizes for HTTP shown in table 1 and shown in table 2 for HTTP with RabbitMQ. Only HTTP has the fastest handling results for 1KB files, task store only type, which is 8268 micro sec. And the highest is the longest time that is on the data size 10000KB with the type of task Base64 Encode is 384996,25 micro sec. For HTTP with RabbitMQ the fastest handling results for 1KB files, task storage type only, which is 8191 micro seconds. And the highest is the longest time, namely the data measuring 10000KB with the type of task Base64 Encode of 192595,125 micro seconds.

Table 1. Performance evaluation (HTTP).

Traditional HTTP	Request Completion Time		(Micro Sec.)		
File Size Each	**Store Only**	**MD5 Encode**	**Base64 Encode**	**Word Count**	**Average**
1 KB	8268	10092	10049	7603	9003
10 KB	8588	11394	13856	9455,25	10823,3125
100 KB	11610,75	26896	19001	12343	17462,6875
1000 KB	16273,75	27175,75	42387,25	79142	41244,6875
10000 KB	23639,25	117956	384996,25	163431,75	172505,813

Table 2. Performance evaluation (HTTP and RabbitMQ).

HTTP + RabbitMQ	Request Completion Time (Micro Sec.)				
File Size Each	**Store Only**	**MD5 Encode**	**Base64 Encode**	**Word Count**	**Average**
1 KB	8191	10111	10111,5	7387,25	4571,5
10 KB	8370	11797	13701	9481,2625	5495,15625
100 KB	5860,375	13540	9550,5	6251,5	8800,59375
1000 KB	8204,875	13645,875	21243,625	39645	20684,8438
10000 KB	11914,625	59072	192595,125	81802,875	86346,1563

4.2. Overall latency

Overall shown in Figure 3 HTTP performance coupled with RabbitMQ shows much better latency than HTTP only. This value can be seen in the block on the far right of the chart where the yellow block is far below the blue block (almost half size). This means that the HTTP latency coupled with RabbitMQ is almost 50% with only HTTP.

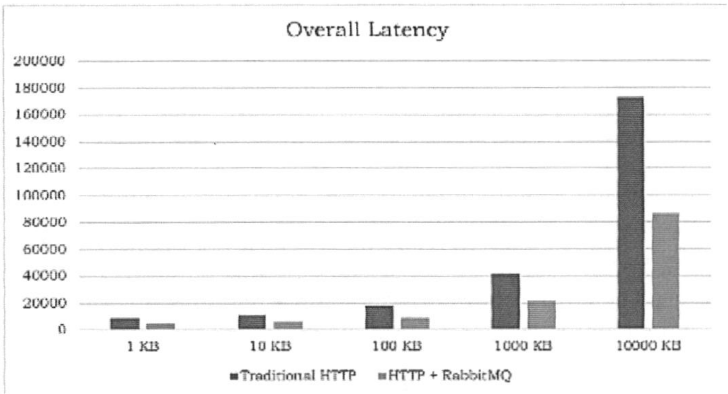

Figure 3. Chart overall latency.

5. Conclusion

Each figure should have a brief caption describing it and, if necessary, a key to interpret the various lines and symbols on the figure. We proposed a prototype of Hadoop based dataset repository for the Polinema's Department of Information Technology. In our proposed architecture, we incorporated RabbitMQ as a message broker to reduce the overall amount of Re-quest Completion Time (RCT). We observed the capability of RabbitMQ enable us to implement concurrency, thus reducing the overall RTC up to 50% compared to traditional HTTP.

References

[1] Fourianalistyawati E, Uswatunnisa A, Arif M M, Paulus S A, Soebandhi S, Reni A and Kurniasih N 2018 Research Development in Indonesia: Ethics Committee in Open Science and Collaboration Era *JPhCS* **1114** 1 012069

[2] Nagy J, Oláh J, Erdei E, Máté D and Popp J 2018 The role and impact of Industry 4.0 and the internet of things on the business strategy of the value chain—the case of Hungary *Sustainability* **10** 10 3491

[3] Jang-Jaccard J and Nepal S 2014 A survey of emerging threats in cybersecurity *Journal of Computer and System Sciences* **80** 5 973-993

[4] Porambage P, Okwuibe J, Liyanage M, Ylianttila M and Taleb T 2018 Survey on multi-access edge computing for internet of things realization *IEEE Communications Surveys & Tutorials* **20** 4 2961-2991

[5] Pan J, Liu W, Ge P, Li F, Shi W, Jia L and Qin H 2020 Real-time segmentation and tracking of excised corneal contour by deep neural networks for DALK surgical navigation *Computer Methods and Programs in Biomedicine* **197** 105679

[6] Ionescu L M, Mazare A, Lita A I and Serban G 2015 Fully integrated artificial intelligence solution for real time route tracking *38th International Spring Seminar on Electronics Technology (ISSE)* pp 536-540 IEEE

[7] Cheung H, Wang S, Zhuang C and Gu J 2018 A simplified power consumption model of information technology (IT) equipment in data centers for energy system real-time dynamic simulation *Applied energy* **222** 329-342

[8] Hillman M F and Subriadi A P 2019 40 Years Journey of Function Point Analysis: Against Real-time and Multimedia Applications *Procedia Computer Science* **161** 266-274

[9] Koppe R and Schäfer A 2015 Enabling central access to marine data: Data portal German marine research *OCEANS 2015-Genova* pp 1-5

[10] Rajkumar S, Abraham S E and Santhi V 2017 Web based portal using a biometrie interface and Android application to assist non-government organizations *International Conference on Intelligent Computing, Instrumentation and Control Technologies (ICICICT)* 898-902 IEEE

[11] Yang C T, Chen C J, Tsan Y T, Liu P Y, Chan Y W and Chan W C 2019 An implementation of real-time air quality and influenza-like illness data storage and processing platform *Computers in Human Behavior* **100** 266-274

[12] Singh R K and Verma H K 2020 Redis Based Messaging Queue and Cache Enabled Parallel Processing Social Media Analytics Framework *Journal of King Saud University-Computer and Information Sciences*

[13] Nahri M, Boulmakoul A, Karim L and Lbath A 2018 IoV distributed architecture for real-time traffic data analytics *Procedia computer science* **130** 480-487

[14] Dobbelaere P and Esmaili K S 2017 Kafka versus RabbitMQ: A comparative study of two industry reference publish/subscribe implementations: Industry Paper *Proceedings of the 11th ACM international conference on distributed and event-based systems* 227-238

[15] Jones B, Luxenberg S, McGrath D, Trampert P and Weldon J 2011 *RabbitMQ performance and scalability analysis. project on CS* **4284**

[16] Fernandes J L, Lopes I C, Rodrigues J J and Ullah S 2013 Performance evaluation of RESTful web services and AMQP protocol *2013 Fifth International Conference on Ubiquitous and*

Future Networks (ICUFN) 810-815 IEEE
[17] Rostanski M, Grochla K and Seman A 2014 Evaluation of highly available and fault-tolerant middleware clustered architectures using RabbitMQ *2014 federated conference on computer science and information systems* 879-884 IEEE
[18] Liu X, Iftikhar N and Xie X 2014 Survey of real-time processing systems for big data *Proceedings of the 18th International Database Engineering & Applications Symposium* 356-361
[19] Ionescu V M 2015 The analysis of the performance of RabbitMQ and ActiveMQ *2015 14th RoEduNet International Conference-Networking in Education and Research (RoEduNet NER)* 132-137 IEEE
[20] Dobbelaere P and Sheykh Esmaili K 2019 *Industry Paper: Kafka versus RabbitMQ A comparative study of two industry reference publish/subscribe implementations*
[21] Tennant M, Stahl F, Rana O and Gomes J B 2017 Scalable real-time classification of data streams with concept drift *Future Generation Computer Systems* **75** 187-199
[22] Hong X J, Yang H S and Kim Y H 2018 Performance analysis of RESTful API and RabbitMQ for microservice web application *International Conference on Information and Communication Technology Convergence (ICTC)* 257-259 IEEE
[23] Nugroho A W and Kusumawardhani S 2020 Distributed Classier for SDGs Topics in OnlineNews using RabbitMQ Message Broker *Journal of Physics: Conference Series. 2nd 2019 ICERA*
[24] Tiropanis T, Hall W, Shadbolt N, De Roure D, Contractor N and Hendler J 2013 The web science observatory *IEEE Intelligent Systems* **28** 2 100-104
[25] Eugster P T, Felber P A, Guerraoui R and Kermarrec A M 2003 The many faces of publish/subscribe *ACM computing surveys (CSUR)* **35** 2 114-131

IOP Conf. Series: Materials Science and Engineering 1073 (2021) 012064 doi:10.1088/1757-899X/1073/1/012064

Enhancing student interest in learning through the development of serious mathematics games

R Rawansyah*, A N Pramudhita and Y Pramitarini

Information Technology Department, Politeknik Negeri Malang, Malang, Indonesia

*rawansyah@polinema.ac.id

Abstract. Learning mathematics in Indonesia today has not given the expected results, it is not better when compared to other countries. If a student is active and able to count quickly and carefully in mathematics, then the student will indirectly be easier to solve problems either in mathematics itself or applied in everyday life. The average educator has not used interesting media to improve children's numeracy skills, there are often difficulties in students' mathematics learning. Many research contributions are directed at taking advantage of the success of video games and using them to benefit from the education domain. The concept of learning in this research is carried out by utilizing the development of information and communication technology through interesting mathematical game applications through web-based learning. The game is developed using scratch by MIT so that the development process is less time consuming and easier. The game is in form of the quiz that makes students utilize their mathematical skills to get rewards. In that way, the competitiveness of the students also increases at the same time with their numeric ability. This learning has an appropriate value to be a breakthrough to balance current technological developments and improve students' numerical ability. Because of the mathematics games has a rewards system, it is interesting for students to learn math.

1. Introduction

Mathematics learning in Indonesia currently has not given the expected results and is not better when compared to other countries, this can be seen from the results of the Program for International Student Assessment (PISA), that the percentage of Indonesian students under level 2 is very large 76, 6% and the percentage of students at levels 4, 5, and 6 are statistically absent [1]. The Organization for Economic Co-operation and Development (OECD) said that students under level 2 were deemed unable to function effectively in the life of the XXI century, therefore students' abilities needed to be further developed in learning mathematics in schools so that the Indonesian nation had the resources human power that can compete in the future life. Students should have cognitive competence after learning mathematics including "Thinking and reasoning, Argumentation, Communication, Modeling, Problem posing and solving, Representation, Using symbolic, formal and technical language and operations, Use of aids and tools"[1], so students can complete various mathematical calculations. Based on the results of the PISA test in 2018 the value of science, reading skills, and mathematics of Indonesian students is still low, number 72 of 79 countries in the world [2]. The PISA test value is shown in figure 1.

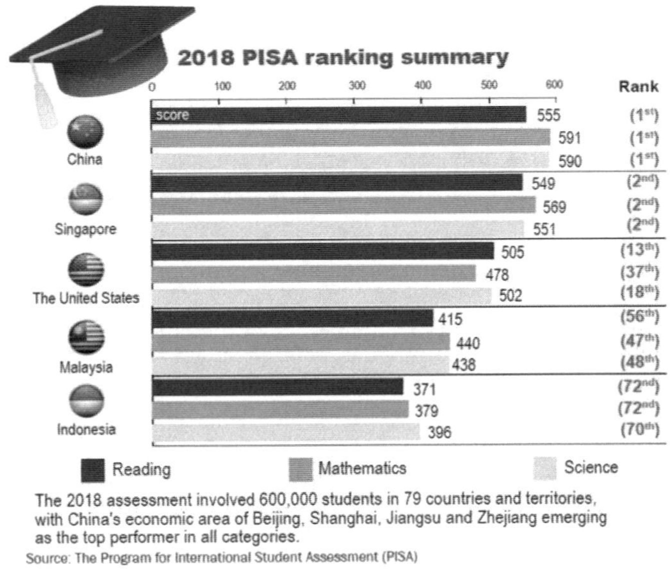

Figure 1. Pisa test results for Indonesia in 2018.

In line with what was conveyed by KEMDIKBUD [3], the creation of the 2013 curriculum aims to make individuals who are capable of productive and creative thinking. The importance of the ability to calculate, not only in mathematical problems but also in the everyday life of children, young people, and even parents [4]. If a child becomes active and can calculate quickly and carefully in learning mathematics, then indirectly the child will be easier to solve problems both in mathematics lesson or applied in everyday life, so that the child can solve problems and look for the right way out according to his ability.

Mathematics is not a difficult thing to learn as long as the delivery strategy is right and by the level of ability to learn it as stated by the Hakim that "in the teaching and learning activities, mathematics is one of the basic science that must be mastered by the student" [5]. Numeracy is the basis of other sciences, especially for the development of science and technology. The practice of learning to count requires learning media or teaching aids that serve to facilitate children to learn, but the average educator has not used attractive media to improve children's numeracy skills. , so that there are often difficulties in learning students' mathematics and cause reluctance to learn mathematics because the material in mathematics is considered difficult to understand because it is too abstract, plus the delivery of teachers who are too monotonous and boring is also one of the reasons why students do not like math lessons [6].

The concept of learning in this research was conducted by utilizing the development of information and communication technology through an interesting math game application through web-based learning. This game learning media has an appropriate value to be a breakthrough to keep pace with current technological developments and improve students' numerical abilities because of the exercises in the form of interesting games in applications that can be used alone as well as competed with friends during learning. so that the use of books that sometimes make students lazy to learn can be reduced.

2. Math and live skill

Mathematics is a subject that needs to be given to all students starting from elementary school to equip students with the ability to think logically, analytically, systematically, critically, and creatively, as well as the ability to work together. Therefore, teaching and learning activities, especially in mathematics, need to be paid more attention to remember how important mathematics is for students to support their lives. Mathematics is a basic subject, in elementary or middle school, learning mathematics is very important because in everyday life we cannot escape the application of mathematics [7]. In solving math problems, which mostly use numbers, numeracy skills are needed. Therefore, numerical skills are particularly important.

3. Serious games

Games are defined as physical and/or mental contests played according to certain rules, with the sole purpose of entertaining or entertaining participants. On the other hand, video games are a special type of game in which they are played on a computer according to certain rules for entertainment, recreation, or winning competitions.

Some perspectives are defining Serious Game or Applied Game as seen from academia and industry. For example, some industry figures believe that the Serious Game should include entertainment elements combined with a practical dimension [8]. The term game can be used to refer to any application that is produced using software developed by the gaming industry, which means that the majority of simulators will be considered as Serious Games [9]. Others argue that the Serious Game is just a computer game and that its classification is nothing more than a marketing technique [10]. The most common definition of a Serious Game is "a game that does not make entertainment or fun the main goal". Following this definition, Serious Games can be distinguished from video games based on their development goals, because Serious Games have a primary purpose other than entertainment.

4. Serious games in education

Many research contributions are directed to take advantage of the success of video games and use them to benefit from the educational domain [11]. Strive to use the addictive nature of games and the players 'obsession with digital games to facilitate the players' learning process. As more and more young people spend hours playing video games, Quest to Learn, a high school that opened in September 2009 in New York City, is based entirely on a game-like learning model. In this school, students spend their day studying by playing games and designing them.

Several educational games were developed for classroom use, and the areas taught by them can vary widely. An example of such a game is Skills Arena which is specially designed for classroom use. The Skill Arena uses the Nintendo Gameboy and teaches students arithmetic skills at various difficulty levels. A study was conducted which involved elementary school students and concluded that playing technology-based games improved students' performance in mathematics [12].

5. Scratch

Scratch is a visual programming language for learning environments that allow beginners (whether students, teachers, learners, or parents) to learn to program without having to think about writing syntax completely wrong. This programming language was created by MIT Media Lab of the Massachusetts Institute of Technology with Scratch, users can create their animations, games, artworks, and more. Scratch users can create programs (called projects) by visually arranging command blocks (called blocks). In this way, students can focus on logic and programming flow (algorithms) without ever or frequently getting error errors due to incorrect syntax writing [13]. Scratch is taught to more than 800 schools and 70 colleges worldwide. In higher education, Scratch is used in Harvard University's CS50 introductory course.

6. Research method

The research approach used in the development research of this learning model is a development research approach. R&D is a process used to develop accountable educational products. The development of the learning model used in this study is the ADDIE (Analysis-Design-Develop-Implement-Evaluate) Learning Design Model with the consideration that ADDIE is suitable for developing learning model products that are right on target, effective and dynamic and are very helpful in learning development [14]. Figure 2 Shows the ADDIE model in general.

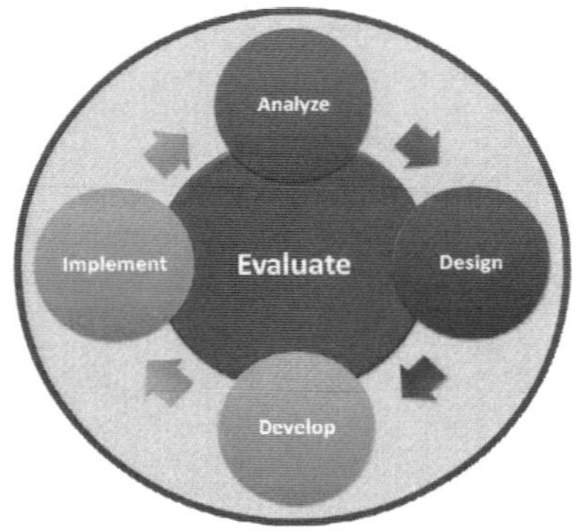

Figure 2. ADDIE Model.

6.1. Analyze
The Analysis stage focuses on the target audience. At the analysis stage, the instructional problems were defined, instructional objectives, learning objectives and identification of the learning environment and student's knowledge

6.2. Design
The design stage is related to targeting, assessment instruments, exercises, content, and analysis related to learning materials, learning plans and media selection. The design phase is carried out systematically and specifically. In this phase, learning media in the form of educational games were chosen which were developed with a scratch platform, this platform was chosen because it is easy to learn and is widely used for various levels of education. And games created using the scratch platform can be played on even low-spec laptops/computers if they have a web browser.

6.3. Develop
In the development stage, the creation and incorporation of the content that has been designed during the design stage are carried out. In this phase, storyboards, content writing and learning media are made.

6.4. Implement
In this stage, a procedure is made for students and facilitators. Training for facilitators includes curriculum material, expected learning outcomes, delivery methods and testing procedures. Other activities that must be done in this stage include distributing material and other supporting materials, as well as preparing for technical problems and discussing alternative plans with students.

6.5. Evaluate
Each stage of the ADDIE process involves formative evaluation. It is multidimensional and an essential component of the ADDIE process. It assumes a formative evaluation form is in the development stage. Evaluation is carried out during the implementation stage with the help of instructors and students. After the implementation of the learning is complete, summative evaluation is carried out to improve learning. The designer of the entire evaluation phase must ascertain whether problems relevant to the training program are resolved and whether the desired objectives are being met.

7. Results
The Created Games is in form of a quiz game, the students can choose the mathematical operation that he wants to practice, whether it is addition, subtraction, multiplication, or division. After students chose the operation students will be taken to a page with series of mathematical problems, there is music that makes students enthusiastic about doing the questions, then, at the end of the game, if there are more

correct answers than wrong answers, the students will get rewards, that is, they can play songs according to the students' wishes. Figure 2 shows some of the screenshots of the game

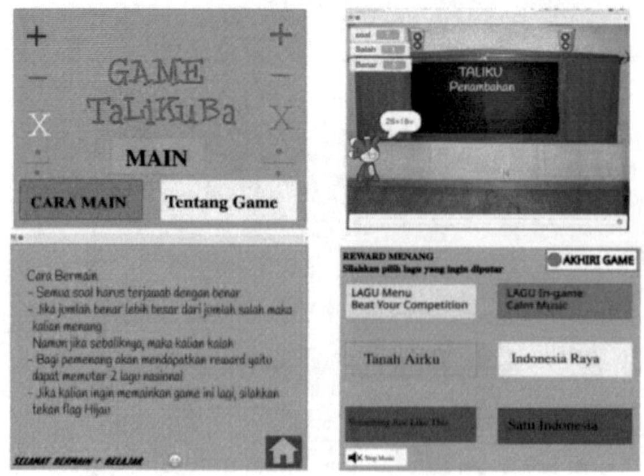

Figure 3. In game screenshots.

Games especially computer games along with learning-teaching activities help the learning of the student and increase its inner motivation and interest by creating challenges and giving control of the game to the player. Training and repetition with custom settings make the memorizing contents easier for the students [15].

NO	Assessed Aspects
Interface Design Aspect	
1	background color mix
2	The combination of writing colors with a comfortable background to the eye
3	Sound and image selection
4	Text color selection
5	font selection
6	the position of the image and character are right
7	game display design
8	Color selection of operator symbols
9	clarity of the rules of the game
10	Music Selection
Game Presentation Aspect	
11	Easy to Use Game
12	Easy to understand Game Instructor
13	Understanding of writing or text in the game
14	use the right language
15	The images presented are appropriate
Aspek Pemrograman	
16	Easy to Operate the symbols presented
17	Game animation settings
18	The suitability of the used operator symbol
Aspek Isi Game	
19	Students can understand the operations of addition, multiplication, subtractions, and division through the game media
20	Match the level of difficulty of the game in accordance with the ability of elementary students

Figure 4. Questionnaires design.

8. Evaluation plan

The Evaluation will be done in SDN Tunjungsekar 1, consist of 4 stages (Validation, Small Group Testing, Large Group testing, Release):

- The Game will be validated by mathematicians, ICT experts and elementary school teacher teachers. If the validation results are good, then we will move to the next stage.
- The Game will then be tested to small groups of students. Each student will be asked to play the games that have been designed. Then The students will be asked to provide assessments through the provided questionnaire sheet and give advice and comment on the game (see figure 4). If the results of the assessments are good, then we will move to the next stage.

- The game will be tested in a large group, and they will also be asked to give the assessment. through the provided questionnaire sheet and give advice and comment on the game. If the Results is Good, then we will move to the next stage.
- The game is ready to be released.

9. Conclusion

The game is developed using scratch by MIT so that the development process is less time consuming and easier. The game is in the form of a quiz game that makes students utilize their mathematical skills to win the quiz and get the rewards. In that way, the competitiveness of the students also increases at the same time with their numeric ability. This learning media has an appropriate value to balance current technological developments and improve student's numerical ability because of the exercises in the form of interesting web-based games. The game can make learning math interesting for the students because of the interface and the reward system. Further Evaluation plan still needs to be done before releasing the game to the public.

References

[1] OECD 2015 *Pisa 2015 Results in Focus* **853**
[2] OECD 2019 PISA 2018 Results. Combined Executive Summaries *J. Chem. Inf. Model.* **53** 9 1689–1699
[3] Kurniasih I and Sani B 2014 *Implementasi Kurikulum 2013 Konsep dan Penerapan*
[4] Hakim D L and Sari R M M 2019 Aplikasi Game Matematika Dalam Meningkatkan Kemampuan Menghitung Matematis *JPPM (Jurnal Penelitian dan Pembelajaran Matematika)* **12** 1 129-141
[5] Hakim D L 2014 Efforts To Improve Student Learning Outcomes By Using Cooperative Learning Type of Student Teams Achievement Division (STAD) *Proceeding of International Conference On Research, Implementation And Education Of Mathematics And Sciences 2014* 18–20
[6] Hakim D L 2017 Penerapan Permainan Saldermath Algebra Dalam Pelajaran Matematika Siswa Kelas Vii Smp Di Karawang *JIPMat* **2** 1
[7] Siagian R E F 2015 Pengaruh minat dan kebiasaan belajar siswa terhadap prestasi belajar matematika *Formatif: Jurnal Ilmiah Pendidikan MIPA* **2** 2
[8] Alvarez J and Michaud L 2008 Serious Games: Advergaming, edugaming, training and more *rapport IDATE*
[9] Sawyer B and Rejeski D 2002 Serious Games: Improving public policy through game-based learning and simulation, Woodrow Wilson International Center for Scholars: 31 *Serious Games* 2002-1
[10] Sawyer B 2009 Foreword: From virtual U to serious game to something bigger *Serious games: Mechanisms and effects*
[11] Laamarti F, Eid M and El Saddik A 2014 An overview of serious games *Int. J. Comput. Games Technol.*
[12] Shin N, Sutherland L M, Norris C A and Soloway E 2012 Effects of game technology on elementary student learning in mathematics *British journal of educational technology* **43** 4 540-560
[13] Maloney J, Resnick M, Rusk N, Silverman B and Eastmond E 2010 The scratch programming language and environment *ACM Transactions on Computing Education (TOCE)* **10** 4 1-15
[14] Anglin G J 2002 Designing effective instruction *Educ. Technol. Res. Dev.* **50** 2 79–83
[15] Chizary F and Farhangi A 2017 Efficiency of educational games on mathematics learning of students at second grade of primary school *Journal of History Culture and Art Research* **6** 1 232-240

IOP Conf. Series: Materials Science and Engineering 1073 (2021) 012065 doi:10.1088/1757-899X/1073/1/012065

Software testing on e-learning Madrasahs using Blackbox testing

S Sutiah[1,*] and S Supriyono[2]

[1]Department of Islamic Education, Faculty of Tarbiyah and Teaching Training, Universitas Islam Negeri (UIN) Maulana Malik Ibrahim Malang, Indonesia
[2]Department of Informatics Engineering, Faculty of Science and Technology, Universitas Islam Negeri (UIN) Maulana Malik Ibrahim Malang, Indonesia

*sutiah@pai.uin-malang.ac.id

Abstract. Distance learning in the pandemic era of COVID 19 must meet the needs of the system. E-Learning Madrasah is an application program developed from the Ministry of Religious Affairs of the Republic of Indonesia. The learning that can be done in the pandemic era of COVID 19 corresponds to government directives. E-Learning Madrasah has an essential role in distance learning at the basic education level up to secondary education. To improve the software functionality required testing system. The following research uses the Blackbox Testing approach to e-learning Madrasah software testing. The testing scenario includes several scenarios, including login form testing, class setting testing, computer-based test testing, student learning evaluation testing. The scenario can be used to test functionality so that the recommendation results can be used to develop e-learning madrasah in the future. The resulting results showed e-learning Madrasah using the BlackBox testing approach to testing software functionality levels with the need to reach 90%. These results show from the experiment of some of the scenarios.

1. Introduction

Distance learning is fundamental in the era of the COVID 19 pandemic. The success of educators in the COVID 19 pandemic can be measured by how effective and efficient the learning method is using e-learning. Related research discusses the use of e-learning at Senior High School in Bengkulu City [1]. Related research uses Moodle to improve performance on e-learning. The development of technology and the world of education today encourages innovation and development to get knowledge quickly without limitation of place and time. One of the good breakthroughs in the world of education is the existence of e-learning or electronic learning media, with the development of the internet and the ease of information, e-learning is now quite common as a means of getting knowledge quickly and easily. The use of Hybrid learning can increase the use of e-learning in improving learning technology [2].

E-learning Madrasah is a free application for Madrasah products that are intended to support the learning process in Madrasas from Raudatul Athfal (RA), Madrasah Ibtidayah (MI), Madrasah Tsanawiyah (MTs) and Madrasah Aliyah (MA), to make it more structured, attractive and interactive. E-learning Madrasah contains content starting from the beginning of the learning process, making competency standards, learning implementation plans, learning materials, teacher journals, daily assessment processing, computer-based exams to processing report cards. Core competencies in student learning can be measured according to the competency level needs of students [3]. To produce more optimal software, appropriate technology that can be used in distance learning is needed. The optimal results can be seen from the results of the system testing that has been done. For more optimal testing, the BlackBox testing method is needed.

The development of learning media is an essential aspect of learning technology. In testing the learning media application, it has a pretty good level of accuracy [4]. Related research uses design-based research (DBR) method in integrating technology in a lot of interactivity in a student's class [5]. Web-based application testing uses a testing scenario with various parameters by evaluating the use of genetic algorithms used to generate testing requirements [6]. Software testing is used to measure the level of complexity of an application program used includes functional and non-functional testing [7]. Software testing has been discussed in research related to using tools that aim to verify software that has been developed [8].

The software testing that is carried out imperfectly will certainly have an adverse effect on the quality of the software produced. Ineffective and incomplete software testing can cause various problems when the software is used by end-users. It can automatically improve the efficiency of the testing process to identify which parts of the software are prone to failure. In the following research, testing the E-Learning Madrasah software with the BlackBox Testing approach. Based on the method used, it will be known weaknesses in the software after testing using the Black Box method and how to find out which output is considered valid.

2. Blackbox testing

System testing aims to see whether the system that has been created is by the original purpose of manufacture and is suitable for use. Testing on the system uses the BlackBox Testing method. The goal is to determine that the parts in the application system correctly display error messages if an error occurs in data input.

2.1. Equivalence partitioning

The following testing techniques are used to reduce the number of test cases. The test parameter is obtained from the input value in each part of the tested domain [9]. Equivalence Partitioning Testing is a method of testing software with the BlackBox Testing approach that breaks or divides the input domain of the program into data classes so that test cases can be obtained. The design of test case equivalence partitioning is based on the evaluation of equivalence classes for input conditions that describe a set of states that are valid or not. The input condition can be a numeric value, a range of values, a related set of values , or a Boolean condition.

2.2. Boundary value analysis / limit testing

The Boundary Value Analysis method is a method of detecting errors by limiting input to a specific limit, a value above or below a limit [10]. Boundary value analysis is one type of testing technique with the Blackbox Testing approach, which tests the upper and lower limits of a value inputted into the application. The main requirement of boundary value analysis is complementing the equivalence class testing technique, which only pays attention to the input value. At the same time, the boundary value analysis also pays attention to the output value. Boundary value analysis examines the input at the upper and lower limits of a valid value.

2.3. Comparison testing

Test each software version with the same data to ensure all versions produce the same output [11].

2.4. Sample testing
Involves several values selected from an equivalent class. Integrate values into test cases. Selected values may be selected in a specific order or at certain intervals.

2.5. Robustness testing
Input data is selected outside of defined specifications. The purpose of this test is to prove that there is no error if the input is invalid.

2.6. Behavior testing
The test results cannot be evaluated if you only test once but can be evaluated if the test is carried out several times, for example, testing the stack data structure.

2.7. Performance testing
Evaluate the program's ability to operate with correct in terms of reference needs, for example, data flow, memory usage size, execution speed. To find out the workload or program configuration conditions. Performance specifications are defined at the specification or design stage. It can be used to test the limits of the application program environment.

2.8. Requirement testing
Requirements specification associated with software (input / output / function / performance) is identified at the requirements specification and design stage. Requirement testing involves creating test cases for each requirement specification associated with the program. Each requirement specification can be traced to a test case using a traceability matrix.

2.9. Endurance testing
Endurance Testing involves a repeated number of test cases to evaluate whether the program conforms to the specification requirements. The test requirements specification is defined at the requirements or design specification stage.

2.10. Cause –effect relationship testing
Software test design techniques cover a wide variety of areas, including identifying input and output cases [12]. The steps taken include determining the input conditions (cause) and the output conditions (effect) and making examples of cases for testing and creating test tables.

3. E-Learning madrasah
E-learning Madrasah is a free application for Madrasah products that are intended to support the learning process in Madrasas from Raudatul Athfal (RA), Madrasah Ibtidayah (MI), Madrasah Tsanawiyah (MTs) and Madrasah Aliyah (MA), to make it more structured, attractive and interactive. Madrasah e-learning has six access roles, including Madrasah Operators (Administrators), Subject Teachers, Counseling Guidance Teachers, Classroom Teachers, Students, and Supervisors (Madrasah Principals and staff). The learning tools used by the teacher are the 2013 Curriculum with a Learning Implementation Plan (RPP) and an assessment rubric that refers to the syllabus. Learning resources use textbooks with material taken from Graduate Competency Standards following Content Competencies and Basic Competencies.

The function of e-learning for learning activities in the classroom is as a supplement, which is optional or optional, complementary, or substitute (substitute). It is said to function as a complementary if the electronic learning material is programmed to complement the learning material received by students in the classroom. As a complement, it means that electronic learning material is programmed to become reinforcement or remedial material for students in following conventional learning activities.

IOP Conf. Series: Materials Science and Engineering 1073 (2021) 012065 doi:10.1088/1757-899X/1073/1/012065

4. Research methodology

The following research includes several stages, including the following Figure 1.

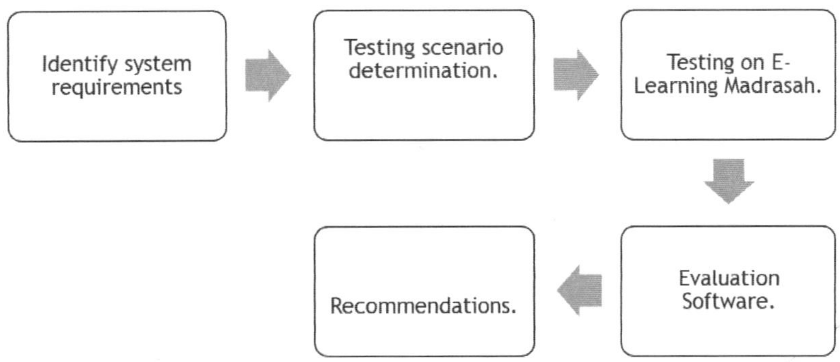

Figure 1. Research steps.

4.1. Identify system requirements

The system requirements phase is the process of understanding the proposed problem and ensuring that it supports the system requirements. Other things that are done include building a solid foundation for system development, system analysis activities, system modeling requirements, and development strategies.

4.2. Testing scenario determination

Testing the madrasah e-learning application based on the data that has been obtained from the application. The testing scenarios include login form testing, class setting testing, computer-based test testing, student learning evaluation testing. A scenario is a story or narrative that is easy to understand to make the application come alive. Stories that are easily accessible help the target of the story to understand the reasons why the problem is being solved by the system to be built. A scenario is not a technical specification description of a software product. A scenario is a narrative or story.

4.3. Testing on e-learning madrasah

Software Testing of e-learning Madrasah based on a scenario that has been made. Testing includes test methods applicable to all levels of software testing: unit, integration, functional, systems, and acceptance. This usually consists of software testing but can also dominate unit testing.

4.4. Evaluation software

Program evaluation is an effort to determine the extent to which the level of success of software which later results from this evaluation can be used by policymakers (Decision maker in making decisions whether the program is appropriate and still appropriate to continue or is still lacking, so it needs revision or even stopped.

4.5. Recommendations

Recommendations are resulting from the results obtained on software testing with the BlackBox Testing approach.

5. Results and discussion

The test scenario includes several things, including the following.

IOP Conf. Series: Materials Science and Engineering 1073 (2021) 012065 doi:10.1088/1757-899X/1073/1/012065

5.1. Login form scenarios

In the login form scenario, it is done by entering the username and password in the application program that has been run. Figure 2 is a display on the login form. The login form is used to find out which system users have obtained access rights from the application program. The test scenario on the login form knows the level of user access, both users by teachers, students, and administrators who have been appointed by the school.

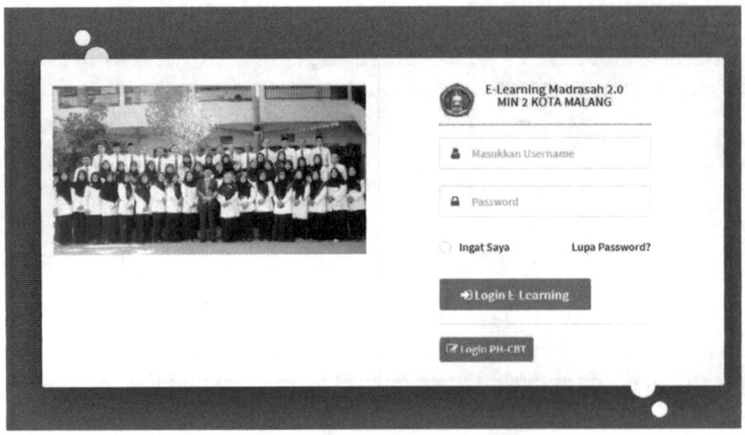

Figure 2. Login form.

5.2. Classroom scenario

Figure 3 shows the form to create a new class that is used for the class management process. From the results of the scenario, input parameters and agendas will be carried out at each meeting conducted based on the class that the teacher is teaching. Scenarios in the classroom setting test are needed to determine the level of validity results of the application function with the BlackBox testing approach.

Figure 3. Form create a new class.

IOP Conf. Series: Materials Science and Engineering 1073 (2021) 012065 doi:10.1088/1757-899X/1073/1/012065

5.3. Computer based test scenarios

The test on the Computer Based Test scenario is shown in Figure 4 From this scenario will measure the functionality of the online exam arrangements carried out by teacher users to students who have been connected in their respective classes. The parameters that were tested included the type of test, randomization of questions, time of the test, and the results obtained when students finished working on the questions that had been given by the teacher. In addition, there will also be a token generation test before the test starts.

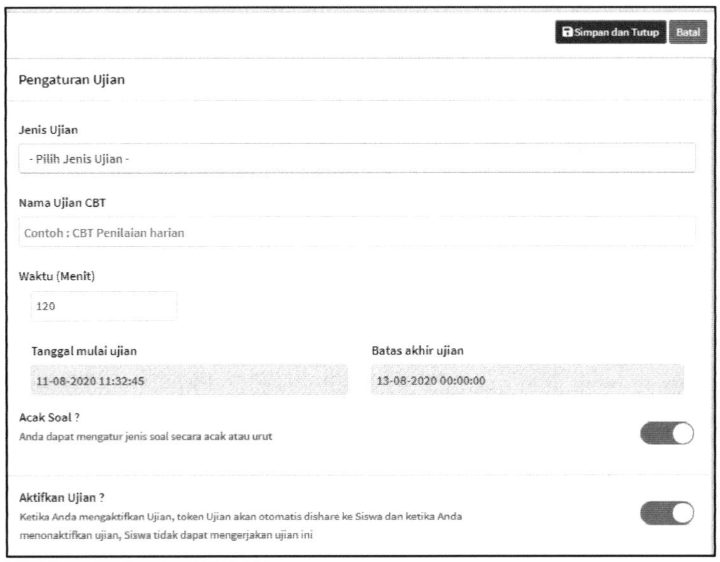

Figure 4. CBT form.

5.4. Student learning evaluation scenarios

The results of the evaluation of student learning make it easier for teachers to measure the level of student achievement in understanding the lessons that have been given. From these results, several things related to the assessment will be obtained. The test scenario in the evaluation of student learning is shown in Figure 5.

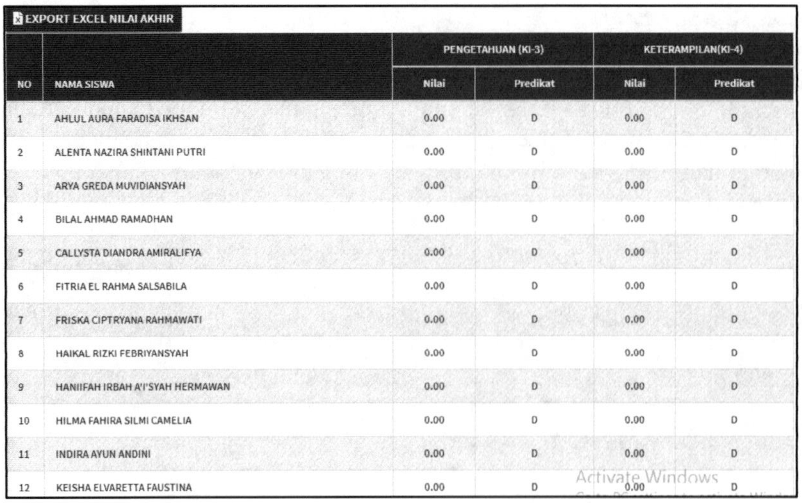

EXPORT EXCEL NILAI AKHIR

NO	NAMA SISWA	PENGETAHUAN (KI-3)		KETERAMPILAN(KI-4)	
		Nilai	Predikat	Nilai	Predikat
1	AHLUL AURA FARADISA IKHSAN	0.00	D	0.00	D
2	ALENTA NAZIRA SHINTANI PUTRI	0.00	D	0.00	D
3	ARYA GREDA MUVIDIANSYAH	0.00	D	0.00	D
4	BILAL AHMAD RAMADHAN	0.00	D	0.00	D
5	CALLYSTA DIANDRA AMIRALIFYA	0.00	D	0.00	D
6	FITRIA EL RAHMA SALSABILA	0.00	D	0.00	D
7	FRISKA CIPTRYANA RAHMAWATI	0.00	D	0.00	D
8	HAIKAL RIZKI FEBRIYANSYAH	0.00	D	0.00	D
9	HANIIFAH IRBAH A'I'SYAH HERMAWAN	0.00	D	0.00	D
10	HILMA FAHIRA SILMI CAMELIA	0.00	D	0.00	D
11	INDIRA AYUN ANDINI	0.00	D	0.00	D
12	KEISHA ELVARETTA FAUSTINA	0.00	D	0.00	D

Figure 5. Student learning evaluation form.

From the results of the scenario, the level of validity of the test was generated, as shown in Table 1. In the scenario that was tested, there were results with different percentage variations.

Table 1. Evaluation result

Scenario	*Valid Test Case (%)*
1	82
2	95
3	85
4	93
5	95

The graph shown in Figure 6 shows the evaluation of software testing with the BlackBox Testing approach in madrasah e-learning. These results follow the requirements of the system applied to a particular school. These results may change according to the level of system requirements that have been identified.

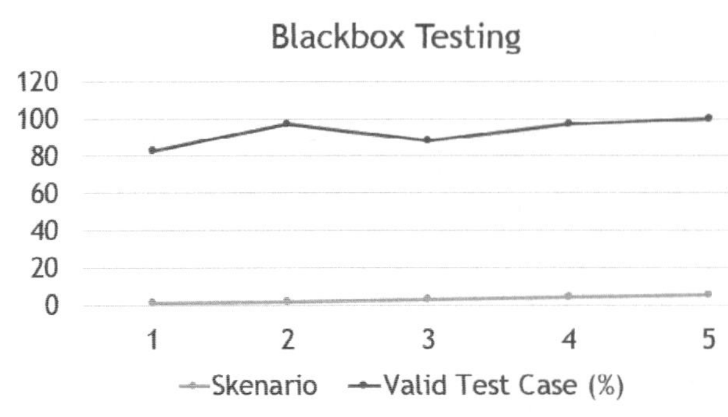

Figure 6. Blacbox testing form.

6. Conclusion

In the application testing scenario that has been tested, there are several things that need to be considered, including the results of the accuracy level of the BlackBox testing examiner that is in accordance with the existing results. The test results in Blackbox testing showed the level of functionality in e-learning Madrasah software shows the results reaching 90%. The recommendation results in the Madrasah e-learning test show are quite optimal results. Blackbox testing is perfect for testing Blackbox testing.

Acknowledgments

The author would like to thank the reviewers and the team who have helped in the process of completing the paper. Thanks to the head of the Department of Informatics Engineering, Faculty of Science and Technology and Department of Islamic Education, Faculty of Tarbiyah and Teaching Training for his support.

References

[1] Puspita Sari A and Setiawan A 2018 The Development of Internet-Based Economic Learning Media using Moodle Approach *Int. J. Act. Learn.* **3** 100–9

[2] Sutiah and Supriyono 2020 Improvement of e-learning based on hybrid learning methods at the university in the era of industrial revolution 4.0 *Int. J. Adv. Sci. Technol.* **29** 2137–42

[3] Sutiah S and Supriyono S 2020 Fuzzy topsis optimization on expert systems for core competency detection and Islamic religious education student learning achievement at Perguruan Tinggi Keagamaan Islam Negeri (PTKIN) *IOP Conf. Ser. Mater. Sci. Eng.* **732**

[4] Sutiah S and Supriyono S 2020 Software Testing on The Learning of Islamic Education Media Based on Information Communication Technology Using Blackbox Testing *IJISTECH (International J. Inf. Syst. Technol.* **3** 254–60

[5] Wang Y H 2020 Design-based research on integrating learning technology tools into higher education classes to achieve active learning *Comput. Educ.* **156** 103935

[6] Boukhris S, Andrews A, Alhaddad A and Dewri R 2017 A case study of black box fail-safe testing in web applications *J. Syst. Softw.* **131** 1339–51

[7] Lonetti F and Marchetti E 2018 *Emerging Software Testing Technologies* vol 108 (Elsevier Inc.)

[8] Siddiqui J H, Rauf A and Ghafoor M A 2018 *Advances in Software Model Checking* vol 108 (Elsevier Inc.)

[9] Irawan Y, Muzid S, Susanti N and Setiawan R 2019 System Testing using Black Box Testing Equivalence Partitioning (Case Study at Garbage Bank Management Information System on Karya Sentosa) 1–7

[10] Prayudha I P A, Hartati R S and Divayana Y 2019 on Learning Management System Applications **4** 25–8

[11] R. FARAH DINI QOYYIMAH, Erfan Rohadi, ST., M. Eng., Ph.D and Rizky Ardiansyah, S.Kom, MT 2020 Pengembangan Sistem Informasi Pemetaan Infrastruktur Sistem Informasi Di Kota Probolinggo *J. Inform. Polinema* **6** 65–70

[12] Babbar H 2017 Software Testing : Techniques and Test Cases *Int. J. Res. Comput. Appl. Robot.* **5** 44–53

IOP Conf. Series: Materials Science and Engineering 1073 (2021) 012066 doi:10.1088/1757-899X/1073/1/012066

3D reconstruction using Structure From Motion (SFM) algorithm and Multi View Stereo (MVS) based on computer vision

M Kholil[*]**, I Ismanto and M N Fu'ad**

Akademi Komunitas Negeri Putra Sang Fajar Blitar, East Java, Indonesia

*moch.kholil89@gmail.com

Abstract. The development of the Information and Computer Technology (ICT) sector, three-dimensional (3D) technology is also growing rapidly. Currently, the need to visualize 3D objects is widely used in animation and graphic applications, architecture, education, cultural recognition and Virtual Reality. 3D modeling of historic buildings has become a concern in recent years. 3D reconstruction is an attempt to document reconstruction or restoration if the building is destroyed. By using the 3D model reconstruction using Structure from Motion (SFM) and Multi View Stereo (MVS) algorithm based on Computer Vision, it is hoped that the results of this 3D modeling can be utilized as an effort to preserve 3D objects in the Penataran Temple cultural heritage area. This research was conducted by taking as many as 61 images of objects in the Blitar Penataran Temple area. The photos obtained were reconstructed into a 3D model using the Structure From Motion algorithm in the meshroom. This research a trial of the original image with a compressed image for reconstruction is used to compare the 3D reconstruction process from the two input data. From 61 images processed using the Structure Form Motion algorithm, 33 poses of camera pose and 3D points were improved, both original and compressed images. The number of iterations compresses 1.4% less than the original image and takes 43.53% faster than the original image.

1. Introduction

Law of the Republic of Indonesia Number 11 Year 2010 article 1 concerning Cultural Heritage explains that cultural preservation is a material cultural heritage in the form of cultural heritage objects on land and / or in water that needs to be preserved because they have important values for history, science, education, religion, and / or culture through the process of determination. In the Law of the Republic of Indonesia Number 11 Year 2010 also stated that what is referred to as preservation is a dynamic effort to maintain the existence of cultural heritage and its value by protecting, developing, and utilizing it.

The development of the Information and Computer Technology (ICT) sector, three-dimensional (3D) technology is also developing rapidly. At present, the need to visualize 3D objects is widely used in animation and graphics applications, architecture, education, cultural recognition and Virtual Reality [1]. 3D modeling of historic buildings has become a concern in recent years. 3D reconstruction is an attempt to document reconstruction or restoration if the building is destroyed.

Law No. 11 Article 5 section 4 also stipulates that the preservation of cultural heritage must be supported by documentation activities before activities that can cause a change in authenticity. The documentation is not only limited to knowing the geometry dimensions of cultural heritage, but also related to how large changes in the dimensions of the geometry that occur within a certain time span. One method of documenting cultural heritage which is currently undergoing development is a three-dimensional modeling method. Utilization of documentation methods by making three-dimensional models of objects or areas of cultural heritage provides many advantages including documentation data that has the shape and dimensions of objects that are precise and easy to store. Therefore, currently making three-dimensional models for the benefit of the documentation of objects and areas of cultural heritage is very necessary in conservation activities so as to be able to maintain the elements of cultural works that are in a state quite complete in such a way that it is still able to provide a complete picture of cultural heritage exists and reflects the important values they contain [1].

The selection of the Penataran Temple area as a research site to model 3D objects in the area is based on the Penataran Temple, one of the grandest and most extensive temples in East Java, located on the western slope of Mount Kelud, north of the city of Blitar with an altitude of 450 meters above sea level. 3D object reconstruction techniques are divided into 2 categories, namely active techniques and passive techniques. Active technique (object scanning) requires control of structured light. Some researchers use a projector or viewer to produce structured light [2]. Other researchers used a laser beam and a video camera. Passive technique is done by taking using two or more images of an object from various positions with the camera [3,4]. This technique is often known as the adoption of photogrammetry or structure from motion. By using the 3D model reconstruction approach based on Multi View Stereo (MVS) using the Structure From Motion (SFM) algorithm, it is hoped that the results of this 3D modeling can be utilized as an effort to preserve 3D objects in the Penataran Temple cultural heritage area.

2. Literature

2.1. Structure Form Motion (SFM)

Structure From Motion (SFM) is the process of estimating the 3D structure of a scene from a series of 2D images. SFM is used in many applications, such as 3D scanning and Augmented Reality [5]. SFM can be calculated in various ways. The way to approach a problem depends on various factors, such as the number and type of cameras used. If the picture is taken with a single calibrated camera, the 3D structure and camera movement can only be restored to an existing scale, changing the structure of the scale, the magnitude of the camera's movement or still maintaining observation. For example, if you place the camera close to an object, you can see the same image as when zooming in and moving the camera away. If you want to calculate the actual scale of structures and movements in the area, additional information is needed, such as:

- The size of the object in the scene.
- Information from other sensors, for example, odometer.

For most applications, such as robots and autonomous drivers, SFM uses more than two displays.

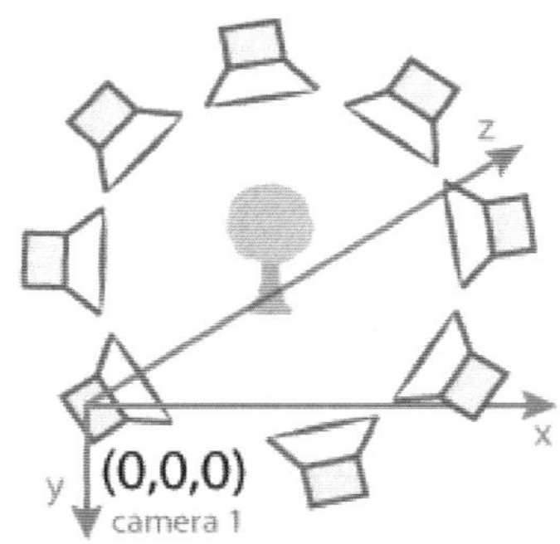

Figure 1. SFM multiple view.

The approach used for SFM from two views can be extended to several views. The set of displays that are used for SFM can be ordered or canceled. The approach taken here assumes a regular sequence of views. SFM from multiple displays requires the corresponding point in several images, called tracks. A typical approach is to calculate tracks from correspondence of paired points.

The SFM algorithm takes as input one set of images and produces two things: the camera parameters of each image, and a set of 3D points seen in images that are often encoded as tracks. Tracks are defined as 3D coordinates from reconstructed 3D Points and a list of corresponding 2D coordinates subset of input images. Most of the cutting-edge SFM algorithms share the same basic processing pipeline.

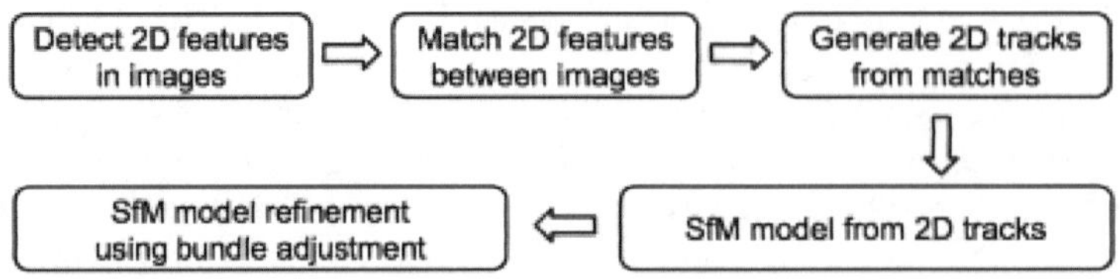

Figure 2. SFM stages.

The effort then focused on two key components of the SFM algorithm:

- Calculate Euclidean reconstructions (up to scale) from multiple cameras, i.e. estimating camera parameters and 3D positions of tracks, and
- Build 2D tracks that are longer.

2.2. Multi View Stereo (MVS)

The origins of stereo multi-view can be traced back to human stereopsis and the first attempt to solve the stereoscopic matching problem as a matter of calculation [6]. To this day, the Multi-View Stereo algorithm has become a very active and fruitful area of research [7]. The Multi-View Stereo version originated as a natural enhancement for the two-display case. Instead of taking two photos from two different viewpoint photos, Multi-View Stereo will capture more angles in between to increase durability, for example for image noise or surface texture [8,9].

Although MVS has the same principles as the classical stereo algorithm, the MVS algorithm is designed to handle images with more varied perspectives, such as sets of images that surround objects, and also handles very large numbers of images, even in millions of orders. The difference in the nature of the MVS problem eventually results in a significantly different algorithm from the classic stereo partner. For example, industrial applications for 3D mapping [6], processing millions of photos over hundreds of kilometers at a time, effectively reconstructing large metropolitan areas, countries and eventually the entire world.

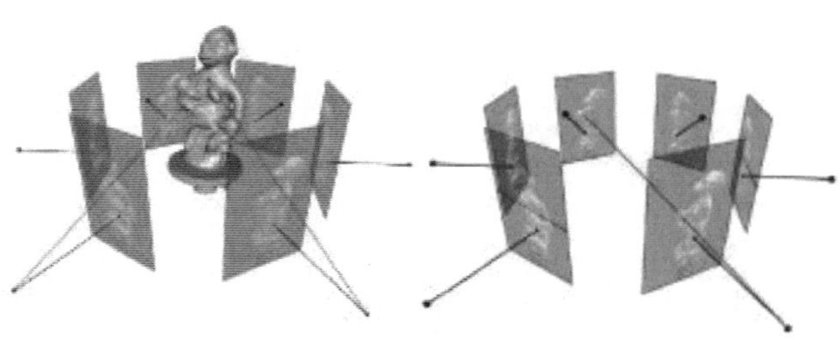

Figure 3. SFM stages.

Matching pixels in an image is a difficult problem unique to stereo or multi-view stereo. In fact, optical flow is another area that is very active in Computer Vision, overcoming the problem of correspondence that is dense across images [10]. The main difference with the existence of MVS is that optical flow is usually a matter of two images (similar to two stereo displays), the camera is not calibrated, and its main application is image interpolation rather than 3D reconstruction.

3. Methodology

This research was carried out sequentially and systematically (sequence) starting from data collection, process to produce output in order to facilitate the process. In this study using the Photogrammetry Pipeline model [11]. The following stages of research are carried out to produce 3D objects from 2D images.

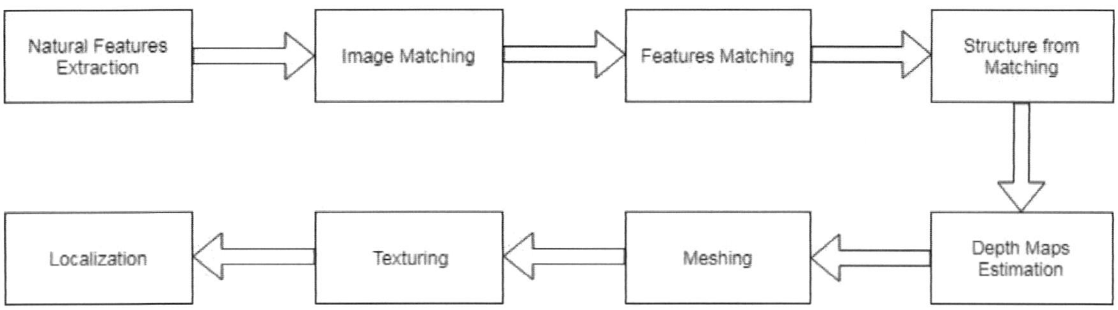

Figure 4. Photogrammetry Pipeline.

3.1. *Natural features extraction*

The purpose of this step is to extract distinct groups of pixels that, to some extent, do not change at the changing camera viewpoints during image acquisition. Therefore, the features in the scene must have similar feature descriptions in all images.

3.2. *Image matching*

The purpose of this section is to find images that are looking into the same area of the scene. To do this, we use shooting techniques to find images that share some content without the cost of completing all of the feature matches in detail. His ambition is to simplify the image in a concise image descriptor which makes it possible to calculate the distance between all image descriptors efficiently.

3.3. *Features matching*

The purpose of this step is to match all features between candidate pairs of images. First, a photometric match was made between the descriptors set of 2 input images. For each feature in figure A, get a list of candidate features in figure B. Because the descriptor space is not linear and well-defined space, it cannot rely on absolute distance values to find out whether the match is valid or not (can only have a higher absolute bound distance). To remove a bad candidate, it is assumed that there is only one valid match in the other image. So for each feature descriptor in the first image, look for the 2 closest descriptors and use the relative threshold between images. This assumption will turn off features in the repetitive structure but has proven to be a strong criterion. It provides a list of candidate matching features based only on photometric criteria. Find the 2 nearest descriptors in the second image for each feature that is computationally intensive with the brute force approach, but there are many optimized algorithms. The most common is the Estimated Nearest Neighbor, but there are alternatives such as, Cascading Hashing.

3.4. *Structure from motion*

The aim of this step is to understand the geometric relationships behind all the observations provided by the input image, and deduce the rigid structure of the scene (3D dots) with poses (position and orientation) and internal calibration of all cameras using the non-linear bundle adjustment method to improve structure and motion as well as minimize reprojection errors.

$$E(P,X) = \sum_{i=1}^{m}\sum_{j=1}^{n} D(X_{ij}, P_i X_j)^2 \tag{1}$$

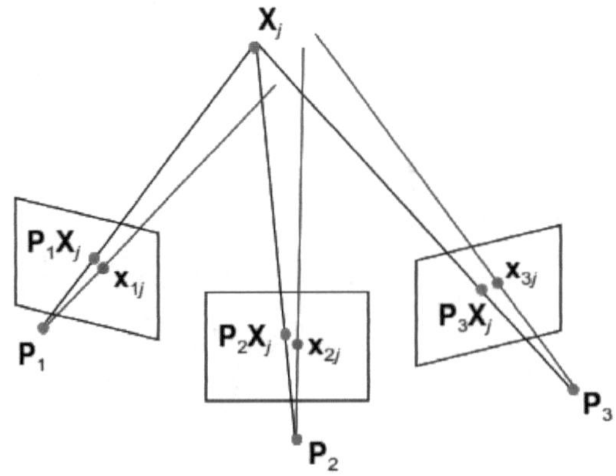

Figure 5. Bundle adjustment process.

Furthermore, the reconstruction step will be calculated automatically from the two initial views which are extended iteratively by adding a new view or what is called Incremental SFM.

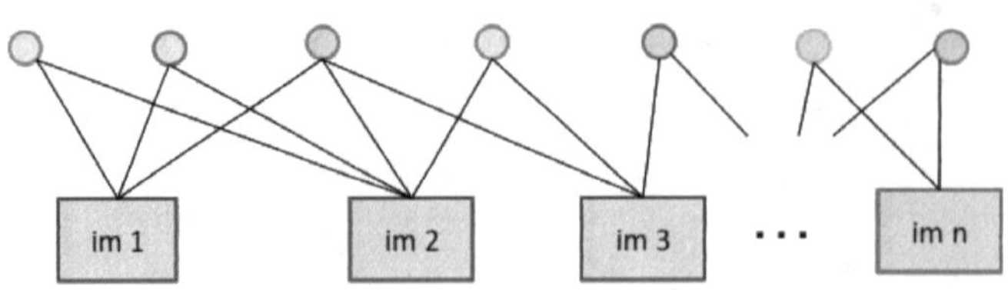

Figure 6. Incremental SFM.

3.5. Depth maps estimation
For all cameras that have been completed by SFM, we want to take the depth value of each pixel using the Semi-Global Matching (SGM) or ADCensus approach applied in AliceVision in Meshroom [11].

3.6. Meshing
The purpose of this step is to make a geometric representation of the dense surface of the scene.

3.7. Texturing
The purpose of this step is for the resulting mesh texture. If the mesh has no associated UVs, it will calculate the UV map automatically. AliceVision implements a basic UV mapping approach to minimize texture space.

3.8. Localization
Based on the results of SFM, camera localization can be performed and take animated camera movements at the 3D reconstruction site.

3.8.1. Camera calibration. The internal camera parameters can be calibrated from several checkered displays. This makes it possible to take parameters of focal length, main point and distortion.

3.8.2. Single camera localization. Use the algorithm presented in the image matching section to localize the nearest image in the SFM result. Then do the matching features with these images as well as with the previous N frames. Then immediately get the 2D-3D association, which is used to localize the camera.

3.8.3. Lots of cameras. If a camera rig is used, it can calibrate the rig. Localize the cameras one at a time in the whole sequence. Then use all valid poses to calculate relative poses between the camera rig and choose a more stable value throughout the image. Then initialize the relative pose of the rig with this value and make a global Bundle Adjustment on all rig cameras. When the rig is calibrated, it can use it to directly localize the pose of the rig from a multi-camera system that is synchronized with the approach.

4. Results and discussion

The 3D object reconstruction research was built from photos as many as 61 images from various angles. Furthermore, the image is reconstructed using the Structure From Motion algorithm in the Meshroom application [11] to produce a complete 3D model with textures. While the photos themselves were taken using the Canon EOS-100M camera. Figure 7 shows the 3D Model Reconstruction using the Multi View Stereo and Structural From Motion method.

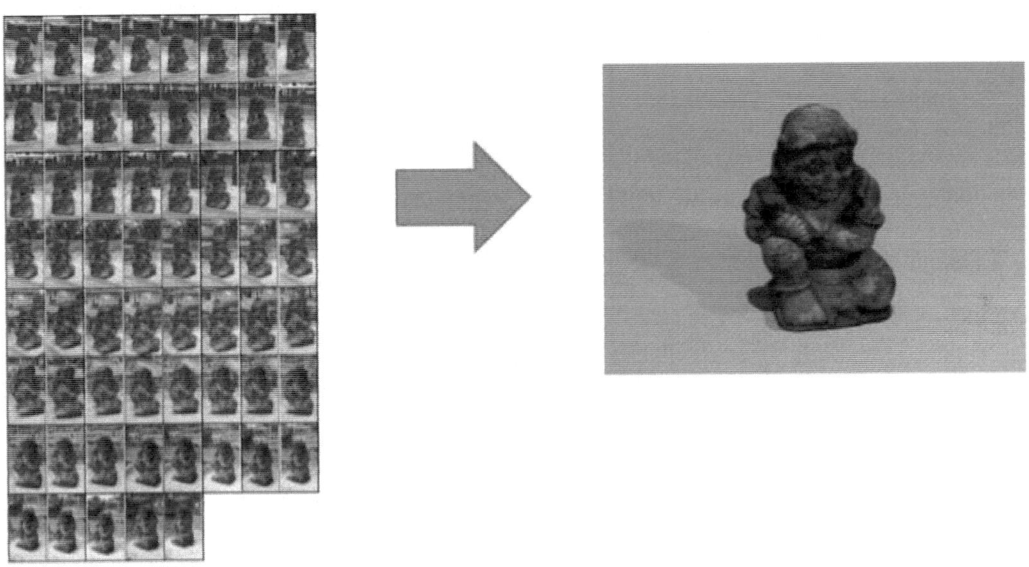

Figure 7. 3D reconstruction of Dwarapala statues in Penataran temple.

Table 1 is a comparison of data processing data from shooting using the Canon EOS-100M camera on the original image or before the data is compressed and after the image is compressed in order to compare the optimal process and results at the Structure Form Motion Budle Adjustment stage.

Table 1. Structure From Motion Bundle Adjustment.

No	SFM Bundle Adjustment	Original Picture	Compress Picture
1	Refine Pose	33	33
2	Iteration	223	220
3	Time (s)	50.30	6.77

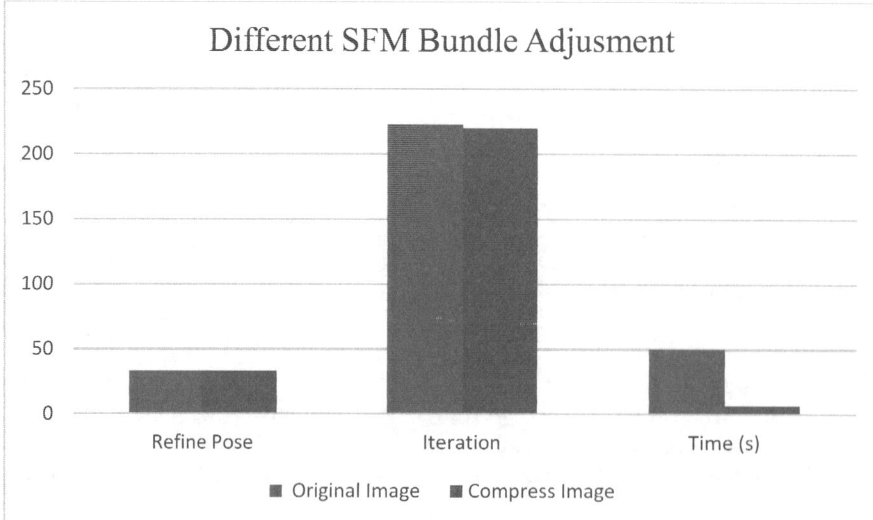

Figure 8. SFM Bundle Adjustment chart.

5. Conclusion

This research was conducted by taking as many as 61 images of objects in the Blitar Penataran Temple area. The photos obtained were reconstructed into a 3D model using the Structure From Motion algorithm in the meshroom. In this study a trial of the original image with a compressed image for reconstruction is used to compare the 3D reconstruction process from the two input data. From 61 images processed using the Structure Form Motion algorithm, 33 poses of camera pose and 3D points were improved, both original and compressed images. The number of iterations compresses 1.4% less than the original image and takes 43.53% faster than the original image.

References

[1] Mulyadi Y 2012 Mengoptimalkan Zonasi Sebagai Upaya Pelestarian Cagar Budaya *Buletin Somba Opu Balai Pelestarian Peninggalan Purbakala Makassar* **15**

[2] Dipanda A and Woo S 2005 Towards a real-time 3D shape reconstruction using a structured light system *Pattern recognition* **38**(10) 1632-1650

[3] Prakoonwit S and Benjamin R 2007 3D surface point and wireframe reconstruction from multiview photographic images *Image and Vision Computing* **25**(9) 1509-1518

[4] Park J S 2005 Interactive 3D reconstruction from multiple images: A primitive-based approach *Pattern recognition letters* **26**(16) 2558-2571

[5] Hwang J T and Ting-Chen C 2016 3D Building Reconstruction By Multiview Images and the Integrated Application With Augmented Reality *The International Archives of Photogrammetry, Remote Sensing and Spatial Information Sciences* **41** 1235

[6] Google Inc. Google maps. http://maps.google.com.

[7] Scharstein D and Szeliski R 2002 A taxonomy and evaluation of dense two-frame stereo correspondence algorithms *International journal of computer vision* **47**(1-3) 7-42

[8] Okutomi M and Kanade T 1993 A multiple-baseline stereo *IEEE Transactions on pattern analysis and machine intelligence* **15**(4) 353-363

[9] Tsai R Y 1983 Multiframe image point matching and 3-d surface reconstruction *IEEE Transactions on Pattern Analysis and Machine Intelligence* (2) 159-174

[10] Baker S, Scharstein D, Lewis J P, Roth S, Black M J and Szeliski R 2011 A database and evaluation methodology for optical flow *International journal of computer vision* **92**(1) 1-31

[11] https://alicevision.org/#photogrammetry (Agustus 1, 2019).

IOP Conf. Series: Materials Science and Engineering 1073 (2021) 012067 doi:10.1088/1757-899X/1073/1/012067

Building Information Modelling for clean water and wastewater system in the medium rise school building

K A Rofi*, R I Hapsari, S S Riskijah, W Harsanti, M A Dharmawan and T Rahman

Department of Civil Engineering, State Polytechnic of Malang, Jl. Soekarno Hatta No. 9, Jatimulyo, Kec. Lowokwaru, Kota Malang, Jawa Timur 65141, Indonesia

*khanifarofi@gmail.com

Abstract. Human activities in a building are closely related to the use of clean water. At this time, the concept of a water installation is demanded to be as effective as possible by minimizing the possibility of water waste caused by leakage problems, broken pipes, blocked pipes, and others. However, some concepts still use traditional techniques by depicting 2D models and manual calculations. The method used is a hydraulic analysis with SNI 03-7065-2005 as standard and using Autodesk Revit MEP as modeling and calculating tools. Necessary data are technical specifications, design drawings, and designed plumbing facilities. The system comprises of Domestic Cold Water, Sanitary Grey Water, Sanitary Black Water, and Vent. Building Information Modelling (BIM) is applied to evaluate the effectiveness of the existing design and design its modification as well as to provide the working drawing. The analysis includes analysis of piping system completeness, system crash, flow, velocity, pressure loss. Based on the report, it can be concluded the flow velocity of Domestic Cold Water is 0.7-6.4 m/s which is out the range of standard value. For the pressure, a manual calculation is performed to validate the report results using the Darcy Weisbach equation that has the same value as the report by Building Information Modeling. The cost of the design conditions has a difference of Rp 8.363.608 or eight million three hundred sixty-three thousand rupiah or 0.4% more efficient.

1. Introduction

Water is one of the primary needs for human life which can be used for several functions, both for daily use and for energy utilization. In the construction of a building, it cannot be separated from the role of the need for clean water. The need for water in a building means water that is used either by the occupants of the building or for other purposes related to building facilities. The need for water in a building is closely related to the function of the building or the activities of building users. This is the basis that each type of building has different needs for clean water.

However, until now the planning of clean water and wastewater systems generally still use conventional methods where the depiction is still limited to 2D media, and planning and control are carried out by manual calculations. With the development of the concept of Building Information Modeling, it is hoped that all projects in Indonesia can be developed and adapt the current, more modern technological concepts in the field of construction which will be easier for all project stakeholders.

Aniendhita [1] and Volk et al. [2] studied the literature review of Building Information Modeling application. However, the application in the modeling was not performed. Further implementation for building object behavior and lean construction performance have been researched [3,4]. In this study, BIM is applied to assist the evaluation of plumbing system efficacy.

This research was conducted to simulate the clean water distribution system along with the sewerage system of the medium rise school building using the necessary hydraulic calculations and analyze the performance of the optimal distribution of clean water and sewage. To simplify the design of clean water distribution networks in buildings, design is assisted by Autodesk Revit MEP software. This is a construction design software in the form of a 3D design that provides an integrated Building Information Modeling. With this technology, the data that is managed will be more accurate and more detailed, starting from modeling, detailing, engineering, drawing, reporting, to scheduling.

2. Methods

At first, the piping system and the dimension are analyzed according to the water demand [5] and volume. The basis for calculation is Indonesian National Standard of Plumbing System Planning Procedure [6]. Afterward, the design is introduced into Building Information Modeling using for the modeling to ensure that it is appropriate, safe, and comfortable to use using Autodesk Revit MEP. The software used is Free Download Autodesk Revit 2019 for Student and Educational Version by creating an account as a student so 1-year license is obtained for free. Control that is carried out in the planning of clean water and wastewater treatment plants includes control of the laying of pipelines whether there are crosses or not. In terms of pressure and flow velocity in the pipe, it is also necessary to control using standards from Indonesian National Standard of Engineering Planning Procedures for Distribution Networks and Clean Water Supply System Service Units [7] regarding the planning procedure for the plumbing system of flow velocity in pipes mentioning that the velocity is 0.9 – 2 (m/s). Meanwhile, according to SNI 7509: 2011 that the maximum flow velocity is 3 m/s. According to PVC pipe standards, the maximum allowed pressure of AW class is 10 kg/cm^2.

Flow control is performed using Revit MEP [8], which can be directly controlled by friction and flow velocity after inputting pipeline dimension data. Control is also carried out on the layout of the pipeline network because if there is an error either in the pipe, connection or other elements, there will be an error in the layout. Flow calculations uses Plumbing Fixture Flow method where this method converts the fixture unit into the volumetric flow using the values listed in Table E103.3 (3) 2012 International Plumbing Code (IPC). The specific flow conversion method is used in the classification of the Domestic Hot/Cold Water System. The results of flow calculations can later be used to determine pipe dimensions. The calculation of pressure drop uses the Colebrook Equation method available in the application. After making sure that the model on Revit MEP matches the expected output, the resource volume calculation can be done automatically. Finally, the comparison of material quantity and budget of existing design and Building Information Modeling design results is discussed.

3. Results and discussion

Figure 1 is the 3D model of the medium rise school building being researched. To explain the colours on the model, the following is the description **Table 1**.

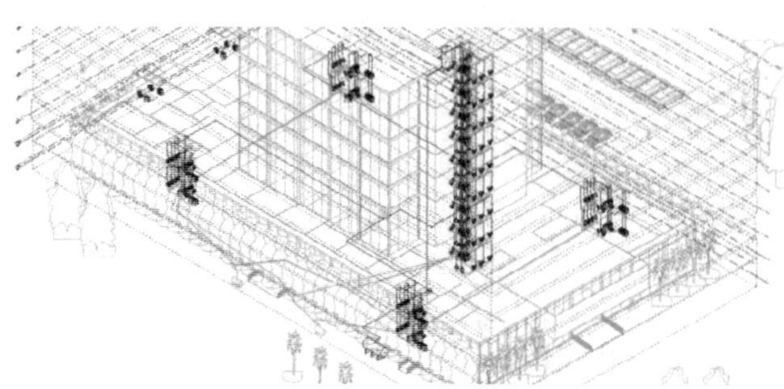

Figure 1. 3D model of the subject.

Table 1. Pipe description.

Pipe Color	Plumbing System Classification	Pipe Type	Slope
Blue	Domestic Cold Water	PVC	0
Red	Sanitary Grey Water	PVC	1%
Green	Sanitary Black Water	PVC	2%
Yellow	Vent	PVC	0

3.1. Layout assessment

Layout control is carried out to ensure that pipelines between the systems do not collide with each other. For a larger scale, usually done for control of one system to another, for example, such as plumbing systems for duct systems and other complex systems. In **Figure 2**, the example of detailed side view of plumbing system is illustrated. Green pipes have a slope of 2% so they appear as tilted line, the same thing as red pipes which have 1% slope. Meanwhile, the blue and yellow pipes appear in straight line because the slope value is 0 or does not have a slope. Because the individual pipes are easily visible, it can be seen that the pipes are not stacked on top of each other.

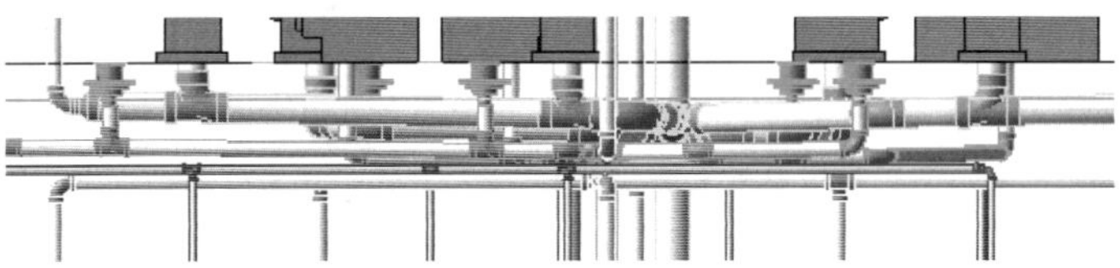

Figure 2. Detailed side view of plumbing system.

3.2. Velocity assessment

It is required to assess the flow velocity of clean water system. **Table 2** provides a summary of the report. From the table, the flow velocity from 5 Domestic Cold Water systems are in the range 0.7 m/s to 6.4 m/s. It can be concluded that for four clean water systems, except Domestic Cold Water 2 system, the velocity is higher than minimum standard, i.e. 0.9 m/s. However, in terms of maximum standard, all systems exceed permissible velocity of 2 m/s. Therefore, if the speed is too high it will cause frequent water blows which can damage plumbing equipment, cause scaly sound from the pipe and cause rapid wear of the pipe surface which later causes pipe leakage.

Table 2. Velocity report summary.

Pipe System	V (m/sec)	
	Min	max
DCW 1	1.7	5.1
DCW 2	0.7	6.4
DCW 3	1.7	5.1
DCW 4	1.7	5.1
DCW 5	1.7	5.1

3.3. Pressure assessment

Verification of the Building Information Modeling results is carried by comparison with manual calculations. Darcy-Weisbach and Bernoulli formulas are used to confirm the head loss and the pressure.

3.3.1. Calculation of head loss with the Darcy-Weisbach formula. To calculate the formula, some values that have been given in the Building Information Modeling need to be stated. **Figure 3**, **Figure 4**, and **Figure 5** show the pipeline properties and the report in Revit MEP. The data that will be used as input data calculations are marked in these figures. The calculation for section 41 on the DCW 2 system using Darcy-Weisbach Formula [9] are as follows:

$$\Delta p = f_D \frac{\rho\, V^2 L}{2\, D_i} = 0{,}017722 \frac{998{,}9114 \times 1{,}63^2 \times 3{,}84}{2 \times 0{,}0779} = 1158{,}4\ N/m^2 \quad (1)$$

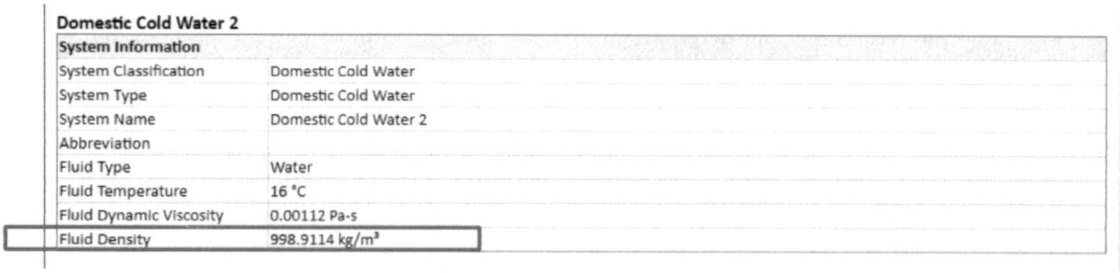

Domestic Cold Water 2

System Information	
System Classification	Domestic Cold Water
System Type	Domestic Cold Water
System Name	Domestic Cold Water 2
Abbreviation	
Fluid Type	Water
Fluid Temperature	16 °C
Fluid Dynamic Viscosity	0.00112 Pa-s
Fluid Density	998.9114 kg/m³

Figure 3. Data for fluid density.

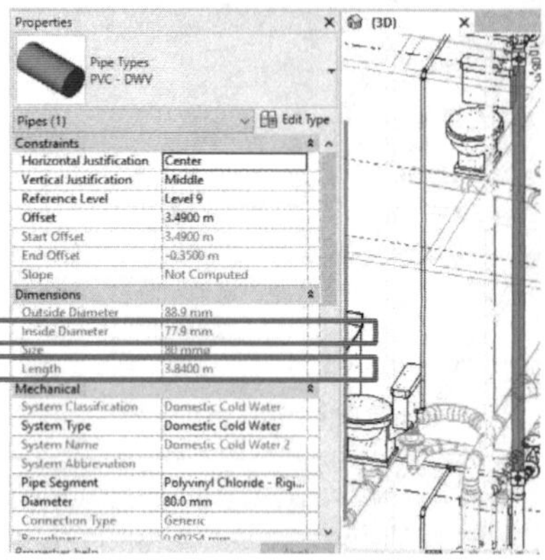

Figure 4. Data for inner diameter, and length.

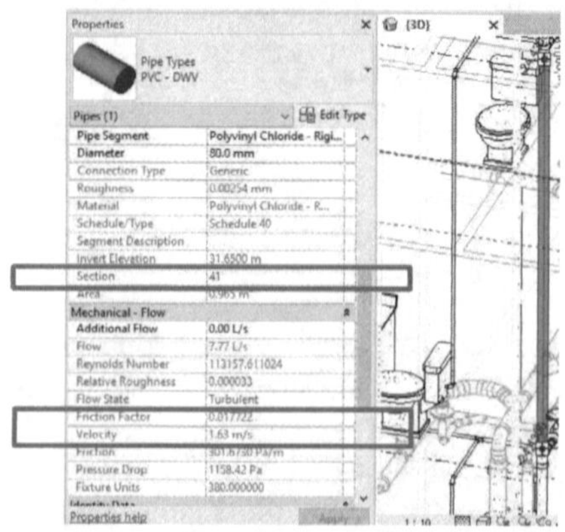

Figure 5. Data for section, Fd and v.

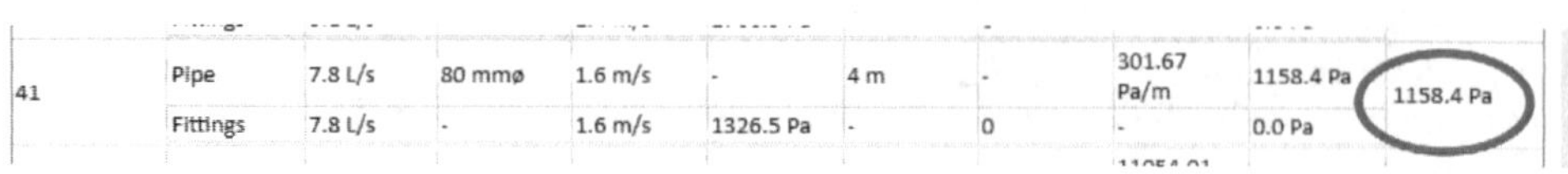

41	Pipe	7.8 L/s	80 mmø	1.6 m/s	-	4 m	-	301.67 Pa/m	1158.4 Pa	1158.4 Pa
	Fittings	7.8 L/s	-	1.6 m/s	1326.5 Pa	-	0	-	0.0 Pa	

Figure 6. Pipe report for section 41.

Based on the manual calculations, the compressive loss in section 41 is 1158.4 N/m^2 or equal to 1158.4 Pa. Compared with Building Information Modeling analysis report shown in **Figure 6**, the pressure provides the same value. It can be proven that the results of the report are valid because they can be proven by using manual calculations using the Darcy - Weisbach equation formula.

3.3.2. Calculation of residual pressure with the Bernoulli formula. In this discussion, the assessment will be carried out on the critical sections according to the report. Figure 7 shows the critical path of Domestic Cold Water 2 system. The critical path is the pipeline that starts from the roof water tank on the roof floor to the pipe at the urinal on the 1st floor. Cumulative Hf at the section should be the same as the static pressure value in the Building Information Modeling properties. Figure 8 is the properties of static pressure for DCW 2 system.

Figure 7. Critical path of Domestic Cold Water 2 system.

Table 3. Residual pressure calculation of Domestic Cold Water 2 system.

Section	Hf		Cumulative	Elevation (m)		ΔH	P		P max control	
	(Pa)	(mH_2O)	(m)	Start	End	(m)	(mH_2O)	(kg/cm^2)	(kg/cm^2)	OK / NO
9	45.7	0.004	0.0046	36	36	0	0.0045	0.000457	10	OK
40	1158.4	0.115	0.1204	36	32	4	4.1204	0.4120	10	OK
71	1010.1	0.101	0.2214	36	28	8	8.2214	0.8221	10	OK
102	882.2	0.088	0.3096	36	24	12	12.3096	1.2310	10	OK
133	771.0	0.077	0.3867	36	20	16	16.3867	1.6387	10	OK
164	629.4	0.062	0.4497	36	16	20	20.4497	2.0450	10	OK
195	503.8	0.050	0.5001	36	12	24	24.5001	2.4500	10	OK
228	377.0	0.037	0.5378	36	8	28	28.5378	2.8538	10	OK
227	44.7	0.005	0.5422	36	4	32	32.5422	3.2542	10	OK
226	64.7	0.007	0.5487	36	4	32	32.5487	3.2549	10	OK
225	360.4	0.036	0.5847	36	4	32	32.5847	3.2585	10	OK
224	61260.5	6.126	6.7108	36	4	32	38.7108	3.8711	10	OK
223	4616.7	0.462	7.1725	36	4	32	39.1725	3.9172	10	OK
222	16114.4	1.611	8.7839	36	4	32	40.7839	4.0784	10	OK
221	5662.2	0.566	9.3501	36	4	32	41.3501	4.1350	10	OK
220	10744.0	1.074	10.4245	36	4	32	42.4245	4.2425	10	OK
219	12260.1	1.226	11.6505	36	4	32	43.6505	4.365	10	OK
229	142694.4	14.269	25.9198	36	1	35	60.9200	6.0920	10	OK

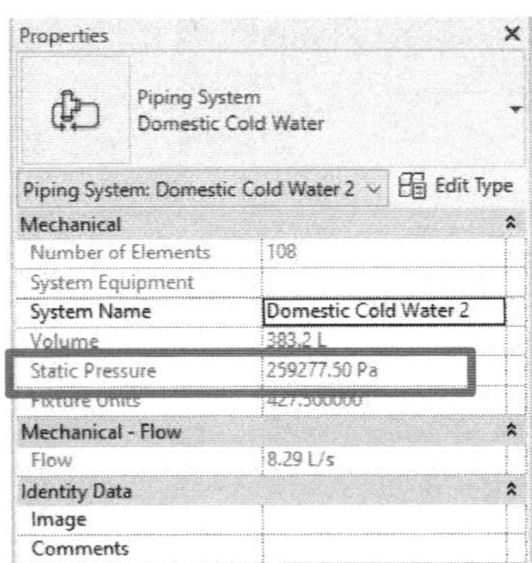

Figure 8. Static pressure of Domestic Cold Water 2 system.

The following is the example of calculation in the section 219. Hf in this section is 12260.1 mH_2O or 12260 Pa. Cumulative pressure at 219 is Hf at section 219 added by cumulative at section 220 or 1.2260 mH_2O added by 10.4245 mH_2O that equals to 11.6505 mH_2O. With start elevation of 36 and end elevation of 4, the height different is 32 m. Applying Bernoulli's formula [10], the remaining pressure in the system is elevation difference added by Hf, that is 32 added by 11.6505, that equals to 43.6506 mH_2O. In kg/cm^2 the pressure is 43.6505 mH_2O divided by 10 or 4.3651 kg/cm^2. The summary of pressure calculations for the entire system are shown in **Table 4.**

Table 4. Residual pressure calculation of Domestic Cold Water 2 system.

Classification System	P (kg/cm²)			P max control	
	min	max	static	(kg/cm²)	OK / NO
Domestic Cold Water 1	0.9972	3.1322	2.4322	10.000	OK
Domestic Cold Water 2	0.4120	6.0920	2.5920	10.000	OK
Domestic Cold Water 3	1.0062	3.1448	2.4445	10.000	OK
Domestic Cold Water 4	1.0221	3.0100	2.3100	10.000	OK
Domestic Cold Water 5	1.0217	2.5630	2.4630	10.000	OK

The analysis shows that the pressure ranges from 0.4120 kg/cm^2 to 6.0920 kg/cm^2 with static pressure ranges from 2.3 kg/cm^2 to 2.6 kg/cm^2. The pressure value is below the maximum permissible pressure value of 10 kg/cm^2. The static pressure value is also quite flat and falls into the standard category for offices, which is between 2.5 to 3.5 kg/cm^2.

3.4. Construction budget

The calculation of the budget plan is intended to determine the total cost required to install the clean water and wastewater system. Construction budget plan is analyzed from unit price and work quantity from automatic calculation by schedule/quantity tool on Revit shown in **Figure 9** as an example. **Table 5** and **Table 6** shows the work items from Building Information Modeling and it cost. The total budget cost for the redesign is IDR 2,481,776,832.00 and for the conventional design is IDR 2,490.140.439.

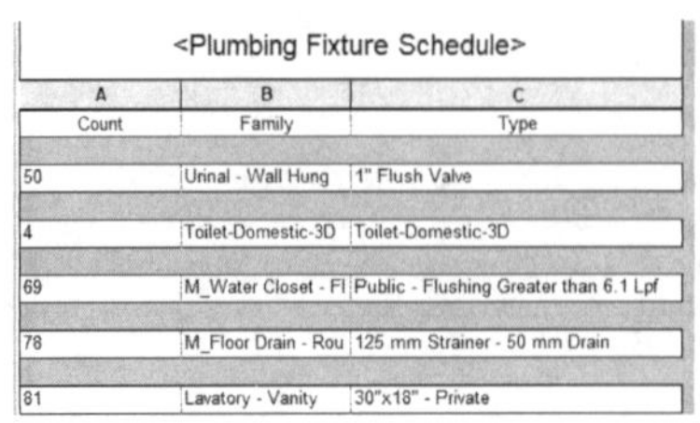

<Plumbing Fixture Schedule>

A	B	C
Count	Family	Type
50	Urinal - Wall Hung	1" Flush Valve
4	Toilet-Domestic-3D	Toilet-Domestic-3D
69	M_Water Closet - Fl	Public - Flushing Greater than 6.1 Lpf
78	M_Floor Drain - Rou	125 mm Strainer - 50 mm Drain
81	Lavatory - Vanity	30"x18" - Private

Figure 9. Quantity calculation on revit.

The difference is not too big because the two designs are simulated in the same way and it lies only in the dimensions of the pipe used. The redesign uses pipe dimensions that have been readjusted to the standard calculations and values, while the conventional is in accordance with the initial design. Overall, the implementation of BIM is beneficial as it allows the construction can be thoroughly analyzed, simulations performed quickly, and performance demonstrated accurately [11].

Table 5. Budget cost of the redesign.

No.	Work Item	Total Price (Rp)
A	Sanitary Tool	665.687.111
B	Clean Water Pipe	437.871.797
C	Wastewater and Vent Pipe	574.998.806
D	Hydrant	204.829.729
E	Complementary	598.389.389
	Sum	2.481.776.832

Table 6. Budget cost of the conventional design.

No.	Work Item	Total Price (Rp)
A	Sanitary Tool	665.687.111
B	Clean Water Pipe	438.592.889
C	Wastewater and Vent Pipe	582.641.322
D	Hydrant	204.829.729
E	Complementary	598.389.389
	Sum	2.490.140.439

4. Conclusions and suggestions

4.1. Conclusions

The simulation of clean water distribution system along with the sewerage system of the medium rise school building and the resources analysis has been presented. By using BIM technique, the pipelines and fixtures are visible. Therefore, it is possible to evaluate the pipe and accessories placement to avoid crash or stacked component. The application in the study building gives flow velocity that ranges from 0.7 m/s to 6.4 m/s. Analysis of the pressure using Darcy-Weisbach and Bernoulli method shows that the BIM report using Revit MEP as the software is accurate. The pressure ranges from 0.4120 kg/cm^2 to 6.0920 kg/cm^2 with static pressure values ranges from 2.3 kg/cm^2 to 2.6 kg/cm^2. The budget cost has a difference IDR 8.363.608, so the redesign is 0.4% more efficient than the conventional. BIM is recommended for plumbing design as the flow parameter as well as the resources can be easily obtained.

4.2. Suggestions
BIM software looks practical, but accuracy is needed to correct the details that potentially interfere when performing analysis and the reports, so the input data and steps used must be precise and appropriate.

Acknowledgments
The authors express sincere gratitude to Komunitas Revit Indonesia for valuable advice in the modeling process.

References
[1] Amalia A R 2016 *Studi literatur tentang program bantu Autodesk Revit Structure* (Skripsi, Surabaya: Institut Teknologi Sepuluh Nopember)
[2] Volk R, Stengel J and Schultmann F 2014 Building information models (BIM) for existing buildings – literature review and future needs *Automation in Construction* **38** 109-127
[3] Lee G, Sacks R and Eastman C A 2006 Specifying parametric building object behavior (BOB) for a building information modeling system *Automation in Construction* **15** 758-776
[4] Sacks R, Koskela L, Dave B A and Owen R 2009 The interaction of lean and building information modeling in construction *Jurnal of Construction Engineering and Management* **136**(9) 968-980
[5] Noerbambang S M and Morimura T 1987 *Planning and Maintenance of Plumbing System* (Jakarta: Pradnya Paramita) *in Indonesian*
[6] Indonesian National Standard of Plumbing System Planning Procedures 03-7065-2005
[7] Indonesian National Standard of Engineering Planning Procedures for Distribution Networks and Clean Water Supply System Service Units 7509:2011
[8] Revit MEP User's Guide 2011 Autodesk
[9] Klaas D K S Y 2009 *Design of Pipe Network* (Bandung: Mandar Maju) *in Indonesian*
[10] Triatmodjo B 2016 *Hidraulic I* (Yogyakarta: Beta Offset) *in Indonesian*
[11] Azhar S 2011 Building Information Modeling (BIM): trends, benefits, risks, and challenges for the AEC industry *Leadership and Management in Engineering* **11**(3) 241-252

IOP Conf. Series: Materials Science and Engineering 1073 (2021) 012068 doi:10.1088/1757-899X/1073/1/012068

Evaluation of educational media of private body parts for down syndrome children

H Pradibta*, S E Sukmana, D S E Ikawati, W I Sabila, A M Hikmah and M A I Ramdani

Information Technology Department, State Polytechnic of Malang, Malang, Indonesia

*hendra.pradibta@polinema.ac.id

Abstract. Educational media of private body parts for Down syndrome children is an educational media in the form of audio-video animation containing information about private body parts. This application developed as part of the supportive learning for Down syndrome children who have entered puberty so that they can avoid sexual abuse. An evaluation is carried out to determine whether the educational media created are according to the needs. The evaluation process is done by interviewing the respondent after conducting a trial on the educational media. The selected respondents were teachers, parents and students. Testing is held at SLB Tk. National Bagian C Malang. The interview questions based on several criteria for assessing multimedia applications, including 1) Ease of use of navigation, 2) Cognition content, 3) Presentation of information, 4) Media integration, 5) Artistic and aesthetics, 6) Learning function. The results of evaluations show that the educational media developed have met the criteria of multimedia assessment. Application of private body parts can provide understanding to Down syndrome children about personal body changes and social contact.

1. Introduction

Educational media for private body parts for Down syndrome children is learning that contains private body parts and the changes that occur when entering puberty. The content created in learning media is in the form of animation and audio. This learning media is intended for children with down syndrome so that they can understand about private body parts and how to protect them in their social environment.

Children with Down syndrome have a level of intelligence below average and have barriers to adjusting to their surroundings. They have a learning delay, have a short memory span, and are less able to think abstractly and complexly [1]. What distinguishes down syndrome children from normal children is in terms of speed, strength, and coordination. Children with Down syndrome tend to be slow to learn new things. They do not understand the changes and development of their physical and biological body parts [2]. This condition can cause problems in the social environment.

Educational Media of Private Body Parts for Down syndrome children developed to support children with mild down syndrome enter puberty, i.e. the early 12-15 years of age, 15-18 years of mid-adolescence, and 18-21 years of late teenage years [3].

2. Literature review

The existence of cases and the urgency of the importance of sexual education for children, has prompted much research on learning about sexual education for children with various methods, problems and different age ranges.

The first study is about the development of media for sex education based on android by presenting it in the form of pictures. The material in the educational media is about sex organs, abuse, educational tips, the origin of the baby, which is focusing on kindergarten children [4]. The second study is about educational media in the form of animation about sex education for kindergarten children [5]. The third study is an understanding of the concept of private body parts in prepubertal autistic children using magnetic board media at SLB Citra Mulia Mandiri Yogyakarta [6]. The fourth study is an educational media in the form of animated videos about the importance of covering private body parts, how to dress, and social boundaries with other people. This media is used by the children with special needs in grade V SLB Kemala Bhayangkari, Tanah Datar Regency [2]. In the current study, the object of the study were children with mild Down syndrome. Then the content created regarding the information on private body parts at puberty is in the form of animation, audio dan video.

3. Research methodology

In this study, several steps are summarised in the research methodology. The steps based on the method used in this research are literature study, data collection, requirement analysis, software development using the MDLC method (Multimedia Development Life Cycle), testing and drawing conclusions (Figure 1).

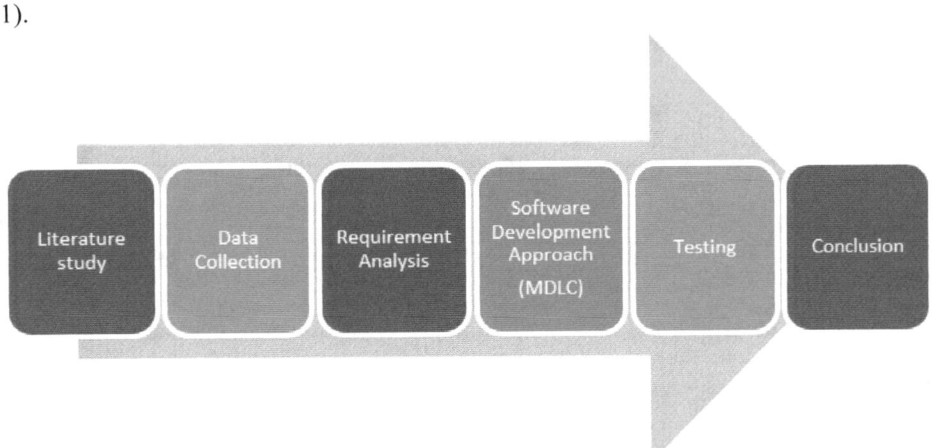

Figure 1. Research methodology.

3.1. Literature study

The literature study was carried out by searching for publications on google scholar. The search is done by entering several keywords that are relevant to the topic. From the search for literature studies that have conducted, publications that discuss sexual education and private body parts are obtained. However, there are not many publications that discuss sexual education and private body parts for children with Down syndrome.

3.2. Data collection

Data were collected by interview and observation methods. The interview method was conducted with teachers, parents and psychologists. Furthermore, the observation method also carried out in order to determine the conditions of learning carried out in SLB.

3.3. MDLC (Multimedia Development Life Cycle)

This study uses the MDLC (Multimedia Development Life Cycle) as a software development approach. In the MDLC method, there are several stages in software development, including concept, design, material collecting, assembly, testing (Figure 2).

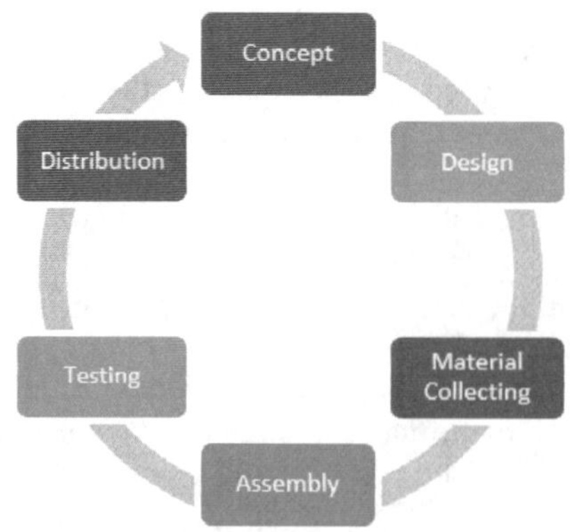

Figure 2. Multimedia development life cycle.

3.3.1. Concept. This educational media was developed for children with mild Down syndrome. The purpose of developing this educational media is to facilitate the learning process regarding private body parts (puberty). These media are used for Down syndrome children aged 12-21 years old, i.e. at the junior and senior high school levels.

3.3.2. Design. Design flowchart in this research can be seen in Figure 3.

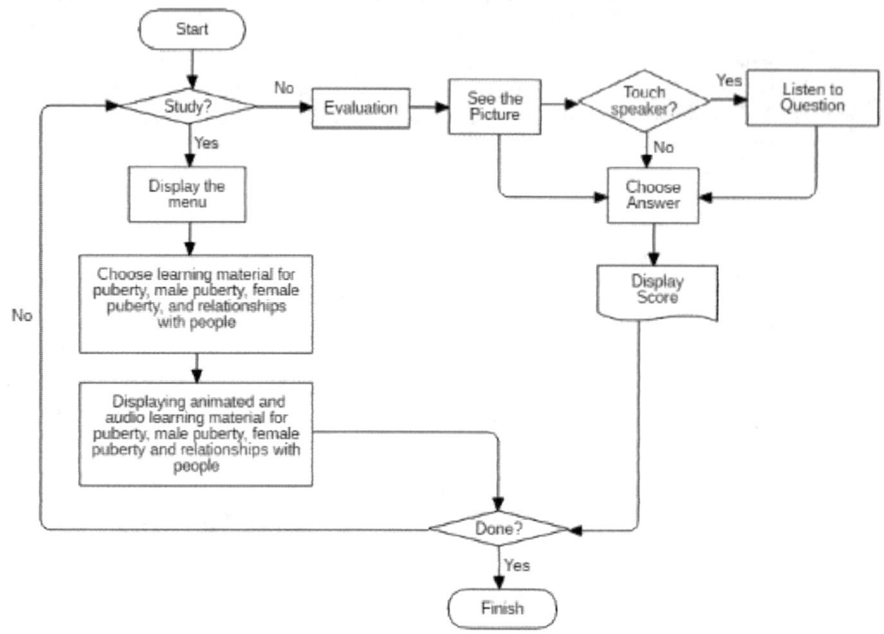

Figure 3. Flowchart.

3.3.3. Material collecting. The next stage after the design is the process of collecting materials used in application development. The material includes material obtained from interviews with psychologists, teachers, textbooks and literature studies. The assets that will be used in the learning media include character assets, background assets, and asset buttons. Then the audio files are used for background music and dubbing.

3.3.4. Assembly. The creation of all objects is based on a flowchart and begins with the creation of 3D objects using a blender application for character creation. The character can be seen in Figure 4.

Figure 4. Character.

Making design assets *User Interface* assisted using Adobe Photoshop CS 6. The process of making educational media uses Unity in 2D and uses the C # programming language (Figure 5).

Figure 5. Creating a user interface.

3.3.5. Testing. Testing and evaluation of educational media of private body parts for children with Down syndrome are carried out using alpha and beta testing. Alpha testing is done using the BlackBox testing method. This method is done by running the application. Then see whether all the functions of the features running well. Beta testing is carried out by involving end-users of children with Down syndrome. At the time of testing, interviews were also conducted following question points related to acceptance, satisfaction and the achievement of the goals of making the educational media.

3.3.6. Distribution. At the distribution stage, the application will be stored on storage media that can be accessed by end-users. This stage also includes an evaluation process for the product so that it can be developed better. The evaluation obtained can be included in the concept and design stages in further development.

4. Results and discussion

Following are the results of learning media that have stored in form.exe.

Figure 6. Educational media of private body parts.

Figure 6 Educational Media of Private Body Parts shows some views of the educational media. Figure 6 shows the home page, learning page with male body part content, female body parts and evaluation page.

The educational media that have been developed are then tested to find out whether the educational media made is in accordance with the needs. Testing was carried out by conducting interviews with teachers, parents and students of SLB C Pembina regarding the educational media that had been built. Before using the application, students, parents and teachers are given an explanation and manual book. This is done to make it easier to test the application. As early information, students who will use the application have never used computer-based learning media before.

The questions in the questionnaire are based on several criteria for assessing multimedia applications, including 1) Ease of use of navigation, so that users can learn easily; 2) Content of cognition, showing the content of knowledge in educational media clearly and easily understood; 3) Presentation of information, explaining the information presented in educational media clearly and precisely; 4) Media integration, namely educational games, there are aspects of knowledge and skills that must be learned; 5) Artistic and aesthetic, namely educational media must have an attractive application appearance and high aesthetic value; 6) The learning function is as an educational medium to provide learning and knowledge desired by users [7].

Figure 7. Educational media testing for mild down syndrome children.

After testing and interviews with teachers, parents and students, the results were as expected. Educational media has a good appearance; the user interface is easy to understand, easy to operate, and in accordance with the learning material. Educational media help students understand both content and evaluation. Students can know about wet dreams, the difference between bedwetting and wet dreams. They also understand how to communicate and interact with family, other people or strangers.

However, there are deficiencies in the learning media. (1) Each learning feature requires facilities to be able to stop animation. Because it is needed for parents, the teacher explains each part of the material repeatedly. (2) Some students cannot use or understand Indonesian, so they must be given an explanation using everyday language.

5. Conclusion

Based on the results of analysis, design, implementation and testing; several conclusions can be drawn; (1) Educational media about the private body parts of Down Syndrome children can provide learning about private body parts that make it easier for children with Down syndrome to learn and understand. (2) Based on the usability test through interviews, satisfactory results were obtained. Users respond well to educational media about body parts so that they are easy to understand and use.

Acknowledgements

The authors would like to thank the State Polytechnic of Malang (Polinema), who have supported this research project.

References

[1] Driya P 2018 *Pola Asuh Ibu Terhadap Kebersihan Pribadi Organ Reproduksi Pada Penyandang Tunagrahita Yang Sudah Mengalami Menstruasi (Studi Kualitatif Pada Siswi SLB-C TPA Jalan Jawa No. 57 Kecamatan Sumbersari Kabupaten Jember)* (Jember: Universitas Jember)

[2] Yuliani T and Armaini A 2019 Media Video Animasi dalam Pendidikan Seks Anak dengan Hambatan Kecerdasan Ringan *J. Penelit. Pendidik. Khusus* **7** 41–6

[3] Prayoga G 2015 *Hubungan Antara Pengetahuan Kesehatan Reproduksi dan Sikap Seksualitas Dengan Perilaku Pacaran Pada Pelajar SLTA di Kota Semarang* (Surakarta: Universitas Muhammadiyah Surakarta)

[4] Padillah R 2018 Pengembangan multimedia pembelajaran sex education berbasis android untuk anak usia dini *BaJET Baturaja J. Educ. Technol.* **2** 117–123

[5] Hanafri M I, Mariana A R and Suryana C 2016 Animasi sex education untuk pembelajaran dan pencegahan pelecehan seksual pada anak usia dini (Studi kasus di TK Kartini) *J. SISFOTEK Glob.* **6** 1-7

[6] Fauzy M Z 2016 Peningkatan Pemahaman Konsep Bagian Tubuh Pribadi pada Anak Autistik Usia Prapubertas Menggunakan Media Papan Magnet di SLB Citra Mulia Mandiri Yogyakarta *WIDIA ORTODIDAKTIKA* **5** 624–633

[7] Munir D and IT M 2009 *Pembelajaran jarak jauh berbasis teknologi informasi dan komunikasi* (Bandung: Alfabeta)

IOP Conf. Series: Materials Science and Engineering 1073 (2021) 012069 doi:10.1088/1757-899X/1073/1/012069

Micro-expression recognition based on motion detection method

U D Rosiani[1,*], P Choirina[2] and M N Shoumi[1]

[1] Department of Information Technology, State Polytechnic of Malang, Malang, Indonesia
[2] Faculty of Science and Technology, Universitas Islam Raden Rahmat Malang, Malang, Indonesia

*rosiani@polinema.ac.id

Abstract. Micro-expressions are emotional representations that occur spontaneously and cannot be controlled by humans. The micro-expression movements are temporary with fast duration and have subtle movements with little intensity. This is difficult to detect with the human eye. Previous studies have shown that micro-expression movements occur in several areas of the face. This study aims to determine the subtle movements in several areas of the face using the motion detection method. We compared the performance of two motion detection methods: the optical flow method and the Block Matching Algorithm (BMA) method. The optical flow method uses the Kanade-Lucas Tomasi (KLT) method and the BMA method uses the Phase Only Correlation (POC) algorithm. Observations were carried out based on region, where the face area was divided into several observation areas: eyebrows, eyes and mouth. Both methods perform motion detection between frames. The KLT method tracks the movement of the observation points on the frame movement. Meanwhile, the POC method matches the blocks between frames. If the two blocks are the same, no motion vector is generated. However, if the two blocks are different, it is assumed that there is a translational motion and a motion vector is generated. Experiments were conducted using a dataset from CASME II with emotional classes of disgust, happiness, surprise, and sadness. The classification accuracy of the POC method is 94% higher than the KLT method of 84.8% which uses the SVM classification.

1. Introduction

Facial expressions are usually considered to be physiological reactions to the emotions within. Human facial expressions are divided into two parts: macro-expression and micro-expression. Macro expression is defined by easily recognizable movements of the facial muscle over many areas of the face which occur between ½ seconds and 4 seconds. Meanwhile, micro-expression movements take the form of smooth muscle movements in some areas of the face that are quicker, from 1/25 seconds to 1/5 seconds [1]. Other people are often not aware and do not see this movement [2]. Similar to macro expressions, micro expressions can be grouped into six basic expression: happy, surprise, anger, sad, disgust, and fear [3]. In many studies on the recognition of micro-expressions, the approach to research is divided into two classes: appearance-based and motion-based [4].

The main approaches to appearance-based methods, such as Gabor wavelet [5] and local binary pattern (LBP) [6], focus on the use of texture representations [7]. However, the texture feature makes it difficult to capture the fast, low-intensity display changes that occur in micro-expressions. Meanwhile, motion features that are usually derived from optical flow are considered to be a robust and accurate approximation to the recognition of micro expressions [8]. This method tries to recognize facial activity by analyzing the movement itself. Since micro-expressions can be well characterized by motion features, the performance of the motion approach is better than that of the appearance-based approach [9].

In this study, the performance of two motion detection methods for recognition of micro-expressions is compared: optical flow method and Block Matching Algorithm (BMA) method. The first method, optical flow method, is part of method in pixel-based motion aims to observe and detect non-rigid motion of facial components [10]. Liu et al., used Main Directional Mean Optical-flow (MDMO) based on robust optical flow in micro-expression video clips by partitioning the area on the face (ROI) [9]. MDMO produces better performance than LBP-TOP and HOOF using SMIC, CASME and CASME II dataset. Another optical flow method is Kanade-Lucas-Tomasi (KLT), capable of detecting subtle motion based on point feature tracking. KLT calculates the number of differences in the square of the intensity between features on two frames. Asmara et al., applied the KLT method for the recognition of micro expressions based on tracking the points formed by the facial landmark method, namely DRMF [11]. This study resulted in an accuracy rate of 79.3% in CASME II with SVM.

The second method, BMA metod, is a part of block-based motion which estimates block-based motion quickly [12]. One of the matching methods in the comparison between blocks is Phase-Only Correlation (POC) which is better than other block matching methods [13]. The POC method can be used to detect translation motion that occur in an image based on the displacement of two images in the cross-spectrum power phase [14]. In this study, we compared the performance of KLT and POC method for micro expression recognition. Observations were made based on region, where the face area is divided into several observation areas: eyebrows, eyes and mouth. Region based observation is better than whole face observation [15].This study aims to obtain the best accuracy with minimal computation time from the comparison of these methods, because the movements generated by micro expressions occur very quickly.

2. Proposed methods

In this study, there are four stages: 1) pre-processing; 2) motion detection; 3) feature extraction; and 4) classification. Figure 1, shows a block diagram of a micro-expression recognition based on motion detection method. In this section, we describe the whole process the comparison of motion detection methods is Phase-only Correlation (POC) and Kanade Lucas Tomasi (KLT) aims to find the best accuracy of the two methods.

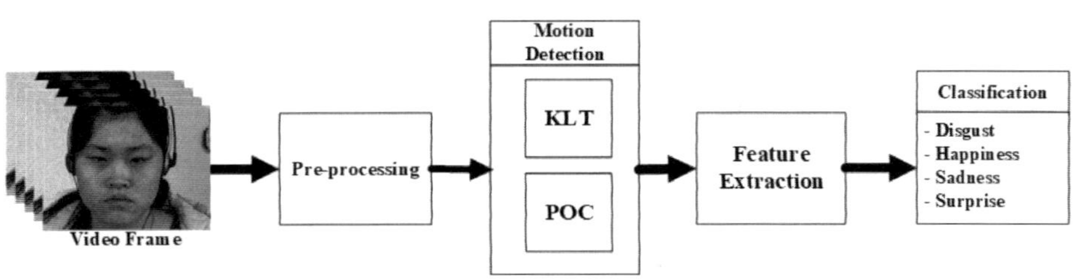

Figure 1. The block diagram for micro-expression recognition based on motion detection method.

2.1. Pre-processing

This section discusses the detection method of the location of components on the face, namely the right eyebrow, left eyebrow, right eye, left eye, and mouth. In the preprocessing stage, we only use the first frame as a reference for the reference frame. This occurs because the changes that result in micro-expressions are very small and can minimize the time in the sequence.

This section discusses the detection method of the location of components on the face, namely the right eyebrow, left eyebrow, right eye, left eye, and mouth. In the preprocessing stage, we only use the first frame as a reference for the reference frame. This occurs because the changes that result in micro-expressions are very subtle and can minimize the time in the sequence.

We used the Discriminative Response Map Fitting (DRMF) method to locate the facial components accurately in a very fast time [16]. The DRMF method can adjust facial shapes with impressive performance and is well suited for handling very large occlusion and illumination conditions [11]. The detection of facial components with DRMF has 3 stages. First, the detection of face location using the Viola-Jones method. Second, computed starting point feature set by extracting patch responses by low dimensional projection. Furthermore, this method repeats the initial feature point by connecting with the target frame.

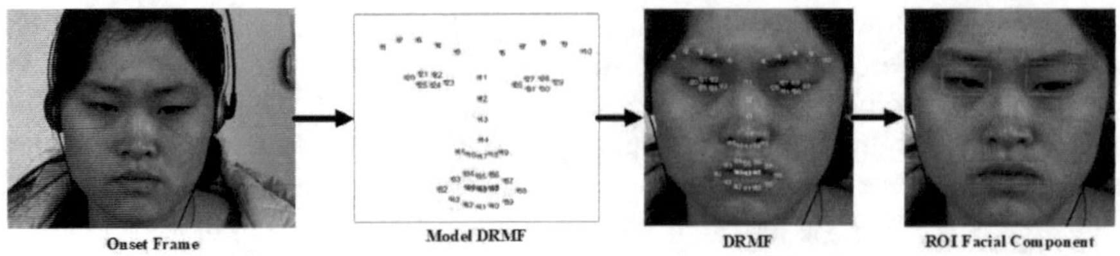

Figure 2. The flow of face component detection.

The result of this method is a set of points totaling 49 and scattered over the area of the face component which has a coordinate value as in Figure 2. Furthermore, we form the ROI (Region of Interest) on the face component and the landmark points of the DRMF which are used as reference points. for other frames. The parameters of the ROI required include the X and Y coordinate values, width, and height for each face component. We used the reference points used in previous studies as in Table 1 [16] for the formation of ROI for each facial component.

Table 1. Reference to the formation of ROI from the DRMF point.

Facial Component	Landmark Points	ROI			
		X	Y	Width	Height
Left Eyebrow	1-5	x (1)	y (2)	110	28
Right Eyebrow	6-10	x (6)	y (9)	110	28
Left Eye	20-25	x (20)	y (22)	81	34
Right Eye	26-31	x (26)	y (30)	81	34
Mouth	32-41	x (32)	y (34)	139	39

The POC method uses the 5x5 block area feature for each ROI as shown in Figure 4. Whereas the KLT method uses the point feature for each ROI component. As in Figure 3, the points generated in each ROI are formed by the function of the Matlab is detectMinEigenFeatures() [17]. This function is used to forming corner points (interest points) on an object based on the minimum eigenvalue algorithm.

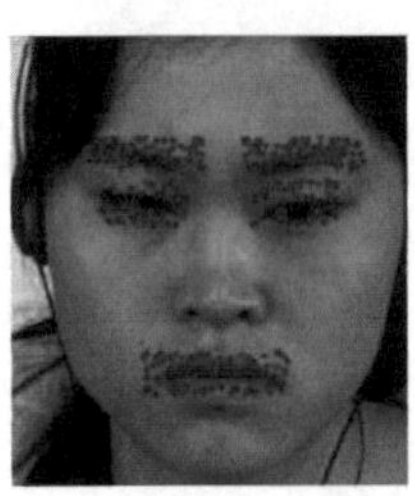

Figure 3. Facial point feature using Eigenvalues for KLT.

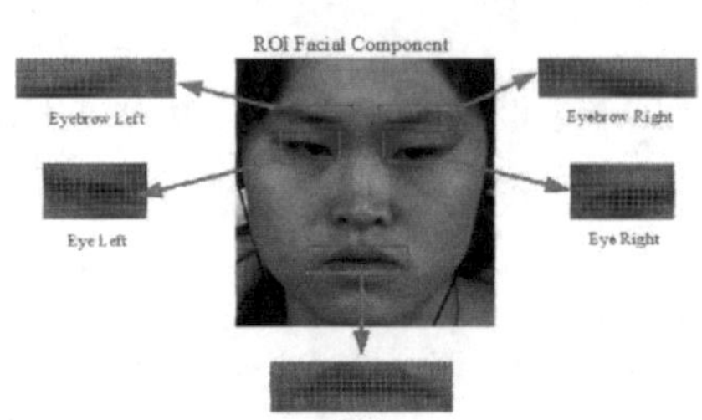

Figure 4. Block area feature for POC.

IOP Conf. Series: Materials Science and Engineering 1073 (2021) 012069 doi:10.1088/1757-899X/1073/1/012069

2.2. *Motion detection for micro-expression recognition*

2.2.1. Kanade Lucas Tomasi (KLT). The optical flow method calculates pixel movement between frames over time. Optical flow can track the motion of each pixel in the image. The calculation used for all pixels is dense optical flow and for some pixels, it uses the sparse optical flow method. KLT is a representative sparse optical flow algorithm for the feature tracking process. We use this method to trace the points that have been formed in each sequence frame. The grayscale value of the same spatial point pixel does not change in all frames. It is called constant grayscale values which are used to assume the optical flow method. Figure 5 explains that when a pixel in position (x, y) at time t moves to the position $(x + dx, y + dy)$ at time $t + dt$, the grayscale values of the two pixels are same.

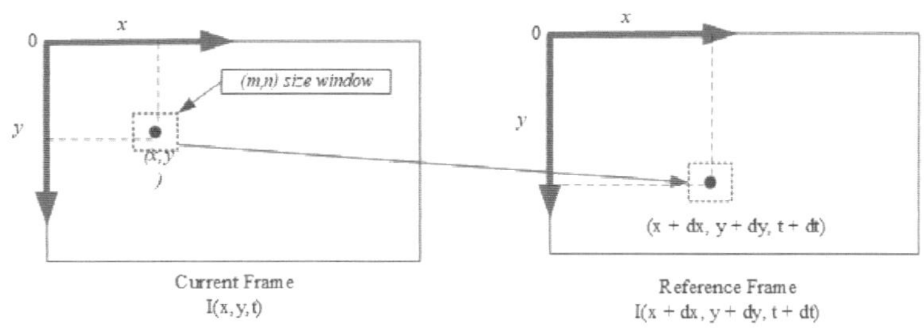

Figure 5. Kanade Lucas Tomasi (KLT) feature tracking algorithm [18].

First, a set of points formed in the previous process (see Figure 3), the first frame is initiated as a starting point before tracking. Furthermore, the KLT method traces the reference frame. KLT calculates the sum of the differences in the squares of intensity between features in the current and previous frames. The motion of a feature point is defined as a displacement that minimizes the amount of difference. This process is repeated in a sequence of images until all feature points can be traced [11]. Motion vectors are formed when the location of the point on the current frame moves with the movement of the reference frame. All point locations in each frame that have been through the tracking process are stored as feature extraction data.

2.2.2. Phase-Only Correlation (POC). Image registration with POC makes it possible to predict movement between images with sub-pixel accuracy. POC functions to measure the similarity between two blocks in the frame, the current block, and the reference block. The function used in POC is the inverse 2D discrete Fourier transformation. The translation and similarity between the two images can be obtained based on the position and height of the correlation peaks. If the two images are the same, the POC graph produces a peak that is 1. On the other hand, if the two images are different, the resulting graph is not 1. The graph of the POC function can be seen in Figure 6.

The main step in calculating the movement of each component of the facial area is dividing each area into blocks. The block size used is 5x5 pixels as shown in Figure 4. Next, match each block between the current frame $I(x, y, t)$ and the reference frame $J(x, y, t + 1)$ using the POC function. If both are equal or there is no shift, the POC value approaches 1 and a motion vector is formed on the block. On the other hand, if the two image blocks are different, no motion vector is formed.

2.3. *Feature extraction*

The output of the two motion detection methods, namely POC and KLT, is the value of the motion vector which consists of the x and y coordinates in each frame comparison. This value will be used as a motion feature for each component of the face area. Motion features consist of horizontal (Rx) and vertical (Ry) components, the magnitude of the vector (M), and the direction of the resulting vector.

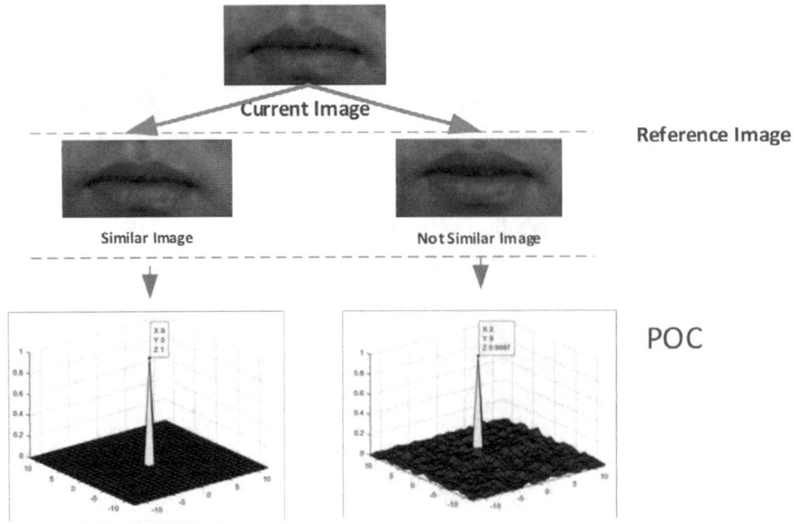

Figure 6. Phase-only correlation between two images.

3. Experimental results

3.1. Database

To verify the effectiveness of the comparison of motion detection methods on the recognition of micro-expressions, we conducted experiments using the Chinese Academy of Sciences Micro-expression (CASME II) dataset. CASME II is a comprehensive database of spontaneous micro expression data and contains 247 video samples with 26 Asian participant subjects with an average age of 22.03 years. Each video in CASME II gives one micro expression to the subject of five categories of micro expression: Happiness, Disgust, Repression, Surprise, Other. In this study, we only used four micro-expressions, namely: Happiness, Disgust, Surprise and Sadness.

3.2. Processing time

The recognition of micro expressions requires real-time processing, it takes fast processing time. Experiments in this study conducted in MATLAB run on Intel Core i5 CPU with RAM size 4 GHz using 64-bit Windows OS. Figure 7 describes the graph of time requirements for processing motion detection methods used in this study, namely POC and KLT. In this graph, it can be seen that the time require for the KLT method is smaller than the POC with an average time of 0.0445 seconds. Whereas the POC method requires time with an average time of 0.2041 seconds.

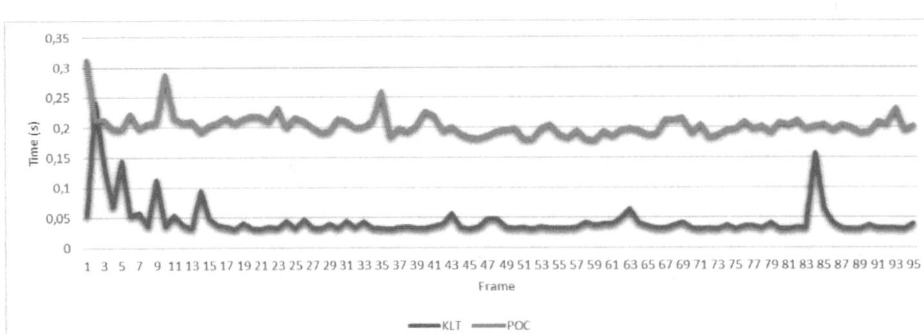

Figure 7. Time comparison of motion detection method on micro expression recognition.

3.3. Comparison accuracy of motion detection methods in micro expression recognition

In this study, the results of feature extraction from the motion detection method will be recognized by machine learning using a Support Vector Machine (SVM) as a measure of the resulting accuracy in classifying emotions in micro expressions. The SVM used is non-linear using the Radial Basis Function (RBF) kernel. RBF is used because it can map the sample to a higher dimensional space. This study resulted in the POC method the resulting accuracy rate of 94.3% was better than the KLT method with an accuracy of 84.8%. This study produced the best accuracy with the KLT (optical flow) method compared to previous studies which are shown in Table 2.

Table 2. Research of motion-based micro-expression detection.

Year	Author	Dataset	Feature	Classifier	Accuracy
2016	Liong et al., [19]	CASME II and SMIC	Optical Strain	SVM	63.41% on CASME II 52.44% on SMIC
2017	Zhang et al., [20]	CASME II	LBP-TOP, Optical Flow	KNN, SVM and RF	Random Forest 62.5%
2018	Zhu et al., [21]	CASME II	LBP-TOP and Optical Flow	SVM	53,3%
2020	**This study**	**CASME II**	**POC, KLT**	**SVM**	**POC 94.3%** **KLT 84.8%**

4. Conclusion

The movements produced by micro-expressions are very subtle and occur in a short period of time. This makes this movement difficult to detect with the human eye. The two methods used in the study, KLT and POC, can detect the movements that occur. The KLT method tracks the motion at the observation points, while the POC method compares the blocks between frames. In the comparison of the performance of the two methods, the classification accuracy with the POC method is higher than the KLT method. This is because more motion features are used. The smaller the block size that is matched by the POC matching method, the more motion features are generated. Meanwhile, for the comparison of processing times, the KLT method is faster than the POC method, this is because less parts are detected in KLT method. Research using a combination of the two methods can be further developed, in order to obtain better accuracy and processing time.

References

[1] Yan W J, Wu Q, Liang J, Chen Y H and Fu X 2013 How fast are the leaked facial expressions: The duration of micro-expressions *Journal of Nonverbal Behavior* **37**(4) 217-230

[2] Liong S T, See J, Wong K, Le Ngo A C, Oh Y H and Phan R 2015 Automatic apex frame spotting in micro-expression database *2015 3rd IAPR Asian conference on pattern recognition (ACPR)* 665-669

[3] Takalkar M A and Xu M 2017 Image based facial micro-expression recognition using deep learning on small datasets *2017 international conference on digital image computing: techniques and applications (DICTA)* 1-7

[4] Lu H, Kpalma K and Ronsin J 2018 Motion descriptors for micro-expression recognition. *Signal Processing: Image Communication* **67** 108-117

[5] Zheng H 2017 Micro-expression recognition based on 2d gabor filter and sparse representation *Journal of Physics: Conference Series* **787** 012013

[6] Ahonen T, Hadid A and Pietikainen M 2006 Face description with local binary patterns: Application to face recognition *IEEE transactions on pattern analysis and machine intelligence* **28**(12) 2037-2041

[7] Wu Q, Shen X and Fu X 2011 The machine knows what you are hiding: an automatic micro-expression recognition system *international conference on affective computing and intelligent*

Interaction (Berlin, Heidelberg: Springer) p 152-162

[8] Liu Y J, Li B J and Lai Y K 2018 Sparse MDMO: Learning a discriminative feature for spontaneous micro-expression recognition *IEEE Transactions on Affective Computing*

[9] Liu Y J, Zhang J K, Yan W J, Wang S J, Zhao G and Fu X 2015 A main directional mean optical flow feature for spontaneous micro-expression recognition *IEEE Transactions on Affective Computing* **7**(4) 299-310

[10] Happy S L and Routray A 2017 Fuzzy histogram of optical flow orientations for micro-expression recognition *IEEE Transactions on Affective Computing* **10**(3) 394-406

[11] Asmara R A, Choirina P, Rahmad C, Setiawan A, Rahutomo F, Yusron R D R and Rosiani U D 2019 Study of DRMF and ASM facial landmark point for micro expression recognition using KLT tracking point feature *Journal of Physics: Conference Series* **1402**(7) 077039

[12] Zhang P, Wei P and Yu H Y 2012 Biogeography-based optimisation search algorithm for block matching motion estimation *IET image processing* **6**(7) 1014-1023

[13] Setiyono B, Hariadi M and Purnomo M H 2012 Investigation of Superresolution using Phase based Image Matching with Function Fitting *Res. J. Eng. Sci.* **2278** 9472

[14] Reddy B S and Chatterji B N 1996 An FFT-based technique for translation, rotation, and scale-invariant image registration *IEEE transactions on image processing* **5**(8) 1266-1271

[15] Duan X, Dai Q, Wang X, Wang Y and Hua Z 2016 Recognizing spontaneous micro-expression from eye region *Neurocomputing* **217** 27-36

[16] Rosiani U D, Ririd A R T H, Choirina P, Sooai A G, Sumpeno S and Purnomo M H 2018 Micro Expression: Comparison of Speed and Marking Accuracy in Facial Component Detection *2018 International Conference on Computer Engineering, Network and Intelligent Multimedia (CENIM)* 221-226

[17] Shi J 1994 Good features to track *1994 Proceedings of IEEE conference on computer vision and pattern recognition* 593-600

[18] Piao J C and Kim S D 2017 Adaptive monocular visual–inertial SLAM for real-time augmented reality applications in mobile devices *Sensors* **17**(11) 2567

[19] Liong S T, See J, Phan R C W, Oh Y H, Le Ngo A C, Wong K and Tan S W 2016 Spontaneous subtle expression detection and recognition based on facial strain *Signal Processing: Image Communication* **47** 170-182

[20] Zhang S, Feng B, Chen Z and Huang X 2017 Micro-expression recognition by aggregating local spatio-temporal patterns *International Conference on Multimedia Modeling* 638-648

[21] Zhu X, Ben X, Liu S, Yan R and Meng W 2018 Coupled source domain targetized with updating tag vectors for micro-expression recognition *Multimedia Tools and Applications* **77**(3) 3105-3124

IOP Conf. Series: Materials Science and Engineering 1073 (2021) 012070 doi:10.1088/1757-899X/1073/1/012070

Optimal multi-layer perceptron parameters for early stage diabetes risk prediction

V N Wijayaningrum[1,*], T H Saragih[2] and N N Putriwijaya[3]

[1] Department of Information Technology, Politeknik Negeri Malang, Malang, Indonesia

[2] Faculty of Mathematics and Natural Sciences, Lambung Mangkurat University, Banjarmasin, Indonesia

[3] Faculty of Information and Communication Technology, Institut Teknologi Sepuluh Nopember, Surabaya, Indonesia

*vivinurw@polinema.ac.id

Abstract. Diabetes is an alarming threat to people around the world because the number of diabetics is increasing every year. Diabetics with other complications have a very high risk of death. Therefore, the use of technology to predict the risk of early diabetes is needed. Neural Network as one part of artificial intelligence plays a role in solving prediction problems with satisfying results. In this study, a multi-layer perceptron neural network is used to predict the risk of early stage diabetes with optimal parameters from the optimization results using Improved Crow Search Algorithm. The test results prove that the multi-layer perceptron with optimal parameters is able to provide better accuracy compared to other algorithms such as J48, PART, Decision Table, Naïve Bayes, AIRS1, AIRS2, and Single Layer Perceptron with the highest accuracy values of 97.69% and 96.92% for one and two hidden layers, respectively. This proves that the proposed solution can be used to predict the early stage diabetes risk.

1. Introduction

Diabetes is a disease caused by a metabolic disorder so that blood sugar levels increase. Diabetes patients show weight loss and frequent urination. In addition, the effects of long-term diabetes can cause sufferers to experience kidney failure, nerve damage, blurred vision, even death [1]. World Health Organization (WHO) states that 422 million people worldwide suffer from diabetes, with the majority of sufferers living in low and middle income countries. The number of diabetes cases has also continued to increase over the last few decades. Each year, 1.6 million deaths are directly related to diabetes. This shows that diabetes is a dangerous disease. Thus, early identification and prediction of diabetes risk is needed to avoid more serious problems.

Currently, the health care system is one of the popular research fields used for early identification of a disease by applying data mining techniques. Data mining can be applied to extract hidden patterns, so that relationships between different patterns in clinical data can be found [2]. Furthermore, machine learning can play a role in teaching computers to carry out the learning process and understanding the given parameters, which are then used for prediction or decision making. This learning process can be done using various classification or clustering algorithms, one of which is the Support Vector Machine. This algorithm performs classification by forming a multidimensional hyperplane that separates the two data clusters by maximizing the margin between the two data clusters. By using the data set on diabetes from the National Health and Nutrition Examination Survey (NHANES) with 14 parameters, SVM performance was measured using the area under the curve (AUC) with a value of 83.47% for the detection of diagnosed or undiagnosed diabetes [3].

In other study, a hybridization mechanism between K-Means and J48 was proposed. This mechanism is carried out using K-Means for data reduction then the classification process is carried out using the J48 decision tree. Data reduction aims to transform raw data into a more useful form, in this case the data set is converted into an ordered and simple form. In this study, PIMA Indian Diabetes Data from UCI repository was used. Missing values in the data set are replaced with mean values according to their attributes, while incorrect classification results are removed from the sample using K-Means algorithm, leaving 532 samples out of 768 samples. Furthermore, J48 classifies the sample data set with an accuracy value of 90.04% [4].

Neural Network is a computational algorithm that is often used as a classification method for high-dimensional data [5]. Several Neural Network algorithms have proven successful for disease diagnosis, such as Multilayer Perceptron for heart disease prediction [6,7], Back Propagation for chronic kidney disease prediction [8], Extreme Learning Machine for Jatropha Curcas disease identification [9], and Learning Vector Quantization for leaf disease detection [10]. However, the learning process on Neural Network algorithms is influenced by the learning rate. Algorithms often take a long time to come up with an optimal solution when the learning rate is too small [11]. To solve this problem, several researchers have performed various hybridization techniques between the Neural Network algorithm and other algorithms, such as genetic algorithms and Back Propagation which are used to determine the initial weights and number of neurons in the hidden layer for the Multi-Layer Perceptron [12]. In other study, Extreme Learning Machine and Modified Simulated Annealing were proposed to optimize Extreme Learning Machine weight values [13]. The use of this hybridization technique aims to increase the accuracy value obtained.

In this study, optimization of Multi-Layer Perceptron parameters using Improved Crow Search Algorithm was proposed. Multi-Layer Perceptron has five main parameters that can affect the given results, namely learning rate, momentum, number of epochs, threshold for errors, and number of neurons in hidden layer. The optimal parameters of the search results using the Improved Crow Search Algorithm are expected to provide high accuracy values for predicting early stage diabetes risk.

2. Research methodology

Multi-Layer Perceptron (MLP) is a class of Neural Network algorithms. It consists of one input layer, one output layer, and one or more hidden layers of nodes that are mapped nonlinearly. The nodes in each layer are connected by weight to all nodes in the next layer. The Perceptron Multi-Layer architecture depends on the choice of the number of layers, the number of nodes in the hidden layer, and the objective functions used [14]. The number of neurons in the input layer is determined based on the number of input variables, while the number of neurons in the output layer is based on the number of output classes [15]. In this study, 16 neurons were used in the input layer and two neurons in the output layer. Meanwhile, the number of hidden layers and the number of neurons in the hidden layer are determined through the optimization process. The architecture of Multi-Layer Perceptron used in this study is shown in Figure 1.

In this study, Multi-Layer Perceptron is used to predict the risk of early stage diabetes. When carrying out the learning process, Multi-Layer Perceptron uses five main parameters, consisting of learning rate, momentum, number of epochs, threshold for errors, and number of neurons in hidden layer. In order to obtain high accuracy, Improved Crow Search Algorithm (ICSA) is used to find optimal values for the Multi-Layer Perceptron parameters. The improvised mechanism of ICSA was adopted from previous research [16]. The parameter optimization mechanism of the Multi-Layer Perceptron using ICSA is illustrated in Figure 2.

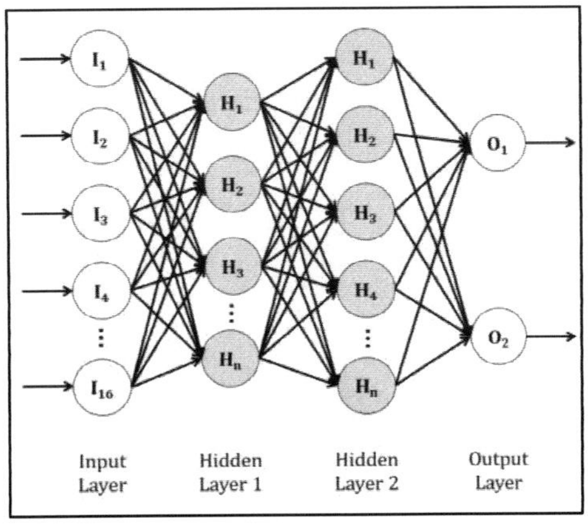

Figure 1. Architecture of multi-layer perceptron.

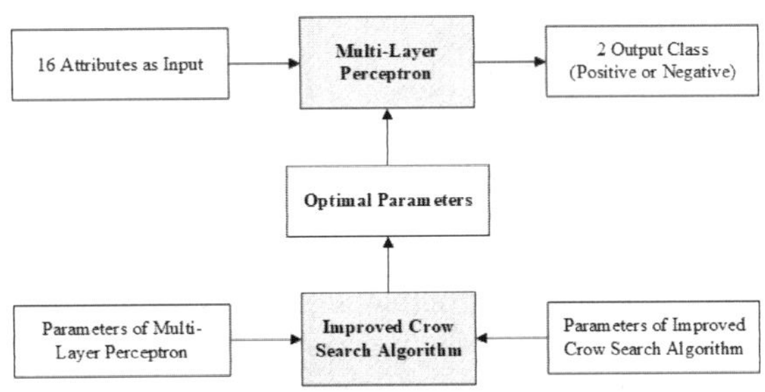

Figure 2. The parameter optimization mechanism.

The parameter optimization mechanism of Multi-Layer Perceptron in Figure 2 begins by inserting the value of each attribute into Multi-Layer Perceptron. To produce a good quality solution, the parameters used for the learning process on Multi-Layer Perceptron are determined based on the search results from Improved Search Crow Algorithm (ICSA). In this search process, ICSA also uses several parameters consisting of flock size, number of iterations, flight length, and awareness probability. The ICSA search results are the optimal parameters used to regulate the learning process on Multi-Layer Perceptron. After carrying out the learning process, Multi-Layer Perceptron classifies the input data to produce a positive or negative class output. From the dataset used, it can be seen the accuracy value obtained at the end.

2.1. Search agent representation

ICSA is a meta-heuristic algorithm that has a population that contains several search agents. Each search agent states a possible solution to the problem. In this study, each agent has six values that represent the parameters of the Multi-Layer Perceptron. Figure 3 shows a representation of a search agent.

Variable	x_1	x_2	x_3	x_4	x_5	x_6
Value range	0 - 1	0 - 1	100 - 1000	1 - 30	1 - 10	1 - 10

Figure 3. Representation of a search agent.

In Figure 3, the variables x_1 to x_6 respectively represent the values for learning rate, momentum, number of epochs, threshold for errors, number of neurons in the first hidden layer, and number of neurons in the second hidden layer. Each search agent has different values but these values are within the range of values specified for each variable.

2.2. Fitness function

The quality of the solutions offered by each search agent can be measured using the fitness function. A search agent who has a high fitness value claims that its solution is good. In this case, the fitness function is calculated using accuracy. To calculate accuracy, (1) is used.

$$\text{Accuracy} = \frac{\text{The number of sample correctly classified}}{\text{The number of all sample}} \tag{1}$$

To evaluate the performance of the algorithm, this study uses *k*-fold cross validation, which is a resampling procedure by separating the dataset into *k* folds. Thus, the calculation of accuracy uses (2) as a fitness function.

$$\text{Accuracy} = \frac{\text{Sum of Accuracy in each fold}}{\text{k}} \tag{2}$$

3. Results and discussion

The dataset used in this study is the early stage diabetes risk prediction dataset derived from the UCI dataset. The dataset consists of 520 instances and 17 attributes consisting of 16 inputs and one output. The output class consists of 'positive' and 'negative' related to early stage diabetes risk prediction.

The performance of the proposed algorithm for identifying early stage diabetes risk is compared with other algorithms such as J48, PART, Decision Table, Naive Bayes, Artificial Immune Recognition System (AIRS1), and AIRS2. In addition, comparisons of various variants of the Perceptron algorithm are also carried out to determine the quality of the given solutions. The ICSA parameter settings used to optimize the Multi-Layer Perceptron parameters are shown in Table 1.

Table 1. Parameter settings used for ICSA.

Parameter	Value	Explanation of Parameter
Flock size	100	The number of agents searching for solutions
Number of iterations	50	The number of searches carried out
Flight length	2.5	How far the search is carried out
Awareness probability	0.1	The tendency of direction in finding a solution

The performance comparison of these algorithms is shown in Table 2. The performance of the proposed algorithm is highlighted in gray. The values shown are the average accuracy using 10-fold cross validation.

Table 2. Comparison of algorithm performance.

Algorithm	Accuracy (%)
J48	92.88
PART	94.42
Decision Table	93.46
Naïve Bayes	86.15
AIRS1	89.62
AIRS2	90.58
Single Layer Perceptron	95.19
Multi-Layer Perceptron without optimization	95.38
Multi-Layer Perceptron with optimization using ICSA (one hidden layer)	92.66 (**Max: 97.69**)
Multi-Layer Perceptron with optimization using ICSA (two hidden layer)	93.18 (**Max: 96.92**)

Because ICSA is a stochastic algorithm, the accuracy given to each execution will be different. Therefore in Table 2, the average accuracy obtained from 10 independent experiments is shown, and the highest accuracy obtained from all these experiments is also shown and symbolized by Max. These results indicate that the Multi-Layer Perceptron with optimization provides the highest accuracy value compared to other algorithms using either one hidden layer or two hidden layers. This is because ICSA is able to find the best parameters on the Multi-Layer Perceptron to get a solution with high accuracy. Meanwhile, the average accuracy given by the Multi-Layer Perceptron with optimization is lower than some other algorithms. This is because the ICSA initialization process is carried out using random initial values which are then improved continuously as the learning process progresses. Thus, the use of a random initial value that is not good enough allows the algorithm to be unable to provide a high enough accuracy.

4. Conclusion

The test results show that the use of Multi-Layer Perceptron with optimal parameters from the optimization results using Improved Crow Search Algorithm can be used to predict the risk of early diabetes with high accuracy. The proposed algorithm is proven to provide the best accuracy compared to other algorithms such as J48, PART, Decision Table, Naive Bayes, AIRS1, AIRS2, Perceptron, and Multi-Layer Perceptron without optimization. By using the optimal parameters of the optimization results, it is known that the accuracy provided by Multi-Layer Perceptron with one hidden layer and two hidden layers has a slight difference. Multi-Layer Perceptron with one hidden layer gives an average accuracy of 92.66% with the highest accuracy of 97.69%, while Multi-Layer Perceptron with two hidden layers gives an average accuracy of 93.18% with the highest accuracy of 96.92%.

The hidden layers on Multi-Layer Perceptron perform computations on input and use activation functions for generating output. However learning with more layers will be easier but more training time is required. The weights of each hidden layer also have a very important role to the output produced by Multi-Layer Perceptron. Weights are used to scale inputs separately. In the next study, the use of optimization mechanisms using Particle Swarm Optimization can be used to optimize the weights used in each hidden layer. Thus, the predicted results are expected to be better so that the algorithm produces higher accuracy.

References

[1] Vijayan V V and Anjali C 2015 Prediction and Diagnosis of Diabetes Mellitus - A Machine Learning Approach *2015 IEEE Recent Advances in Intelligent Computational Systems (RAICS)* 122–7
[2] Kumari S and Singh A 2012 A Data Mining Approach for the Diagnosis of Diabetes Mellitus *2013 7th International Conference on Intelligent Systems and Control (ISCO)* (IEEE) 373–5
[3] Yu W, Liu T, Valdez R, Gwinn M and Khoury M J 2010 Application of support vector machine modeling for prediction of common diseases: the case of diabetes and pre-diabetes *BMC Med. Inform. Decis. Mak.* **10** 16
[4] Chen W, Chen S, Zhang H and Wu T 2017 A Hybrid Prediction Model for Type 2 Diabetes Using K-means and Decision Tree *2017 8th IEEE International Conference on Software Engineering and Service Science (ICSESS)* 386–90
[5] Wójcik P I and Kurdziel M 2019 Training neural networks on high-dimensional data using random projection *Pattern Anal. Appl.* **22** 1221–1231
[6] Yan H, Jiang Y, Zheng J, Peng C and Li Q 2006 A multilayer perceptron-based medical decision support system for heart disease diagnosis *Expert Syst. Appl.* **30** 272–81
[7] Sonawane J S and Patil D R 2014 Prediction of Heart Disease Using Multilayer Perceptron Neural Network *International Conference on Information Communication and Embedded Systems* (IEEE) 1–6
[8] Borisagar N, Barad D and Raval P 2017 Chronic Kidney Disease Prediction Using Back Propagation Neural Network Algorithm *Proceedings of International Conference on*

Communication and Networks (Singapore: Springer) p 295–303
[9] Saragih T H, Fajri D M N, Mahmudy W F, Abadi A L and Anggodo Y P 2018 Jatropha Curcas Disease Identification with Extreme Learning Machine *Indones. J. Electr. Eng. Comput. Sci.* **12** 883–888
[10] Upadhyay A, Singh J and Shinde R 2019 Learning Vector Quantization Based Leaf Disease Detection *International Conference on Recent Developments in Science, Engineering and Technology* (Singapore: Springer) p 246–53
[11] Rahmi A, Wijayaningrum V N, Mahmudy W F and Parewe A M A K 2016 Offline Signature Recognition using Back Propagation Neural Network *Indones. J. Electr. Eng. Comput. Sci.* **4** 678–83
[12] Castillo P A, Merelo J J, Prieto A, Rivas V and Romero G 2000 G-Prop: Global optimization of multilayer perceptrons using GAs *Neurocomputing* **35** 149–63
[13] Saragih T H, Mahmudy W F, Abadi A L and Anggodo Y P 2018 Application of extreme learning machine and modified simulated annealing for jatropha curcas disease identification *Int. J. Adv. Soft Comput. its Appl.* **10** 108–19
[14] Ramchoun H, Idrissi M A J, Ghanou Y and Ettaouil M 2017 New Modeling of Multilayer Perceptron Architecture Optimization with Regularization: An Application to Pattern Classification *IAENG Int. J. Comput. Sci.* **44** 261–9
[15] Taşkıran M, Çam Z G and Kahraman N 2015 An Efficient Method to Optimize Multi-Layer Perceptron for Classification of Human Activities *Int. J. Comput. Commun. Instrum. Eng.* **2** 191–5
[16] Wijayaningrum V N and Putriwijaya N N 2020 An Improved Crow Search Algorithm for Data Clustering *Emit. Int. J. Eng. Technol.* **8** 86–101

IOP Conf. Series: Materials Science and Engineering 1073 (2021) 012071 doi:10.1088/1757-899X/1073/1/012071

Object recognition using cognitive artificial intelligence's knowledge growing system: A preliminary study

A D W Sumari[1,2,*], I N Syamsiana[1], D R H Putra[1,3] and R A Asmara[1,3]

[1] Cognitive Artificial Intelligence Research Group (CAIRG), Department of Electrical Engineering, State Polytechnic of Malang, Jl. Soekarno-Hatta No. 9, Malang 65141, Indonesia

[2] Faculty of Defense Technology, Indonesia Defense University, IPSC, Sentul, Bogor, Indonesia

[3] Department of Information Technology, State Polytechnic of Malang, Jl. Soekarno-Hatta No. 9, Malang 65141, Indonesia

*arwin.sumari@polinema.ac.id

Abstract. Object recognition has been a challenge for an intelligent system. There have been various approaches to develop such a system by utilizing machine learning especially which are based on neuron that is, neural network and deep learning. Common problems when using those approaches are the first one is dataset availability and the second one is the number of data. Lack of data causes neural-based approaches cannot be well operated, while a small number of data causes low accuracy results on the system. From another point of view, a considered-new technology from Cognitive Artificial Intelligence (CAI) perspective called as Knowledge Growing System (KGS) which may cope with such problems. With the capability to build its own knowledge from nothing, KGS is able to carry out recognition while developing its knowledge regarding the phenomenon it is trying to recognize. In this research, we showed KGS capability to perform object recognition as it is developing knowledge when interacting with such object directly. We did a benchmark on face recognition use-case with some common machine learning methods to show their performance on a small number of data, and KGS showed good results. With 100 feature-set from 5 persons' face images, KGS achieves Degree of Certainty (DoC) as much as 80% which is the system's prediction accuracy that enables it recognizing the person based on that-moment data. Even though it is still lower compared to machine learning methods, but KGS shows advantages that it does not require high computational cost because it requires no training and no model development. In this research, we also showed that KGS enables the fast-deployment light-operated object recognition system.

1. Introduction

Object detection and recognition is a very easy task for humans since they were born. Humans can do this task every second, every day in their life, They can detect an object continued with recognizing it in just a very short time. Object detection and recognition can be carried out by using whether one of their sensory organs or the combination of two or more sensory organs. One of the easiest tasks that is the hardest one to an intelligent system is visual object detection and recognition which utilizes eyes. With eyes, humans can see, watch, and observe images and videos. They can easily extract information from such data and use the extracted information to build knowledge. They then can use the obtained

IOP Conf. Series: Materials Science and Engineering 1073 (2021) 012071 doi:10.1088/1757-899X/1073/1/012071

knowledge for making a decision and action. Detection, Recognition, and Identification (DRI) are parts of computer or machine vision, techniques that are utilized to enable computer or machine having humans' vision capabilities, such as detecting, recognizing, and identifying using their vision sensory organ.

Identification is the highest level of human's vision capability in discriminating or differentiating one object to others. From his original paper [1], Johnson divided it into four criteria, namely, Detection, Orientation, Recognition, and Identification (DORI). Other researcher uses Detection, Classification, Recognition, and Identification (DCRI) [2]. Humans can do all those activities easily, but the machine has to work very hard to do the same ones. Humans' vision sensory organ identifies an object through features such as size, shape, weight, smell, texture, and color. In some cases, by using only one feature, human is able to obtain an inference based on his identification. In some other case, the human needs more features in order to get better inference so as to minimize error in making a decision. Detection, recognition, and identification are actually the steps that have to be carried out to get the most accurate inference. The machine is only an entity that works if a human gives it instructions and works on behalf of the human to carry out works. To make the machine performs human-like vision capability, it has to be fed with appropriate inputs that have to be well processed using a correctly-selected image or video processing. Therefore, the primary things that have to be prepared are entities that represent image or video called features as mentioned previously.

In the real world, sometimes we find new objects which we have no knowledge about them before. If we put ourselves as a machine learning-based system, we will try to find the most possible match with the objects that we have ever learnt before. But if we put ourselves as a system that has no knowledge about any object, how can we know what objects are those? What will we be doing to recognize them and how do we do it? For a machine learning-based system to obtain very good classification accuracy, it needs a considerable amount of storage and computation power [3] because huge past data is required for training before used. When such a system is tasked to recognize an object in that-moment without training in the first place, it will not be possible that it can work without knowledge that it should develop during training. To answer this question, there has been a new perspective in AI known as Knowledge Growing System (KGS) [4] that has capability to develop its own knowledge while interacting with the objects. In this paper, we share our preliminary study on the use of KGS for object recognition.

In order to deliver an easy-to-understand presentation, this paper is arranged as follow. A brief introduction regarding the aim of this paper has been already given in Section I. In Section II, we deliver a brief on image processing and feature extraction methods which can deliver appropriated inputs to KGS, and also a brief on KGS. An example of the use of KGS for object recognition with a use-case in face recognition will be delivered in Section III. The paper is concluded in Section IV with some concluding remarks and also ways forward to enhance KGS for object recognition.

2. A brief on related theories

2.1. Recognition as the result of prediction

Recognition in a very simple definition is the ability to acknowledge the type of a detected object. Humans cannot recognize an object if it does not exist. The existence of an object in an image or video can only be known if it is detected by eyes. Humans can recognize what the object is if he already has knowledge about it. But if not, he will try to approximate what the object is. What he is doing is to predict what type of object is. Therefore, it can be said that recognition is the result of the prediction. The quality of recognition depends on the accuracy of the prediction. Prediction aims at forecasting unobserved outcomes or future behavior [5] and can be performed in two types, namely, quantitatively and qualitatively [6]. In the AI world, prediction can be using various methods whether neuron-based ones such as Neural Networks (NN) and deep learning, or statistical-based ones such as Support Vector Machine (SVM) and K-Nearest Neighbors (KNN). On the other hand, there is also an alternative approach based on cognitive psychology called Knowledge Growing System (KGS) [7]. This is the primary reason that KGS is called as Cognitive AI [8].

2.2. KGS as the recognition engine

In AI learning, knowledge of objects can be obtained from two means. The first one is by giving labels to objects in supervised learning, and the second one is by creating clusters of objects which are differentiated by their attributes or characteristics in unsupervised learning. These kinds of learning require data at high volume in the first place to be used to train the AI methods. The less the volume of data, the less the accuracy of prediction. On the other side, KGS is designed to have the ability to use the data it acquires at that moment to detect and recognize objects. KGS adopts constructivism theory especially the theory of learning by interaction to enable brain generating knowledge. Interaction is performed by using its sensors. The quantitative value of the obtained knowledge will show the detected and recognized object. The detailed explanation of KGS can be found in Sumari and Ahmad [4] and Sumari et al. [9]. There are four main formulas for KGS as shown in (1), (2), (3), and (4).

$$\lambda = (2^{\delta} - \delta) - 1 \tag{1}$$

where $\delta = 1, \ldots, i, \ldots, n$ is the number of sensors, or we can generalize it as the number of information sources, and $\lambda = 1, \ldots, j, \ldots m$ is a collection of knowledge generated as the result of the fusion of information from all relevant corresponding sensors.

$$P(\psi_1^j) = \frac{\sum_{i=1}^{\delta} P(\vartheta_i^j)}{\delta} \tag{2}$$

$P(\psi_1^j) \in \Psi$ and it is called as New Knowledge Probability Distribution (NKPD) which is the fusion result of the relevant corresponding sensors, $P(\vartheta_i^j)$. This is a collection of information that can be furthered extracted to obtain inference or new knowledge. The inference or new knowledge at this point can be obtained by applying (3) which becomes the quantitative value of the grown knowledge. The accuracy of prediction can be measured with Degree of Certainty (DoC) by using (4).

$$P(\psi_1^j)_{estimate} = \odot \left[P(\psi_1^j)\right] \tag{3}$$

$$DoC = \left|P(\theta_j)_{estimate} - \phi_1^j\right| \times 100\% \tag{4}$$

2.3. Image processing and feature extraction

Image processing is a method for processing images into digital forms for specific purposes. The inputs to image processing are images and the outputs are images with improved quality. For example, images that are less sharp in color, noise (such as white spots), and blurry (blurred) require a process to improve their quality. Therefore, images that have been performed with image processing will have better information [10,11]. Feature extraction is a way to get a value from the image, and aims to look for areas of significant features in an image depending on the intrinsic characteristics and applications. The region can be defined in a global or local environment and is distinguished by shape, color, texture, size, intensity, geometry, statistical properties, and so on [12-14].

One of the objects which has long been researched is the face, that is, how to recognize a person from his/her attributes. Before an object can be recognized, it has to go to image detection in the first place. There has been various model for image detection for human face. There is one model that has been a leading player in face detection arena, namely Viola-Jones which provides a fast and accurate object detection system like face region and facial features like eye, nose, and mouth [14]. After the face area and the face parts are successfully detected by the Viola-Jones model, the midpoint of the area is taken by using (5) [15]. This process is followed by the calculation of the distance of the face parts by using Euclidean Distance (ED) in (6). The results of all processes are the features of faces that become the inputs to KGS. The feature consists of the distance from the right eye to the left eye, the distance from the right eye to the nose, the distance from the left eye to the nose, the distance from the right eye with the mouth, left eye with nose and nose with the mouth.

$$(x,y) = \left(x_1 + \frac{width}{2}, y_1 + \frac{height}{2}\right) \tag{5}$$

$$ED = \sqrt{(x_1 - x_2)^2 + (y_1 - y_2)^2} \tag{6}$$

3. Object recognition using knowledge growing system

Object recognition using an already-trained AI system has been a common thing. Recognizing an object is done by carrying out the prediction. The prediction is obtained after comparing it with the knowledge about objects the system had been trained. It is not a hard task because the hard challenge is already handled during data collection for system training. For KGS, the challenge occurs during the interaction with the object. The more the attributes it can get, the more accurate the result will be. To observe KGS performance to recognize objects from that-moment data, data of faces from five different persons will be used. The object recognition using KGS is illustrated in Figure 1.

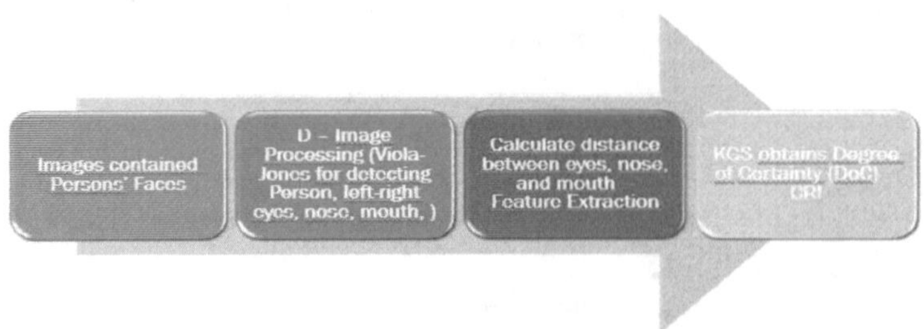

Figure 1. The illustration of object recognition with KGS.

3.1. Data preparation and processing

We prepared objects, that is, face images from 5 different persons numbered 0 to 4. By using Viola-Jones, we obtained extracted features of 5 different person's face which consist of facial features namely, distances between eyes and nose, distances between eyes and mouth, and distance between nose and mouth. Each individual person has 20 facial images taken from the same distance of 1.5 meters with the conditions as follows: the same lighting, the same taken time, almost the same pose, and almost the same expression. In this case, images with index numbered 1 to 20 belong to person 0, 21 to 40 belong to person 1, and so on. The extracted data is stored into a table using a spreadsheet application and then is converted to CSV form for further processing using KGS. A partial result of the five individual person faces feature-set is shown in Figure 3. As be seen in the two tables, there are no label for each person. An example of feature extraction for person 4 at index 2 is depicted in Figure 2.

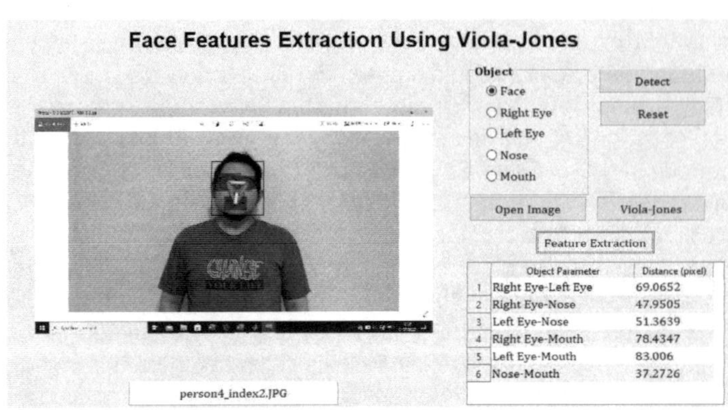

Figure 2. An example of face feature extraction by using Viola-Jones model for person 4.

No	Attributes/Indeks	1	2	3	4	5	6	7	8	9	10	11	12	13	14	15	16	17	18	19	20
1	right eye-left eye	87.5728	86.9166	88.5664	89.5056	86.8692	87.135	86.4696	87.0201	87.7568	85.4239	86.6386	86.6386	87.367	87.0761	87.9616	88.1887	88.5664	88.9073	88.7482	86.5231
2	right eye-nose	59.0953	58.6941	59.7599	59.4012	60.8769	59.4012	58.346	59.7599	58	58.6941	57.6823	58.7069	60.8153	59.0529	60.1082	57.2756	59.4012	58.6941	58.7963	59.397
3	left eye-nose	66.2741	66.562	68.3904	69.0163	64.2125	68.1175	66.2288	67.0317	68.3904	65.5687	65.8976	67.0317	66.6521	67.3573	67.6849	69.7227	68.7204	67.9154	66.9795	67.7366
4	right eye-mouth	96.729	96.3846	96.831	97.0979	96.0417	95.9544	96.3846	95.9544	95.6151	95.1972	93.7243	95.5249	95.5249	95.0289	96.0417	94.1541	96.4689	95.9544	94.6678	95.1893
5	left eye-mouth	99.4409	98.8863	98.976	98.6978	97.1198	100.276	99.1615	99.3491	100.09	98.4238	98.6154	99.6255	97.6934	97.3974	98.4238	100.121	100.001	99.2031	100.386	97.3974
6	nose-mouth	43.0726	42.5793	41.3824	41.5993	41.3038	41.512	43.0726	41.4367	42.2641	41.8629	41.8629	42.006	39.8811	40.302	40.6079	41.1886	41.6803	42.0862	42.2966	39.8121

Figure 3. Attributes for person 0.

3.2. Recognition

We have developed KGS software for another purpose but because it is multi-purpose, so we used it for this experiment. According to the mechanism illustrated in Figure 2, we have obtained the results from KGS as depicted in Figure 4. The results show that indices 1 to 40 have DoC = 100%, indices 41 to 60 have DoC around 31%, indices 61 to 80 have DoC = 0%, and indices 81 to 100 have DoC around 88%. From these results, KGS recognizes that there are two same persons, that is, person 0 dan person 1 (indices 1 to 20 and indices 21 to 40), and three other different persons.

3.3. Analysis

KGS is able to recognize 4 out of 5 persons by finding DoC for each group of features in feature-set. KGS recognizes person 0 dan person 1 as the same person, while person 2 and person 3 as other ones. But for person 4 even though KGS is able to recognize, there are many anomalies and some indices have the same values. If the images were not informed at the first place came from 5 different persons where each person has 20 images and was inputted to the system in a series, then KGS may recognize the person 4's pattern of NKPD as many persons. The pattern shown by indices 1 to 40 is consistent even though there is one index that is not consistent. The same thing occurs to indices 41 to 60. The most consistent pattern is shown by indices 61 to 80, and it is warranted to this pattern represents different person than the previous ones.

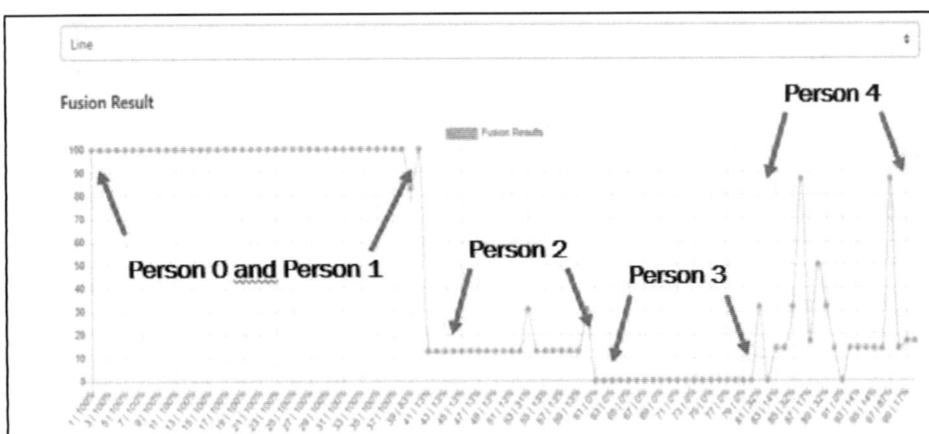

Figure 4. Persons recognition resulted from KGS.

A benchmark with machine learning methods such as NN, K-NN, Bayes, and SVM. For machine learning methods, the benchmark was carried out by splitting the feature-set into training set and testing set with 80:20 ratio, where 80 features were used as the training set and 20 features were used as the testing set. For KGS, because it does not need training phase so the feature-set which contains 100 features was inputted to the system. The benchmark results are given in Figure 5. The ability of KGS to recognize 4 out of 5 persons without training can be considered as a good result. The problem why KGS recognizes person 0 as the same as person 1 or vice versa, is because the features of the two persons' face images are very similar. This is a challenge to have much better feature extractor to be able to obtain different features for very similar images.

Object	Machine Learning					KGS
	Trained Label	Recognition after Training				That-Moment Recognition without Training
		NN	SVM	1-NN	Bayes	
Person 0's Face	0	0	0	0	0	0/1
Person 1's Face	1	1	1	1	1	1/0
Person 2's Face	2	2	2	2	2	2
Person 3's Face	3	3	3	3	3	3
Person 4's Face	4	4	4	4	4	4
Prediction Accuracy ML		90%	95%	90%	95%	
Prediction Accuracy (DoC KGS)						80%

Figure 5. Benchmark of the recognition results of KGS and machine learning.

4. Concluding remarks

4.1. Conclusions

Object recognition is easy for human even he/she has not had knowledge about it in the first place. He/she easily recognizes it by having knowledge of its attributes. If this mechanism can be emulated in the machine, then we can have a human-like system which is fast-deployed and light-operated. In this study we utilized and used KGS as such system with person face images as the objects to be recognized. The experiment results as well as its benchmark results to machine learning methods show a good prospect for KGS as an object recognition engine in a recognition system. As shown in Figure 5, KGS without training achieves 80% accuracy results. This results however, is lower compared to machine learning methods. On the other hand, KGS does not require high computational cost because it does not need training and require to develop the model that has to be done for machine learning. The advantages of KGS over machine learning are it does not need past data and automatically cuts the training phase which impacts on fast-deployment. It also does need huge storage to store past data which impacts to light-operation in the context of computing infrastructure.

4.2. Ways forward

KGS is a new technology and is a new perspective [16] as well as a novel field in AI [17]. However, there are still many things that can be explored to make it much better and mature for object recognition. There are things that will be carried out further as follows.

- Enhancing the recognition mechanism in order to cope with features with very small differences.
- Using different objects for next studies such as air, land, and sea vehicles.
- Developing the mechanism for object identification which is the highest level in DCRI level.

References

[1] Sjaardema T A, Smith C S and Birch G C 2015 History and Evolution of the Johnson Criteria *SANDIA Report SAND2015-6368* (Albuquerque, New Mexico: Sandia National Laboratories)
[2] Israel S A and Irvine J M 2012 Heartbeat biometrics: a sensing system perspective *International Journal of Cognitive Biometrics* **1**(1) 39-65
[3] Davies E R 2018 Computer Vision Principles, Algorithms, Applications, Learning Fifth Edition (London: Elsevier) p 371
[4] Sumari A D W and Ahmad A S 2018 Cognitive Artificial Intelligence: Concept and Applications for Humankind *Intelligent System* **23** 55

[5] Bzdok D, Altman N and Krzywinski M 2018 Statistics versus machine learning *Nat Methods* **15** 233–234
[6] Dobson A J 2013 *Introduction to statistical modelling* (New York: Springer)
[7] Sumari A D W, Ahmad A S, Wuryandari A I and Sembiring J 2010 Constructing brain-inspired knowledge-growing system: a review and a design concept *Second International Conference on Distributed Framework and Applications 2010* 95-102
[8] Sumari A D W and Ahmad A S 2017 Knowledge-Growing System: the origin of the Cognitive Artificial Intelligence *Proceedings of the 6th International Conference on Electrical Engineering and Informatics 2017* 1-7
[9] Sumari A D W, Sereati C O, Syamsiana I N, Wibisono M N, Abdulharis R and Putra D H P 2020 Cognitive artificial intelligence decision support system based on knowledge growing system *International Journal of Advanced Science and Technology (IJAST)* **29** 3734-3743
[10] Kang S, Choi B and Jo D 2016 Faces detection method based on skin color modeling *J. Syst. Archit.* **64** 100–109
[11] Akakin H Ç, Salah A A, Akarun L and Sankur B 2006 2D/3D facial feature extraction *Image Processing: Algorithms and Systems, Neural Networks, and Machine Learning* **6064** 60641D
[12] Hidayati K, Utomo A B and Wibawanto H 2018 Pengenalan wajah menggunakan metode local binary pattern (LBP) dan principal component analysis (PCA) untuk citra berkualias buruk *e-Proceeding Eng.* **5** 7928–7938
[13] Pongakkasira K and Bindemann M 2015 The shape of the face template: Geometric distortions of faces and their detection in natural scenes *Vision Res.* **109** 99–106
[14] C Rahmad, R A Asmara, D R H Putra, I Dharma, H Darmono and I Muhiqqin 2020 Comparison of Viola-Jones Haar Cascade Classifier and Histogram of Oriented Gradients ({HOG}) for face detection *IOP Conference Series: Materials Science and Engineering* **732**
[15] Sabri N, Henry J, Ibrahim Z, Ghazali N, Mangshor N N A, Johari N F M and Ibrahim S 2018 A Comparison of Face Detection Classifier using Facial Geometry Distance Measure *2018 9th IEEE Control and System Graduate Research Colloquium (ICSGRC)* 116-120
[16] Sumari A D W, Ahmad A S, Wuryandari A I and Sembiring J 2009 Knowledge Growing System: a new perspective on Artificial Intelligence, *Proc. of the 5th Int. Conf. on Information & Communication Technology and System 2009* 217-222
[17] Sumari A D W, Ahmad A S, Wuryandari A I, Sembiring J and Sahlan F 2010 An introduction to Knowledge-Growing System: a novel field in Artificial Intelligence *Jurnal Ilmiah Teknologi Informasi* (JUTI) **8** 11-20

IOP Conf. Series: Materials Science and Engineering 1073 (2021) 012072 doi:10.1088/1757-899X/1073/1/012072

Control position of the double nozzles on the Y (+) and Y (-) axis of 3D symmetric bilateral printing using G-Code

V U Audiana[1*], B Setiawan [1], A D W Sumari[1,2] and S Wibowo[1]

[1]Departement of Electrical Engineering, State Polytechnic of Malang, Jl. Soekarno-Hatta No. 9, Malang 65141, Indonesia.
[2]Faculty of Defense Technology, Indonesia Defense University, IPSC, Sentul, Bogor, West Jawa, Indonesia

*virnaumro03@gmail.com

Abstract. The most widely used universal Cartesian 3-Dimensional (3D) Printing today is not fast enough. However, by applying double nozzles, the process for making symmetric bilateral product can be done in half of the time. A symmetric bilateral product is an object that has the same size on the right and left sides. So, the method can be implemented by having double actuators that are mounted by a nozzle on each side. Both nozzles, called as Y and -Y nozzle, move on the same time and distance but in different direction. Besides that, an appropriate mechanical support is needed to be prepared, so that the motors can move in different direction at the same time. Therefore, a design simulation for two Y actuators was carried out. The simulation and the data taking were using actuator simulation software. The simulation results shows that the G-code can control two Y actuators with 99.64% accuracy with average error of 0.216% for revolution speed, and 100% accuracy with average error 0% for distance. The motor can move in different directions with the same type of screw.

1. Introduction

The 3-Dimensional (3D) printing is one of the main technologies that can sustain the development of the industrial system 4.0 [1]. It can print complex components and make the production process short and simple. 3D printer is a machine that can create real objects from 3D designs using computer-aided design (CAD) software [2]. Therefore, a program is needed to translate images or designs which have been made by CAD into a G-Code file. The G-Code, in this case, will be processed by the microcontroller to then be translated into motion on stepper motors.

The most widely used universal Cartesian 3-Dimensional (3D) Printing. It is not fast enough because only using one instrument (nozzle) for making products especially of 3D symmetric bilateral products [3]. A symmetric bilateral product is an object that has the same size on the right and left sides. So, it is very important to develop a printing method or system that can serve bilateral symmetric product formation faster than the universal 3D. By applying double nozzles at Y axis as Y and -Y nozzle, the process for making symmetric bilateral products can be done in half of the time.

Double nozzles can be implemented by having double actuators that are mounted by a nozzle on each side and it has moved at the same time and distance but in a different direction. So that the movement of the double nozzles is coordinated, it is necessary a design simulation for control two Y actuators at a distance, movement, direction of movement, and speed of motion of the two stepper motors.

IOP Conf. Series: Materials Science and Engineering 1073 (2021) 012072 doi:10.1088/1757-899X/1073/1/012072

The purpose of this research is to accelerate the manufacture of bilateral symmetrical products and to support government programs in accelerating industrial infrastructure 4.0.

2. Literature review

2.1. 3D printing symmetric bilateral

3D printer is a machine that can create real objects from 3D designs using CAD software. CAD is the use of computers to assist in the modification, analysis, creation and optimization of a designs [4] and outputs in file format for printing, machining, or other manufacturing operations. CAD software can be used to design curves and drawings in two-dimensional (2D) space or in 3D space [5]. The 3D design is saved in the stereolithography (STL) format and then sent to 3D Printing to make the 3D design layer by layer until it becomes a real 3D shaped object [2]. The 3D printing symmetric bilateral machines can form bilateral symmetrical products with the same size, angle, and shape on the right and the left sides.

2.2. 3D printing slicer software

The slicer software is a computer software used in the 3D printing processes for the conversion of a 3D object model to a specific instruction for the printer. The conversion is carried out to convert a model in STL format to printer commands in G-code [6-8]. G-code uses computer numerical control (CNC) programming language. It is used in computer-aided manufacturing to control the automated machine tools. G-code instructions provided to a machine controller tells the motors where to move, what path to follow, and how fast to move [9].

2.3. Firmware 3D printer

The firmware manages all of the machine's real-time activities including sending movement coordinates to the stepper motors through the stepper drivers, controlling the sensors, the lights, the heater elements, and tracking the bed levelling [10]. Interfaces for 3D Printers and CNC software suite handles the printing process and allows direct control of the 3D printers.

2.4. Actuator simulation software and screw

Actuator simulation software is used to show the movement of actuator and uses mathematical models to replicate the behavior of an actual electronic device or circuit. Simulation software allows for modeling of the circuit operation and is an invaluable analysis tool [11]. The series animation of actuator simulation can be used to help understanding the series so that it have educational value. A screw is a mechanism that converts a rotational motion to a linear one, and a rotational force (torque) to a linear one [12].

3. Methods

The method can be implemented by having double actuators that are mounted by a nozzle on each side. Both nozzles move on the same time and distance but in different direction. Therefore, a design of simulation for two Y actuators has to be carried out.

3.1. Simulation procedure

Figure 1 shows the simulation procedure. CAD software is used to design a half of the bilateral symmetrical products. 3D objects from the STL format are converted in the G-code format using 3D printing slicer software. It is slicing objects into a layer by layer section. After the completion of the slicing, the slicing software needs to make information of slice of the object shape, size, and other related one into the G-code instruction which 3D printer is able to identify for finishing the 3D printing task. The G-code instructions provide a machine controller that tells the motors where to move. The firmware that is included in the microcontroller software will manage all real-time machine activities including sending coordinate movements to the stepper motor through the stepper driver. The interface for 3D printers and CNC will handle the printing process according to the entered G-code. The simulation for the two Y actuators uses actuator simulation software.

The stepping motor used is capable of doing 1.8 °/step turn and it needs 200 steps per revolution (360°). The screw is used and requires one revolution to cover a distance of 2 mm. So the motor must be rotated in 360° for a distance of 2 mm. The steps of the motor must be set at 200 steps, but the settings on the step motor driver are at least 400 steps. Therefore, synchronizing the program is needed.

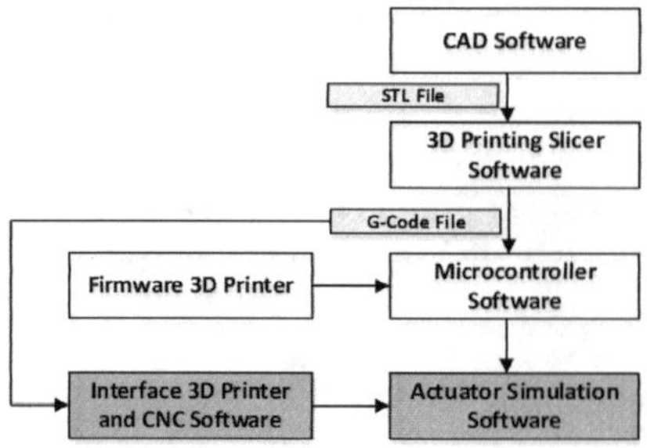

Figure 1. Block diagram of the simulation.

The calibartion of the firmware 3D printer, where microstep driver is 400 pulse/rev and p is pitch of screw using equation (1).

$$Firmware\ Kalibration = \frac{Microstep\ Driver}{p} \quad (1)$$

Step settings in the program is as follows.

```
//#define DISTINCT_E_FACTORS
/**
 * Default Axis Steps Per Unit (steps/mm)
 * Override with M92
 * X, Y, Z, E0 [, E1[, E2[, E3[, E4]]]]
 */
#define DEFAULT_AXIS_STEPS_PER_UNIT {160, 200, 160, 500}
```

3.2. Schematic of two Y stepping motors

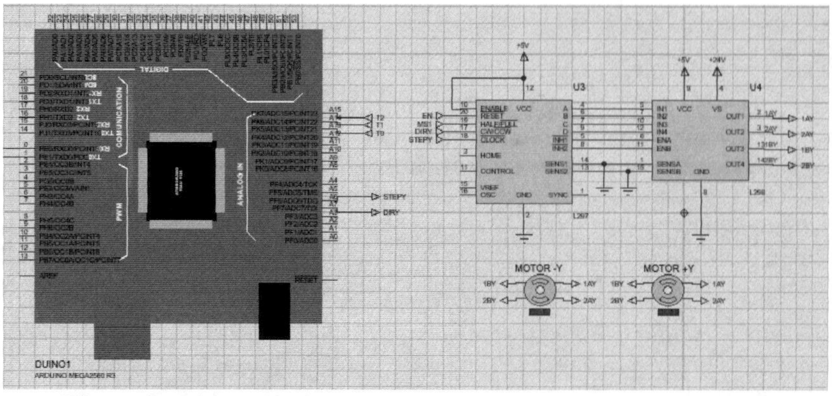

Figure 2. Schematic of two Y stepping motors in actuator simulation.

Figure 2 shows the simulation circuit for the two Y motors using actuator simulation software. The controller uses Arduino Mega 2560, which has been inputted with the firmware software. The output of the controller is used as input for the motor driver L297 and L298, which will regulate the movement of the two stepper motors. The output of actuator simulation is frequency and pulse. So the revolutions speed, that is Rotation per Minute (RPM) can be computed using equation (2).

$$RPM = \frac{\Theta s^\circ}{360^\circ} \, x \, f \, x \, 60 \tag{2}$$

(2) is used to ensure the revolution speed generated in the simulation is appropriate or not. The revolution speed is influenced by frequency. The greater the frequency, the faster the revolution speed of the motor produced. The pulse obtained using equation (3) and can be used as a basis for calculating the travelled distance. The setting step is 200 steps for 2 mm. Therefore, the setting is 100 step per mm.

$$d = \frac{pulse}{100} \tag{3}$$

3.3. The model of screws

Appropriate mechanical support needs to be prepared so that the nozzles can move in a different direction at the same time. Figure 3 shows the model of screws. When the shaft of the screw is rotated relative to the stationary threads, the screw moves along its axis relative to the medium surrounding it. So, the nozzles can move in different directions with the same type of screw. When the motors rotate Clock Wise (CW), the nozzles will move forward and head toward the center. Meanwhile, when the motors turn Counter-Clock Wise (CCW) the nozzles will move backward. The linear distance l, a screw shaft moves when it is rotated through an angle of α degrees can be found by using equation (4) where p, is the pitch of the screw.

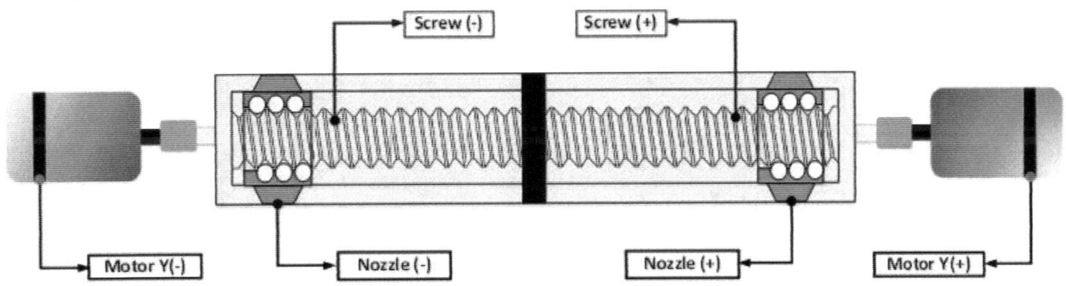

Figure 3. The model of screws.

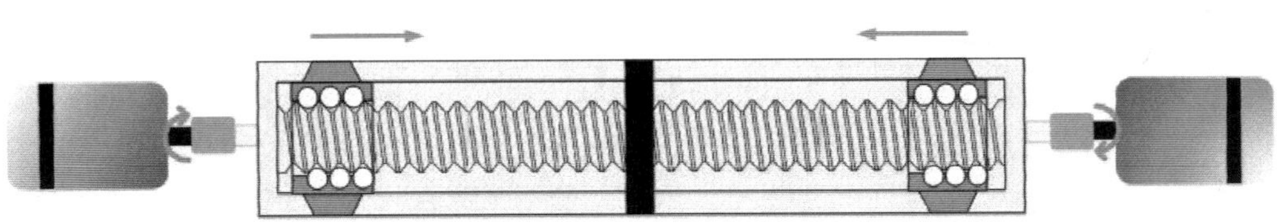

Figure 4. The movement of the nozzles.

$$l = \frac{\alpha^\circ x \, p}{360^\circ} \tag{4}$$

3.4. Planning motor torque in Y axis

The Y-axis weight is 2.95 kg (an HDPE pellets extruder), where Friction Coefficient of bearing (μ) is 0.61, the outer diameter of the screw (*dl*) is 12 mm, the effective diameter of the screw (*de*) is 10 mm, the pitch of screw (*p*) is 2 mm, and the workload (*W*) is 2.95kg or equal to 28.2 N. So, minimum torque is found by using equation (5).

$$T = \frac{F\ x\ de}{2} + \left(\frac{1 + \pi\ x\ de}{\pi\ x\ de - \mu\ x\ p}\right) \tag{5}$$

$$T = \frac{F\ x\ de}{2} + \left(\frac{1+\pi\ x\ de}{\pi\ x\ de - \mu\ x\ p}\right)$$
$$= \frac{47,6492\ x\ 10}{2} + \left(\frac{1+3,14\ x\ 10}{3,14\ x\ 10 - 0,61\ x\ 2}\right)$$

= 255.771 N.mm

= **0.25577 Nm**

Based on these calculations, the torque chosen for the Y-axis stepper motor specifications that is 3.6 Nm or National Electrical Manufacturers Association (NEMA) 23, because it has met the minimum torque required.

4. Results and discussions

To show the performance of the two Y stepping motors, a simulation has been performed. The results are shown in Table 1 and Table 2 as well as in Figure 5. Table 1 shows the results of rotation speed simulation. The distance travel of the motors is set 100 mm and the revolution speed is set 500-1500 RPM. The greater the frequency, the faster the revolution speed of the motor is produced.

Table 1. Rotation speeds simulation.

Input				Output			
Distance Travel	Linear Speed Set for CNC (mm/min)	Revolution Speed (RPM)	Frequency (Hz)	Revolution Speed (RPM)	RPM Error	Angle (°) Y+	Angle (°) Y-
100	1000	500	1665	499.5	0.1%	305	305
	1500	750	2491	747.3	0.36%	24.3	24.3
	2000	1000	3328	998.4	0.16%	269	269
	2500	1250	4158	1247.4	0.2%	341	341
	3000	1500	4987	1496	0.26	226	226

RPM error is used to determine the accuracy rate of the revolution speed and is computed using equation (6).

$$RPM\ Error = \frac{Input\ (RPM) - Output\ (RPM)}{Input\ (RPM)}\ x\ 100\% \tag{6}$$

$$Revolution\ speed\ Accuracy = 100\% - Highest\ RPM\ Error\ (\%) \tag{7}$$

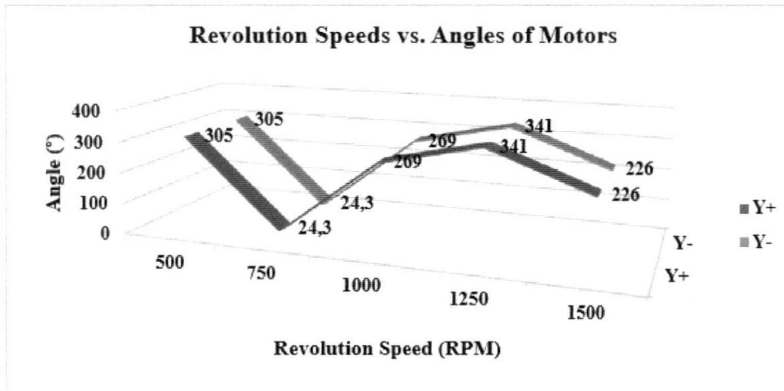

Figure 5. Graph revolution speed and angles of motors.

So, the revolution speed set and Y-Axis motor simulation results have 99.64% accuracy with average error of 0.216%. These errors can still be tolerated because the software has not been used to represent the actual conditions of the circuit, especially those relate to the timing. Therefore, the error does not interfere the system performance. Figure 5 shows Y+ and Y- motors rotate CW with the same rotation angle. Based on these test results, it is certain that the Y+ and Y- motors can rotate together with the same frequency. Based on Table 2, when inputting the G-code programs G1 X9.554 Y2.00 E2.9917, it means to move at the Y-axis is 2mm. The distance sets and Y-Axis motor simulation results have an average error of 0%. The G-code can control two actuators with 100% accuracy, and the nozzles can move according to the G-code coordinates. Distance error is used to determine the accuracy rate of distance. Distance error is computed using equation (8).

Table 2. Distance accuracy.

G Code Program Examples	Angle (°)		Pulse	d (mm)	Distance Error
	Y+	Y-			
G1 X9.554 Y2.00 E2.9917	138	138	200	2	0%
G1 X99.010 Y5.69 E10.5261	297	297	569	5.69	0%
G1 X989.554 Y36.11 E23.9917	93.6	93.6	3611	36.11	0%
G1 X993.161 Y56.18 E76.2003	294	294	5618	56.18	0%
G1 X966.733 Y102.74 E215.9811	194	194	10274	102.74	0%
G1 X938.460 Y150.11 F4800	108	108	15011	150.11	0%
G1 X886.465 Y274.88 E337.4778	163	163	27488	274.88	0%
G1 X870.314 Y319.10 E456.4987	31.5	31.5	31910	319.10	0%
G1 X791.066 Y581.57 E937.7580	103	103	58157	581.57	0%
G1 X745.768 Y948.41 E1561.2789	12.6	12.6	94841	948.41	0%

$$Distance\ Error = \frac{Gcode\ (mm) - d\ (mm)}{Gcode(mm)}\ x\ 100\% \quad (8)$$

$$Distance\ \ Accuracy\ = \ 100\% - \ Highest\ Distance\ Error\ (\%) \quad (9)$$

5. Conclusion

The simulation results prove that the G-code programs can control two Y actuators with 99.64% accuracy, with average error as much as 0.216% for revolution speed, and 100% accuracy with average error of 0% for distance. The nozzles can move in different directions with the same type of screw.

Acknowledgment
The authors would like to thank to the State Polytechnic of Malang for funding to the research through 2020 Thesis Magister Research Scheme.

References

[1] Dinas Kementrian Perindustrian No Title
[2] More M P 2013 *3D printing making the digital real*
[3] Luo X, Zhang L, Ren L and Lali Y 2020 A dynamic and static data based matching method for cloud 3D printing *Robot. Comput. Integr. Manuf.* **61** 101858
[4] Sarcar M M M, Rao K M and Narayan K L 2008 *Computer aided design and manufacturing* (PHI Learning Pvt. Ltd.)
[5] Farin G, Hoschek J and Kim M-S 2002 *Handbook of computer aided geometric design* (Elsevier)
[6] Evans B 2012 *Practical 3D printers: The science and art of 3D printing* (Apress)
[7] Keon A B 2018 *3D Printing at School and Makerspaces: Project Learning with 3D Printing. Cavendish Square*
[8] Kloski L W and Kloski N 2016 *Getting Started with 3D Printing: A Hands-on Guide to the Hardware, Software, and Services Behind the New Manufacturing Revolution* (Maker Media, Inc.)
[9] Apro K 2008 *Secrets of 5-axis machining* (Industrial Press Inc.)
[10] Hoque M M, Jony M M H, Hasan M M and Kabir M H 2019 Design and Implementation of an FDM Based 3D Printer *2019 International Conference on Computer, Communication, Chemical, Materials and Electronic Engineering (IC4ME2)* (IEEE) pp 1–5
[11] Archived copy. Archived from the original on 2010-12-16. Retrieved 2011-03-11
[12] Young J F 2000 Basic Mechanics". ELEC 201: Introduction to Engineering Design *Electrical and Computer Engineering Dept., Rice Univ.*

HDPE plastic extruder design and control for 3D printing with plastic pellet feeding method

B Setiawan[1]*, D C Permatasari[1], A D W Sumari[1,2], T Winarno[1], V U Audiana[1] and A M Damayanti[1]

[1] Department of Electrical Engineering, State Polytechnic of Malang, Jl. Soekarno-Hatta No. 9, Malang 65141, East Java, Indonesia
[2] Faculty of Defense Technology, Indonesia Defense University, IPSC, Sentul, Bogor, West Java, Indonesia

*budhy.setiawan@polinema.ac.id

Abstract. 3-Dimensional (3D) Printing Technology (3D printer) has become a Standard Acceleration Tool for Industry 4.0. Until now, 3D Printing system material melting technology has been using Fused Deposition Modelling (FDM) method. FDM is the melting of plastic materials through melting temperature and the formation of 3D objects in layer by layer manner. Meanwhile, the feeding for melting material is in the form of rolling filament. Currently, in this research FDM uses plastic pellete feeding method. Its exstrusion is called extruder, that is, plastic pellet processed through heater zone and pushed out by screw to the nozzle where the nozzle is a Computer Numerical Control (CNC) drives 3-Degree-of-Freedom (3DOF) applying cartesian axes. In this research, the printing uses high density polyethylene plastic pellete (HDPE). Temperature control in melting heater using proportional-integral-derivative (PID) method is controlled by microntroller and programmed through marlin firmware. The design and implementation result showed that the extruder system is capable to melt HDPE pellet at 200ºC with heater power 300Watt and the screw can rotate to push out melted plastic at 2mm of nozzle.

1. Introduction

3-Dimensional (3D) printer is the most popular issue in recent years. Actually, 3D printing and rapid prototyping can be traces back to 1980. 3D printing brings mass production of customized products to life. Furthermore, not only it can produce models that traditional manufacture technology cannot produce, but also it can save much time in developing and making new products. On the comparison of traditional model manufacturing, this new technology can print out delicate components, and the cycle and producing process are short and simple. Furthermore, it has diverse choices of raw materials and can generate models which are complicated. These advantages make 3D printing stands out markedly and becomes a trend nowadays [1].

More recent contributions considered is screw extrusion as a useful technology for 3D printing application allowing to manufacture a wider variety of materials than FDM, while using polymer granules as raw material [2]. The polymer is fed into the hooper and passed through the temperature zones where it is heated and melted. The melted polymer material is pushed forward by a powerful screw and passed through the extruder mechanism to form the die by nozzle [3]. In order to produce a good quality of filament plastic, the temperature in melting zone must be appropriately set and precisely

IOP Conf. Series: Materials Science and Engineering 1073 (2021) 012073 doi:10.1088/1757-899X/1073/1/012073

controlled [3]. To print a 3D object, this research uses plastic pellet polymer that is directly formed from the nozzle of extruder. The plastic pellet is used in this research are high-density polyethylene (HDPE).

2. Literature review

2.1. 3D printing

3D printing is a new manufacturing technology, related to materials, computing mathematics, and other disciplines. 3D printing has shown a huge advantage for the automotive, aerospace manufacturing industry and brought a lot of conveniences. With the development of 3D printing technology, variety materials printing methods have been developed, such as the most commonly used fused deposition modelling (FDM) technology. FDM is printing technology that uses computer aided design of layer by layer manufacturing print and has the advantages of low cost, easy maintenance, and so on [4]. Meanwhile, the technology of FDM for melting material use filament. But in this research FDM uses plastic pellet feeding method. The process of plastic melted become an object is extruder.

2.2. Screw Extruder and heater extruder

Extruder is a part of 3d printing and process used for manufacturing objects with fixed shapes and specific properties. A screw extruder is divided into one or several conveying zones (transport zones), melting zones (for material fusion), and mixing zones in which the extruded melt is submitted to high pressure, before its eviction through the nozzle. The pumping effect of the rotating screw moves the material from the feeding zone to the nozzle which influences the transport phenomena by having a restrictive effect, namely, the melt is accumulated behind the nozzle and fills completely the available volume in this region. The net flow rate at the nozzle exit is mainly influenced by the material flow in the longitidunal direction if one neglects the clearance between the screw and the extruder barrel [5]. Heater extruder is an extruder heating block consist of an aluminum mounting block and a 40W heating element where is rated at 12V. The power dissipated by this heating element is controlled by varying the duty cycle of the PWM applied by the microcontroller.

2.3. Stepper motor

Stepper motor is an electromechanical actuator which converts the input pulse train into a precisely defined increment in the shaft position. Each pulse moves the shaft through a fixed angle,called step angle. Stepper motors have emerged as cost-effective alternatives for direct-current (DC) servo motors in high-speed, motion control applications, where the high torque is not required, with the improvements in permanent magnets and the incorporation of solid-state circuitry and logic devices in their drive systems [6].

2.4. Marlin firmware

Marlin firmware is structured to utilize a repetitive loop method that periodically checks for serial input. To determine the command process flow,a manual code review was performed tracing the path of incoming G-code commands from reception to execution. It was found that while the firmware uses sequential numbering and check sums to validate incoming commands from the desktop application before processing, it is possible to bypass this functionality by directly updating private variables with desired hardware parameters values [7].

3. The proposed method extruder

In this research, extruder uses plastic pellete feeding method. Plastic pellet processed through melting zone and pushed out by screw to the nozzle. The advantages of extruder uses plastic pellet are low cost because plastic pellet is cheaper than rolling of filament and do not require a lot of time to print an object with diameter of nozzle that is used 2mm. On figure 1 is the mechanical design of the extruder for 3D Printing. Parts of extruder such as inlet as the pipe hole to insert HDPE plastic pellet, a stepper motor to rotate a screw, a screw rotates to push HDPE plastic pellet through melted process by heater and a nozzle with diameter of 2mm are used to form the HDPE filament.

IOP Conf. Series: Materials Science and Engineering 1073 (2021) 012073 doi:10.1088/1757-899X/1073/1/012073

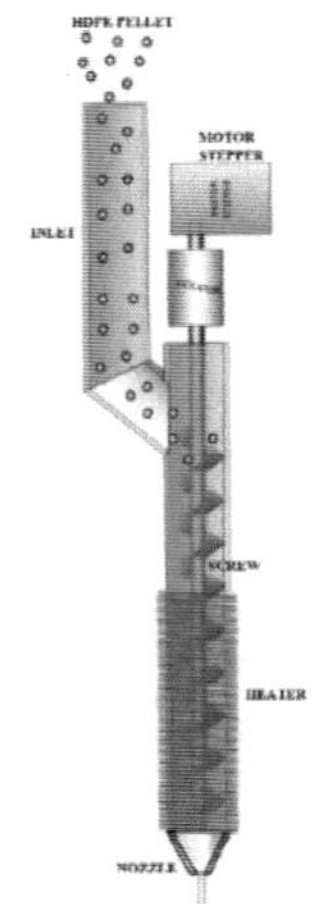

Figure 1. The mechanic design of the extruder.

3.1. Design of heater power

Design of heater to determine heater power. Heater is used to melt HDPE plastic pellet with melting temperature at 200^{0}C where discharge speed of the filament is 100mm/s with nozzle diameter of 2 mm. To calculate the maximum discharge of the filament or debit of filament, use (1)

$$Q = A \text{ x } V = (314 \text{ x } 1 \text{ x}1) \text{ x } 100 \quad (1)$$
$$Q = 314$$

A = Area of nozzle with diameter 2mm (mm^2)
V = discharge speed of filament (mm/s)
Q = debit of filament (mm^3/s)

From the above result, the weigth of HDPE filament can be calculated by using (2)

$$W = \rho \text{ x } Q = 0.96 \ x \ 0.314 \quad (2)$$
$$W = 0.301$$

P = density (g/cm^3)
Q = debit of filament (cm^3/s)
W = Weight or mass of filament (g/s)

The next step is to calculate the energy of HDPE to produce filament with speed 100 mm/s by using (3)

$$E = m \text{ x } c_{hdpe} \text{ x } (t_2 - t_1) = 0.000301 \ x \ 2.25 \ x \ 447.15 \quad (3)$$
$$E = 0.302$$

m = mass of filament (kg/s)
c = specific heat of HDPE (kJ/kg^{0}K)
t_2 - t_1 = (final temperature 200°C – initial temperatur 26°C) then convert celcius to kelvin (+273.15)
E = Energy (kJ)

IOP Conf. Series: Materials Science and Engineering 1073 (2021) 012073 doi:10.1088/1757-899X/1073/1/012073

Then convert the energy to power in watthour using (4)

$$P = E\,x\,Wh = 0.302\;x\,0.278 \tag{4}$$

$$P = 0.084$$

E = Energy (kJ)
Wh = Watthour
P = Power (Watthour)

From the above result, then convert the heater power from watthour to wattsecond by using (5)

$$P = 0.084 \text{ x } Ws = 0.084 \text{ x } 3600 \tag{5}$$

$$P = 302.4$$

Ws = Wattsecond
P = Power (Wattsecond)

So heater power needed to melt HDPE pellet with the speed of filament 100mm/s is 302.4 Wattsecond.

3.2. Torque design of screw

The calculation design of screw torque to determine the stepper motor specification is used for this system. The angle of screw (Θ) is 34.3°, the slide angle of screw (α) is 55.7, and the friction force of screw (f) is 45.75 Newton. Before calculating the torque, the thrust force of screw has to be calculated by using (6).

$$F = \eta + f\cos\alpha + \frac{f\sin\alpha}{\sin\alpha} = 75 + 45.75\cos 55.7 + \frac{45.75\sin 55.7}{\sin 55.7} \tag{6}$$

$$F = 146.53$$

f = friction force (N)
α = slide angle (°)
η = viscosity of HDPE (N)
F = thrust force (N)

Then calculate the toque using (7)

$$T = \frac{F\,x\,de}{2}\left(\frac{1+\pi\,x\,de}{\pi\,x\,de-\,\mu\,x\,l}\right) = \frac{146.531\,x\,4.4}{2}\left(\frac{1+3.14\,x\,4.4}{3.14\,x\,4.4-\,0.61\,x\,15}\right) \tag{7}$$

$$T = 1.3$$

l = pitch screw (mm)
de = effective diameter of screw (mm)
μ = friction coefficient of bearing (N)
F = thrust force (N)
T = Torque (Nm)

So the torque of screw for this system is 1.3 Nm

3.3 Speed design of stepper motor

In this system, speed design filament with 2mm diameter is 100mm/s and extruder system use stepper motor to rotate the screw. The speed of stepper motor can calculated by using (8)

$$n = \frac{v \; x \; 60}{s} = \frac{0.1 \; x \; 60}{0.015} \tag{8}$$

$$n = 400$$

n = speed of stepper motor (RPM)
v = speed of filament (mm/s)
s = screw pitch (m)

So the speed design for the stepper motor for the speed of filament for 100 mm/s is 400 RPM.

4. Methods

Figure 2 shows how the system works. The extruder divides the system into two parts, namely, the temperature control for melting temperature of HDPE pellet and the speed of the stepper motor for rotating the screw. The process of temperature control is carried out by microcontroller with proportional-integral-derivative (PID) method. Temperature controller uses heater to melt HDPE pellet and temperature sensor is used for sensing the temperature setpoint of the melted HDPE. This system uses a stepper motor with Nema 23 type for rotating the screw and pushing out the melted pellet to the nozzle. The output of 2mm nozzle is discharge of plastic melted.

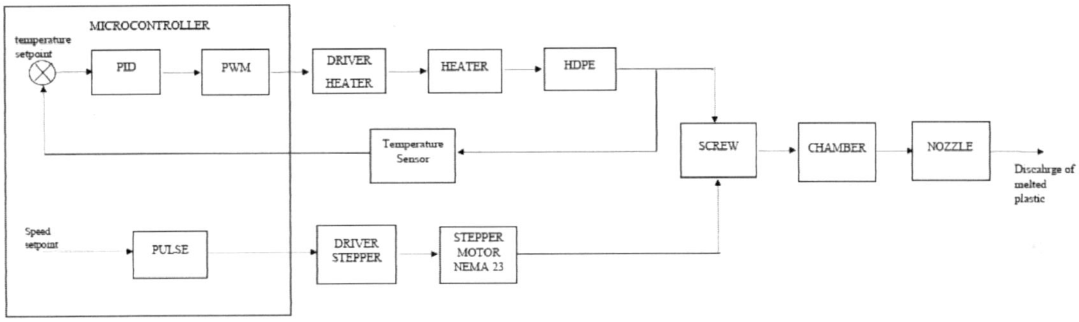

Figure 2. Block diagram of the extruder system.

5. Results and discussion

The results for extruder system consists of temperature testing and stepper motor testing.

5.1. Temperature testing

Temperature testing was conducted to know if the temperature control system worked well or not. Temperature control used heater to melt plastic pellet. Based on design heater power in equation (5), heater power needed to melt HDPE pellet is 300Watt. In this case, the temperature control system is used to stabilize the temperature on the melting process. The temperature set point is entered on system with melting temperature of HDPE pellet at 200^{0}C. The result of the temperature response toward the time on heating process is displayed in Figure 3.

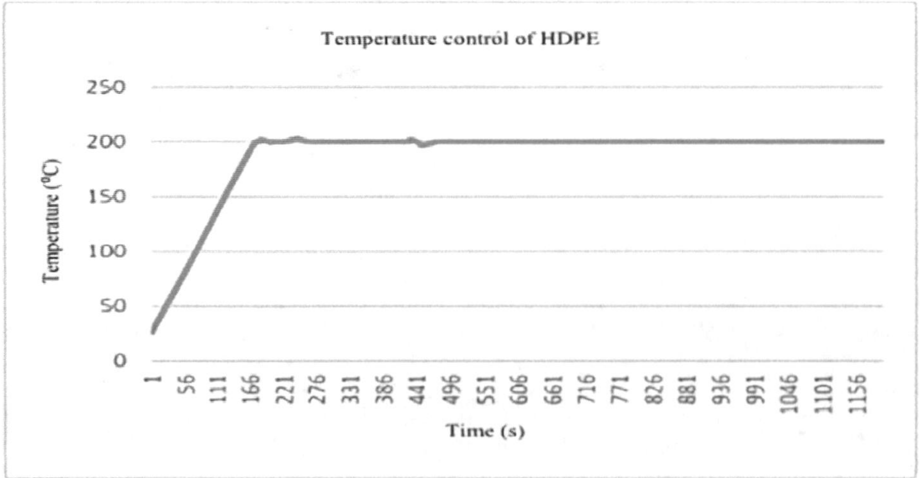

Figure 3. Melting temperature control response.

Figure 3 shows the temperature test with PID control method. The result of melting temperature control for HDPE pellet reached stability at 200°C.

5.2. Stepper motor testing

Table 1 is the data of testing result for speed of stepper toward length of the filament. The stepper motor was set on 200 pulse then changed the values of the stepper speed to rotate the screw in marlin firmware to know the length of HDPE filament that was come out. The result shown in Table 1, the speed of the stepper motor of 100mm/min produced 81 mm filament length, the speed of the stepper motor 90 mm/min produced 83 mm filament length, the speed of the stepper motor 80 mm/min produced 80 mm filament length, and the speed of the stepper motor 70 mm/min produced 84 mm filament length. The lower the speed of the stepper motor, the lower of HDPE filament discharged from the nozzle and also the lower the frequency of stepper motor is.

Table 1. Testing result speed of the stepper.

Speed of the stepper (mm/min) Marlin	Time (s)	Length of Filament (mm)	Frequency (Hz)
100	71	81	331.1
90	82	83	297.6
80	92	80	265.2
70	99	84	232

6. Conclusion

The design and testing result showed that the extruder system is capable to melt HDPE pellet at 200°C with heater power 300Watt. Temperature control of heater by using a PID method can reached stability at 200°C and speed of stepper motor to rotate the screw can produced melted plastic at 2mm of nozzle.

References

[1] Lin K-H, Shen C-Y, Du J-L, Wang G-Y, Chen H-M and Tseng J-D 2016 A design of constant temperature control system in 3D printer *2016 IEEE International Conference on Consumer Electronics-Taiwan (ICCE-TW)* (IEEE) pp 1–2

[2] Koga S, Straub D, Diagne M and Krstic M 2018 Thermodynamic Modeling and Control of Screw Extruder for 3D Printing *2018 Annual American Control Conference (ACC)* pp 2551–6

[3] Ravi S and Balakrishnan P A 2010 Modelling and control of an anfis temperature controller for plastic extrusion process *2010 International Conference On Communication Control And Computing Technologies* (IEEE) pp 314–20

[4] Xiaoyong S, Liangcheng C, Honglin M, Peng G, Zhanwei B and Cheng L 2017 Experimental analysis of high temperature PEEK materials on 3D printing test *2017 9th International conference on measuring technology and mechatronics automation (ICMTMA)* (IEEE) pp 13–6

[5] Diagne M and Krstic M 2015 State-dependent input delay-compensated bang-bang control: Application to 3d printing based on screw-extruder *2015 American Control Conference (ACC)* (IEEE) pp 5653–8

[6] Morar A 2015 The modelling and simulation of bipolar hybrid stepping motor by Matlab/Simulink *Procedia Technol.* **19** 576–83

[7] Moore S B, Glisson W B and Yampolskiy M 2017 Implications of malicious 3D printer firmware

IOP Conf. Series: Materials Science and Engineering 1073 (2021) 012074 doi:10.1088/1757-899X/1073/1/012074

The role of the blades number and pitch angles to the capability of self-start the Darrieus wind turbine

S Adiwidodo*, E Yudiyanto, B Wahyudi and S H Susilo

Department of Mechanical Engineering, State Polytechnic of Malang, Jl. Soekarno-Hatta No. 9, Malang 65141, Indonesia

*satworo.adiwidodo@polinema.ac.id

Abstract. Darrieus is classified as a vertical axis wind turbine that has a simple construction. It is widely used for small-scale domestic needs in rural or aquaculture areas. As with other types of vertical axis wind turbines, Darrieus have low efficiency. There is resistance on one side of the blade which aggravates the shaft rotation. Another problem is the low ability to self-starting. Some even need help from an external source to start. This is certainly detrimental in terms of energy conversion. The purpose of this study is to obtain information on the wind speed needed for a Darrieus to self-starting on variations in the number of blades and pitch angles. The blade profile used is NACA 6412. Variations in the number of blades are 2,3,4 and 5 and pitch angles of -10°, -5°, 0°, 5°, and 10°. Data retrieval was carried out experimentally using the wind tunnel. It is shown that the self-start wind speed was strongly influenced by the number of turbine blades. The number of blades 5 requires the lowest average self-start wind speed. The pitch angle has a different effect on the wind self-starting speed depending on the number of blades.

1. Introduction

The energy that is widely used today comes from very limited fossil fuel. Various studies have been conducted to utilize renewable energy. One of the renewable energies that are being researched at this time is wind energy. As with the horizontal axis wind turbine, there have been many developments in the vertical axis wind turbine [1,2]. The most widely applied and researched vertical axis wind turbine is the Darrieus type.

As with other types of vertical axis wind turbine (VAWT), Darrieus usually have low efficiency when compared to horizontal axis wind turbines, however, this type of turbine has the advantage of being able to receive fluid flow from all directions. The low efficiency of VAWT caused by the wind flow in the turbine is split into two sides. One side of the blade is hindering the other side of the turbine blades. There are several studies to improve the performance of the Darrieus wind turbine. Among them by changing the pitch angle [3-5]. It has been designed pitch-controlled H-Darrieus VAWT with asymmetric NACA 0021 airfoil for maximum lift drag ratio [3]. The numerical simulation shows that the pitch angle has a significant effect on the aerodynamic characteristics of the wind turbine. The airfoil used is NACA0021 with the number of blades 2. The simulation results show that the maximum pressure difference on the blade surface was obtained at the blade pitch angle of $\beta = 6°$ in the upstream region. However, the maximum pressure coefficient was shown at the blade pitch angle of $\beta = 8°$ in the downstream region. The torque coefficient acting on a single blade reached its maximum value at the blade pitch angle of $\beta = 6°$ [4]. The simulation of the effect of blade pitch angle on the aerodynamic performance of the NACA 63415 H-Darrieus VAWT blade at a low wind speed of 6 m/s shows a positive pitch angle of +5° improves turbine performance in the upwind position, while a negative pitch angle of -5° increases the performance of the turbine in the upwind position. downwind which reduces wake [5].

Another problem with the Darrieus wind turbine is its low ability to self-starting. Some even need an external source to start. Several studies have been conducted to solve this problem. The simulation of NACA 0012 shown that a lightly loaded, three-bladed rotor always has the potential to self start under steady wind conditions, whereas the starting of a two-bladed device is dependent upon its initial starting orientation [6]. It has been investigated a numerical model is used to simulate the starting of an H-rotor Darrieus turbine under steady wind conditions demonstrating that modeling remains constrained by the quality of data on aerofoil characteristics [7]. An innovative blade profile design for Darrieus VAWT is presented and named EN0005, offering the ability to self-start without the need for extra components or external electricity feed-in and without compromising a good performance at high TSR [8]. Mathematical modeling has been investigated to control the angle of attack of the blade profile, which allows a self-start to wind speeds as low as possible is presented for the straight blade of Darrieus turbine [9]. Systematic synthesis on the self-starting aerodynamic characteristics of vertical axis wind turbines based on the numerical analysis shows that the solidity and fixed pitch angle influence the aerodynamic characteristics with β = -2.5° has better self-starting aerodynamic characteristics [10]. A 2D numerical model base it is found that the incident flow velocity and the moment of inertia of the rotor have little effect on the averaged values of tip-speed ratios in the equilibrium stage under no-load conditions. In the system load calculations, four modes of self-starting were found: stable equilibrium mode, unstable equilibrium mode, switch mode and halt mode [11]. A numerical method was developed and validated to accurately simulate the interaction between the wind and VAWT indicated that the VAWT with S-1046 blade had a better overall performance as compared to the one with the NACA0018 blade, where the reduced self-starting time [12].

From previous research, it is known that changes in pitch angle greatly affect the aerodynamic characteristics and performance of the Darrieus wind turbine. Changes in the ability of self-star are rarely discussed concerning changes in pitch angle. This research was conducted experimentally using asymmetric airfoils with variations in the number of blades and pitch angle. The purpose of this study is to obtain information on changes in the self-starting wind speed of the Darrieus wind turbine on variations in blade number and blade pitch angle.

2. Material and methods

The effect of the number of blades and the pitch angle on the Darrieus wind turbine was carried out experimentally. The wind turbine prototype has a rotor diameter of 1 m with an airfoil span of 1 m. The blade profile used is an asymmetrical airfoil NACA 6412 (Figure 1).

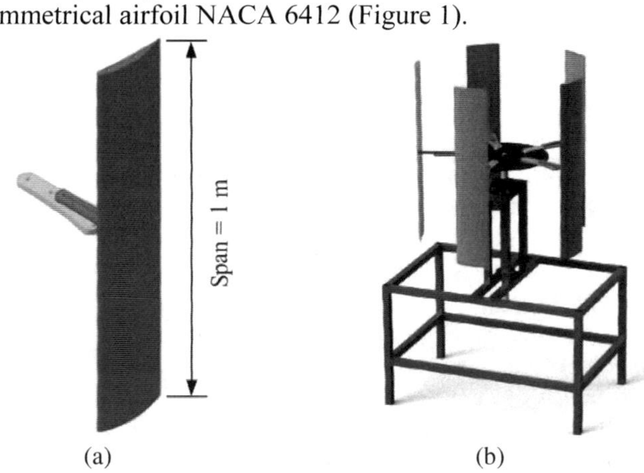

Figure 1. Darrieus wind turbine (a) airfoil NACA 6412; (b) prototype complete design.

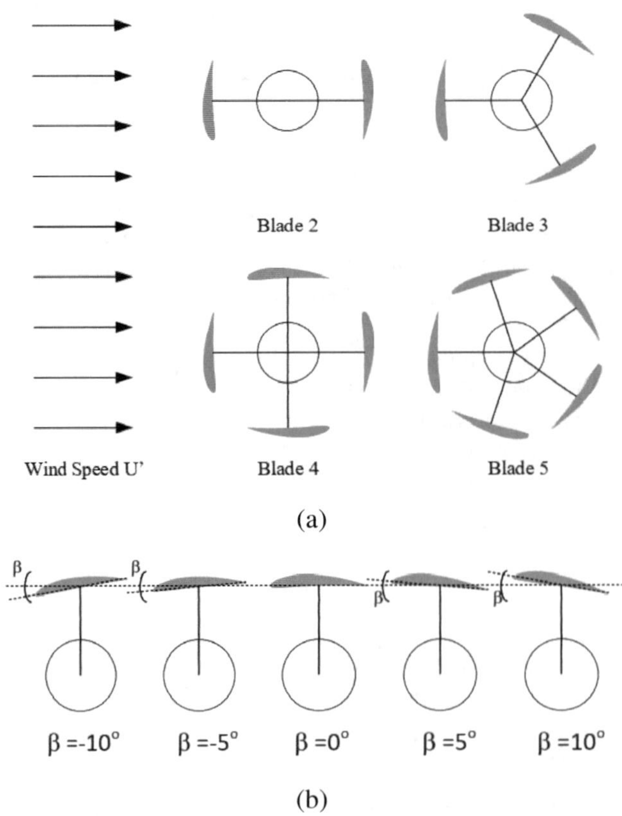

Figure 2. Experiment variables (a) blade number with initial orientation; (b) blade pitch angle (β).

Variations in the number of blades are 2,3,4 and 5 and pitch angles of β = -10°, -5°, 0°, 5°, and 10° (Figure 2). Data retrieval was carried out experimentally using the wind tunnel. At initial orientation, there is a blade radius in the direction of the flow of wind on the front side. The wind speed in the control from zero is increased slowly until the turbine starts moving. When the turbine starts moving, the wind speed is recorded as the turbine's self-starting speed. Fan motor speed control with variable speed drive. Wind speed measurement is done using a portable digital anemometer. Data were recorded and plotted using Excel.

3. Results and discussion

3.1. Effect of blade number to self-starting wind speed

The number of blades is very influential on the wind speed needed to have self-starting as shown in Figure 3. The fewer number of blades, the higher the wind speed needed to have self-starting, conversely the greater number of blades the lower the wind speed needed to do self-starting. It can be understood that the greater the number of blades, the more wind energy can be converted to torque. The number of blades will increase the area that receives wind energy. Figure 3 shows the blade number 5 has an average low self-starting speed at all pitch angles. The pitch angle affects the tendency of self-starting wind turbine speed. At zero-degree pitch angle, self-starting speed occurs with a linear tendency towards the number of blades. Meanwhile, the minus or positive angle tends to change in the polynomial of the blade count. For the negative pitch angle, the change in velocity tends to decrease as the number of blades increases, until the number of blades 4 reaches the lowest speed then rises. Conversely, for a positive pitch angle, there is a tendency to rise at low amounts to 3, then drop to the lowest speed at 4.17 m/s at the pitch angle of β =10° with the number of blades 5.

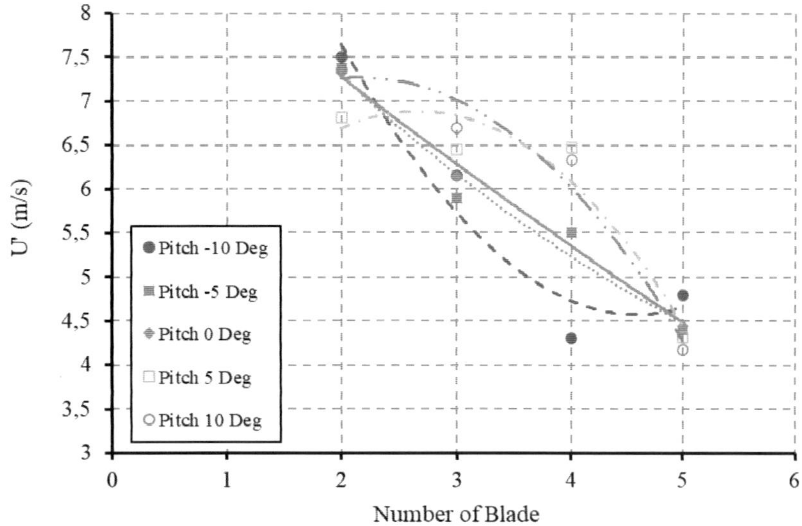

Figure 3. Wind self-starting speed with various number of blade.

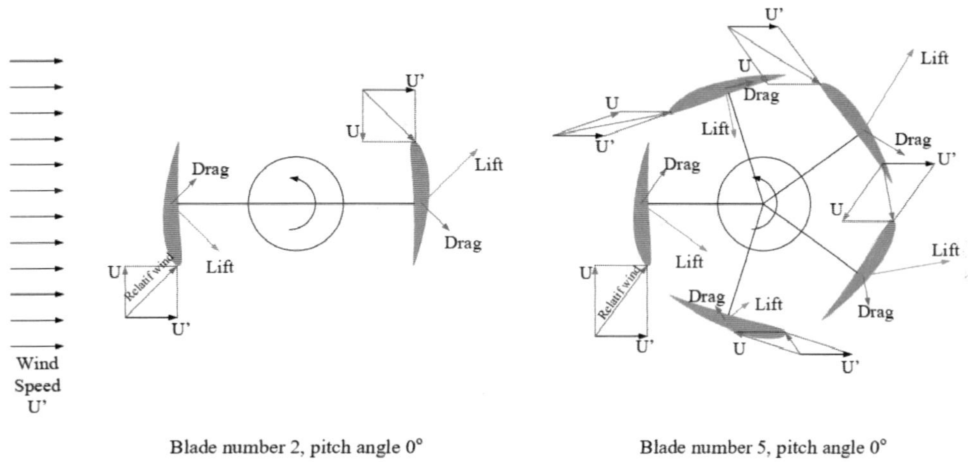

Figure 4. Force vector (unscale) for various number of blade.

Figure 4 shows the force vector works on the airfoil blade. Increasing the number of blades increases the area that can convert wind energy. The addition of the force vector to the turbine blade increases the torque required by the turbine to rotate so that the number of blades 5 has a lower self-starting wind speed.

3.2. Effect of pitch angle to self-starting wind speed

Figure 5 shows that for the number of blades 2 and 5, the change in pitch angle from negative to positive values reduces the self-starting wind speed, while the number of blades 3 and 4 shows the opposite trend. The turbine with the number of blades 2 achieves the lowest self-starting wind speed at a pitch angle of β = 5°, the number of blades 3 at a pitch angle of β = -5°, the number of blades 4 with a pitch angle of β = -10°, and the number of blades 5 at a pitch angle of β = 10°.

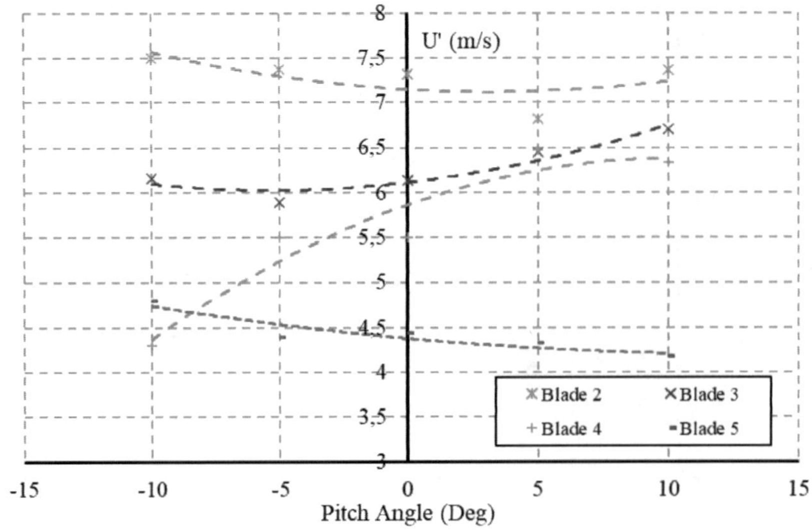

Figure 5. Wind self-starting speed with various pitch angle.

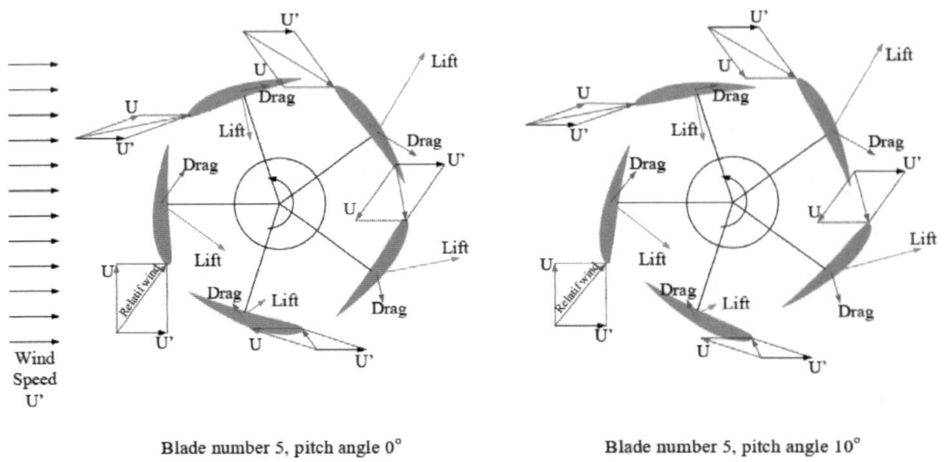

Figure 6. Force vector (unscale) for various pitch angle

Figure 6 shows the number of blades 5 with a pitch angle of $\beta = 10°$ which increase the airfoil's angle of attack. The angle of attack is the angle formed by the chord line with the relative wind. There is an additional angle of attack on the main blade on the downwind section. Increasing the angle of attack increases the lift produced. The increase in lift is of course accompanied by an increase in drag, but the ratio between lift and drag is still high, giving additional torque to the turbine shaft. That is why at low wind speeds the turbine is capable of self-starting. These results confirm the simulation study [4] for the type of symmetrical NACA optimum pitch angle $\beta = 6°\text{-}8°$. At higher pitch angles, the airfoil has a large angle of attack which can cause a stall, which is counterproductive to the performance of wind turbines. This research was conducted on one type of asymmetrical airfoil (NACA 6412). Future work was to compare the self-starting performance of various forms of NACA, both symmetrical and asymmetrical types.

4. Conclusions

From the results experiment of a Darrieus wind turbine with a various number of blade and pitch can be concluded as follows:

- It is shown that the self-start wind speed was strongly influenced by the number of turbine blades. The number of blades 5 requires the lowest average self-start wind speed.
- The pitch angle has a different effect on the wind self-starting speed depending on the number of blades.

Acknowledgment

This research was supported by the Applied Research Fund of DIPA State Polytechnic of Malang Number: SP DIPA-023.18.2.677606/2020. This research is a collaboration with students in completing their final projects. Thanks to Ade Yafi Reyhan, Bagas Mahardika, Jauhari Ulfi, M. Ilham Wildani, and M. Rizky Fadillah who helped prepare the prototype and retrieve the data.

References

[1] Möllerström E, Gipe P, Beurskens J, and Ottermo F 2019 A Historical Review of Vertical Axis Wind Turbines Rated 100 KW and Above *Renewable and Sustainable Energy Reviews* **105** (2019) 1-13
[2] Adiwidodo S, Wahyudi B, Yudiyanto E, Subagiyo S, Hartono M, and Baananto F 2020 Simulation Study of Savonius Tandem Blade Wind Turbine Using an Adjustable Deflector *IOP Conference Series: Materials Science and Engineering* **732** (1)
[3] Mauri M, Bayati I, and Belloli M 2014 Design and Realisation of a High-Performance Active Pitch-Controlled H-Darrieus VAWT for Urban Installations *In 3rd Renewable Power Generation Conference*
[4] Yang Y, Guo Z, Song Q, Zhang Y, and Li Q 2018 Effect of Blade Pitch Angle on the Aerodynamic Characteristics of a Straight-Bladed Vertical Axis Wind Turbine Based on Experiments and Simulations *Energies* **11**(6)
[5] Mazarbhuiya H M S M, Biswas A, and Sharma K K 2020 Low Wind Speed Aerodynamics of Asymmetric Blade H-Darrieus Wind Turbine-Its Desired Blade Pitch for Performance Improvement in the Built Environment *Journal of the Brazilian Society of Mechanical Sciences and Engineering* **42**(6)
[6] Dominy R G, Lunt P, Bickerdyke A, and Dominy J 2007 Self-Starting Capability of a Darrieus Turbine *Proceedings of the Institution of Mechanical Engineers, Part A: Journal of Power and Energy* **221**(1) 111-210
[7] Hill N, Dominy R, Ingram G, and Dominy J 2009 Darrieus Turbines: The Physics of Self-Starting *Proceedings of the Institution of Mechanical Engineers, Part A: Journal of Power and Energy* **223**(1) 21-29
[8] Batista N C, Melício R, Mendes V M F, Calderón M, and Ramiro A 2015 On a Self-Start Darrieus Wind Turbine: Blade Design and Field Tests *Renewable and Sustainable Energy Reviews* **5** 508-522
[9] Douak M and Aouachria Z 2015 Starting Torque Study of Darrieus Wind Turbine *International Journal of Mathematical, Computational, Physical, Electrical, and Computer Engineering* **9**(8) 472-477
[10] Zhu J, Huang H, and Shen H 2015 Self-Starting Aerodynamics Analysis of Vertical Axis Wind Turbine *Advances in Mechanical Engineering* **7**(12) 1-12
[11] Liu Z, Qu H, and Shi H 2016 Numerical Study on Self-Starting Performance of Darrieus Vertical Axis Turbine for Tidal Stream Energy Conversion *Energies* **9**(10)
[12] Sun X, Zhu J, Hanif A, Li Z, and Sun G 2020 Effects of Blade Shape and Its Corresponding Moment of Inertia on Self-Starting and Power Extraction Performance of the Novel Bowl-Shaped Floating Straight-Bladed Vertical Axis Wind Turbine *Sustainable Energy Technologies and Assessments* **38**(2020) 100648

IOP Conf. Series: Materials Science and Engineering 1073 (2021) 012075 doi:10.1088/1757-899X/1073/1/012075

Testing of black chromium coating on low carbon steel

S Hadi*, H P Buwono, A Setiawan, S S M Valeria and H Setyawan

Department of Mechanical Engineering, State Polytechnic of Malang, Jl. Soekarno-Hatta 9, Malang 65141, Indonesia

*syamsul.hadi@polinema.ac.id

Abstract. The problem lies in the colour resistance of the black chromium electroplating results that are not yet known for its effective test methods. The purpose of this research is to find an effective method for testing colour resistance. The methods used include preparing the specimen from motorcycle rim, black chromium electroplating, heating the specimens in Muffle furnace for different temperatures, cooling the specimen to room temperature, measure and calculate the colour state in order to decide the fading level, corroding the specimen in several acidity solutions from chloric acid, calculate the corrosion rate and analysis of the results. The results of the study in the form of (1) measurements with Colourflex showed that heating the low carbon steel rim for 2 hours at 200°C the colour remained dark black, at 250, 300, 350, and 400°C the colours faded 20, 40, 45, and 60% respectively. (2) corrosion test shows that at pH 3, 4, 5, and 6 the corrosion rate is 0.166, 0.136, 0.122 and 0.107 mm/year which the colour begins to fade at pH 4 and increase for more acid.

1. Introduction

Chromium electroplating has been widely used for the purpose of wear resistance of a component, corrosion resistance, and a decorative product. Electroplating with the aim of hardening is generally done by coating in stages with copper coating as a base layer, followed by Nickel coating, and finally coated with chromium. For decorative products made from low carbon steel, it is done by initial coating with Nickel to get a better coating resistant to use, then chromium coating is carried out under certain conditions at the current density and duration of the process with a certain solution, in order to obtain black chromium staining.

Previous experiments had obtained preliminary results indicating shiny and fading colours for certain electroplating conditions. It is not yet known how good the results of the colouring of black chromium and how much colour durability requires testing with certain methods.

The colour resistance of black chromium electroplating has been carried out with various methods including heating test and corrosion test. The real object used is a motorcycle rim made from low carbon steel that has been coated with nickel, then followed by electroplating black chromium at optimum conditions. Colour fading test is done by heating the specimen in an electric furnace for 2 hours for a certain temperature, then measuring its fading with Colourflex. Corrosion test is done by immersion specimen in a solution at a certain acidity and duration of time.

Low carbon steel is used for motorcycle rims because it has a higher strength compared to aluminium rim material, but has the disadvantage of being rusty unless it is coated with nickel or nickel and then chromium.

The development of market aesthesis in white rim nickel coating results tend to decrease and switch to black rims that are increasingly attracting users/teenagers. Low carbon steel is a metal with the main element Fe in group VIII in the periodic element table and the melting temperature of iron is 1538°C, the density of 7.8 g/cm^3 about 3 times the density of aluminium, 2.7 g/cm^3. Iron in a pure state is rarely obtained because pure iron easily reacts with oxygen in the air and in the process of fusion binds to carbon so that carbon iron is formed.

Iron and carbon steel are distinguished by their carbon content, for pure iron only has a small amount of carbon called ferrite which is relatively pure iron, whereas in most metals iron is steel with a carbon content between 0.01 and about 1.9% and the rest is called cast iron containing carbon between 1.9% and 6.5% maximum which in fact cast iron is produced around 4.3% due to lower melting temperatures whose production requires lower energy. Steel is divided into 3 categories are low carbon steel (LCS), medium carbon steel (MCS) and high carbon steel (HCS) with carbon content in a row between 0.01% and 0.025, 0.025 and 0.55, and 0.55 and 1.9%. Low carbon steel is also known as ST37 [1].

Black motorcycle rims are produced from electroplating that uses black chromium (CrO_3) with the conditions of the process on the current density and duration of immersion in a particular solution. The principle of electroplating based on Michael Faraday's Law states that electrolysis cells can be used to determine the amount of reacting substances which is proportional to the electric charge used in a certain period. In both Voltaic cells and electrolysis cells, there is a quantitative relationship between the number of substances that react and the electrical charge involved in the reaction and its oxidation.

The principles of Faraday's Law include: (1) the mass of the substance deposited in an electrode is proportional to the electric current (electron flow) involved in the electrochemical cell, and (2) the equivalent mass of the substance deposited on the electrode is proportional to the electric current flowed in the cell electrochemistry.

The Faraday formula is $\mathbf{w = (e\,i\,t) / F}$ (1)

where w: mass of substance (g), e: equivalent mass or (M/valence), i: electric current (A), t: time (second), F: Faraday's constant = 96,485 C/mol [2].

2. Related works

Chromium which is very resistant to corrosion, scratches and wears [3] is widely used as a decorative metal coating material.

A colourimetry shows the value of L* a* b* for the colour state where if L is 0 (zero) means jet black and if 100 is white, if "a" is negative (-) means green and if "a" positive value (+) means red, and if "b" is negative (-) means blue and if "b" is positive (+) means yellow [4].

The results of the study revealed that high-speed oxy-fuel facilities can superimpose NiCrBSi superalloy powder and $Cr_3C_2NiCr75/25$ alloy powder on machine components that can reach hardness around 850 $HV_{0.3}$ [5] which both alloy elements contain chromium.

Black chromium coatings are widely used in many applications as solar panels, optical instruments and decorative coatings of a product [6]. Nickel electroplating with 0.31 g/cm^3 concentration produces the best nickel deposit with a thickness of 5.13 μm at 0.5V for 15 minutes [7]. The current density of 50 A/dm^2 in electroplating produces the best black homogeneity. In addition to the chromium oxide and hydroxide layers, there is also cobalt oxide, hydroxide and Co_3O_4 plays the most dominant role in affecting the level of blackness [8]. Before electrodeposition of black chromium, bright nickel coating is carried out on steel and copper plates to get the best corrosion resistance and good optical properties in the absorption of solar energy [9]. Trivalent chromium and cobalt coating annealed at 300°C showed results with an increase in hardness from 380 to 460 VHN, and increased corrosion resistance from 41.3 to 77.9%, which offers good absorption [10].

SEM results show that film coatings from chromic acid, acetic acid and barium acetate baths are denser with nano-sized granules than chromic acid, fluorosilicic acid and barium carbonate baths, which are porous, with micro-sized granules [11]. Trivalent chromium coating replaces hexavalent chromium coating because hexavalent chromium coating is toxic and pollutes the environment and health of workers even though it provides a hard and wear-resistant and corrosion-resistant coating [12]. The electrical conductivity of the coating results increases with increasing carbon content which has been distributed rather uniformly at all layer thicknesses [13]. Baths of trivalent chromium sulfate-containing formic acid and carbamide as deposits have a nanocrystalline type (3-5 nm) and carbon present as chromium carbide in the layer is evenly distributed with better wear characteristics than hexavalent chromium bath [14]. Cr (III) from the Cr-C layer causes cracking due to increased internal stress by excessive carbon content in the main salt concentration and after heat treatment at 400° C, Cr_7C_3 and $Cr_{23}C_6$ that occur in the layer can increase hardness and show better wear resistance [15]. By using pulse currents, the hardness and wear parameters of the electrodeposit can be increased sufficiently for example, in $t_{on} = t_{off} = 1$ s, the hardness reaches a value of around 1250 HVN, whereas in steady-state electrolysis the hardness is around 900 HVN [16]. The carbon content of Cr-C alloy coatings depends on one of them at a lower pH value and a higher depositional current density resulting in a higher amount of carbon that can enter the alloy [17]. The solar selective coating made by electrodeposition technique from black chromium has the highest absorption rate at 94.6-96% with a voltage of 5V, the distance between the anode and cathode 5 cm and the immersion time of 5 minutes in an electrolytic bath [18].

3. Materials and methods

The research specimen was chosen in the form of nickel coated low carbon steel motorcycle rim with an arc length of 70 mm, 50 mm width, 1.4 mm thickness that was available and bought in trade for the 178 mm (17 inches) rim diameter is shown in Figure 1.

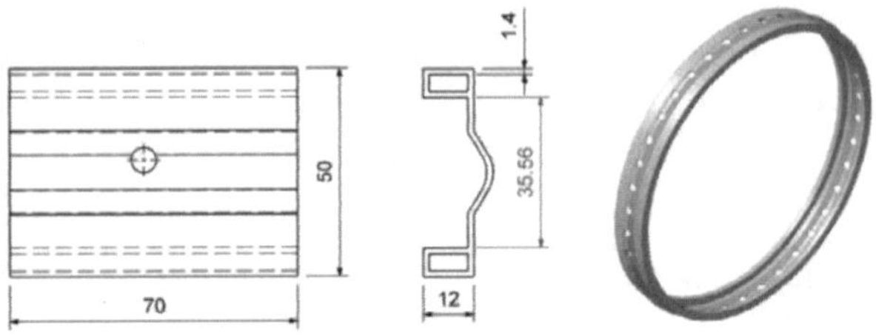

Figure 1. Motorcycle rim with an arc length of 70 mm, 50 mm width, 1.4 mm thickness.

The electroplating procedure is carried out as follows: (1) Cutting of motorcycle rim with an arc length of 70 mm and making holes for specimen hanging, (2) Cleaning the specimen with a metal cleaner (NaOH 20%), (3) Soaking the specimen in the activation solution (50% HCl: 50% Aquadest) for 3 minutes, (4) Preparation of electrolyte solution for electroplating black chromium in a breaker glass with a concentration of chromic acid (CrO_3) 40% and sodium nitrite ($NaNO_3$) 0.5% in 1-litre aqua dest, (5) The electroplating is carried out with the principle shown in Figure 2 with the best conditions at a 5V, a current density of 50-60 A/dm^2, a duration of 8 minutes and at 50°C, (6) Clean specimens with aqua dest, and (7) Specimens cutting for heating and corrosion testing.

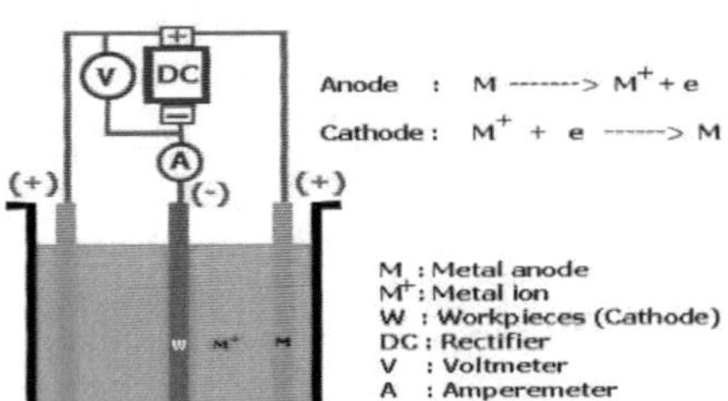

Figure 2. The electroplating principle involves the measurement of voltage and current [19].

The black Chromium was measured by Colorflex machine on the surface before and after heating at 200, 250, 300, 350, 400°C for 2 hours and then calculate the colour fading. The black chromium corrosion test on motorcycle rim was carried out by immersion in the solution with a certain acidity level for 1 week so that the known difference in weight (weight lost) dissolved into the acid solution which was then calculated as a corrosion rate. The flow diagram of the colour resistance research is shown in Figure 3.

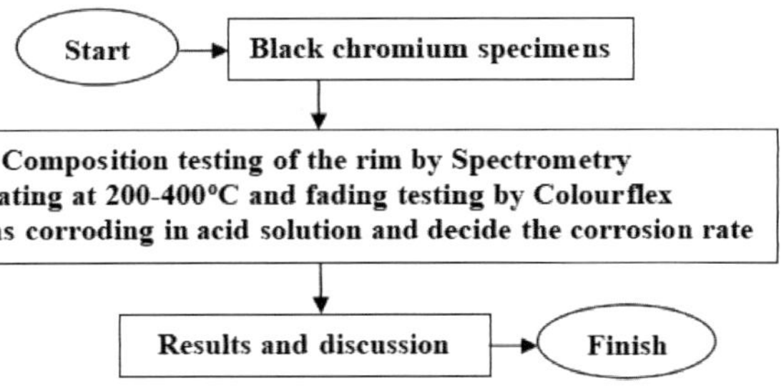

Figure 3. The flow diagram of the colour resistance research of the black chromium.

The methods studied were colour fading (after being heated at a certain temperature and duration of time) and weight loss after corroding.

The results of the composition test of the motorcycle rim specimens are shown in Table 1.

Table 1. The composition of motorcycle rim specimens.

No	Element	%	No.	Element	%
1	C	0.058	10	Al	0.033
2	Si	0.013	11	Co	<0.005
3	Mn	0.254	12	Mg	<0.005
4	P	0.13	13	Nb	<0.005
5	S	<0.010	14	Ti	<0.005
6	Cr	<0.010	15	V	<0.005
7	Mo	<0.010	16	W	0.100
8	Ni	0.185	17	Fe	99.42
9	Cu	<0.005			

4. Results and discussion

The data in column ΔE in Table 2 is obtained from the formula $\Delta E = \sqrt{(L^*)^2+(a^*)^2+(b^*)^2}$. For the L* value obtained by reducing the standard L* value according to https:/colorlib.com/etc/metro-colors/ which is L* = 10.767 with the value obtained from the reading on the Colorflex machine that is 9.24 then squared, 10.767-9.24=1.527 which then squared is obtained 2.332, likewise for a* and b* obtained in the same way. Value of $\Delta E = \sqrt{(L^*)^2+(a^*)^2+(b^*)^2} = \sqrt{(2.332 + 0.3481 + 0.545)} = 1.7959$. The value of ΔE as a resultant value of a combination of 3 colour elements is, in fact, a rim specimen that has been coated with black chromium that has been heated for 2 hours at 200°C which in reality still does not change at all with the appearance as at room temperature, namely 100% black or jet black.

Table 2. Colorflex reading data for L*, a*, and b* black chromium motorcycle rims.

Specimen code	L*	a*	b*	ΔE	Remarks
	The table of black color standard, https:/colorlib.com/etc/metro-colors/				
	10,767	-0	-0.002		
A (Without heating)	9,24	-0.59	0,74		
$(L^*)^2$ or $(a^*)^2$ or $(b^*)^2$	2,332	0,3481	0,545	1,7959	100% black
B (heating 2 h at 200° C	14,34	-0,29	-1,32		
$(L^*)^2$ or $(a^*)^2$ or $(b^*)^2$	12,7663	0,0841	1,7371	3,819	20% faded
C (heating 2 h at 250° C	15,97	-0,72	1,64		
$(L^*)^2$ or $(a^*)^2$ or $(b^*)^2$	27,0712	0,5184	2,6830	5,5021	40% faded
D (heating 2 h at 300° C	16,29	-0,80	-2,08		
$(L^*)^2$ or $(a^*)^2$ or $(b^*)^2$	30,5035	0,64	4,3180	5,9549	45% faded
E (heating 2 h at 350° C	18,04	0,85	1,57		
$(L^*)^2$ or $(a^*)^2$ or $(b^*)^2$	52,8965	0,7225	2,4711	7,489	60% faded
F (heating 2 h at 400° C	16,61	-0,22	-0,68		
$(L^*)^2$ or $(a^*)^2$ or $(b^*)^2$	34,1406	0,0484	0,4596	6,269	80% faded

Table 2 shows that heating the motorcycle rim after coated with black chromium at 200 to 400°C after reading with Colorflex shows a fading of 20 to 80%. Specifically, the ΔE value for the F code specimen Colorflex reading values are no longer appropriate, because of the possibility of overheating which has caused the black chromium layer to burn which may occur on the surface of the rim there is a thin layer of charcoal.

Specimens with code A is specimen without heating, and specimens that are heating at 200, 250, 300, 350 and 400°C have codes B, C, D, E, and F is shown in Figure 4. The faded color causes the absorption of solar energy to decrease as it has been stated that the highest absorption rate of 94.6-96% was achieved at the electroplating process at 5V which has been the same carried out in this study and previous researchers [18].

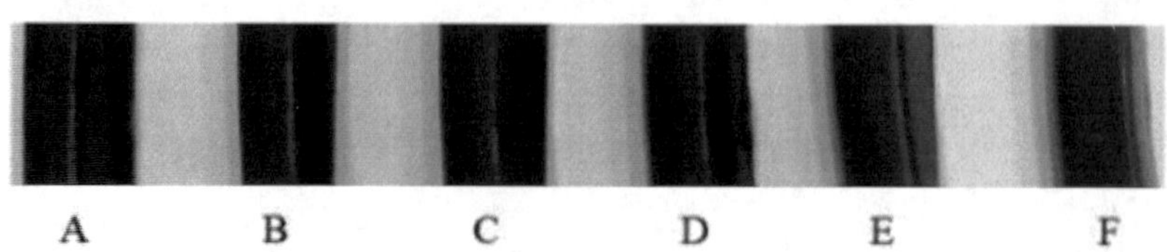

Figure 4. Specimens: A, without heating, and B-F, heating at 200, 250, 300, 350 and 400°C.

The corrosion rate of black chromium coating on the rim in one liter Aqua dest with 0.015 ml HCl for pH 3, with 0.01 ml HCl for pH 4, with 0.001 ml HCl for pH 5 and with 0.0001 ml HCl for pH 6, which is immersed for 168 hours (1 week) whose results are shown in Figure 5.

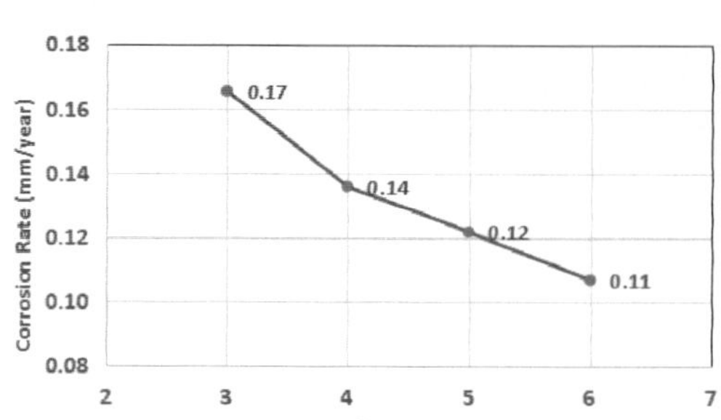

Figure 5. The corrosion rate of the black chromium coating on the rim in the HCl solution.

The corrosion resistance of black chrome coatings at the highest 95.9% was achieved at annealing at 300°C but decreased to 77.9% at 400°C which indicates that the colour resistance of more than 300°C has decreased [10]. Similar to heating at 400°C on black chromium on motorcycle rims experience high colour fading up to 80% is shown in Figure 6. In this study, a current density of 50 A/dm^2 was selected in the same electroplating process carried out by other researchers which produced the best black homogeneity where in addition to chromium oxide and hydroxide layers, there were also cobalt oxides, hydroxides and Co_3O_4 which played a dominant role in producing the dark level [8].

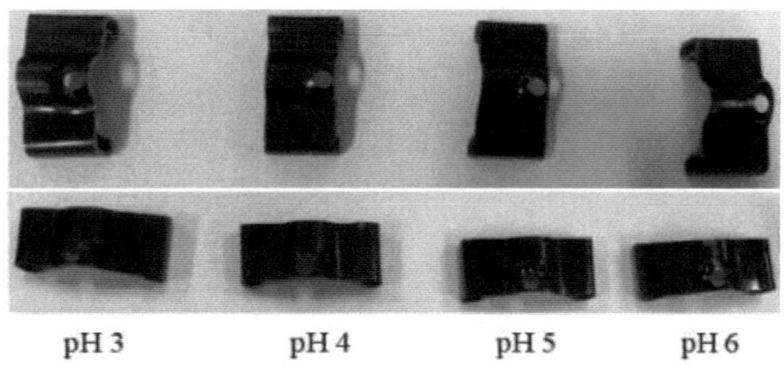

Figure 6. Corrosion test specimen of black chromium on motorcycle rims before (top side) and after (bottom side) immersed in corrosive solutions.

Testing the colour resistance of the 2 methods from the results of the black chromium electroplating shows that the heating test provides a simpler implementation and immediately results in fading degradation, while the other test for corrosion test requires long time because it awaits the process of corroding layer until the colour fades.

5. Conclusions

The conclusion of colour resistance testing of the black chromium rim are (1) The resistance of black chromium still shows a dark black by heating to 200°C for 2 hours, the rest at 250, 300, 350, 350, and 400°C experience fading successively at 20, 40, 45, 60, and 80%, and (2) Corrosion resistance decreases with increasing acidity which the colour begins to fade at pH 4 and more acid with corrosion rates respectively at 0.166, 0.136, 0.122 and 0.107 mm/year at pH 3, 4, 5, and 6 which the colour begins to fade at pH 4 and increase for more acid. For further works it is necessary to carry out hardness, scratch and compression tests of motorcycle rims so that it can be proven that the coating remains firmly attached and nothing is peeled off.

Acknowledgement

This research is supported by the State Polytechnic of Malang by the contract number SP DIPA-042.01.2.401004/2020 with Agreement Letter Number of 5198/PL2.1/HK/2020.

References

[1] Hadi S 2016 *Materials Technology* Andi Offset Yogyakarta Indonesia
[2] Anonymuous 2020a Expl Faraday's law using inexpen screen-printed elec a simple electrop exp for the general chemistry laboratory Pine Res Instr Doc #: DRL10009
[3] Snyder D L *Decorative Chromium Plating Engelhard Corp Beachwood Ohio* [online] retrieved from https://p2infohouse.org/ref/29/28168.pdf accessed on 10 July 2020
[4] Macary R and Tailor D 2019 Black Chromium Finishing-Beauty and the Beast *NASF Surf Tech White Papers* **83**(12) 1-7
[5] Karuppasamy S, Sivan V, Natarajan S, Babu S P K, Duraiselvam M, and Dhanuskodi R 2018 Est of wear resist HVOF coat *J Inst. E. India*
[6] Takadoum J 2010 Black coatings: a review *J Appl Phy E Sci* **52**(3)

[7] Oloruntoba D, Eghwubare O, and Oluwole O 2011 Effect of some process variables on nickel elec Leo *Elec J of Pract and Tech* **18** 79-94
[8] Sheu H H, Lu C E, Lee H B, Pu N W, Wu P F, Hsieh S H, and Ger M D 2016 Elect of black chrom-cobalt alloy based on triv sulfate *J of theTaiwan Inst of Chem E* **59** 496-505
[9] Surviliene S, Cesuniene A, Juskenas R, Selskiene A, Bucinskiene D, Kalinauskas P, and Jureviciute I 2014 The use of trivalent chromium bath *Appl Surf Sci* **305** 492-497
[10] Jeeva P A, Karthikeyan S, and Narayanan S 2013 Performance characteristics of corrosion resisteance black coatings *Proc Eng* **64** 491-496
[11] Jafari S and Rozati S M 2011 Characterization of black chrome films prepared by electroplating technique *Solar Thermal Appl World Renew Energy Congr Sweden*
[12] Kagajwala B, Hall T D, Inman M, Taylor E J, Griffin B, Cushnie G, Taylor R, Jaworowski M, and Bonivel J 2012 Functional trivt chrom elect *NASF SUR/FIN Conf in Nevada*
[13] Lu C E, Pu N W, Hou K H, Tseng C C, and Ger M D 2013 The effect of formic acid concentration on the conduct and corr resistance *Appl Surf Sci* **282** 544-551
[14] Danilov F I, Protsenko V S, Gordiienko V O, Kwon S C, Lee J Y, and Kim M 2011 Nanocrystalline hard chrom electrodep… *Appl Surf Sci* **257**(18) 8048-8053
[15] Liao C W, Hou K H, Wang H C, Lu C E, and Ger M D 2015 Corrosion and wear resistance of electroplating trivalent chromium-carbon coating *Key Eng Mat* **642** 45-49
[16] Protsenko V S, Danilov F I, Gordiienko V O, Baskevich A S, and Artemchuk V V 2012 Improv hard and tribo char Int'l J of Ref Met and Hard Mat **31** 281-283
[17] Ghaziof S, Golozar M A, and Raeissi K 2010 Charact of as-deposited and annealed Cr-C alloy coat produced from a trivalent chro bath *J of Allo and Comp* **496**(1-2) 164-168
[18] Medeiros I D M, Gomes K C, Gonçalves R P N, and Galvão G O 2019 Selective solar surface solar based on black chromium *Mat Res* **22**(2)
[19] Anonymuous 2020 [online] retrieved from http://digilib.polban.ac.id/files/disk1/145/jbptppolban-gdl-adrianluth-7226-3-bab2--3.pdf accessed in 10 July 2020

IOP Conf. Series: Materials Science and Engineering 1073 (2021) 012076 doi:10.1088/1757-899X/1073/1/012076

Quality by design of optimum parameter to minimize the weight of plastic products

M Hartono*, G Gumono, W Wirawan, N Qosim and H Wicaksono
Mechanical Engineering, State Polytechnic of Malang, Malang, Indonesia

*moh.hartono@polinema.ac.id

Abstract. The research aims to control product quality off-line (quality by design) in producing lightweight plastic products for food packaging. The lightweight product is very profitable for the producers due to the efficient production costs. Quality by design was done by designing controlled parameters in the plastic molding process which includes clamping force, holding time, injection pressure, injection speed and injection temperature and their interactions. The plastic material that is made observable is Polypropylene (PP). The response parameter in this study is the weight of the product. The parameter design method used was Factorial Experiment Design with 5 independent variables with different levels of variation. The results showed that the combination of injection temperature and holding time had a significant effect on product weight. To produce heavier products that are in the area above the holding time of 0.4 seconds at any injection temperature. Meanwhile, to produce a lightweight product (under 43 grams) is at a combination of injection temperature below 263 C and holding time of about 0.3 seconds. The standard weight of the product on the plain bucket food packaging for this yogurt container is 42.5 gram.

1. Introduction

Nowadays, objects made of plastic are increasingly being encountered, starting from household appliances, food wrappers, electronic equipment and others. Without realizing it, this plastic object has replaced products made of other materials such as wood, metal, bamboo, glass and so on. In general, the properties of plastics are less stiff and strong when compared to metals, however, the ratio of strength to weight and stiffness to weight ratio is better than metal [1].

According to the industry ministry, the consumption of plastic products is still quite high in various manufacturing sectors, such as the food and beverage product packaging industry, the cosmetics industry, the electronics industry and the automotive industry. Throughout 2018, the plastic industry grew by 6.92%, an increase from 2017's growth of 2.47%. The number of plastic industries currently reaches 925 companies with the ability to produce various kinds of plastic products. The total production of this sector in 2018 reached 7.23 million tons. Meanwhile, the demand for plastic products has increased by an average of 5% in the last five years [2]. One example of a demand for plastic products is for packaging food containers. But on the market there are many unhygienic food container products because they are made of non-food grade plastic material.

In an effort to maintain and improve the quality of plastic products, companies face the problem of selecting and determining the physical attributes of the products that affect quality, which are better known as quality characteristics. Recently, quality control methods have shifted to quality by design methods [3]. Initially, the quality control method was carried out by quality control which was carried out after the product was finished. Then the method is improved by conducting an inspection while the production process is in progress. In this quality control method, it is known as online quality control because it is carried out simultaneously both during the actual production process and shortly after being produced, many losses are borne by the company, including losses in scrap costs, products and materials dispose and loss of processing time. Until finally the quality control method was refined again into a quality by design method, which is an off-line quality control and implemented when the product has not been made and has not entered the actual production process, because it is still a prototype. However, there are many advantages, including the efficiency of raw material costs, the existence of a mature production process planning and anticipating all possibilities that can improve product quality [4].

Plastics are generally classified into 3 (three) types, namely: thermoplastics, thermo-setting and elastomers. Thermoplastics (thermo-plastics) are a type of plastic that will soften when heated and harden when cooled. Examples of thermoplastic materials include: polyethylene, polypropylene, and PVC (polyvinyl chloride). Thermosetting plastic will harden when heated and cannot be recycled (recycled). Examples of thermo-setting plastics are: bakelit, silicon, epoxy and others, while elastic (elastomer) is a very elastic material. an example of an elastic material is synthetic rubber [5].

Injection molding is the process of forming an object or product from a plastic material with a certain shape and size that requires a mold or mold. In principle, plastic molding is a tool used to make components from plastic materials by means of a plastic molding machine. The product forming process begins with inserting plastic material in the form of granules into the machine hopper. Then the material will enter the barrel where the barrel is hot then pushed by a screw mechanism through the engine nozzle and the sprue bushing enters the mold cavity which is already closed. After a few moments of cooling, the mold will be opened and the product will be ejected by an ejector mechanism. In the injection molding process, there are several parameters that affect the quality of injection molding products and can affect the optimal amount of production, including injection temperature, injection pressure, injection time, holding time, cooling time, injection speed, and clamping force [6].

2. Methods

In this research, there are 5 (five) parameters observed in this experiment, namely injection temperature (temperature zone-1 (C)), injection pressure (bar), injection speed (% of cm/sec), holding time (sec) and clamping force (KgF). While the response variable is product weight (grams). The aims of the experiment are make a product made from Polypropylene that is as light as possible. The product made is a plain bucket, which is a yogurt container made of plastic cans, with an outer diameter of 115 mm and an inner diameter of 112 mm and a white bottom with a diameter of 93 and 135 mm. In this experiment, the weight of the product is designed to be as light as possible in order to increase efficiency in product delivery to consumers. For this purpose, this experiment will observe the effect of each parameter of the plastic molding process which contributes to the weight of the product. In this experiment, using a factorial experimental design with 5 factors and each with 3 factor levels and 3 replications each so that the number of experimental runs was 729 runs. The production process and experiments are carried out at the Production Workshop of the Department of Mechanical Engineering, State Polytechnic of Malang, which also serves orders from outside the campus. The machine used is still Maya Jieli MJM-JLB198-HQ Plastic Injection Molding Machine and the standard of the specified product weight is 42.5 ± 0.5 cm as in Figure 1.

IOP Conf. Series: Materials Science and Engineering 1073 (2021) 012076 doi:10.1088/1757-899X/1073/1/012076

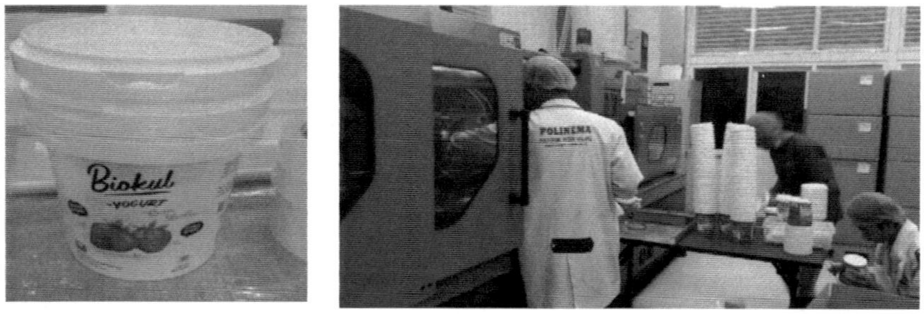

Figure 1. The product and production process for plain buckets.

3. Results and discussion

Table 1. Anova and model summary.

Source	DF	Adj SS	Adj MS	F-Value	P-Value
Model	8	349,15	43,644	770,08	0,000
Linear	4	275,70	68,924	1216,13	0,000
TEMP.ZONA 1(C)	2	29,06	14,529	256,35	0,000
HOLD.TIME(S)	2	246,64	123,320	2175,91	0,000
2-Way Interactions	4	73,46	18,364	324,02	0,000
TEMP.ZONA 1(C)*HOLD.TIME(S)	4	73,46	18,364	324,02	0,000
Error	720	40,81	0,057		
Lack-of-Fit	234	10,01	0,043	0,68	1,000
Pure Error	486	30,79	0,063		
Total	728	389,96			

Model summary

S	R-sq	R-sq(adj)	R-sq(pred)
0,238066	89,54%	89,42%	89,27%

From Tables 1 and 2, it turns out that of the 5 independent variable parameters observed, there are only 2 independent variables that have a significant effect on product weight. The two variables are injection temperature (Zone-1) and holding time (Holding time). Likewise, the interaction or combination of injection temperature and holding time also has a significant effect on product weight. It can be seen from the P-value whose value is <□ of 0.05. From the summary model (Model Summary), it is found that the coefficient of determination is 89.54%. This value shows that the model of the influence of the variable injection temperature, injection pressure, injection speed, holding time and clamping force has an effect on the variation of product weight by 89.54%. While the remaining 10.46% is influenced by other factors that have not been observed and experimental errors.

Table 2 below shows the regression equation coefficients generated from the RSM Box Behnken method. To display the coefficient of this regression equation using the Stepwise method so that only those shown in the table have a significant effect (P-value <0.05). The choice of the Stepwise method is to be efficient in displaying tables.

Table 2. The regression equation model influence of independent variables (injection temperature, injection pressure, injection speed, holding time and clamping force) on product weight variables.

Weight (g) =	43,6556 - 0,1770 TEMP.ZONA 1(C)_260 - 0,1021 TEMP.ZONA 1(C)_265
	+ 0,2790 TEMP.ZONA 1(C)_270 - 0,7597 HOLD.TIME(S)_0,3 + 0,1066 HOLD.TIME(S)_0,4
	+ 0,6531 HOLD.TIME(S)_0,5 - 0,5239 TEMP.ZONA 1(C)*HOLD.TIME(S)_260 0,3
	+ 0,2444 TEMP.ZONA 1(C)*HOLD.TIME(S)_260 0,4 + 0,2794 TEMP.ZONA 1(C)*HOLD.TIME(S)_260 0,5
	- 0,0481 TEMP.ZONA 1(C)*HOLD.TIME(S)_265 0,3 + 0,0313 TEMP.ZONA 1(C)*HOLD.TIME(S)_265 0,4
	+ 0,0169 TEMP.ZONA 1(C)*HOLD.TIME(S)_265 0,5 + 0,5720 TEMP.ZONA 1(C)*HOLD.TIME(S)_270 0,3
	- 0,2757 TEMP.ZONA 1(C)*HOLD.TIME(S)_270 0,4 - 0,2963 TEMP.ZONA 1(C)*HOLD.TIME(S)_270 0,5

The above equation is used to make a prediction or weight estimation model of the product using the parameters of the equation above. To make predictions, the following parameters are converted:
Factor level: Injection Temperature 260,265 and 270 are replaced by level -1, 0 and 1
Holding time 0,3, 0,4, and 0,5 are replaced by level -1, 0 and 1
So that if the regression equation above is replaced with the following numbers:

```
Weight = 43,6556 - 0,1770*(-1) - 0,1021*(0) + 0,270 * (1) - 0, 7597*(-1)
        + 0,1066*(0)+0,6531*(1) - 0,5239*(-1) * (-1)+ 0,2444*(-1)*(0)
        + 0,2794*(-1)*(1) - 0,0481*(0)*(-1) + 0,0313 *(0)*(0)
        + 0,0169 *(0)*(1)+ 0,5720 *(1)*(-1) - 0,2757*(1)*(0)
        - 0,2963*(1)*(1)
```

Weight = 43,8528 gram
It appears that the prediction results are still close to the expected weight.

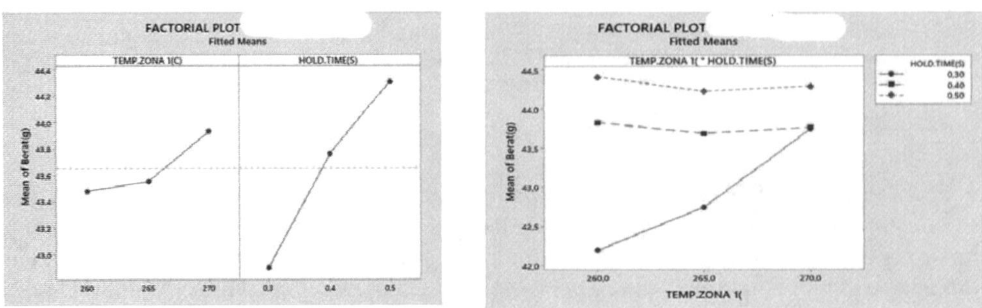

Figure 2. Main effect plot and interaction plot of injection temperature and holding time against product weight.

In accordance with Table 2, the graph in Figure 2 reinforces the significant effect of injection temperature and holding time on product weight. In this experiment, the increase in injection temperature and holding time can increase the weight of the product

In the graph above, it also can be seen that the combination of injection temperature and holding time has a significant effect on product weight, especially the combination of level 1 (temperature 270) injection temperature and level -1 holding time 03 which gives an increase in product weight. This is also in accordance with the magnitude of the regression coefficient above with a contribution of 0.5720 which is the largest contribution than the other coefficients. In the graph above, it can also be shown that the holding time level 0 and level 1 (0.4 and 0.5) has a greater effect on product weight. However, the combination with the increase in injection temperature is not very significant.

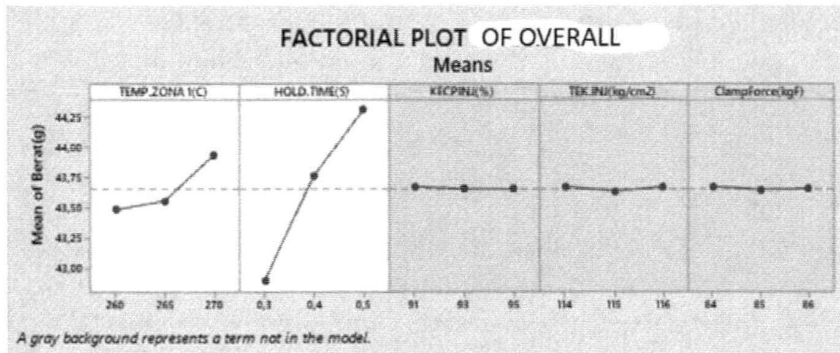

Figure 3. Effect of other factors on product weight.

In the graph above, it can be seen that for the factors of velocity, injection pressure and clamping force the three of them do not have a significant effect on product weight.

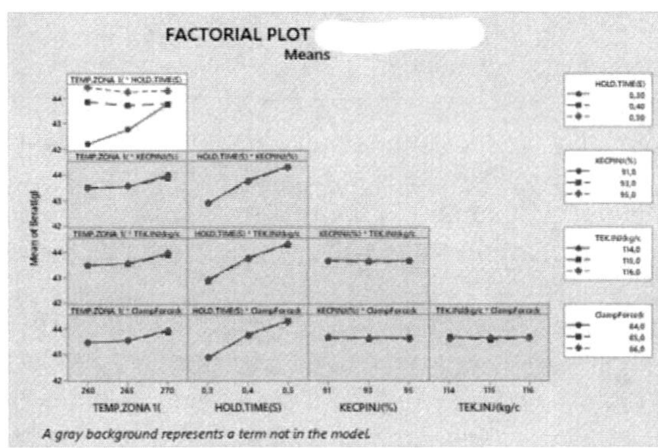

Figure 4. Effect of various interactions of independent variables on product weight.

From the graph above, it can be seen that the interaction of various factors that shows a significant effect is only those related to the holding time and injection temperature. For a combination involving injection temperature and holding time, it always results in an increase in product weight, from insignificant to significant increases.

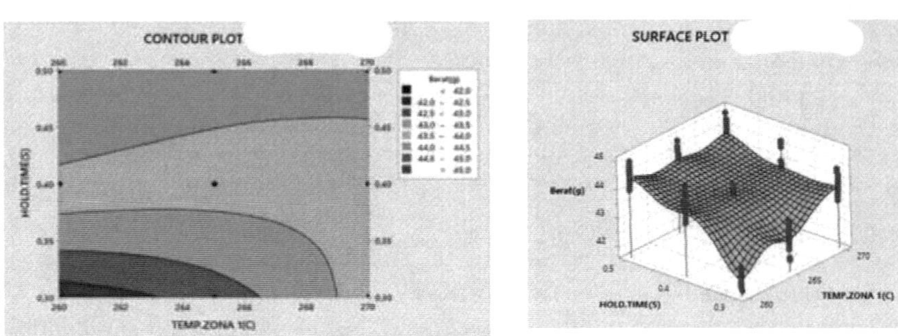

Figure 5. Contour plot and surface plot for the effect of injection and holding temperature time to product weight.

Figure 5 shows a contour plot for the combination of injection temperature and holding time for product weight. To produce a heavier product be in the area above the holding time of 0.4 seconds at any injection temperature. Meanwhile, to produce a light product (under 42 grams), it is at a combination of injection temperature below 263 C and holding time of about 0.3 seconds.

In Figure above, we also can see the surface plot for the combination of injection temperature and holding time for product weight. To produce a heavier product be in the area above the holding time of 0.4 seconds at any injection temperature. Meanwhile, to produce a light product (under 42 grams), it is at a combination of injection temperature below 263 C and holding time of about 0.3 seconds.

The results of this study are consistent with what has been done by Oktem et al [7] who have examined the plastic molding process for producing thin products. From the results of this research, it turns out that the holding time factor is also a determining factor in producing a thin product. Also the injection pressure factor contributes to the production of quality product characteristics for these thin dimensions. The difference with this study is in the aspect of injection temperature. In this research, the target is to produce a lightweight product so that the parameters that determine are the holding time and injection temperature parameters.

The results of this study are relevant to the research conducted by Fitriyanto et al [8] with the title "Analysis of PS 135 Rearview Products with Injection Time Parameters for Polypropylene Plastic Material (PP) in the Injection Molding Process". In this study, using the black PP material type Globalene 7533 with injection time independent variables and various variations. And the dependent variable is a mold temperature of 50 ° C, a melt temperature of 230 ° C, a maximum machine injection pressure of 85 MPa, a machine clamp open time of 7 s. With the dependent variable shot volume, total part weight, confidence of fill, and quality prediction. The results of this study indicate that the increase in injection time variation causes an increase in the total part weight. The increase in total part weight is due to the long injection time duration so that the plastic molecules injected into the mold become solid so that the product weight increases [8].

In another similar research conducted by Hassan [9] with the title "An Experimental Work on The Effect of Injection Molding Parameters on The Cavity Pressure and Product Weight". This study used 2 types of materials, namely LDPE and PS. And there are 4 independent variables in this study, namely, cooling time, injection pressure, packing pressure, and injection temperature. With 2 dependent variables, namely cavity pressure and product weight. The results of this study indicate that injection pressure and injection temperature have a significant effect on product weight compared to the other two independent variables [9].

Research on plastics regarding product weight is closely related to product defects because improper composition of independent variables can cause product defects. At low temperatures there will be a short shoot defect and too high a temperature there will be a flash defect. This is evidenced by previous research, namely [10]. On the independent variable holding time has a significant effect. Significant results are obtained because the longer the time to inject the plain bucket packaging product will increase the product weight and reduce shrinkage product defects because the product is stronger and denser. So, defects in packaged products can be minimized. This is evidenced by Wahyudi [11]. So it can be explained that the higher the injection temperature, the decrease in the weight of the plain bucket product and the increasing the holding time, the product weight increases.

In addition, the longer the holding time, the denser the plastic molecules that are injected into the cavity, this happens because the air trapped in the cavity is small due to the large injection pressure during the product manufacturing process. And if the holding time is shorter, the density of plastic molecules injected into the cavity will also decrease. This happens because there is air trapped in the cavity which is caused by too short a holding time so that there is still air trapped causing product defects.

4. Conclusion

To be able to minimize the product weight (bucket plain) which is also a quality product criterion, the significant parameter settings are holding time and injection temperature. The longer the holding time, the heavier the product. Likewise, the injection temperature. For food packaging products (bucket plain), the holding time is set at 0.3 seconds and the injection temperature is at 260-262.5 C.

Acknowledgment
The author would like to thank P2M State Polytechnic of Malang who has commissioned research on this plastic experiment study to the researcher and the team. Thanks to all team members who helped carry out this research.

References
[1] Bryce D M 1996 *Plastic injection molding: manufacturing process fundamentals* (Society of Manufacturing Engineers)
[2] Departemen Perindustrian dan Perdagangan 2018 *Wajah Industri Plastik Indonesia tahun 2018*
[3] Hartono M and Irianto D 2000 Pengembangan Response Surface Methodology pada Karakteristrik Dinamis untuk Meningkatkan Kualitas Produk *Jurnal Teknik dan Managemen Industri ITB* **2** 2
[4] Myers R H, Khuri A I and Vining G G 1992 Response Surface Alternatives to the Taguchi Robust Parameter Design Approarch *The American Statistician* **46** pp 131-139
[5] Schey J A 2009 2009 *Proses Manufaktur: Introduction to Manufacturing Processes* (Rines, Ed. Andi)
[6] Hartono M and Pratikto P B S 2020 Optimization on the Injection Molding Propypopylene Parameters Using Central Composite Design for Minimizing Defects *Journal of Southwest Jiaotong University* **55** 2
[7] Oktem H, Erzurumlu T and Uzman I 2007 Application of Taguchi optimization technique in determining plastic injection molding process parameters for a thin-shell part *Materials & design* **28** 4 1271-1278
[8] Fitriyanto S 2013 *Analisis Produk Spion PS 135 Dengan Pengaturan Parameter Injection Time Material Plastik Polypropylene (PP) Pada Proses Injection Molding (Studi Kasus PT. Sinar Agung Selalu Sukses Karanganyar Jawa Tengah)*
[9] Hamdy H 2012 An Experimental Work on The Effect of Injection Molding Paramters on The Cavity Pressure and Product Weight *International Journal of Advance Manufacture Technology* **67** 675-686
[10] Yanto H, Saputra I and Satoto S W 2018 Analisa Pengaruh Temperatur dan Tekanan Injeksi Moulding terhadap Cacat Produk *Jurnal Integrasi* **10** 1 pp 1-6
[11] Wahyudi U 2015 Pengaruh Injection TIME dan Backpressure terhadap Cacat Penyusutan pada Produk Kemasan Toples dengan Injection Molding Menggunakan Material Polistyrene *Jurnal Teknik Mesin Mercu Buana* **4** 3 81-90

IOP Conf. Series: Materials Science and Engineering 1073 (2021) 012077 doi:10.1088/1757-899X/1073/1/012077

The effect of variation of extra fan condenser and engine speed to COP of mobile air conditioners

N Nurhadi*, L Listiyono, M Maskuri and B R Prasetyo

Department of Mechanical Engineering, State Polytechnic of Malang, Malang, Indonesia

*nurhadi@polinema.ac.id

Abstract. Efforts to increase coefficient of performance (COP) of Mobile Air Conditioners (MAC) still continue to be done, including by increasing the extra fan condenser speed. The problem is whether it can improve the COP, it needs to be tested further. This study focuses to examine the effect of extra fan condenser and engine speed variations to COP. This study was conducted on a MAC engine stand which able to measure the pressure and temperature of MAC system. The data measured at inlet compressor (P1, T1), outlet compressor (P2, T2), outlet condenser (P3, T3), and outlet expansion valve (P4, T4), at engine speed of 1.000, 1.500, 2.000, 2.500 and 3.000 rotation per minute (RPM), respectively with variations extra fan condenser speed of 0.15, 1.74, and 2.22 m/sec. The data was plotted on p-h diagram to obtain enthalpy value, so that obtained COP value by calculation. The major result obtained are the variations of extra fan condenser and engine speed had an effect on the COP. The highest COP (4.48) was achieved at an engine speed of 2,000 RPM with extra fan condenser speed 2.22 m/sec. Increasing extra fan condenser speed causes better cooling at condenser and cooling effect on the evaporator.

1. Introduction

The problem of Mobile Air Conditioners (MAC) that is not cold or not cold at all can be caused by many factors, one of which is the poor cooling performance of the condenser. For example, the condenser is dirty, the extra fan condenser rotates slowly or not at all. To overcome this, many people make modifications by changing the speed of the extra fan condenser in the hope that it can reduce the temperature of the mobile cabin so that it is cooler and more comfortable or the performance increases. The cooling performance of MAC is defined as Coefficient of Performance (COP). COP air conditioning system is the effect of refrigeration divided by compression work [1].

Efforts to increase COP of Mobile Air Conditioners still continue to be done, for example using condensed water [2], increasing the extra fan condenser speed [3]. Whether the change in the speed of the extra condenser fan is really able to increase the cooling effect on the evaporator, needs to be tested further.

Therefore, in this study defined the problem formulation as how is the effect of the extra fan condenser speed variations and engine speed on the COP MAC? The purpose of study to examine the effect of variations in the extra fan condenser and engine speed to COP of MAC.

2. Method

The study was conducted on a modified engine stand of MAC and was able to display the results of measurements of the pressure and temperature of the MAC system, as shown at **Figure 1-2.**

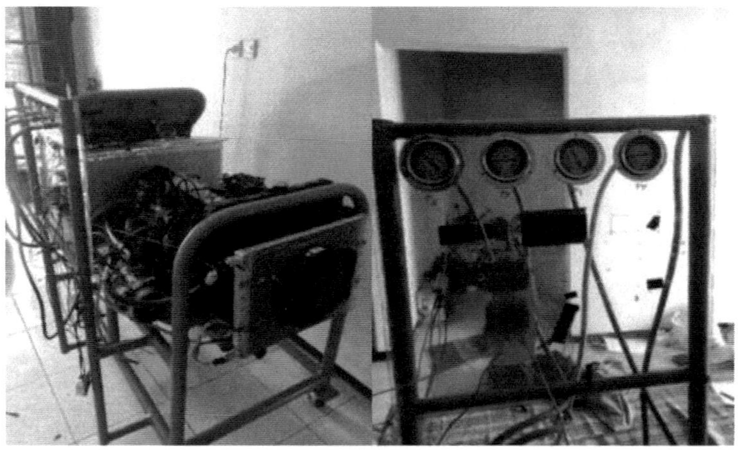

Figure 1. Modified engine stand of MAC.

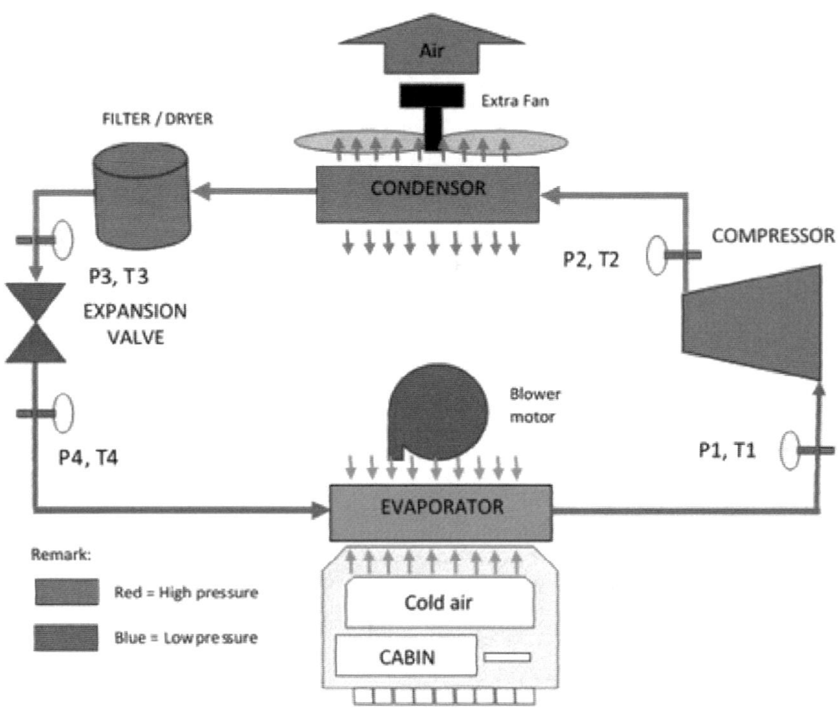

Figure 2. The schematic diagram of MAC [4].

Data is collected by measuring the pressure and temperature at inlet compressor (P1, T1), outlet compressor (P2, T2), outlet condenser (P3, T3), and outlet the expansion valve (inlet evaporator) (P4, T4), at engine speed of 1,000, 1,500, 2,000 2,500 and 3,000 RPM, respectively with variations in the condenser extra fan speed of the 0.15; 1.74; and 2.22 m/sec. The data obtained is then plotted on the ph diagram to obtain the enthalpy value at each point of pressure and temperature, so that the COP value can be calculated, according to equation 1 [4].

$$COP = \frac{h1-h3}{h2-h1} = \frac{h1-h3}{h2-h1} \quad \ldots\ldots (1)$$

2.1. Variable

In this study, there are two variables, including independent variable and dependent variable The Independent variables consist of variations in extra fan condenser speed (0.15; 1.74; 2.22 m/s) and variations in engine speed (1.000, 1.500, 2.000, 2.500, and 3.000 RPM). The dependent variable is COP of MAC

2.2. Hypothesis

Hypothesis in this study defined as:
H0: There is no effect between the extra fan condenser speed and engine speed to the COP of MAC
H1: There is an effect between the extra fan condenser speed and engine speed to the COP of MAC

2.3. Data analysis

The COP data that has been obtained is then graphed on the relationship between the variations of extra fan condenser speed and engine speed to the COP. Data analysis used SPSS software with correlation and regression test to test the effect of variations in the speed of extra fan condenser speed and engine speed to COP of MAC [5].

3. Results and discussion

3.1. Data

The data obtained from the results of this study are in the form of pressure and temperature at various points, various engine speeds and various extra speed of the condenser fan, which are written with notations (P1, P2, P3, P4) for pressure, and (T1, T2, T3, T4) for temperature, as Table 1.

The pressure and temperature data that have been obtained are then plotted on the p-H diagram to obtain the enthalpy value. The enthalpy value plotting the p-h diagram is then used to calculate the COP value based on COP formula on equation 1.

Table 1. The data of pressure, temperature, enthalpy and COP at various points, various engine speeds and various extra speed of the condenser fan.

Engine speed (RPM)	Extra Fan speed m/s	P1 (bar)	P2 (bar)	P3 (bar)	P4 (bar)	T1 (°C)	T2 (°C)	T3 (°C)	T4 (°C)	h1 (kj/kg)	h2 (kj/kg)	h3 (kj/kg)	h4 (kj/kg)	COP
1.000	0.15	3,63	13,00	13,00	2,50	10,50	85,25	45,45	9,07	400	470	267	267	1,90
	1.74	3,23	10,50	10,50	2,40	8,75	83,20	43,36	7,13	400	469	256	256	2,09
	2.22	3,37	12,75	12,75	2,40	9,25	80,85	39,55	8,28	401	464	257	257	2,29
1.500	0.15	3,10	11,50	11,50	2,10	8,15	81,20	43,40	7,43	402	466	264	264	2,16
	1.74	3,17	12,13	12,13	2,07	7,53	82,30	42,75	7,07	404	468	260	260	2,25
	2.22	3,30	11,20	11,20	2,40	9,65	82,45	39,57	8,58	405	465	250	250	2,58
2.000	0.15	3,30	14,50	14,50	2,13	9,80	70,15	36,75	6,70	406	452	246	246	3,48
	1.74	3,00	13,00	13,00	1,97	9,87	70,15	37,95	5,72	407	451	233	233	3,95
	2.22	3,13	11,80	11,80	2,13	10,05	61,35	43,50	8,10	408	441	260	260	4,48
2.500	0.15	3,00	14,20	14,20	1,80	7,55	80,15	44,25	6,75	403	460	269	269	2,35
	1.74	3,00	13,90	13,90	2,03	8,12	73,25	46,72	6,58	404	456	266	266	2,65
	2.22	3,10	12,20	12,20	1,90	8,35	68,85	44,20	6,20	405	450	260	260	3,22
3.000	0.15	3,23	13,50	13,50	1,57	6,85	82,25	52,15	7,72	400	464	275	275	1,95
	1.74	3,03	12,15	12,15	2,00	7,15	83,15	44,35	4,55	402	465	260	260	2,25
	2.22	2,73	10,78	10,70	1,50	7,45	78,25	37,50	6,98	404	460	256	256	2,64

Remarks: P1: inlet compressor pressure, P2: outlet compressor pressure, P3: outlet condenser pressure, P4: outlet expansion valve pressure, T1: inlet compressor temperature, T2: outlet compressor temperature, T3: outlet condenser temperature, T4: outlet expansion valve temperature, h1: outlet evaporator enthalpy, h2: inlet condenser enthalpy, h3: outlet condenser enthalpy, h4: inlet evaporator enthalpy, COP: Coefficient of Performance, kj/kg = kilo joule/kilo gram.

Furthermore, COP data at various of extra fan condenser speeds and various engine speeds are shown at **Figure 3.**

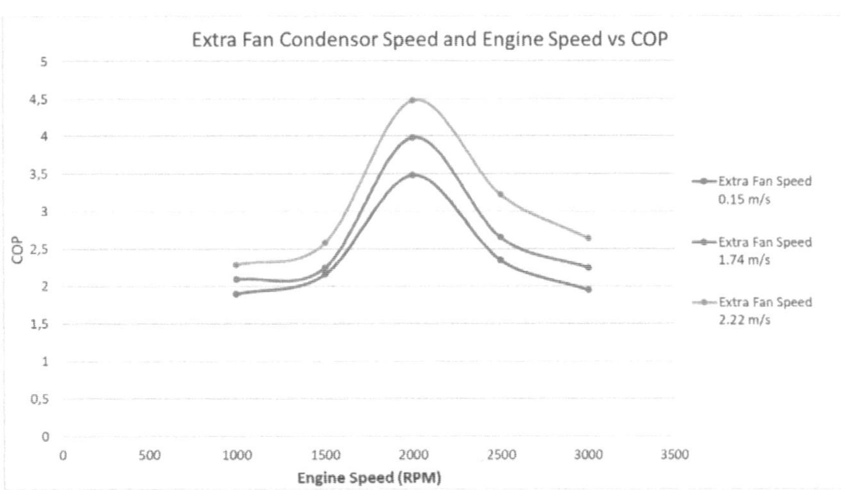

Figure 3. Graphic of COP at various extra fan condenser speed and engine speed.

IOP Conf. Series: Materials Science and Engineering 1073 (2021) 012077 doi:10.1088/1757-899X/1073/1/012077

3.2. Discussion

In the graph as shown in **Figure 3**, it can be seen that at the extra fan condenser speed of 0.15 m/s, the COP value is 1.90 at engine speed 1,000 RPM; 2.16 at engine speed 1,500 RPM; 3.48 at engine speed 2.000 RPM; 2.35 at engine speed 2.500 RPM; and 1.95 at engine speed 3.000 RPM. At engine speed 1.000 RPM, the COP value is still low, and at the higher engine speed (2000 RPM), the COP value was increase, and then decrease again after the engine speed is above 2.000 RPM.

At the extra fan condenser speed 1.74 m/s, COP value is 2.09, at engine speed 1,000 RPM; 2.25 at engine speed 1,500 RPM; 3.95 at engine speed 2.000 RPM; 2.65 at engine speed 2.500 RPM; and 2.25 at engine speed 3.000 RPM. At the engine speed 1.000 RPM, the COP value is still low, and at the higher speed (2.000 RPM) the COP value was increase, and decrease again after the speed is above 2.000 RPM.

At the extra fan condenser speed 2.22 m/s, the COP value is 2.29, at the engine speed 1,000 RPM; 2.58 at engine speed 1,500 RPM; 4.48 at engine speed 2.000 RPM; 3.22 at engine speed 2.500 RPM; and 2.64 at engine speed 3.000 RPM; At the engine speed 1.000 RPM, the COP value is still low, and at the higher speed (2.000 RPM) the COP value was increase, and decrease again after the speed is above 2.000 RPM.

At low engine speed (1,000 RPM) the COP value is still low, and the COP value increases as the engine speed increases and decreases again after 2,000 RPM. This happens because on the higher engine speed, the compressor shaft speed is higher, so the refrigerant pressure increases and the COP value is also increase. However, at the engine speed above 2.000 RPM, even though the compressor shaft speed is higher, the compressor is quite hot, so the refrigeration effect decreases, and the COP value was also decrease.

The COP value at the extra fan speed 2.22 m/s is 4.48, better than at the extra fan speed 0.15 m/s and 1.74 m / s (3.48 and 3.95) as shown in Figure 3. The lowest COP value at the extra fan speed 0.15 m/s and engine speed 2.000 RPM is 3.48, and the highest COP value at an extra fan speed 2.22 m/s and engine speed 2.000 RPM is 4.48, because the higher engine speed (above 2000 RPM) the refrigerant flow rate was decreases. Increasing the extra fan speed on the condenser causes an increasing the refrigeration effect, so the COP was increase also. This result is related to study of Effendy [6] and Tjahjono [7]. If the speed of the extra fan increases, the COP was reached to the optimum certain conditions (2,000 RPM). At engine speed 1.000 RPM, the COP value is not optimal yet, because at engine speed 1.000 RPM the MAC just started working. The MAC compressor already works, but is not optimal because the engine requires a lot of power to be able to move the compressor. Therefore, it can be interpreted that the load from the engine is too large at engine speed 1.000 RPM. At the engine speed 2.000 RPM the function of MAC components is optimal because the compressor speed flow off the refrigerant in accordance the needed of other components, and if the engine speed is increased, the MAC performance will decrease. In the engine speed above 2.000 RPM the COP value decreases again because at that speed, the compressor is quite hot, so that the refrigerant pressure is increases and the refrigeration effect is decreases, so the COP value also decreases.

4. Conclusion

The extra fan condenser speed variation greatly affects the COP of the Mobile Air Conditioner. At the extra fan condenser speed 0.15 m/s, the highest COP value 3.48 is achieved at 2,000 RPM engine speed; at the extra fan condenser speed 1.74 m/s, the highest COP value 3.95 is achieved at 2,000 RPM engine speed; at the extra fan condenser speed 2.22 m/s, the highest COP value 4.48 is achieved at 2,000 RPM engine speed. The increasing of the the extra fan condenser speed causes an increasing the refrigeration effect, so that the COP increases. If the extra fan condenser speed increases, the COP will reach the optimum in certain conditions (2,000 RPM engine speed). At 2.000 RPM engine speed, the performance of Mobile Air Conditioner components is optimal because the compressor is able to flow refrigerant as needed. For further action in order to obtain the best COP value on mobile air conditioner, the extra fan condenser must be ensured that it can rotate at maximum rotation and engine speed is around 2000 RPM.

Acknowledgement
This study was carried out within the statutory fund of the UPT P2M State Polytechnic of Malang.

References
[1] Stoecker W F and Jones J W 1982 *Refrigeration and Air Conditioning* (Jakarta: Erlangga)
[2] Candela L S and Grummy W 2014 Increased COP (Coefficient of Performance) Mobile Air Conditioning System Using Condensed Water *Journal of Mechanical Engineering* **02** 02 162-171
[3] Poernomo H 2015 Analysis of the Performance Characteristics of a Cooling System (Air Conditioning) Using Freon R-22 Based on Variations of Condenser Cooling Fan Rotations *Ship's Journal* **12** 1 1
[4] Marwani M 2010 Effect of Changes in Condenser Fan Rotations on Performance of Air Conditioning Machine *Proceeding of 9th Annual National Seminar on Mechanical Engineering (SNTTM) Palembang* 541-544
[5] Sugiyono S 2010 *Quantitative Research Methods, Qualitative and R & D* (Bandung: Alfabeta)
[6] Effendy M 2005 The Effect of Condenser Cooling Speed on Coefficient of Performance of Air Conditioning *Engineering Journal GELAGAR* **16** 01 51–58
[7] Tjahjono T and Arna D 2014 Optimization of air velocity in the condenser on the performance of mobile air conditioners with Freon 12 working fluid *Journal of MEDIA MESIN* **15** 2 55-67

IOP Conf. Series: Materials Science and Engineering 1073 (2021) 012078 doi:10.1088/1757-899X/1073/1/012078

Control of rotation speed using angle change or effective cross-sectional area of blades on wind turbines

Y Yulianto*, A Pracoyo and B Priyadi

Department of Electrical Engineering, Malang State Polytechnic, Malang, Indonesia

*yulianto_poltek@yahoo.com

Abstract. To cope with wind changes when the speed is excessive, the turbine needs to be equipped with a speed regulator. To control the turbine rotational speed can be done by changing the variable angle or blade cross-sectional area. Changes in blade angle are automatically followed by changes in blade cross-sectional area. To get a more stable rotational speed, the cross-sectional area of the blades is changed inversely with the wind speed. Experiments were carried out to obtain an effective blade cross-sectional area, in order to obtain a characteristic curve of the change in angle or the effective cross-sectional area of the blade when wind speed changes. Based on the results of this test, a wind turbine prototype has been designed using 3 blades with a maximum area of 0.22 m2 each. The change in angle is automatically carried out by adding a centrifugal pendulum at a position half the radius of the blade length. Control of rotational speed using a centrifugal pendulum is quite effective as a control for wind turbine rotational speed. Although the rotating speed is not very constant, it shows better stability.

1. Introduction

Wind speed changes all the time. These changes are not only directional but form a profile due to the stability of the atmosphere. Wind energy with changes in speed and direction when used to rotate wind turbines can also provide various results for the turbine rotational speed [1]. If a fluctuating wind turbine rotational speed is not preferred, a control device is needed to maintain the wind turbine rotational speed so that its rotation is more stable [2]. The variables that affect the wind turbine rotational speed are the wind speed and the effective cross-sectional area of the blades. The purpose of the effective cross-sectional area is the cross-sectional area of the blade which is hit by the wind vertically. One option for controlling wind turbine rotational speed is to change the effective cross-sectional area of the blades. If the wind speed is too large, the effective cross-sectional area of the blade must be reduced, and vice versa. While the effective cross-sectional area of the blade depends on the angle of the blade to the wind direction [3,4]. One of the easiest ways to change the effective cross-sectional area of the blade is to change the blade angle or sweep area [5] to the wind direction. Changing the blade angle can be done with an active or passive control system. Active control system is a control system that uses electric or pneumatic energy. But in this paper, it is focused using the centrifugal force of the pendulum as a blade angle controller which is classified as a passive control system. A speed control model utilizing centrifugal force has been designed in Chena et al [6]. The paper also explains the success of a system at high wind speeds and has good performance at low wind speeds.

2. Literature review

2.1. *Wind energy*

There are three renewable energies that are very easy and cheap to obtain, namely: (1) solar energy, (2) wind energy, and (3) water potential energy. Besides, there are several other energy sources that are more expensive to use, such as geothermal energy, sea wave energy, and others. The required wind speed is at least in the range of 5.5 m/s and in reality very few areas have continuous winds. But in coastal areas or areas at high altitude, there are quite a lot of winds with a more stable continuity and speed [7]. The advantage of wind energy is that the potential for wind is sufficient to provide a supply of its energy needs. The disadvantage of using wind energy is the need to use moving turbines so that intensive maintenance is required.

The uneven heating of the earth's surface triggers winds. Warmer air will rise and its position is replaced by cooler air, this movement of air flow is called wind. If examined in more detail, it turns out that the direction of this wind flow is very uneven, moving turbulence, there are ripples and eddies moving in directions. However, the wind flow still has an average flow velocity in the horizontal direction which is the dominant wind direction that occurs which can be used as a source of energy to drive wind turbines. The irregular wind direction causes the wind energy to be not entirely potential to generate energy to rotate the wind turbine, and for the calculation it is necessary to assume a constant known as the power coefficient. The power extractable curve at a particular wind speed has been tested [8]. The energy obtained from wind power is the kinetic energy. If A_{ef} represents the swept cross-sectional area and V represents the wind speed, then the available power (P) is expressed by equation (1) [7]. If l represents the length of the blade and r represents the length of the blade arm, the swept cross-sectional area can be expressed as in equation 2 [3,7,9].

$$P = \frac{1}{2}\rho A_{ef} V^3 \tag{1}$$

$$A_{ef} = \pi l(l + 2r) \tag{2}$$

In converting the kinetic energy from the wind into mechanical energy, it is greatly influenced by the ability of the blades to capture wind energy. The ability of the blades to capture wind energy from the available resources is called the power coefficient which is closely related to the efficiency of the turbine, as stated in equation 3 [3,10].

$$C_p = \frac{P_{mec}}{P_{Wind}} \tag{3}$$

To control the rotational speed can be done by lowering the C_p value or reducing the efficiency of the wind turbine when the availability of wind energy is excessive. This reduction in efficiency must pay attention to the availability of excessive wind energy. One way to lower the turbine coefficient is to provide interference by turning the blades at a smaller angle. The conversion of energy at the turbine shaft will have high efficiency if the entire surface of the blade is subjected to maximum pressure from the wind. The rotational speed and blade angle greatly affect the formation of the relative speed of the wind so that they also affect the value of turbine efficiency. Error in blade angle or rotation of the turbine too fast can reduce the cross-sectional area of the blades that are gusted by wind. Wind turbine efficiency can be measured through the tip speed ratio (λ) [3,11].

$$\lambda = \frac{\omega r}{v_w} \tag{4}$$

2.2. *Centrifugal force on the pendulum in energy*

A three-blade turbine model is selected, each 120° mechanical apart, on one side of the blade with a certain distance from the blade shaft (in the middle along the blade) a pendulum is installed as shown in Figure 1. One side of the blade is held using a stopper and the other side is held using a spring, then the pendulum is fixed at a certain distance from the turbine shaft on the rear side equipped with an arm to determine the center of gravity of the blade in rotating conditions.

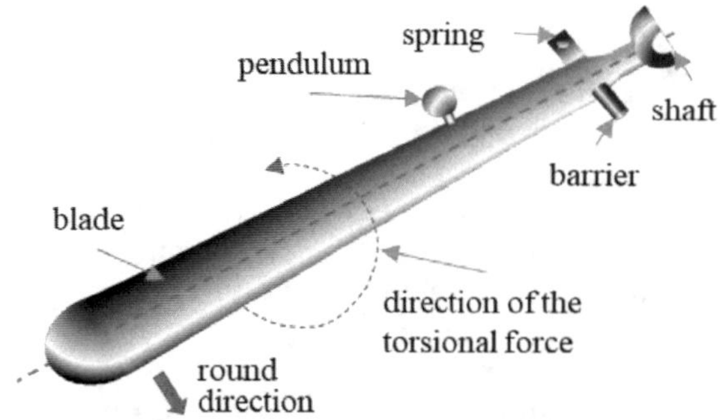

Figure 1. Mounting the pendulum, stopper, and spring on the blade.

The choice of distance between the pendulum and the rotating axis affects the mass of the pendulum, but does not affect the need for spring strength. For ease of implementation, the spring is installed near the rotary axis. Meanwhile, to obtain a larger centrifugal force with a smaller pendulum size, it is installed some distance from the rotary axis. The installation of the pendulum farther from the rotating axis causes the blade to be subjected to more torsional force and may result in damage. The mass and the mounting distance of the pendulum must be the same in order to obtain a steady balance. It should be done to measure the balance of all blades and pendulums in the attached condition. Figure 2 shows the concept of analyzing the force generated by a pendulum in a wind turbine.

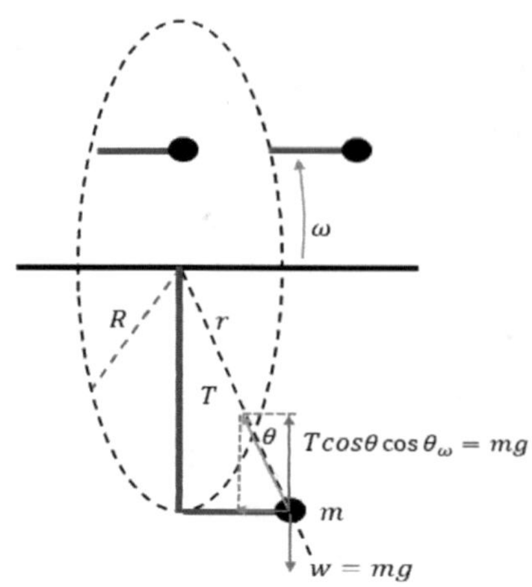

Figure 2. Pendulum centrifugal force on wind turbine blades.

The motion of the pendulum is a vertical circle with radius R and angular velocity ω. The amount of torque caused by the mass of the pendulum is $W=mg$. By referring to Figure 2, the force generated by the pendulum mass can be reflected towards the center of rotation. Then, θ_ω starts from 0° to 360°, therefore the value moves from +1 to -1. And the force towards r that is generated because of the rotation. And the total force held towards the center of the axis is:

$$T_w = \frac{m\omega^2 R}{\sin\theta} + \frac{mg}{\cos\theta \cos\theta_\omega} \quad (5)$$

From equation 3, it appears that the magnitude of the force towards the center of the axis changes along its rotation. But in practice three pendulums are used with 120° separate installations, so that the resultant becomes zero. With the effect of angle θ_ω the resultant total gravity caused by the pendulum is always equal to zero, so that it still provides balance to each blade. The torsional force that occurs on each pure blade only remains the centrifugal force of:

$$T_W = \frac{m\omega^2 R}{\sin\theta} \tag{6}$$

The force of the spring $T_k = k_s\, x, = k\, L\, sin\ \theta_s$, k_s is the spring constant and x is the linear torsion distance, L is half the width of the blade and θ_s is the angle of torsion. In equilibrium conditions, for θ=θ_s, then:

$$\theta = sin^{-1}\sqrt{\frac{m\omega^2 R}{k_s}} \tag{7}$$

$$\theta = sin^{-1}\omega\sqrt{k} \quad \text{for } 0<\theta<90^o \tag{8}$$

If the amount of torsion angle on the blade is expected to be 15° at a rotating speed of 60 rpm or ω = 6.28, for radius (R) the placement of the pendulum is 40 cm from the center axis, and the blade width (L) is 20 cm then θ = *tan-1* (0.5 x 20/40) = 14° and the mass required is 3.4 k (grams).

3. Methodology and data

3.1. Methodology

A wind tunnel was constructed equipped with an anemometer and a wind turbine to be tested [11]. The anemometer in the form of the windmill and wind turbine being tested is equipped with sensors to record the rotational speed. Each rotating speed is displayed for direct observation. The sensor is also connected to a computer equipped with the PLX-DAQ Excel application program to record data. There are two types of wind turbines tested. The first is a windmill with four blades whose blade angle can be adjusted manually and the second is a wind turbine equipped with a pendulum. Tests are carried out several times at different angles to obtain the characteristics of the blades against changes in wind speed. The wind tunnel which has been installed with a wind speed measuring device and test equipment is operated at a certain speed, and data is taken every 2 seconds for 30 minutes. The data from the wind speed gauge are added up so that they are proportional to the area of the curve. Then the sums are compared. By collecting data, equation 4 can be modified into a comparison between the wind turbine rotational speed and the reference turbine rotational speed. The test result data that has been processed in the form of a curve is shown in Figure 3. From Figure 3 it can be seen that the greater the blade angle, the greater the wind energy extract. Of course this does not happen at a 90° blade angle, this incident is described in Figure 4.

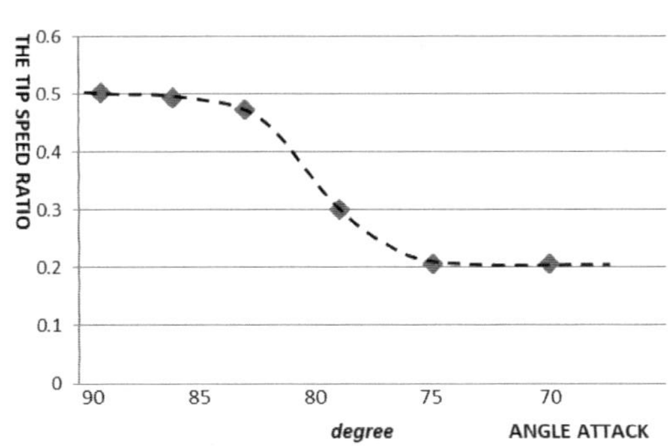

Figure 3. Comparison of turbine rotation speed to reference windmill speed.

For Figure 4, the value of this data is also shown in the curve obtained in terms of the duration and angle of a certain blade added and compared between the two. Measurements are made in constant wind speed and natural wind speed (not constant). The purpose of this treatment is to avoid errors in the anemometer measuring instrument, as well as to anticipate changes in wind which are always unstable. The comparison results of the measurement results are shown in Figure 4 in the form of a curve.

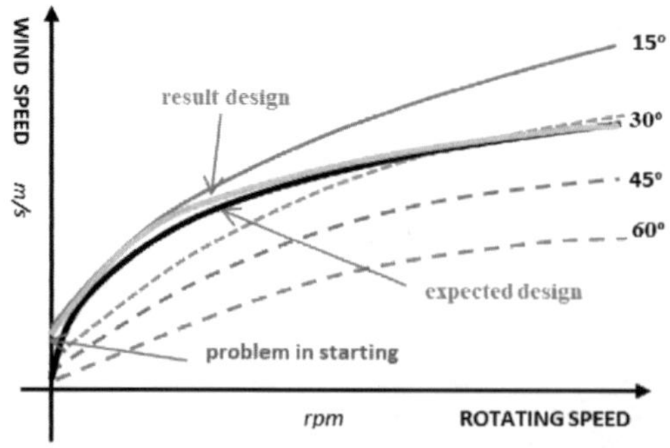

Figure 4. Curve design result of blade angle shift.

4. Results and discussions

Giving the pendulum the same mass and the correct position will not cause imbalance and also will not increase the shaking when it rotates. The gravitational force generated by the pendulum will cancel each other out in obtaining balance at its center of gravity. To reduce the mass of the pendulum and maintain the required centrifugal force, the pendulum can be shifted towards the end of the blade. Selection of a pendulum mass that is too large will be very dangerous for the strength of the blade. The pendulum will not be able to twist the blade to zero degrees; the angle of twist depends on the length of the pendulum arm. This is not a problem because following the characteristic curve in Figure 7, it states that the blade does not need to be twisted to zero degrees. At a twist angle of 45°, the rotating speed has dropped dramatically. The initial setting of the blade angle should be at 83° this is to overcome for easier initial starting. Because it requires a stopper at an angle of 83° and held by the spring on the other side. The strength of the spring can be regulated and compressed by the centrifugal force generated by the pendulum. The centrifugal force will twist the blades when the blades rotate, the higher the rotating speed will cause the twist angle to also increase or the attack angle will decrease and if the rotational speed decreases, the spring will push towards its initial position.

In Figure 3, the characteristic curve of the pit specification ratio is shown, which means that the initial setting angle is at 4 degrees and the greater the angle it can reduce the efficiency as well as reduce the rotational speed at a certain wind speed. Figure 4 shows a characteristic curve compared to the design characteristic curve after being equipped with a pendulum. With the addition of speed control using this pendulum, it is necessary to adjust the strength of the spring to get a curve that matches the design results.

5. Conclusions

Even though the maximum power occurs at the blade angle of 83°, the initial setting value is selected at the angle of the blade of 75° to make it easier in starting. This angle decreases with increasing wind speed. Although the relationship of changes in blade angle to wind speed is not linear, it is sufficient to provide good speed limiting stability. This stability value also depends on the length of the arm of the pendulum. In excessive wind speed, a certain amount of energy is wasted by narrowing the effective cross-sectional area or decreasing the value of the tip speed ratio (λ) so that the output stress and rotational speed are not excessive. The limit of wind speed that can damage the wind turbine can be taken care of by adjusting the setting value of the spring release automatically.

Acknowledgments

The author would like to thank Director and UPT P2M, Malang State Polytechnic who have given us the opportunity to carry out this research.

References

[1] Shankar R N, Kumar L R and Ramana M V 2019 Design and Fabrication of Horizontal Axis Wind Turbine *International Journal of Research and Analytical Reviews (IJRAR)* **06** 1 545-549

[2] Wharton S and Lundquist J K 2012 Atmospheric stability affects wind turbine power collection *Environmental Research Letters* **7** 1 014005

[3] Peter J S and Richard J C 2012 Wind Turbine Blade Design *Energies* **5** 3425-3449

[4] Chen Y J, Tsai Y F, Huang C C, Li M H and Hsiao F B 2014 The design and analysis of passive pitch control for horizontal axis wind turbine *Energy Procedia* **61** 683-686

[5] Chirca M, Dranca M, Oprea C A, Teodosescu P D, Pacuraru A M, Neamtu C and Breban S 2020 Electronically Controlled Actuators for a Micro Wind Turbine Furling Mechanism *Energies* **13** 16 4207

[6] Chen Y J, Tsai Y F, Huang C C, Li M H and Hsiao F B 2014 The design and analysis of passive pitch control for horizontal axis wind turbine *Energy Procedia* **61** 683-686

[7] Tong W 2010 Fundamentals of wind energy *Wind power generation and wind turbine design* 1-44

[8] Lundquist J and Clifton A 2012 How turbulence can impact power performance *North American Windpower* **9** 1

[9] Schubel P J and Crossley R J 2012 Wind turbine blade design *Energies* **5** 9 3425-3449

[10] Herráez I, Akay B, van Bussel G J, Peinke J and Stoevesandt B 2016 Detailed analysis of the blade root flow of a horizontal axis wind turbine *Wind Energ. Sci* **1** 89-100

[11] Çetin N S, Yurdusev M A, Ata R and Özdamar A 2005 Assessment of optimum tip speed ratio of wind turbines *Mathematical and Computational Applications* **10** 1 147-154

The effect of strain hardening on the limit forming ratio on the single-step incremental hole flanging process on aluminium plates

I Mashudi*, M Fakhruddin and A Hardjito

Department of Mechanical Engineering, State Polytechnic of Malang, Jl. Soekarno Hatta, No. 9, Malang 65141, Indonesia

*imam.mashudi.polinema@gmail.com

Abstract. Hole-flanging is the process of forming a plate to produce a collar around the hole. This formed collar has several important functions to increase the bonding surface area, as a supporting medium for other components, as a guide, increase construction rigidity, etc. This collar is generally formed through conventional hole flanging which involves making the initial hole in the plate, then punch with a certain profile and a larger diameter pressing the plate in one step of the forming process. This forming process will cause the deformation of the plate so that a collar is formed around the hole. In recent studies, it can be found that the collar can be formed not only with one stage of the conventional hole flanging process, but can also be achieved by a multi-stage incremental hole flanging process using a punch with a multi-purpose general profile. In this research, an experimental study was carried out on the effect of 'strain hardening' on the limit forming ratio (LFR) in the single-step incremental hole flanging process on aluminium plates using conical punch. The research hypothesis shows the influence of 'strain hardening' on the limit forming ratio (LFR) in the single-step incremental hole flanging process.

1. Introduction

In the industrial and automotive sectors, connections between pipes are often carried out in production processes. Generally, in the process of connecting pipes, binding is needed by sheet plates in the form of holes with a collar / flange that matches the pipe profile. As shown in Figure 1 the process of making holes with this collar is commonly done by hole-flanging [1].

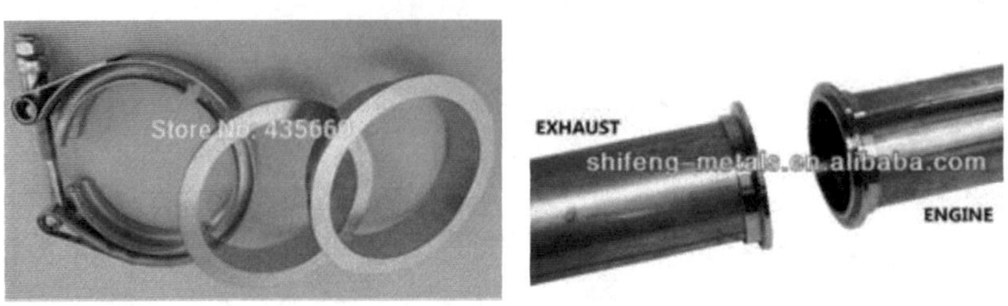

Figure 1. Exhaust flange and clamp.

IOP Conf. Series: Materials Science and Engineering 1073 (2021) 012079 doi:10.1088/1757-899X/1073/1/012079

Conventionally, the hole-flanging process is commonly used to produce a hole with a collar, as shown in Figure 1. In this process, plates with pre-cut holes are gripped and formed plasticly using a combination of bending and circumferential stretching forces. In the industrial world, many applications can be done through this process such as strengthening the connection to the piping [2]. But as the shape of the holes and collars will be produced, the investment needed will also increase. Because in the conventional hole-flanging process, when there is a change in geometry, the punch and dies also change their shape [3]. This causes, conventional hole-flanging processes are only suitable for mass production. In contrast to conventional hole flanging, this incremental hole flanging method offers a solution for making collars with holes on a small production scale as well as prototyping, and offers a lower investment for dynamic and complex flange designs [4], as shown in Figure 2.

Some of the research that has been done by previous researchers is incremental backward by Petek A. and single-stage incremental by Borrego M. Where in the research conducted by Petek A., it was found that on the DC 05 sheet, using the incremental backward method, data collection and data processing using the Response Surface method at the best variation to produce the optimal collar hole [5]. Meanwhile, Borrego M. researched the AL-7075, using the single-stage incremental sheet forming method, where the focus of his research was getting the limit forming ratio (LFR). At various precut holes, the max LFR is 1.62 [6,7].

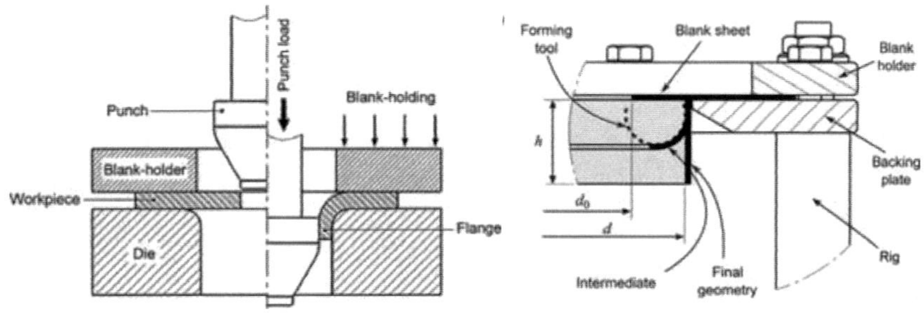

Figure 2. Conventional hole flanging vs incremental hole flanging [8].

Here we offer a combination of previous research with the incremental backward single stage sheet metal forming method. Where the hole flanging process combines upward movement of the collar formation process with the spiral motion of the forming tool.

The contribution of this research is to analyze and determine the hardening strain that occurs during the formation process using the single-stage incremental backward hole flanging method. In addition, the researchers also analyzed the effect on the limit forming ratio associated with the strain rate that occurred. So that the optimal formation process can be obtained.

2. Methods

In this work, the forming method used is single-stage incremental hole flanging, using aluminium plates, conical forming tools, and 3 axis CNC machines. Retrieval of hardness data using a micro Vickers hardness tester. The data were processed using tabulation and graphic methods, to get the effect of strain hardening.

2.1. Single-stage incremental hole flanging

Single-stage incremental backward hole flanging is the process of forming a flange towards the Z + coordinate. This method includes a flexible metal forming. In this work, a plate chuck without a contour is used, which makes it a die less fixture. forming tools that have a conical profile are used during this forming process. The dies are then placed and positioned on the CNC machine table. The workpiece, in this case, aluminium, is positioned on the dies. In one forming motion, the tool moves closer to the aluminium plate that was previously perforated, then performs a spiral motion as a form of the forming process. As shown in Figure 3. Single-stage Incremental Backward Hole Flanging Process [9].

IOP Conf. Series: Materials Science and Engineering 1073 (2021) 012079 doi:10.1088/1757-899X/1073/1/012079

Figure 3. Proposed single-stage incremental backward hole flanging process.

Table 1. Forming setup.

Specification	Units
Dies Material	ST 37
Holding Dimension	150 mm x150 mm x1 mm
Forming Tool Material	HSS
Dimension	Dia. 20 mm x 150 mm (45° Conical)
Workpiece Material	Aluminium
Dimension	150 mm x150 mm x1 mm

2.2. Forming parameter
The process parameters in this study are as follows, 20 mm diameter of precut hole, which is produced by milling process. Up to 80 mm diameter which is produced by forming process. 50 rpm of Spindle speed, constant step over by 1 mm. 4 variation of feed rate, 500, 1000, 1500, and 2000 mm/min where the variation in the speed of formation or strain rate is represented by the feed rate. The results that have been obtained will be carried out an observation process on the cross section of the collar for changes in hardness caused by strain hardening during the forming process.

Table 2. Forming parameter.

Specification	Units
Type of steel	Al 1050-0
Hardness	50 HVN
Pre-cut Hole	20 mm
Finish Hole	Up to 80 mm
Spindle Speed	50 rpm
Forming Step Over	1 mm
Forming Rate	500, 1000, 1500 & 2000 $^{mm}/_{min}$

2.3. Hardness observation
Hardness testing in this study used a micro Vickers hardness tester. As shown in Figure 4, where the use of this tool is because the observed workpiece cross-section is very small (below 5 mm). the cross-section of the forming result is then cut to produce the cross-section as shown in Figure 5 below. Then the observation points are divided from the edge of the section to the original material.

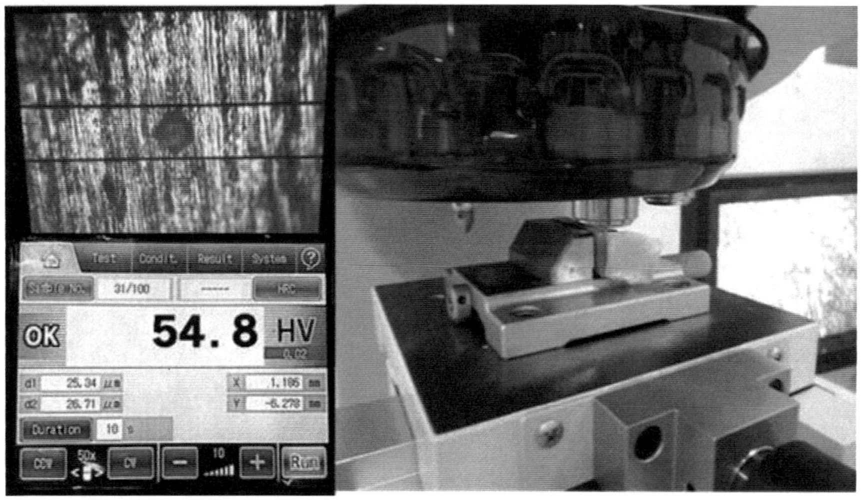

Figure 4. Micro Vickers hardness tester.

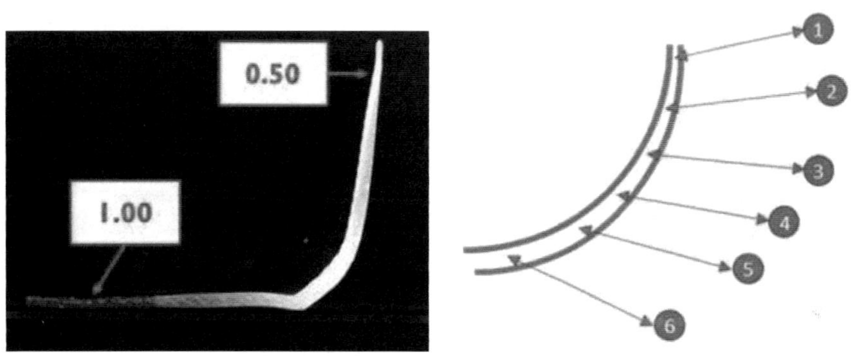

Figure 5. Cross section view of workpiece after forming process and point of hardness observation.

Table 3. Micro Vickers specification.

Specification	Units
Model	HM-210 Type A
Fixed test Force	10-1000 gf
Loading Rate	60 μ/sec
Load Dwell Time	0-999 sec
Indenter	Vickers
Objective Lens	10x, 50x
Filar Eye Piece	Dual Line, 10x, 0.01 μ min

3. Results and discussions

The process of observing changes in the hardness of the cross section of the collar is done using the micro Vickers hardness tester method, As shown in Figure 5. Where in the cross section of the collar is divided into five points of observation, where from the upper end of the collar experiencing thinning, point number one, to the parent material, point number five. As a reference, the average hardness value of the parent material is 50.5 HVN. In this study it was found that, the highest hardness value was found in observation at point number one and in all variations, As shown in Figure 6.

At this point the material experiences the most extreme thinning. This happens due to the strain hardening process during the formation movement. The strain rate was represented as feed rate during forming process. The lowest hardness value at observation point one is found at the highest feed rate or strain rate variation, which is at 2000 mm / min. This happens because at a high feed rate the material experiences less friction time, so the heat generated is lower which causes the material to be affected by changes in hardness due to the minimal impact of heat treatment during the forming process [10].

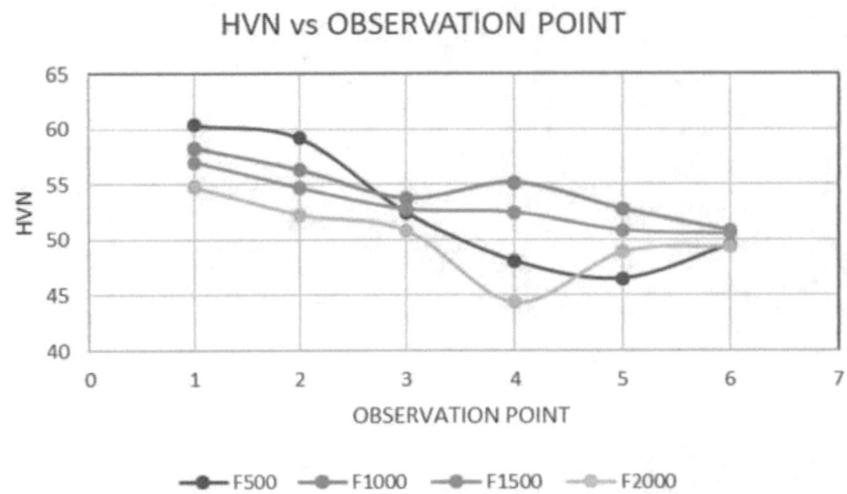

Figure 6. Hardness Vickers Number (HVN) vs Hardness observation point.

At the maximum feed rate or strain rate of 2000 mm/min, aluminium started to crack. At this point, we conclude that maximum forming rate was 2000 mm/min at finish hole of 80 mm. Therefore, the Limit Forming Ratio (LFR) on this study is 80 mm divided by 20 mm equal to 4.0. At forming rate of 2000 mm/min, strain hardening process was at lower point, because of insufficient of friction during the forming process, therefore, the material begins to crack eventually. These phenomes can be shown in figure 7.

Table 4. Hardness value for each observation points.

F (mm/min)	1	2	3	4	5	6	
500	60.4	59.2	52.5	48.1	46.5	49.5	
1000	58.3	56.3	53.8	55.2	52.8	50.9	**HVN**
1500	57.0	54.7	52.8	52.5	50.9	50.6	
2000	54.8	49.5	50.9	44.4	48.9	49.4	

Figure 7. (Left) Flange produced at 1500 mm/min of forming speed & (Right) Flange produced at 2000 mm/min of forming speed.

4. Conclusions

The conclusion of the research we have done is highest strain hardening was occurred at the edge of the flange, at forming speed of 2000 mm/min at 1mm forming step over and at 80 mm finish hole, aluminum start to crack, below 2000 mm/min at same parameters, there is no sign of crack in the aluminum. Therefore, in this study maximum limit forming ratio (LFR) is achieved at 4.0, at forming speed of 2000 mm/min. In the future, this research needs to consider lubrication during the forming process. lubrication during the forming process is expected to be able to suppress cracks on the plate surface during the forming process.

References

[1] Huang Y M and Chien K H 2001 The formability limitation of the hole-flanging process *Journal of Materials Processing Technology* **117**(1-2) 43-51

[2] Cristino V A, Montanari L, Silva M B, Atkins A G and Martins P A F 2014 Fracture in hole-flanging produced by single point incremental forming *International Journal of Mechanical Sciences* **83** 146-154

[3] Dewang Y, Purohit R and Tenguria N 2017 A study on sheet metal hole-flanging process *Materials Today: Proceedings* **4**(4) 5421-5428

[4] Centeno G, Silva M B, Cristino V A M, Vallellano C and Martins P A F 2012 Hole-flanging by incremental sheet forming *International Journal of Machine Tools and Manufacture* **59** 46-54

[5] Petek A and Kuzman K 2012 Backward hole-flanging technology using an incremental approach *Strojniški vestnik-Journal of Mechanical Engineering* **58**(2) 73-80

[6] Borrego M, Morales-Palma D, Martínez-Donaire A J, Centeno G and Vallellano C 2015 On the study of the single-stage hole-flanging process by SPIF *Procedia engineering* **132**(1) 290-297

[7] Borrego M, Morales-Palma D, Martínez-Donaire A J, Centeno G and Vallellano C 2016 Experimental study of hole-flanging by single-stage incremental sheet forming *Journal of Materials Processing Technology* **237** 320-330

[8] Hirt G, Ames J, Bambach M and Kopp R 2004 Forming strategies and process modelling for CNC incremental sheet forming *CIRP Annals* **53**(1) 203-206

[9] Cui Z and Gao L 2010 Studies on hole-flanging process using multistage incremental forming *CIRP Journal of Manufacturing Science and Technology* **2**(2) 124-128

[10] Fakhruddin M, Choiron M A and Purnowidodo A 2017 The Influence of Process Parameters toward Collar Height on Incremental Backward Hole-Flanging Process *JEMMME (Journal of Energy, Mechanical, Material, and Manufacturing Engineering)* **2**(1) 23-30

IOP Conf. Series: Materials Science and Engineering 1073 (2021) 012080 doi:10.1088/1757-899X/1073/1/012080

Web innovation in horizontal wind pipe turbine propeler

B Wahyudi*, S Sudarmadji, S Sarjiyana and M Maskuri

Mechanical Department, State Polytechnic of Malang, Malang, Indonesia

*bagus.wahyudi@polinema.ac.id

Abstract. The efficiency limitations of wind turbines provided by Betz's Theorem prompted researchers to develop new geometric shapes of wind turbines propeller that split pipe spiral slice by placing the web between turbine blades. The idea of developing the shape of the turbine was inspired by the shape of a duck's or swan's leg which has a web between its fingers that functions to obtain the thrust when swimming in water. The purpose of this study is to obtain the strength of material with optimal geometry and aerodynamic parameters in producing wind turbine efficiency. The method of obtaining research data is using solid work software for design, Ansys software for CFD simulations and Wind Tunnel laboratory equipment for experiments and also using Minitab 19 software for data processing. The results of the CFD simulation and experiment based research show that the turbine propeller turbine with using the web has higher efficiency than without the web.

1. Introduction

The current problem is designing wind turbines from materials that are easily found in the market, one of which is pipes. The actual function of the pipe is to drain a fluid. Thus it is necessary to analyze the working stress that occurs in the construction and simulation of CFD, so that this designed pipe propeller wind turbine can be applied in the community.

Unlike natural resources that come from fossils such as oil and gas, wind is a natural resource that will never run out. Indonesia is a tropical country that has a lot of wind potential, but until now wind power is still rarely used as an energy source [1]. The potential for wind flow in Indonesia with an average speed of between 9 m / s to 12 m / s is an opportunity to be developed as a wind power plant [2].

A very influential factor in the performance of wind turbines is wind speed, but a very large wind speed will make the wind turbine damaged and not work efficiently. Therefore, before designing and building a wind turbine, we must first consider the potential wind speed in the area [3]. Similar research has been conducted by Monteiro et al. [4], Murugan et al. [5], and Abdelwaly et al. [6], entitled the effect of wind speed to obtain the maximum electrical power efficiency in wind turbines.

Aerodynamic design principles for wind turbine blades must take into account the careful blade design, airfoil selection, and optimal angle of attack [7]. The aerodynamic performance of horizontal axis wind turbines is highly dependent on many parameters, including airfoil type and blade geometry [8]. The method used to optimize the geometry of small wind turbine blades is obtained from a spiral split circular pipe with an optimal distribution and airfoil sweep can be obtained with the right cutting path. The geometry of the blade is obtained by folding the shape of the blade surface into the pipe, it is very difficult to describe the shape of the blade in the pipe. Moreover, the surface is located in three-

IOP Conf. Series: Materials Science and Engineering 1073 (2021) 012080 doi:10.1088/1757-899X/1073/1/012080

dimensional space. In Figure 1 there are three sections of the blade (tip, root, and intermediate) and their relative positions on the longitudinal axis of the blade [9].

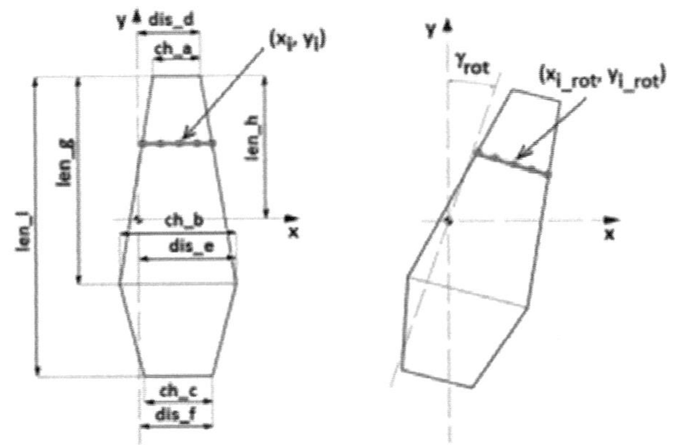

Figure 1. Blade shape geometric parameters [9].

The geometry of the blade is obtained by folding the shape of the blade surface into the pipe, it is very difficult to describe the shape of the blade in the pipe. Moreover, the surface is located in three-dimensional space. The shape of the surface parameters shown in Figure 1. are the three sections of the blade (tip, root, and intermediate) and their relative positions on the longitudinal axis of the blade. The shape of the blade surface can also be rotated with respect to the origin of the axis (x, y, z) from the angle γ_{rot}, can be seen in Figure 1.

Although the measurement data (experiment-based) is able to provide precise information on the proper wind turbine performance parameters of the flow around the blade, the development of more effective designs is still going to be a slow process based on trial and error. Various Computational Fluid Dynamic (CFD) techniques have been used to predict the field flow around wind turbine blades and their efficiency [10]. Hence to overcome the limitations of the experimental methodology, this study used CFD simulations to investigate the aerodynamics of the flow field around the blade.

1.1. Problem and basic principle

The wind kinetic energy of an object with mass m and velocity v is $E = 0.5 \times m \times v2$, assuming that the velocity v does not approach the speed of light. Since mass can be replaced by air density ρ, area A, and velocity v, it can be written: $m = \rho.A.v$.

The formulation of kinetic energy in a wind turbine is

$$EK = \frac{1}{2} m v^2 \tag{1}$$

The mass flow rate equation is:

$$m = \rho A v \tag{2}$$

By changing the air mass at eq (1) with eq (2), the formulation of the potential power calculated from the wind velocity is:

$$Pa = \frac{1}{2} \rho A v^3 \tag{3}$$

Coefficient of power is ratio output and input

$$\mathrm{Cp} = \frac{\mathrm{P_m}}{\mathrm{P_a}} \tag{4}$$

The mechanical power of wind turbine

$$\mathrm{Pm} = \frac{1}{2}\boldsymbol{\rho A v^3 Cp} \tag{5}$$

The tip speed ratio is the ratio of the blade tip tangential speed to the free wind speed. For certain wind speeds, the tip speed ratio will affect the rotation of the rotor. The lift type wind turbine will have a relatively larger tip speed ratio compared to the drag wind turbine [11-13].

$$\boldsymbol{\lambda} = \frac{\mathbf{2.\pi.n.r}}{\mathbf{60 \times v}} \tag{6}$$

The maximum power coefficient that can be achieved by the wind turbine according to the Betz limit is 59.26%. However, in practice the value obtained from the center of the power coefficient is around 45%. Figure 2 is a graph depicting the Betz ideal constant and the actual turbine power coefficient as a function of TSR.

Based on the empirical approach, when practitioners build a horizontal wind turbine, it is usually only guided by the relationship between the parameters of the turbine power, diameter and TSR as shown in figure 3.

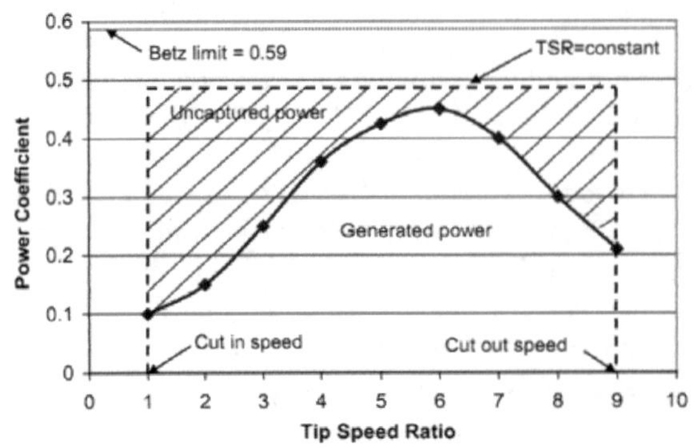

Figure 2. Wind turbine performance diagram.

Power (watts)	Diameter (metres)	TSR=4	TSR=6	TSR=8	TSR=10
10	0.4	2032	3047	4063	5079
50	0.8	909	1363	1817	2271
100	1.2	642	964	1285	1606
250	1.9	406	609	813	1016
500	2.7	287	431	575	718
1000	3.8	203	305	406	508
2000	5.3	144	215	287	359
5000	8.4	91	136	182	227

Figure 3. Design parameter relationship between Power, Diameter and TSR [4].

2. Methodology

This research was conducted using two methods, namely experimental tests on the wind tunnel and CFD simulation, from the two methods obtained data and the value of the effect (P-value) of wind speed and angle of attack on wind turbine shaft torque. Tests conducted on the web-less pipe propeller wind turbine model and the pipe propeller wind turbine model using the web.

The idea of developing the shape of the pipe propeller wind turbine using the web was inspired by the shape of a duck's or swan's leg which has a web between its fingers that functions to obtain the thrust when swimming in water. The web as shown in Figure 4 is made of plates that are assembled on the corner hub between two pipe blades made of the same material and thickness as the pipe material. The web plane is also set to have a certain slope to function as a guide for the flow of wind to enter the cavity in the pipe. The wind will be caught by the web and flowed through the cavity in the pipe towards the radial end.

The idea of developing the shape of the pipe propeller wind turbine using the web was inspired by the shape of a duck's or swan's leg which has a web between its fingers that functions to obtain the thrust when swimming in water. The web as shown in Figure 4 is made of plates that are assembled on the corner hub between two pipe blades made of the same material and thickness as the pipe material. The web plane is also set to have a certain slope to function as a guide for the flow of wind to enter the cavity in the pipe. The wind will be caught by the web and flowed through the cavity in the pipe towards the radial end.

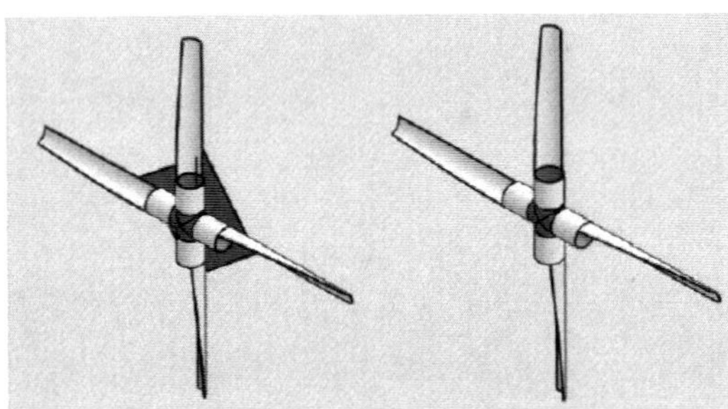

Figure 4. The pipe propeller wind turbine model using the web and without web'.

Figure 5. The pipe propeller wind turbine without web model on position testing.

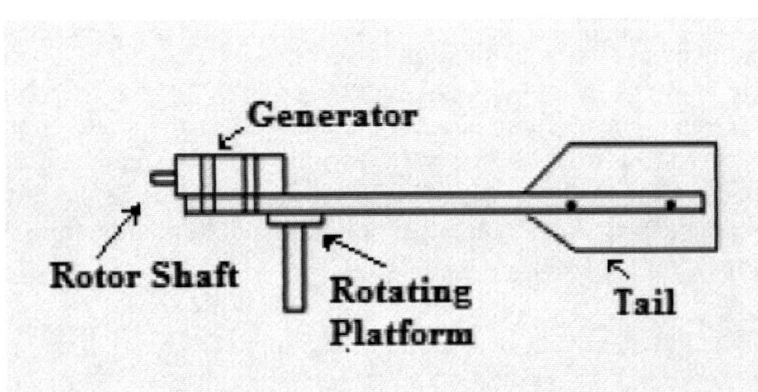

Figure 6. Sketch of wind turbine arrangement.

DOE (design of experiment) is comprised of statistical and mathematical techniques for optimization process development and improvement, which utilizes the experiment design, regression analysis, and variance analysis. In this case, the response variable Efficiency "Cp" (y) is affected by two independent variables: Angle of attack "α" or (x1) and wind speed "V" or (x2). The significance effect of those independent variables (x1, x2) that lead to maximum power (and working stress) can be obtained from an appropriate model formulation. Variance and regression analyses can be used to estimate regression coefficients in the quadratic polynomial model and to generate an uncertainty measure in the coefficients [14].

3. Results and discussion

The geometry design of the propeller pipe wind turbine was tested using CFD simulations to produce data to be analyzed using Minitab 19 software. This working stress study used the Regression and ANOVA analysis method. It has two independent variables consisting of wind speed and angle of attack. Likewise, the aerodynamic study of turbine performance also uses variable wind speed and angle of attack. With the DOE method, it can be seen the effect of the independent variable on the dependent variable, as well as the effect of the interaction between the two independent variables on the dependent variable or the response in the form of working stress on the turbine body (σ) as well as the turbine power coefficient (CP).

Table 1. Regression analysis of working stress using minitab.

Analysis of Variance					
Source	**DF**	**Adj SS**	**Adj MS**	**F-Value**	**P-Value**
Regression	2	2,42492E+12	1,21246E+12	77,50	0,000
Wind speed	1	2,27797E+12	2,27797E+12	145,61	0,000
Angel of Attack	1	1,46953E+11	1,46953E+11	9,39	0,022
Error	6	93868333333	15644722222		
Total	8	2,51879E+12			
Coefficients					
Term	**Coef**	**SE Coef**	**T-Value**	**P-Value**	**VIF**
Constant	-1749333	309203	-5,66	0,001	
Wind speed	616167	51063	12,07	0,000	1,00
Angel of Attack	12227	3989	3,06	0,022	1,00
Regression Equation					
Working stress =	-1749333 + 616167 wind speed + 12227 Angle of attack				

The maximum working stress value in the wind turbine simulation test with the addition of a web was obtained at a wind speed of 7 m / s and a blade angle of attack -12.8 °, that is 1.736E + 08 N / m² or 173600 MPa, while in the wind turbine simulation test without additional web obtained at a wind speed of 7 m / s and an angle of attack of 12.8 °, that is 2.893E + 06 N / m² or 2893 MPa. In a wind turbine with an additional web, the higher the wind speed and the smaller the angle of attack of the blade, the greater the value of the resulting working stress. In a wind turbine without additional web, if the wind speed and angle of attack of the blade are greater, the greater the value of the resulting working stress. Based on coefficient value of the working stress regression equation on table 1, the wind speed giving more dominant effect then angle of attack.

Figure 7. Stream line on the pipe propeller wind turbine with addition web and without web.

Figure 7 shows that in a turbine with additional web there is a perfect vortex behind the blade, whereas in a turbine without a web there is no vortex or turbulence. From the two images above, it can be seen that a relatively large vortex is obtained by the wind turbine with the addition of a web. From the simulation results, it shows that the wind turbine using the web addition can rotate optimally compared to the wind turbine without a web.

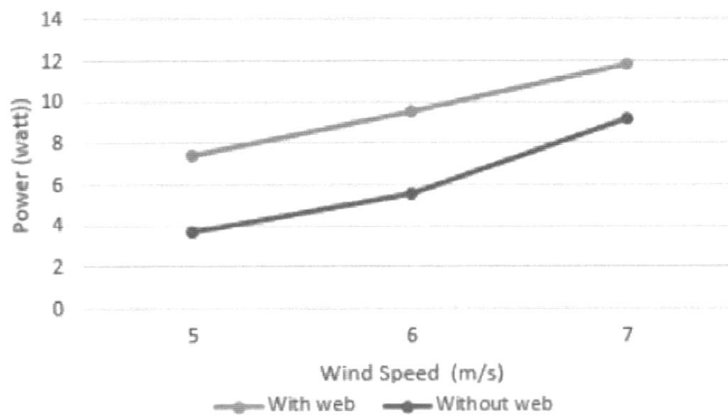

Figure 8. Comparison of power generation between pipe propeller wind turbine using web and without web resulted from model testing wind tunnel laboratory.

From the results of model testing in Figure 8, it shows a large power difference between the turbines using the web and without using the web. At a wind speed of 7 m / s, it is shown that the wind turbine using the web produces a power of 11.84 watts, while the wind turbine without using the web is only 9.20 watts.

4. Conclusion

The simulation test using *Solidwork* software is now equipped with the Computational Fluid Dynamics (CFD) feature, the test results show that the wind turbine with the addition of the web has a wider vector distribution than the wind turbine without using the web so that the power generated by the wind turbine is also greater and more stable. However, this type of wind turbine with the addition of a web produces a larger vortex.

The use of PVC pipe material to build propeller wind turbines has proven to be quite strong in simulation tests using *Solidwork* software and experimental tests in wind tunnel laboratories up to a speed of 8 m / s.

Acknowledgments

Authors wish acknowledge for fund assistance from Director of State Polytechnic of Malang, financial support from DIPA 2020 very helpful to achieve goal of this research. This research was funded by Directorate General of High Education - State Polytechnic of Malang through SP DIPA No. 023.18.2.677606/2020.

References

[1] Sanusi A, Soeparman S, Wahyudi S and Yuliati L 2016 Experimental study of combined blade savonius wind turbine *International Journal of Renewable Energy Research (IJRER)* **6**(2) 614-619

[2] Multazam T, Putri R I, Pujiantara M, Priyadi A and Hery P M 2016 Wind farm site selection base on fuzzy analytic hierarchy process method; Case study area Nganjuk *2016 International Seminar on Intelligent Technology and Its Applications (ISITIA)* 545-550

[3] Tavner P, Edwards C, Brinkman A and Spinato F 2006 Influence of wind speed on wind turbine

reliability *Wind Engineering* **30**(1) 55-72

[4] Monteiro J P, Silvestre M R, Piggott H and Andre J C 2013 Wind tunnel testing of a horizontal axis wind turbine rotor and comparison with simulations from two Blade Element Momentum codes *Journal of wind engineering and industrial aerodynamics* **123** 99-106

[5] Murugan N, Umamaheswari M, Vimal S I and Sivashanmugam P 2012 Experimental investigation on power output in aged wind turbines *Advances in Mechanical Engineering* **4** 380986

[6] Abdelwaly M, El-Batsh H and Hanna M B 2019 Numerical study for the flow field and power augmentation in a horizontal axis wind turbine *Sustainable Energy Technologies and Assessments* **31** 245-253

[7] Schubel P J and Crossley R J 2012 Wind turbine blade design review *Wind engineering* **36**(4) 365-388

[8] Khalil Y, Tenghiri L, Abdi F and Bentamy A 2020 Improvement of aerodynamic performance of a small wind turbine *Wind Engineering* **44**(1) 21-32

[9] Ceruti A 2019 Meta-heuristic multidisciplinary design optimization of wind turbine blades obtained from circular pipes *Engineering with Computers* **35**(2) 363-379

[10] Cai X, Gu R, Pan P and Zhu J 2016 Unsteady aerodynamics simulation of a full-scale horizontal axis wind turbine using CFD methodology *Energy Conversion and Management* **112** 146-156

[11] Rathod V K and Kamdi S Y Design of PVC Bladed Horizontal Axis Wind Turbine for Low Wind Speed Region *Int. Journal of Engineering Research and Applications* 2248-9622

[12] Schubel P J and Crossley R J 2012 Wind turbine blade design *Energies* **5**(9) 3425-3449

[13] Ragheb M and Ragheb A M 2011 Wind turbines theory-the betz equation and optimal rotor tip speed ratio *Fundamental and advanced topics in wind power* **1**(1) 19-38

[14] Wahyudi B and Adiwidodo S 2017 The influence of moving deflector angle to positive torque on the hydrokinetic cross flow savonius vertical axis turbine *International Energy Journal* **17**(1)

IOP Conf. Series: Materials Science and Engineering 1073 (2021) 012081 doi:10.1088/1757-899X/1073/1/012081

Analysis of fuel pressure on the performance of motorcycle engine with ethanol fuel

B Irawan*, Y A Winoko, E Puspitasari and T Dwiyono

Malang State Polytechnic, Malang, Indonesia

*bambang.irawan@polinema.ac.id

Abstract. Gasoline engines fuel cannot be directly replaced with pure ethanol, moreover ethanol contains a lot of water. The purpose of the study was to analyze the effect of fuel pump pressure on performance of 4-stroke engine motor cycle by ethanol fuel. The independent variable is pressure of the fuel and the engine speed. The dependent variable is the engine power, torque, and specific fuel consumption. The power and torque test use inertia dynamometer. The results showed the addition of fuel pump pressure, at 3.5 bar pressure can produce higher power at each engine speed with a maximum power of 6 HP at 8000 rpm. As for the best results torque obtained at a pressure of 3.5 bar which is 5,5 Nm at 7500 rpm, besides that the lowest specific fuel consumption SFC at a pressure of 3,5 bar is 0.409 kg/HPh. The results of this study can be concluded that gasoline vehicles can use ethanol fuel by increasing the fuel pressure up to 3.5 bar.

1. Introduction

The fossil fuel is dwindling rapidly due to increasing application of fossil fuel for the generation of energy and gasoline engine use. Therefore, it is necessary to increase the use of renewable energy and alternative fuels. One of options is the alternative energy and renewable energy is ethanol. Ethanol presently are being used for fuel blending with gasoline or can use alone. Ethanol is a fairly extensive alternative energy to replace petroleum. In Indonesia, the sugar mill produces white sugar, bagasse and molasses. the ethanol is produced from molasses fermentation.

There have been many studies on ethanol being used as fuel for motor vehicles, but the research is that alcohol is mixed with gasoline. There is very little research on the use of pure ethanol for fuel. It clearly confirmed that E100, 95.5 vol % ethanol, can be effectively used as an alternative fuel in used 4-stroke motorcycles whose carburetor were purposely designed for the use of gasoline91 [1].

This research is about ethanol used for motorcycle fuel, pure ethanol (E100) without being mixed with gasoline. Pure 100% ethanol without water is expensive and difficult to find. This study used ethanol with a content of 95% and 5% water.

Vehicle engines without direct modification use pure ethanol (E100) fuel is difficult to live. Can live but cannot be normal, if the engine speed is accelerated it be halting. Analysis of using gasoline-corncob waste bioethanol blends in motorcycle engine onto compression ratio [2,3]. Optimization of the octane response of gasoline and ethanol mixtures [4].

The problem is how to make a vehicle engine using E100 fuel, the engine can start normally. One of them is by increasing the amount of ethanol fuel into the engine. To increase the amount of fuel, one way is to increase the fuel pressure.

IOP Conf. Series: Materials Science and Engineering 1073 (2021) 012081 doi:10.1088/1757-899X/1073/1/012081

There have been many studies conducted by several researchers such as research. Ethanol–gasoline and methanol–gasoline fuel for spark-ignition engines [5]. Fuel effect of 5% ethanol in 88, 92, and 98 RON gasoline on motorcycle engine performance [6]. Make use of the benefits of ethanol in advanced engine-fuel system [7]. The benefits of hight octane fuels in spark-ignition engines [8]. Optimal use of ethanol in dual fuel application [9]. Effect of the compression ratio and duration injection on performance of engine [10]. Estimate the octane ratings of ethanol-gasoline mixtures using base fuel properties [11]. Performance enhancement of used carburetor motorcycles fueled by flex-fuel gasoline-ethanol blends [12]. Performance of bio ethanol petrol blends in a spark ignition engine [13,14]. Experimental on twin spark ignition gasoline engine at different gasoline-methanol blends [15]. Bioethanol and gasoline mixture to power on motor vehicle [16]. Performance of a spark-ignition automobile engine while using ethanol blends as fuel [17]. Review on bioethanol as alternative fuel for gasoline engines [18]. Summary of ethanol-fuel engine [19].

2. Basic theory

Properties of gasoline and ethanol [20].

Table 1. Properties of gasoline and ethanol.

	Gasoline	**Ethanol**
Air fuel ratio	14,6	9
Density (kg/l)	0.74	0.785
O_2 (%)		34.73
RVP (kPa)		250
RON	91-99	111
MON	81-89	104
Molecular weight	111.21	46.07
Carbon (%)	86,3	52.2
Hydrogen (%)	14.8	13.1
Lower Heating (MJ/kg)	44.3	27
Heat of Evaporation (MJ/kg)	305	840
Normal boiling (^{0}C)	38	78
Autoignition Temp. (^{0}C)	246-280	365

Reaction of ethanol fuel [21].

Perfect reaction:

$$a\ C_2H_5OH + b\ H_2O + c\ O_2 \rightarrow d\ CO_2 + e\ H_2O + \text{Energy}$$

Not perfect reaction:

$$a\ C_2H_5OH + b\ H_2O + c\ O_2 \rightarrow d\ CO_2 + e\ H_2O + f\ CH + g\ CO + h\ O_2 + \text{Energy}$$

Mass flow rate of fuel:

$$m = A.(2.\rho.dp)^{0.5}.c_f$$

Where:

m = mass flow rate (kg/s)
A = area of injector hole (m^2)
dp = different of fuel pressure (Pa)
ρ = density (kg/m^3)
c_f = correction factor (viscosity, shape, etc)

The fuel pressure changes so that the mass flow rate of fuel will change.

3. Materials and methods

3.1. Materials

3.1.1. The engine. The engine used for this investigation is an motor cycle spark ignition engine whose main characteristics are:

Type	
No of cylinder	: 1
power	: max. 7.5 HP
Bore and Stroke	: 50 mm and 58 mm
Compression ratio	: 9.1:1
Type of cooling	: air cooling
Type of transmission	: automatic
Normal fuel pressure	: 2.6 bar
Duration of injection	: plus 30 % from normal

3.1.2. Apparatus test
- Dynamometer, inertia type
- Flowmeter
- Manometer
- Thermometer
- Tachometer

3.1.3. Raw materials
- Ethanol (ethanol 95 % +water 5 %)
- Gasoline RON 92

3.2. Methods
Gasoline fuel is not used to mix with ethanol but is used only to compare the test results with ethanol. The first test uses pure gasoline with the engine condition as it is without changing specifications. Furthermore, using pure ethanol fuel without mixture with gasoline. Testing with ethanol fuel injector opening duration is increased by 30%, without additional injection time, the engine cannot start normally. Testing with ethanol fuel in addition to the opening time of the injectors plus the pressure is increased. The fuel pressure for ethanol is made at 3.5 bar, 4 bar and 4.5 bar.

Figure 1. Implementation of data collection.

IOP Conf. Series: Materials Science and Engineering 1073 (2021) 012081 doi:10.1088/1757-899X/1073/1/012081

4. Results and discussion

Ethanol fuel with normal original standard pressure can cause the engine to not start normally. When the engine accelerates, there will be a abnormal and even when the engine is running, the engine may die. This can happen because the amount of ethanol fuel that enters the cylinder is not fulfiiled.

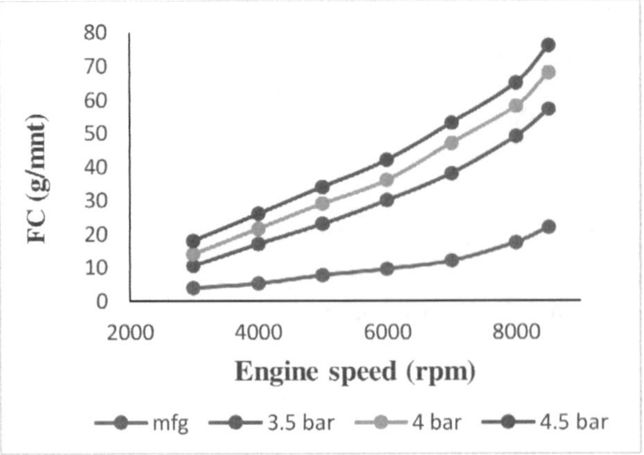

Figure 2. Engine speed vs fuel consumption.

Figure 2 shows a graph of the measurement results of the fuel consumption of gasoline and ethanol. It can be seen that gasoline fuel is much less than ethanol, this happens because gasoline does not change the engine while ethanol has increased fuel pressure and increased injection time.

As the fuel pressure entering the injector increases, the amount of fuel entering the intake manifold will increase according to the equation. From Figure 2 the total fuel consumption rates increases as the engine speed. These fuel consumption rates show how it may be much more economical to power the engine with gasoline than with the investigated of ethanol. This behavior is attributed to the low energy value of the ethanol fuel, which is distinctly lower than that of the gasoline fuel. The maximum fuel consumption value is ethanol with a pressure 4.5 bar, 4 bar, 3.5 bar and the minimum is gasoline.

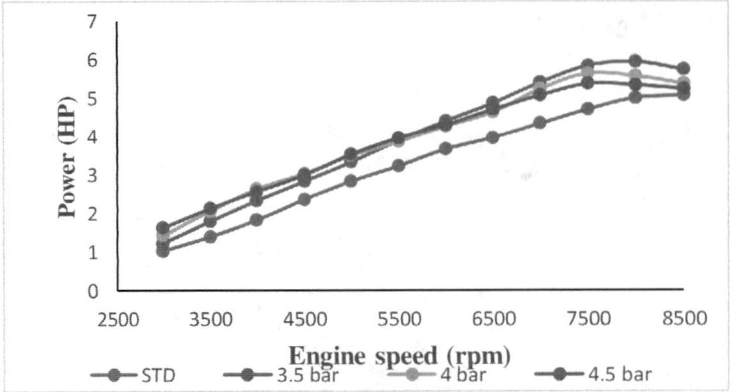

Figure 3. Engine speed vs power.

The effect of ethanol and gasoline on engine power can be observed in Figure 3. Maximum power obtained with gasoline is 5.13 HP at 8500 rpm. The maximum power obtained for ethanol with pressure of 3.5 bar is 6.0 HP at 8000 rpm, pressure of 4 bar is 5.6 HP at 7500 rpm, and pressure 4.5 bar is 5.3 HP at 7500 rpm. The power produced by ethanol fuel is always higher than that of gasoline, whereas based on theory shows that the calorific energy value of gasoline is 1.7 times the calorific of ethanol. This is because the amount of ethanol that enters the cylinder is far more than gasoline, so the power is large.

The picture shows that the power produced is different. Based on the theory that the more fuel that is put into the cylinder, the greater the power produced. The graph shows that for pressures ranging from 3.5 bar until 4.5 bar the power generated does not coincide. At first it shows that the high power pressure is high but at higher revolutions the power can be different. At a pressure of 3.5 bar, the power is actually higher than at a higher pressure. This can happen because the engine speed, the higher the air entering the cylinder, the less so that the fuel cannot burn properly due to excess fuel.

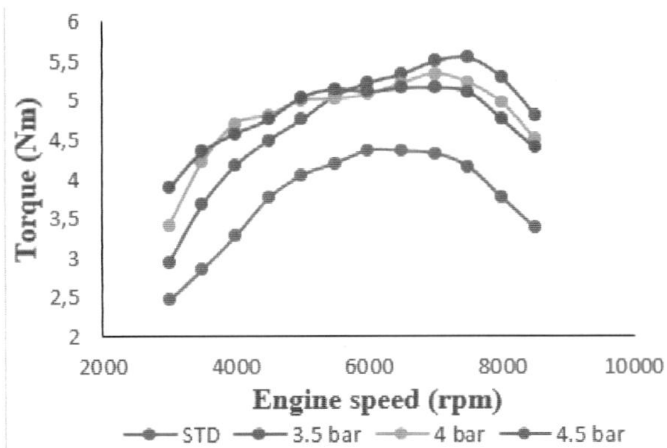

Figure 4. Engine speed vs torque.

The effect of ethanol and gasoline on engine torque can be observed in Figure 4. Maximum torque obtained with gasoline is 4.3 Nm at 6000 rpm. The maximum torque obtained for ethanol with pressure of 3.5 bar is 5.5 Nm at 7500 rpm, pressure of 4 bar is 5.3 Nm at 7000 rpm and pressure 4.5 bar is 5.1 Nm at 7000 rpm. The torque produced by ethanol fuel is always above the torque of gasoline. This is because the amount of ethanol that enters the cylinder is far more than gasoline so that the torque is greater.

Based on the theory that the more fuel that is put into the cylinder, the greater the torque produced. The graph shows that for pressures ranging from 3.5 bar to 4.5 bar the resulting torque is not the same. At first it shows that the high torsion pressure is high but at higher revolutions the torque can be different. At a pressure of 3.5 bar, the torque is actually higher than at a higher pressure of ethanol. This can happen because the engine speed, the higher the air entering the cylinder, the less so that the fuel cannot burn properly due to excess fuel. It can also be caused by greater engine friction so that the torque decreases.

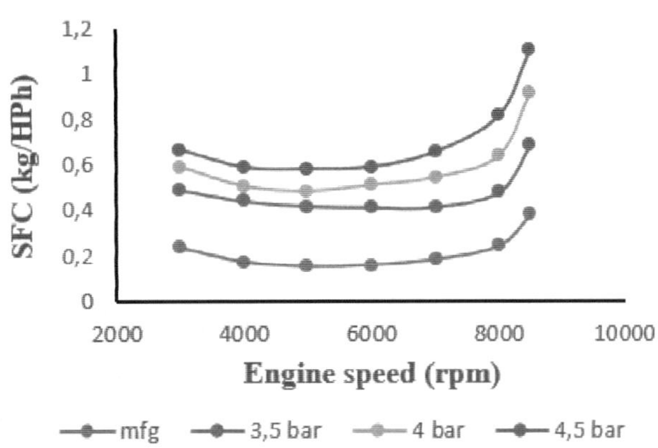

Figure 5. Engine speed vs SFC.

Figure 5 shows a graph of the measurement results and calculations of the specific fuel consumption of gasoline and ethanol. It can be seen that the SFC for gasoline is much less than that of ethanol, this happens because there is no change in the engine for gasoline, while the ethanol fuel pressure is increased and the injection time is increased.

As the engine speed increases, the injector will be opened for longer as a result, the fuel increases. With the increase in fuel, SFC does not necessarily increase because there is a power factor generated by the engine. Figure 5 shows that the initial SFC is high and with increasing rotation it will decrease but until a certain engine speed the SFC will increase again. This incident is due to the fact that the power produced does not increase significantly with the fuel being entered, due to incomplete combustion.

The minimum SFC value of ethanol for 3.5 bar is 0.409 (kg/HPh) at 6000 rpm, 4 bar is 0.48 (kg/HPh) at 5000 rpm and 4.5 bar is 0.577 (kg/HPh) at 5000 rpm. The gasoline SFC smaller then ethanol, for the normal pressure of fuel is 0.148 (kg/HPh) at 5000 rpm.

5. Conclusion

By increasing the ethanol fuel pressure and increasing the injector opening time, the engine can run normally. However, ethanol fuel consumption is more wasteful than gasoline. The maximum power obtained for ethanol with a pressure of 3.5 bar is 6.0 HP at 8000 rpm. The maximum torque obtained for ethanol with a pressure of 3.5 bar is 5.5 Nm at 7500 rpm. The fuel pressure for ethanol is made at 3.5 bar, it can make the vehicle's engine run normally up to 9,000 rpm.

Acknowledgments

The authors are grateful to the P2M Malang State Polytechnic for Research Grant Program 2020 for financial support and VEDC Malang for facilitating research laboratory.

References

[1] Suthisripok T 2012 The use of E100 to fuel a used 4-stroke motorcycle *American Journal of Applied Sciences* **9**(5) 647

[2] Hasan H, Maksum H, Gunawan S, Eswanto E and Januariyansah S 2019 Analysis of Using Gasoline-Corncob Waste Bioethanol Blends in Four-Stroke Motorcycle Engine onto Exhaust Emission and Compression Ratio *ACEIVE 2019: Proceedings of the the 3rd Annual Conference of Engineering and Implementation on Vocational Education, ACEIVE 2019, 16 November 2019, Universitas Negeri Medan, North Sumatra, Indonesia* 172

[3] Hatte P and Bhalerao Y 2019 Influence of Ethanol-Gasoline Fuel Fractions on Variable Compression Ratio Engine *International Journal of Recent Technology and Engineering* **8**(3) 2929-2936

[4] Badra J, AlRamadan A S and Sarathy S M 2017 Optimization of the octane response of gasoline/ethanol blends *Applied Energy* **203** 778-793

[5] Yusuf A A and Inambao F L 2018 Progress in alcohol-gasoline blends and their effects on the performance and emissions in SI engines under different operating conditions *International Journal of Ambient Energy* 1-17

[6] Adian F, Sugiarto B, Wibowo C S, Zikra A and Mulya T 2019 The effect of 5% ethanol in 88, 92, and 98 RON gasoline on motorcycle engine performance *AIP Conference Proceedings* **2114** 020018

[7] Morganti K, Almansour M, Khan A, Kalghatgi G and Przesmitzki S 2018 Leveraging the benefits of ethanol in advanced engine-fuel systems *Energy conversion and management* **157** 480-497

[8] Morganti K, Viollet Y, Head R, Kalghatgi G, Al-Abdullah M and Alzubail A 2017 Maximizing the benefits of high octane fuels in spark-ignition engines *Fuel* **207** 470-487

[9] Jo Y S, Bromberg L and Heywood J 2016 Optimal use of ethanol in dual fuel applications: effects of engine downsizing, spark retard, and compression ratio on fuel economy *SAE International Journal of Engines* **9**(2) 1087-1101

[10] Hasan H, Sudarmanta B and Paloboran M 2017 Influence of the Compression Ratio and Duration Injection on Performance and Emission of Sinjai Engine Type Flexible Fuel Engine 150 cc Fueled Bioethanol E70 *IPTEK The Journal of Engineering* **3**(3)

[11] Anderson J E and Wallington T J 2020 Novel Method to Estimate the Octane Ratings of Ethanol–Gasoline Mixtures Using Base Fuel Properties *Energy and Fuels* **34**(4) 4632-4642

[12] Dinh T, Nguyen K, Pham T and Nguyen V 2020 Study on performance enhancement and emission reduction of used carburetor motorcycles fueled by flex-fuel gasoline-ethanol blends *Journal of the Chinese Institute of Engineers* 1-12

[13] Nair P A, Sarath S, Kumar A S, George A J and Das A G 2018 Experimental Study on the Effects of Bio Ethanol Petrol Blends in a Spark Ignition Engine: Performance Analysis *Advanced Manufacturing and Materials Science* 327-336

[14] Saikrishnan V, Karthikeyan A and Jayaprabakar J 2018 Analysis of ethanol blends on spark ignition engines *International journal of ambient energy* **39**(2) 103-107

[15] AS D S and Antony A J 2017 Experimental study on digital twin spark ignition gasoline engine at different gasoline-methanol blends *International Journal of Applied Engineering Research* **12**(13) 3817-3821

[16] Koç M, Sekmen Y, Topgül T and Yücesu H S 2009 The effects of ethanol–unleaded gasoline blends on engine performance and exhaust emissions in a spark-ignition engine *Renewable energy* **34**(10) 2101-2106

[17] Solomon W C, Bonet M U and Mohammed S U 2018 An Analytical Performance Investigation of A Spark-Ignition Automobile Engine While Using Ethanol Blends As Fuel *American Journal of Engineering Research (AJER)* **7** 288-297

[18] Thangavelu S K, Ahmed A S and Ani F N 2016 Review on bioethanol as alternative fuel for spark ignition engines *Renewable and Sustainable Energy Reviews* **56** 820-835

[19] Choudhary A and Suryavanshi A 2018 Ethanol-Fueled Engine *International Research Journal of Engineering and Technology (IRJET)* **5**

[20] Raja A S, Arasu A V and Sornakumar T 2015 Effect of gasoline-ethanol blends on performance and emission characteristics of a single cylinder air cooled motor bike SI engine *Journal of Engineering Science and Technology* **10**(12) 3

[21] Irawan B 2017 Perhitungan Energi Pembakaran Bahan Bakar di Dalam Silinder Mesin Bensin *Seminar Nasional Teknologi Terapan (MESIN)* **3** 13-16

AUTHOR INDEX